Quantum Mechanics

The important changes quantum mechanics has undergone in recent years are reflected in this new approach for students.

A strong narrative and over 300 worked problems lead the student from experiment, through general principles of the theory, to modern applications. Stepping through results allows students to gain a thorough understanding. Starting with basic quantum mechanics, the book moves on to more advanced theory, followed by applications, perturbation methods and special fields, and ending with new developments in the field. Historical, mathematical, and philosophical boxes guide the student through the theory. Unique to this textbook are chapters on measurement and quantum optics, both at the forefront of current research. Advanced undergraduate and graduate students will benefit from this new perspective on the fundamental physical paradigm and its applications.

Online resources including solutions to selected problems and 200 figures, with color versions of some figures, are available at www.cambridge.org/Auletta.

Gennaro Auletta is Scientific Director of Science and Philosophy at the Pontifical Gregorian University, Rome. His main areas of research are quantum mechanics, logic, cognitive sciences, information theory, and applications to biological systems.

Mauro Fortunato is a Structurer at Cassa depositi e prestiti S.p.A., Rome. He is involved in financial engineering, applying mathematical methods of quantum physics to the pricing of complex financial derivatives and the construction of structured products.

Giorgio Parisi is Professor of Quantum Theories at the University of Rome "La Sapienza." He has won several prizes, notably the Boltzmann Medal, the Dirac Medal and Prize, and the Daniel Heineman prize. His main research activity deals with elementary particles, theory of phase transitions and statistical mechanics, disordered systems, computers and very large scale simulations, non-equilibrium statistical physics, optimization, and animal behavior.

Quantum Mechanics

GENNARO AULETTA
Pontifical Gregorian University, Rome

MAURO FORTUNATO
Cassa Depositi e Prestiti S.p.A., Rome

GIORGIO PARISI
''La Sapienza'' University, Rome

CAMBRIDGE
UNIVERSITY PRESS

CAMBRIDGE
UNIVERSITY PRESS

University Printing House, Cambridge CB2 8BS, United Kingdom

One Liberty Plaza, 20th Floor, New York, NY 10006, USA

477 Williamstown Road, Port Melbourne, VIC 3207, Australia

314-321, 3rd Floor, Plot 3, Splendor Forum, Jasola District Centre, New Delhi - 110025, India

103 Penang Road, #05-06/07, Visioncrest Commercial, Singapore 238467

Cambridge University Press is part of the University of Cambridge.

It furthers the University's mission by disseminating knowledge in the pursuit of education, learning and research at the highest international levels of excellence.

www.cambridge.org
Information on this title: www.cambridge.org/9781107665897

First published 2009
First paperback edition 2013

A catalogue record for this publication is available from the British Library

Library of Congress Cataloging in Publication data
Auletta, Gennaro, 1957–
Quantum mechanics : into a modern perspective / Gennaro Auletta, Mauro Fortunato, Giorgio Parisi.
p. cm.
Includes bibliographical references and index.
ISBN 978-0-521-86963-8
1. Quantum theory. I. Fortunato, Mauro. II. Parisi, Giorgio. III. Title.
QC174.12.A854 2009
530.12–dc22
2009004303

ISBN 978-0-521-86963-8 Hardback
ISBN 978-1-107-66589-7 Paperback

Contents

Figures

Tables

Definitions, principles, etc.

Corollaries

Definitions

Lemmas

Principles

Theorems

Boxes

Symbols

Latin letters

a	proposition, number
$\hat{a} = \sqrt{\frac{m}{2\hbar\omega}}\left(\omega\hat{x} + \imath\hat{\dot{x}}\right)$	annihilation operator
$\hat{a}_{\mathbf{k}}$	annihilation operator of the **k**-th mode of the electromagnetic field
$\hat{a}^{\dagger} = \sqrt{\frac{m}{2\hbar\omega}}\left(\omega\hat{x} - \imath\hat{\dot{x}}\right)$	creation operator
$\hat{a}^{\dagger}_{\mathbf{k}}$	creation operator of the **k**-th mode of the electromagnetic field
$\vert a\rangle$	(polarization) state vector (along the direction **a**)
$\vert a_j\rangle$	element of a discrete vector basis $\{\vert a_j\rangle\}$
A	number
$\mathrm{A}(\zeta)$	Airy function
$\mathbf{A}$	vector potential
$\hat{\mathbf{A}}(\mathbf{r},t) = \sum_{\mathbf{k}} c_{\mathbf{k}} \times \left[\hat{a}_{\mathbf{k}}\mathbf{u}_{\mathbf{k}}(\mathbf{r})e^{-\imath\omega_{\mathbf{k}}t} + \hat{a}^{\dagger}_{\mathbf{k}}\mathbf{u}^{*}_{\mathbf{k}}(\mathbf{r})e^{\imath\omega_{\mathbf{k}}t}\right]$	vector potential operator
$\mathcal{A}$	apparatus
$\vert\mathcal{A}\rangle$	ket describing a generic state of the apparatus
b	proposition, number
$\vert b\rangle$	(polarization) state vector (along the direction **b**)
$\vert b_j\rangle$	element vector of a discrete basis $\{\vert b_j\rangle\}$
B	number, intensity of the magnetic field
$\mathrm{B} = h/8\pi^2\mathrm{I}$	rotational constant of the rigid rotator
$\mathbf{B} = \nabla \times \mathbf{A}$	classical magnetic field
$\hat{\mathbf{B}}(\mathbf{r},t) = \imath\sum_{\mathbf{k}}\left(\frac{\hbar k}{2cL^3\epsilon_0}\right)^{\frac{1}{2}} \times \left[\hat{a}_{\mathbf{k}}e^{\imath(\mathbf{k}\cdot\mathbf{r}-\omega_{\mathbf{k}}t)} - \hat{a}^{\dagger}_{\mathbf{k}}e^{-\imath(\mathbf{k}\cdot\mathbf{r}-\omega_{\mathbf{k}}t)}\right]\mathbf{b}_{\lambda}$	magnetic field operator
c	speed of light, proposition
c_j, c'_j	generic coefficients of the j-th element of a given discrete expansion
c_{a_j}	coefficient of the basis element $\vert a_j\rangle$
c_{b_j}	coefficient of the basis element $\vert b_j\rangle$
$c_n^{(0)}$	coefficients of the expansion of a state vector in stationary state at an initial moment $t_0 = 0$

$c(\eta), c(\xi)$	coefficient of the eigenkets of continuous observables $\hat{\eta}$ and $\hat{\xi}$, respectively
$\vert c\rangle$	polarization state vector (along the direction **c**)
C, C'	constants
C	coulomb charge unit, correlation function
$\mathcal{C}$	cost function
C_{jk}	cost incurred by choosing the j-th hypothesis when the k-th hypothesis is true
$\mathbb{C}$	field of complex numbers
d	electric dipole
d	distance
$\mathcal{D}$	decoherence functional
$\hat{D}(\alpha) = e^{\alpha\hat{a}^\dagger - \alpha^*\hat{a}}$	displacement operator
e	exponential function
e	electric charge
$\mathbf{e} = (e_x, e_y, e_z)$	vector orthogonal to the propagation direction of the electromagnetic field
$\vert e\rangle$	excited state
$\vert e_k\rangle$	k-th ket of the environment's eigenbasis $\{\vert e_j\rangle\}$
E	energy
E_n	n-th energy level, energy eigenvalue
E_0	energy value of the ground state
E	one-dimensional electric field
$\mathbf{E} = -\nabla V_e - \frac{\partial}{\partial t}\mathbf{A}$	classical electric field
$\hat{\mathbf{E}}(\mathbf{r}, t) = \imath \sum_{\mathbf{k}} \left(\frac{\hbar\omega_{\mathbf{k}}}{2\epsilon_0}\right)^{\frac{1}{2}} \times \left[\hat{a}_{\mathbf{k}}\mathbf{u}_{\mathbf{k}}(\mathbf{r})e^{-\imath\omega_{\mathbf{k}}t} - \hat{a}^\dagger_{\mathbf{k}}\mathbf{u}^*_{\mathbf{k}}(\mathbf{r})e^{\imath\omega_{\mathbf{k}}t}\right]$	electric field operator
$\mathcal{E}$	environment
$\hat{E}$	effect
$\vert\mathcal{E}\rangle$	ket describing a generic state of the environment
f	arbitrary function
$\mathbf{f}, \mathbf{f}'$	arbitrary vectors
$\vert f\rangle$	final state vector
F	force, arbitrary classical physical quantity
$\mathbf{F}_e$	classically electrical force
$\mathbf{F}_m$	classically magnetic force
$\mathrm{F}_m(\phi)$	eigenfunctions of $\hat{l}_z$
$\mathcal{F}(x) = \wp(\xi < x)$	distribution function of a random variable that can take values $< x$
g	arbitrary function, gravitational acceleration
$\vert g\rangle$	ground state
$G^{(n)}$	coherence of the n-th order

G	Green function
$G_0(\mathbf{r}', t'; \mathbf{r}, t) = -\imath \left[\frac{m}{2\pi\imath\hbar(t'-t)}\right]^{\frac{3}{2}} e^{\frac{\imath m\lvert\mathbf{r}'-\mathbf{r}\rvert^2}{2\hbar(t'-t)}}$	free Green function
$\mathcal{G}$	group
$\hat{G}$	generator of a a continuos transformation or of a group
$h = 6.626069 \times 10^{-34}\mathrm{J} \cdot \mathrm{s}$	Planck constant
$\hbar = h/2\pi$	
$\lvert h\rangle$	state of horizontal polarization
$\mathcal{H}$	Hilbert space
$\mathcal{H}_{\mathcal{A}}$	Hilbert space of the apparatus
$\mathcal{H}_{\mathcal{S}}$	Hilbert space of the system
$\hat{H}$	Hamiltonian operator
$\hat{H}_0$	unperturbed Hamiltonian
$\hat{H}_A$	Hamiltonian of a free atom
$\hat{H}_F$	field Hamiltonian
$\hat{H}_{\mathrm{I}}$	interaction Hamiltonian
$\hat{H}_{\mathrm{I}}^{\mathrm{I}}$	interaction Hamiltonian in Dirac picture
$\hat{H}_{JC}$	Jaynes–Cummings Hamiltonian
$\hat{H}_r$	planar part of the Hamiltonian
$\mathrm{H}_n(\zeta) = (-1)^n e^{\zeta^2} \frac{d^n}{d\zeta^n} e^{-\zeta^2}$ n-th	Hermite polynomial, for all $n \neq 0$
$\imath$	imaginary unity
$\boldsymbol{\imath}$ x	Cartesian versor
$\lvert i\rangle$	initial state vector
I	intensity (of radiation)
I	moment of inertia
$\hat{I}$	identity operator
$\Im(z) = \frac{z-z^*}{2\imath}$	imaginary part of a complex number z
$\boldsymbol{\jmath}$ y	Cartesian versor
$\hat{\boldsymbol{\jmath}} = \hat{\boldsymbol{J}}/\hbar = (\hat{\jmath}_x, \hat{\jmath}_y, \hat{\jmath}_z)$	
$\lvert j\rangle$	arbitrary ket, element of a continuous or discrete basis
$\lvert j, m\rangle$	eigenket of $\hat{J}_z$
$\mathbf{J}$	density of the probability current
J_{I}	incidental current density
J_R	reflected current density
J_{T}	transmitted current density
$\hat{\mathbf{J}} = \hat{\mathbf{L}} + \hat{\mathbf{S}} = (\hat{J}_x, \hat{J}_y, \hat{J}_z)$	total angular momentum
$\hat{\hat{\mathcal{J}}}$	jump superoperator
$k, \mathbf{k}$	wave vector
$\boldsymbol{k}$ z	Cartesian versor
k_{B}	Boltzmann constant
$\lvert k\rangle$	generic ket, element of a continuous or discrete basis

l	arbitrary length
$\hat{\mathbf{l}} = (\hat{l}_x, \hat{l}_y, \hat{l}_z) = \hat{\mathbf{L}}/\hbar$	
$\hat{l}_\pm = \hat{l}_x \pm \imath \hat{l}_y$	raising and lowering operators for the levels of the angular momentum
$\|l\rangle$	generic ket
$\|l, m_l\rangle$	eigenket of $\hat{l}_z$
$\mathrm{L}(q_1, \ldots, q_n; \dot{q}_1, \ldots, \dot{q}_n)$	classical Lagrangian function
$\hat{\mathrm{L}}$	Lagrangian multiplier operator
$\hat{\mathbf{L}} = (L_x, L_y, L_z)$	orbital angular momentum
$\hat{\hat{\mathcal{L}}}$	Lindblad superoperator
m	mass of a particle
m_e	mass of the electron
m_n	mass of the nucleus
m_p	mass of the proton
m_l	magnetic quantum number
m_j	eigenvalue of $\hat{j}_z$
m_s	spin magnetic quantum number or secondary spin quantum number
$\|m\rangle$	generic ket, eigenket of the energy
M	measure of purity
$\mathcal{M}$	meter
$\hat{M}$	arbitrary matrix
$\mathbf{n}, \mathbf{n}'$	direction vectors
$\|n\rangle$	eigenvector of the harmonic oscillator Hamiltonian or of the number operator
N	number of elements of a given set
$\mathcal{N}$	normalization constant
$\hat{N} = \hat{a}^\dagger \hat{a}$	number operator
$\hat{N}_\mathbf{k} = \hat{a}^\dagger_\mathbf{k} \hat{a}_\mathbf{k}$	
o_j	j-th eigenvalue of the observable $\hat{O}$
$\|o\rangle$	eigenket of the observable $\hat{O}$
$\|o_j\rangle$	j-th eigenket of the observable $\hat{O}$
$\hat{O}, \hat{O}', \hat{O}''$	generic operators, generic observables
$\hat{O}^{\mathrm{H}}$	observable in the Heisenberg picture
$\hat{O}^{\mathrm{I}}$	observable in the Dirac picture
$\hat{O}^{\mathrm{S}}$	observable in the Schrödinger picture
$\hat{O}_{\mathcal{A}}$	apparatus' pointer
$\hat{O}_{\mathcal{S}}$	observable of the object system
$\hat{O}_{\mathrm{ND}}$	non-demolition observable
$\left\|\hat{O}\right\}$	super-ket (or S-ket)
p_k	classical generalized momentum component
$\hat{\mathbf{p}} = (\hat{p}_x, \hat{p}_y, \hat{p}_z)$	three-dimensional momentum operator
$\hat{p}_x$	one-dimensional momentum operator

$\hat{\dot{p}}_x$	time derivative of $\hat{p}_x$
$\hat{p}_r = -\imath\hbar\frac{1}{r}\frac{\partial}{\partial r}r$	radial part of the momentum operator
$\mathrm{P}(\alpha, \alpha^*)$	P-function
$\hat{P}_j$	projection onto the state $\mid j\rangle$ or $\mid b_j\rangle$
$\mathcal{P}$	path predictability
$\hat{\mathcal{P}}$	path predicability operator
$\wp_j$ or $\wp(j)$	probability of the event j
$\wp_k(\mathbf{D}) = \wp(\mathbf{D}\vert\mathrm{H}_k)$	probability density function that the particular set $\mathbf{D}$ of data is observed when the system is actually in state k
$\wp(\mathrm{H}_j\vert\mathrm{H}_k) = \mathrm{Tr}\left(\hat{\rho}_k\hat{E}_{\mathrm{H}_j}\right)$	conditional probability that one chooses the hypothesis H_j when H_k is true
q_k	classical generalized position component
Q	charge density
$\mathrm{Q}(\alpha, \alpha^*) = \frac{1}{\pi}\langle\alpha\vert\hat{\rho}\vert\alpha\rangle$	Q-function
$\mathcal{Q}$	quantum algebra
$\mathbb{Q}$	field of rational numbers
r	spherical coordinate
$\mathbf{r}\cdot\mathbf{r}'$	scalar product between vectors $\mathbf{r}$ and $\mathbf{r}'$
r_k	k-th eigenvalue of a density matrix
$r_0 = \frac{\hbar^2}{me^2}$	Bohr's radius
$\hat{\mathbf{r}} = (\hat{x}, \hat{y}, \hat{z})$	three-dimensional position operator
R	reflection coefficient
$\mathrm{R}(r)$	radial part of the eigenfunctions of $\hat{l}_z$ in speherical coordinates
$\mathcal{R}, \mathcal{R}'$	reference frames
$\mathcal{R}$	reservoir
$\mathbb{R}$	field of real numbers
$\Re(z) = \frac{z+z^*}{2}$	real part of a complex quantity z
$\hat{R}, \hat{\mathbf{R}}(\beta, \phi, \theta)$	rotation operator, generator of rotations
$\hat{\mathrm{R}}_{\hat{O}}$	resolvent of the operator $\hat{O}$
$\hat{\mathcal{R}}_j = \sum_{k=1}^{N}\wp_k^{\mathrm{A}}C_{jk}\hat{\rho}_k$	risk operator for the j-th hypothesis
$\mid R\rangle$	initial state of the reservoir
s	spin quantum number
$\hat{\mathbf{s}} = (\hat{s}_x, \hat{s}_y, \hat{s}_z) = \hat{\mathbf{S}}/\hbar$	spin vector operator
$\hat{s}_\pm = \hat{s}_x \pm \imath\hat{s}_y$	raising and lowering spin operators
S	action
$\mathcal{S}$	generic quantum system
S	entropy
$\hat{\mathbf{S}} = (\hat{S}_x, \hat{S}_y, \hat{S}_z)$	spin observable
t	time
$\hat{t}$	time operator
$\mid t\rangle$	eigenket of the time operator

T	transmission coefficient
T	temperature, classical kinetic energy
$\hat{T}$	kinetic energy operator
$\hat{\mathcal{T}}$	time reversal operator
$\hat{\hat{\mathcal{T}}}, \mathcal{T}$	generic transformation
$u(\nu, T)$	energy density
$\mathbf{u_k}(\mathbf{r}) = \frac{\mathbf{e}}{L^{\frac{3}{2}}} e^{\imath \mathbf{k} \cdot \mathbf{r}}$	**k**-th mode function of the electromagnetic field
U	scalar potential
$\hat{U}$	unitary operator
$\hat{U}_{\mathrm{BS}}$	beam splitting unitary operator
$\hat{U}_{\mathrm{PBS}}$	polarization beam-splitting unitary operator
$\hat{U}_{\mathrm{CNOT}}$	unitary controlled-not operator
$\hat{U}_f$	Boolean unitary transformation
$\hat{U}_F$	Fourier unitary transformation
$\hat{U}_H$	unitary Hadamard operator
$\hat{U}_{\mathbf{p}}(\mathbf{v})$	unitary momentum translation
$\hat{U}_P$	permutation operator
$\hat{U}_{\mathbf{R}}(\phi)$	rotation operator
$\hat{U}_{\mathcal{R}}$	space-reflection operator
$\hat{U}_t$	time translation unitary operator
$\hat{U}_x(a)$	one-dimensional space translation unitary operator
$\hat{U}_{\mathbf{r}}(\mathbf{a})$	three-dimensional space translation unitary operator
$\hat{U}_\theta$	unitary rotation operator
$\hat{U}_\phi$	unitary phase operator
$\hat{U}_\tau^{\mathcal{SA}} = e^{-\frac{\imath}{\hbar}\int_0^\tau dt \hat{H}_{\mathcal{SA}}(t)}$	unitary operator coupling system and apparatus for time interval τ
$\hat{U}_t^{\mathcal{SA},\mathcal{E}} = e^{-\frac{\imath}{\hbar}t\hat{H}_{\mathcal{SA},\mathcal{E}}}$	unitary operator which couples the environment $\mathcal{E}$ to the system and apparatus $\mathcal{S}+\mathcal{A}$ at time t
$\tilde{\hat{U}}$	antiunitary operator
$\tilde{\hat{U}}_{\mathcal{T}}$	time reversal
$\hat{\mathcal{U}}$	generic transformation that can be either unitary or antiunitary
$\vert v\rangle$	state of vertical polarization
$\vert v_n\rangle$	element of a discrete basis
V	potential energy
V_e	scalar potential of the electromagnetic field
$V_c(r) = \frac{\hbar^2 l(l+1)}{2mr^2}$	centrifugal-barrier potential energy
V_{c}	classical potential energy
$\hat{V}$	potential energy operator

V	volume
$\mathbf{V}$	generic vector
$\mathcal{V}$	visibility of interference, generic vectorial space
$\hat{\mathcal{V}}$	visibility of interference operator
w_k	k-th probability weight
$\vert w_n\rangle$	element of a discrete basis vector
$W(\alpha,\alpha^*) = \frac{1}{\pi^2}\int d^2\alpha e^{-\eta\alpha^*+\eta^*\alpha}\chi_W(\eta,\eta^*)$	Wigner function
x	first Cartesian axis, coordinate
$\vert x\rangle$	eigenket of $\hat{x}$
$\hat{x}$	one-dimensional position operator
$\dot{\hat{x}}$	time derivative of $\hat{x}$
$\hat{X}_1 = \frac{1}{\sqrt{2}}(\hat{a}^\dagger + \hat{a})$	quadrature
$\hat{X}_2 = \frac{\imath}{\sqrt{2}}(\hat{a}^\dagger - \hat{a})$	quadrature
$\mathcal{X}$	set
y	second Cartesian axis, coordinate
$Y_{lm}(\theta,\phi)$	spherical harmonics
z	third Cartesian axis, coordinate
Z	atomic number
$Z(\beta) = \mathrm{Tr}\left(e^{-\beta\hat{H}}\right)$	partition function
$\mathcal{Z}$	parameter space
$\mathbb{Z}$	field of integer numbers

Greek letters

α	angle, (complex) number
$\vert\alpha\rangle = e^{-\frac{\vert\alpha\vert^2}{2}}\sum_{n=0}^{\infty}\frac{\alpha^n}{\sqrt{n!}}\vert n\rangle$	coherent state
β	angle, (complex) number, thermodynamic variable = $(k_{\mathrm{B}}T)^{-1}$
$\vert\beta\rangle$	coherent state
γ	damping constant
Γ	Euler gamma function
Γ	phase space
$\hat{\Gamma}_k$	reservoir operator
δ_{jk}	Kronecker symbol
$\delta(x)$	Dirac delta function
$\Delta = \nabla^2 = \frac{\partial^2}{\partial x^2} + \frac{\partial^2}{\partial y^2} + \frac{\partial^2}{\partial z^2}$	Laplacian
Δ_ψ	uncertainty in the state $\vert\psi\rangle$
ϵ	small quantity
ϵ_{jkn}	Levi–Civita tensor
ε	coupling constant
$\varepsilon_0 = \left(\frac{\omega}{2\epsilon_0\hbar l^3}\right)^{\frac{1}{2}}\vert\mathbf{d}\cdot\mathbf{e}\vert$	vacuum Rabi frequency

$\varepsilon_n = \varepsilon_0\sqrt{n+1}$	Rabi frequency
$\varepsilon_{\mathcal{SA}}$	coupling between object system and apparatus
$\varepsilon_{\mathcal{SM}}$	coupling between object system and meter
ζ	arbitrary variable, arbitrary (wave) function
ζ_S, ζ_A	number of possible configurations of bosons and fermions, respectively
η	arbitrary variable, arbitrary (wave) function
$\hat{\eta}$	arbitrary (continuous) observable
$\vert\eta\rangle$	eigenkets of $\hat{\eta}$
θ	angle, spherical coordinate
ϑ	generic amplitude
$\hat{\vartheta}_k(m) = \left\langle m \left\vert \hat{U}_t \right\vert k \right\rangle$	amplitude operator connecting a premeasurement ($\vert k\rangle$), a unitary evolution ($\hat{U}_t$), and a measurement ($\vert m\rangle$)
$\Theta_{lm}(\theta)$	theta component of the spherical harmonics
$\Theta(\theta)$	part of the spherical harmonics depending on the polar coordinate θ
$\hat{\Theta}, \hat{\hat{\Theta}}$	arbitrary transformation (superoperator)
ι	constant, parameter
$\vert\iota\rangle$	internal state of a system
κ	parameter
λ	wavelength
$\lambda_c = h/mc$	Compton wavelength of the electron
$\lambda_T = \frac{\hbar}{\sqrt{2mk_{\mathrm{B}}T}}$	thermal wavelength
$\Lambda = \mu_B B_{\mathrm{ext}}$	constant used in the Paschen–Bach effect
$\hat{\Lambda}_j$	Lindblad operator
$\boldsymbol{\mu}$	classically magnetic dipole momentum
$\hat{\boldsymbol{\mu}}_l = \frac{e\hbar}{2m}\hat{\mathbf{l}}$	orbital magnetic momentum of a massive particle
$\hat{\boldsymbol{\mu}}_s = Q\frac{e\hbar}{2m}\hat{\mathbf{s}}$	spin magnetic momentum
$\mu_B = \frac{e\hbar}{2m}$	Bohr magneton
μ_0	magnetic permeability
ν	frequency
ξ	random variable, variable
$\xi(r) = R(r)r$	change of variable for the radial part of the wave function
$\hat{\xi}$	arbitrary (continuous) observable
$\vert\xi\rangle$	eigenkets of $\hat{\xi}$
$\Xi(x)$	Heaviside step function
$\hat{\Pi}$	parity operator
ρ	(classical) probability density
$\hat{\rho}$	density matrix (pure state)
$\dot{\hat{\rho}}$	time-evolved density matrix

$\hat{\hat{\rho}}$	mixed density matrix
$\hat{\rho}_f$	density matrix for the final state of a system
$\hat{\rho}_i$	density matrix for the initial state of a system
$\hat{\varrho}_j$	reduced density matrix of the j-th subsystem
$\hat{\rho}_{\mathcal{SA}}$	density matrix of the system plus apparatus
$\hat{\rho}_{\mathcal{SAE}}$	density matrix of the system plus apparatus plus environment
$\sigma_x^2 = \langle \hat{x}^2 \rangle - \langle \hat{x} \rangle^2$	variance of $\hat{x}$
σ_x	standard deviation (square root of the variance) of $\hat{x}$
$\sigma_p^2 = \langle \hat{p}_x^2 \rangle - \langle \hat{p}_x \rangle^2$	variance of $\hat{p}_x$
σ_p	standard deviation (square root of the variance) of $\hat{p}_x$
$\hat{\sigma}_+ = \| e \rangle \langle g \|$	raising operator
$\hat{\sigma}_- = \| g \rangle \langle e \|$	lowering operator
$\hat{\boldsymbol{\sigma}} = (\hat{\sigma}_x, \hat{\sigma}_y, \hat{\sigma}_z)$	Pauli (two-dimensional) spin matrices
$\varsigma(s)$	wave component of the spin
$\| \varsigma \rangle$	ket of the object system
τ	time interval, interaction time between two or more systems
$\tau_d \simeq \gamma^{-1} \left(\frac{\lambda_T}{\Delta x} \right)^2$	decoherence time
ϕ	angle, spherical coordinate
$\hat{\phi}$	angle operator
$\| \phi \rangle$	eigenket of the angle operator
$\| \varphi \rangle, \| \varphi' \rangle$	state vectors
$\varphi(\xi)$	eigenfunctions of the observable with eigenvector $\| \xi \rangle$
$\varphi_k(x)$	plane waves
$\varphi_{\mathbf{k}}(\mathbf{r})$	spherical waves
$\varphi_p(x)$	momentum eigenfunctions in the position representation
$\varphi_{x_0}(x)$	position eigenfunctions in the position representation
$\tilde{\varphi}_{p_0}(p_x)$	momentum eigenfunctions in the momentum representation
$\tilde{\varphi}_x(p_x)$	position eigenfunctions in the momentum representation
$\varphi_\xi(x)$	scalar product $\langle x \| \xi \rangle$
$\varphi_\eta(\xi)$	scalar product $\langle \xi \| \eta \rangle$
Φ	flux of electric current
Φ_M	magnetic flux
$\| \Phi \rangle$	generic ket for compound systems
$\chi_\xi(\eta) = \int d\mathcal{F}(x) e^{\imath \eta x}$	classical characteristic function of a random variable ξ

$\chi(\eta, \eta^*) = e^{\|\eta\|^2} \int d^2\alpha e^{\eta\alpha^* - \alpha\eta^*} Q(\alpha, \alpha^*)$	characteristic function
$\chi_W(\eta, \eta^*) = e^{-\frac{1}{2}\|\eta\|^2} \chi(\eta, \eta^*)$	Wigner characteristic function
$\| \psi \rangle, \| \psi' \rangle$	state vectors
$\| \psi(t) \rangle$	time-evolved or time-dependent state vector
$\| \psi_E \rangle$	Eigenket of energy corresponding to eigenvalue E (in the continuous case)
$\| \Psi_F \rangle$	quantum state of the electromagnetic field
$\| \psi_n \rangle$	n-th stationary state
$\| \psi \rangle_{\rm H}$	state vector in the Heisenberg picture
$\| \psi \rangle_{\rm I}$	state vector in the Dirac picture
$\| \psi \rangle_{\rm S}$	state vector in the Schrödinger picture
$\psi(x), \psi(\mathbf{r})$	wave functions in the position representation
$\psi(\eta), \psi(\xi)$	wave functions of two arbitrary continuous observables, η and ξ, respectively
$\tilde{\psi}(p_x), \tilde{\psi}(\mathbf{p})$	Fourier transform of the wave functions
$\psi(\mathbf{r}, s)$	wave function with a spinor component
$\psi(r, \theta, \phi)$	eigenfunctions of $\hat{l}_z$ in spherical coordinates
$\psi_p(x), \psi_{\mathbf{p}}(\mathbf{r}), \psi_k(x), \psi_{\mathbf{k}}(\mathbf{r})$	momentum eigenfunctions in the position representation
$\psi_E(x)$	energy eigenfucntion in the position representation
ψ_S, ψ_A	symmetric and antisymmetric wavefucntions, respectively
$\| \Psi \rangle$	ket of a compund system
$\| \Psi \rangle_{\mathcal{SA}}$	ket describing an objects system plus apparatus compound system
$\| \Psi \rangle_{\mathcal{SM}}$	ket describing an objects system plus meter compound system
$\| \Psi \rangle_{\mathcal{SAE}}$	ket describing an objects system plus apparatus plus environment compound system
$\Psi(x), \Psi(\mathbf{r})$	wave function of a compound system
$\omega = 2\pi\nu$	angular frequency
$\omega_B = \frac{eB}{m}$	electron cyclotron frequency
ω_{jk}	ratio between energy levels $E_k - E_j$ and $\hbar$
Ω	space

Other Symbols

∇	Nabla operator
$\langle \cdot \| \cdot \cdot \rangle$	scalar product
$\langle j_1, j_2, m_1, m_2 \| j, m \rangle$	Clebsch–Gordan coefficient
$\| \cdot \rangle \langle \cdot \|$	external product
$\langle \cdot \rangle$	mean value

$\mathrm{Tr}(\hat{O})$	trace of the operator $\hat{O}$
$\otimes$	direct product
$\oplus$	direct sum
$\forall$	for all ...
$\exists$	there is at least one ... such that
$a \in X$	the element a pertains to the set X
$X \subset Y$	X is a proper subset of Y
$a \Longrightarrow b$	a is sufficient condition of b
$\vee$	inclusive disjunction (OR)
$\wedge$	conjunction (AND)
$a \mapsto b$	a maps to b
$\rightarrow$	tends to ...
$\vert 0\rangle, \vert 1\rangle$	arbitrary basis for a two-level system, qubits
$\vert 1\rangle, \vert 2\rangle, \vert 3\rangle, \vert 4\rangle$	set of eigenstates of a path observable
$\vert \mathbf{0}\rangle = \vert 0,0,0\rangle$	vacuum state
$\vert \uparrow\rangle, \vert \downarrow\rangle$	arbitrary basis for a two-level system, eigenstates of the spin observable (in the z-direction)
$\vert \leftrightarrow\rangle$	state of horizontal polarization
$\vert \updownarrow\rangle$	state of vertical polarization
$\vert \nearrow\rangle$	state of 45° polarization
$\vert \nwarrow\rangle$	state of 135° polarization
$\vert \smile\rangle_c, \vert \frown\rangle_c$	living- and dead-cat states, respectively
$[\cdot, \cdot\cdot]_- = [\cdot, \cdot\cdot]$	commutator
$[\cdot, \cdot\cdot]_+$	anticommutator
$\{\cdot, \cdot\cdot\}$	Poisson brackets
$\partial_t = \frac{\partial}{\partial t}$	partial derivatives
$\partial_j = \frac{\partial}{\partial j}$, with $j = x, y, z$	

Abbreviations

AB	Aharonov–Bohm
BS	beam splitter
Ch.	chapter
CH	Clauser and Horne
CHSH	Clauser, Horne, Shimony, and Holt
Cor.	corollary
cw	continuous wave
Def.	definition
EPR	Einstein, Podoloski, and Rosen
EPRB	Einstein, Podoloski, Rosen, and Bohm
Fig.	figure
GHSZ	Greenberger, Horne, Shimony, and Zeilinger
GHZ	Greenberger, Horne, and Zeilinger
iff	if and only if
HV	hidden variable
LASER	light amplification by stimulated emission of radiation
LCAO	linear combination of atomic orbitals
lhs	left-hand side
MWI	many world interpretation
p.	page
PBS	polarization beam splitter
POSet	partially ordered set
Post.	postulate
POVM	positive operator valued measure
Pr.	principle
Prob.	problem
PVM	projector valued measure
rhs	right-hand side
Sec.	section
SGM	Stern–Gerlach magnet
SPDC	spontaneous parametric down conversion
SQUID	superconducting quantum interference device
Subsec.	subsection
Tab.	table
Th.	theorem
VBM	valence bond method

Acknowledgements for the revised edition

We are indebted to our colleague Shang–Yung Wang of the Tamkang University, Taiwan for a thorough critical reading of our book, the enlightening correspondence and the very helpful suggestions. His precious work has made this revised edition possible.

Introduction

Why yet another book on quantum mechanics? Quantum mechanics was born in the first quarter of the twentieth century and has received an enormous number of theoretical and experimental confirmations over the years. It is considered to be the fundamental physical paradigm, and has a wide range of applications, from cosmology to chemistry, and from biology to information sciences. It is one of the greatest intellectual achievements of the past century. As an effect of its invention, the very concept of physical reality was changed, and "observation," "measurement," "prediction," and "state of the system" acquired a new and deeper meaning.

Probability was not unknown in physics: it was introduced by Boltzmann in order to control the behavior of a system with a very large number of particles. It was the missing concept in order to understand the thermodynamics of macroscopic bodies, but the structure of the physical laws remained still deterministic. The introduction of probability was needed as a consequence of our lack of knowledge of the initial conditions of the system and of our inability to solve an enormous number of coupled non-linear differential equations.

In quantum mechanics, the tune is different: if we have 10^6 radioactive atoms *no* intrinsic unknown variables decide which of them will decay first. What we observe experimentally seems to be an irreducible random process. The original explanation of this phenomenon in quantum mechanics was rather unexpected. All atoms have the same probability of having decayed: only when we observe the system do we select which atoms have decayed in the past. In spite of the fact that this solution seems to be in contrast with common sense, it is the only possible one in the framework of the conventional interpretation of quantum mechanics. Heisenberg, de Broglie, Pauli, Dirac, and many others invented a formalism that was able to explain and predict the experimental data and this formalism led, beyond the very intention of the men who constructed it, to this conceptual revolution. Then, the old problem of the relations among the observer and the observed object, discussed for centuries by philosophers, had a unexpected evolution and now it must be seen from a new, completely different perspective.

Once established, quantum mechanics became a wonderful and extremely powerful tool. The properties of the different materials, the whole chemistry, became for the first time objects that could be predicted from the theory and not only phenomenological rules deduced from experiments. The technological discovery that shaped the second half of last century, the transistor (i.e. the basis of all the modern electronics and computers) could not have been invented without a deep command of quantum mechanics.

The advances of recent years have not only concentrated on the problems of interpretation that could be (wrongly) dismissed as metaphysical by some people, considering them to be beyond experimental tests. In the last 30 years, the whole complex of problems connected to quantum mechanics and the meaning of measurements started to be studied from a new perspective. Real, not only *Gedanken* experiments began to be done on some of the most elusive properties of quantum mechanics, i.e. the existence of correlations among spatially separated systems that could not be explained using the traditional concept of probability. The precise quantum mechanical meaning of measurements started to be analyzed in a more refined way (e.g. quantum non-demolition measurements were introduced) and various concepts from statistical mechanics and other fields of physics began to be used.

This is not only an academic or philosophical problem. The possibility of constructing a quantum computer, which would improve the speed of present day computers by an incredible factor, is deeply rooted in these achievements. It is now clear that a quantum computer can solve problems, which on conventional computers take a time exploding as exponent of some parameter (e.g. the factorization into primes of a number of length N), in a time which is only a polynomial in N. The technical problems to be overcome in constructing a quantum computer are not easy to solve, but this result has a high conceptual status, telling us how deeply quantum mechanics differs from classical mechanics. Another quantum-information puzzling phenomenon, i.e. teleportation, has been recently proved experimentally to exist and it is a very active area of experimental research.

The arguments above explain why this new situation imposes the necessity to treat this field in a new way. The idea of writing this book came to one of us in 2000; it has taken more than eight years to accomplish this challenge.

Outline

The book is divided into four parts:

I *Basic features of quantum mechanics*
Part I deals with the basic framework of the theory and the reasons for its birth. Furthermore, starting from the fundamental principles, it explains the nature of quantum observables and states, and presents the dynamics of quantum systems and its main examples.

II *More advanced topics*
In Part II we introduce angular momentum, spin, identical particles, and symmetries. Moreover, we give a special emphasis to the quantum theory of measurement.

III *Matter and light*
We devote Part III to some of the most important applications of quantum theory: approximation methods and perturbation theory, the hydrogen atom, simple molecules, and quantum optics.

IV *Quantum information: state and correlations*
Finally, we deal with the most recent topics: the quantum theory of open systems, state

measurement, quantum correlations and non-locality, and quantum information and computation.

In this book there is material for four one-semester courses. It may also serve as a guide for short courses or tutorials on specific and more advanced topics.

Methodology

(1) In our exposition we have tried to follow a "logical" order, starting from the principles of classical mechanics, the need of quantum mechanics with its fundamental assumptions (superposition, complementarity, and uncertainty principles). Then, we present the main features of observables and states, before going forward to the dynamics and to more sophisticated stuff, applications, and special areas.

(2) We have made an effort to use a pedagogical style. In particular:

  (i) We prove or let the reader prove (through problems that are solved on the book's website) practically all our results: we try to lead the reader to reach them step by step from previous ones.

  (ii) We have made the choice to present Dirac algebra and operatorial formalism from the very beginning, instead of starting with the wave-function formalism. The latter is obtained naturally as a particular representation of the former. This approach has the advantage that we are not obliged to repeat the fundamental mathematical tools of the theory.

  (iii) We present our main principles and results in a pragmatic way, trying to introduce new concepts on the basis of experimental evidence, rather than in an axiomatic way, which may result cumbersome for readers who are learning quantum mechanics.

  (iv) We have made an effort to pay particular attention to cross-references in order to help the (inexpert) reader to quickly find the necessary background and related problems.

(3) We have taken into account some of the most recent developments at theoretical and experimental level, as well as with respect to technological applications: quantum optics, quantum information, quantum non-locality, state measurement, etc.

(4) We believe that measurement theory constitutes a fundamental part of quantum mechanics. As a consequence we have devoted an entire chapter to this issue.

(5) When necessary, we have emphasized interpretational as well as historical issues, such as complementarity, measurement, nature of quantum states, and so on.

(6) We propose to the reader a large number of problems (more than 300), and the less trivial ones (about half of them) are solved in a pedagogical way.

(7) From time to time, we have chosen to treat special topics in "boxes."

Apparatus

Besides a large number of cross-references, we also list the following tools:

(1) The book contains 200 figures among drawings, photographs, and graphs, distributed in all chapters (a sample of color figures can be found on the book's website). We consider this graphic support a very important aspect of our exposition. In this context, figure captions are particularly accurate and often self-contained.
(2) The book contains an extensive bibliography (almost 600 entries, most of which are quoted in the text) and a "Further reading" section at the end of each chapter. Name of authors in italics in citations refer to books, those in roman text refer to journals, papers, and other publications.
(3) The book contains full, accurate, and comprehensive indices (table of contents, subject index, author index, list of figures, list of tables, list of abbreviations, list of symbols, list of boxes, list of theorems, definitions, and so on) together with a summary of the main concepts at the end of each chapter.

Readers

This book is addressed to people who want to learn quantum mechanics or deepen their knowledge of the subject. The requirement for understanding the book is a knowledge of calculus, vectorial analysis, operator algebra, and classical mechanics.

The book is primarily intended for third- and fourth-year undergraduate students in physics. However, it may also be used for other curricula (such as mathematics, engineering, chemistry, computer sciences, etc.). Furthermore, it may well be used as a reference book for graduate students, researchers, and practitioners, who want a rapid access to specific topics. To this purpose the extensive indices and lists are of great help. It may even serve as an introduction to specific areas (quantum optics, entanglement, quantum information, measurement theory) for experienced professionals from different fields of physics. Finally, the book may prove useful for scientists of other disciplines who want to learn something about quantum mechanics.

Acknowledgements

We would like to thank Enrico Beltrametti, Michael Heller, Artur Ekert, and Willem de Muynck for a critical reading of part of the manuscript. We would also like to warmly thank all those persons – friends, colleagues, teachers, students – who helped and influenced us in the writing of the book.

Finally, we dedicate this book to our families for their continual support and love, and for tolerating our many absences during the completion of the book.

PART I

BASIC FEATURES OF QUANTUM MECHANICS

1 From classical mechanics to quantum mechanics

In this chapter we shall first summarize some conceptual and formal features of classical mechanics (Sec. 1.1). Modern physics started with the works of Galileo Galilei and Isaac Newton from which classical mechanics, one of the most beautiful and solid intellectual buildings of the human history, came out. The architecture of classical mechanics was developed between the end of the eighteenth century and the second half of the nineteenth century, and its present form is largely due to Lagrange, Jacobi, and Hamilton. As we shall see in this chapter, classical mechanics is built upon the requirement of determinism, a rather complex assumption which is far from being obvious. In Sec. 1.2 we shall present the two main conceptual features of quantum mechanics on the basis of an ideal interferometry experiment: the superposition principle and the principle of complementarity. In Sec. 1.3 a first formal treatment of quantum-mechanical states is developed: quantum states are represented by vectors in a space that turns out to be a Hilbert space. In Sec. 1.4 the significance of probability for quantum mechanics is explained briefly: we will show that probability is not just an ingredient of quantum mechanics, but is rather an intrinsic feature of the theory. Furthermore, we shall see that quantum probability is not ruled by Kolmogorov axioms of classical probability. Finally, we discuss the main evidences which have historically revealed the necessity of a departure from classical mechanics. Our task then is to briefly present the principles upon which quantum mechanics is built (in Secs. 1.2–1.4) and to summarize in Sec. 1.5 the main evidences for this new mechanics.

1.1 Review of the foundations of classical mechanics

Classical mechanics is founded upon several principles and postulates, sometimes only implicitly assumed. In the following we summarize and critically review such assumptions.[1]

First of all, in classical mechanics a *principle of perfect determination* is assumed: all properties of a physical system $\mathcal{S}$ are perfectly determined at any time. Here, we define a *physical system* as an object or a collection of objects (somehow interrelated) that can be (directly or indirectly) experienced through human senses, and a *property* as the value that can be assigned to a physical variable or observable describing $\mathcal{S}$. *Perfectly determined* means then that each (observable) variable describing $\mathcal{S}$ has at all times a definite value.

[1] See [Auletta 2004].

Some of these properties will have a value that is a real number, e.g. the position of a particle, others an integer value, e.g. the number of particles that constitute a compound system.

It is also assumed that all properties can be in principle perfectly known, e.g. they can be perfectly measured. In other terms, the measurement errors can be – at least in principle – always reduced below an arbitrarily small quantity. This is not in contrast with the everyday experimental evidence that any measurement is affected by a finite resolution. Hence, this assumption can be called the postulate of *reduction to zero of the measurement error*. We should emphasize that this postulate is not a direct consequence of the principle of perfect determination because we could imagine the case of a system that is objectively determined but cannot be perfectly known.

Moreover, the variables associated with a system $\mathcal{S}$ are in general supposed to be continuous, e.g. given two arbitrary values of a physical variable, all intermediate possible real values are also allowed. This assumption is known as the *principle of continuity*.

At this point we can state the first consequence of the three assumptions above: If the state of a system $\mathcal{S}$ is perfectly determined at a certain time t_0 and its dynamical variables are continuous and known, then, knowing also the conditions (i.e. the forces that act on the system), it should be possible (at least in principle) to predict with certainty (i.e. with probability equal to one) the future evolution of $\mathcal{S}$ for all times $t > t_0$. This in turn means that the future of a classical system is unique. Similarly, since the classical equations of motion (as we shall see below) are invariant under time reversal (the operation which transforms t into $-t$) also the past behavior of the system for all times $t < t_0$ is perfectly determined and knowable once its present state is known. Such a consequence is usually called *determinism*. Determinism is implemented by assuming that the system satisfies a set of first-order differential equations of the form

$$\frac{d}{dt}\mathbf{S} = F[\mathbf{S}(t)], \tag{1.1}$$

where $\mathbf{S}$ is a vector describing the state of the system. It is also assumed that these equations (called equations of motion) have one and only one solution, and this situation is usual if the functional transformation F is not too nasty.

Another very important principle, implicitly assumed since the early days of classical mechanics but brought into the scientific debate only in the 1930s, is the *principle of separability*: given two non-interacting physical systems $\mathcal{S}_1$ and $\mathcal{S}_2$, all their physical properties are separately determined. Stated in other terms, the outcome of a measurement on $\mathcal{S}_1$ cannot depend on a measurement performed on $\mathcal{S}_2$.

We are now in the position to define what a *state* in classical mechanics is. Let us first consider for the sake of simplicity a particle moving in one dimension. Its initial state is well defined by the position x_0 and momentum p_0 of the particle at time t_0. The knowledge of the equations of motion of the particle would then allow us to infer the position $x(t)$ and the momentum $p(t)$ of the particle at all times t.

It is straightforward to generalize this definition to systems with n degrees of freedom. For such a system we distinguish a coordinate configuration space $\{q_1, q_2, \ldots, q_n\} \in \mathbb{R}^n$ and a momentum configuration space $\{p_1, p_2, \ldots, p_n\} \in \mathbb{R}^n$, where the q_j's ($j = 1, \ldots, n$) are the generalized coordinates and the p_j's ($j = 1, \ldots, n$) the generalized momenta. On the other hand, the phase space Γ is the set $\{q_1, q_2, \ldots, q_n; p_1, p_2, \ldots, p_n\} \in \mathbb{R}^{2n}$. The state of a system with n degrees of freedom is then represented by a point in the $2n$-dimensional phase space Γ.

Let us consider what happens by making use of the Lagrangian approach. Here, the equations of motion can be derived from the knowledge of a Lagrangian function. Given a generalized coordinate q_j, we define its *canonically conjugate variable* or *generalized momentum* p_j as the quantity

$$p_j = \frac{\partial}{\partial \dot{q}_j} \mathrm{L}(q_1, \ldots, q_n; \dot{q}_1, \ldots, \dot{q}_n), \tag{1.2}$$

where the $\dot{q}_k$ are the generalized velocities. In the simplest case (position-independent kinetic energy and velocity-independent potential) we have

$$\mathrm{L}(q_1, \ldots, q_n, \dot{q}_1, \ldots : \dot{q}_n) = T(\dot{q}_1, \ldots, \dot{q}_n) - V(q_1, \ldots, q_n), \tag{1.3}$$

where L is the *Lagrangian function* and T and V are the kinetic and potential energy, respectively. The kinetic energy is a function of the generalized velocities $\dot{q}_j$ $(j = 1, \ldots, n)$ and may also be written as

$$T = \sum_j \frac{p_j^2}{2m_j}, \tag{1.4}$$

i.e. as a function of the generalized momenta p_j $(j = 1, \ldots, n)$, where m_j is the mass associated with the j-th degree of freedom.

In an alternative approach, a classical system is defined by the function

$$H = T(p_1, p_2, \ldots, p_n) + V(q_1, q_2, \ldots, q_n), \tag{1.5}$$

which is known as the Hamiltonian or the energy function, simply given by the sum of kinetic and potential energy. Differently from the Lagrangian function, H is directly observable because it represents the energy of the system. The relationship between Lagrangian and Hamiltonian functions is given by

$$H = \sum_j \dot{q}_j p_j - \mathrm{L}(q_1, \ldots, q_n, \dot{q}_1, \ldots : \dot{q}_n) \tag{1.6}$$

in conjunction with (1.2).

For the sake of simplicity we have assumed that the Lagrangian and the Hamiltonian functions are not explicitly time-dependent. The coordinate q_k and momentum p_k, together with their time derivatives $\dot{q}_k$, $\dot{p}_k$, are linked – through the Hamiltonian – by the Hamilton canonical equations of motion

$$\dot{q}_k = \frac{\partial H}{\partial p_k}, \quad \dot{p}_k = -\frac{\partial H}{\partial q_k}, \tag{1.7}$$

which can also be written in terms of the Poisson brackets as

$$\dot{q}_k = \{q_k, H\}, \quad \dot{p}_k = \{p_k, H\}. \tag{1.8}$$

The Poisson brackets for two arbitrary functions f and g are defined as

$$\{f,g\} = \sum_j \left(\frac{\partial f}{\partial q_j}\frac{\partial g}{\partial p_j} - \frac{\partial f}{\partial p_j}\frac{\partial g}{\partial q_j} \right), \tag{1.9}$$

and have the following properties:

$$\{f,g\} = -\{g,f\}, \tag{1.10a}$$

$$\{f,C\} = 0, \tag{1.10b}$$

$$\{Cf + C'g, h\} = C\{f,h\} + C'\{g,h\}, \tag{1.10c}$$

$$0 = \{f,\{g,h\}\} + \{g,\{h,f\}\} + \{h,\{f,g\}\}, \tag{1.10d}$$

$$\frac{\partial}{\partial t}\{f,g\} = \left\{\frac{\partial f}{\partial t}, g\right\} + \left\{f, \frac{\partial g}{\partial t}\right\}, \tag{1.10e}$$

where C, C' are constants and h is a third function. Equation (1.10d) is known as the Jacobi identity. The advantage of this notation is that, for any function f of q and p, we can write

$$\frac{d}{dt} f = \{f, H\}. \tag{1.11}$$

It is easy to see that Newton's second law can be derived from Hamilton's equations. In fact, from Eq. (1.8) we have

$$\dot{q}_k = \{q_k, H\} = \frac{p_k}{m_k}, \tag{1.12a}$$

$$\dot{p}_k = \{p_k, H\} = -\frac{\partial V}{\partial q_k}. \tag{1.12b}$$

From Eq. (1.12a) one obtains $p_k = m_k \dot{q}_k$ (the definition of generalized momentum), which, substituted into Eq. (1.12b), gives

$$m_k \ddot{q}_k = -\frac{\partial V}{\partial q_k}. \tag{1.13}$$

Since $F_k = -\partial V/\partial q_k$ is the generalized force relative to the k-th degree of freedom, Eq. (1.13) can be regarded as Newton's second law. As a consequence, Newton's second law can be written in terms of a first-order differential equation (as anticipated above). However, in this case we need both the knowledge of position and of momentum for describing a system.

In classical mechanics the equations of motion may also be determined by imposing that the action

$$\mathrm{S} = \int_{t_1}^{t_2} dt \mathrm{L}(q_1, \ldots, q_n, \dot{q}_1, \ldots, \dot{q}_n) \tag{1.14}$$

has an extreme value. This is known as the *Principle of least action* or Maupertuis–Hamilton principle.

The application of this principle yields the *Lagrange equations*

$$\frac{d}{dt}\left(\frac{\partial \mathrm{L}}{\partial \dot{q}_k}\right) - \frac{\partial \mathrm{L}}{\partial q_k} = 0, \tag{1.15}$$

which, as Hamilton equations, are equivalent to Newton's second law. In fact, we have

$$\frac{\partial L}{\partial \dot{q}_k} = m_k \dot{q}_k \quad \text{and} \quad \frac{\partial L}{\partial q_j} = -\frac{\partial V}{\partial q_k}, \tag{1.16}$$

from which it again follows that

$$\frac{d}{dt}(m_k \dot{q}_k) = -\frac{\partial V}{\partial q_k}. \tag{1.17}$$

For this reason, the Lagrange and Hamilton equations are equivalent. The main difference is that the former is a system of n second-order equations in the generalized coordinates, whereas Hamilton equations constitute a system of $2n$ first-order equations in the generalized coordinates q_k and momenta p_k.

From the discussion above it turns out that any state in classical mechanics can be represented by a point in the phase space, i.e. it is fully determined given the values of position and momentum (see also Subsec. 2.3.3) – when these values cannot be determined with arbitrary precision we have to turn to probabilities. As a consequence, in a probabilistic approach the system is described by a distribution of points in the phase space, whose density ρ at a certain point $(q_1, \ldots, q_n; p_1, \ldots, p_n)$ measures the probability of finding the system in the state defined by that point. It follows that ρ is a real and positive quantity for which

$$\int d^n q \int d^n p \rho(q_1, \ldots, q_n; p_1, \ldots, p_n) = 1, \tag{1.18}$$

i.e. the probability of finding the system in the entire phase space Γ is equal to one. The density ρ allows us to calculate, at any given time, the mean value of any given physical quantity F, i.e. of a function $F(q_1, \ldots, q_n; p_1, \ldots, p_n)$ of the canonical variables thanks to the relation

$$\bar{F}(\{q\}, \{p\}) = \int d^n q \int d^n p \rho(\{q\}, \{p\}) F(\{q\}, \{p\}), \tag{1.19}$$

where $\{q\}$ and $\{p\}$) stand for $(q_1, \ldots, q_n)$ and $(p_1, \ldots, p_n)$, respectively.

The dynamics of a statistical ensemble of classical systems is subjected to the *Liouville equation* (or continuity equation). Let us denote with $\rho(q, p; t)$ the density of representative points that at time t are contained in the infinitesimal volume element $d^n p \, d^n q$ in Γ around q and p. Then it is possible to show that we have

$$\frac{d\rho}{dt} = \{\rho, H\} + \frac{\partial \rho}{\partial t} = 0 \tag{1.20}$$

or

$$\frac{\partial \rho}{\partial t} = \{H, \rho\}. \tag{1.21}$$

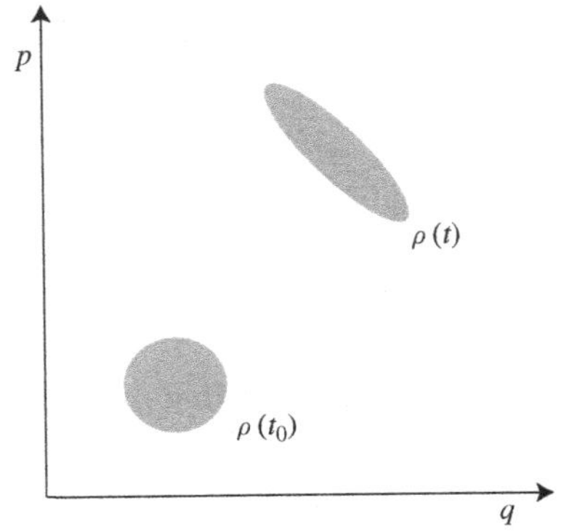

Figure 1.1 Graphical representation of the Liouville theorem. The area that defines the density of representative points must remain the same even if its shape may change with time.

Then, the Liouville theorem states that the density of representative points in the phase space Γ is constant (see Fig. 1.1).

From what we have seen above, we can finally define the basic feature of a *state* that we would have in *any* mechanics: it is the collection of all the properties of a system that can be simultaneously known – an issue that will be discussed extensively later (see also Sec. 15.5). In classical mechanics, for the principle of perfect determination, such a collection is also *complete*. This means that, according to this principle, a definite value is assigned to every physically sensible variable. In quantum mechanics, as we shall see in the following, this is not the case.

1.2 An interferometry experiment and its consequences

In this section we aim to draw, with the help of an ideal experimental setup, a series of consequences that will allow us to introduce some basic features of quantum mechanics. To some extent, this setup will become the guiding tool for many of the discussions that will follow in the present and next chapters. This experimental setup is essentially an optical interferometer. We therefore wish to discuss first the basic features of the photon – the quantum of light.

1.2.1 The quantum of light and the photoelectric effect

The hypothesis of the existence of the quantum of light was introduced by Albert Einstein in 1905 starting from Planck's solution to the black-body problem. In this way, he was able to explain the photoelectric effect, i.e. the emission of electrons by a metal surface

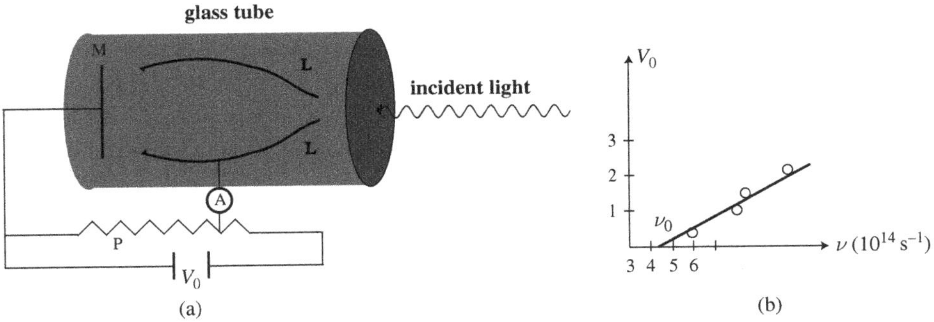

Figure 1.2 (a) Schematic representation of the experimental setup for detecting the photoelectric effect. The two layers L, whose potential difference with respect to the metal surface M may be changed through the potentiometer P, collect the electrons emitted by M, which in turn is illuminated by the incident light. The resulting electric current is measured by the amperometer A. The maximum kinetic energy of the emitted electrons is measured by the (inverted) potential V_0 necessary to make the current vanish. This quantity turns out to be independent of the intensity of the incident light. (b) The experimental results are shown in terms of the behavior of V_0 as a function of the frequency ν of the incident light.

when it is illuminated by light.[2] In fact, in classical physics light is treated as a wave and as such it is delocalized. It turns out that this classical picture is unable to account for the photoelectric effect: for a wave, a very long period of time would be needed in order to deliver the energy required for an atom to emit an electron. However, it is experimentally known that the effect is almost instantaneous. This is understandable only if one admits that light is made up of localized energy packets. If we denote by T_e the kinetic energy of the emitted electron and by U its binding energy (i.e. the minimum energy which is required to extract the electron from the metal), then they are related to the energy of the photon responsible for the photoelectric effect by the following expression (see Fig. 1.2):

$$E = T_e + U. \tag{1.22}$$

According to Einstein's proposal, there is a relation between the energy of the photon and the frequency ν of the electromagnetic radiation that is given by

$$E = h\nu, \tag{1.23}$$

where

$$h = 6.626069 \times 10^{-34}\,\mathrm{J\,s} \tag{1.24}$$

is the Planck constant. First, it is important to note that this phenomenon has a threshold: for photons with frequencies smaller than $\nu_0 = U/h$, the photoelectric effect is not

[2] See [Einstein 1905]. For a historical reconstruction see [*Mehra/Rechenberg* 1982–2001, I, 72–83]. Actually the term "photon" is due to G. N. Lewis [Lewis 1926].

observed at all, no matter how great the intensity of the light beam. However, above threshold the number of emitted electrons is proportional to the intensity of radiation. Second, the kinetic energy of the outgoing electrons is proportional to the frequency of the electromagnetic radiation. This relationship is surprising because classically the energy of a wave is proportional to the intensity, i.e. to the square of its amplitude and does not depend on the frequency.

It results then that the energy of photons occurs in quantized amounts. The quantization of energy (of the matter) was proposed for the first time by Planck in 1900 as a solution to the black-body radiation problem (as we shall see in Subsec. 1.5.1). This assumption is traditionally known as the *quantum postulate* and, after Einstein's contribution, can be reformulated as: the energy of an electromagnetic radiation with frequency ν can only assume discrete values, i.e.

$$E = nh\nu, \tag{1.25}$$

where $n = 1, 2, \ldots$. As we shall see in Subsec. 1.5.4 and in Ch. 11, this assumption was also applied by Niels Bohr to the atomic model.

Being the (energy) quantum of light, the photon can be absorbed and emitted by single atoms. As a consequence, photons can be detected by certain apparata (called *photodetectors*) in well defined positions exactly as it happens for particles. It is worth mentioning that in optimal conditions a single photoreceptor (rod) of a human eye is able to detect a single photon[3] and therefore to function as a photodetector (even though with a small efficiency).[4]

1.2.2 The Mach–Zehnder interferometer

Let us now describe an experiment from which some basic aspects of quantum mechanics can be inferred. The set up is shown in Fig. 1.3 and is known as a Mach–Zehnder *interferometer*. Let us first describe it using classical optics. It essentially consists of two beam splitters, i.e. two half-silvered mirrors which partly reflect and partly transmit an input light beam, two mirrors, and two photodetectors. A light beam coming from the source is split by the first beam splitter into the upper and lower paths. These are reflected at the mirrors and recombined at the second beam splitter before the photodetectors D1 and D2, which we assume to be ideal, i.e. with 100% efficiency. In the upper path a phase shifter is inserted in order to create a relative phase difference ϕ between the two component light beams. A phase shift which is a multiple of 2π brings the situation back to the original one, while a phase shift $\phi = \pi$ corresponds to the complete out-of-phase situation. At BS2 the two beams interfere and such interference may be destructive ($\phi = \pi$) or constructive

[3] See [*Hubel* 1988].

[4] Since photons travel at a relativistic speed, one may be surprised that they are introduced in a textbook about non-relativistic quantum mechanics. However, for our purposes, photons are very useful tools and we do not need to consider their relativistic nature.

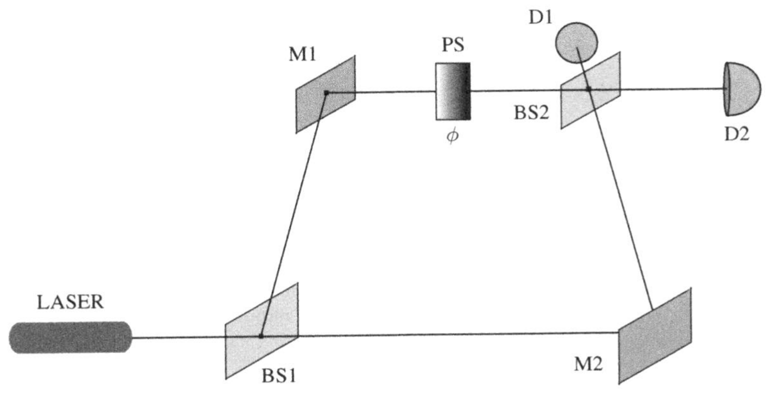

Figure 1.3 Schematic setup of the Mach–Zehnder interferometer (lateral-downward view). The source beam coming from the laser is split at the first beam splitter BS1. After reflections at the two mirrors M1 and M2, the upper and lower paths are recombined at the second beam splitter BS2 and then detected at the photodetectors D1 and D2. PS denotes a phase shifter which causes a relative shift ϕ of the upper beam.

($\phi = 0$). For example, destructive interference at D2 means that the observed intensity at such photodetector is equal to zero (*dark output*). This in turn means that D1 will certainly click (constructive interference). The transmission and reflection coefficients T and R of the beam splitters can vary between 0 and 1, with $R^2 + T^2 = 1$. When $T = R = 1/\sqrt{2}$, we have a 50%–50% beam splitter. All the devices present in this setup are linear, i.e. such that the output is proportional to the input.

Up to now the description is purely classical, and the light has wave-like properties – for instance, a phase. Therefore, having already considered the photoelectric effect, we see that light may display both wave-like and corpuscular features. We face here a new and surprising situation that appears even paradoxical from a classical viewpoint. In the next subsections we shall try to shed some "light" on this state of affairs and draw the necessary consequences.

Box 1.1 **Interferometry**

Interferometry is a widely used technique for detecting "waves" of different nature. There are many different forms of interferometry depending on the nature of the "objects" to be detected and on the configurations of the mirrors. One of the first interferometers was that of Michelson and Morley who used it to demonstrate the invariance of the speed of light (see Fig. 1.4). The Michelson–Morley experiment was performed in 1887 at what is now Case Western Reserve University, and is considered to be the first strong evidence against the theory of a luminiferous aether. Figure 1.5 shows an interferometer for photons that is useful for indirectly detecting gravitational waves.

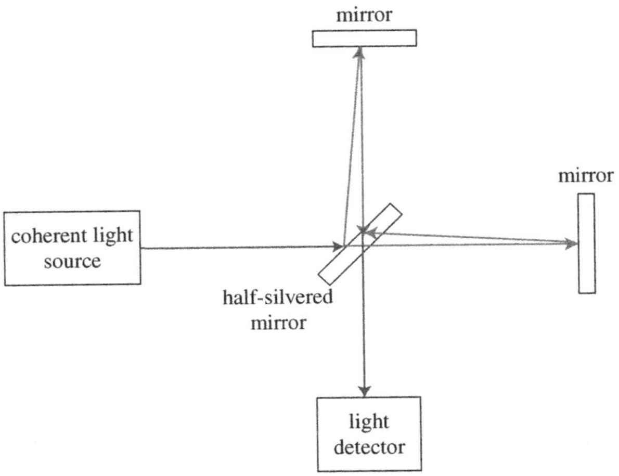

Figure 1.4 The Michelson–Morley interferometer. A single source of monochromatic light is sent through a half-silvered mirror that is used to split it into two beams travelling at right angles to one another (top view). After leaving the splitter, the beams travel out to the ends of long arms where they are reflected back into the middle on small mirrors. They are then recombined on the far side of the splitter in an eyepiece, producing a pattern of constructive and destructive interference based on the length of the arms. Any slight change in the amount of time the beams spent in transit would then be observed as a shift in the positions of the interference fringes. If the aether were stationary relative to the sun, then the Earth's motion would produce a shift of about 0.04 fringes.

1.2.3 First consequence: superposition principle

Let us now imagine what happens when a single photon at a time (i.e. the time interval between the arrival of two successive photons is much larger than the time resolution of the detector) is sent through the Mach–Zehnder interferometer. The number of photons per second can easily be calculated by knowing the intensity of light and the energy of the photons. It is an experimental fact that at each time a single photon is detected either at D1 or at D2, and never at both detectors. However, after $N \gg 1$ runs, which are required in order to obtain a good statistics, we experimentally observe that the detector D1 will click $N(1 - \cos\phi)/2$ times and detector D2 $N(1 + \cos\phi)/2$ times.[5] Again, if $\phi = \pi$, D2 does not click. Repeating the same experiment for a large number of times with different values of ϕ, we would obtain the plots shown in Fig. 1.6. This behavior is typical of an interference phenomenon. Since at most one photon at a time is present within the apparatus, then one can speak of *self-interference* of the photon.[6]

Self-interference has been experimentally verified for the first time by Pfleegor and Mandel.[7] Successively, further confirmations have come, among many others, from the experiments performed by Grangier, Roger, and Aspect[8] and by Franson and Potocki.[9]

[5] For a formal derivation of these formulas in term of probabilities see Subsec. 2.3.4.
[6] This concept was introduced for the first time by Dirac [*Dirac* 1930, 9].
[7] See [Pfleegor/Mandel 1967].
[8] See [Grangier *et al.* 1986a, Grangier *et al.* 1986b].
[9] See [Franson/Potocki 1988].

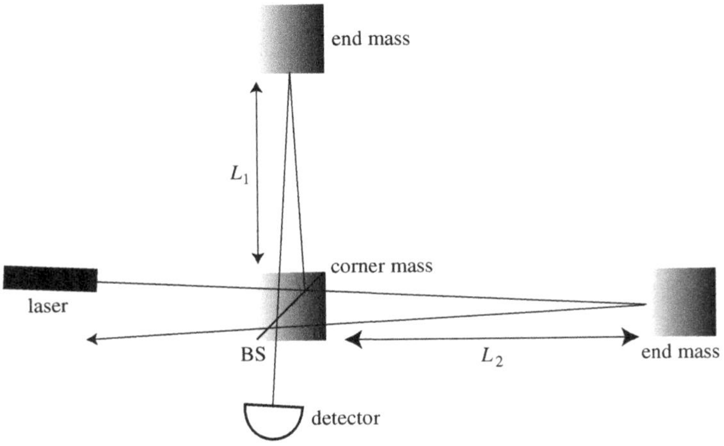

Figure 1.5 A Michelson–Morley-type interferometer for detecting gravitational waves (top view). Three masses hang by wires from the overhead supports at the corner and ends of the interferometer. When the first crest of a gravitational wave enters the laboratory, its tidal forces (gravitational forces producing stretching along the direction of a falling body and squeezing along the orthogonal direction) should stretch the masses apart along the L_1 arm while squeezing them together along L_2. When the wave's first crest has passed and its first trough arrives, the directions of stretch and squeeze will be changed. By monitoring the difference $L_1 - L_2$, one may look for gravitational waves. This is provided by a laser beam which shines onto a symmetric BS on the corner mass. The two outgoing beams go down the two arms and bounce off mirrors at the end of the arms and then return to the BS. The beams will be combined and split so that one part of each beam goes back to the laser and another part goes toward the photodetector. When no gravitational wave is present, the contributions from the two beams interfere in such a way that all the light goes back to the laser. See also [*Thorne* 1994, 383–85].

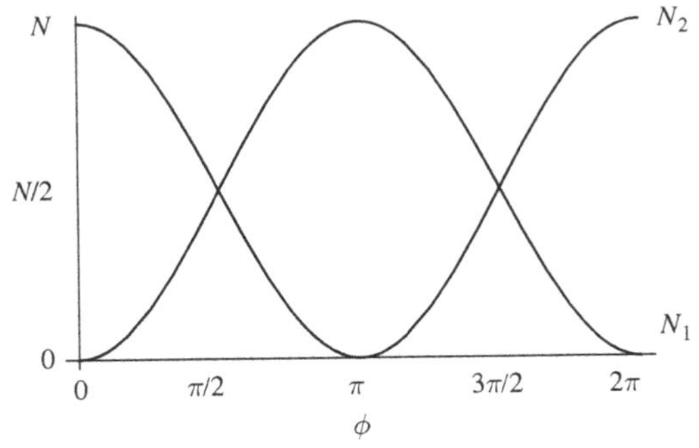

Figure 1.6 The two curves show the statistical results of photon counting at detectors D1 and D2. N_1 and N_2 denote the number of photons counted at detectors D1 and D2, respectively. It should be noted that, for each value of ϕ, $N_1(\phi) + N_2(\phi) = N$.

In Fig. 1.7 we report the experimental results obtained by Grangier, Roger and Aspect which confirm the expectations of Fig. 1.6.

Self-interference forces us to admit that the photon is not localized in either of the two arms. Now, let us suppose that we remove BS1. Then, the photon will certainly travel along the lower path (it is fully transmitted). We can label the "state" of the photon in such a case

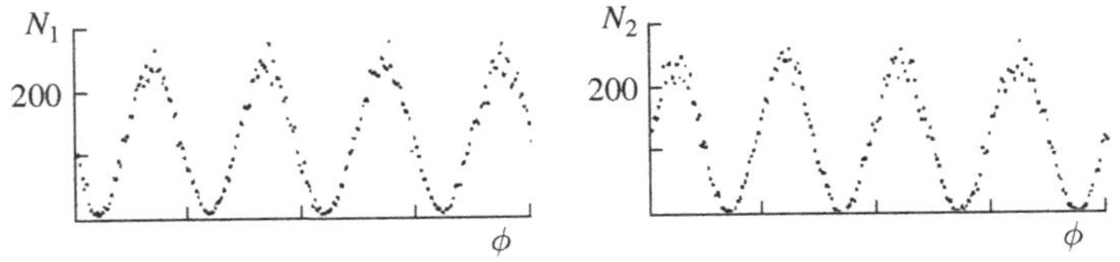

Figure 1.7 Results of the experiment performed by Grangier, Roger, and Aspect. As before, N_1 and N_2 denote the number of photons counted at detectors D1 and D2, respectively, as functions of the phase shift ϕ. Adapted from [Grangier *et al.* 1986b].

by the symbol $|\,\psi_l\rangle$, where the subscript l denotes the lower path. On the other hand, if BS1 is replaced by a 100% reflecting mirror, the photon will take with certainty the upper path and its state may then be denoted in this case by the symbol $|\,\psi_u\rangle$, where the subscript u refers to the upper path. As a consequence, when the partial reflecting mirror BS1 is put in its place, we are led to the conclusion that the state of the photon should be a combination (a *superposition*) of both the states $|\,\psi_l\rangle$ and $|\,\psi_u\rangle$ associated with the two arms of the interferometer. Therefore, we state in general terms the first theoretical suggestion of our ideal experiment:

Principle 1.1 (Superposition principle) *If a quantum system $\mathcal{S}$ can be in either of two states, then it can also be in any linear combination (superposition) of them.*

In the example above, the state $|\,\psi\rangle$ of the photon after BS1 can be expressed as the superposition

$$|\,\psi\rangle = c_u\,|\,\psi_u\rangle + c_l\,|\,\psi_l\rangle, \tag{1.26}$$

where c_u and c_l are some coefficients whose meaning will be explained below. Equation (1.26) represents the fact that it is not possible to assign a well-defined path to the photon: the state is a combination of the contribution of the two paths, i.e. it is delocalized. We should emphasize that this state of affairs is a clear violation of the classical principle of perfect determination (see Sec. 1.1) according to which the photon should be *either* in the upper path *or* in the lower path. In other words, Eq. (1.26) – describing a superposition *of states* – cannot be interpreted as a classical superposition of waves. In fact, in the latter case the components of the superposition would be *two spatial waves*, whereas in the case of Eq. (1.26) the components $|\,\psi_l\rangle$ and $|\,\psi_u\rangle$ represent possible states of the *same system*. Therefore, the wave-like properties of the photon discussed in Subsec. 1.2.2 cannot be taken in a classical sense.

We finally stress that the superposition principle is not just a consequence of our ignorance of the path *actually* taken by the photon, as we shall see in the following subsection.

1.2.4 Second consequence: complementarity principle

It is clear from the preceding analysis that, for $\phi = 0$ (π), detector D1 (D2) will never click. This *dark output* may even be used in order to detect the presence of an obstacle in one of the two paths without directly interacting with it. Let us place an object in the lower arm of the interferometer and set $\phi = 0$. Then the presence of this object will prevent the interference and allow, at least with some probability different from zero, that the photon will actually be detected at D1. This phenomenon is known as *interaction-free measurement* and shall be analyzed in greater details in Sec. 9.6. Turning the argument around, we can state that a detection event in D1 tells us with certainty that an object is in the lower arm and that the photon has taken the upper arm to the detector, i.e. it was localized in one of the two arms. It should be noted that in some cases the photon is not detected at all because it is absorbed by the object. Still in those cases when detector D1 clicks, we have learned about the position of the object without directly interacting with it, something which classically would be evidently not possible.[10] As we wrote above, it is evident that interference cannot be a manifestation of subjective ignorance. If this were the case, its presence or absence would not allow us to acquire objective information.

A second consequence of the experiments discussed above is that every time the photon is localized (i.e. we know with certainty that it has taken either the upper or the lower arm), interference is actually destroyed since it is a direct consequence of the superposition of $| \psi_u \rangle$ and $| \psi_l \rangle$. In other words, we cannot acquire information about the path actually taken by the photon without disturbing the detected interference and consequently change the state of the photon itself. This consequence can be generalized by the following principle:

Principle 1.2 (Complementarity principle) *Complete knowledge of the path is not compatible with the presence of interference.*

Principle 1.2 states that the knowledge of the path is complementary to the interference. It should be stressed, however, that the term "path" does not necessarily refer to the spatial path, and in the following chapters we shall consider several and different instances of this concept. Moreover, complementarity is here expressed in such a way that it does not necessarily imply a sharp yes/no alternative. As we shall see (in Subsec. 2.3.4), it rather consists of a trade-off between the partial gain of information and partial interference. In other words, an increase in the knowledge of the path occurs at the "expense" of the interference and *vice versa*, so that full localization (particle-like behavior) and full interference ("superposition-affected" or wave-like behavior) are only limiting cases of a continuous range of behaviors. Therefore, quantum systems can neither be considered as classical particles, nor as classical waves.

10 This has far-reaching consequences, if one thinks that the 1971 Nobel prizewinner in physics, Dennis Gabor, supported the wrong idea that one cannot acquire information about an object system if at least one photon does not interact with it [Kwiat *et al.* 1996].

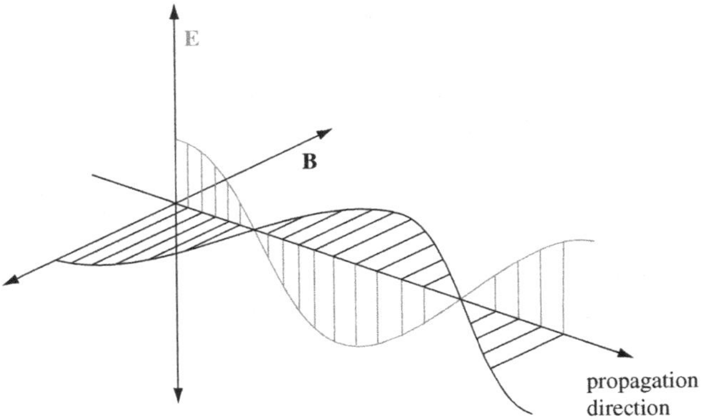

Figure 1.8 **Oscillation of electric (E, in grey) and magnetic (B, in black) fields associated with electromagnetic waves.**

The complementarity principle was first formulated by Niels Bohr at the Como Conference in 1927 and communicated to a large audience in an article in *Nature* [Bohr 1928].[11] Bohr intended it as a generalization of what was at that time known as the uncertainty principle. As we shall see (in Sec. 2.3), the uncertainty relation is another main point of departure from classical mechanics to the extent in which it states that it is impossible to jointly know with arbitrary precision a pair of conjugate variables, as for example the position and the momentum are.

1.3 State as vector

In the previous section we have seen that the state in quantum mechanics has "strange" characteristics, namely it violates at least the classical principle of perfect determination. We shall now consider an important property of light – polarization – as a tool for analyzing some further features of the quantum state.

1.3.1 Polarization states

Just as classical light can be, photons can be (linearly, circularly, or elliptically) polarized.[12] Classically, light polarization refers to the direction of oscillation of the electric (or magnetic) field associated with the electromagnetic wave (see Fig. 1.8). Normal light (e.g. that from a light bulb) is unpolarized, i.e. the electric field oscillates in all possible directions orthogonal to the propagation direction (see Fig. 1.9). However, if we insert a polarizing filter P1 on the light path, i.e. a filter which only allows the transmission of light polarized

[11] For a historical reconstruction see [*Mehra/Rechenberg* 1982–2001, VI, 163–99].

[12] See [*Jackson* 1962, 273–78].

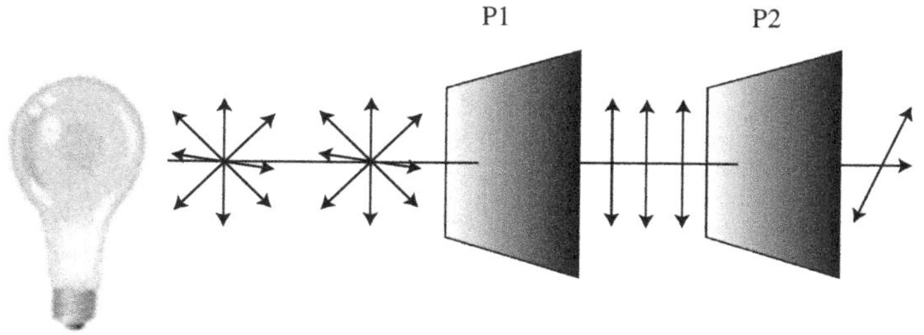

Figure 1.9 The light from a bulb is unpolarized – the electric field oscillates in all possible directions. After it passes through the vertical polarizing filter P1, only the vertically polarized component survives. A second filter (P2), with an orientation **b** which makes an angle θ with the orientation **a** of P1 (here vertical) may be inserted.

along a certain direction **a**, say vertical, we may produce a beam of polarized light but with a lower intensity than that before the filter. In fact, some photons will pass and some will not. We may then assign a state of vertical polarization $|v\rangle$ to each photon that has passed. Similarly, if P1 were rotated 90° about its axis, as output we would have photons in the state of horizontal polarization $|h\rangle$. The superposition principle implies that we should interpret $|v\rangle$ and $|h\rangle$ as vectors in a linear vector space $\mathcal{V}$.[13] In fact, if $|v\rangle$ and $|h\rangle$ are represented by two vectors in $\mathcal{V}$, then also $c_v|v\rangle + c_h|h\rangle$ (where c_v, c_h are some coefficients) will be in $\mathcal{V}$. Since the polarization directions lie in a plane, the vector space $\mathcal{V}$ has dimension 2. Therefore $|v\rangle$ and $|h\rangle$ can be thought of as an orthogonal basis in $\mathcal{V}$, and, as we shall see later, any photon polarization state can be written as a linear combination of them, as in Eq. (1.26). *State vectors*, i.e. vectors as representative of quantum states, were first introduced by Dirac [Dirac 1926a].[14]

1.3.2 Projectors and Hilbert space

Suppose now that we insert a second polarizing filter P2 with a different polarization axis **b** which makes an angle θ with the orientation **a** of P1 (see again Fig. 1.9). We know from classical physics that the transmitted beam will be polarized along the **b** direction and its intensity will be $I_2 = I_1 \cos^2\theta$, where I_1 is the intensity of the beam after P1. If **a** and **b** are orthogonal directions, i.e. $\theta = \pi/2$, obviously $I_2 = 0$.

Let us now observe what happens when we send single photons one at a time through the apparatus. As we said above, after P1 we only have photons polarized along **a**. Since the photon cannot be divided, the observer will see that – even though all the photons are in the same state – some photons will pass through P2 and some will not. We see here that we can only speak of a certain *probability* that a particular photon will pass through the apparatus. In fact, if we repeat the experiment many times and for different values of

[13] A complete list of properties of linear vector spaces can be found in a good handbook about linear spaces, for instance in [*Byron/Fuller* 1969–70, Ch. 3]. For a rigorous treatment of the problem see also Subsec. 8.4.3.
[14] See also [*Mehra/Rechenberg* 1982–2001, IV, 141–47].

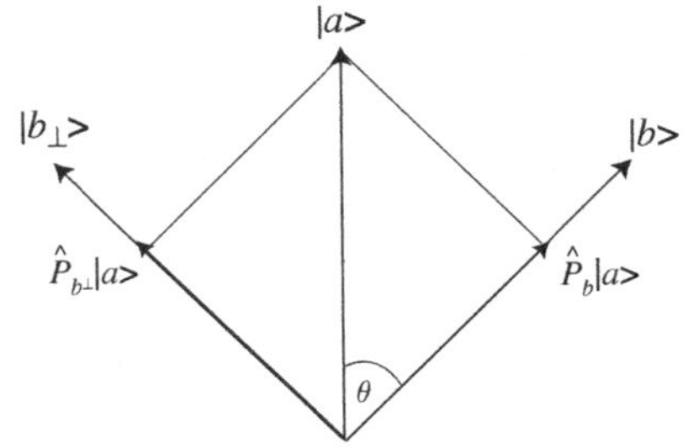

Figure 1.10 Decomposition of an arbitrary vector $|a\rangle$. The pair $|b\rangle$ and $|b_\perp\rangle$ form an orthonormal basis in the polarization vector space $\mathcal{V}$.

the angle θ, we shall be able to reconstruct the probability that a photon polarized along **a** passes through P2, which is given by $\wp(\theta) = \cos^2\theta$. Quantum mechanics has to account for such an experimental result. In general terms, let us denote by $|a\rangle$ the state vector corresponding to a photon polarized along direction **a** and similarly $|b\rangle$ for **b**. As it is clear from Fig. 1.10, $|a\rangle$ can be decomposed as

$$|a\rangle = \cos\theta\,|b\rangle + \sin\theta\,|b_\perp\rangle, \tag{1.27}$$

where $|b_\perp\rangle$ is the vector orthogonal to $|b\rangle$ (representing a photon in a state with a polarization orthogonal to the **b** direction). The space of states is here just a two-dimensional vector space, as previously suggested. Then, after P2 the state of the photon will be $|b\rangle$ with probability $\wp(\theta)$, whereas with probability $\sin^2\theta = 1 - \wp(\theta)$ the photon will be absorbed by P2. In other words, the state of the photon which has passed P2 is *projected* onto $|b\rangle$. Mathematically, the operator which performs such projection is called a *projector* and describes the selection performed by the filter P2.[15] The projector can be written as[16]

$$\hat{P}_b = |b\rangle\,\langle b|. \tag{1.28}$$

This way of writing projectors is justified by the fact that it is possible to associate to any vector $|b\rangle \in \mathcal{V}$ a vector $\langle b|$ which, if complex numbers are introduced, belongs to the isomorphic space $\mathcal{V}^*$. Following Dirac[17] we call the $|b\rangle$ vector *ket* and the $\langle b|$ vector *bra*: if $|b\rangle$ is a column vector, $\langle b|$ is the corresponding complex conjugate row vector. This terminology expresses the fact that the scalar product of two arbitrary vectors $|c\rangle$ and $|d\rangle$, which is often written as (d, c), can be written as $\langle d \mid c\rangle$ (*bra–ket* or *bracket*), where the two adjacent vertical lines have been contracted for brevity.

Then, by inspecting Fig. 1.10, we may rewrite Eq. (1.27) as

$$|a\rangle = \cos\theta \begin{pmatrix} 1 \\ 0 \end{pmatrix} + \sin\theta \begin{pmatrix} 0 \\ 1 \end{pmatrix} = \begin{pmatrix} \cos\theta \\ \sin\theta \end{pmatrix}. \tag{1.29}$$

15 Projectors were first introduced in quantum mechanics by von Neumann [*von Neumann* 1932, *von Neumann* 1955].

16 We shall always write operators with a hat to distinguish them from usual numbers, also called classical numbers or *c*-numbers.

17 See [*Dirac* 1930, 18].

Let us now generalize the previous example and consider the ket $| b\rangle$ as a vector pertaining to the n-dimensional vector space $\mathcal{V}^n$, as

$$| b\rangle = \begin{pmatrix} c_1 \\ c_2 \\ \dots \\ c_n \end{pmatrix}, \tag{1.30}$$

where $c_1, c_2, \dots, c_n$ are numbers: they turn out to be the coefficients of the expansion of $| b\rangle$ on the orthonormal set $\{| j\rangle\}$ in the n-dimensional vector space $\mathcal{V}^n$. This expansion (which is a superposition) can be written as

$$| b\rangle = \sum_{j=1}^{n} c_j \, | j\rangle. \tag{1.31}$$

We recall that a set of vectors is called *orthonormal* if and only if (iff):

- the scalar product between two different vectors in the set is equal to zero;
- the norm of each vector in the set is equal to one.

Moreover, any complet set of n vectors in a n-dimensional vector space is called a *basis*.

The bra corresponding to $| b\rangle$ is $\langle b |$:

$$\langle b | = \begin{pmatrix} c_1^* & c_2^* & \dots & c_n^* \end{pmatrix}, \tag{1.32}$$

where $c_1^*, c_2^*, \dots, c_n^*$ are the complex conjugates of the coefficients above.[18]

As we have said, the projector (1.28) is an operator, which in the finite n-dimensional case of our example, is mathematically expressed by the $n \times n$ matrix given by the row–times–column product

$$\begin{aligned} \hat{P}_b &= \begin{pmatrix} c_1 \\ c_2 \\ \dots \\ c_n \end{pmatrix} \begin{pmatrix} c_1^* & c_2^* & \dots & c_n^* \end{pmatrix} \\ &= \begin{bmatrix} |c_1|^2 & c_1 c_2^* & \dots & c_1 c_n^* \\ c_2 c_1^* & |c_2|^2 & \dots & c_2 c_n^* \\ \dots & \dots & \dots & \dots \\ c_n c_1^* & c_n c_2^* & \dots & |c_n|^2 \end{bmatrix}. \end{aligned} \tag{1.33}$$

Let us now consider some properties of kets and bras. The multiplication of kets and bras by a scalar, i.e. the multiplication of all components by the same number, is a linear operation, i.e.

$$\alpha \left(| b\rangle + | c\rangle\right) = \alpha \, | b\rangle + \alpha \, | c\rangle, \tag{1.34a}$$

$$\alpha \left(\langle b | + \langle c |\right) = \alpha \, \langle b | + \alpha \, \langle c |, \tag{1.34b}$$

18 There are several reasons why it is necessary to introduce complex numbers (pertaining to $\mathbb{C}$) in quantum mechanics. Let us mention here that, in order to account for interference, the coefficients in the superposition (1.26) need to be complex numbers.

where α is any (in general complex) number. Given the one-to-one correspondence between bra and ket vectors, one also has that the bra corresponding to the ket $|b\rangle + |b'\rangle$ is $\langle b| + \langle b'|$, while the bra corresponding to $\alpha\,|b\rangle$ is $\alpha^*\,\langle b|$, where α^* is the complex conjugate of α. It should be stressed that the ket $|a\rangle$ and the bra $\langle b|$ are of different nature (they belong to different spaces) and therefore cannot be added to each other.

Let us consider two (ket) vectors $|a\rangle$ and $|b\rangle$ and again the orthonormal basis $\{|\,j\rangle\}$ in an n-dimensional space $\mathcal{V}^n$. The kets $|a\rangle$ and $|b\rangle$ can both be expanded in the same basis (see also Eq. (1.31))

$$|a\rangle = \sum_{j=1}^{n} c'_j\,|\,j\rangle \quad \text{and} \quad |b\rangle = \sum_{j=1}^{n} c_j\,|\,j\rangle, \tag{1.35}$$

where the c'_j's and c_j's $\in \mathbb{C}$. Then, the scalar product $\langle b\,|\,a\rangle$ can be defined as

$$\begin{aligned}\langle b\,|\,a\rangle &= \begin{pmatrix} c_1^* & c_2^* & \dots & c_n^* \end{pmatrix} \begin{pmatrix} c'_1 \\ c'_2 \\ \dots \\ c'_n \end{pmatrix} \\ &= c_1^*c'_1 + c_2^*c'_2 + \ldots + c_n^*c'_n \\ &= \sum_{j=1}^{n} c_j^*c'_j. \end{aligned} \tag{1.36}$$

The following properties of the scalar product follow from the above definition:

$$(\langle b| + \langle c|)\,|a\rangle = \langle b\,|\,a\rangle + \langle c\,|\,a\rangle, \tag{1.37a}$$
$$\langle a|\,(|b\rangle + |c\rangle) = \langle a\,|\,b\rangle + \langle a\,|\,c\rangle, \tag{1.37b}$$
$$\langle b|\,(\alpha\,|a\rangle) = \alpha\,\langle b\,|\,a\rangle, \tag{1.37c}$$
$$(\alpha\,\langle b|)\,|a\rangle = \alpha\,\langle b\,|\,a\rangle, \tag{1.37d}$$
$$\langle b\,|\,a\rangle = \langle a\,|\,b\rangle^*, \tag{1.37e}$$
$$\langle a\,|\,a\rangle = 0 \quad \text{iff} \quad |a\rangle = 0. \tag{1.37f}$$

The definition and properties of the scalar product also allows us to introduce in a natural way the *norm* of a vector through the relation

$$\parallel a \parallel = (\langle a\,|\,a\rangle)^{\frac{1}{2}}. \tag{1.38}$$

Summarizing, the following operations are allowed:

$$|\cdot\rangle + |\cdot\rangle \quad \text{(sum of kets)}, \tag{1.39a}$$
$$\langle\cdot| + \langle\cdot| \quad \text{(sum of bras)}, \tag{1.39b}$$
$$\langle\cdot\,|\,\cdot\rangle \quad \text{(scalar product)}, \tag{1.39c}$$
$$|\cdot\rangle\,\langle\cdot| \quad \text{(external product)}, \tag{1.39d}$$

whereas, as we have said above, the sum of a bra and a ket $\langle\cdot| + |\cdot\rangle$ is not. Finally, we recall that the expression $|\cdot\rangle\,\langle\cdot|$ always denotes an operator.

A linear vector space endowed with a scalar product and which is complete and separable (having a countable dense subset, as it happens for the Euclidean space $\mathbb{R}^n$) is called a *Hilbert space* and is symbolized by $\mathcal{H}$.[19] Hilbert spaces are the natural framework for state vectors in quantum mechanics and can have a finite or infinite dimension. In the following, bras and kets describe state vectors in a Hilbert space, whereas, as we have seen, vectors in a configuration space are symbolized by bold letters. In a Hilbert space $\mathcal{H}$ it is always possible to find a complete orthonormal set of vectors, that is, an orthonormal *basis* on $\mathcal{H}$.

Let us now discuss the properties of projectors. Since in our example the projector (1.28) describes the projection onto the state $|b\rangle$ of the photon which has passed P2, then, if we let the projector $\hat{P}_b$ act on the state $|b\rangle$, we should obtain the state $|b\rangle$ again, i.e.

$$\hat{P}_b\,|b\rangle = |b\rangle\,\langle b\,|\,b\rangle = |b\rangle, \tag{1.40}$$

since $\langle b\,|\,b\rangle = 1$. It is also evident that such a projector has value 1 when a photon has actually passed P2 (i.e. has positively passed the test represented by P2, which means a "yes") and value 0 when a photon has been absorbed (i.e. has not passed the test represented by P2, which means a "no"). Then, projectors in a bidimensional space have a binary form and can be understood as propositions expressing a physical state of affairs ("the photon has passed P2") that can be evaluated as true (1) or false (0).

Being $\{|\,j\rangle\}$ an orthonormal basis in an n-dimensional Hilbert space, the corresponding projectors have the following properties:

$$\sum_j \hat{P}_j = \sum_j |\,j\rangle\,\langle j\,| = \hat{I}, \tag{1.41a}$$

$$\hat{P}_j\hat{P}_k = |\,j\rangle\,\langle j\,|\,k\rangle\,\langle k\,| = \delta_{jk}\hat{P}_k, \tag{1.41b}$$

where in Eq. (1.41a) the sum is extended over all possible j's, $\hat{I}$ is the identity operator, and in Eq. (1.41b) δ_{jk} is the Kronecker symbol:

$$\delta_{jk} = 0, \quad \forall j \neq k \quad \text{and} \quad \delta_{jj} = 1. \tag{1.42}$$

Property (1.41a) expresses the fact that a projection over the entire space does not affect the state (is not a selection). Property (1.41b) expresses the fact that the product of mutually exclusive selections is zero, and $\hat{P}_j^2 = \hat{P}_j$ (*idempotency*) (see Prob. 1.6).

It should be emphasized that projectors (and obviously any linear combination of projectors) are *linear operators*. A generic operator $\hat{O}$ acting on a Hilbert space is said to be linear when

$$\hat{O}\,(\alpha\,|a\rangle + \beta\,|b\rangle) = \alpha\hat{O}\,|a\rangle + \beta\hat{O}\,|b\rangle, \tag{1.43}$$

for all vectors $|a\rangle$ and $|b\rangle$, where α and β are (complex) numbers. It can be shown (see Prob. 1.8) that any operator $\hat{O}$ that is a linear combination of projectors maps $\mathcal{H}$ into $\mathcal{H}$, that is if a vector $|a\rangle \in \mathcal{H}$, then also $\hat{O}\,|a\rangle \in \mathcal{H}$. Projectors are only a first example of operators that act on quantum mechanical states. In the next chapter we shall see other kinds of operators.

[19] See [*Halmos* 1951, 16–17]. For a rigorous treatment see also Subsec. 8.4.3.

Let us now turn back again to the polarization state example (1.27). We are now in the position to see how the projector $\hat{P}_b$ defined by Eq. (1.28) acts on the state (1.27)

$$\begin{aligned}\hat{P}_b \,|\, a\rangle &= |\, b\rangle \langle b \,|\, (\cos\theta \,|\, b\rangle + \sin\theta \,|\, b_\perp\rangle) \\ &= \cos\theta \,|\, b\rangle \langle b \mid b\rangle + \sin\theta \,|\, b\rangle \langle b \mid b_\perp\rangle \\ &= \cos\theta \,|\, b\rangle, \end{aligned} \tag{1.44}$$

since $\langle b \mid b\rangle = 1$ and $\langle b \mid b_\perp\rangle = 0$. In fact, $\{|\, b\rangle, \,|\, b_\perp\rangle\}$ is an orthonormal basis on the polarization Hilbert space of dimension 2. The vector $\cos\theta \,|\, b\rangle$ has the same direction of vector $|\, b\rangle$ but a smaller length (its norm is equal to $|\cos\theta| < 1$) (see Fig. 1.10). However, in quantum mechanics vectors with the same direction and different lengths are taken to describe the same state. In other words, the same state is represented by the equivalence class of all vectors that can be constructed from one another by multiplication times a (complex) number. This is justified by the fact that quantum state vectors are usually taken to be normalized (meaning, as we shall see in the next subsection, that global phase factors have no relevance).

The reduction of the norm of the state vector after the application of $\hat{P}_b$ describes the fact that only a fraction $\cos^2\theta$ of the photons in state $|\, a\rangle$ has passed the test represented by P2. We also emphasize that the resulting state vector $\hat{P}_b \,|\, a\rangle$ should *not* be regarded as the state of the photon after interaction with the filter, while the information about the chance that the photon is absorbed by the filter is coded in the normalization of the state vector.

From Fig. 1.10 it is evident that $\cos\theta = \langle b \mid a\rangle$ and $\sin\theta = \langle b_\perp \mid a\rangle$. In other words, Eq. (1.27) may be rewritten as

$$\begin{aligned}|\, a\rangle &= \langle b \mid a\rangle \,|\, b\rangle + \langle b_\perp \mid a\rangle \,|\, b_\perp\rangle \\ &= |\, b\rangle \langle b \mid a\rangle + |\, b_\perp\rangle \langle b_\perp \mid a\rangle \\ &= \left(\hat{P}_b + \hat{P}_{b_\perp}\right) |\, a\rangle, \end{aligned} \tag{1.45}$$

which is in agreement with Eq. (1.41a). We have made use of the fact that

$$\langle b \mid a\rangle \,|\, b\rangle = |\, b\rangle \langle b \,|\, a\rangle \quad \text{and} \quad \langle b_\perp \mid a\rangle \,|\, b_\perp\rangle = |\, b_\perp\rangle \langle b_\perp \,|\, a\rangle, \tag{1.46}$$

since the scalar products $\langle b \mid a\rangle$ and $\langle b_\perp \mid a\rangle$ are c-numbers. We are able now to show in a pictorial way the superposition principle. Take again Eq. (1.27). It is evident that, by varying θ, $|\, a\rangle$ will span all the range from $|\, b\rangle$ (when $\theta = 0°$) to $|\, b_\perp\rangle$ (when $\theta = 90°$) (see Fig. 1.10). In other words, if a system can be in a state $|\, b\rangle$ and in a state $|\, b_\perp\rangle$, it can be in *any* linear combination of $|\, b\rangle$ and $|\, b_\perp\rangle$, where any possible superposition $|\, a\rangle$ is determined by the coefficients $c_b = \cos\theta = \langle b \mid a\rangle$ and $c_{b_\perp} = \sin\theta = \langle b_\perp \mid a\rangle$.

1.3.3 Poincaré sphere representation of quantum states

It is very interesting to note that there exists a useful graphic representation, known as the Poincaré sphere, of a generic quantum state of a two-level system. A *two-level system* is a system that possesses two orthogonal states, generally denoted by $|\uparrow\rangle$ and $|\downarrow\rangle$.

In general terms, the quantum state of a two-level system may be written as

$$|\, \psi\rangle = a \,|\uparrow\rangle + b \,|\downarrow\rangle. \tag{1.47}$$

Given $| \uparrow \rangle$ and $| \downarrow \rangle$, therefore, the state $| \psi \rangle$ is completely defined by a set of two complex coefficients $a = |a|e^{\imath\alpha}$ and $b = |b|e^{\imath\beta}$, i.e. by four real parameters. However, for normalization reasons (see Sec. 1.4) we must have $|a|^2 + |b|^2 = 1$, which reduces the number of linearly independent real parameters to 3. Moreover, the immaterial global phase factor of any quantum state allows us to reduce this number to 2. We shall return to this point,[20] but, intuitively speaking, only *relative* phase factors between different components of a superposition are relevant since they determine the interference behavior of the system (see Subsec. 1.2.3). Then, we may write $| \psi \rangle$ as

$$| \psi \rangle = e^{\imath\alpha} \left(|a| \, | \uparrow \rangle + |b| e^{\imath(\beta - \alpha)} \, | \downarrow \rangle \right), \tag{1.48}$$

which, up to the global phase α, reduces to

$$| \psi \rangle = |a| \, | \uparrow \rangle + |b| e^{\imath(\beta - \alpha)} \, | \downarrow \rangle . \tag{1.49}$$

Now, without any loss of generality, we may define

$$|a| = \cos \frac{\theta}{2}, \quad |b| = \sin \frac{\theta}{2}, \quad \beta - \alpha = \phi, \tag{1.50}$$

so that Eq. (1.47) can be finally rewritten as

$$| \psi \rangle = \cos \frac{\theta}{2} \, | \uparrow \rangle + e^{\imath\phi} \sin \frac{\theta}{2} \, | \downarrow \rangle , \tag{1.51}$$

which may be considered as a generalization of Eq. (1.27). It is now clear that, in order to completely define $| \psi \rangle$, it is sufficient to know the values of the two angles θ and ϕ, with $0 \leq \theta \leq \pi$ and $0 \leq \phi < 2\pi$. These, in turn, may be interpreted as the polar and azimuthal angles, respectively, of the spherical coordinates $\{r, \theta, \phi\}$. We may therefore make a one-to-one correspondence between points on a spherical surface and quantum states of a two-level system. In particular, given the normalization condition, we have that $r = 1$. The correspondence between quantum states and points on the spherical surface of unit radius is schematically illustrated in Fig. 1.11. In the polarization framework[21] (where $| \uparrow \rangle$ and $| \downarrow \rangle$ are replaced by $| v \rangle$ and $| h \rangle$, respectively), states of the type (1.51) may be considered as states of elliptical polarization.

For example, the vector pointing to the north pole ($\theta = 0$) will represent the state $| \uparrow \rangle$ while the vector pointing to the south pole ($\theta = \pi$) the state $| \downarrow \rangle$. On the other hand, all vectors lying on the equatorial plane (where $\theta = \pi/2$) represent all states of symmetric superposition of the basis states (states of circular or linear polarization) of the type

$$| \psi_{\text{sym}} \rangle = \frac{1}{\sqrt{2}} \left(| \uparrow \rangle + e^{\imath\phi} \, | \downarrow \rangle \right), \tag{1.52}$$

for any $\phi \in [0, 2\pi]$.

[20] See the discussion at the end of Subsec. 2.1.3.

[21] Originally, Poincaré introduced the homonimous sphere for representing polarization [*Poincaré* 1892].

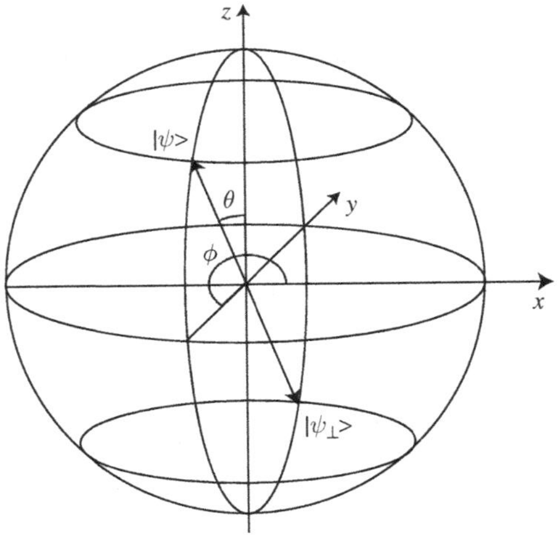

Figure 1.11 Poincaré sphere representation of states of a two-level quantum system. The state $|\psi\rangle$ is represented by a vector having an angle θ with the polar axis (here z) and a projection onto the x–y plane which forms an angle ϕ with the x-axis. The orthogonal state $|\psi_\perp\rangle$ is represented as the point of the surface of the sphere that is diametrically opposed to $|\psi\rangle$ and is defined by the angles $\theta' = \pi - \theta$ and $\phi' = \phi + \pi$. The states $|\psi\rangle$ and $|\psi_\perp\rangle$ are on two parallels at the same angular distance to their respective poles.

1.4 Quantum probability

In the polarization example of the previous section we have introduced the probability of some events. After the polarizing filter P1 (see Fig. 1.9) all photons have been prepared in the same state, namely $|a\rangle$. It can then appear quite strange, from a classical point of view, to find that they behave in different ways (some will pass P2 and some will be absorbed). This is a common state of affairs in quantum mechanics. A further evidence of this situation can be found, for instance, in the phenomenon of radioactive decay. Let us consider a piece of radioactive material. If at time t_0 we have N_0 non-decayed atoms, then at a successive time t the number of undecayed atoms will be $N_t = N_0 e^{-\gamma t}$, where $\gamma > 0$ is the decay constant that is characteristic of the particular material. However, it is not possible to predict which atom will decay at which time, even though all the atoms can be thought of as being in the same state. To the best of our knowledge, there is no experiment that can be performed in order to predict which atom will decay next and at what time. We can only speak of a certain probability that a particular atom will decay in a given time interval. As we shall see, the use of probability in quantum mechanics is not a consequence of subjective ignorance that could be reduced by some improvement of knowledge. Instead, it should be taken as an irreducible property of quantum systems. Thus, in contrast to classical mechanics, quantum mechanics has an *intrinsically* probabilistic character.

At this point, one might also think that quantum mechanics faces the same situation as thermodynamics does. However, the statistical character of thermodynamics is due to the large number of particles (atoms or molecules) present in a macroscopic piece of (fluid or solid) matter: it is due to the fact that it is practically (for our human means, technically) impossible to measure the position and the momentum of an ensemble of particles whose number is of the same order of Avogadro number $N_A = 6.02 \times 10^{23}$. Instead, in quantum mechanics, the probabilistic character is intrinsic to the behavior of *each* single particle.

We have seen that the coefficients of the two "alternatives" (photon passing through P2 in state $| b\rangle$ and absorbed photon in state $|b_\perp\rangle$) in Eq. (1.27) are $\cos\theta$ and $\sin\theta$, respectively. On the other hand, from the previous section we also know that the probabilities of the corresponding events are $\cos^2\theta$ and $\sin^2\theta$. Then, we see that there is a clear relationship between those coefficients and these probabilities:[22] these are the square of the coefficients. We may conclude that the coefficients $\cos\theta$ and $\sin\theta$ in Eq. (1.27) can be interpreted as *probability amplitudes* of the alternatives associated with the corresponding state vectors. By *probability amplitude* we mean *prima facie* a quantity whose square gives the probability of the associated events.

In general, however, the coefficients preceding kets and bras are complex numbers whereas probabilities must be real and non-negative numbers in the interval $[0, 1]$. In the general case, therefore, in order to obtain the probability of a certain event, one has to compute the square modulus of the amplitude associated with the corresponding measurement outcome. For instance, if the probability amplitudes for the photon in the state $| a\rangle$ to pass P2 or to be absorbed are $\langle b \mid a\rangle$ and $\langle b_\perp \mid a\rangle$ [see Eq. (1.45)], respectively, then the corresponding probabilities are $| \langle b \mid a\rangle |^2$ and $| \langle b_\perp \mid a\rangle |^2$. It is also evident that

$$\langle a \mid a\rangle = | \langle b \mid a\rangle |^2 + | \langle b_\perp \mid a\rangle |^2 = 1. \tag{1.53}$$

This amounts to the requirement that the sum of the probabilities of all disjoint events of a given set is equal to one (this is the well-known Kolmogorov's probability axiom). This is connected to the problem of *normalization* of the states in quantum mechanics (Subsec. 2.2.1). States that satisfy a condition of the type (1.53) are said to be *normalized*.

It is also important to note a further difference between probabilities in classical and quantum mechanics. To display these differences in a most effective way, let us go back to the Mach–Zehnder experiment that we have treated in Subsec. 1.2.2. From the discussion in Subsec. 1.2.3 we can deduce that the probabilities for a single photon to be detected at detectors D1 and D2 are

$$\wp_{u+l}(\mathrm{D1}) = \frac{(1-\cos\phi)}{2} \quad \text{and} \quad \wp_{u+l}(\mathrm{D2}) = \frac{(1+\cos\phi)}{2}, \tag{1.54}$$

respectively. As we have seen, this result is due to interference between two "alternatives": photon taking the upper path and photon taking the lower path. This justifies the notation above where the subscript $u + l$ means that the photon can take both paths. If we block the lower path by inserting a screen S between BS1 and M2 (see Fig. 1.12), then it is clear that the photon will be absorbed by the screen with probability $1/2$. With probability $1/2$,

[22] As pointed out by Born in [Born 1926, Born 1927a, Born 1927b]. For a history see [*Mehra/Rechenberg* 1982–2001, VI, 36–55].

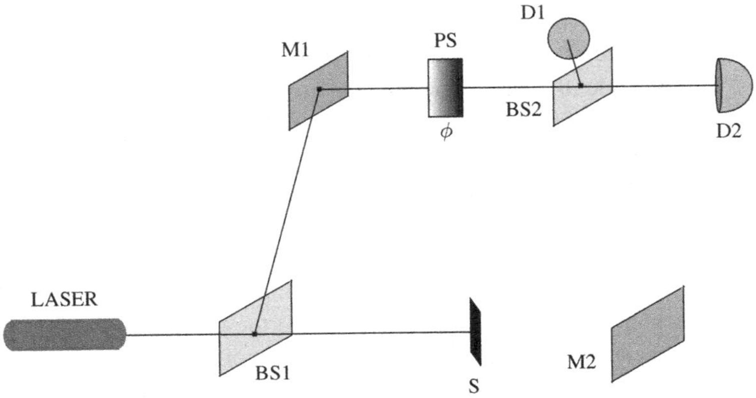

Figure 1.12 Mach–Zehnder interferometer with the lower path blocked by the screen S.

instead, the photon will take the upper path and will be reflected by M1. At the symmetric beam splitter BS2 it will have a 50% chance of going to detector D1 and a 50% chance of going to detector D2. The overall probabilities for a photon impinging on BS1 to be detected at D1 or D2 will be therefore $\wp_u(\mathrm{D1}) = 1/4$ and $\wp_u(\mathrm{D2}) = 1/4$, respectively. A similar analysis can be performed in the case where we insert the screen S in the upper path between BS1 and M1: again the probabilities for detection at D1 or D2 will be $\wp_l(\mathrm{D1}) = \wp_l(\mathrm{D2}) = 1/4$. We immediately see that the probability for detection at, say, detector D1 when both paths are open is in general *not* equal to the sum of the probabilities of being detected at D1 after taking the two paths separately, i.e.

$$\wp_{u+l}(\mathrm{D1}) \neq \wp_u(\mathrm{D1}) + \wp_l(\mathrm{D1}), \tag{1.55}$$

except in the cases $\phi = \pi/2$ and $\phi = 3\pi/2$. This is due to the fact that, when both paths are open, the state of the photon after BS1 is not a mere addition of the two alternatives but rather a quantum superposition of them. This contradicts the basic structure of classical probability. In classical probability theory,[23] given two events A and B, we have that

$$\wp(A + B) \leq \wp(A) + \wp(B). \tag{1.56}$$

In the example of Fig. 1.12 this inequality is violated for all values of $\phi \neq \{\pi/2, 3\pi/2\}$ either for D1 or for D2.

Let us compare this result with that obtained in the case where the experimental setup shown in Fig. 1.12 is replaced by its classical analogue. In this classical device, photons are replaced by bullets and the beam splitters by random mechanisms that send each bullet in one of the two paths, with equal probability over many runs. Then, if both paths are open, the probability of detection at both D1 and D2 is equal to 0.5, i.e. $\wp_{u+l}(D1) = \wp_{u+l}(D2) = 1/2$. On the other hand, if one of the two paths is blocked, we have $\wp_u(D1) = \wp_u(D2) = \wp_l(D1) = \wp_l(D2) = 1/4$. It clearly results that, in this

[23] See [*Gnedenko* 1969, 48–49].

classical example, Eq. (1.55) becomes an equality and therefore the requirement (1.56) is obviously satisfied.

The result (1.55) is strictly related to the fact that quantum probabilities are calculated as square moduli of the corresponding amplitudes and that, therefore, in quantum mechanics amplitudes and not the corresponding probabilities sum linearly. In particular, when more than one "alternative" (or "path") lead to the same measurement outcome, one has first to sum the amplitudes corresponding to the different "alternatives" and then to calculate its square modulus in order to obtain the probability of that measurement outcome.

1.5 The historical need of a new mechanics

In this section we wish to enumerate and briefly discuss the major problems that physicists had to face at the end of the nineteenth and the beginning of the twentieth centuries. As we shall see, there was at that time a number of experimental facts which simply could not be explained in the framework of classical physics. These experimental facts are the playground in which quantum mechanics was built.

1.5.1 The black-body radiation problem

We have seen in Subsec. 1.2.1 that Einstein's interpretation of the photoelectric effect forces us to assume that electromagnetic radiation is made out of quanta of energy $h\nu$ called photons.

Einstein took his starting point from Max Planck's work of 1900. The problem faced by Planck was the emission of a black body. Let us consider a hollow body with internal surface at constant and uniform temperature T. Electromagnetic waves are produced from the different elements $d\mathbf{S}$ of the internal surface $\mathbf{S}$. These waves are also absorbed by the different surface elements $d\mathbf{S}$. One might expect that this mutual energy exchange between all $d\mathbf{S}$ reach an equilibrium. Experimentally, this is exactly what happens. However, classical physics is not able to correctly predict the spectral properties of the black body. The *spectrum* of the black-body radiation is given by the function $f(\nu)$ such that $f(\nu)d\nu$ represents the energy of the electromagnetic field contained in the unit volume at a frequency between ν and $\nu + d\nu$. Then,

$$f(\nu) = \frac{\partial u(\nu, T)}{\partial \nu}, \tag{1.57}$$

where $u(\nu, T)$ is the energy density at temperature T and frequency ν. Computing the energy density by using the classical energy equipartition law, yields the well-known Rayleigh–Jeans formula

$$f(\nu) = \frac{8\pi}{c^3} k_B T \nu^2, \tag{1.58}$$

where $k_B = 1.3807 \times 10^{-23}$J/K is the so-called Boltzmann constant, c is the speed of light, and $k_B T$ is the equipartition energy associated with each oscillator of frequency ν. This

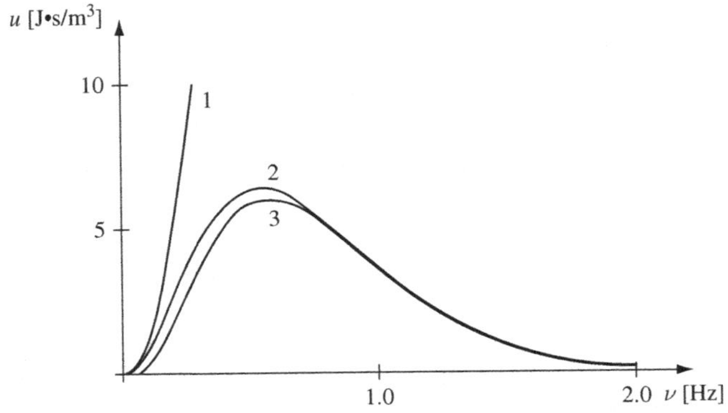

Figure 1.13 **Black-body radiation intensity corresponding to the formula of Rayleigh–Jeans (1), Planck (2), and Wien (3). Adapted from [*Bialynicki-B. et al.* 1992, 7].**

formula does not agree well with experimental data, and, above all, paradoxically predicts an infinite total intensity I_c of the emitted radiation

$$I_c = \frac{c}{4}\int_0^\infty d\nu f(\nu) = \frac{2\pi}{c^2}k_B T\int_0^\infty d\nu \nu^2 = \infty. \tag{1.59}$$

This situation is called *ultraviolet catastrophe* and is illustrated in Fig. 1.13. This is what goes under the name of the *black-body radiation problem.*

Planck proposed to consider the black-body internal surface as a collection of N linear harmonic resonators.[24] If

$$S_B^N = k_B \ln w_E \tag{1.60}$$

represents the Boltzmann entropy of the total system, and E_N its total energy, then the quantity w_E comes to represent here the number of different ways in which E_N may be distributed among the resonators. On the contrary, Planck treated [Planck 1900a, Planck 1900b] E_N as consisting of a finite number n_ϵ of discrete energy elements ϵ, each of them having a definite value for each frequency ν

$$E_N = n_\epsilon \epsilon(\nu). \tag{1.61}$$

If we indicate by $\bar{E}$ the average energy of the oscillators, we have

$$\frac{n_\epsilon}{N} = \frac{\bar{E}}{\epsilon}, \tag{1.62}$$

and, after some calculations (see Prob. 1.10), the entropy takes the final form

$$S_B^N = k_B N\left[\left(1+\frac{\bar{E}}{\epsilon}\right)\ln\left(1+\frac{\bar{E}}{\epsilon}\right) - \frac{\bar{E}}{\epsilon}\ln\frac{\bar{E}}{\epsilon}\right]. \tag{1.63}$$

[24] A more complete historical reconstruction of what follows can be found in several books [*Jammer* 1966, 7–16] [*Kuhn* 1978, 97–110] [*Mehra/Rechenberg* 1982–2001, I, 24–59].

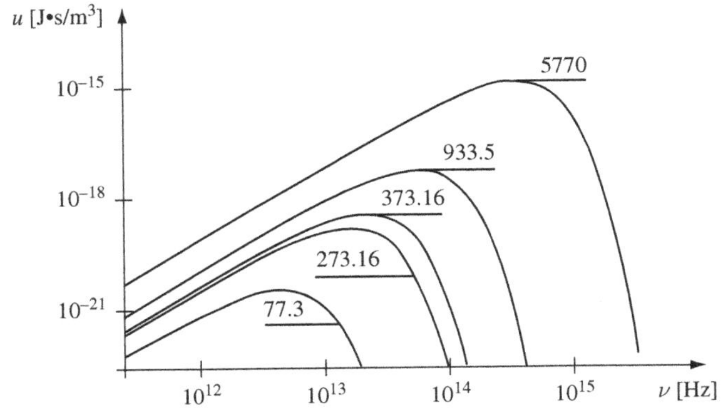

Figure 1.14 Planck's radiation curves in logarithmic scale for the increasing temperatures of liquid nitrogen, melting ice, boiling water, melting aluminium, and the solar surface. Adapted from [*Bialynicki-B. et al.* 1992, 8].

Since

$$\frac{\partial S_{\mathrm{B}}^{N}}{\partial U} = \frac{1}{T}, \tag{1.64}$$

where the internal energy $U = N\bar{E}$, then (see Fig. 1.14 and Prob. 1.11)

$$\bar{E} = \frac{\epsilon}{e^{\epsilon/k_{\mathrm{B}}T} - 1}, \tag{1.65}$$

where $\epsilon = h\nu$ (see Subsec. 1.2.1). This formula agrees very well with experimental data. By making use of the classical equations of motion it can be proved that the average of the black body can be obtained. By substituting the rhs of Eq. (1.65) in place of $k_{\mathrm{B}}T$ in the Rayleigh–Jeans formula (1.58) and by making use of the notation $\beta = (k_{\mathrm{B}}T)^{-1}$, we finally obtain the correct expression for the spectrum

$$f(\nu) = \frac{8\pi\nu^3}{c^3}\frac{h}{e^{h\nu\beta} - 1}. \tag{1.66}$$

As we see in Fig. 1.13, the Rayleigh–Jeans formula is correct at small frequencies but diverges at larger frequencies, whereas the Planck formula (1.66) reaches a maximum and then decreases as ν goes to infinity.

Before ending this subsection, a historical remark is in order. It is worth emphasizing that in Planck's view matter can be modelled as a collection of resonators, and in this sense its energy is quantized, although Planck never assumed that the energy of the oscillators is actually a multiple of ϵ.[25] He only pointed out that, as far as the computation of entropy is concerned, the quantization hypothesis gives the correct results. Moreover, according to Planck, the formula $\epsilon = h\nu$ only applies to matter quantization and is not at all a manifestation of light quantization. In later papers Planck made clear that the energy could be emitted by resonators in a quantized form, but is still absorbed in a continuous way.

[25] See [Parisi 2005b].

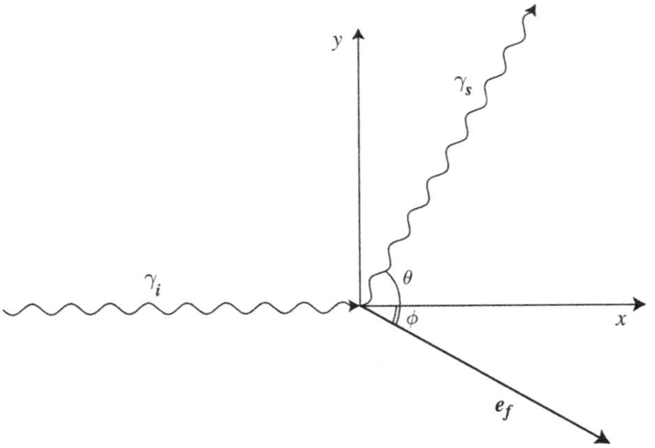

Figure 1.15 Schematic representation of the Compton effect. Here γ_i is the incident photon, impinging with momentum $h\nu_i/c$ on the electron located at the origin and initially at rest. The photon scattered at an angle θ and with momentum $h\nu_s/c$ is represented by γ_s, while the final electron with momentum mv at an angle ϕ is indicated as e_f.

1.5.2 Photoelectric and Compton effects

We have already seen (in Subsec. 1.2.1) that the photoelectric effect can be correctly interpreted if one admits that light is made of energy quanta. This was the first evidence of the quantization of the electromagnetic field. A further confirmation of the quantization of electromagnetic radiation was found by Arthur Compton, who investigated the scattering of x-rays after their collision with electrons [Compton 1923].[26] The wavelength of scattered x-rays λ_s was slightly longer than that of incident x-rays λ_i – this is the essence of the *Compton effect*. The change of wavelength is a function of the angle θ (see Fig. 1.15) at which the scattered radiation is observed according to the formula

$$\Delta\lambda = 2\lambda_c \sin^2 \frac{\theta}{2}, \tag{1.67}$$

where $\Delta\lambda = \lambda_s - \lambda_i$ and

$$\lambda_c = \frac{h}{mc} = 2.42 \times 10^{-12}\text{m} \tag{1.68}$$

is the so-called Compton wavelength of the electron. This discontinuous change cannot be explained in terms of the classical electromagnetic theory of light. On the contrary, it may be accounted for by assuming that photons of incident energy $E_i = h\nu_i$ and incident momentum $p_i = h/\lambda_i$ collide with the electrons of the target (which may be supposed to be at rest) and are successively deflected with reduced energy $E_s = h\nu_s$ and reduced momentum $p_s = h/\lambda_s$. This collision may be thought of as a two-step

[26] For a historical reconstruction see [*Mehra/Rechenberg* 1982–2001, I, 520–32].

process: an absorption of the photon by the electron followed by the successive emission of a photon of different energy. As a consequence, there is a transfer of energy $\Delta E = E_i - E_s$ and of momentum $\Delta \mathbf{p} = \mathbf{p}_i - \mathbf{p}_s$ to the electron so that the energy and momentum conservation laws are satisfied for the total system (photon + electron) in each collision.

Following Fig. 1.15 we have that

$$\frac{h\nu_i}{c} = \frac{h\nu_s}{c}\cos\theta + mv\cos\phi, \tag{1.69a}$$

$$\frac{h\nu_s}{c}\sin\theta = mv\sin\phi \tag{1.69b}$$

express momentum conservation along the x- and y-axes, respectively, while

$$h\nu_s + \frac{1}{2}mv^2 = h\nu_i \tag{1.70}$$

expresses energy conservation during the scattering process. From Eqs. (1.69)–(1.70) one obtains Eq. (1.67) (see Prob. 1.12).

It is worth mentioning that measuring the Compton wavelength of the electron yields the value of Planck constant h. when the speed of light and the mass of electron are known.

We would like to end this subsection with a historical remark. Einstein's interpretation of the photoelectric effect involving the corpuscular nature of light had not completely convinced the scientific community about the quantization of the electromagnetic field. In this respect, the Compton effect, where energy and momentum are conserved in each single collision, played the role of a definitive experimental evidence of radiation quantization and convinced even the most skeptical physicists.

1.5.3 Specific heat

At the beginning of the 19th century is was already experimentally known that the specific heat per mole of monatomic, diatomic, and multiatomic ideal gases is given by $\frac{3}{2}R, \frac{5}{2}R, 3R$, respectively, where $R = N_A k_B$ and N_A is the Avogadro number. According to the classical equipartition law, an equilibrium system has an average energy of $k_B T/2$ for each harmonic degree of freedom (i.e quadratic term appearing in its Hamiltonian). As a consequence, the specific heat per mole should be equal to $\nu R/2$, where ν is here the number of the degrees of freedom of the molecule. While this correctly explains the experimental value of the specific heat of monoatomic ideal gases, it cannot account for the above experimental values in the case of diatomic and multiatomic ideal gases. This disagreement between the classical equipartition prediction and the experimental value of the molar heat capacity cannot be explained by using a more complex model of the molecule, since adding more degrees of freedom can only increase the predicted specific heat, not decrease it. This discrepancy can be solved by taking into account that energy levels are quantized according to prediction of quantum mechanics [see Sec. 4.4 on the quantum harmonic oscillator].

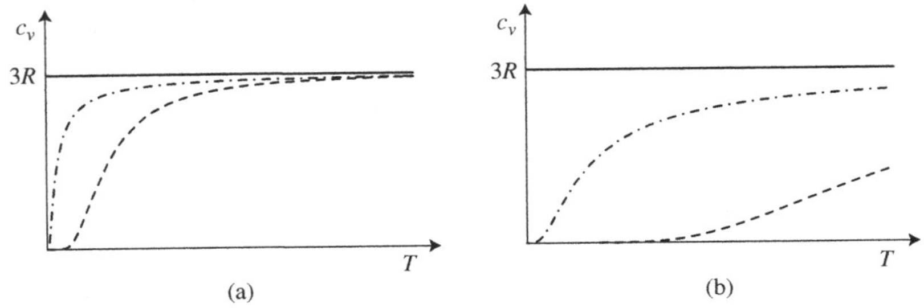

Figure 1.16 (a) Dulong–Petit's (full horizontal line $c_v = 3R$), Einstein's (dashed line), and Debye's (dot-dashed line) predictions for the specific heat of solids: all converge to the same limit $c_v = 3R$ for large temperatures. (b) Same as (a) but magnified around the origin (at low temperatures): Einstein's curve decreases exponentially as $T \to 0$, whereas Debye's curve correctly predicts the T^3 behavior.

In the case of solids, assuming small vibrations of the N atoms around their equilibrium positions, the classical energy equipartition law ensures that the internal energy U is equal to $3Nk_BT$, or $3RT$ per mole. Therefore, the specific heat per mole for all solids should be given by

$$c_v = \left(\frac{\partial U}{\partial T}\right)_V = 3R, \tag{1.71}$$

which is known as the Dulong–Petit law. Classical physics then predicts a specific heat which is constant at all temperatures, a fact which agrees well with experimental data at high temperatures. For low temperatures, however, it is experimentally found that the specific heat goes to zero as soon as $T \to 0$ (see Fig. 1.16). Also this behavior is totally incomprehensible in the classical context.

In 1906 Einstein[27] tried to solve this problem by assuming that the average energy of the oscillating (non-interacting) atoms is given by Planck's Eq. (1.65) so that the total internal energy may be written as

$$U = \frac{3Nh\nu}{e^{h\nu\beta} - 1}, \tag{1.72}$$

from which he derived the specific heat per mole as

$$c_v = \frac{3R\,(h\nu\beta)^2\, e^{\frac{h\nu}{k_BT}}}{\left(e^{h\nu\beta} - 1\right)^2}. \tag{1.73}$$

According to this equation, c_v decreases with T at low temperatures but is equal to $3R$ at high temperature. However, Eq. (1.73) decays exponentially for $T \to 0$, which is faster than the observed behavior. In the 1912 Debye showed that the correct behavior of the specific heat can be recovered both at small and at large temperatures when one takes into account the simultaneous oscillation of interacting atoms (see Fig. 1.16), and uses information about the sound velocity. The computation of the energy of the block radiation and of the internal energy of the solid can be done in a parallel way.

[27] For a historical reconstruction see [*Mehra/Rechenberg* 1982–2001, I, 113–44].

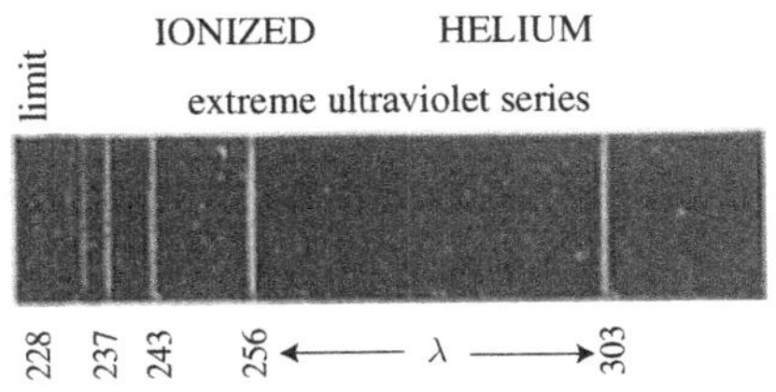

Figure 1.17 Lyman series for ionized helium. Adapted from [*White* 1934, 32].

1.5.4 Atomic spectra and stability of atoms

It has been known since the second half of the nineteenth century that the spectrum of the electromagnetic radiation emitted by diluted gases is not a continuous function of the wavelength. Instead, the intensity of the emitted radiation is a collection of sharp peaks which are located at wavelengths characteristic of the different elements in the periodic table. Moreover, no radiation is emitted at wavelengths in between peaks – an example of such a spectrum is shown in Fig. 1.17. This fact is difficult to understand in the framework of classical physics, since it would mean that electrons orbit around the nucleus with selected frequencies, and tends to force us to admit that electrons in an atom possess discrete "stationary" energy levels. We shall deal with atomic models in Ch. 11. It is worth mentioning that a similar phenomenon (i.e. the presence of spectral lines) shows up also in the case of absorption spectra.

Furthermore, it was known from the very beginning that Rutherford's planetary atom-model (1911) was affected by the great problem of instability. In fact, due to the centripetal acceleration, a negatively charged particle (the electron) revolving around a positively charged nucleus should continuously radiate energy and rapidly fall, following a spiral trajectory, onto the nucleus. Bohr searched for a solution to the stability problem in extending Planck's postulate. He then postulated that, for an atomic system, there exists a discrete set of permissible (stationary) stable orbits characterized by energy values $E_1, E_2, \ldots$, and that these are governed by the ordinary laws of classical mechanics [Bohr 1913, 874].[28] As long as the electron remains in one of these orbits, no energy is radiated. The energy of stationary states can be obtained from the quantization rule and the mechanical equilibrium condition (that the electromagnetic force is equal to the centripetal force), and, for the hydrogen atom, is given by the formula

$$E_n = -\frac{2\pi^2 me^4}{h^2 n^2} \quad (n = 1, 2, 3, \ldots). \tag{1.74}$$

The energy is emitted (or absorbed) during the transition from one stationary state to the other in a discontinuous way – an electron is said to jump from one level to the other – so that the amount of energy emitted (absorbed) is quantized in accordance with Planck's Law (1.25) and Sommerfeld's hypothesis.[29] For arbitrary transition from the level k to the level

[28] For a historical reconstruction see [*Mehra/Rechenberg* 1982–2001, I, 155–257].

[29] See [Sommerfeld 1912].

j $(k > j)$, we have that the angular frequency (the frequency ν times 2π) of the emitted radiation (the so-called *Bohr frequency*) is given by

$$\omega_{kj} = -\frac{\Delta E_{kj}}{\hbar}, \tag{1.75}$$

where

$$\Delta E_{kj} = E_j - E_k \tag{1.76}$$

and

$$\hbar = \frac{h}{2\pi} = 1.054571 \times 10^{-34}\,\text{J s}. \tag{1.77}$$

The opposite transition (from a level j to a level k with higher energy) is possible only in the presence of *absorption* of the same quantized amount of energy $E_k - E_j$. It is worth mentioning that Bohr's solution did not really solve the problem of atomic instability: he was not able to show why some orbits are stationary, because he still assumed that the electron's trajectories were classical. We shall return later on this problem (see Ch. 11). In spite of this weakness, Bohr's contribution opened the way to the understanding of the periodic table of elements and of chemical bonds.

1.5.5 Electron diffraction

The phenomena of black-body radiation, of Compton, and of photoelectric effects suggest that light – traditionally understood in wave–like terms – may have particle-like features. Conversely, in the 1920s, matter – which traditionally was understood in particle-like terms – was shown to have wave-like properties. Davisson and Germer [Davisson/Germer 1927] performed an experiment in which a beam of electrons was diffracted by a crystal. They observed a regular diffraction figure only at certain incidence angles – exactly as happens for light.[30] This result was an experimental confirmation of de Broglie's hypothesis that matter can be treated as a wave. In fact in 1924–25 de Broglie [de Broglie 1924, de Broglie 1925][31] had already stated that to a particle of momentum p can be associated a wavelength λ given by

$$\lambda = \frac{h}{p}. \tag{1.78}$$

Since $\lambda = 2\pi k^{-1}$, where k is the propagation vector, it is possible to rewrite Eq. (1.78) as $p = \hbar k$. As a consequence, we are led to the conclusion that, according to the complementarity principle (see Subsec. 1.2.4), both light and matter have wave-like and particle-like features.

[30] More recently, wave-like effects of atoms have also been observed, such as a shift in phase during interaction with a surface [Perreault/Cronin 2005].

[31] For a history of de Broglie's contribution see [*Mehra/Rechenberg* 1982–2001, I, 595–604].

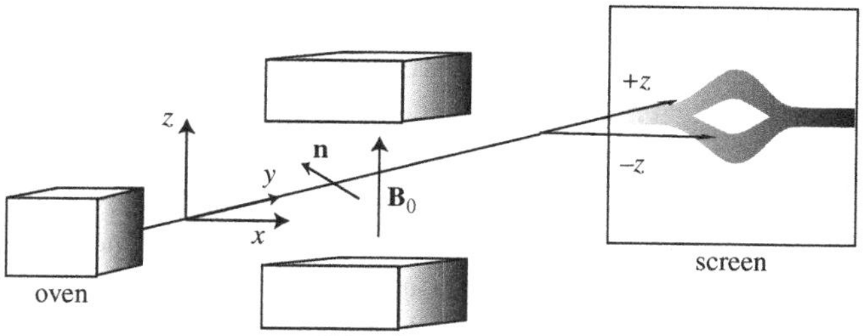

Figure 1.18 The Stern–Gerlach experiment. A typical explanation states that due to the magnetic field an initial polarization **n** is changed into "spin-up" ($+z$) or "spin-down" ($-z$), relative to the main field direction $\mathbf{B}_0$ (of an inhomogeneous magnetic field) if the particle is found in the upper or lower part of the deflected beam, respectively.

1.5.6 Intrinsic magnetic moment

We owe the discovery of the intrinsic magnetic momentum of microentities to a series of experiments carried out by Stern and Gerlach and Uhlenbeck and Goudsmit.[32] In these experiments, a beam of identically prepared silver atoms is sent through a magnetic field oriented in such a way that the gradient of the field is constant and perpendicular to the beam axis. The emerging silver atoms are captured by a screen whose plane is perpendicular to the initial beam axis (see Fig. 1.18). The result shows that the atoms accumulate in two separate "spots." This is another aspect of quantization, leading to the conclusion that the atoms have an intrinsic angular momentum – the spin – that can assume only discrete values. We shall treat the quantum mechanical theory of angular momentum in Ch. 6.

1.5.7 Final considerations

The path taken by quantum mechanics from its first appearance (1900) to a precise formulation of the theory (1925–27) was very long and difficult. In the first 20 years of the twentieth century the majority of physicists still believed that classical mechanics would have been able – sooner or later – to explain the "quantum anomalies" as effects of some forces acting at a microlevel. Gradually, however, it became clear that it was not possible to eliminate these anomalies and that they were not completely compatible with the classical framework. This growing awareness of the inadequacy of classical physics did not result in a new satisfactory formulation until in 1925–27 Heisenberg with his *matrix mechanics* and Schrödinger with his *wave mechanics* were able to by the foundations of a new theory that was thereafter called quantum mechanics[33] [see Chs. 2–3]. Later, it was

[32] See [Gerlach/Stern 1922a, Gerlach/Stern 1922b, Gerlach/Stern 1922c] [Uhlenbeck/Goudsmit 1925, Uhlenbeck/Goudsmit 1926]. For a historical reconstruction of the Stern–Gerlach experiment see [*Mehra/Rechenberg* 1982–2001, I, 422–45] and for a history of the theory of the spin see [*Mehra/Rechenberg* 1982–2001, I, 684–709].

[33] See [Heisenberg 1925, Heisenberg 1927] and [Schrödinger 1926].

proved[34] that Heisenberg's and Schrödinger's formulations are just two different representation of the same theory. As we shall see, this new mechanics has some homologies with classical mechanics. Notwithstanding, there are also many important differences both at technical and conceptual levels.

One of the most difficult tasks was the interpretation of this new formalism. As we shall see some questions remained open for many years (for instance, the measurement problem (see Ch. 9)). However, already in 1927–28 Bohr was able to provide a general framework, founded on the complementarity principle, which, although needing integrations and corrections, still provides a good structure for an understanding of the theory. This framework is known as the *Copenhagen interpretation*.

In the remaining chapters of parts I and II, we shall study Heisenberg's and Schrödinger's contributions and see how they can be understood as two different representations of quantum theory. We shall also show the necessary corrections to be introduced into the Copenhagen interpretation.

Summary

In this chapter we have briefly reviewed classical mechanics and compared it with quantum mechanics. We may summarize the main results as follows:

- Quantum mechanical states are represented by *vectors* in a (complex) Hilbert space.
- If two states are allowed, every linear combination of them is also allowed (*superposition principle*), with the consequence that quantum mechanics violates the principle of perfect determination (and therefore determinism of properties, which characterizes classical mechanics).
- There is a fundamental (and smooth) *complementarity* between particle-like behavior and wave-like behavior (the complementarity principle).
- As a consequence of these two principles, the use of *probability* in quantum mechanics is not due to subjective ignorance but is an intrinsic feature of the theory concerning individual systems (and this is a further evidence of the violation of determinism, because in the general case only probabilistic predictions about properties are possible).
- The structure of quantum probability is deeply different from that of classical probability theory, in the sense that it is not ruled by Kolmogorov axioms and is rooted in the concept of the *probability amplitude*.
- Finally, we have collected the most important historical *experimental evidence* for the departure from classical mechanics: black-body radiation, Compton and photoelectric effects, atomic spectra, specific heat, quantization of atomic levels, electron diffraction, and spin.

[34] The proof was first sketched by Pauli and then built into a new mathematical framework by von Neumann in [von Neumann 1927a, von Neumann 1927b, von Neumann 1927c, von Neumann 1929].

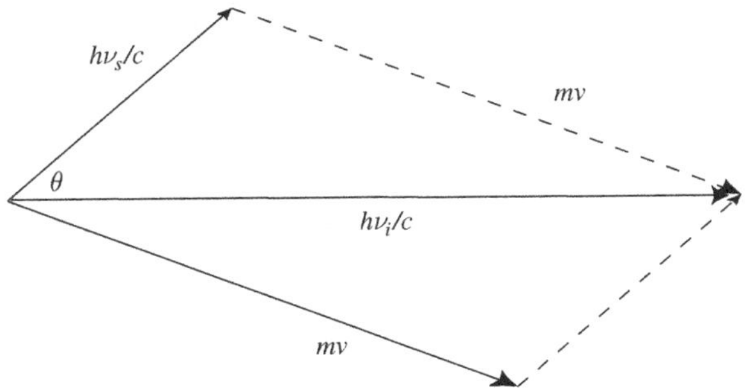

Figure 1.19 Momentum conservation in the Compton effect.

Problems

1.1 Assume that the principle of perfect determination fails. Does then determinism fail as well?

1.2 Assume that the principle of continuity fails. Does then determinism fail as well?

1.3 Making use of one of the properties (1.10), show that we have $\{f, f\} = 0$.

1.4 Show that, if f and g are two constants of motion (i.e. $\{f, H\} = \{g, H\} = 0$) then also their Poisson bracket $\{f, g\}$ is a constant of motion – for the quantum-mechanical counterpart of this problem see also Prob. 3.16.

1.5 Consider the polarization of a photon in the state

$$|\nearrow\rangle = \frac{1}{\sqrt{2}}(|v\rangle + |h\rangle). \tag{1.79}$$

What is the probability $\wp(45°)$ that the photon will pass a filter oriented at 45° relative to the horizontal axis? And the probability $\wp(135°)$?

1.6 Prove Eq. (1.41b).

1.7 Take an orthonormal basis $\{|j\rangle\}$, $1 \le j \le n$ on a Hilbert space of dimension n. Compute the result of the action of the projector $\hat{P}_k = |k\rangle\langle k|$ on a state $|\psi\rangle = \sum_{j=1}^{n} c_j |j\rangle$, where the c_j's are arbitrary complex numbers with $\sum_{j=1}^{n} |c_j|^2 = 1$.

1.8 Taking advantage of Prob. 1.7, prove that, for any $\mathcal{H}$, if $|a\rangle \in \mathcal{H}$, also $\hat{O} \in \mathcal{H}$, where $\hat{O}$ is a linear combination of projectors acting on $\mathcal{H}$.

1.9 Compute the norm of the result of Prob. 1.7. Is this norm larger or smaller than one? Why? Explain the physical meaning of this result.

1.10 Derive Eq. (1.63).

1.11 Starting from the expression (1.63) of the entropy of the black body and taking into account Eq. (1.64), derive Eq. (1.65).

1.12 Derive Eq. (1.67) from energy and momentum conservation in the Compton effect (Eqs. (1.69)–(1.70)).

(*Hint*: Take advantage of the scheme in Fig. 1.19 to rewrite momentum conservation in a single equation, not involving the angle ϕ. Also take into account the fact that $\nu_i \simeq \nu_s$.)

Further reading

Classical mechanics

Arnold, Vladimir I., *Mathematical Methods of Classical Mechanics* (tr. from russ.), New York: Springer, 1978; 2nd edn., 1989.
A wide and very technical exposition of classical mechanics.

Auletta, Gennaro, Critical examination of the conceptual foundations of classical mechanics in the light of quantum physics. *Epistemologia*, **27** (2004), 55–82.

Goldstein, Herbert, *Classical Mechanics*, Cambridge, MA: Addison-Wesley, 1950, 1965.

Landau, L. D. and Lifshitz, E. M., *Mechanics* (Engl. trans.), *The Course of Theoretical Physics*, vol. I, Oxford: Pergamon, 1976.

Superposition and complementarity

Bohr, Niels, The quantum postulate and the recent development of atomic theory. *Nature*, **121** (1928), 580–90.

Vectors and Hilbert spaces

Byron, F. W. Jr. and Fuller, R. W., *Mathematics of Classical and Quantum Physics*, 1969–70; New York: Dover, 1992.

Quantum probability

Gudder, Stanley P., *Quantum Probability*, Boston, MA: Academic Press, 1988.

Early developments of quantum theory

Jammer, Max, *The Conceptual Development of Quantum Mechanics*, New York: McGraw-Hill, 1966; 2nd edn., Thomas Publications, 1989.

Mehra, J. and Rechenberg, H., *The Historical Development of Quantum Theory*, New York: Springer, 1982–2001.

2 Quantum observables and states

In this chapter we shall mainly present the basic formalism that was initially developed by Heisenberg,[1] also known as matrix mechanics (see Subsec. 1.5.7). We will first introduce in Sec. 2.1 the concept of quantum observables. Then, the problem of discrete and continuous spectra will be discussed and the basic non-commutability of quantum-mechanical observables will be deduced. While in Sec. 2.1 we discuss observables on a general formal level, in Sec. 2.2 some basic quantum-mechanical observables will be defined, and then different representations discussed and commutation relations derived. In Sec. 2.3 a basic uncertainty relation is derived. In the same section the relationship between uncertainty, superposition, and complementarity will be discussed. Finally, in Sec. 2.4 complete subsets of commuting observables will be shown to be Boolean subalgebras pertaining to a quantum algebra which is not Boolean.

2.1 Basic features of quantum observables

This section is devoted to a general and formal exposition of quantum observables. In Subsec. 2.1.1 we shall learn how one can mathematically represent quantum observables as Hermitian operators. In Subsec. 2.1.2 we shall see how to change a basis, while in Subsec. 2.1.3 we shall find the relationship between eigenvalues of the observables and probabilities and learn how to calculate mean values. In Subsec. 2.1.4 we shall deal with an operator diagonalization. Finally, in Subsec. 2.1.5, the basic non-commutability of quantum observables will be presented by means of an example.

2.1.1 Variables and operators

The lesson we have learnt from the experimental evidences reported in Sec. 1.5 is that – in both the cases of light and matter – energy may have a discrete spectrum. This has important consequences in the definition of physical quantities in quantum mechanics. In fact, in classical mechanics physical quantities are represented by real variables and functions of real variables. For example, the coordinates q_k's and momenta p_k's in Sec. 1.1 are real variables, whereas the Hamiltonian H, which represents the energy, is a real function of

[1] See [Heisenberg 1925, Heisenberg 1927].

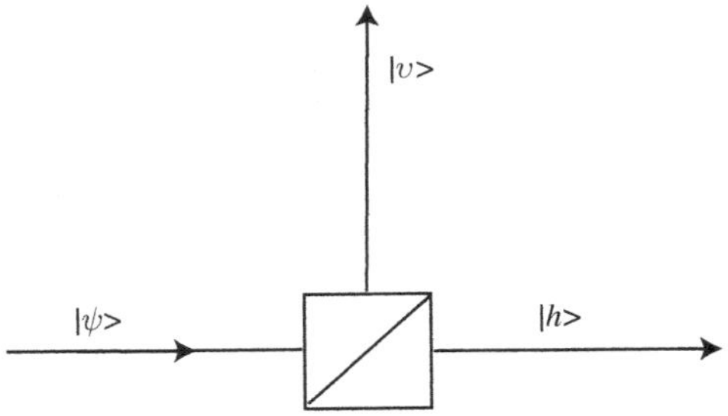

Figure 2.1 Schematic representation of a polarization beam splitter (PBS). An incoming photon from the left in an arbitrary polarization state $|\psi\rangle$ is split by the PBS in such a way that either the photon is found in the lower path with horizontal polarization or in the upper path with vertical polarization. Of course, the state of the photon after the PBS is a quantum superposition of the two alternatives.

the q_k's and of the p_k's. In all these cases, classical variables are continuous. On the other hand, quantum mechanics has to face a situation where physical quantities may have a continuous spectrum, a discrete spectrum, or a combination of both. Mathematically, real variables are not a natural tool for a mechanics facing such a situation. On the other hand, as we shall see later, the spectrum of operators on an infinite-dimensional Hilbert space may have both a continous and a discreate component. Inspired by this consideration, we propose the following principle:

Principle 2.1 (Quantization principle) *Observables in quantum mechanics are represented by operators on a Hilbert space.*

In order to find an adequate representation of quantum observables in terms of operators, let us first consider the discrete case and turn back to the example of polarization already discussed in Secs. 1.3 and 1.4. Suppose that we want to measure the polarization of a system. This can be effected, for instance, with the help of a polarization beam splitter (PBS), which is a particular type of beam splitter that separates photons with vertical and horizontal polarization (see Fig. 2.1). This device is particularly interesting as it somehow "combines" the two experimental setups proposed in the previous chapter: the Mach–Zehnder interferometer (see Fig. 1.3) and the polarization-filter experiment (see Fig. 1.9). Since we can only obtain either vertical or horizontal polarization as outcome, we may make use of the two projectors associated with the two polarization vectors $|v\rangle$ and $|h\rangle$, i.e. $\hat{P}_v$ and $\hat{P}_h$, respectively. In other words, independently of what the state before the measurement was (in general a superposition of $|v\rangle$ and $|h\rangle$), the projectors $\hat{P}_v$ and $\hat{P}_h$ describe the two situations where the measured photon can be found as a result of the measurement. A similar situation would have been obtained if, instead of $\hat{P}_v$ and $\hat{P}_h$, we had considered the polarization along the 45° and 135° orientations. In this case, we would have used an alternative set of projectors, e.g. $\hat{P}_{45}$, $\hat{P}_{135}$, describing again the two possible states in which the system can be found after measurement. As in any kind of measurement, the possible outcomes may be represented as ticks on a reading scale and may be represented by real numbers. Therefore, we may associate to each projector a real number which is the observed outcome itself. In other words, in the discrete case, we can understand the reading scale as a partition of the space of the possible outcome values. Suppose for the sake of simplicity that, for the example chosen here, the two mutually exclusive results have values $+1, -1$, so that, when measuring, if the photon is detected in the upper path (vertical polarization) the outcome is $+1$, whereas if the photon is found in the lower path (horizontal polarization) the outcome is -1. The "polarization observable"[2] can be intuitively conceived as a combination of these two possible outcomes, i.e. something like $(+1)\hat{P}_v + (-1)\hat{P}_h$. In other words, the polarization observable is defined as a combination of its possible values and of the associated projectors. Since a linear combination of operators is itself an operator, we see that a quantum mechanical observable is, in the general case,[3] an operator. In our example we can write the polarization observable as $\hat{O}_P = (+1)\hat{P}_v + (-1)\hat{P}_h$. Therefore, quantum mechanical observables can be represented by operators which act on the state vectors belonging to the Hilbert space $\mathcal{H}$ of a given system. We may generalize the previous result to any discrete spectrum (either finite- or infinite-dimensional) by stating that a generic quantum observable $\hat{O}$ can be written as

$$\hat{O} = \sum_j o_j \hat{P}_j, \tag{2.1}$$

where the o_j's are the *eigenvalues*[4] of $\hat{O}$, the $|o_j\rangle$'s its *eigenvectors*, and the $\hat{P}_j = |o_j\rangle\langle o_j|$'s the corresponding projectors. The sum is extended over all the possible measurement outcomes. The set of all eigenvalues is called *spectrum* of the observable and Eq. (2.1) is called the *spectral representation* of the operator $\hat{O}$. Since $\hat{O}$ is a sum of linear operators (see Eq. (1.43)), therefore it is itself a linear operator.

From Eq. (2.1) it immediately follows that

$$\begin{aligned} \hat{O}\,|o_k\rangle &= \sum_j o_j \hat{P}_j\,|o_k\rangle \\ &= \sum_j o_j\,|o_j\rangle\,\langle o_j\,|\,o_k\rangle = \sum_j o_j\,|o_j\rangle\,\delta_{jk} \\ &= o_k\,|o_k\rangle. \end{aligned} \tag{2.2}$$

The relation $\hat{O}\,|o_k\rangle = o_k\,|o_k\rangle$ can be regarded as the *eigenvalue equation* of the observable $\hat{O}$.[5] The eigenvectors $|o_k\rangle$'s associated with the eigenvalues o_k's in Eq. (2.2) are also

[2] We shall call any physical quantity in quantum mechanics an *observable*.

[3] We shall consider some problems of this generalization in Sec. 3.9.

[4] See [*Byron/Fuller* 1969–70, 120–21].

[5] Strictly speaking, one should distinguish between the observable, which is a physical quantity, and the associated operator, which is a mathematical entity. But, for brevity, we will often write "the observable $\hat{O}$."

called the *eigenkets* (i.e. the eigenstates) of the observable $\hat{O}$. As an example, consider the action of the polarization observable $\hat{O}_P$ on the horizontal-polarization state $|h\rangle$, i.e.

$$\begin{aligned}\hat{O}_P|h\rangle &= (+1\,|v\rangle\langle v| - 1\,|h\rangle\langle h|)\,|h\rangle \\ &= +|v\rangle\langle v\,|\,h\rangle - |h\rangle\langle h\,|\,h\rangle \\ &= -|h\rangle.\end{aligned} \tag{2.3}$$

Since quantum-mechanical observables are represented by operators, the possible values of an observable are the eigenvalues of the corresponding operator. However, not all mathematical operators are suitable for representing observables. In fact, since the values of an observable have to represent physical quantities and therefore must be real, the operator associated with a quantum observable must be a Hermitian operator.[6] An operator $\hat{O}$ is said to be *Hermitian* or self-adjoint when $\hat{O} = \hat{O}^{\dagger}$, where $\hat{O}^{\dagger} = (\hat{O}^{*})^{\mathrm{T}}$ is the transposed conjugate or adjoint of $\hat{O}$ (see Box 2.1), and the transposed matrix $\hat{O}^{\mathrm{T}}$ is obtained by interchanging rows and columns of the matrix $\hat{O}$. In other words, for any vectors $|\varphi\rangle, |\psi\rangle$,

$$\left\langle \varphi \middle| \hat{O} \middle| \psi \right\rangle = \left(\left\langle \psi \middle| \hat{O}^{\dagger} \middle| \varphi \right\rangle \right)^{*}, \tag{2.4}$$

where $\left\langle \varphi \middle| \hat{O} \middle| \psi \right\rangle$ must be interpreted as the scalar product between the bra $\langle\varphi|$ and the ket $\hat{O}|\psi\rangle$, resulting from the action of the operator $\hat{O}$ onto the ket $|\psi\rangle$.[7] This result can be summarized by the following theorem:

Theorem 2.1 (Hermitian operators) *Any quantum mechanical observable can be represented by a Hermitian operator.*

We also note that projectors are observables, since an arbitrary vector $|\psi\rangle$ can be put in the form of a sum of eigenvectors of an arbitrary projector $\hat{P}$, i.e.

$$|\psi\rangle = \hat{P}|\psi\rangle + (\hat{I} - \hat{P})|\psi\rangle. \tag{2.5}$$

In fact, $\hat{P}|\psi\rangle$ is the eigenket of $\hat{P}$ corresponding to the eigenvalue $+1$, since one has (see also Eq. (1.41b)) $\hat{P}(\hat{P}|\psi\rangle) = \hat{P}|\psi\rangle$. Instead, the vector $(\hat{I} - \hat{P})|\psi\rangle$ is the eigenvector of $\hat{P}$ with the eigenvalue 0, i.e.

$$\hat{P}(\hat{I} - \hat{P})|\psi\rangle = (\hat{P} - \hat{P}^2)|\psi\rangle = 0. \tag{2.6}$$

[6] This is a consequence of a theorem which states that a linear transformation $\hat{O}$, on a Hilbert space, is Hermitian if and only if (iff) $\langle\psi|\hat{O}\psi\rangle$ is real for all vectors $|\psi\rangle$ [*Byron/Fuller* 1969–70, I,154].

[7] In the real and finite case self-adjoint operators are represented by symmetric matrices.

Box 2.1 **Hermitian and bounded operators**

Strictly speaking, our definition of a Hermitian operator is valid only in the case of a finite-dimensional Hilbert space. If the dimension is infinite, there may be problems with the definition above if the operator is not bounded. An operator $\hat{O}$ on a Hilbert space $\mathcal{H}$ is said to be *bounded* if there exists a real constant C such that, for any non-zero vector $|\psi\rangle \in \mathcal{H}$, we have

$$\frac{\left\langle \psi \left| \hat{O}^{\dagger}\hat{O} \right| \psi \right\rangle}{\langle \psi \mid \psi \rangle} \leq C^2. \tag{2.7}$$

The lowest possible value of C is the norm of the operator $\hat{O}$.

If the operator is not bounded there are some vectors $|\psi\rangle$ in $\mathcal{H}$ such that $\hat{O}|\psi\rangle$ is not defined (it is formally infinite). In this case the operator is defined only in a dense subspace of the Hilbert space $\mathcal{H}$. Then, the definition of the adjoint or Hermitian conjugate of $\hat{O}$ becomes more subtle. Indeed, in some cases the operator is symmetric, i.e.

$$\left\langle \varphi \mid \hat{O}\psi \right\rangle = \left\langle \hat{O}\varphi \mid \psi \right\rangle, \tag{2.8}$$

where $\left|\hat{O}\psi\right\rangle = \hat{O}|\psi\rangle$, $\left\langle \hat{O}\varphi\right|$ is the bra corresponding to $\left|\hat{O}\varphi\right\rangle$, and $|\varphi\rangle$, $|\psi\rangle$ are vectors in the Hilbert space such that $\left|\hat{O}\varphi\right\rangle$ and $\left|\hat{O}\psi\right\rangle$ are finite. If we define the operator $\hat{O}^{\dagger}$ by the relation

$$\left\langle \varphi \mid \hat{O}\psi \right\rangle = \left\langle \hat{O}^{\dagger}\varphi \mid \psi \right\rangle, \tag{2.9}$$

symmetry implies that $\hat{O}$ and $\hat{O}^{\dagger}$ do coincide if both vectors belong to the domain of the operator $\hat{O}$. However, the operator $\hat{O}^{\dagger}$ may have a larger domain. Often, a symmetric non-bounded operator that is naturally defined in a given subspace of $\mathcal{H}$ may be extended in a non-unique way to a larger subspace so that it becomes Hermitian; in other words, there may be many self-adjoint extensions of the symmetric operator $\hat{O}$.

Generally speaking, the study of self-adjoint extensions of operators is a rather complex subject in functional analysis – the interested reader may refer to [*Fano* 1971, 279–86, 330–54]. However, there is a simple theorem that states that, if the symmetric observable $\hat{O}$ is bounded from below, there is a natural self-adjoint extension and one can forget all problems concerning the uniqueness of self-adjoint extensions.

Moreover, let $\hat{P} = |\varphi\rangle\langle\varphi|$ and consider two arbitrary states $|a\rangle$ and $|b\rangle$. Then,

$$\left\langle a \left| \hat{P}^{\dagger} \right| b \right\rangle = \left\langle b \left| \hat{P} \right| a \right\rangle^{*} = (\langle b \mid \varphi\rangle\langle\varphi \mid a\rangle)^{*} = \langle a \mid \varphi\rangle\langle\varphi \mid b\rangle = \left\langle a \left| \hat{P} \right| b \right\rangle. \tag{2.10}$$

Stated in simple terms, any projector is Hermitian. In the finite-dimensional case we can state the following fundamental spectral theorem:

Theorem 2.2 (Finite-dimensional spectrum) *The eigenvectors $\{|o_j\rangle\}$ of any Hermitian operator $\hat{O}$ on $\mathcal{H}$ span the Hilbert space and can be chosen to be an orthonormal basis for $\mathcal{H}$.*

Box 2.2 **Example of a Hermitian operator**

Going back to the example of the photon polarization observable $\hat{O}_P$ (see p. 45), we may write the state vectors $|v\rangle$ and $|h\rangle$, corresponding to vertical and horizontal polarization, respectively, as (see also Eq. (1.29))

$$|v\rangle = \begin{pmatrix} 1 \\ 0 \end{pmatrix}, \quad |h\rangle = \begin{pmatrix} 0 \\ 1 \end{pmatrix} \tag{2.11}$$

in the two-dimensional polarization Hilbert space $\mathcal{H}_P$. It follows that

$$\langle v \mid h\rangle = \langle h \mid v\rangle = 0, \quad \langle v \mid v\rangle = \langle h \mid h\rangle = 1, \tag{2.12}$$

i.e. the vectors $|v\rangle$ and $|h\rangle$ form an orthonormal basis on $\mathcal{H}_P$. Then, the projectors associated with $|v\rangle$ and $|h\rangle$ can be written as

$$\hat{P}_v = |v\rangle\langle v| = \begin{pmatrix} 1 \\ 0 \end{pmatrix}\begin{pmatrix} 1 & 0 \end{pmatrix} = \begin{bmatrix} 1 & 0 \\ 0 & 0 \end{bmatrix}, \tag{2.13a}$$

$$\hat{P}_h = |h\rangle\langle h| = \begin{pmatrix} 0 \\ 1 \end{pmatrix}\begin{pmatrix} 0 & 1 \end{pmatrix} = \begin{bmatrix} 0 & 0 \\ 0 & 1 \end{bmatrix}, \tag{2.13b}$$

from which it easily follows that $\hat{P}_v + \hat{P}_h = \hat{I}$. As a consequence, the polarization observable can be constructed as

$$\begin{aligned} \hat{O}_P &= (+1)\,\hat{P}_v + (-1)\,\hat{P}_h \\ &= \begin{bmatrix} 1 & 0 \\ 0 & 0 \end{bmatrix} - \begin{bmatrix} 0 & 0 \\ 0 & 1 \end{bmatrix} \\ &= \begin{bmatrix} 1 & 0 \\ 0 & -1 \end{bmatrix}. \end{aligned} \tag{2.14}$$

From Eq. (2.14) it easily follows that $\hat{O}_P = \hat{O}_P^\dagger$, i.e. that $\hat{O}_P$ is a Hermitian operator. Needless to say, the number of linearly independent projectors in a spectral representation of an observable (see Eq. (2.1)) must be equal to the dimension of the Hilbert space of the system.

In fact, if the eigenvalues of $\hat{O}$ are all distinct, then its eigenvectors are orthonormal and indeed form an orthonormal basis on $\mathcal{H}$, as is well known from linear algebra. But it can also be that different eigenvectors correspond to the same eigenvalue. In this case such eigenvalue is said to be *degenerate*. When one (or more) eigenvalue is degenerate with multiplicity k, it is always possible to find, in the k-dimensional subspace of $\mathcal{H}$ spanned by the eigenvectors corresponding to the degenerate eigenvalue, k linearly independent (but not necessarily orthogonal) eigenvectors[8] (see also Subsec. 3.1.4).

[8] Then, one can apply the Gram–Schmidt orthonormalization procedure to find a complete orthonormal set – see [*Byron/Fuller* 1969–70, 159–60].

Let us consider a generic state $| \psi \rangle$ of a system $\mathcal{S}$. Following Th. 2.2, $| \psi \rangle$ can be expanded in the basis $\{| o_j \rangle\}$ (see also Eqs. (1.35)), i.e.

$$| \psi \rangle = \sum_j c_j \, | o_j \rangle, \tag{2.15}$$

where $c_k = \langle o_k | \psi \rangle$ are complex numbers which represent probability amplitudes (see Sec. 1.4). In fact, multiplying both sides of Eq. (2.15) by $\langle o_k |$ from the left, we have

$$\langle o_k | \psi \rangle = \sum_j c_j \langle o_k | o_j \rangle = \sum_j c_j \delta_{jk} = c_k. \tag{2.16}$$

Note that an arbitrary observable $\hat{O}$ can always be expanded as

$$\hat{O} = \sum_j | b_j \rangle \langle b_j | \hat{O} = \sum_{j,n} | b_j \rangle \langle b_j | \hat{O} | b_n \rangle \langle b_n |, \tag{2.17}$$

where (see Eq. (1.41a))

$$\sum_j | b_j \rangle \langle b_j | = \hat{I}, \tag{2.18}$$

and $\{| b_k \rangle\}$ is an orthonormal basis on the Hilbert space $\mathcal{H}$ of the system. From Eq. (2.17) it is clear that, if one knows all the matrix elements

$$O_{jn} = \langle b_j | \hat{O} | b_n \rangle \tag{2.19}$$

of $\hat{O}$ on a given basis, then $\hat{O}$ is fully determined.

Needless to say, given an arbitrary state vector $| \psi \rangle$, it is always possible to construct a Hermitian operator $\hat{O}$ which has $| \psi \rangle$ among its eigenvectors (actually, there are several Hermitian operators having $| \psi \rangle$ among their eigenvectors). In fact, we can take, e.g., $\hat{O} = | \psi \rangle \langle \psi |$, which is a projector and also a Hermitian operator. In this case $| \psi \rangle$ would be an eigenvector of $\hat{O}$ with eigenvalue 1.

We have said that the spectrum of an observable can be continuous, discrete, or a combination of both. We have already examined in the previous subsection the discrete case. Now, what happens if we have a continuous observable such as the position? In this case we have an infinite-dimensional Hilbert space.[9] In an infinite-dimensional Hilbert space we can use an equation of the type of Eq. (2.1) only for compact operators. In a more general case we should use the continuous counterpart of Eq. (2.1). This is the content of the *spectral theorem*, which may summarized as follows:[10] for any Hermitian operator $\hat{O}$ there always exists a spectral representation given by

$$\hat{O} = \int do \, o \, \hat{P}(o), \tag{2.20}$$

where $\hat{P}(o) = | o \rangle \langle o |$. In this representation an arbitrary ket $| \psi \rangle \in \mathcal{H}$ may be expanded as (see also Eq. (2.15))

[9] Also for discrete spectra one can have infinite-dimensional Hilbert spaces. For example – as we shall see in Sec. 4.4 – the energy of a simple harmonic oscillator has a discrete (though infinite) number of eigenvalues.

[10] A formal proof can be found in specialized textbooks [*Prugovečki* 1971, 250–51] [*Holevo* 1982, 52–64].

$$| \psi \rangle = \int do\, c(o) | o \rangle, \tag{2.21}$$

where $c(o) = \langle o | \psi \rangle$ is a (complex) function of the eigenvalues o of the observable $\hat{O}$.

Similarly to the the discrete case (see Eqs. (1.41)), the continuous projectors have the properties

$$\int do \hat{P}(o) = \int do | o \rangle \langle o | = \hat{I}, \tag{2.22a}$$

$$\hat{P}(o)\hat{P}(o') = \delta(o - o')\hat{P}(o), \tag{2.22b}$$

where $\delta(x)$ is the Dirac delta function, which has the following properties:

$$\delta(x) = 0, \quad \forall x \neq 0, \tag{2.23a}$$

$$\int_{-\infty}^{+\infty} dx \delta(x) = 1, \tag{2.23b}$$

$$\int_{-\infty}^{+\infty} dx \delta(x) f(x) = f(0), \quad \text{[The integral is defined iff } f(x) \text{ is a continuous function in } x = 0], \tag{2.23c}$$

$$\delta(ax) = \frac{1}{|a|}\delta(x), \tag{2.23d}$$

$$\delta\left[f(x)\right] = \sum_{j=1}^{N} \frac{1}{|f'(x_j)|}\delta(x - x_j), \quad [x_j \text{ are the zeroes of } f(x)]. \tag{2.23e}$$

Actually, the Dirac δ-function is not a proper function. Rather, it is a more complex object, namely a distribution. For our practical purposes, however, it may be considered as a function.

In the following subsections we shall discuss some properties of observables, limiting ourselves to the discrete case. We shall return to the discussion of the properties of continuous observables before introducing momentum and position operators (in Sec. 2.2).

2.1.2 Change of basis

It is interesting to note that a generic state vector can be expanded in different bases. It is then natural to ask what is the relation between the representations of the state vector in the two different bases. To answer this question, let us go back once again to our polarization example (see Sec. 1.3). In this case, a generic state vector $| \psi \rangle$ can be expanded in the basis $\{| h \rangle, | v \rangle\}$ as

$$| \psi \rangle = c_h | h \rangle + c_v | v \rangle, \tag{2.24}$$

where

$$c_h = \langle h | \psi \rangle \quad \text{and} \quad c_v = \langle v | \psi \rangle. \tag{2.25}$$

However, the vector $|\psi\rangle$ can be also expanded in a different polarization basis $\{|b\rangle, |b_\perp\rangle\}$ as

$$|\psi\rangle = c_b |b\rangle + c_{b_\perp} |b_\perp\rangle, \tag{2.26}$$

where

$$c_b = \langle b \mid \psi\rangle \quad \text{and} \quad c_{b_\perp} = \langle b_\perp \mid \psi\rangle. \tag{2.27}$$

Moreover, $|b\rangle$ and $|b_\perp\rangle$ may be in turn expanded in the basis $\{|h\rangle, |v\rangle\}$ (see Th. 2.2: p. 47) as

$$|b\rangle = \langle h \mid b\rangle |h\rangle + \langle v \mid b\rangle |v\rangle, \tag{2.28a}$$

$$|b_\perp\rangle = \langle h \mid b_\perp\rangle |h\rangle + \langle v \mid b_\perp\rangle |v\rangle, \tag{2.28b}$$

where we have taken advantage of the usual relation

$$|h\rangle\langle h| + |v\rangle\langle v| = \hat{I}. \tag{2.29}$$

Using Eqs. (2.28), Eq. (2.26) may be rewritten as

$$\begin{aligned} |\psi\rangle &= c_b \left(\langle h \mid b\rangle |h\rangle + \langle v \mid b\rangle |v\rangle\right) + c_{b_\perp} \left(\langle h \mid b_\perp\rangle |h\rangle + \langle v \mid b_\perp\rangle |v\rangle\right) \\ &= \left(c_b \langle h \mid b\rangle + c_{b_\perp} \langle h \mid b_\perp\rangle\right) |h\rangle + \left(c_b \langle v \mid b\rangle + c_{b_\perp} \langle v \mid b_\perp\rangle\right) |v\rangle. \end{aligned} \tag{2.30}$$

The last expression has to be equal to the rhs of Eq. (2.24), i.e.

$$c_h = \langle h \mid b\rangle c_b + \langle h \mid b_\perp\rangle c_{b_\perp}, \tag{2.31a}$$

$$c_v = \langle v \mid b\rangle c_b + \langle v \mid b_\perp\rangle c_{b_\perp}, \tag{2.31b}$$

which is the desired relation between the sets of coefficients $\{c_h, c_v\}$ and $\{c_b, c_{b_\perp}\}$.

In matrix notation, Eqs. (2.31) may be cast in the more compact form

$$\begin{pmatrix} c_h \\ c_v \end{pmatrix} = \hat{U} \begin{pmatrix} c_b \\ c_{b_\perp} \end{pmatrix}, \tag{2.32}$$

where (see Prob. 2.2)

$$\hat{U} = \begin{bmatrix} \langle h \mid b\rangle & \langle h \mid b_\perp\rangle \\ \langle v \mid b\rangle & \langle v \mid b_\perp\rangle \end{bmatrix}. \tag{2.33}$$

The matrix $\hat{U}$ is unitary (see Box 2.3), i.e. (see also Prob. 2.3)

$$\hat{U}\hat{U}^\dagger = \hat{U}^\dagger\hat{U} = \hat{I} \tag{2.34}$$

or $\hat{U}^{-1} = \hat{U}^\dagger$.

Box 2.3 **Unitary operators**

Unitary operators are a special class of the *normal* operators, i.e. of the operators $\hat{O}$ which commute with their adjoint: $\hat{O}\hat{O}^\dagger = \hat{O}^\dagger\hat{O}$. It is interesting to note that any linear operator $\hat{O}$ may be decomposed as $\hat{O} = \hat{O}' + \imath\hat{O}''$, where $\hat{O}'$ and $\hat{O}''$ are Hermitian operators (see Prob. 2.4). In this respect, we may establish an analogy between linear operators and complex numbers, where the two Hermitian operators $\hat{O}'$ and $\hat{O}''$ play the "role" of the real and imaginary parts of a complex number. Now, the necessary and sufficient condition for $\hat{O}$ to be normal is that $\hat{O}''\hat{O}' = \hat{O}'\hat{O}''$ [*Halmos* 1951, 42–43]. Unitary operators have to fulfill the further condition (see Prob. 2.5)

$$\left(\hat{O}'\right)^2 + \left(\hat{O}''\right)^2 = \hat{I}, \tag{2.35}$$

which corresponds to the condition for a complex number $z = x + \imath y$ to have modulus 1, i.e. $x^2 + y^2 = 1$. This analogy justifies the use of the term "unitary."

Notice also that unitary operators are not necessarily Hermitian. In fact, for a normal operator $\hat{O} = \hat{O}' + \imath\hat{O}''$ to be Hermitian, the additional condition $\hat{O}'' = 0$ is required. Furthermore, if $\hat{O}$ is also unitary, it must satisfy $\hat{O} = \hat{O}'$ with (see Prob. 2.6)

$$\hat{O}' = \left(\hat{O}'\right)^{\dagger} \quad \text{and} \quad \left(\hat{O}'\right)^2 = \hat{I}. \tag{2.36}$$

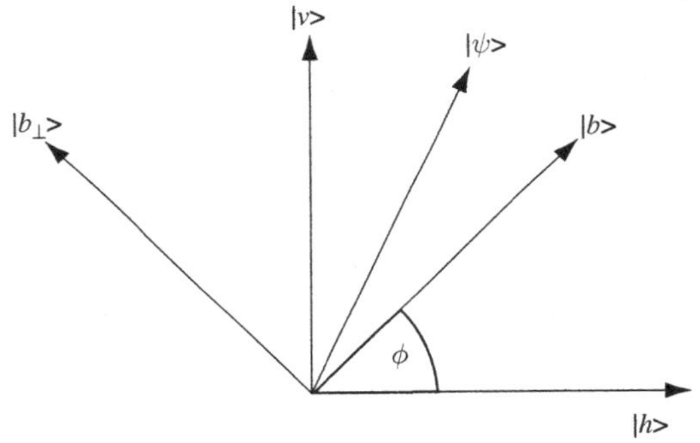

Figure 2.2 **Change of basis. The basis $\{|b\rangle, |b_\perp\rangle\}$ is obtained from the original basis $\{|h\rangle, |v\rangle\}$ by a counterclockwise rotation of 45°.**

As we shall see, the unitary character is a distinctive feature of several basic quantum transformations and, depending on the context, it may be expressed in operatorial terms, time reversibility or spatial invariance. In general, it is a signature of the existence of a symmetry (see Ch. 8).

We can choose, as a particular instance of $|b\rangle$ and $|b_\perp\rangle$, the polarization vectors at 45° and 135°, respectively, which means that the angle ϕ between $|b\rangle$ and $|h\rangle$ and between $|b_\perp\rangle$ and $|v\rangle$ is 45° (see Fig. 2.2). Then, the matrix $\hat{U}$ is given by

$$\hat{U} = \frac{1}{\sqrt{2}}\begin{bmatrix} 1 & -1 \\ 1 & 1 \end{bmatrix}, \tag{2.37}$$

which is a particular instance of the rotation matrix for a two-dimensional system

$$\hat{U}(\phi) = \begin{bmatrix} \cos\phi & -\sin\phi \\ \sin\phi & \cos\phi \end{bmatrix}, \tag{2.38}$$

for $\phi = \pi/4$. In other words, in the finite-dimensional case a change of basis can be represented by a *rotation of the axes in the Hilbert space*. As we shall see in Ch. 8, any unitary operator describes a kind of rotation.

It is possible to generalize the derivation above to any state vector $|\psi\rangle$. Let us expand it in terms of an orthonormal basis $\{|b_k\rangle\}$ as

$$|\psi\rangle = \sum_j c_{b_j} |b_j\rangle, \tag{2.39}$$

with some complex coefficients $c_{b_k} = \langle b_k | \psi\rangle$. We may always choose to expand the state vector $|\psi\rangle$ in a different orthonormal basis $\{|a_k\rangle\}$. In this case, we have

$$|\psi\rangle = \sum_n c_{a_n} |a_n\rangle, \tag{2.40}$$

where $c_{a_k} = \langle a_k | \psi\rangle$. It is therefore interesting to look for the relationship between the sets of coefficients $\{c_{b_k}\}$ and $\{c_{a_k}\}$. For this purpose, we may insert the expression $\sum_n |a_n\rangle\langle a_n| = \hat{I}$ (see Eq. (1.41a)) into Eq. (2.39), so as to obtain

$$\begin{aligned} |\psi\rangle &= \sum_{n,j} c_{b_j} |a_n\rangle \langle a_n | b_j\rangle \\ &= \sum_{n,j} U_{n,j} c_{b_j} |a_n\rangle, \end{aligned} \tag{2.41}$$

with the matrix elements $U_{n,j} = \langle a_n | b_j\rangle$. Since the rhs of Eqs. (2.40) and (2.41) have to be identical due to the uniqueness of the expansion in terms of an orthonormal basis, we must conclude that

$$c_{a_k} = \sum_j U_{k,j} c_{b_j}, \tag{2.42}$$

which is the desired relation. If we interpret the sets of coefficients $\{c_{b_j}\}$ and $\{c_{a_j}\}$ as column vectors $\mathbf{c}_b$ and $\mathbf{c}_a$, respectively, we can write Eq. (2.42) in matrix form as

$$\mathbf{c}_a = \hat{U}\mathbf{c}_b, \tag{2.43}$$

or

$$\begin{pmatrix} c_{a_1} \\ c_{a_2} \\ \dots \\ c_{a_n} \end{pmatrix} = \begin{bmatrix} \langle a_1 | b_1\rangle & \langle a_1 | b_2\rangle & \dots & \langle a_1 | b_n\rangle \\ \langle a_2 | b_1\rangle & \langle a_2 | b_2\rangle & \dots & \langle a_2 | b_n\rangle \\ \dots & \dots & \dots & \dots \\ \langle a_n | b_1\rangle & \langle a_n | b_2\rangle & \dots & \langle a_n | b_n\rangle \end{bmatrix} \begin{pmatrix} c_{b_1} \\ c_{b_2} \\ \dots \\ c_{b_n} \end{pmatrix}. \tag{2.44}$$

It should now appear evident at this point that superposition (see p. 18) is basis-dependent: a state vector which appears to be a superposition in a certain basis (relatively to an observable) may well not be such in a different basis (relatively to a different observable). For instance the vectors $|v\rangle$ and $|h\rangle$ in Eq. (2.11) are an eigenbasis of the observable (2.14).

It is not very difficult to translate change of basis in a continuous "language." Let us consider a generic state vector $|\psi\rangle$ that can be expanded as

$$|\psi\rangle = \int d\xi c(\xi)\,|\xi\rangle. \tag{2.45}$$

The vector $|\psi\rangle$ could also have been expanded in another basis, say $\{|\eta\rangle\}$ as follows:

$$\begin{aligned} |\psi\rangle &= \int d\eta c'(\eta)\,|\eta\rangle \\ &= \int d\eta c'(\eta) \int d\xi\,|\xi\rangle\,\langle\xi\mid\eta\rangle \\ &= \int\int d\eta d\xi c'(\eta)\,\langle\xi\mid\eta\rangle\,|\xi\rangle, \end{aligned} \tag{2.46}$$

from which we find the desired relation

$$c(\xi) = \int d\eta\,\langle\xi\mid\eta\rangle\,c'(\eta). \tag{2.47}$$

We prove in the following three important properties of the change of basis (for the discrete case).

- First, it is easy to show that the matrix $\hat{U}$ in Eq. (2.42) is unitary.

Proof

We have that

$$(U_{n,j})^* = \langle b_j \mid a_n\rangle \quad \text{and} \quad \left(U^\dagger\right)_{n,j} = \langle b_n \mid a_j\rangle. \tag{2.48}$$

Then we have that

$$\left(UU^\dagger\right)_{n,j} = \sum_k U_{n,k}\left(U^\dagger\right)_{k,j} = \sum_k \langle a_n \mid b_k\rangle\langle b_k \mid a_j\rangle = \langle a_n \mid a_j\rangle = \delta_{nj}, \tag{2.49}$$

which means that $\hat{U}\hat{U}^\dagger = \hat{I}$.

Similarly, we have

$$\left(U^\dagger U\right)_{n,j} = \sum_k \left(U^\dagger\right)_{n,k} U_{k,j} = \sum_k \langle b_n \mid a_k\rangle\langle a_k \mid b_j\rangle = \delta_{nj}. \tag{2.50}$$

Q.E.D

- Second, the unitary transformation that instantiates a change of basis preserves the scalar product between two arbitrary kets $|\psi\rangle$ and $|\psi'\rangle$. This result may appear evident since the scalar product should be independent from the basis chosen to compute it. Nevertheless, as we shall see later (in Subsec. 3.5.1 and Ch. 8; see also Eqs. (1.35)–(1.36)), scalar-product conservation is an important property of unitary transformations.

Proof

Let us decompose the two kets $|\psi\rangle$ and $|\psi'\rangle$ into the two basis $\{|b_k\rangle\}$ and $\{|a_k\rangle\}$:

$$|\psi\rangle = \sum_j c_{b_j} |b_j\rangle, \quad |\psi'\rangle = \sum_j c'_{b_j} |b_j\rangle, \tag{2.51a}$$

$$|\psi\rangle = \sum_j c_{a_j} |a_j\rangle, \quad |\psi'\rangle = \sum_j c'_{a_j} |a_j\rangle. \tag{2.51b}$$

The scalar product of the two vectors in the basis $\{|b_k\rangle\}$ is easily calculated:

$$\langle\psi' \mid \psi\rangle = \sum_{k,j} \langle b_k \mid b_j\rangle c'^*_{b_k} c_{b_j} = \sum_j c'^*_{b_j} c_{b_j}, \tag{2.52}$$

since $\langle b_k \mid b_j\rangle = \delta_{kj}$. Next, we calculate the same scalar product in the $\{|a_k\rangle\}$ basis and finally show that it is the same as Eq. (2.52):

$$\langle\psi' \mid \psi\rangle = \sum_{k,j} c'^*_{a_k} c_{a_j} \langle a_k \mid a_j\rangle = \sum_j c'^*_{a_j} c_{a_j}. \tag{2.53}$$

But $c_{a_j} = \sum_n U_{j,n} c_{b_n}$ (see Eq. (2.42)) and, analogously, $c'^*_{a_j} = \sum_n U^*_{j,n} c'^*_{b_n} = \sum_n c'^*_{b_n} U^\dagger_{n,j}$. Substituting these expressions into Eq. (2.53), we obtain

$$\begin{aligned}\langle\psi' \mid \psi\rangle &= \sum_j \sum_n c'^*_{b_n} U^\dagger_{n,j} \sum_k U_{j,k} c_{b_k} \\ &= \sum_{n,k} c'^*_{b_n} \left(\sum_j U^\dagger_{n,j} U_{j,k} \right) c_{b_k} = \sum_k c'^*_{b_k} c_{b_k},\end{aligned} \tag{2.54}$$

since

$$\sum_j U^\dagger_{n,j} U_{j,k} = \left(U^\dagger U\right)_{n,k} = \hat{I}_{n,k} = \delta_{nk}. \tag{2.55}$$

Q.E.D

- Finally, it can be shown that the trace of an operator does not change under change of basis. With trace of an operator we intend here the sum of the diagonal elements of the corresponding matrix, i.e.

$$\mathrm{Tr}(\hat{O}) = \sum_j \langle b_j \mid \hat{O} \mid b_j\rangle, \tag{2.56}$$

where, as usual, $\{|b_k\rangle\}$ is an orthonormal basis on the Hilbert space of the system.

Proof

Let us first write the trace of $\hat{O}$ in the basis $\{|a_k\rangle\}$, i.e. $\mathrm{Tr}(\hat{O}) = \sum_j \langle a_j \mid \hat{O} \mid a_j\rangle$. Now the trace of $\hat{O}$ in the basis $\{|b_k\rangle\}$ can be written as

$$\begin{aligned}\mathrm{Tr}(\hat{O}) &= \sum_{j,n,k} \langle b_j \mid a_n\rangle \langle a_n \mid \hat{O} \mid a_k\rangle \langle a_k \mid b_j\rangle \\ &= \sum_{j,n,k} \langle a_k \mid b_j\rangle \langle b_j \mid a_n\rangle \langle a_n \mid \hat{O} \mid a_k\rangle \\ &= \sum_{n,k} \langle a_k \mid a_n\rangle \langle a_n \mid \hat{O} \mid a_k\rangle = \sum_k \langle a_k \mid \hat{O} \mid a_k\rangle, \end{aligned} \tag{2.57}$$

where $\sum_j \langle b_j \mid b_j\rangle = \sum_n \langle a_n \mid a_n\rangle = \hat{I}$.

Q.E.D

2.1.3 Values of observables

So far we have discussed a few basic properties of quantum observables. It is now natural to ask what the "effect" of the action of an observable on a state vector is. One might think that the action on a given system's state of the operator $\hat{O}$ corresponding to a generic observable describes the "effect" of a measurement of that observable on the system. This is not the case. In fact, if we have an arbitrary polarization state $|\, b\rangle$ (see Subsec. 2.1.1), the action of the polarization observable

$$\hat{O}_P = (+1)\,|v\,\rangle\,\langle v\,| + (-1)\,|h\,\rangle\,\langle h\,| \tag{2.58}$$

does not produce a state corresponding to any measured outcome. In other words

$$\hat{O}_P\,|\,b\rangle \neq (+1)\,|\,v\rangle \quad \text{and} \quad \hat{O}_P\,|\,b\rangle \neq (-1)|\,h\rangle, \tag{2.59}$$

in general, since $|\,b\rangle$ may well be a superposition state of the form

$$|\,b\rangle = c_v\,|\,v\rangle + c_h|\,h\rangle\,. \tag{2.60}$$

In this case, we clearly have

$$\begin{aligned}\hat{O}_P\,|\,b\rangle &= \hat{O}_P\,(c_v\,|\,v\rangle + c_h|\,h\rangle\,) \\ &= (|v\,\rangle\,\langle v\,| - |h\,\rangle\,\langle h\,|)\,(c_v\,|\,v\rangle + c_h|\,h\rangle\,) \\ &= c_v\,|v\,\rangle\,\langle v\,|\,v\rangle - c_h\,|h\,\rangle\,\langle h\,|\,h\rangle = c_v\,|\,v\rangle - c_h|\,h\rangle, \end{aligned} \tag{2.61}$$

which shows that an observable induces a "transformation" on a given state that in general does not yield one of its eigenvectors as output. The only exception is when the initial state is already an eigenvector of the observable (see Eq. (2.3)).

Quantum measurement theory is a complex aspect of quantum mechanics and will be the object of later examination (in Ch. 9). For the time being, let us say that the measurement process requires that the object system interact at least with an apparatus and should lead to a "change" in the state of the object system from the initial state (which may be a superposition relatively to the measured observable) to a final state described by the eigenvector corresponding to the measured eigenvalue – this "change" is the heart of the measurement problem in quantum mechanics. Then, when actually performing a measurement of a given observable $\hat{O}$, we must obtain as outcome one of the possible values (eigenvalues) of $\hat{O}$.[11] In particular, it is not possible to find intermediate values between eigenvalues. Since the measured outcome is certainly one of the possible eigenvalues of $\hat{O}$ and since we have seen (in Sec. 1.4) that quantum probabilities must be expressed as square moduli of some amplitudes, we can then postulate what follows.[12]

Principle 2.2 (Statistical algorithm) *Given that a quantum system is completely defined by a vector* $|\,\psi\rangle$ (Eq. (2.15)), *the probability of having a determinate measurement result – an eigenvalue* o_k *of the measured observable* $\hat{O}$ *– is given by*

$$\wp(o_k, \psi) = |c_k|^2, \tag{2.62}$$

where the complex coefficient c_k *is the amplitude* $c_k = \langle o_k \mid \psi\rangle$, *and the eigenvector* $|\,o_k\rangle$ *of* $\hat{O}$ *corresponds to the eigenvalue* o_k.

This algorithm is of particular relevance because it provides the general mathematical connection between the coefficients of the expansion of the system state onto a given basis and the probabilities of the corresponding outcomes of a measurement process. It is also evident that we must have

$$\sum_j |c_j|^2 = 1 \tag{2.63}$$

in Eqs. (2.15) and (2.62). It is then evident that $c_j \in l^{(2)}$, where $l^{(2)}$ is the space of successions for which the sum of square moduli is finite. This condition reflects, on one hand, the *normalization* of the state vectors (see also Eq. (2.108)), and, on the other hand, the requirement that the sum of the probabilities of all disjoint events of a given set is equal to one, i.e. Kolmogorov's probability axiom (see p. 29).

It is straightforward to extend the statistical algorithm to the continuous spectrum case, where we recall that Eq. (2.15) may be rewritten as (see Eq. (2.21))

$$|\,\psi\rangle = \int do\; c(o)\,|\,o\rangle, \tag{2.64}$$

where the coefficient $c(o) = \langle o \mid \psi\rangle$ is a continuous function of the eigenvalue o and the $|\,o\rangle$ are the eigenkets of the observable $\hat{O}$, i.e.

$$\hat{O}\,|\,o\rangle = o\,|\,o\rangle. \tag{2.65}$$

11 This provides a first evidence that the state vector cannot be measured with a single measurement. This subject will be discussed extensively in Ch. 15.

12 See [Born 1926].

As a consequence, Eq. (2.62) translates into

$$\wp(o, \psi) = |c(o)|^2 \tag{2.66}$$

where $\wp(o, \psi)do$ has here to be interpreted as the probability that the eigenvalue be in the interval $(o, o + do)$. In the continuous case, the normalization condition (2.63) translates into

$$\int do\ |c(o)|^2 = 1. \tag{2.67}$$

It may well happen that an observable $\hat{O}$ presents simultaneously a continuous and a discrete spectrum. For instance, let us assume that the discrete spectrum ranges between $-\infty$ and $\tilde{o}$, above which value the spectrum becomes continuous up to $+\infty$. Then, if

$$\hat{P}_D = \sum_j |\, o_j \,\rangle \langle\, o_j \,| \quad \text{and} \quad \hat{P}_C = \int_{\tilde{o}}^{+\infty} do\, |\, o \,\rangle \langle\, o \,|, \tag{2.68}$$

where the sum is extended over the discrete eigenvalues, we must also have

$$\hat{P}_C + \hat{P}_D = \hat{I}. \tag{2.69}$$

Principle 2.2 provides the connection quantum theory and experimental measurement statistics. In other words, if we perform a large number of observations, we expect that the statistics of different possible outcomes will tend to their corresponding probabilities as the number of measurement runs grows. Principle 2.2 also allows us to define the useful concept of the mean or *expectation value* $\langle \hat{O} \rangle_\psi$ of an observable $\hat{O}$ on a certain state $|\,\psi\rangle$. We start from the usual definition of mean value for a probability distribution.

Classically, the expectation value of a random variable ξ is defined as[13]

$$\overline{\xi} = \int dx\ x\ \wp(x), \tag{2.70}$$

where $\wp(x)$ is called the probability density function and is such that $\wp(x)dx$ is the probability that the random variable ξ takes on a value in the interval $(x, x + d\xi)$. Equation (2.70) is valid when the probability distribution is continuous and the integral

$$\int dx |x| \wp(x) \tag{2.71}$$

exists. In the most general case, the expectation value of the random variable ξ may be expressed in terms of the distribution function

$$\mathcal{F}(x) = \wp(\xi < x), \tag{2.72}$$

[13] See [*Gnedenko* 1969, 125–32, 165–86, 219, 227] [*Gudder* 1988, 30].

which gives the probability that ξ will take on a value less than x, i.e.

$$\bar{\xi} = \int x d\mathcal{F}(x). \tag{2.73}$$

For the continuous case we then have

$$\mathcal{F}(x) = \int_{-\infty}^{x} dz \wp(z). \tag{2.74}$$

Another useful concept is that of *characteristic function*. The characteristic function of a classical random variable ξ is defined as the expectation of the random variable $e^{\imath\eta\xi}$, where η stands for a real parameter. If $\mathcal{F}(x)$ is the distribution function of ξ (see Eq. (2.72)), the characteristic function is given by

$$\chi_\xi(\eta) = \int d\mathcal{F}(x) e^{\imath\eta x}. \tag{2.75}$$

A distribution function is uniquely determined by its characteristic function. If x is a point of continuity of $\mathcal{F}(x)$, then

$$\mathcal{F}(x) = \frac{1}{2\pi} \lim_{y \longrightarrow -\infty} \lim_{c \longrightarrow \infty} \int_{-c}^{+c} d\eta \frac{e^{-\imath\eta y} - e^{-\imath\eta x}}{\imath\eta} \chi_\xi(\eta), \tag{2.76}$$

where the limit in y is evaluated with respect to any set of points y that are points of continuity for the function $\mathcal{F}(x)$. The n-th derivative of the characteristic function, calculated at $\eta = 0$ gives – apart from a multiplicative factor – the n-th *moment* of the random variable

$$\chi_\xi^{(n)}(0) = \imath^n \bar{\xi}^n, \tag{2.77}$$

so that the first derivative is the expectation value of ξ times the imaginary unity, the second derivative gives the opposite of the second moment of ξ, and so on.

Let us now come back to our original problem of defining the expectation values of a quantum observable. In order to calculate the mean value of an observable $\hat{O}$, we write the analogy of Eq. (2.70) making use of the quantum probability density $\wp(o_j, \psi)$. For the discrete case, we expand the definition in several steps as

$$\begin{aligned}
\left\langle \hat{O} \right\rangle_\psi &= \sum_j \wp(j, \psi) o_j = \sum_j |c_j|^2 o_j = \sum_j c_j^* c_j o_j \\
&= \sum_j \langle o_j \mid \psi \rangle \langle \psi \mid o_j \rangle o_j = \sum_j o_j \left\langle o_j \left| \hat{P}_\psi \right| o_j \right\rangle \\
&= \sum_j \langle o_j \mid \psi \rangle \langle \psi \mid \hat{O} \mid o_j \rangle \\
&= \sum_j \langle \psi \mid \hat{O} \mid o_j \rangle \langle o_j \mid \psi \rangle \\
&= \left\langle \psi \left| \hat{O} \right| \psi \right\rangle,
\end{aligned} \tag{2.78}$$

where $\hat{P}_\psi = |\psi\rangle\langle\psi|$, and we have made use of the eigenvalue equation $\hat{O}|o_k\rangle = o_k|o_k\rangle$ (Eq. (2.2)) and of the property $\sum_j |o_j\rangle\langle o_j| = \hat{I}$ (Eq. (1.41a)). Since the eigenvalues of a Hermitian operator are real, the expectation value of an observable in any state $|\psi\rangle$ must also be real.

Box 2.4 **Example of mean value**

Let us again consider the example of the polarization observable $\hat{O}_P$ and its eigenbasis given in Box 2.2 (p. 48). Then, the mean value of $\hat{O}_P$ on the state $|b\rangle = c_v|v\rangle + c_h|h\rangle$ is

$$\begin{aligned}\langle b|\hat{O}_P|b\rangle &= \left(\langle v|c_v^* + \langle h|c_h^*\right)\hat{O}\left(c_v|v\rangle + c_h|h\rangle\right)\\ &= \left[c_v^*\begin{pmatrix}1 & 0\end{pmatrix} + c_h^*\begin{pmatrix}0 & 1\end{pmatrix}\right]\begin{bmatrix}1 & 0\\ 0 & -1\end{bmatrix}\left[c_v\begin{pmatrix}1\\0\end{pmatrix} + c_h\begin{pmatrix}0\\1\end{pmatrix}\right]\\ &= \begin{pmatrix}c_v^* & c_h^*\end{pmatrix}\begin{pmatrix}c_v\\ -c_h\end{pmatrix}\\ &= |c_v|^2 - |c_h|^2.\end{aligned} \tag{2.79}$$

It should be stressed that all the relevant physical quantities are related to expressions of the type of Eq. (2.78), i.e. to the mean value of some operator. In fact, we have:[14]

Theorem 2.3 (Observables' equality) *Two observables $\hat{O}$ and $\hat{O}'$ are equal if and only if*

$$\langle\psi|\hat{O}|\psi\rangle = \langle\psi|\hat{O}'|\psi\rangle, \quad \forall|\psi\rangle. \tag{2.80}$$

The proof is immediate. One can also see that any global phase factor $e^{\imath\phi}$ which might pertain to a state vector $|\psi\rangle$ cancels out when calculating the mean value of an observable onto $|\psi\rangle$. This means that the relevant physical quantities do not depend on the global phase factor of the state vector, which is therefore irrelevant (see p. 26). On the other hand, relative phase factors between different components of a superposition state are physically relevant since they determine the interference behavior of the corresponding quantum system (see Subsec. 1.2.3).

2.1.4 Diagonalization of operators

Finding the eigenvalues of a observable $\hat{O}$ on a finite Hilbert space is equivalent to diagonalizing the matrix corresponding to the operator $\hat{O}$. In quantum mechanics it is often useful to put an operator in a diagonal form since then its diagonal elements are its

[14] See [*Messiah* 1958, 633–36].

eigenvalues and therefore the possible outcomes of a measurement process. The procedure with which one may diagonalize a matrix corresponds to the solution of the characteristic polynomial pertaining to the matrix. It is well known[15] that, if the o_j's are the eigenvalues of $\hat{O}$ corresponding to the eigenvectors $|o_j\rangle$'s, the matrix whose columns are given by the eigenvectors is a diagonalizing matrix for $\hat{O}$. Then, in the non-degenerate case, the $n \times n$ matrix $\hat{U}$ formed by the eigenvectors will be the diagonalizing matrix for $\hat{O}$. This means that, if $o_j^k = \langle b_k \mid o_j\rangle$ is the k-th component of the eigenvector $|o_j\rangle$ in a certain basis $\{|b_k\rangle\}$ such that

$$|o_j\rangle = \sum_k o_j^k \, |b_k\rangle, \tag{2.81}$$

the diagonalizing matrix $\hat{U}$ will be written as

$$\hat{U} = \begin{bmatrix} o_1^1 & o_2^1 & \cdots & o_n^1 \\ o_1^2 & o_2^2 & \cdots & o_n^2 \\ \cdots & \cdots & \cdots & \cdots \\ o_1^n & o_2^n & \cdots & o_n^n \end{bmatrix}, \tag{2.82}$$

i.e. $U_{jk} = o_k^j$. It is easy to see that the matrix $\hat{U}$ is unitary. In fact, since the $|o_j\rangle$ are (orthonormal) eigenvectors, then

$$\sum_n (o_k^n)^* o_j^n = \langle o_k \mid o_j\rangle = \delta_{kj}. \tag{2.83}$$

On the other hand,

$$(U^\dagger U)_{kj} = \sum_n (U^\dagger)_{kn} U_{nj} = \sum_k U_{nk}^* U_{nj} = \sum_n (o_k^n)^* o_j^n = \delta_{kj}, \tag{2.84}$$

where use has been made of Eqs. (2.82)–(2.83) and which proves that $\hat{U}$ is unitary.

2.1.5 Non-commutativity

Let us now come back to our polarization example (see Sec. 1.3). We have supposed a polarization filter P1 with direction **a** and have inserted another polarizing filter P2 with a polarization axis **b** which makes an angle θ with the orientation **a** of P1. Suppose now that directions **a** and **b** are orthogonal. We have already seen that in this case there is no output light after P2. The question is: what happens if, between the two orthogonal polarizing filters, we now insert a third filter P3 (see Fig. 2.3) with a polarization axis **c** at an angle $\phi \neq 0, \pi/2$ relative to the first polarization axis? The final observed intensity (after P2) is $I_2 = I_1 \cos^2\phi \cos^2(\theta - \phi) = I_1 \cos^2\phi \sin^2\phi$, which, for $\phi \neq 0, \pi/2$, is not zero. On the other hand, after the beam has passed the first filter, the component parallel to **b** must be zero. This is difficult to understand intuitively, because it appears that the addition of the intermediate filter P3 should not increase the output intensity. The only way

[15] See [*Byron/Fuller* 1969–70, 123].

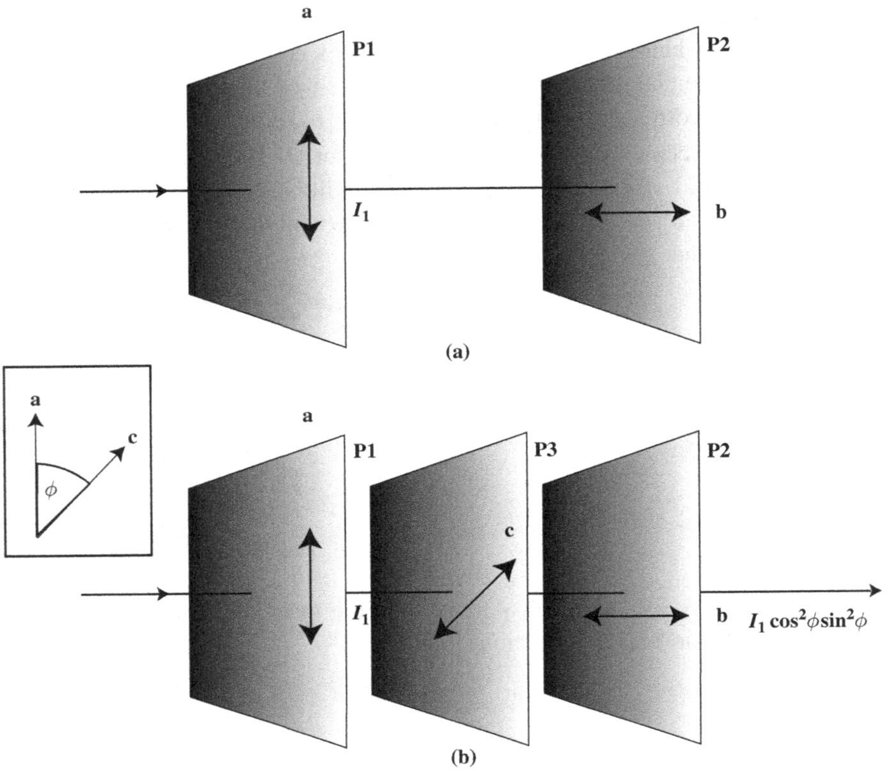

Figure 2.3 (a) An initial (unpolarized) light beam passes the filter P1 with polarization direction **a** (here supposed to be vertical). After P1, the light is polarized (with a reduced intensity I_1) along **a**. Then, no photon can pass the filter P2 whose polarization direction **b** is orthogonal to **a**. (b) If we insert a third polarization filter P3 between P1 and P2 with a polarization direction **c** at an angle $\phi \neq 0, \pi/2$ relative to **a**, then the final output intensity will be $I_2 = I_1 \cos^2 \phi \sin^2 \phi$.

to overcome this difficulty is to admit that after P3 the state of the photon is described by a superposition (see Subsec. 1.2.3) of the two states corresponding to a polarization parallel to **a** and parallel to **b**, respectively.

If we invert the order of P2 and P3 the output intensity must be clearly zero since, as we know, in this case it must already be zero after P2. Here we see that the order of such filters is crucial for determining what the output is. We know from the previous chapter that the action of a filter can be interpreted as an operation on the state of the photon during its travel along the apparatus. Generalizing this simple result, we may state that in quantum mechanics different operations may not commute – i.e. the order of the operations determines the possible outcome. In other words, the fundamental difference between classical and quantum-mechanical physical quantities (see also Subsec. 2.1.1) is that the former are mathematically represented by classical numbers (c-numbers), and therefore commute, whereas the latter are represented by quantum numbers (q-numbers), i.e. operators, and *do not necessarily commute*.

If we indicate with $\hat{P}_2$ the projection performed by the filter P2, and with $\hat{P}_3$ the operation performed by the filter P3, then we can formulate the above statement as

$$\hat{P}_2\hat{P}_3 - \hat{P}_3\hat{P}_2 = \left[\hat{P}_2, \hat{P}_3\right]_- \neq 0, \tag{2.85}$$

where the expression

$$\left[\hat{O}, \hat{O}'\right]_- = \hat{O}\hat{O}' - \hat{O}'\hat{O} \tag{2.86}$$

is called the *commutator* of the operators $\hat{O}$, $\hat{O}'$. In the rest of the book we shall omit the minus sign in $[\cdot, \cdot]_-$ for the sake of simplifying notation. Let us briefly prove Eq. (2.85).

Proof

Suppose that filter P1 selects states of vertical polarization and P2 states of horizontal polarization described by the vectors (2.11). The projectors associated with P1 and P2 can be described as $\hat{P}_1 = \hat{P}_v$ and $\hat{P}_2 = \hat{P}_h$, where $\hat{P}_v$ and $\hat{P}_h$ are given by Eqs. (2.13a). It is then evident that the successive operations performed by P1 and P2 give a zero output

$$\hat{P}_1\hat{P}_2 = \begin{bmatrix} 1 & 0 \\ 0 & 0 \end{bmatrix}\begin{bmatrix} 0 & 0 \\ 0 & 1 \end{bmatrix} = \begin{bmatrix} 0 & 0 \\ 0 & 0 \end{bmatrix} = 0; \tag{2.87a}$$

$$\hat{P}_2\hat{P}_1 = \begin{bmatrix} 0 & 0 \\ 0 & 1 \end{bmatrix}\begin{bmatrix} 1 & 0 \\ 0 & 0 \end{bmatrix} = \begin{bmatrix} 0 & 0 \\ 0 & 0 \end{bmatrix} = 0, \tag{2.87b}$$

from which it follows that $\hat{P}_1$ and $\hat{P}_2$ trivially commute, i.e. $[\hat{P}_1, \hat{P}_2] = 0$. On the other hand, the state selected by $\hat{P}_3$ can be described by the superposition

$$| c\rangle = \cos\phi \,| v\rangle + \sin\phi | h\rangle. \tag{2.88}$$

Then, $\hat{P}_3 = | c\rangle \langle c |$ is

$$\begin{aligned} \hat{P}_3 &= \left(\cos^2\phi \,| v\rangle \langle v | + \cos\phi \sin\phi \,| v\rangle \langle h | + \sin\phi\cos\phi \,| h\rangle \langle v | + \sin^2\phi \,| h\rangle \langle h |\right) \\ &= \begin{bmatrix} \cos^2\phi & \cos\phi\sin\phi \\ \sin\phi\cos\phi & \sin^2\phi \end{bmatrix}. \end{aligned} \tag{2.89}$$

Therefore, we can now prove the result (2.85) as follows:

$$\begin{aligned} \left[\hat{P}_2, \hat{P}_3\right] &= \begin{bmatrix} 0 & 0 \\ 0 & 1 \end{bmatrix}\begin{bmatrix} \cos^2\phi & \sin\phi\cos\phi \\ \sin\phi\cos\phi & \sin^2\phi \end{bmatrix} - \begin{bmatrix} \cos^2\phi & \sin\phi\cos\phi \\ \sin\phi\cos\phi & \sin^2\phi \end{bmatrix}\begin{bmatrix} 0 & 0 \\ 0 & 1 \end{bmatrix} \\ &= \begin{bmatrix} 0 & 0 \\ \sin\phi\cos\phi & \sin^2\phi \end{bmatrix} - \begin{bmatrix} 0 & \sin\phi\cos\phi \\ 0 & \sin^2\phi \end{bmatrix} \\ &= \begin{bmatrix} 0 & -\sin\phi\cos\phi \\ \sin\phi\cos\phi & 0 \end{bmatrix} \neq 0. \end{aligned} \tag{2.90}$$

Q.E.D

Box 2.5 **Commutation and product of Hermitian operators**

Note that, although projectors are Hermitian operators, their product is not necessarily a Hermitian operator [*Halmos* 1951, 41–42]. The necessary and sufficient condition for this is precisely that the two projectors (or, in more general terms, the two Hermitian operators) commute. In fact, if

$$\hat{O}_1 = \hat{O}_1^\dagger \quad \text{and} \quad \hat{O}_2 = \hat{O}_2^\dagger, \tag{2.91}$$

we have

$$\left(\hat{O}_1\hat{O}_2\right)^\dagger = \hat{O}_2^\dagger\hat{O}_1^\dagger = \hat{O}_2\hat{O}_1, \tag{2.92}$$

but this does not necessarily imply that we also have

$$\hat{O}_2\hat{O}_1 = \hat{O}_1\hat{O}_2. \tag{2.93}$$

It should be noted that non-commuting operations are not specific to quantum mechanics. In fact, it is well known that rotations in the three-dimensional space in general do not commute. For instance, a $\pi/2$ rotation about the z-axis followed by a $\pi/2$ rotation about the y-axis is not equivalent to a $\pi/2$ rotation about the y-axis followed by a $\pi/2$ rotation about the z-axis (see Fig. 2.4).

What we have seen teaches us that the operators representing quantum-mechanical observables in general do not commute. The concept of non-commutability was first introduced in quantum theory by Heisenberg and formally refined by Born and Jordan and Heisenberg himself, and it is often taken to represent the very birth of quantum mechanics.[16] Later (see Subsec. 2.2.7), we shall discuss the commutation relations between concrete quantum mechanical observables. Here we only want to establish some general properties of commutators. The most fundamental among these properties are the following[17] (notice the analogy with the properties of Poisson brackets (1.10); see also Sec. 3.7):

- Commutators are antisymmetric, i.e.

$$\left[\hat{O}, \hat{O}'\right] = -\left[\hat{O}', \hat{O}\right]. \tag{2.94}$$

- Commutators are bilinear, that is, given two (complex) scalars α and β, we have

$$\left[\alpha\hat{O} + \beta\hat{O}', \hat{O}''\right] = \alpha\left[\hat{O}, \hat{O}''\right] + \beta\left[\hat{O}', \hat{O}''\right]. \tag{2.95}$$

- Commutators (as well as classical Poisson brackets) satisfy the Jacobi identity, i.e. (see Prob. 2.11)

$$\left[\hat{O}, \left[\hat{O}', \hat{O}''\right]\right] + \left[\hat{O}', \left[\hat{O}'', \hat{O}\right]\right] + \left[\hat{O}'', \left[\hat{O}, \hat{O}'\right]\right] = 0. \tag{2.96}$$

[16] See [Heisenberg 1925] [Born/Jordan 1925] [Born *et al.* 1926]. For a history of Heisenberg's contribution see [*Mehra/Rechenberg* 1982–2001, II, 261–312]. For a historical reconstruction of the matrix – mechanics formulation see the third volume of [*Mehra/Rechenberg* 1982–2001].

[17] See [*Weyl* 1936, 260].

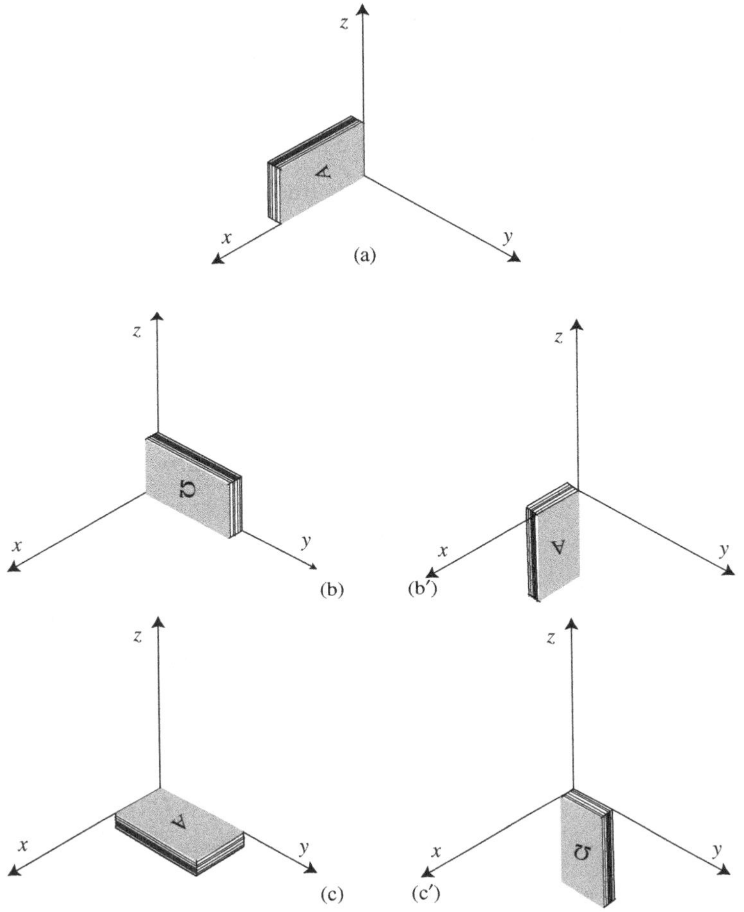

Figure 2.4 Two sequences of two rotations of a book (A is on the first cover while Ω is on the last cover) are shown. In the first sequence, i.e. (a)–(b)–(c), a rotation about the z-axis has been first applied, followed by a rotation about the y-axis. In the second sequence, i.e. (a)–(b′)–(c′), the rotation about the y-axis follows the rotation about the z-axis. All rotations are of 90° and anticlockwise. Since the two final configurations (c) and (c′) are different, rotations in the three-dimensional space in general do not commute.

In addition, it is possible to verify that commutators also satisfy the following properties:

- Any operator $\hat{O}$ commutes with itself, that is

$$\left[\hat{O}, \hat{O}\right] = 0. \tag{2.97}$$

- Any operator $\hat{O}$ commutes with any other operator that is only a function of $\hat{O}$ (see Prob. 2.12)

$$\left[\hat{O}, f\left(\hat{O}\right)\right] = 0. \tag{2.98}$$

- Given three operators $\hat{O}$, $\hat{O}'$ and $\hat{O}''$, we have (see Prob. 2.13)

$$\left[\hat{O}\hat{O}', \hat{O}''\right] = \left[\hat{O}, \hat{O}''\right]\hat{O}' + \hat{O}\left[\hat{O}', \hat{O}''\right]. \tag{2.99}$$

We finally note that, given three observables $\hat{O}$, $\hat{O}'$ and $\hat{O}''$, for which $\left[\hat{O}, \hat{O}'\right] = 0$ and $\left[\hat{O}, \hat{O}''\right] = 0$, it is not necessarily true that $\left[\hat{O}', \hat{O}''\right] = 0$ (see Prob. 2.14).

We have seen (Th. 2.2: p. 47) that the eigenvectors of a Hermitian operator constitute a basis on the underlying Hilbert space. What is then the relationship between this fundamental property and commutability? This question is answered by the following important theorem:

Theorem 2.4 (Commuting observables) *Two observables $\hat{O}$ and $\hat{O}'$ commute if and only if they admit a common basis of eigenvectors.*

Proof

First we prove that, if the observables $\hat{O}$ and $\hat{O}'$ admit a common basis of eigenvectors, then they commute. The hypothesis can be stated as follows:

$$\hat{O} \mid b_k\rangle = o_k \mid b_k\rangle\,, \quad \hat{O}' \mid b_k\rangle = o'_k \mid b_k\rangle, \tag{2.100}$$

where $\{\mid b_k\rangle\}$ is the common basis. Then, we have

$$\hat{O}\hat{O}' \mid b_k\rangle = \hat{O}o'_k \mid b_k\rangle = o'_k\hat{O} \mid b_k\rangle = o'_k o_k \mid b_k\rangle; \tag{2.101a}$$

$$\hat{O}'\hat{O} \mid b_k\rangle = \hat{O}'o_k \mid b_k\rangle = o_k\hat{O}' \mid b_k\rangle = o_k o'_k \mid b_k\rangle. \tag{2.101b}$$

Since o'_k and o_k are c-numbers, and Eqs. (2.101) hold for any k, it is evident that $\hat{O}$ and $\hat{O}'$ commute.

Now we prove that, if the observables $\hat{O}$ and $\hat{O}'$ commute, then they admit a common basis of eigenvectors. In fact, if $\{\mid a_k\rangle\}$ is a certain basis on the Hilbert space, we have

$$\left(OO'\right)_{jk} = \langle a_j \mid \hat{O}\hat{O}' \mid a_k\rangle = \sum_n \langle a_j \mid \hat{O} \mid a_n\rangle\,\langle a_n \mid \hat{O}' \mid a_k\rangle = \sum_n O_{jn}O'_{nk}. \tag{2.102}$$

Assuming now that $\{\mid a_k\rangle\}$ is a basis of eigenkets of $\hat{O}$, or $\hat{O} \mid a_k\rangle = o_k \mid a_k\rangle$, we can rewrite Eq. (2.102) as follows:

$$\left(OO'\right)_{jk} = \sum_n \langle a_j \mid \hat{O} \mid a_n\rangle\,\langle a_n \mid \hat{O}' \mid a_k\rangle = \sum_n o_n\delta_{jn}O'_{nk} = o_j O'_{jk}. \tag{2.103a}$$

Inverting the order of the operators, we have instead

$$\left(O'O\right)_{jk} = \sum_n \langle a_j \mid \hat{O}' \mid a_n\rangle\,\langle a_n \mid \hat{O} \mid a_k\rangle = \sum_n O'_{jn}o_k\delta_{nk} = o_k O'_{jk}. \tag{2.103b}$$

Now, since $[\hat{O}, \hat{O}'] = 0$, then we must have $(o_j - o_k)O'_{jk} = 0$. This condition is trivially satisfied for $j = k$. If, for $j \neq k$, we had $o_j \neq o_k$ (the non-degenerate case), we would have already proved the result. In fact, in such a case we would have $O'_{jk} = 0$ for $j \neq k$, i.e.

$O'_{jk} = O'_{jj}\delta_{jk}$. Then, the operator $\hat{O}'$ would be diagonal in the basis of the eigenvectors of $\hat{O}$. In general, however, there can be several eigenvectors for the same eigenvalue (degeneracy of the eigenvalues) (see Th. 2.2, p. 47, and comments). In this case we could have $o_j = o_k$ even if $j \neq k$. In this circumstance, if we take a certain degenerate eigenvalue, we may consider the subspace spanned by the eigenvectors which pertain to that eigenvalue. In this subspace $\hat{O}'$ can be diagonalized because it is Hermitian and also $\hat{O} = o_k\hat{I}$ (where $\hat{I}$ is the identity in this subspace). Then, even in this case $\hat{O}$ and $\hat{O}'$ can be jointly diagonalized and therefore share a common basis of eigenvectors.

Q.E.D

As a consequence of Th. 2.4, the following corollary can be proved:

Corollary 2.1 (Simultaneous measurability) *The necessary and sufficient condition for two observables $\hat{O}$ and $\hat{O}'$ to be simultaneously measurable with arbitrary precision is that they commute.*

Proof

In fact, if $\hat{O}$ and $\hat{O}'$ commute, then we have a common eigenbasis $\{|o_k, o'_k\rangle\}$ such that

$$\hat{O}\,|o_k, o'_k\rangle = o_k\,|o_k, o'_k\rangle, \tag{2.104a}$$

$$\hat{O}'\,|o_k, o'_k\rangle = o'_k\,|o_k, o'_k\rangle. \tag{2.104b}$$

This means that there is a common basis in which both observables are perfectly determined. In other words, for each state $|o_k, o'_k\rangle$ of the basis, the observable $\hat{O}$, if measured, gives with certainty the eigenvalue o_k as outcome as well as the observable $\hat{O}'$, if measured, gives with certainty the eigenvalue o'_k as outcome.

Q.E.D

As an immediate consequence of Cor. 2.1 we may state that non-commuting observables cannot be simultaneously measurable with arbitrary precision. In general, therefore, given the set of observables of a physical system $\mathcal{S}$, it will be possible to divide them into separate subsets of reciprocally commuting observables. These are called *complete sets* and represent the maximum number of properties of $\mathcal{S}$ that can be jointly known (see also Subsec. 2.2.7). We see here that, while in classical mechanics it is possible to know jointly all the properties of a system, in quantum mechanics by a *complete description* we mean the knowledge of all the observables in certain complete (but not necessarily disjoint) sets.

As we shall see in the following, the non-commutability between quantum mechanical observables has extraordinary implications in the very foundations of the theory and in the corresponding interpretation of its physical reality (see e.g. Subsec. 2.3.3).

2.2 Wave function and basic observables

In this section we shall apply (and develop) the formalism of the previous section to concrete quantum observables. After having introduced the concept of wave function (Subsec. 2.2.1), we will discuss the difficult problem of normalization (Subsec. 2.2.2). In Subsec. 2.2.3 we introduce the position operator whereas in Subsec. 2.2.4 we introduce the momentum observable. In Subsec. 2.2.5 we analyze the relationship between position and momentum representations. In Subsec. 2.2.6 the energy observable is shortly introduced – further developments can be found in Ch. 3. Finally, in Subsec. 2.2.7, the basic commutation relation between position and momentum is deduced.

2.2.1 Wave function

So far we have discussed examples of quantum systems in terms of photons and their polarization. However, quantum mechanics has to apply to a generic microscopic system as well. In particular, it should enable us to deal with a particle (or an ensemble of particles) moving in a configuration space. In order to investigate the quantum mechanical behavior of such a system we need to introduce the position observable $\hat{\mathbf{r}} = (\hat{x}, \hat{y}, \hat{z})$. For the sake of simplicity let us first treat the one-dimensional case. Let us denote the eigenvectors of the position operator $\hat{x}$ by $| x \rangle$. Since $\hat{x}$ is a continuous operator, we have to use the generalization given at the end of Subsec. 2.1.1. Given a generic state vector $| \psi \rangle$, we may try to expand this vector in terms of the position eigenvectors, and make use of the identity (see Eq. (2.22a))

$$\int dx \, | x \rangle \langle x | = \hat{I} \tag{2.105}$$

in order to obtain (see Eq. (2.21))

$$\begin{aligned} | \psi \rangle &= \int dx \, | x \rangle \langle x \mid \psi \rangle \\ &= \int dx \psi(x) \, | x \rangle, \end{aligned} \tag{2.106}$$

so that we may interpret the scalar product (see also Eq. (2.16))

$$\langle x \mid \psi \rangle = \psi(x) \tag{2.107}$$

as the continuous probability amplitude of finding the particle in the interval $(x, x + dx)$, will be the probability of finding the particle between x and $x + dx$, and $\wp(x) = |\psi(x)|^2$ will be the corresponding probability density (see Pr. 2.2: p. 57, and also Sec. 1.4). We then expect that the probability of finding the particle anywhere in the whole configuration space be equal to one; that is (see Eq. (2.67))

$$\int_{-\infty}^{+\infty} dx|\psi(x)|^2 = \int_{-\infty}^{+\infty} dx\psi^*(x)\psi(x) = 1. \tag{2.108}$$

This is also called the *normalization condition* and the function $\psi(x)$ is called the *wave function* of the particle.

Notice that, since a normalized state vector is defined up to a phase factor (see p. 60), for the wave function a global phase factor is irrelevant.

From what we have seen above, it turns out that there is an obvious relation between the *bracket* Dirac notation and integrals of products of wave functions in the configuration space: they both represent scalar products. Since the integral (2.108) may be written as $\int dx\psi^*(x)\psi(x)$, it can be also expanded as the scalar product $(\psi(x), \psi(x))$. More specifically, let us consider two state vectors $|\psi\rangle$ and $\left|\psi'\right\rangle$. These may be also represented by the corresponding wave functions $\psi(x) = \langle x \mid \psi\rangle$ and $\psi'(x) = \left\langle x \mid \psi'\right\rangle$. It is then straightforward to write

$$\begin{aligned}\left\langle \psi \mid \psi' \right\rangle &= \int_{-\infty}^{+\infty} dx \, \langle \psi \mid x\rangle \left\langle x \mid \psi'\right\rangle \\ &= \int_{-\infty}^{+\infty} dx\psi^*(x)\psi'(x).\end{aligned} \tag{2.109}$$

It may happen that a wave function is not normalized, so that we have

$$\int_{-\infty}^{+\infty} dx|\psi(x)|^2 = N, \tag{2.110}$$

where N is finite and different from 1. Then, in order to normalize the wave function $\psi(x)$ it is sufficient to consider a wave function $\psi_{\rm norm}(x)$ such that

$$\psi_{\rm norm}(x) = \frac{1}{\sqrt{N}}\psi(x). \tag{2.111}$$

Note that the eigenfuctions of the continuous spectrum are not normalizable. In the next subsection we shall investigate this problem.

2.2.2 Normalization

Let us consider a one-dimensional observable $\hat{\xi}$ with a continuous spectrum. The state vector of the particle can be expanded as (see Eqs. (1.35) and (2.21))

$$|\psi\rangle = \int d\xi c(\xi) |\xi\rangle, \tag{2.112}$$

where the vectors $|\xi\rangle$ are the eigenkets of the observable $\hat{\xi}$ and $|c(\xi)|^2 \, d\xi$ represent the probability that the value of $\hat{\xi}$ can be found in the interval $(\xi, \xi + d\xi)$. We can then write the following identity:

$$\int d\xi\, |c(\xi)|^2 = \int dx\, |\psi(x)|^2. \tag{2.113}$$

Substituting $\psi^*(x) = \langle \psi \mid x \rangle$ as expanded in Eq. (2.112) into the rhs of Eq. (2.113), we obtain

$$\int d\xi c^*(\xi)c(\xi) = \int d\xi c^*(\xi)\left[\int dx \psi(x)\varphi^*_\xi(x)\right], \tag{2.114}$$

where $\varphi_\xi(x) = \langle x \mid \xi \rangle$, which yields

$$c(\xi) = \langle \xi \mid \psi \rangle = \int dx \psi(x)\varphi^*_\xi(x). \tag{2.115}$$

Back-substituting $\psi(x) = \langle x \mid \psi \rangle$ as expanded in Eq. (2.112) into Eq. (2.115), we obtain

$$c(\xi) = \int d\xi' c\left(\xi'\right)\left[\int dx \varphi_{\xi'}(x)\varphi^*_\xi(x)\right], \tag{2.116}$$

from which we must conclude that (see Eqs. (2.23))

$$\int dx \varphi_{\xi'}(x)\varphi^*_\xi(x) = \delta(\xi - \xi'). \tag{2.117}$$

In other words, the eigenfunctions of an observable with a continuous spectrum are not normalizable.

Summing up, concerning normalization we have three possible cases:

- The wave function is normalized:

$$\int dx\, |\psi(x)|^2 = 1. \tag{2.118}$$

- The wave function is not normalized but is normalizable:

$$\int dx\, |\psi(x)|^2 = N \neq 1, \quad |N| < \infty. \tag{2.119}$$

- The wave function is not normalizable:

$$\int dx\, |\psi(x)|^2 = \infty. \tag{2.120}$$

In the latter case, $|\psi(x)|^2\, dx$ cannot represent the probability of finding the particle in the interval $(x,\ x + dx)$. However, the ratio

$$\frac{\wp(x')}{\wp(x'')} = \frac{|\psi(x')|^2}{|\psi(x'')|^2} \tag{2.121}$$

still determines the relative probabilities pertaining to two different values x' and x'' of the position.

2.2.3 Position operator

The wave function $\psi(x)$, viewed as a function of the position $\hat{x}$, is a particular representation of the state vector $\mid \psi \rangle$. In this representation the operator $\hat{x}$ takes a very simple form. The eigenvectors $\mid x \rangle$'s of $\hat{x}$ represent state vectors for which the position has a determined

Table 2.1 Different cases and ways of expressing the basic quantum formalism

Discrete case		Continuous case	
Dirac algebra	Wave-function	Dirac algebra	Wave-function
$\vert\psi\rangle = \sum_j c_j \vert\varphi_j\rangle$	$\psi(x) = \sum_j c_j \varphi_j(x)$	$\vert\psi\rangle = \int d\xi \vert\xi\rangle \langle\xi \vert \psi\rangle$	$\psi(x) = \int d\xi c(\xi)\varphi_\xi(x)$
$c_j = \langle\varphi_j \vert \psi\rangle$	$c_j = \int dx\varphi_j^*(x)\psi(x)$	$c(\xi) = \langle\xi \vert \psi\rangle$	$c(\xi) = \int dx\varphi_\xi^*(x)\psi(x)$
$\hat{O} = \sum_j o_j \hat{P}_j$		$\hat{\xi} = \int d\xi \hat{P}(\xi)$	

value, i.e. the eigenvalue x associated with $|x\rangle$. In other words, the eigenvalue equation (see also Eq. (2.2)) of the observable $\hat{x}$ may be written as

$$\hat{x}\,|x\rangle = x\,|x\rangle. \tag{2.122}$$

Writing the Hermitian conjugate of Eq. (2.122)

$$\langle x\,|\,\hat{x} = x\,\langle x\,|\,. \tag{2.123}$$

where the eigenvalues x are obviously real (see Th. 2.1), and taking the scalar product with a generic state vector $|\,\psi\rangle$, we obtain

$$\hat{x}\psi(x) = x\psi(x), \tag{2.124}$$

which means that the one-dimensional position operator $\hat{x}$ in the position representation acts both on its eigenvector $|x\rangle$ and on the wave function $\psi(x)$ simply as a multiplication by the scalar x. This may be seen equally well by using the concept of mean value (see Eq. (2.78)). Let us write

$$\langle\hat{x}\rangle_\psi = \langle\psi\,|\hat{x}|\,\psi\rangle = \int_{-\infty}^{+\infty} dx\,\langle\psi\,|\,x\rangle\langle x\,|\hat{x}|\,\psi\rangle, \tag{2.125}$$

since integration over the set of projectors $|x\rangle\,\langle x\,|$ yields the identity (see Eq. (2.22a)). On the other hand, given the meaning of $|\psi(x)|^2$, we may also write (see Eq. (2.78))

$$\langle x\rangle_\psi = \int_{-\infty}^{+\infty} dx x|\psi(x)|^2 = \int_{-\infty}^{+\infty} dx\psi^*(x)\psi(x)x = \int_{-\infty}^{+\infty} dx\,\langle\psi\,|\,x\rangle\,\langle x\,|\,\psi\rangle\,x. \tag{2.126}$$

Comparing the rhs of Eqs. (2.125) and (2.126), we find

$$\langle x\,|\hat{x}|\,\psi\rangle = x\,\langle x\,|\,\psi\rangle, \tag{2.127}$$

which is equivalent to Eq. (2.124).

Finally, it is convenient to determine the eigenfunction $\varphi_{x_0}(x) = \langle x\,|\,x_0\rangle$, where the position operator takes on the determined value x_0. To this end, we first note that, for any x, we have (see Eqs. (2.122) and (2.124))

$$\hat{x}\varphi_{x_0}(x) = x\varphi_{x_0}(x) = x_0\varphi_{x_0}(x). \tag{2.128}$$

The second equality of Eq. (2.128) is automatically satisfied for $x = x_0$. For $x \neq x_0$ the equality is satisfied only if $\varphi_{x_0}(x) = 0$. Therefore, it follows that

$$\varphi_{x_0}(x) = \delta(x - x_0). \tag{2.129}$$

2.2.4 Momentum operator

The aim of the present and of the next two subsections is to introduce two additional physical quantities in quantum mechanics: momentum and energy. In order to accomplish this task, we need a link with classical mechanics where energy and momentum were first defined in a natural way. This link is provided by the *correspondence principle*, first formulated in [Bohr 1920], which may be stated as follows:[18]

Principle 2.3 (Correspondence principle) *The quantum-mechanical physical quantities should tend to the classical-mechanical counterparts in the macroscopic limit.*

With *macroscopic limit* we mean a physical scale where the action (1.14) is much larger than Planck's constant. In this situation h (see Subsec. 1.2.1 and Sec. 1.5), is negligible and quantum effects (such as superposition, interference, etc.) are very small. For this reason, sometimes – though in a improper way – the classical limit is referred to as the physical situation in which $h \to 0$. This limit should not frighten the reader. Although h is a constant and as such cannot change, the limit should be understood as expressing the relative weight of this quantity with respect to the system's action.

In classical mechanics, *momentum* is defined as the quantity which is conserved under global spatial translations or, alternatively, as the generator of spatial translations. Let us then consider a one-dimensional particles described by the wave function $\psi(x)$. A rigid translation by a quantity a of this system will change $\psi(x)$ into[19]

$$\begin{aligned}\psi(x+a) &= \psi(x) + a\frac{\partial}{\partial x}\psi + \frac{a^2}{2}\frac{\partial^2}{\partial x^2}\psi + \frac{a^3}{6}\frac{\partial^3}{\partial x^3}\psi + \cdots + \frac{a^n}{n!}\frac{\partial^n}{\partial x^n}\psi + \cdots \\ &= \sum_{j=0}^{\infty}\frac{a^j}{j!}\frac{\partial^j}{\partial x^j}\psi(x)) \\ &= \hat{U}_a\psi(x).\end{aligned} \tag{2.130}$$

[18] For a historical reconstruction see [*Mehra/Rechenberg* 1982–2001, I, 246–57].

[19] In the one-dimensional case we could use the total derivative instead of the partial one. However, we prefer to use partial derivatives throughout the book.

The unitary operator $\hat{U}_a$ [see Eq. (2.34)], translating the wave function of the system by an amount a, which is called *translation operator*, can be written as

$$\hat{U}_a = e^{\imath a \hat{G}_T}. \tag{2.131}$$

As a consequence, the generator $\hat{G}_T$ of the spatial displacements can be identified as

$$\hat{G}_T = -\imath \frac{\partial}{\partial x} \tag{2.132}$$

up to a constant factor and must represent the quantum-mechanical *momentum operator* of the particle (see also Subsec. 3.5.4), i.e.

$$\hat{p}_x = -\imath \frac{\partial}{\partial x} \tag{2.133}$$

up to a constant factor. For dimensional reasons, and following the correspondence principle, we take this constant to be $\hbar = h/2\pi$. In the three-dimensional case, the momentum operator is then given by

$$\hat{\mathbf{p}} = -\imath \hbar \nabla. \tag{2.134}$$

Once we have determined the form of the momentum operator, it is necessary to find its *eigenfunctions*. In other words, we have to solve the eigenvalue equation

$$\hat{p}_x \varphi_p(x) = p_x \varphi_p(x) \tag{2.135}$$

for the unknown functions $\varphi_p(x) = \langle x \mid p_x \rangle$ in the position representation, where the $\mid p_x \rangle$'s are the one-dimensional eigenkets of the momentum. This amounts to solving the differential equation

$$\frac{\partial}{\partial x} \varphi_p(x) = \frac{\imath}{\hbar} p_x \varphi_p(x), \tag{2.136}$$

which has been obtained upon substitution of

$$\hat{p}_x = -\imath \hbar \frac{\partial}{\partial x} \tag{2.137}$$

into Eq. (2.135). The solutions of Eq. (2.136) can be immediately written as

$$\varphi_p(x) = C e^{\frac{\imath}{\hbar} p_x x}, \tag{2.138}$$

where C is some integration constant.

It is convenient to express the momentum p_x in terms of the *wave* or *propagation vector*

$$k_x = \frac{2\pi}{\lambda}. \tag{2.139}$$

The link between these two quantities comes from the de Broglie relationship (1.78), from which it can easily be derived that

$$p_x = \frac{h}{\lambda} = \hbar k_x. \tag{2.140}$$

In terms of the wave vector the momentum eigenfunctions (2.138) may be written as

$$\varphi_k(x) = Ce^{\imath k_x x}. \tag{2.141}$$

These eigenfunctions are often called *plane waves*.

It can easily be seen that these functions, being the eigenfunctions of an observable with a continuous spectrum, are not normalizable (see Subsec. 2.2.2). In fact,

$$\int_{-\infty}^{+\infty} dx |\varphi_p(x)|^2 = \begin{cases} 0 & \text{for} \quad p_x \neq 0 \\ \int_{-\infty}^{+\infty} |C|^2 dx = \infty & \text{for} \quad p_x = 0 \end{cases}. \tag{2.142}$$

As we have seen, the only function which satisfies the normalization requirements of Eq. (2.142) is the Dirac delta function. The orthonormality condition for the momentum eigenfunctions will then be given by

$$\int_{-\infty}^{+\infty} dx \varphi_p^*(x) \varphi_{p'}(x) = |C|^2 \int_{-\infty}^{+\infty} dx e^{\frac{\imath}{\hbar}(p'_x - p_x)x} = 2\pi\hbar\, |C|^2\, \delta(p'_x - p_x), \tag{2.143}$$

where we have made use of the formula[20]

$$\frac{1}{2\pi} \int_{-\infty}^{+\infty} dx e^{\imath \alpha x} = \delta(\alpha). \tag{2.144}$$

Therefore the constant C of Eq. (2.138) or Eq. (2.141) may be taken as equal to $(2\pi\hbar)^{-\frac{1}{2}}$ and the momentum eigenfunctions can be finally written as

$$\varphi_k(x) = \frac{1}{\sqrt{2\pi\hbar}} e^{\imath k_x x} \tag{2.145}$$

or

$$\varphi_p(x) = \frac{1}{\sqrt{2\pi\hbar}} e^{\frac{\imath}{\hbar} p_x x}. \tag{2.146}$$

In the three-dimensional case it is straightforward to generalize the result (2.145) into

$$\varphi_{\mathbf{k}}(\mathbf{r}) = \frac{1}{\sqrt{8\hbar^3 \pi^3}} e^{\imath \mathbf{k}\cdot\mathbf{r}}, \tag{2.147}$$

where $\mathbf{k} = (k_x, k_y, k_z)$, and which, in contrast to Eq. (2.141), are called *spherical waves*.

[20] See [*Byron/Fuller* 1969–70, 246–53].

2.2.5 Momentum representation

So far we have worked in the position representation. As we have said, this means that the wave function ψ is considered as a function of the position x (or $\mathbf{r}$ in the three-dimensional case). As we have seen in the previous subsection, in this representation the momentum acts on $\psi(x)$ as a differential operator, whereas the position observable is simply a multiplication operator. However, we may as well consider a different representation according to which the state vector $| \psi\rangle$ is projected onto the eigenbra $\langle p_x |$ of the momentum operator, i.e. $\tilde{\psi}(p_x) = \langle p_x | \psi\rangle$ can be viewed as the wave function in the momentum representation. We have used the superscript tilde in order to emphasize that the functional dependence of $\tilde{\psi}$ on p_x is in general obviously different from the functional dependence of ψ on x.[21] In the momentum representation it is obviously the momentum that acts as a multiplication operator, that is

$$\hat{p}_x \tilde{\psi}(p_x) = p_x \tilde{\psi}(p_x). \tag{2.148}$$

What is the connection between the momentum and the position representations of the wave function? In order to answer this question, we note that any wave function $\psi(x) = \langle x | \psi\rangle$ may be rewritten as

$$\psi(x) = \int dp_x \langle x | p_x\rangle \langle p_x | \psi\rangle, \tag{2.149}$$

where we have taken advantage of the fact that integration over the set of projectors $| p_x\rangle \langle p_x |$ gives the identity operator (see Eq. (2.22a)). Now we recall that $\langle x | p_x\rangle = \varphi_p(x)$ is the momentum eigenfunction in the position representation and $\langle p_x | \psi\rangle = \tilde{\psi}(p_x)$, so that, making use of Eq. (2.146), Eq. (2.149) becomes

$$\psi(x) = \int dp_x \varphi_p(x) \tilde{\psi}(p_x) = \frac{1}{\sqrt{2\pi\hbar}} \int dp_x e^{\frac{i}{\hbar} p_x x} \tilde{\psi}(p_x). \tag{2.150a}$$

Equation (2.150a) shows that the inverse Fourier transform[22] of the wave function in the momentum representation gives the wave function in the position representation. Therefore, by inverting Eq. (2.150a), one obtains that the wave function in the momentum representation

$$\tilde{\psi}(p_x) = \frac{1}{\sqrt{2\pi\hbar}} \int dx e^{-\frac{i}{\hbar} p_x x} \psi(x) \tag{2.150b}$$

[21] It is a historical contingency depending on the development of quantum mechanics that, when the first wave function was introduced, it was written $\psi(x)$. If it had been written $\psi_x(x)$, one would have written $\psi_p(p_x)$ for indicating the wave function of the momentum in the momentum representation. Since it did not happen, we are obliged to choose forms like $\tilde{\psi}(p_x)$ in order to indicate both the different dependence and the different representation.

[22] See [*Byron/Fuller* 1969–70, 246–53].

is simply given by the Fourier transform of the corresponding wave function in the position representation. Using Dirac formalism (as in Eq. (2.149)), Eq. (2.150b) may be also written as

$$\tilde{\psi}(p_x) = \langle p_x \mid \psi\rangle = \int dx \, \langle p_x \mid x\rangle \langle x \mid \psi\rangle = \int dx \, \langle p_x \mid x\rangle \, \psi(x). \tag{2.151}$$

Then, changing the representation corresponds to projecting the state vector $\mid \psi\rangle$ onto different basis eigenvectors: position eigenvectors in the case of position representation and momentum eigenvectors for momentum representation.

It can be shown (see Prob. 2.17) that, if $\psi(x)$ is normalized, i.e.

$$\text{if} \quad \int_{-\infty}^{+\infty} dx |\psi(x)|^2 = 1, \quad \text{then also} \quad \int_{-\infty}^{+\infty} dp_x |\tilde{\psi}(p_x)|^2 = 1. \tag{2.152}$$

This is called the *Bessel–Parseval relationship.*

We have seen that the momentum acts as a differential operator in the position representation. Conversely, one may ask what is the form of the position observable in the momentum representation. It can be proved that (see Prob. 2.18)

$$\hat{x}\tilde{\psi}(p_x) = \imath\hbar \frac{\partial}{\partial p_x} \tilde{\psi}(p_x). \tag{2.153}$$

We should emphasize that writing a wave function in different representations is a special instance of the change of basis (see Subsec. 2.1.2), and in particular it corresponds to a change from the $|x\rangle$ to the $\mid p_x\rangle$ basis and *vice versa.* In Subsec. 2.1.2 we have mainly considered the problem of a change of basis in the discrete case. Let us now address this problem under general terms in the continuous case. Any state vector $\mid \psi\rangle$ can be expanded in an arbitrary orthonormal basis $\{\mid \xi\rangle\}$ – given by the eigenvectors of a continuous one-dimensional observable $\hat{\xi}$ – as (see Eq. (2.112))

$$\mid \psi\rangle = \int d\xi \mid \xi\rangle \langle \xi \mid \psi\rangle = \int d\xi \, \psi_\xi(\xi) \mid \xi\rangle . \tag{2.154a}$$

Similarly, the state vector $\mid \psi\rangle$ can be expanded in a different basis $\{\mid \eta\rangle\}$ – given by the eigenvectors of another one-dimensional observable $\hat{\eta}$, not necessarily conjugate to $\hat{\xi}$ – as (see also Eq. (2.46))

$$\mid \psi\rangle = \int d\eta \mid \eta\rangle \langle \eta \mid \psi\rangle = \int d\eta \psi_\eta(\eta) \mid \eta\rangle, \tag{2.154b}$$

where the functions $\psi(\eta)$ and $\psi_\xi(\xi)$ take the role of continuous coefficients in two different expansions of $\mid \psi\rangle$ (see Subsec. 2.2.2).

The relation between $\psi(\eta)$ and $\psi_\xi(\xi)$ is easily derived as

$$\begin{aligned} \psi_\xi(\xi) = \langle \xi \mid \psi\rangle &= \int d\eta \, \langle \xi \mid \eta\rangle \langle \eta \mid \psi\rangle \\ &= \int d\eta \varphi_\eta(\xi)\psi(\eta) \\ &= \hat{U}(\eta, \xi)\psi(\eta), \end{aligned} \tag{2.155}$$

where $\varphi_\eta(\xi) = \langle \xi \mid \eta \rangle$ are the eigenfunctions of the observable $\hat{\eta}$ in the ξ-representation. In the following we show that the transformation

$$\psi(\eta) \longrightarrow \psi_\xi(\xi) = \hat{U}(\eta, \xi)\psi(\eta) \tag{2.156}$$

conserves the scalar product and is therefore unitary (see p. 54 and also Ch. 8).[23] In fact, let us take two generic wave functions $\psi(\eta)$ and $\psi'(\eta)$ in the η–representation, with the scalar product

$$(\psi, \psi') = \int d\eta \psi^*(\eta)\psi'(\eta). \tag{2.157}$$

The application of the U-transformation on $\psi(\eta)$ and $\psi'(\eta)$ gives straightforwardly

$$\hat{U}(\eta, \xi)\psi(\eta) = \int d\eta \varphi_\eta(\xi)\psi(\eta), \tag{2.158a}$$

$$\hat{U}(\eta, \xi)\psi'(\eta) = \int d\eta \varphi_\eta(\xi)\psi'(\eta), \tag{2.158b}$$

whose scalar product yields

$$\begin{aligned}\left(\hat{U}\psi, \hat{U}\psi'\right) &= \int d\xi \left[\int d\eta \varphi_\eta^*(\xi)\psi^*(\eta)\right]\left[\int d\eta' \varphi_{\eta'}(\xi)\psi'(\eta')\right] \\ &= \int d\eta d\eta' \psi^*(\eta)\psi'(\eta') \int d\xi \varphi_\eta^*(\xi)\varphi_{\eta'}(\xi).\end{aligned} \tag{2.159}$$

Now, we know that $\varphi_{\eta'}(\xi) = \langle \xi \mid \eta' \rangle$ and $\varphi_\eta(\xi) = \langle \xi \mid \eta \rangle$, and therefore

$$\begin{aligned}\int d\xi \varphi_\eta^*(\xi)\varphi_{\eta'}(\xi) &= \int d\xi \langle \eta \mid \xi \rangle \langle \xi \mid \eta' \rangle \\ &= \langle \eta \mid \eta' \rangle = \delta(\eta - \eta').\end{aligned} \tag{2.160}$$

Substituting this result into Eq. (2.159), we obtain

$$\begin{aligned}\left(\hat{U}\psi, \hat{U}\psi'\right) &= \int d\eta d\eta' \psi^*(\eta)\psi'(\eta')\delta(\eta - \eta') \\ &= \int d\eta \psi^*(\eta)\psi'(\eta) \\ &= (\psi, \psi').\end{aligned} \tag{2.161}$$

It is worth noticing that in the special case of position $\hat{x}$ and momentum $\hat{p}_x$ – i.e. in general, of conjugate observables – the unitary transformation $\hat{U}$ is represented by the Fourier transform $\hat{U}_F$, so that we have

$$\tilde{\psi}(p_x) = \hat{U}_F\psi(x), \quad \psi(x) = \hat{U}_F^\dagger\tilde{\psi}(p_x), \tag{2.162}$$

where

$$\hat{U}_F f(x) = \frac{1}{\sqrt{2\pi\hbar}} \int dx e^{-\frac{i}{\hbar}p_x x} f(x), \quad \hat{U}_F^\dagger g(p_x) = \frac{1}{\sqrt{2\pi\hbar}} \int dp_x e^{\frac{i}{\hbar}p_x x} g(p_x). \tag{2.163}$$

[23] See also [*Fano* 1971, 75].

The main difference with respect to the discrete case treated in Subsec. 2.1.2 is that here, in the continuous and therefore infinite-dimensional case, we cannot express this unitary operation by means of a matrix.

2.2.6 Energy

As we have already seen in Sec. 1.1, a classical system is well defined by its energy, in the sense that the knowledge of the Hamiltonian function H allows us to derive the equations of motion. The Hamiltonian is given by the sum of the kinetic energy T and the potential energy V. In quantum mechanics this should hold true as well (see Pr. 2.3: p. 72). However, following Th. 2.1 (p. 46) we know that the energy must also be represented by a Hermitian operator. Classical mechanics helps us in determining such an operator. In fact in order to "quantize" the Hamiltonian it is sufficient first to replace in the classical formula $H = T + V$ the corresponding expressions in terms of momentum and position and finally to consider these physical quantities as operators. Thus, one immediately obtains an expression of the Hermitian operator $\hat{H}$, called the Hamiltonian operator in terms of the operators $\hat{x}$ and $\hat{p}_x$ which we have discussed in the previous subsections.

For instance, in the case of a one-dimensional particle subject to the potential energy $V(x)$ one has

$$\hat{H} = \hat{T} + \hat{V} = \frac{\hat{p}_x^2}{2m} + V(\hat{x}), \tag{2.164}$$

where m is the mass of the particle. As we know from Subsecs. 2.2.1–2.2.4, in the position representation – in which we shall usually work if not otherwise stated – the position operator acts as a simple multiplication, whereas the momentum acts as a differential operator. In such a representation, therefore, we shall write

$$\hat{H} = -\frac{\hbar^2}{2m}\frac{\partial^2}{\partial x^2} + V(x). \tag{2.165}$$

We have already seen that, as particular instances of the general formula (2.2), there are a position eigenvalue equation (2.122) and a momentum eigenvalue equation (2.135). Also for the energy it is possible to write an eigenvalue equation

$$\hat{H}\,|\,\psi_E\rangle = E\,|\,\psi_E\rangle, \tag{2.166}$$

where $|\,\psi_E\rangle$ are the eigenkets of the energy and the E's the corresponding eigenvalues. For reasons which shall become clear in Ch. 3, the states $|\,\psi_E\rangle$ in Eq. (2.166) are called *stationary states*. In the same chapter we shall see that the energy plays a fundamental role in the dynamics of a quantum system.

It is clear from Eqs. (2.165) and (2.166) that the eigenfunctions and eigenvalues of the energy will depend on the potential V and therefore on the particular kind of system we are facing. In the simplest case of a unidimensional free particle, the eigenvalue equation of the energy in the position representation becomes

$$\hat{H}\psi_E(x) = -\frac{\hbar^2}{2m}\frac{\partial^2}{\partial x^2}\psi_E(x) = E_p\psi_E(x), \tag{2.167}$$

where E_p is the eigenvalue of the energy corresponding to the momentum p. Eq. (2.167) may be rewritten as

$$\psi''_E(x) = -\frac{2mE_p}{\hbar^2}\psi_E(x), \tag{2.168}$$

where $\psi''_E(x)$ is the second derivative of $\psi_E(x)$ with respect to x. Its general solution is given by

$$\psi_E(x) = C_1 e^{\imath kx} + C_2 e^{-\imath kx}, \tag{2.169}$$

where C_1 and C_2 are integration constants and $k = \sqrt{2mE_p}/\hbar$. Therefore, in the case of a free particle, the energy eigenfunctions have a form similar to that of the momentum eigenfunctions (2.141).[24] This fact is not surprising. Indeed, in this case, we have $\hat{H} = \hat{p}_x^2/2m$ and, as a consequence, $\hat{p}_x$ commutes with $\hat{H}$ (see Eq. (2.98)), or $[\hat{H}, \hat{p}_x] = 0$. One can easily deduce from the classical expression of the energy of a free particle,

$$E_p = \frac{p_x^2}{2m}, \tag{2.170}$$

that $k = \sqrt{p_x^2}/\hbar$. The only difference between momentum and energy eigenfunctions, in the case of a one-dimensional free particle, is that we may distinguish the case of a particle moving from the left to the right ($e^{\imath kx}$ with momentum $p_x = \hbar k$) from the case of a particle moving from the right to the left ($e^{-\imath kx}$ with momentum $-p_x = -\hbar k$). In both cases, however, the particle has the same energy $E = p_x^2/2m$.

2.2.7 Commutation relations for position and momentum

So far we have seen that in general quantum mechanical observables may not commute. It is interesting to consider the case of momentum and position operators which we have discussed in the previous subsections. Our aim then is to compute the commutator

$$[\hat{x}, \hat{p}_x] = \hat{x}\hat{p}_x - \hat{p}_x\hat{x}. \tag{2.171}$$

To this end we apply such a commutator to an arbitrary wave function $\psi(x)$:

$$\begin{aligned}[\hat{x}, \hat{p}_x]\psi(x) &= x\left(-\imath\hbar\frac{\partial}{\partial x}\right)\psi(x) + \imath\hbar\frac{\partial}{\partial x}\left[x\psi(x)\right] \\ &= -\imath\hbar x\psi'(x) + \imath\hbar\psi(x) + \imath\hbar x\psi'(x) \\ &= \imath\hbar\psi(x),\end{aligned} \tag{2.172}$$

24 Being the energy of a free particle a continuous observable, its eigenfunctions are not normalizable (see Subsec. 2.2.2).

where use has been made of Eqs. (2.124) and (2.133). Since the wave function $\psi(x)$ in Eq. (2.172) must hold for arbitrary wavefunctions $\psi(x)$, we may write

$$\left[\hat{x}, \hat{p}_x\right] = \imath\hbar\hat{I}. \tag{2.173a}$$

Similar expressions can be derived (see Prob. 2.21) in the same way for the other components of momentum and position

$$\left[\hat{y}, \hat{p}_y\right] = \left[\hat{z}, \hat{p}_z\right] = \imath\hbar\hat{I}. \tag{2.173b}$$

On the other hand, we have

$$\begin{aligned}\left[\hat{x}, \hat{p}_y\right] &= \left[\hat{x}, \hat{p}_z\right] = 0,\\ \left[\hat{y}, \hat{p}_x\right] &= \left[\hat{y}, \hat{p}_z\right] = 0,\\ \left[\hat{z}, \hat{p}_x\right] &= \left[\hat{z}, \hat{p}_y\right] = 0,\end{aligned} \tag{2.173c}$$

as one can immediately verify by applying a similar procedure as in Eq. (2.172) to the corresponding commutators.

Equations (2.173) can be unified through the relation

$$\left[\hat{r}_j, \hat{p}_k\right] = \imath\hbar\delta_{jk}, \tag{2.174}$$

where $j, k = (x, y, z)$ and $\hat{r}_x = \hat{x}, \hat{r}_y = \hat{y}, v_z = \hat{z}$.

It is then clear that $\hat{x}$ and $\hat{p}_x$ (as well as $\hat{y}$ and $\hat{p}_y$ or $\hat{z}$ and $\hat{p}_z$) cannot have a common basis of eigenvectors (see Th. 2.4). As a consequence they are not simultaneously measurable with arbitrary precision (see Cor. 2.1). In this case we shall say that $\hat{x}$ and $\hat{p}_x$ (as well as $\hat{y}$ and $\hat{p}_y$ or $\hat{z}$ and $\hat{p}_z$) are *incompatible* observables. This feature characterizes the state in quantum mechanics and distinguishes it radically from the classical state. On the other hand, a component of the position along a certain axis can be determined with arbitrary precision simultaneously with the component of the momentum along any of the two other axes.

We can now go back to the definition of complete sets (see p. 67). For a one-dimensional particle (a single degree of freedom), for instance, we only have two independent observables, the position $\hat{x}$ and the momentum $\hat{p}_x$, since any

Box 2.6 **Wave packet**

In the classical limit, matter waves (see Subsec. 1.5.5) become classical particles, that is we suppose that the wave be confined in a sufficiently small region to be approximated as a point-like entity.

In the one-dimensional case, the simplest type of wave is a plane and monochromatic wave

$$\psi(x,t) = e^{\imath(kx - \omega t)}. \tag{2.175}$$

In the classical approximation, we should relate k to p_x. To this end, as we have said, we should associate to the particle a wave confined in a small region. Although Eq. (2.175) does not have this character, we may write a superposition of waves with neighboring wave vectors, namely a *wave packet*, such that

$$\Psi(x,t) = \int dk A(k) e^{\imath(kx-\omega t)}, \tag{2.176}$$

where $A(k)$ is a function that has support in a small region of extension Δk around a given value k_0 of k. For the sake of simplicity we take $A(k)$ to be real. If the factor $e^{\imath(kx-\omega t)}$ oscillates many times in this region, the integral will be negligible. On the contrary, values of $\Psi(x, t)$ that are significantly different from zero are obtained when the phase $kx - \omega t$ stays approximately constant in that region, i.e. when

$$\left|\frac{d}{dk}(kx - \omega t)\right| \cdot \Delta k < 1. \tag{2.177}$$

This means that the wave packet $\Psi(x,t)$ is mainly confined in a spatial region of width $(\Delta k)^{-1}$ around the center

$$x_0 = t \cdot \frac{d\omega}{dk}. \tag{2.178}$$

From Eq. (2.178) it is clear that the center of the wave packet moves with constant velocity

$$v_g = \frac{d\omega}{dk}, \tag{2.179}$$

which is called *group velocity*. This has to be contrasted to the phase velocity

$$v_\phi = \frac{\omega}{k}, \tag{2.180}$$

which corresponds to the velocity of propagation of the plane wave (Eq. (2.175)). When taking the classical limit, it is the group velocity and not the phase velocity that should be considered as the the particle velocity

$$v = \frac{dE}{dp} = \frac{p}{m}. \tag{2.181}$$

other observable (e.g. the Hamiltonian $\hat{H}$) can be understood as a function of $\hat{x}$ and $\hat{p}_x$. It is then evident that complete sets are given by $\{\hat{x}\}$ and $\{\hat{p}_x\}$. For the three-dimensional case the problem is a little more complicated. We have here three components for the momentum and three for the position. In order to build a complete set we have to write a triple of observables with the prescription that any of its elements must commute with the other two and there is no additional element that commutes with all elements of the triple. Any element of the triple must be taken from one of the following pairs (each one composed of mutually exclusive elements) $\{\hat{x}, \hat{p}_x\}$, $\{\hat{y}, \hat{p}_y\}$, and $\{\hat{z}, \hat{p}_z\}$. We have then eight possible complete sets

$$\begin{array}{llll} \{\hat{x}, \hat{y}, \hat{z}\}, & \{\hat{x}, \hat{p}_y, \hat{p}_z\}, & \{\hat{x}, \hat{p}_y, \hat{z}\}, & \{\hat{x}, \hat{y}, \hat{p}_z\}, \\ \{\hat{p}_x, \hat{p}_y, \hat{p}_z\}, & \{\hat{p}_x, \hat{y}, \hat{z}\}, & \{\hat{p}_x, \hat{y}, \hat{p}_z\}, & \{\hat{p}_x, \hat{p}_y, \hat{z}\}. \end{array} \tag{2.182}$$

It is also evident that every quantum observable commutes with the identity operator $\hat{I}$ (whose mention is therefore omitted in the above sets).

2.3 Uncertainty relation

In Subsec. 2.1.1 we have formulated a basic principle of quantum theory: the quantization principle. Now, we derive the uncertainty relation between position and momentum, which is a direct consequence of the operatorial character of the quantum observables. In fact, we have already seen (in Subsec. 2.2.7) that two quantum-mechanical observables may not commute and that in such case they are not simultaneously measurable with arbitrary precision. As a consequence, the observables of the system can be divided into complete subsets of commuting observables (see Subsec. 2.2.7). The uncertainty relation is then the quantitative formulation of the impossibility of simultaneously measuring a pair of non-commuting observables. In Subsec. 2.3.1 we derive the uncertainty relation for the pair $(\hat{x}, \hat{p}_x)$, while in Subsec. 2.3.2 we generalize the uncertainty relation to any pair of observables. In Subsec. 2.3.3 we analyze the consequences of the uncertainty relation on the phase space representation for quantum mechanical systems. Finally, in Subsec. 2.3.4 we briefly discuss the relationship between the uncertainty relation and the complementarity and superposition principles.

2.3.1 Derivation of the uncertainty relation

The uncertainty relation, derived for the first time by Heisenberg,[25] formally defines the minimum value of the product of the uncertainties of two canonically conjugate variables (see Sec. 1.1). In the following we shall derive the uncertainty relation in the case of position and momentum for the one-dimensional case.[26]

Let us take a normalized wave function $\psi(x)$ for which both the position and momentum mean values (see Eq. (2.78)) are zero, i.e.

$$\langle \hat{x} \rangle_\psi = \langle \psi \mid \hat{x} \mid \psi \rangle = 0 \quad \text{and} \quad \langle \hat{p}_x \rangle_\psi = \langle \psi \mid \hat{p}_x \mid \psi \rangle = 0. \tag{2.183}$$

The normalization condition excludes the case of eigenfunctions of position and momentum (see Subsec. 2.2.2). However, this does not represent a loss of generality, as we shall see below.

25 See [Heisenberg 1927]. For a history see [*Mehra/Rechenberg* 1982–2001, VI, 130–63].

26 On this point we follow the formulation of Landau [*Landau/Lifshitz* 1976b] who in turn follows the derivations by Pauli [*Pauli* 1980, 21] and Weyl [*Weyl* 1950, 77, 393–94]. We proceed in this way for pedagogical reasons even though – from a formal standpoint – we could have started from the commutation relations, derived the result of the next subsection, and finally introduced the uncertainty relation between position and momentum as a special application of this general result. For the same reasons in the present subsection we make use of the wave-function formalism introduced previously.

We define now the *uncertainties*

$$\Delta_\psi x = \left\langle \left(\hat{x} - \langle \hat{x} \rangle_\psi\right)^2 \right\rangle_\psi^{\frac{1}{2}} = \left[\langle \psi \mid \left(\hat{x} - \langle \psi \mid \hat{x} \mid \psi \rangle\right)^2 \mid \psi \rangle \right]^{\frac{1}{2}}, \tag{2.184a}$$

$$\Delta_\psi p_x = \left\langle \left(\hat{p}_x - \langle \hat{p}_x \rangle_\psi\right)^2 \right\rangle_\psi^{\frac{1}{2}} = \left[\langle \psi \mid \left(\hat{p}_x - \langle \psi \mid \hat{p}_x \mid \psi \rangle\right)^2 \mid \psi \rangle \right]^{\frac{1}{2}}, \tag{2.184b}$$

as the *standard deviations*, i.e. the square roots of the variances (or dispersions) of the position and momentum operators $\hat{x}$, $\hat{p}_x$, calculated on the state $\mid \psi \rangle$, which we assume here to be finite.[27] For any wave function $\psi(x)$, we must have

$$\int_{-\infty}^{+\infty} dx \left| ax\psi(x) + \frac{d\psi(x)}{dx} \right|^2 \geq 0, \tag{2.185}$$

where a is an arbitrary real constant. The square modulus inside the integral in Eq. (2.185) is explicitly equal to

$$a^2x^2|\psi(x)|^2 + ax\left(\psi(x)\frac{d\psi^*(x)}{dx} + \psi^*(x)\frac{d\psi(x)}{dx}\right) + \frac{d\psi(x)}{dx}\frac{d\psi^*(x)}{dx}. \tag{2.186}$$

Now the integration of the first term in Eq. (2.186) gives

$$a^2 \int_{-\infty}^{+\infty} dx x^2 |\psi(x)|^2 = a^2 \left(\Delta_\psi x\right)^2. \tag{2.187a}$$

Integrating by parts the second term in Eq. (2.186) yields

$$\begin{aligned} a\int_{-\infty}^{+\infty} dx \left(\psi(x)\frac{d\psi^*(x)}{dx} + \psi^*(x)\frac{d\psi(x)}{dx}\right) x &= a\int_{-\infty}^{+\infty} dx \frac{d|\psi(x)|^2}{dx} x \\ &= ax|\psi(x)|^2\Big|_{-\infty}^{+\infty} - a\int_{-\infty}^{+\infty} dx |\psi(x)|^2 \\ &= -a\int_{-\infty}^{+\infty} dx |\psi(x)|^2 = -a, \end{aligned} \tag{2.187b}$$

where we have made use of the fact that

$$\lim_{x \to \pm\infty} x|\psi(x)|^2 = 0, \tag{2.187c}$$

due to the normalization of $\psi(x)$.[28] Finally, integrating by parts the last term in Eq. (2.186), we have

[27] This definition is not completely free of problems [Hilgevoord/Uffink 1983, Hilgevoord/Uffink 1988]. However, for our needs, it works.

[28] The condition $\int_{-\infty}^{+\infty} dx\, |\psi(x)|^2 = 1$ implies that, for sufficiently well-behaved functions, $|\psi(x)|^2$ tends to zero for $x \to \pm\infty$ more rapidly than $1/x$, from which the consequence (2.187c) follows.

$$\int_{-\infty}^{+\infty} dx \frac{d\psi(x)}{dx}\frac{d\psi^*(x)}{dx} = \psi^*(x)\frac{d\psi(x)}{dx}\bigg|_{-\infty}^{+\infty} - \int_{-\infty}^{+\infty} dx\psi^*(x)\frac{d^2\psi(x)}{dx^2}$$

$$= \hbar^{-2}\int_{-\infty}^{+\infty} dx\psi^*(x)\hat{p}_x^2\psi(x)$$

$$= \hbar^{-2}\left(\Delta_\psi p_x\right)^2, \tag{2.187d}$$

where we have taken advantage of the fact that

$$\lim_{x\to\pm\infty} \psi^*(x)\frac{d\psi(x)}{dx} = 0, \tag{2.187e}$$

because $\langle \hat{p}_x \rangle_\psi = 0$, and

$$\frac{d}{dx} = \frac{\imath}{\hbar}\hat{p}_x. \tag{2.187f}$$

Therefore, Eq. (2.185) may be rewritten as

$$a^2\left(\Delta_\psi x\right)^2 - a + \hbar^{-2}\left(\Delta_\psi p_x\right)^2 \geq 0. \tag{2.188}$$

In order to satisfy this condition, the discriminant of the quadratic expression in the lhs of Eq. (2.188) should be negative, i.e.

$$1 - 4\hbar^{-2}\left(\Delta_\psi x\right)^2\left(\Delta_\psi p_x\right)^2 \leq 0, \tag{2.189}$$

or

$$\Delta x \Delta p_x \geq \frac{\hbar}{2}, \tag{2.190}$$

where we have dropped any reference to the state $|\psi\rangle$ because this relation holds for any quantum state. Equation (2.190) represents the *uncertainty relation* for position and momentum.[29] Needless to say, similar expressions hold for the y and z components. Moreover, as we shall see (in Secs. 3.8, 6.5, 13.3, and Subsec. 13.4.2), analogous uncertainty relations can be written also for other pairs of conjugate observables. The value $\hbar/2$ represents then the maximum attainable certainty, i.e. the *minimum uncertainty* product allowed by the uncertainty relations, a value that can be attained under special physical circumstances, a fact that has a particular relevance in the case of the harmonic oscillator (see Sec. 4.4 and Prob. 4.20). The relation (2.190) states that when one tries to reduce the uncertainty of one of the two conjugate observables, then necessarily the uncertainty of the other increases. The argument we have used to derive Eq. (2.190) started from the assumption that $\psi(x)$ is a normalized wave function. As we have said, this excludes explicitly

[29] This relation is also called the "Heisenberg inequality." From a historical point of view, this relation has been considered as a founding principle of quantum mechanics, known as the "uncertainty principle." In some textbooks this principle is introduced as such, whereas we have chosen to derive the uncertainty relation from basic principles.

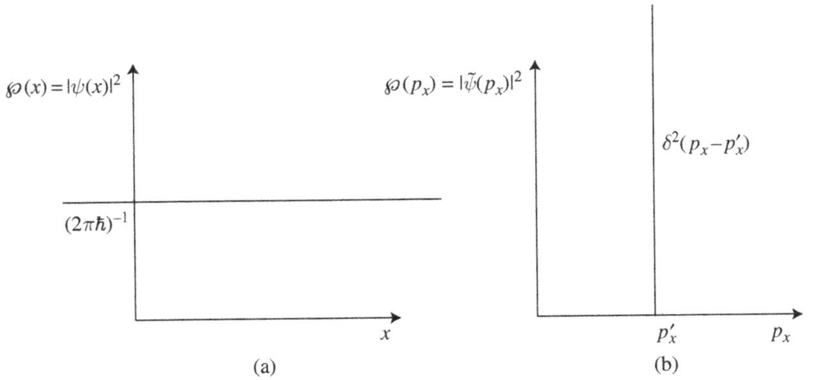

Figure 2.5 Probability distributions of (a) position and (b) momentum for a momentum eigenfunction.

the eigenfunctions of momentum and position. However, it is possible to have an infinitely precise determination of one of the two observables, say the momentum ($\Delta_\psi p_x = 0$). In this case, the wave function of the system would be given by Eq. (2.146), i.e.

$$\psi(x) = \varphi_p(x) = \frac{1}{\sqrt{2\pi\hbar}} e^{\frac{i}{\hbar} p_x' x}. \tag{2.191}$$

The Fourier transform of Eq. (2.191) is

$$\begin{aligned}\tilde{\psi}(p_x) &= \frac{1}{2\pi} \int_{-\infty}^{+\infty} e^{\frac{i}{\hbar} x p_x'} e^{-\frac{i}{\hbar} x p_x} \\ &= \frac{1}{2\pi} \int_{-\infty}^{+\infty} e^{-\frac{i}{\hbar} x (p_x - p_x')}, \\ &= \delta(p_x - p_x').\end{aligned} \tag{2.192}$$

so that the probability distribution of momentum is (see Prob. 2.22)

$$\wp(p_x) = \left|\tilde{\psi}(p_x)\right|^2 = \delta^2(p_x - p_x'). \tag{2.193}$$

The square modulus of the wave function (2.191) is $|\psi(x)|^2 = (2\pi\hbar)^{-1}$: the probability distribution of the position is uniform, which means that all position values are equiprobable, i.e. $\Delta_\psi x = \infty$ (see Fig. 2.5).

Similarly, if one takes an eigenfunction of the position observable (with eigenvalue x_0), i.e. $\psi(x) = \delta(x - x_0)$, one has (see Prob. 2.16)

$$\tilde{\psi}(p_x) = \tilde{\varphi}_{x_0}(p_x) = \frac{1}{\sqrt{2\pi\hbar}} e^{-\frac{i}{\hbar} p_x x_0}, \tag{2.194}$$

and all values of momentum are equally probable (see Fig. 2.6).

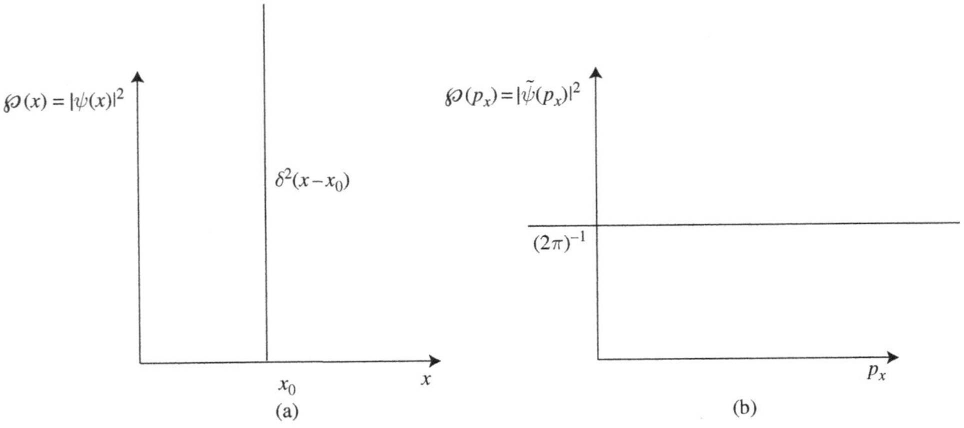

Figure 2.6 **Probability distributions of (a) position and (b) momentum for a position eigenfunction.**

2.3.2 Generalized uncertainty relation

The result of Subsec. 2.3.1 is valid for any pair of canonically conjugate observables, and can be generalized to two arbitrary observables (not necessarily conjugate) $\hat{O}$ and $\hat{O}'$.[30] In other words, as we shall see below, it is possible to write an uncertainty relation for any pair of non-commuting observables. Given an arbitrary state vector $|\,\psi\rangle$ on which, without loss of generality,[31] $\hat{O}$ and $\hat{O}'$ have zero expectation values (i.e. $\langle\psi\,|\,\hat{O}|\,\psi\rangle = \langle\psi\,|\,\hat{O}'|\,\psi\rangle = 0$), let us consider the vectors

$$|\,\varphi\rangle = \hat{O}|\,\psi\rangle \quad \text{and} \quad |\,\varphi'\rangle = \hat{O}'|\,\psi\rangle. \tag{2.195}$$

The Cauchy–Schwarz inequality[32] ensures that

$$\left|\langle\varphi\,|\,\varphi'\rangle\right| \leq \left(\langle\varphi\,|\,\varphi\rangle\right)^{\frac{1}{2}} \left(\langle\varphi'\,|\,\varphi'\rangle\right)^{\frac{1}{2}}. \tag{2.196}$$

Substituting the definitions of $|\,\varphi\rangle$ and $|\,\varphi'\rangle$ into Eq. (2.196), we obtain

$$|\,\langle\psi\,|\,\hat{O}\hat{O}'|\,\psi\rangle\,| \leq \left(\langle\psi\,|\,\hat{O}^2|\,\psi\rangle\right)^{\frac{1}{2}} \left(\langle\psi\,|\,\hat{O}'^2|\,\psi\rangle\right)^{\frac{1}{2}}, \tag{2.197a}$$

since operators $\hat{O}$ and $\hat{O}'$ are Hermitian. Interchanging the role of $|\,\varphi\rangle$ and $|\,\varphi'\rangle$, we also have

$$|\,\langle\psi\,|\,\hat{O}'\hat{O}|\,\psi\rangle\,| \leq \left(\langle\psi\,|\,\hat{O}^2|\,\psi\rangle\right)^{\frac{1}{2}} \left(\langle\psi\,|\,\hat{O}'^2|\,\psi\rangle\right)^{\frac{1}{2}}. \tag{2.197b}$$

30 The derivation of the following result is in [Robertson 1929]. Robertson follows Weyl's derivation of uncertainty relations [*Weyl* 1950, 77, 393–94] and therefore applies a general and abstract mathematical formalism. Our own derivation, however, is based on state vectors rather than on wavefunctions and is slightly different.

31 If $\langle\hat{O}\rangle_\psi = a \neq 0$, then one may always redefine $\hat{O}$ as $\hat{O}'' = \hat{O} - a$, so that $\langle\hat{O}''\rangle_\psi = 0$.

32 See [*Byron/Fuller* 1969–70, 148].

It is well known that, for any complex numbers a and b and a real positive number c, we have that $|a| \leq c$ and $|b| \leq c$ imply $|a - b| \leq 2c$.[33] Since the rhs of Eqs. (2.197) is real and positive (see comments following Eq. (2.78)), we also have

$$| \langle \psi \mid \hat{O}\hat{O}' | \psi \rangle - \langle \psi \mid \hat{O}'\hat{O} | \psi \rangle | \leq 2 \left(\langle \psi \mid \hat{O}^2 | \psi \rangle \right)^{\frac{1}{2}} \left(\langle \psi \mid \hat{O}'^2 | \psi \rangle \right)^{\frac{1}{2}}. \quad (2.198)$$

From Eq. (2.198) it easily follows that

$$| \langle \psi \mid \left[\hat{O}, \hat{O}' \right] | \psi \rangle | \leq 2\Delta_\psi O \cdot \Delta_\psi O', \quad (2.199)$$

where by $\Delta_\psi O$ we mean, as usual, the standard deviation of the values of the observable $\hat{O}$ in the state $| \psi \rangle$. This finally gives

$$(\Delta_\psi O) \cdot (\Delta_\psi O') \geq \frac{1}{2} \left| \langle \psi \left| \left[\hat{O}, \hat{O}' \right] \right| \psi \rangle \right|. \quad (2.200)$$

This result is particularly interesting because it shows that the uncertainty relation is a direct consequence of the non-commutability between quantum observables. Moreover, it is a general result which deals with any pair of arbitrary observables (not necessarily conjugate). We know that conjugate observables do not commute. In this case an uncertainty relation of the type of Eq. (2.190) can be derived from Eq. (2.200). On the other hand, for commuting observables there is no limit (at least in principle) on the precision of simultaneous measurements. However, there are observables which do not commute but neither are conjugate. In this case, Eq. (2.200) generates an uncertainty relation which is "less strict" than that for conjugate observables since the quantity in the rhs of Eq. (2.200) is the absolute value of the mean of the commutator of the two observables.[34] This expectation value can therefore assume different values depending on the "degree of commutativity." Such a situation is related to the concept of smooth complementarity which we have already introduced in Subsec. 1.2.4 and which we will discuss in greater detail in the next subsections.

As an example of application of the central result of Eq. (2.200), we may derive the already known result (2.190). In this case we have $[\hat{x}, \hat{p}_x] = \imath\hbar\hat{I}$ and $\langle \psi \mid [\hat{x}, \hat{p}_x] \mid \psi \rangle = \imath\hbar$ for any state $| \psi \rangle$. Substituting this result in Eq. (2.200) immediately provides the desired uncertainty relation (2.190).

2.3.3 Quantum state and quantum phase space

We have seen (in Sec. 1.3 and Subsec. 2.2.1) that the state vector $| \psi \rangle$ (or the corresponding wave function $\psi(x)$) describes the state of a quantum-mechanical system. We want to

[33] The triangular inequality ensures indeed that $|a \pm b| \leq |a| + |b| \leq 2c$.

[34] It is worth emphasizing that, even for two non-commuting observables, the rhs of Eq. (2.200) may be zero when the commutator between these two observables is not a number but an operator whose mean value on the state under consideration is zero.

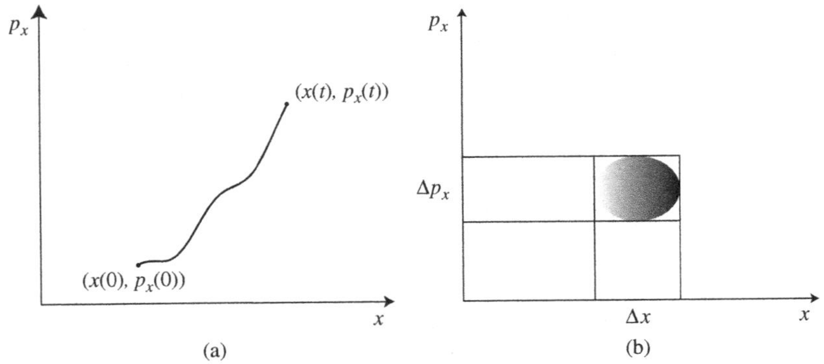

Figure 2.7 (a) Time evolution of a classical degree of freedom in phase space: at any time t, the state of the system is described by a point. (b) Graphical representation of a state in the quantum-mechanical phase space. According to the uncertainty relation, a single degree of freedom should be represented by an elliptical spot whose minimal area $\pi(\Delta x/2) \cdot (\Delta p_x/2)$ is $h/16$.

stress here that the vector $|\psi\rangle$ (as well as the wave function in any possible representation) contains the *whole* information that we may in principle acquire about the system, i.e. it represents a complete description of the state of the system. As we know, however, the knowledge of $|\psi\rangle$ does not allow us to predict with certainty the result of the measurement of any observable. The best we can do is to calculate the probability distribution of the outcomes of a given measurement or experiment (see Pr. 2.2: p. 57). While classical mechanics is ruled by the principle of perfect determination and therefore a classical state is characterized by the collection of all its physical properties (see Subsec. 1.1), the quantum-mechanical state is intrinsically probabilistic (see Sec. 1.4) and *affected by uncertainty*, i.e. not all observables can be completely determined at the same time. This finds expression in the fact that quantum observables can be cast into separate complete sets (see Subsec. 2.1.5 and 2.2.7). In other terms, there are probabilistic features in quantum mechanics that are not expression of subjective ignorance, but rather of the intrinsic nature of microscopic systems.

The concept of quantum state has profound implications in the phase-space representation of a system. A classical-state representation in phase space is necessarily pointlike. In fact, due to the principle of perfect determination, momentum and position may both have a perfectly determined value, and, as a consequence, the state of a classical system can be represented by a point in the phase space. If one considers the time evolution of the system, then this point will trace a well-defined trajectory in the phase space in the form of a curve (see Fig. 2.7(a)).

On the contrary, due to the uncertainty relation (which is a consequence of the non-commutability between observables that are at least conjugate), a phase-space representation for a quantum system at a given instant cannot be pointlike: Such a representation must reflect the fact that the uncertainties in position and momentum are both finite and that their product cannot be smaller than $\hbar/2$. Therefore, we may depict this circumstance by an elliptical spot in the phase space whose horizontal and vertical dimensions are equal to the position and momentum uncertainties, respectively, and whose minimal area is $h/16$ (see

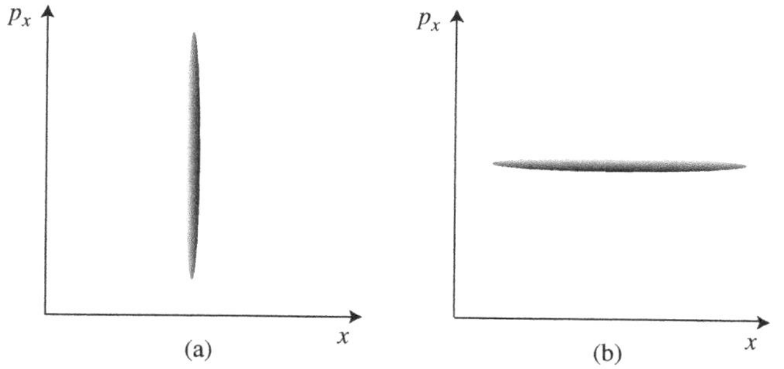

Figure 2.8 **Inverse proportionality between momentum and position uncertainties. When the position is accurately determined, the momentum becomes highly uncertain (a), and *vice versa* (b), since the product $\Delta x \Delta p_x$ has to remain equal to or larger than $\hbar/2$.**

Fig. 2.7(b)). Moreover, any improvement in the determination of momentum will be paid in terms of a proportional increase in uncertainty for position and *vice versa* (see Fig. 2.8).

This has important methodological and philosophical consequences. In fact, since we cannot have simultaneously perfect determination of two conjugate observables, if we wish to know with great accuracy one of the two, then we are obliged to *choose* between position and momentum. In any case, it is clear that quantum mechanics forces us to consider knowledge as a matter of choice rather than of a mirror image of a *datum*.[35] This also has a relevance for the measurement problem (see Ch. 9). In this sense the uncertainty relation is not only the quantum-mechanical counterpart of the principle of perfect determination but also of the principle of perfect knowledge (see Sec. 1.1).

Another consequence of this situation is that trajectories *do not exist* in quantum mechanics (see also Subsec. 1.2.3). This is true both in the phase space (for what we have said above) and in the configuration space. In fact, if one could define a trajectory, say, of a one-dimensional particle, i.e. a curve $x(t)$, then it would also be possible to determine the velocity and therefore the momentum of the particle, violating the uncertainty relations. At first sight, this might apparently contradict what is observed in experiments with Wilson chambers (and similar particle detectors) where particle's tracks are recorded as a series of bubbles.[36] However, this is not the case, since what is observed in these devices is not a true trajectory: the size of the bubbles and the momentum uncertainty taken simultaneously do not violate the uncertainty relation (Eq. (2.190)).

2.3.4 Superposition, uncertainty, and complementarity

What is the relationship between the uncertainty relation and the basic principles of quantum mechanics? As we have seen, the uncertainty relation is a consequence of the quantization principle (p. 44). Notice that the superposition principle is a positive statement – it enriches the space of states of a physical system – whereas the uncertainty relation is rather limiting: it imposes constraints on the maximal amount of knowledge

[35] On this point see [*Weyl* 1950, 76].
[36] See [*Segrè* 1964, Ch. 2].

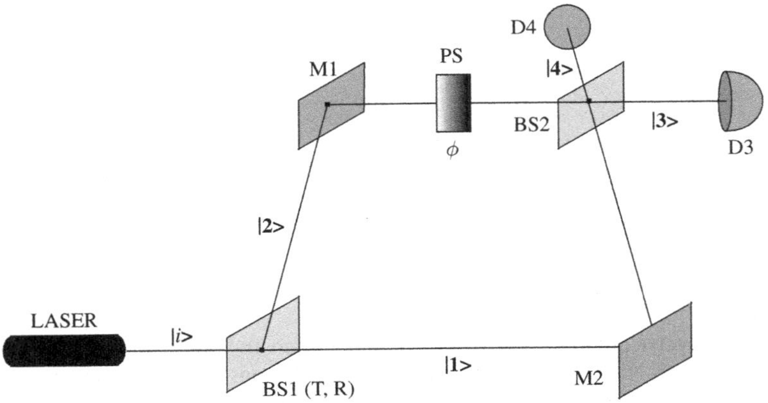

Figure 2.9 Smooth complementarity between wave and particle shown by an interferometry experiment.

that one can in principle extract from a system, and therefore on its measurability. For this reason, the relationship between measurement and uncertainty relation should not be understood in the sense that these constraints on the knowledge are caused by some form of perturbation. On the contrary, it is the uncertainty relation itself that poses constraints on what a measurement can perform, independently from the type of measurement one performs and therefore also from the perturbation it causes.[37]

The complementarity principle (Subsec. 1.2.4) deserves a deeper discussion. As a matter of fact, not only the uncertainty principle but also the complementarity principle concerns the interpretation of the quantum-mechanical entities, i.e. the ontological nature of the entities the theory refers to, and provides a bridge between dynamics and measurements, as we shall see in Ch. 9. The concepts of particle and wave are of classical origin and represent the two extreme cases of a spectrum of behaviors of quantum systems. In classical mechanics, the sharp distinction between these two concepts is justified by the fact that matter and waves are considered and treated in completely different ways. On the contrary, according to quantum mechanics, matter and waves are two sides of the same coin. To address this point in finer detail, let us discuss again the ideal experiment of Subsec. 1.2.2. Let us write the initial state as $|\,i\rangle$, which represents a photon impinging on BS1 from the left (see Fig. 2.9). We assume that BS1 has (in general, complex) transmission (T) and reflection (R) coefficients that for the sake of simplicity we impose here to be real. These coefficients may be changed, still satisfying the relation $\mathrm{T}^2 + \mathrm{R}^2 = 1$. The BS1 transformation can be described as

$$|\,i\rangle \overset{\mathrm{BS1}}{\mapsto} \mathrm{T}\,|\,1\rangle + \mathrm{R}\,|\,2\rangle, \tag{2.201}$$

that is, a quantum superposition of the states "photon in lower path" and "photon in upper path". After the two mirrors and the phase shifter the state becomes

$$\mathrm{T}\,|\,1\rangle + e^{\imath\phi}\mathrm{R}\,|\,2\rangle\,. \tag{2.202}$$

[37] This statement will show all its richness and importance in the light of the discussions about the interaction-free measurement (see Subsec. 1.2.4 and Sec. 9.6) and the quantum non-demolition measurement (see Sec. 9.11).

describing the fact that only the upper path acquires a phase factor $e^{\imath\phi}$. The second beam splitter is assumed to be a 50%–50% one. Its action may be described by the transformations

$$| 1\rangle \overset{BS2}{\mapsto} \frac{1}{\sqrt{2}}(| 3\rangle - | 4\rangle), \quad | 2\rangle \overset{BS2}{\mapsto} \frac{1}{\sqrt{2}}(| 3\rangle + | 4\rangle), \tag{2.203}$$

so that we can write the outgoing state as

$$T\,| 1\rangle + e^{\imath\phi}R\,| 2\rangle \overset{BS2}{\mapsto} \frac{1}{\sqrt{2}}\left[T\,(| 3\rangle - | 4\rangle) + e^{\imath\phi}\,(| 3\rangle + | 4\rangle)\right]. \tag{2.204}$$

Ordering the terms, the final state is

$$| f\rangle = \frac{1}{\sqrt{2}}\left[\left(\mathrm{T} + e^{\imath\phi}\mathrm{R}\right)| 3\rangle - \left(\mathrm{T} - e^{\imath\phi}\mathrm{R}\right)| 4\rangle\right]. \tag{2.205}$$

It is easy to calculate the final detection probabilities at the two detectors D3 and D4, which are given by the square moduli of the amplitudes of the states $| 3\rangle$ and $| 4\rangle$ in the above expression, respectively, i.e.

$$\begin{aligned} \wp_3 &= \frac{1}{2}\left(\mathrm{T} + e^{\imath\phi}\mathrm{R}\right)\left(\mathrm{T} + e^{-\imath\phi}\mathrm{R}\right) = \frac{1}{2} + \mathrm{TR}\cos\phi, \\ \wp_4 &= \frac{1}{2}\left(\mathrm{T} - e^{\imath\phi}\mathrm{R}\right)\left(\mathrm{T} - e^{-\imath\phi}\mathrm{R}\right) = \frac{1}{2} - \mathrm{TR}\cos\phi. \end{aligned} \tag{2.206}$$

As it should be, we have that $\wp_3 + \wp_4 = 1$. We now see that if the transmission coefficient is $\mathrm{T} = 0, 1$ (and correspondingly $\mathrm{R} = 1, 0$), then we have a perfectly determined path (if $\mathrm{T} = 0$ the photon will always take the path 2, whereas if $\mathrm{T} = 1$ the photon will always take the path 1) and the interference term in $\wp_3$ and $\wp_4$ vanishes: in this case we have $\wp_3 = \wp_4 = 1/2$. On the contrary, if $\mathrm{R}^2 = \mathrm{T}^2 = 1/2$, then we have maximal interference and also maximal indetermination of the path, since in this case the photon has equal probability to take path 1 and path 2. However, when $1/2 < \mathrm{T}^2 < 1$ or $0 < \mathrm{T}^2 < 1/2$, we have a range of possibilities where partial path information and partial interference are simultaneously present.

This state of affairs can be quantitatively formulated as a relationship between the visibility $\mathcal{V}$ of interference and the predictability $\mathcal{P}$ of the path.[38] In fact if one repeats the same experiment a large number of times and for different values of ϕ, one will obtain pictures similar to Fig. 1.6, where the profile may be viewed as the light intensity I detected at detectors D3 and D4. Such intensities will then be proportional to the probabilities of detecting the photons at D3 and D4, respectively. The interference visibility $\mathcal{V}$ may then be defined by

$$\mathcal{V} = \frac{I_{\max} - I_{\min}}{I_{\max} + I_{\min}} = 2\mathrm{TR}. \tag{2.207}$$

Similarly, the path predictability $\mathcal{P}$ may be seen as the probability of correctly predicting the path taken by the photon. It is clear that $\mathcal{P}$ will be equal to zero for a symmetric beam

[38] This analysis was performed by Greenberger and Yasin [Greenberger/Yasin 1988] on the basis of a paper by Wootters and Zurek [Wootters/Zurek 1979]. A first experimental evidence of smooth complementarity can be found in [Badurek *et al.* 1983]. See also [Mittelstaedt *et al.* 1987] [Englert 1996].

splitter BS1 while it will be equal to one for $T = 0, 1$. If we limit ourselves to the range $1/2 < T < 1$, then we may define $\mathcal{P}$ as

$$\mathcal{P} = \frac{T^2 - R^2}{T^2 + R^2} = T^2 - R^2. \tag{2.208}$$

It is easy to see that we have

$$\mathcal{V}^2 + \mathcal{P}^2 = 1, \tag{2.209}$$

which is known as *Greenberger–Yasin equality*. This means that, besides the two limiting cases $\mathcal{P} = 1, \mathcal{V} = 0$ (i.e. particle-like behavior), and $\mathcal{P} = 0, \mathcal{V} = 1$ (i.e. wave-like behavior), all possible intermediate values are also allowed. If we take the Poincaré-sphere representation of (two-level) quantum states, the two possible paths (1 and 2, given by $T = 1$ and $T = 0$, respectively), can be represented as north and south poles of the sphere (see Fig. 1.11: p. 28), respectively, whereas the equator represents all states of maximal interference and therefore of highest visibility, which are distinguished from one another by the phase difference ϕ. Given a certain angle ϕ, all points but the extremes (i.e. for all $1/2 < T^2 < 1$) of the arc joining the north pole (where $\theta = 0$) with the point on the equator (where $\theta = \pi/2$) characterized by ϕ represent states which are intermediate between a wave-like and a corpuscular behavior. All points but the extremes (i.e. for all $0 < T^2 < 1/2$) on the arc joining the same point on the equator and the south pole (where $\theta = \pi$), again represent intermediate states between a wave-like and a corpuscular behavior with the same phase ϕ. Summing up, for $\theta = 0$ we have $\mathcal{P} = 1$ and $\mathcal{V} = 0$, for $\theta = \pi/2$, we have $\mathcal{P} = 0$ and $\mathcal{V} = 1$, and finally for $\theta = \pi$ we have $\mathcal{P} = -1$ and $\mathcal{V} = 0$. This suggests that we may take

$$\mathcal{P} = \cos\theta, \quad \mathcal{V} = \sin\theta. \tag{2.210}$$

The result (2.209) is the essence of what we have called *smooth* wave–particle complementarity (see also the conclusions of Subsec. 2.3.2). In other words, the complementarity principle states that quantum–mechanical entities display any possible intermediate behavior between these two extreme forms (particle-like and wave-like).

It is worth emphasizing that one might be tempted to consider the complementarity principle as a consequence of the uncertainty relation. As a matter of fact, a lively debate on this issue developed during the 1990s within the scientific community (see Subsec. 9.5.2.) The output of such a debate would seem to establish the foundational character of the complementarity principle.

2.4 Quantum algebra and quantum logic

The quantum-mechanical formalism we have introduced so far poses serious questions concerning the algebraic structure of the theory. In classical mechanics we can build propositions about the world that are true or false. These propositions are statements about

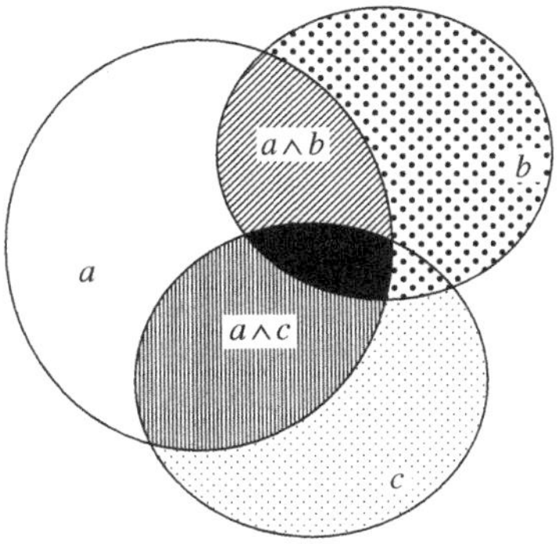

Figure 2.10 **Illustration of the distributive law $a \wedge (b \vee c) = (a \wedge b) \vee (a \wedge c)$. The three sets a, b, and c represent propositions.**

the properties of the physical system under consideration, that is, about the values of observables (see Sec. 1.1). It is possible to perform several operations on these propositions, that is, we can establish different relations among them. For instance, we can add these propositions, and this corresponds to the logical addition, that is to the inclusive disjunction OR, or we can multiply propositions, which corresponds to the conjuction AND. Once defined such operations, they constitute, together with the atomic propositions, a propositional algebra.[39]

This algebra is called a Boolean algebra, and it is an algebra for which distributivity yields, which is, as we shall see, strictly related with commutativity. Distributivity consists in the relationships[40]

$$a \wedge (b \vee c) = (a \wedge b) \vee (a \wedge c), \tag{2.211a}$$

$$a \vee (b \wedge c) = (a \vee b) \wedge (a \vee c), \tag{2.211b}$$

which are valid for any elements a, b, c of the algebra and where $\wedge$ and $\vee$ are the symbols for conjunction (AND) and inclusive disjunction (OR), respectively. The conjunction of two propositions is defined as true if and only if both propositions are true, while the inclusive disjunction is defined as true if and only if at least one of the two propositions is true. An example of Eq. (2.211a) is given by Fig. 2.10. The elements of a Boolean alegbra, i.e. propositions, have a formal analogy with projectors, as far as they are idem potent.

That quantum mechanics violates distributivity can be seen by the following example. Consider the usual Mach–Zehnder setup shown in Fig. 2.9. Suppose that the state of the photon is in a superposition of path $|1\rangle$ and path $|2\rangle$ before BS2. Then, the proposition a, "The photon is in a superposition state of $|1\rangle$ and $|2\rangle$," may be represented by the disjunction of two propositions, say a' ("The photon takes path 1") and a'' ("The photon takes the path 2"). Now, the proposition a can be true even if the system is neither in path 1 nor in path 2. Furthermore, suppose that the relative phase ϕ is tuned in such a way

[39] On the abstract concept of algebra see also Subsecs. 8.4.3 and 8.4.4.

[40] See [*Bocheński* 1970], especially part II, about Boolean algebras.

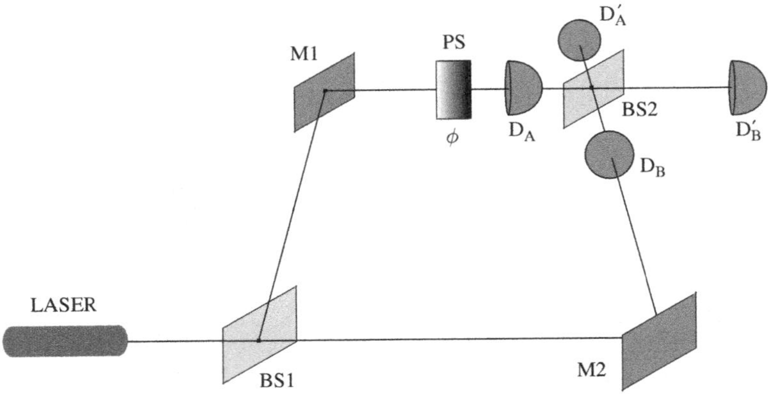

Figure 2.11 **Experimental arrangement in order to show that one can generate the two Boolean sub-algebras $\{1, 0, D_A, D_B\}$ and $\{1, 0, D'_A, D'_B\}$ by considering the two mutual exclusive arrangements given by D_A and D_B and by D'_A and D'_B, respectively. $\{1, 0, D_A, D'_A, D_B, D'_B\}$ is instead a non-Boolean algebra.**

that $|4\rangle$ is a dark output of the interferometer – detector D4 never clicks. In this case the proposition b

$$b = a \wedge c = \left(a' \vee a''\right) \wedge c \tag{2.212}$$

amounts to the assertion that $c =$ "The detector D3 clicks" is true. In fact, there are only three possibilities: the photon takes path 1, the photon takes path 2, or the photon takes both paths 1 and 2. In each of these three cases, the proposition $a = a' \vee a''$ is true. On the other hand, the proposition d,

$$d = \left(a' \wedge c\right) \vee \left(a'' \wedge c\right), \tag{2.213}$$

means something different, namely that we have two possibilities, that the photon passes through path 1 and detector D3 clicks or that the photon passes through path 2 and detector D3 clicks. These propositions are in general both false, so that d is also false: if the photon had already a determined path before BS2, it would be split at BS2 and would have a non-zero probability to be detected by D4 – so that D4 could not be considered a dark output. This means that in quantum mechanics we are forced to write

$$\left(a' \vee a''\right) \wedge c \neq \left(a' \wedge c\right) \vee \left(a'' \wedge c\right), \tag{2.214}$$

and to reject classical distributivity. Now, there is a close relationship between commutability and distributivity.[41]

In other words, a quantum algebra is not Boolean in itself but can be decomposed in Boolean subalgebras, and any subalgebra is the counterpart – on an algebraic level – of a complete set of observables (see p. 67). Let us analyze this point in greater detail by means of an example.[42] Take the arrangement shown in Fig. 2.11, which represents a Mach–Zehnder interferometer (see Subsec. 1.2.2), where only one photon at the time is

[41] See [*Beltrametti/Cassinelli* 1981, 126–27].

[42] See [Quadt 1989, 1030].

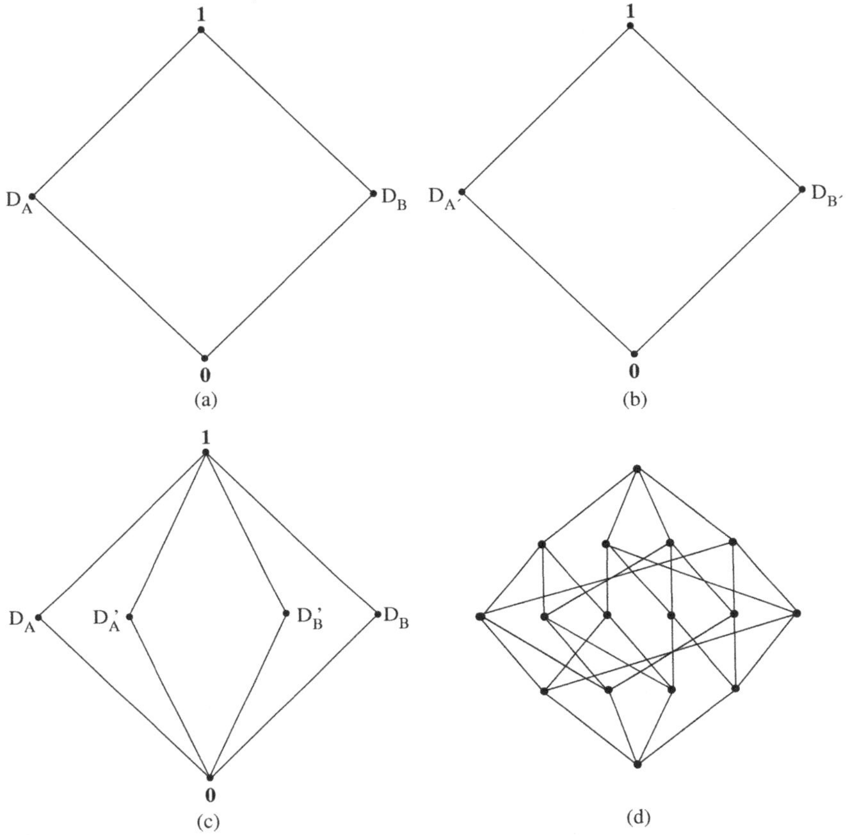

Figure 2.12 Hasse diagrams of several Boolean and non-Boolean algebras. A Boolean algebra is a collection of sets that includes the sets **1**, **0**, the complement of any set, the intersections of all pairs of sets and the sums of all pairs of sets. (a) Hasse diagram of the Boolean subalgebra $\{\mathbf{1}, \mathbf{0}, D_A, D_B\}$. (b) Hasse diagram of the Boolean subalgebra $\{\mathbf{1}, \mathbf{0}, D'_A, D'_B\}$. (c) Non-Boolean algebra $\{\mathbf{1}, \mathbf{0}, D_A, D_B\}, D'_A, D'_B\}$. It is easy to see that the subalgebras $\{\mathbf{1}, \mathbf{0}, D_A, D_B\}$ and $\{\mathbf{1}, \mathbf{0}, D'_A, D'_B\}$ are Boolean but the algebra $\{\mathbf{1}, \mathbf{0}, D_A, D_B\}, D'_A, D'_B\}$ is not. In (d) the Hasse diagram of a Boolean algebra with four elements is shown.

sent through the apparatus. We have two possible alternative (mutually exclusive) settings: either we choose to measure the path of the photon by placing the detectors at positions D_A and D_B or we choose to detect the interference by placing the detectors at positions D'_A and D'_B (after the second beam splitter). It is clear that in both settings one and only one of the two detectors will click in absence of losses – which we do not consider here. Then the two settings are *incompatible* and generate two Boolean subalgebras given by the elements $\{\mathbf{1}, \mathbf{0}, D_A, D_B\}$ and $\{\mathbf{1}, \mathbf{0}, D'_A, D'_B\}$, respectively, where $\mathbf{1}, \mathbf{0}$ are the identity (the sum or disjunction of all elements) and the null element (the intersection or conjunction of all elements) of the set, respectively. The element D_j $(j = A, B)$ corresponds to the proposition "Detector D_j has clicked." However, the total algebra is not Boolean.[43] The two subalgebras and the total algebra may be graphically depicted as Hasse diagrams (see Fig. 2.12). A Hasse diagram is a graphical representation of a partially ordered set (or POSet), which is ordered by a relation $\leq$, with an implied upward orientation. A point is drawn for each element of the set and line segments are drawn between these points according to the following two rules:

- If $a < b$ in the Poset then the point corresponding to a appears lower in the drawing then the point corresponding to b.
- The line segment between the points corresponding to any two elements a and b of the poset is included in the drawing if and only if a covers b or b covers a.

Summary

In this chapter we have developed the basic formalism of quantum observables and different representations of quantum states. We may summarize the main results as follows:

- Quantum-mechanical observables can have a continuous spectrum, a discrete one or a combination of both, and therefore cannot be represented by variables, as in classical mechanics, but must be represented by *operators*, as stated by the quantization principle.
- These operators must be *Hermitian*, since possible measurement results are represented by the eigenvalues of the operator and a physical quantity must always take on real values.
- Quantum mechanical observables do not necessarily *commute*. In this case they do not share a common eigenbasis, with the consequence that it is possible to introduce different representations (e.g. position and momentum representations) associated with the relative observable.
- Due to non-commutability there are intrinsic limitations on the maximal amount of information one can extract from a system. This is the content of the *uncertainty relation* and has the consequence of limiting the quantum mechanical states which are allowed.
- We have examined the relationships between superposition, uncertainty, and complementarity. While the superposition and complementarity principles increase the spectrum of the possible states relative to classical mechanics and are fundamental principles of quantum mechanics, uncertainty relations pose constraints on the possible measurements and are consequence of the quantization principle.
- The elements (states) of the *quantum phase space* cannot be pointlike as the classical one is, but must be rather represented by spots whose minimal area is given by the uncertainty relation between position and momentum.
- Another consequence of non-commutability is that one can partition the set of observables of a physical system into *subsets* of commuting observables. These subsets represent Boolean sub-algebras, while distributivity is not valid on the whole quantum algebra.

[43] See also the discussion of quantum probability in Sec. 1.4.

Problems

2.1 Given an arbitrary state vector $|\psi\rangle$, it is always possible to construct a Hermitian operator $\hat{O}$ which has $|\psi\rangle$ among its eigenvectors (see p. 49). Show that $\hat{O} = |\psi\rangle\langle\psi|$ is not the only Hermitian operator whose eigenvectors contain $|\psi\rangle$.

2.2 Prove that Eq. (2.32) reduces to an identity when replacing the coefficients by the explicit forms given by Eqs. (2.25) and (2.27).

2.3 Prove that the matrix $\hat{U}$ of Eq. (2.33) is unitary.

2.4 Show that any linear operator $\hat{O}$ can be decomposed as $\hat{O} = \hat{O}' + \imath\hat{O}''$, where $\hat{O}'$ and $\hat{O}''$ are Hermitian operators.

2.5 Show that the necessary and sufficient conditions for an operator to be unitary is that it is normal and satisfies condition (2.35).

2.6 Consider a generic two-dimensional matrix

$$\hat{O} = \begin{bmatrix} a & b \\ c & d \end{bmatrix}.$$

Prove that in order for it to be unitary and Hermitian, it has to be of the form

$$\hat{O} = \begin{bmatrix} a & b \\ b^* & -a \end{bmatrix},$$

where a is real.

2.7 Calculate the expectation value in Box 2.4 by making use of the properties of $\hat{O}_P$ and of the scalar product without employing the explicit matricial form.

2.8 Eq. (2.1) has been introduced in a heuristic way, as a generalization of an intuitive result. Prove it in a rigorous way.

2.9 Take the 2×2 matrix

$$\hat{O} = \begin{bmatrix} 0 & -\imath \\ \imath & 0 \end{bmatrix}.$$

(a) Find its eigenvalues and eigenvectors.
(b) Derive the diagonalizing matrix of $\hat{O}$ and its diagonal form.

2.10 Consider a system of three polarization filters as in Subsec. 2.1.5 (see Fig. 2.3) with $\phi = \pi/6$ and $\theta = \pi/3$. If N is the number of photons passing filter P1, how many photons will pass on average through the entire apparatus?

2.11 Verify Eq. (2.96).

2.12 Prove Eq. (2.98).
(*Hint*: Take the Taylor expansion of $f(\hat{O})$ in powers of $\hat{O}$.)

2.13 Prove Eq. (2.99).

2.14 Prove that, given three observables $\hat{O}$, $\hat{O}'$, and $\hat{O}''$, for which $\left[\hat{O}, \hat{O}'\right] = 0$ and $\left[\hat{O}, \hat{O}''\right] = 0$, then it is not necessarily true that $\left[\hat{O}', \hat{O}''\right] = 0$.
(*Hint*: Take $\hat{O}$ as the identity operator $\hat{I}$.)

2.15 Derive the momentum eigenfunctions in the momentum representation.
(*Hint*: There are two independent ways to derive this result. Either one closely follows the derivation of the position eigenfunction in the position representation (see Subsec. 2.2.1), or one applies Eq. (2.150b) with $\psi(x) = \varphi_p(x)$ (see Eq. (2.146)). It is instructive to verify that these two methods lead to the same result.)

2.16 Derive the position eigenfunctions in the momentum representation.
(*Hint*: See Prob. 2.15.)

2.17 Prove that, if $\psi(x)$ is normalized, i.e. if $\int_{-\infty}^{+\infty} dx |\psi(x)|^2 = 1$, then also $\int_{-\infty}^{+\infty} dpx |\tilde{\psi}(p_x)|^2 = 1$.

2.18 Prove Eq. (2.153).
(*Hint*: Calculate the expectation value of the position operator in the momentum representation and take advantage of the Fourier transform.)

2.19 Prove that

$$\hat{P}(x) \mid p_x\rangle = \frac{1}{2\pi\hbar} e^{\frac{\imath}{\hbar} p_x x} \mid x\rangle, \tag{2.215}$$

where $\hat{P}(x)$ projects onto the position eigenvector $\mid x\rangle$.

2.20 Prove the result (2.173a) in the momentum representation.

2.21 Derive Eqs. (2.173b) and (2.173c).

2.22 Prove that $\int_{-\infty}^{+\infty} dx \delta^2(x) = +\infty$.

2.23 Prove that $[\hat{x}, \hat{p}_x^2] = 2\imath\hbar \hat{p}_x$.

2.24 Consider a one-dimensional free particle in an eigenstate of the momentum operator with eigenvalue p_0. Derive the uncertainty relation between its energy and its position following Eq. (2.200).
(*Hint*: Take advantage of the result of Prob. 2.23.)

2.25 Generalize the result of Prob. 2.23 to $[\hat{x}, \hat{p}_x^n] = n\imath\hbar \hat{p}_x^{n-1}$.

2.26 Prove that to $[\hat{p}_x, f(\hat{x})] = -\imath\hbar f'(\hat{x})$.

2.27 Derive the commutation relation $[\hat{x}, f(\hat{p}_x)] = \imath\hbar f'(\hat{p}_x)$, where $f(\hat{p}_x)$ is an arbitrary function of the momentum operator and $f'(\hat{p}_x)$ is its first derivative made with respect to $\hat{p}_x$.
(*Hint*: Write the Taylor expansion of $f(\hat{p}_x)$ and use the result of Prob. 2.25.)

2.28 Calculate the probabilities $\wp_3$ and $\wp_4$ of Subsec. 2.3.4 for a symmetric beam splitter and verify that they are equal to the corresponding probabilities of Subsec. 1.2.3.

2.29 Verify Eq. (1.55) by explicitly calculating the involved amplitudes and probabilities.
(*Hint*: See the formalism used in Subsec. 2.3.4.)

2.30 Compute the state (2.205) in the case R = 0, T = 1 and denote it by $\mid f_0\rangle$. Repeat the same procedure with R = 1, T = 0, and denote the resulting state by $\mid f_1\rangle$. Show that, for $\mathrm{R} = \mathrm{T} = 1/\sqrt{2}$ the resulting state $\left| f_{1/2}\right\rangle$ may be expressed as a linear superposition of $\mid f_0\rangle$ and $\mid f_1\rangle$. Make a comparison with the Poincaré sphere formalism.

2.31 Starting from the example of Sec. 2.4, in which it is shown that we may have $\left(a' \vee a''\right) \wedge c$ true but $\left(a' \wedge c\right) \vee \left(a'' \wedge c\right)$ false, show that we may have $\neg(a' \wedge a'')$ true even if neither a' nor a'' is false.

Further reading

Basic features of generic quantum observables

Byron, F. W. Jr. and Fuller, R. W., *Mathematics of Classical and Quantum Physics*, 1969–70; New York: Dover, 1992.

Fano, Guido, *Mathematical Methods of Quantum Mechanics*, New York: McGraw-Hill, 1971.

Prugovečki, Eduard, *Quantum Mechanics in Hilbert Space*, 1971; 2nd edn., New York: Academic Press, 1981.

Uncertainty, complementarity, and superposition

Hilgevoord, J. and Uffink, J. B. M., Overall width, mean peak width and the uncertainty principle. *Physics Letters*, **95A** (1983), 474–76.

Greenberger, D. M. and Yasin, A., Simultaneous wave and particle knowledge in a neutron interferometer. *Physics Letters* **128A**, (1988), 391–94.

Quantum logic

Beltrametti, E. and Cassinelli, G., *The Logic of Quantum Mechanics*, Redwood City, CA: Addison-Wesley, 1981.

3 Quantum dynamics

In the first two chapters we have examined the basic principles – superposition (p. 18), complementarity (p. 19), quantization (p. 44), statistical algorithm (p. 57), and correspondence (p. 72) (see also Subsec. 2.3.4) – and the basic entities, observables and states, of quantum mechanics, as well as the main differences with respect to classical mechanics. While what we have discussed so far is rather a static picture of observables and states, in this chapter we shall deal with quantum dynamics, i.e. with the time evolution of quantum-mechanical systems.

Historically, after Bohr had provided a quantized description of the atom (see Subsec. 1.5.4), Einstein showed the quantized nature of photons (see Subsec. 1.2.1), and de Broglie hypothized the wave-like nature of matter (see Subsec. 1.5.5), the first building block of quantum mechanics was provided by the commutation relations, proposed by Heisenberg in 1925,[1] whose consequence is represented by the uncertainty relation (see Subsec. 2.2.7 and Sec. 2.3). This was the subject of the previous chapters. The dynamical part of the theory was proposed by Schrödinger in 1926,[2] and is known as the Schrödinger equation. It is also known as wave mechanics (see Subsec. 1.5.7). In this chapter we shall show that Heisenberg's and Schrödinger's formulations are only two different aspects of the same theory. We shall also come back to this point in Sec. 8.1.1. Here, first we shall derive the fundamental equation which rules quantum dynamics (Sec. 3.1), and, in Sec. 3.2, we shall summarize the main properties of the Schrödinger equation. Moreover, in Sec. 3.3, we shall show that the Schrödinger equation is invariant under Galilei transformations. In Sec. 3.4 we shall discuss a first example of the Schrödinger equation (a one-dimensional particle in a box). Then in Sec. 3.5, we shall introduce the unitary transformations in a general form, and in Sec. 3.6 the Heisenberg and the Dirac pictures, which are representations of the quantum evolution equivalent to the Schrödinger picture. In Sec. 3.7 the fundamental Ehrenfest theorem will be presented, while in Sec. 3.8 we shall derive the uncertainty relation between energy and time. Finally, in Sec. 3.9 we shall discuss the difficult problem of finding a self-adjoint operatorial representation of time and present some possible solutions.

[1] See [Heisenberg 1925].
[2] See [Schrödinger 1926].

3.1 The Schrödinger equation

In this section we derive the Schrödinger equation (Subsec 3.1.1). Then, we briefly return to the difference between classical and quantum mechanics concerning determinism (Subsec. 3.1.2). In Subsec. 3.1.3 we shall learn how to represent the time evolution of an arbitrary initial state in the basis of energy eigenvectors (the stationary states). Finally, in Subsec. 3.1.4 we shall briefly discuss the problem of degenerate eigenvalues.

3.1.1 Derivation of the Schrödinger equation

We have already seen (in Subsec. 2.3.3) that the state vector (or the wave function) contains complete information about the state of a quantum system at a given time. We now need to determine how the system (and therefore its state) evolves with time. In classical mechanics, the knowledge of position and momentum of a particle (i.e. the knowledge of its state) allows, together with the knowledge of the forces which act on the particle, the univocal determination of position and momentum of the particle at any future (and past) time through Newton's second law (see Sec. 1.1).

In order to build the equation which gives the time evolution of a quantum state, it is useful to take into account the requirements imposed by the mathematical formalism and the conceptual aspects developed in the previous chapters. Let us assume that the equation is deterministic. First, the evolution equation must only contain the first time derivative of the state vector (see Eq. (1.1)). If it were not so, then the knowledge of the state at the initial time t_0 would not be sufficient for determining its evolution at future times, since the solution of a n-th order differential equation requires the knowledge of the first $n-1$ derivatives at time t_0. This would contradict the assumption that the state vector contains complete information about the state of a quantum system. On the most general grounds, we can therefore write the evolution equation as

$$\frac{\partial}{\partial t} \mid \psi \rangle = \hat{O} \mid \psi \rangle, \tag{3.1}$$

where $\hat{O}$ must be a linear operator to be determined.[3] This requirement directly follows from the superposition principle (see p. 18) and the consequent linearity of quantum mechanics. In other words, the evolution equation has to be linear and homogeneous.

Moreover, the operator $\hat{O}$ in Eq. (3.1) must evidently represent the generator of time translations or, equivalently, the quantity which is conserved under time translations. We know from classical mechanics that such a quantity is represented by the energy, or the Hamiltonian function of the system. Therefore, following the correspondence principle (see Subsec. 2.2.4) we take the operator $\hat{O}$ to be a function of the Hamiltonian operator $\hat{H}$ only and rewrite Eq. (3.1) as

[3] For the time being we limit ourselves to the case in which the operators $\hat{O}$ does not explicitly depend on time. We shall consider time dependency in Secs. 10.3 and 14.2.

$$\frac{\partial}{\partial t} | \psi \rangle = f\left(\hat{H}\right) | \psi \rangle. \tag{3.2}$$

Moreover, in the case of a composite system made of two subsystems with Hamiltonians $\hat{H}_1$ and $\hat{H}_2$, respectively, the generator of time translations must satisfy

$$f\left(\hat{H}\right) = f\left(\hat{H}_1\right) + f\left(\hat{H}_2\right), \tag{3.3}$$

in order to respect the linearity requirement. Since we must have $\hat{H} = \hat{H}_1 + \hat{H}_2$, the only possibility is that

$$\hat{O} = f\left(\hat{H}\right) = a\hat{H}, \tag{3.4}$$

where a is a (complex) constant to be determined. In order to determine a, we take advantage of the fact that $\langle \psi \mid \psi \rangle = 1$ for any state $| \psi \rangle$ and therefore also at any time (see p. 26). As a consequence, the norm of the state must be conserved and

$$\frac{\partial}{\partial t} \langle \psi \mid \psi \rangle = \frac{\partial \langle \psi |}{\partial t} | \psi \rangle + \langle \psi | \frac{\partial | \psi \rangle}{\partial t} = 0. \tag{3.5}$$

Substituting Eq. (3.2) and its Hermitian conjugate

$$\frac{\partial}{\partial t} \langle \psi | = \langle \psi | a^* \hat{H}, \tag{3.6}$$

into Eq. (3.5) and using Eq. (3.4), we obtain

$$\langle \psi | \left(a^* \hat{H} + a\hat{H}\right) | \psi \rangle = 0, \tag{3.7}$$

since $\hat{H} = \hat{H}^\dagger$, the Hamiltonian being a Hermitian operator. In order for Eq. (3.7) to be valid for any state $| \psi \rangle$, we must have $a^* = -a$, or, in other words, a must be a pure imaginary number. Such a quantity (see Eq. (3.2)) has the dimension of the inverse action and, therefore, its inverse, for convenience and in agreement with the correspondence principle, can be expressed as $1/a = \imath\hbar$. The quantum-mechanical evolution equation then takes the final form

$$\imath\hbar \frac{\partial}{\partial t} | \psi \rangle = \hat{H} | \psi \rangle, \tag{3.8}$$

which is known as the *Schrödinger equation*.[4]

If the equations of motion of the system explicitly depend on time, then the Hamiltonian operator $\hat{H}$ in Eq. (3.8) will also depend on time. In general, as we have seen in Subsec. 2.2.6, the Hamiltonian $\hat{H}$ will be given by the sum of the kinetic and potential energy operators, $\hat{H} = \hat{T} + \hat{V}$.

For a one-dimensional particle and in the position representation, one may rewrite Eq. (3.8) as a partial differential equation for the wave function (see Subsec. 2.2.1, Eq. (2.165), and Prob. 3.1)

[4] See [Schrödinger 1926]. For a historical reconstruction of Schrödinger's great contribution see [*Mehra/Rechenberg* 1982–2001, V, 404–576].

$$\imath\hbar\frac{\partial\psi(x,t)}{\partial t}=\left[\frac{\hat{p}_x^2}{2m}+V(x,t)\right]\psi(x,t), \tag{3.9}$$

where $V(x,t)$ is the (in general time-dependent) potential energy. Using the definition $\hat{p}_x=-\imath\hbar(\partial/\partial x)$ of the momentum operator in the position representation (see Subsec. 2.2.4), we obtain

$$\imath\hbar\frac{\partial\psi(x,t)}{\partial t}=\left[-\frac{\hbar^2}{2m}\frac{\partial^2}{\partial x^2}+V(x,t)\right]\psi(x,t). \tag{3.10}$$

In particular, the Schrödinger equation for a free particle (when $V(x,t)=0$) in the position representation can be written as[5]

$$\frac{\partial\psi(x,t)}{\partial t}=\frac{\imath\hbar}{2m}\frac{\partial^2\psi(x,t)}{\partial x^2}. \tag{3.11}$$

For an initial wave function represented by a plane wave $\psi(t_0)\propto e^{\imath kx}$, which is also an energy eigenfunction (see Subsec. 2.2.6), the time-dependent wave function may then be written as

$$\psi(t)\propto e^{\imath(kx-\omega_k t)}, \tag{3.12}$$

where

$$\omega_k=\frac{E}{\hbar}=\hbar\frac{k^2}{2m}. \tag{3.13}$$

It is also useful to write the Schrödinger equation in the position representation for the three-dimensional case

$$\imath\hbar\frac{\partial\psi(\mathbf{r},t)}{\partial t}=\left[-\frac{\hbar^2}{2m}\Delta+V(\mathbf{r},t)\right]\psi(\mathbf{r},t), \tag{3.14}$$

where

$$\Delta=\nabla^2=\frac{\partial^2}{\partial x^2}+\frac{\partial^2}{\partial y^2}+\frac{\partial^2}{\partial z^2} \tag{3.15}$$

is the Laplacian in Cartesian coordinates.

3.1.2 Determinism and probabilism

From Eq. (3.8) it is clear that, consistently with our assumptions, given a certain Hamiltonian operator, any initial state vector $|\psi(t_0)\rangle$ will evolve in a deterministic way. This means that the knowledge of the state vector at an initial time t_0 and of the Hamiltonian allows the univocal determination of the state at any future (and past) times. However, this is deeply different from what happens in classical mechanics (see Sec. 1.1, and in particular Eq. (1.1)). There, the deterministic evolution concerns *all the properties* of the system,

[5] Note that Eq. (3.11) is formally identical to a classical diffusion equation.

i.e. at any time t the value of every observable (and therefore the state itself) is perfectly determined.

In quantum mechanics, on the contrary, the state has an intrinsically probabilistic nature (see Subsec. 2.3.3). This does not mean, however, that it is ontologically defective: it is exactly as determined as it should be, given the superposition principle and uncertainty relations (see Sec. 2.3). On the other hand, as we have said, the wave function (or the corresponding state vector) expresses it completely, so that any attempt at improving the knowledge of the state beyond the quantum formalism has failed (see Sec. 16.3).

The Schrödinger equation concerns the deterministic evolution of a probability amplitude and this circumstance affects the probability distributions of observables, too. In fact, after a certain time evolution there will still be a certain probability for an observable to assume a given value (see Pr. 2.2: p. 57). In conclusion, even though the fundamental evolution equation is deterministic, the structure of the theory remains intrinsically probabilistic (see also Sec. 1.4).

3.1.3 Stationary states

In the case where the potential does not explictly depend on time,[6] the formal solution of the Schrödinger equation is easily obtained by integration of Eq. (3.8), and is given by

$$| \psi(t)\rangle = e^{-\frac{i}{\hbar}\hat{H}t} | \psi(0)\rangle, \tag{3.16}$$

where $| \psi(0)\rangle$ is the state vector at time $t_0 = 0$. In general, however, it is not trivial to determine the action of the (unitary) operator $e^{-\frac{i}{\hbar}\hat{H}t}$ onto the state vector $| \psi(0)\rangle$. It appears now clear that the eigenvectors and eigenvalues of $\hat{H}$ play a central role in the determination of the time evolution of a quantum system. In fact, let us assume that the initial state vector of the system be an eigenstate of the Hamiltonian operator, i.e.

$$\hat{H} | \psi(0)\rangle = E | \psi(0)\rangle, \tag{3.17}$$

where E is the corresponding eigenvalue of $\hat{H}$ (see Eq. (2.166)). Then, the action of the operator $e^{-\frac{i}{\hbar}\hat{H}t}$ onto the state vector $| \psi(0)\rangle$ becomes trivial (a multiplication by the phase factor $e^{-\frac{i}{\hbar}Et}$) and the state vector at time t can be simply written as

$$| \psi(t)\rangle = e^{-\frac{i}{\hbar}Et} | \psi(0)\rangle . \tag{3.18}$$

As a consequence, an energy eigenstate *does not evolve with time* since a phase factor is irrelevant for the determination of a state (see end of Subsec. 2.1.3).

In the general case, when the initial state $| \psi(0)\rangle$ is not an energy eigenstate, things get a bit harder. There is, however, a general procedure which may be employed in order to find

[6] In general, the solution of the Schrödinger equation with a time-dependent Hamiltonian is a complex problem. In certain cases the problem can be addressed by making use of some approximation methods, for instance assuming that the potential changes very slowly in time (see Sec. 10.3).

the time-evolved state vector $| \psi(t)\rangle$. First, one has to solve the eigenvalue equation for the Hamiltonian (see Subsec. 2.2.6). For a discrete spectrum, one has

$$\hat{H} \mid \psi_n\rangle = E_n \mid \psi_n\rangle, \tag{3.19}$$

where the eigenvectors $| \psi_n\rangle$ are called the *stationary states* and the E_n are the corresponding eigenvalues, namely the energy levels of the system. Second, one has to expand the initial state vector $| \psi(0)\rangle$ onto the basis $\{| \psi_n\rangle\}$, i.e. to determine the complex coefficients $c_n^{(0)}$ such that (see Subsec. 2.1.2)

$$| \psi(0)\rangle = \sum_n c_n^{(0)} \mid \psi_n\rangle. \tag{3.20}$$

Finally, one is able to explicitly evaluate the rhs of Eq. (3.16) as

$$| \psi(t)\rangle = \sum_n e^{-\frac{i}{\hbar}E_n t} c_n^{(0)} \mid \psi_n\rangle, \tag{3.21}$$

which represents the time evolution of *any* initial state given its expansion in the basis of energy eigenvectors.

In the case of a continuous spectrum, the sums in Eqs. (3.19)–(3.21) have to be replaced by the corresponding integration signs. Therefore, Eq. (3.19) becomes

$$\hat{H} \left| \psi_E \right\rangle = E \left| \psi_E \right\rangle, \tag{3.22}$$

while the expansion (3.20) can be written as

$$| \psi(0)\rangle = \int dE\, c^{(0)}(E) \left| \psi_E \right\rangle. \tag{3.23}$$

As a consequence, the time evolution of any initial state $| \psi(0)\rangle$ for the continuous case can be formulated as

$$| \psi(t)\rangle = \int dE\, c^{(0)}(E) e^{-\frac{i}{\hbar}Et} \left| \psi_E \right\rangle. \tag{3.24}$$

It is evident from the procedure above that the solution of the energy eigenvalue equation is a *necessary* step for the determination of the time evolution of any system. Therefore, Eq. (3.19) or Eq. (3.22) is often called the *stationary Schrödinger equation*, while Eq. (3.8) is called the *time-dependent Schrödinger equation*.

We also note that the time-dependent Schrödinger equation for the wave function $\psi(x,t)$ is the wave function at time t (see Eq. (3.9)) can be formally solved in analogy with Eq. (3.16) as

$$\psi(x,t) = e^{\frac{i}{\hbar}\hat{H}t} \psi(x,0). \tag{3.25}$$

As for Eq. (3.21), also in this case the actual solution requires an expansion of $\psi(x,0)$ in terms of the eigenfunctions $\psi_n(x) = \langle x \mid \psi_n\rangle$ of $\hat{H}$, and reads

$$\psi(x,t) = \sum_n e^{-\frac{i}{\hbar}E_n t} c_n^{(0)} \psi_n(x). \tag{3.26}$$

Note that, since the integral $\int_{-\infty}^{+\infty} dx |\psi(x,t)|^2$ does not depend on time, a normalized wave function will remain normalized (see Prob. 3.5).

We have already seen that the energy spectrum of a system is given by the set of all possible energy eigenvalues and that stationary states correspond to these eigenvalues. In particular, if the Hamiltonian is bounded from below (as in physically interesting cases) (see Box 2.1), there is a state corresponding to the minimum eigenvalue that is called *ground state*, as, for example, the lowest energy level of one electron in an atom (see Ch. 11). As we know, such a spectrum can be continuous, discrete, or a combination of both these possibilities (see Subsecs. 1.5.4 and 2.1.1). This is in close relationship with the normalization of the wave function describing the system. In fact, eigenfunctions corresponding to discrete eigenvalues are normalizable, whereas eigenfunctions corresponding to continuous eigenvalues are not (see Subsec. 2.2.2). In the former case $\int dx\, |\psi_n(x)|^2 < \infty$ (see Eq. (2.108)), which means that $\psi_n(x)$ tends to zero sufficiently fast for $x \to \pm\infty$ so that the integral converges. As a consequence, the probability of finding the particle at large distances goes rapidly to zero. This kind of states are called *bound states*. Instead, if we consider a non-normalized wave function, i.e. corresponding to an eigenvalue belonging to the continuous spectrum, we have that $\int dx\, |\psi_n(x)|^2 = \infty$, which means that the system extends to infinity and the stationary states are called *unbound*. In this case, the eigenfunctions may not vanish at infinity. For instance, when $x \to \infty$, their absolute value may oscillate indefinitely. Wave functions of this type are useful for collision problems. It may also happen, of course, that the eigenfunctions tend to zero not sufficiently fast as x goes to infinity.

3.1.4 Degenerate eigenvalues

It is interesting to note that energy eigenvalues can be degenerate (see Th. 2.2: p. 47 and comments, and also the proof of Th. 2.4: p. 66), i.e. it can be the case that two or more eigenvectors (or eigenfunctions) share the same eigenvalue. This has a relevant physical meaning.

We may establish a necessary and sufficient condition in order to have energy degeneracy. Suppose that the observable $\hat{O}$ commutes with the Hamiltonian and is not the identity operator, i.e. $[\hat{H}, \hat{O}] = 0$, and that $\hat{O}$ is not a function of $\hat{H}$ only. Then,

$$\hat{H}\hat{O} \left| \psi_j \right\rangle = \hat{O}\hat{H} \left| \psi_j \right\rangle = E_j \hat{O} \left| \psi_j \right\rangle, \tag{3.27}$$

which shows that the ket $\hat{O}\left|\psi_j\right\rangle$ is an eigenket of $\hat{H}$ with eigenvalue E_j just as $\left|\psi_j\right\rangle$ is. Moreover, the ket $\hat{O}\left|\psi_j\right\rangle$ cannot be proportional to $\left|\psi_j\right\rangle$, i.e. it cannot be written in the form $\hat{O}\left|\psi_j\right\rangle = f(E_j)\left|\psi_j\right\rangle$, where $f(E_j)$ is a function of the j-th energy eigenvalue. In fact, if this were the case, $\hat{O}$ would be equal to $f(\hat{H})$, which would contradict one of the assumptions.

Let us now prove the equivalence in the other sense, that is, that degeneracy implies commutativity. If we have degeneracy, we may always partition the whole Hilbert space $\mathcal{H}$ into subspaces $\mathcal{H}_1, \mathcal{H}_2, \ldots$, in each of which the energy eigenvalue is constant, i.e.

$$\mathcal{H} = \mathcal{H}_1 \oplus \mathcal{H}_2 \oplus \cdots , \tag{3.28}$$

where the symbol $\oplus$ denotes the direct sum. This is technically true only if the spectrum is discrete. In the case of a continuous spectrum, a similar formula holds, but it is more complex. Then,

$$\hat{H} \mid \varphi\rangle = E_k \mid \varphi\rangle, \tag{3.29}$$

where $\mid \varphi\rangle \in \mathcal{H}_k$. This means that the Hamiltonian $\hat{H}_k$ in each subspace $\mathcal{H}_k$ is a multiple of the identity. Here, $\hat{H}_k$ is the block of $\hat{H}$ pertaining to $\mathcal{H}_k$, i.e.

$$\hat{H}_k = \begin{bmatrix} E_k & 0 & 0 & \cdots \\ 0 & E_k & 0 & \cdots \\ & & \ddots & \\ 0 & \cdots & \cdots & E_k \end{bmatrix}, \tag{3.30}$$

where the dimension of $\mathcal{H}_k$ is equal to the degree of degeneracy of the eigenvalue E_k. We can then build an Hermitian operator

$$\hat{O} = \hat{O}_1 \otimes \hat{O}_2 \otimes \cdots , \tag{3.31}$$

where $\otimes$ denotes the direct product between operators belonging to different subspaces of the total Hilbert space, and $\hat{O}_k$ is an arbitrary operator that takes a vector on $\mathcal{H}_k$ into a vector of $\mathcal{H}_k$, and obviously commutes with $\hat{H}_k$. This implies that $[\hat{O}, \hat{H}] = 0$.

3.2 Properties of the Schrödinger equation

Before going into the details of the solution of the (stationary and time-dependent) Schrödinger equation, it is very useful to look for fundamental properties of the solutions to these equations which can be derived a priori at an abstract mathematical level.[7]

3.2.1 Regularity

For what concerns *continuity* it is possible to state the following properties of the wave function $\psi(\mathbf{r})$ (we refer here to the Schrödinger equation in the form (3.14)):

(i) $\psi(\mathbf{r})$ has to be single-valued and continuous. It is single-valued because there cannot be two different probability amplitudes for the same position, and it must be continuous because the Schrödinger equation (3.14) requires it to be differentiable.

[7] In this section we somewhat follow the arguments by Landau and Lifshitz [*Landau/Lifshitz* 1976b, §§ 18 and 21]. See also [*Messiah* 1958, 98–114].

(ii) For the same reasons the continuity of $\psi(\mathbf{r})$ and of its first derivatives must hold true even when $V(\mathbf{r})$ is discontinuous but finite. On the contrary, if in some regions $V(\mathbf{r}) = \infty$, then, while $\psi(\mathbf{r})$ must still be continuous, its first derivatives need not necessarily be so (see Sec. 3.4 and Subsec. 4.2.2).

(iii) Since a particle cannot penetrate an infinite potential wall, it is clear that, in the open regions where $V(\mathbf{r}) = \infty$, $\psi(\mathbf{r})$ has to be equal to zero. Therefore, for continuity, we must have $\psi(\mathbf{r}) = 0$ at the border of such regions. However, in this case, the derivatives will be discontinuous. When the potentially is infinite only in one point (i.e. a δ-function), the wave function need not vanish at this point but it still has to be continuous. Again, the first derivative of the wave function at this point need not be continuous. On the contrary, in the regions where $V(\mathbf{r}) < \infty$ then $\psi(\mathbf{r})$ cannot be identically zero. As we shall see (in Sec. 4.3), this means that a quantum particle may have a non-vanishing probability amplitude to be found in a classically forbidden region. In other words, we may have $\psi(\mathbf{r}) \neq 0$ in regions where the energy E of the particle is smaller than the potential energy V. Unlike classical mechanics, in quantum mechanics this is not a contradiction.

(iv) For any dimension, the wave function of the ground state never vanishes (where we have assumed the absence of a magnetic field, i.e. that the Schrödinger equation is real (see Subsec. 11.3.3 and Sec. 11.4)).

3.2.2 Energy eigenvalues

Concerning the *energy eigenvalues* we have the following three properties of the Schrödinger equation :

(i) Since the Hamiltonian is given by the sum of the kinetic and potential energies, we have, also for the mean values, $\langle E \rangle = \left\langle \hat{H} \right\rangle = \left\langle \hat{T} \right\rangle + \left\langle \hat{V} \right\rangle$. Now, the kinetic energy is always positive and therefore $\langle \hat{T} \rangle > 0$. When the potential energy has a minimum value $V_{\min}$, it is clear that $\left\langle \hat{V} \right\rangle \geq V_{\min}$ and $\left\langle \hat{H} \right\rangle > V_{\min}$. Since this has to hold true for any state, also any energy eigenvalue has to be larger than $V_{\min}$, i.e.

$$\langle E_n \rangle > V_{\min}. \tag{3.32}$$

(ii) If $V \to 0$ for $r \to \infty$,[8] then the negative eigenvalues of the energy are discrete and therefore the corresponding eigenstates are bound states (see Subsec. 3.1.3), while the positive eigenvalues correspond to the continuous spectrum (infinite motion). This is so because at large distances the potential energy is negligible and therefore the motion is almost free: a free motion, however, can only correspond to positive eigenvalues. In particular, if the potential energy is positive everywhere and tends to zero at the infinity (as in Fig. 3.1), then the discrete spectrum is absent and the only possible motion of the particle is infinite. In fact, in this case, we must have

[8] It is always possible to define $V \to 0$ for $r \to \infty$ when the force vanishes sufficiently fast at infinity.

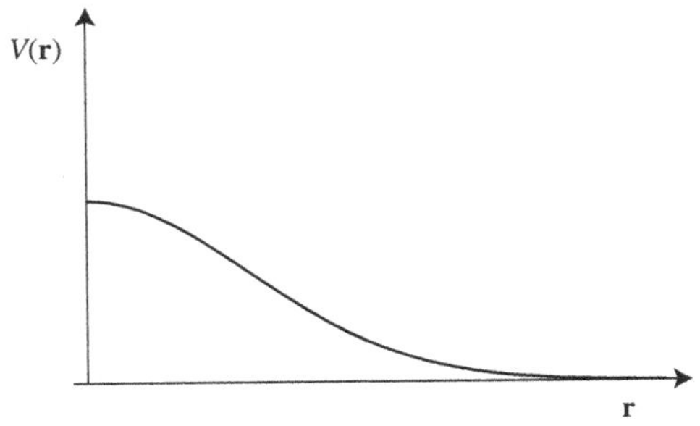

Figure 3.1 Positive potential vanishing at infinity.

$E_n > 0$ (see Eq. (3.32)), and we have already established that for positive energies the spectrum has to be continuous.

(iii) If $V(\infty) = \infty$, then the whole spectrum is discrete.

(iv) The Schrödinger equation for the wave functions of the stationary states (i.e. the energy eigenfunctions), in the absence of a magnetic field (which is not an external potential), does not contain complex terms, and therefore its solutions may be chosen to be real. For non-degenerate eigenvalues of the spectrum, this is straightforward, since any eigenfunction and its complex conjugate must both satisfy the same equation, and may differ at most by an irrelevant phase factor. In the case of degenerate energy eigenvalues, the corresponding eigenfunctions may be complex. However, by suitably choosing appropriate linear combinations of them, it is always possible to establish a set of real eigenfunctions.

3.2.3 One-dimensional case

For the one-dimensional case we can establish the following additional properties (we refer here to the Schrödinger equation in the form (3.10)):

(i) In a discrete spectrum there are no degenerate eigenvalues (let us for semplicity consider the case where $V(x) < \infty$). In fact, let us assume that ψ_1 and ψ_2 be two different eigenfunctions corresponding to the same energy eigenvalue E. They both have to satisfy the Schrödinger equation (see Subsec. 3.1.4)

$$\psi''(x) = \frac{2m(V - E)}{\hbar^2}\psi(x), \tag{3.33}$$

where $\psi''(x)$ denotes the second derivative of $\psi(x)$ with respect to x. Therefore we have

$$\frac{\psi_1''}{\psi_1} = \frac{2m(V - E)}{\hbar^2} = \frac{\psi_2''}{\psi_2}, \tag{3.34}$$

from which it follows that $\psi_1''\psi_2 - \psi_2''\psi_1 = 0$. Integrating this relation we obtain $\psi_1'\psi_2 - \psi_2'\psi_1 = C$, where C is a constant. Since ψ_1 and ψ_2 tend to zero for $x \to \infty$, the constant C must be zero and therefore $\psi_1'/\psi_1 = \psi_2'/\psi_2$. Integrating this relation

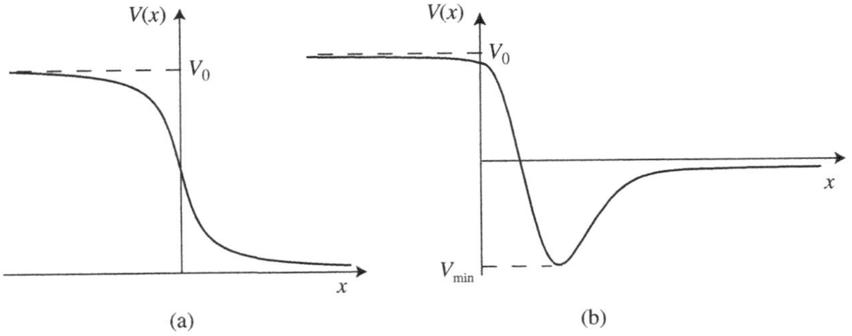

Figure 3.2

Potential function tending to finite values as $x \to \pm\infty$.

once more, we conclude that $\psi_1 = C'\psi_2$, where C' is again a constant. Therefore ψ_1 and ψ_2 are not linearly independent. We should emphasize that this is not true in larger dimensions. In fact, in this case there exist non-trivial observables which commute with the Hamiltonian, and therefore there is degeneracy (see Subsec. 3.1.4).

(ii) For the wave functions of the discrete spectrum it is possible to state the following theorem, which we shall not prove: the eigenfunction $\psi_n(x)$ corresponding to the $(n+1)$-th energy eigenvalue E_n vanishes n times for finite values of x (see the examples in Secs. 3.4, 4.1, and 4.4).

(iii) Let us assume that, for $x \to \pm\infty$, the potential energy $V(x)$ tends to finite values (as in Fig. 3.2(a) and (b)). We take $V(+\infty) = 0$, $V(-\infty) = V_0$, and assume $V_0 > 0$. The presence of the discrete spectrum is possible only for values of the energy that do not allow the particle to escape to infinity, that is, for negative values of the energy. Moreover, the energy has also to be larger than the minimum value of $V(x)$, i.e. the potential energy must have at least one minimum with $V_{\min} < 0$ (as in Fig. 3.2(b)). In the range $V_0 > E > 0$ the spectrum is continuous and the motion of the particle is infinite, since it can escape to arbitrary large positive x values. This is why the spectrum in the case of Fig. 3.2(a) is only continuous. For $V_0 > E > 0$ all eigenvalues are also not degenerate – in order to prove this, it is possible to apply the same proof as that discussed in the context of discrete spectrum of property (i), since here both functions ψ_1 and ψ_2 vanish for $x \to \pm\infty$. Finally, for $E > V_0$, the spectrum is continuous and the motion of the particle is infinite in both directions ($x \to \pm\infty$).

(iv) If $V(x)$ is even, then the wave functions $\psi(x)$ of the stationary states are either even or odd. In fact, if we have $V(x) = V(-x)$, the Schrödinger equation does not change under the transformation $x \to -x$, and if $\psi(x)$ is a solution of the Schrödinger equation, also $\psi(-x)$ is a solution which has to coincide with $\psi(x)$ up to a constant factor, that is $\psi(x) = C\psi(-x)$. Changing sign once more we obtain $\psi(x) = C^2\psi(x)$, hence $C = \pm 1$, which proves the result (see also Prob. 3.6). This property may be extended to the three-dimensional case.

(v) For a potential well of the type shown in Fig. 3.3 (when we have $V(x) \leq V(\infty)$, $\forall x$ and $V(x) < V(\infty)$ for some x), there exists a bound state independently of the height of the well.

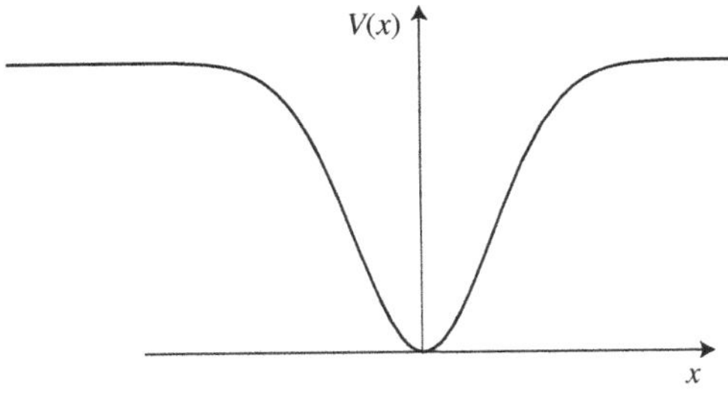

Figure 3.3 Potential well.

3.3 Schrödinger equation and Galilei transformations

A further property of the Schrödinger equation concerns its behavior under relativistic transformations. In general terms, a relativity theory tells us how physical quantities are transformed under change of the inertial reference frame.[9] Necessarily, the overall physical picture must not depend on the reference frame. As a consequence, any mechanics must be invariant with respect to the underlying relativity theory. For instance, Galilean relativity is the relativity attached to classical mechanics and quantum mechanics, whereas special relativity underlines relativistic quantum mechanics and quantum eld theory.[10] Here we are dealing with microscopic phenomena, however, which occur at a speed much smaller than the speed of light. Therefore, our non-relativistic quantum mechanics has to be invariant under Galilei transformations. In the following we shall test whether the Schrödinger equation is invariant and how the wave function changes under these transformations. For the sake of simplicity we shall restrict ourselves to the free motion of a particle in the three-dimensional case.

Let us take two reference frames $\mathcal{R}$ and $\mathcal{R}'$ such that $\mathcal{R}'$ moves with respect to $\mathcal{R}$ with constant velocity $\mathbf{V}$. We assume that at time $t = 0$ the origins of the two frames coincide, i.e. $O = O'$. The relation between the space–time coordinates in $\mathcal{R}$ and $\mathcal{R}'$ may then be written as

$$\mathbf{r} = \mathbf{r}' + \mathbf{V}t, \tag{3.35a}$$

$$t = t', \tag{3.35b}$$

where the vectors $\mathbf{r}$ and $\mathbf{r}'$ represent the position of a point particle P in $\mathcal{R}$ and $\mathcal{R}'$, respectively (see Fig. 3.4). Eqs. (3.35) are known as *Galilei transformations*. As a consequence

[9] Einstein's general relativity theory – which is not a subject of this book – also includes gravitation and therefore accelerated frames. See [*Hartle* 2003].

[10] Again, quantum field theory goes beyond the scope of this book. The interested reader is referred to specialized handbooks, for instance to [*Mandl/Shaw* 1984].

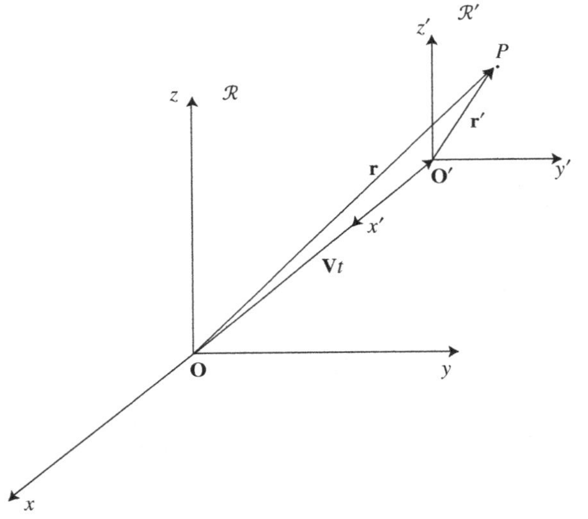

Figure 3.4 **Relationship between two different inertial reference frames $\mathcal{R}$ and $\mathcal{R}'$ under Galilei transformations. $\mathcal{R}'$ is in motion with respect to $\mathcal{R}$ with constant velocity $\mathbf{V}$.**

of Eqs. (3.35), the quantities which characterize the free motion of the particle (i.e. momentum and energy) transform according to the relations

$$\mathbf{p} = \mathbf{p}' + m\mathbf{V}, \tag{3.36a}$$

$$E = E' + \mathbf{V} \cdot \mathbf{p}' + \frac{m\mathbf{V}^2}{2}, \tag{3.36b}$$

where m is the mass of the particle. In order to find the relation between the wave functions written in the two inertial reference frames $\mathcal{R}$ and $\mathcal{R}'$, we need to derive the transformation rule for the plane-wave form of the wave function. In other words, the wave function in the frames $\mathcal{R}$ and $\mathcal{R}'$ can be written as (see Subsecs. 2.2.4 and 3.1.3, and Prob. 3.3)

$$\psi(\mathbf{r}, t) = e^{\frac{\imath}{\hbar}(\mathbf{p}\cdot\mathbf{r} - Et)}, \tag{3.37a}$$

$$\psi'(\mathbf{r}', t') = e^{\frac{\imath}{\hbar}(\mathbf{p}'\cdot\mathbf{r}' - E't')}, \tag{3.37b}$$

up to a normalization factor. Substituting the relations (3.35) and (3.36) into Eq. (3.37b) yields

$$\begin{aligned} \psi'(\mathbf{r}', t') &= \exp\left\{\frac{\imath}{\hbar}\left[(\mathbf{p} - m\mathbf{V}) \cdot (\mathbf{r} - \mathbf{V}t) - \left(E - \mathbf{V} \cdot \mathbf{p}' - \frac{m\mathbf{V}^2}{2}\right) t\right]\right\} \\ &= e^{\frac{\imath}{\hbar}(\mathbf{p}\cdot\mathbf{r} - Et)} e^{\frac{\imath}{\hbar}(\frac{1}{2}m\mathbf{V}^2 t - m\mathbf{V}\cdot\mathbf{r})}. \end{aligned} \tag{3.38a}$$

Using Eq. (3.37a) we finally have

$$\psi(\mathbf{r}, t) = \psi'(\mathbf{r} - \mathbf{V}t, t) e^{\frac{\imath}{\hbar}(m\mathbf{V}\cdot\mathbf{r} - \frac{1}{2}m\mathbf{V}^2 t)x}, \tag{3.39}$$

which is the transformation rule we were looking for. Since the exponential in the rhs of Eq. (3.39) is just a phase factor which does not contain the relevant quantities of the free

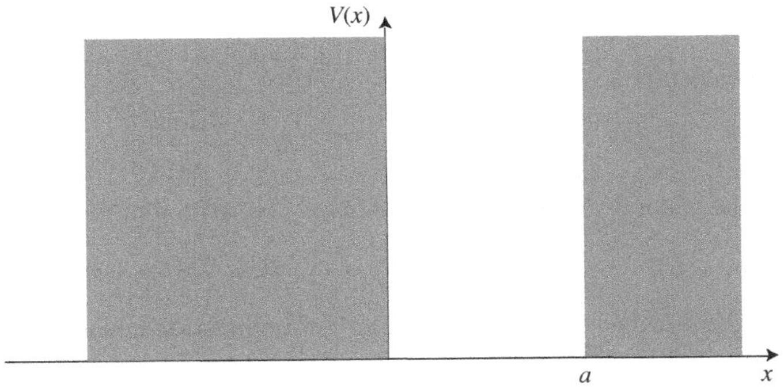

Figure 3.5 Schematic representation of the potential $V(x)$ for a particle in a box. The particle is confined in the segment $(0, a)$ by the presence of the two infinitely high potential walls at $x = 0$ and $x = a$, as described in Eq. (3.40). The left and right walls extend to $x \rightarrow -\infty$ and to $x \rightarrow +\infty$, respectively.

motion of the particle, we can state that the Schrödinger equation is invariant under Galilei transformations.

It should be noted that we have performed this calculation for the case of a plane wave. A generic wave function, however, can always be expanded into a series or integral of plane waves and, therefore, the general result may easily be derived from Eq. (3.39) (see Prob. 3.7). Moreover, in the case of a system of particles, the exponent in the rhs of Eq. (3.39) should evidently contain a sum over all the particles.

3.4 One-dimensional free particle in a box

As a first example of a quantum-mechanical system, let us consider a simple model: a one-dimensional free particle constrained between infinite potential walls located at $x = 0$ and $x = a$ (see Fig. 3.5) so that we have

$$V(x) = \begin{cases} +\infty & \text{if} \quad x > a \quad \text{or } x < 0 \\ 0 & \text{if} \quad 0 \leq x \leq a \end{cases}. \tag{3.40}$$

Therefore, for $0 \leq x \leq a$, the particle is free and the Schrödinger equation in the position representation can be written as

$$-\frac{\hbar^2}{2m}\frac{\partial^2}{\partial x^2}\psi(x) = E\psi(x). \tag{3.41}$$

Since the particle cannot penetrate into infinite walls, it is clear that $\psi(x) = 0$ for $x > a$ or $x < 0$ and, for continuity, we must have $\psi(0) = \psi(a) = 0$.[11] Equation (3.41) may be rewritten as

[11] See Property iii) in Subsec. 3.2.1. Moreover, it will be evident in the following that in this case (i.e. when the potential walls are infinitely high) the first derivative of the wave function is not continuous. For the reason of this behavior see the discussion at the end of Sec. 4.1 (p. 144).

$$\psi''(x) = -k^2\psi(x), \tag{3.42}$$

where

$$k = \frac{\sqrt{2mE}}{\hbar}. \tag{3.43}$$

The general solution of Eq. (3.42) may be written in the form

$$\psi(x) = \mathcal{N} \sin(kx + \phi), \tag{3.44}$$

where $\mathcal{N}$ is a (in general complex) normalization constant and ϕ a phase which has to be determined. Since

$$\psi(0) = \mathcal{N} \sin\phi = 0, \tag{3.45a}$$

we have $\phi = 0$. Moreover,

$$\psi(a) = \mathcal{N} \sin(ka) = 0 \tag{3.45b}$$

implies that[12] $ka = n\pi$ $(n = 1, 2, \ldots)$, i.e.

$$k = k_n = \frac{n\pi}{a}. \tag{3.46}$$

This means that the possible values of k are discrete, or *quantized*. This in turn has the consequence that also the energy levels have to be discrete. Given Eq. (3.43), the energy levels turn out to be

$$E = E_n = \frac{\pi^2\hbar^2}{2ma^2}n^2. \tag{3.47}$$

A schematic drawing of the energy levels (3.47) is given in Fig. 3.6. The corresponding eigenfunctions may then be written as

$$\psi_n(x) = \mathcal{N} \sin\left(\frac{n\pi}{a}x\right). \tag{3.48}$$

In order to determine the coefficient $\mathcal{N}$, we may take advantage of the fact that, being the spectrum discrete, the $\psi_n(x)$ must be normalized, i.e.

$$\int_0^a dx |\psi_n(x)|^2 = 1. \tag{3.49}$$

From this it follows that $|\mathcal{N}|^2 a/2 = 1$. Since an overall phase factor is irrelevant from the point of view of the wave function (see Subsec. 2.2.1), we are allowed to take a real value of $\mathcal{N}$, i.e. (see Prob. 3.9)

$$\mathcal{N} = \sqrt{\frac{2}{a}}, \tag{3.50}$$

and can finally write the energy eigenfunctions as

$$\psi_n(x) = \sqrt{\frac{2}{a}} \sin\left(\frac{n\pi}{a}x\right). \tag{3.51}$$

[12] Note that we cannot have $n = 0$ because in this case ψ would identically be zero.

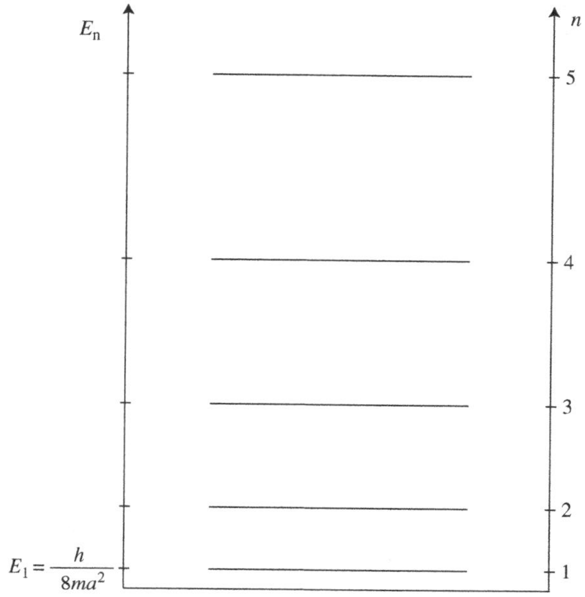

Figure 3.6 Schematic diagram of the first five energy levels for a one-dimensional particle of mass m confined in a box of dimension a.

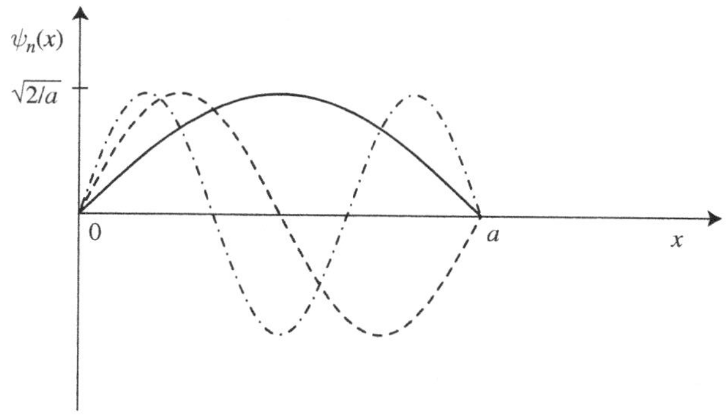

Figure 3.7 First three energy eigenfunctions for a one-dimensional particle confined in a box of dimension a: $\psi_1(x)$ is shown in solid line, $\psi_2(x)$ in dashed line, and $\psi_3(x)$ in dot-dashed line. It is interesting to note that the n-th eigenfunction has $n-1$ nodes inside the interval $(0, a)$ (see property (ii) in Subsec. 3.2.3). Moreover, in each of the n intervals between two successive eigenvalues there is a node of the $(n+1)$-th eigenfunction (and of all the following ones). This is rather a general property of the one-dimensional Schrödinger equation.

The first three of such eigenfunctions are displayed in Fig. 3.7.

It is interesting to note that the state with $E = 0$ is not allowed. In fact, if this were not the case, we would also have $p_x = 0$, which in turn would mean that the particle is at rest. This is not possible, however, in quantum mechanics, since it would imply an obvious violation of the uncertainty relation ($\Delta x = \Delta p_x = 0$). On the other hand, if $E > 0$, then there is an ambiguity in the sign of p_x, which may point rightward or leftward (see the discussion at the end of Subsec. 2.2.6: p. 79). This implies that

$$\Delta p_x \simeq 2p_x = 2\hbar k = 2\hbar\frac{n\pi}{a}, \tag{3.52}$$

and, since $\Delta x \simeq a$ (the particle is spread over the allowed region $(0, a)$), we have

$$\Delta x \Delta p_x \simeq 2\pi\hbar n = nh. \tag{3.53}$$

Even for the state of minimal uncertainty – the ground state ($n = 1$) – we would then have

$$\Delta x \Delta p_x \simeq h > \frac{\hbar}{2}. \tag{3.54}$$

Furthermore, it should be noted that the density of the energy levels (3.47) increases with m and a. This means that, when m and a are large (which is the case for macroscopic objects), the quantum levels become approximately continuous.

At this point we are in the position to apply the general procedure of Subsec. 3.1.3 and find the time evolution of *any* initial wave function $\psi(x, 0)$ of a particle in a box. In fact, we only have to expand $\psi(x, 0)$ into a series of the energy eigenfunctions $\psi_n(x)$ (see Eq. (3.51)), i.e.

$$\psi(x, 0) = \sum_n c_n(0)\psi_n(x), \tag{3.55}$$

where

$$\sum_n |c_n(0)|^2 = 1 \tag{3.56}$$

for normalization reasons. Then, according to Eqs. (3.25) and (3.26), we finally have

$$\begin{aligned}\psi(x, t) &= e^{-\frac{\imath}{\hbar}\hat{H}t}\psi(x, 0)\\ &= \sum_n c_n(0)e^{-\frac{\imath}{\hbar}E_n t}\psi_n(x)\\ &= \sum_n c_n(t)\psi_n(x)\\ &= \sqrt{\frac{2}{a}}\sum_n c_n(0)e^{-\imath\frac{\pi^2\hbar}{2ma^2}n^2 t}\sin\left(\frac{n\pi}{a}x\right),\end{aligned} \tag{3.57}$$

where

$$c_n(t) = c_n(0)e^{-\imath\frac{\pi^2\hbar}{2ma^2}n^2 t}. \tag{3.58}$$

Box 3.1 Cyclic property of the trace

This is a crucial property that we shall use on several occasions throughout the book. It states that, in the case of the product of n operators (or square matrices), we have

$$\mathrm{Tr}\left[\hat{O}_1 \cdot \hat{O}_2 \cdots \hat{O}_n\right] = \mathrm{Tr}\left[\hat{O}_n \hat{O}_1 \cdot \hat{O}_2 \cdots \hat{O}_{n-1}\right]. \tag{3.59}$$

For square matrices of finite dimensions it is straightforward to prove Eq. (3.59). In particular, we prove that it is true for $n = 2$, i.e.

$$\mathrm{Tr}\left[\hat{O}\hat{O}'\right] = \mathrm{Tr}\left[\hat{O}'\hat{O}\right]. \tag{3.60}$$

In fact, we have

$$\mathrm{Tr}\left[\hat{O}\hat{O}'\right] = \sum_n \sum_j o_{nj} o'_{jn}, \tag{3.61}$$

since $(\hat{O}\hat{O}')_{ik} = \sum_j o_{ij} o'_{jk}$, and

$$\begin{aligned}\mathrm{Tr}\left[\hat{O}'\hat{O}\right] &= \sum_n \sum_j o'_{nj} o_{jn} \\ &= \sum_n \sum_j o'_{jn} o_{nj} \\ &= \mathrm{Tr}\left[\hat{O}\hat{O}'\right],\end{aligned} \tag{3.62}$$

where we have interchanged the role of the indices n and j. This result means that the trace of any commutator is zero in the finite case. However, this does not hold true in the case of infinite dimensions, where restrictive conditions on the trace must be considered. By induction, it is trivial to generalize to the case of the product of n operators and obtain Eq. (3.59).

3.5 Unitary transformations

3.5.1 General properties of unitary operators

We have seen in Subsec. 3.1.1 how the state of a quantum-mechanical system at times $t > t_0$ is related to the initial state at $t = t_0$. In particular, following Eq. (3.16), we may write

$$|\,\psi(t)\rangle = e^{-\frac{\imath}{\hbar}\hat{H}(t-t_0)}\,|\,\psi(t_0)\rangle = \hat{U}_{t-t_0}\,|\,\psi(t_0)\rangle\,. \tag{3.63}$$

Therefore, the states at times t and t_0 are connected by a class of transformations induced by the Hamiltonian operator $\hat{H}$,

$$|\,\psi\rangle \mapsto e^{-\frac{\imath}{\hbar}\hat{H}t}\,|\,\psi\rangle, \tag{3.64}$$

and these transformations are of the form $\hat{U} = e^{\imath a\hat{O}}$, where a is a real constant and $\hat{O}$ a Hermitian operator. They are *unitary transformations*. We have already met them in the last chapter, and, as we know (see Eq. (2.34)), a unitary transformation is characterized by the property that $\hat{U}\hat{U}^\dagger = \hat{U}^\dagger\hat{U} = \hat{I}$.

Now, we wish to show that unitary transformations possess some important properties, which can be summarized as follows:

- First, unitary transformations preserve the scalar product between state vectors (see Subsecs. 2.1.2 and 2.2.5). This can be easily shown as follows. Let us take two state vectors $|\psi\rangle$ and $|\varphi\rangle$ and a unitary transformation $\hat{U}$ such that

$$\left|\psi'\right\rangle = \hat{U}\,|\psi\rangle \quad \text{and} \quad \left|\varphi'\right\rangle = \hat{U}\,|\varphi\rangle\,. \tag{3.65}$$

Then, we have

$$\left\langle\psi'\mid\varphi'\right\rangle = \left\langle\psi\left|\hat{U}^\dagger\hat{U}\right|\varphi\right\rangle = \langle\psi\mid\varphi\rangle. \tag{3.66}$$

- Second, unitary transformations preserve the trace of an operator. We recall that any normal operator can be written as (see Eq. (2.1) and Box 2.3: p. 51)

$$\hat{O} = \sum_j o_j \left|o_j\right\rangle\!\left\langle o_j\right|, \tag{3.67}$$

for some basis $\{\left|o_j\right\rangle\}$ and where the numbers o_j are not necessarily real. From this it immediately follows that *any* transformation $\hat{\mathcal{U}}$ on an operator $\hat{O}$ (see also Subsec. 8.1.1) will act as

$$\hat{O} \mapsto \hat{O}_{\mathcal{U}} = \sum_j o_j\hat{\mathcal{U}}\,|o_j\rangle\,\langle o_j|\,\hat{\mathcal{U}}^\dagger = \hat{\mathcal{U}}\hat{O}\hat{\mathcal{U}}^\dagger. \tag{3.68}$$

As a consequence, we have

$$\mathrm{Tr}\left(\hat{O}_U\right) = \mathrm{Tr}\left(\hat{U}\hat{O}\hat{U}^\dagger\right) = \mathrm{Tr}\left(\hat{U}^\dagger\hat{U}\hat{O}\right) = \mathrm{Tr}\left(\hat{O}\right), \tag{3.69}$$

where we have made use of the cyclic property of the trace (see Box 3.1).

- Finally, unitary transformations may play the same role of canonical transformations in classical mechanics, since they leave the canonical commutation relations (we call the commutation relations between observables corresponding to canonically conjugate variables *canonical commutation relations*) invariant as – in classical mechanics – canonical transformations leave Poisson brackets invariant (see Sec. 1.1, as well as also Sec. 3.7 and Ch. 8). In fact, let

$$\hat{q}' = \hat{U}\hat{q}\hat{U}^\dagger \quad \text{and} \quad \hat{p}' = \hat{U}\hat{p}\hat{U}^\dagger \tag{3.70}$$

be unitary transformations of the canonical observables $\hat{q}$ and $\hat{p}$. Then,

$$\begin{aligned}\left[\hat{q}', \hat{p}'\right] &= \hat{U}\hat{q}\hat{U}^\dagger\hat{U}\hat{p}\hat{U}^\dagger - \hat{U}\hat{p}\hat{U}^\dagger\hat{U}\hat{q}\hat{U}^\dagger \\ &= \hat{U}\hat{q}\hat{p}\hat{U}^\dagger - \hat{U}\hat{p}\hat{q}\hat{U}^\dagger \\ &= \hat{U}\left[\hat{q}, \hat{p}\right]\hat{U}^\dagger \\ &= \imath\hbar\hat{I}.\end{aligned} \tag{3.71}$$

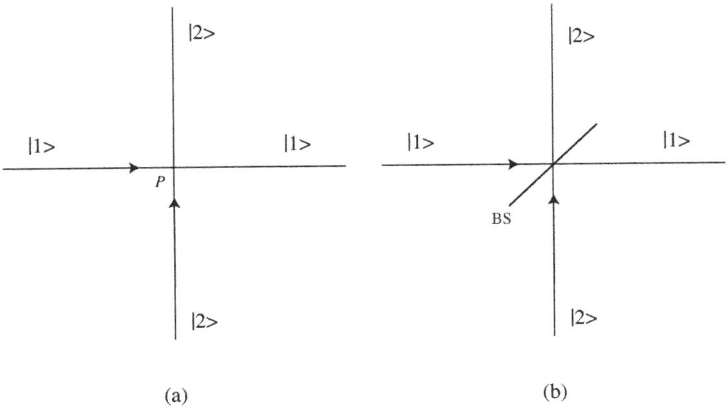

Figure 3.8 (a) Scheme of input–output formalism. (b) Beam Splitters as unitary operators.

3.5.2 Beam splitters as unitary transformations

In this book we have largely made use of beam-splitting as a tool for introducing new concepts. Generally speaking, the action of a beam splitter can be described by an input–output relation of the type

$$|\,\mathrm{out}\rangle = \hat{U}_{\mathrm{BS}}\,|\,\mathrm{in}\rangle\,,$$

where the matrix $\hat{U}$ is given by

$$\hat{U}_{\mathrm{BS}} = \begin{bmatrix} \mathrm{T} & \mathrm{R}' \\ \mathrm{R} & \mathrm{T}' \end{bmatrix},$$

T and R being the (in general, complex) transmission and reflection coefficients, respectively. By imposing the requirement of unitarity of the matrix $\hat{U}_{\mathrm{BS}}$, we must have[13]

$$\mathrm{R} = \mathrm{R}'^{*}, \quad \mathrm{T} = -\mathrm{T}'^{*}$$

which implies $|\mathrm{R}| = |\mathrm{R}'|$ and $|\mathrm{T}| = |\mathrm{T}'|$. For the common case of a beam splitter that has the same effect on a beam incident through port 1 as on a beam incident through port 2, that is, a symmetric beam splitter, we have $\mathrm{R} = \mathrm{R}'$ and $\mathrm{T} = \mathrm{T}'$. By imposing this condition together with that of a fifty–fifty beam splitter ($|\mathrm{R}| = |\mathrm{T}|$, which together with the previous one implies $|\mathrm{R}| = |\mathrm{R}'| = |\mathrm{T}| = |\mathrm{T}'|$), it is possible to prove that the unitary matrix above turns into

$$\hat{U}_{\mathrm{BS}} = \frac{1}{\sqrt{2}} \begin{bmatrix} 1 & \imath \\ \imath & 1 \end{bmatrix}. \tag{3.72}$$

Let us consider the example represented in Fig. 3.8. In Fig. 3.8(a) we consider a photon in the initial state

$$|\,1\rangle = \begin{pmatrix} 1 \\ 0 \end{pmatrix}. \tag{3.73}$$

13 See Holbrow, C. H., Galvez, E., and Parks, M. E., "Photon Quantum Mechanics and Beam Splitters", *American Journal of Physics* **70** (2002): 260–65.

It is clear that, after the point P, the state will still be $|1\rangle$. The same is if the incoming photon is $|2\rangle$, given by

$$|2\rangle = \begin{pmatrix} 0 \\ 1 \end{pmatrix}. \tag{3.74}$$

Also in this case, the state will be the same after the geometrical point P. Consider now a symmetric beam splitter, as in Fig. 3.8(b). It is clear that in this case we should have the transformations

$$|1\rangle \mapsto \frac{1}{\sqrt{2}}(|1\rangle + \imath|2\rangle), \quad |2\rangle \mapsto \frac{1}{\sqrt{2}}(\imath|1\rangle + |2\rangle). \tag{3.75}$$

It is not difficult to see that this is a unitary transformation. Its transposed conjugate is given by

$$\hat{U}^{\dagger}_{\mathrm{BS}} = \frac{1}{\sqrt{2}}\begin{bmatrix} 1 & -\imath \\ -\imath & 1 \end{bmatrix}. \tag{3.76}$$

We then have

$$\hat{U}_{\mathrm{BS}}\hat{U}^{\dagger}_{\mathrm{BS}} = \frac{1}{2}\begin{bmatrix} 2 & 0 \\ 0 & 2 \end{bmatrix} = \begin{bmatrix} 1 & 0 \\ 0 & 1 \end{bmatrix}, \tag{3.77a}$$

$$\hat{U}^{\dagger}_{\mathrm{BS}}\hat{U}_{\mathrm{BS}} = \frac{1}{2}\begin{bmatrix} 2 & 0 \\ 0 & 2 \end{bmatrix} = \begin{bmatrix} 1 & 0 \\ 0 & 1 \end{bmatrix}. \tag{3.77b}$$

We remark that in the derivation of Subsec. 2.3.4 we have made use of two beam-splitter matrices with real transmission and reflection coefficients, that is, of the matrix

$$\begin{bmatrix} \mathrm{T} & \mathrm{R} \\ \mathrm{R} & -\mathrm{T} \end{bmatrix},$$

for BS1 and of the matrix

$$\frac{1}{\sqrt{2}}\begin{bmatrix} 1 & 1 \\ 1 & -1 \end{bmatrix}$$

for BS2. These two matrices can derived form the general case by imposing the reality of T and R but cannot account for symmetric beam-splitting. Moreover, the latter describes a 50%–50% beam splitter. A slightly more complex example is represented by the polarization beam splitter (PBS) (see Fig. 2.1). In this case, we have another degree of freedom, namely polarization. The distinctive feature of a PBS is, e.g., that it reflects incoming photons of vertical polarization and transmits those of horizontal polarization, independently from the incoming path. As a consequence, there are four possibilities for the incoming and outgoing vectors, that is, the four-dimensional system's Hilbert space is spanned by the four kets

$$|1, \leftrightarrow\rangle = \begin{pmatrix} 1 \\ 0 \\ 0 \\ 0 \end{pmatrix}, \quad |1, \updownarrow\rangle = \begin{pmatrix} 0 \\ 1 \\ 0 \\ 0 \end{pmatrix}, \tag{3.78a}$$

$$|2, \leftrightarrow\rangle = \begin{pmatrix} 0 \\ 0 \\ 1 \\ 0 \end{pmatrix}, \quad |2, \updownarrow\rangle = \begin{pmatrix} 0 \\ 0 \\ 0 \\ 1 \end{pmatrix}. \tag{3.78b}$$

The effect of the PBS can be represented by the following transformations on the basis vectors

$$|1, \leftrightarrow\rangle \mapsto |1, \leftrightarrow\rangle\,, \quad |1, \updownarrow\rangle \mapsto \imath\, |2, \updownarrow\rangle\,, \tag{3.79a}$$

$$|2, \leftrightarrow\rangle \mapsto |2, \leftrightarrow\rangle\,, \quad |2, \updownarrow\rangle \mapsto |1, \updownarrow\rangle\,. \tag{3.79b}$$

These transformations can be expressed by the 4×4 matrix

$$\hat{U}_{\mathrm{PBS}} = \begin{bmatrix} 1 & 0 & 0 & 0 \\ 0 & 0 & 0 & \imath \\ 0 & 0 & 1 & 0 \\ 0 & \imath & 0 & 0 \end{bmatrix}, \tag{3.80}$$

whose transposed conjugate is

$$\hat{U}^{\dagger}_{\mathrm{PBS}} = \begin{bmatrix} 1 & 0 & 0 & 0 \\ 0 & 0 & 0 & -\imath \\ 0 & 0 & 1 & 0 \\ 0 & -\imath & 0 & 0 \end{bmatrix}. \tag{3.81}$$

It is straightforward to verify that

$$\hat{U}_{\mathrm{PBS}}\hat{U}^{\dagger}_{\mathrm{PBS}} = \hat{U}^{\dagger}_{\mathrm{PBS}}\hat{U}_{\mathrm{PBS}} = \hat{I}, \tag{3.82}$$

and that therefore $\hat{U}_{\mathrm{PBS}}$ is unitary.

3.5.3 Time translations

As we have seen, the time translation operator (see Eq. (3.16))

$$\hat{U}_t = e^{-\frac{\imath}{\hbar}\hat{H}(t-t_0)} \tag{3.83}$$

is the operator which describes the transformation from the state vector (or the wave function) at time t_0 to the state vector (or the wave function) at time t. The inverse of the unitary transformation $\hat{U}_t$ describes a backward time translation. So, for example (for $t_0 = 0$),

$$|\psi(0)\rangle = \hat{U}_t^{-1}\,|\psi(t)\rangle = \hat{U}_{-t}\,|\psi(t)\rangle\,. \tag{3.84}$$

Since $\hat{U}_t^{-1} = \hat{U}_t^{\dagger}$, we also have

$$\hat{U}_t^{\dagger} = \hat{U}_{-t}, \tag{3.85}$$

which shows the forward–backward time symmetry in quantum mechanics. This implies that the Schrödinger equation is invariant under time-reversal transformations if, together with the substitution $t \mapsto -t$, one also applies the replacement $|\,\psi\rangle \mapsto \langle\psi\,|$. In fact, taking the Hermitian conjugate of Eq. (3.8) one has

$$\imath\hbar\frac{\partial}{\partial(-t)}\langle\psi\,| = \langle\psi\,|\,\hat{H}. \tag{3.86}$$

We have shown (see Subsec. 2.1.1, in particular Eq. (2.16)) that scalar products between state vectors represent probability amplitudes. For example, the expression $\langle\psi\,|\,\hat{U}_t\,|\,\varphi\rangle$ is the probability amplitude that, given an initial state $|\,\varphi\rangle$, measures how close it evolves unitarily to a final state $|\,\psi\rangle$ at time t. If we take the complex conjugate of this expression we obtain

$$\left(\langle\psi\,|\,\hat{U}_t\,|\,\varphi\rangle\right)^* = \langle\varphi\,|\,\hat{U}_{-t}\,|\,\psi\rangle. \tag{3.87}$$

The rhs of Eq. (3.87) represents the probability amplitude that a final state $|\,\psi\rangle$ (at time t) evolves unitarily backwards to an initial state $|\,\varphi\rangle$ (at time $t_0 = 0$). In this context, we see that kets may be thought of as input states, whereas bras as output states of a certain physical evolution or process. In particular, the expression $\langle\psi\,|\,\hat{U}_t\,|\,\psi\rangle$ may be understood as the probability amplitude that an initial state $|\,\psi\rangle$ remains unaltered after the unitary evolution for a time t and its square modulus is sometimes called the *autocorrelation function*.

From what we have seen above, given the Hamiltonian $\hat{H}$ of a quantum-mechanical system and its initial state vector $|\,\psi(0)\rangle$, the unitary operator $\hat{U}_t$ allows us to determine the state vector at subsequent times t. However, in Subsecs. 2.1.1 and 2.1.3 we stated that the measurement of any observable always gives as an outcome one (and only one) of the eigenvalues of the associated operator with a certain probability. This seems to be in contradiction with the unitary evolution of Eq. (3.63). Once again, this is the so-called measurement problem in quantum mechanics, as no unitary evolution will ever be able to account for an abrupt change of the state vector from an arbitrary superposition to one of its components (see Probs. 3.12 and 3.13). Of course, given a certain superposition state, it is always possible to build an unitary operator that brings it to one of its components – an example for this has been given in the previous subsection with the (polarization) beam-splitter unitary transformations. However, this assumes that one already knows a priori which superposition the system is in, i.e. that one already knows the initial state. In other words, there is no way to find a unitary transformation which provides the observer with the information that represents the final outcome of the measurement process (as it will be shown in Subsecs. 15.2.2 and 15.3.2).

We can also approach this problem from a slightly different point of view. As we have shown in Subsec. 2.1.2, finite-dimensional unitary transformations can be represented by rotations of a given state vector (in the Hilbert space of the system) by a certain angle (e.g. the BS unitary transformations considered in the previous subsection are by an angle of 90°). Now, given a certain superposition state, it is always possible to bring it to coincide with one of its components with a suitable rotation by a given angle, but it is impossible to bring *any* state vector to coincide with a certain component with a rotation by the *same* angle (that is with the same unitary transformation).

As we shall see in Ch. 9, the above apparent contradiction can be approached by considering the non-unitary evolution represented by a measurement as a process occurring in a subsystem which is part of a larger system whose evolution is nevertheless unitary.

3.5.4 Stone theorem

The fact that a unitary transformation can be cast into the form $\hat{U} = e^{\imath a \hat{O}}$ is rather general. Let us consider following theorem:

Theorem 3.1 (Stone) *Given a family $\hat{U}(a)$ of unitary operators, where $a \geq 0$ is a real parameter, satisfying the semigroup property* (see Subsec. 8.4.2)

$$\hat{U}(a)\hat{U}(a') = \hat{U}(a + a'), \tag{3.88}$$

then it is possible to write

$$\hat{U} = e^{\imath a \hat{O}}, \tag{3.89}$$

where $\hat{O}$ is a Hermitian operator.

The Stone theorem ensures the existence of a Hermitian infinitesimal generator for an Abelian group of unitary transformations. We have explicitly derived the unitary operator $\hat{U}_x$ for spatial translations (see Eq. (2.130)), which can be considered a specific instance of Eq. (3.89), where the momentum operator is identified as the generator of spatial translations, a point which will be the subject of Ch. 8.

We have already seen some further examples of unitary transformations when we have dealt with the matrix $\hat{U}$ occurring in the change of basis (see Subsec. 2.1.2; see also Subsec. 2.2.5) and with the diagonalization of an operator (in Subsec. 2.1.4). Other examples of unitary transformations have been given by the beam splitter and polarization beam splitter transformations, and the time translations, as we have seen in the previous subsections.

It is also interesting to observe that any unitary operation on a two-level system corresponds to a rotation of the Poincaré sphere (see Subsec. 1.3.3).

3.5.5 Green's function

The elements of the time-translation unitary matrix can be written as

$$\left\langle k \left| e^{-\frac{\imath}{\hbar}\hat{H}(t-t_0)} \right| j \right\rangle = \imath \mathrm{G}(k, t; j, t_0), \tag{3.90}$$

where $| j\rangle$ and $| k\rangle$ are some state vectors. The functions G are called *Green's functions*. In order to appreciate their importance, let us start from Eq. (3.63), that is

$$| \psi(t')\rangle = e^{-\frac{\imath}{\hbar}\hat{H}(t'-t)} | \psi(t)\rangle, \tag{3.91}$$

which relates the state vector at time t' to the state vector at time t. If we multiply both sides of Eq. (3.91) times $\left\langle \mathbf{r}' \right|$ from the left and make use of the resolution of the identity in the form $\hat{I} = \int d\mathbf{r} \,|\, \mathbf{r}\rangle \langle \mathbf{r} \,|$, we obtain

$$\psi(\mathbf{r}',t') = \imath \int d\mathbf{r} G(\mathbf{r}',t';\mathbf{r},t)\psi(\mathbf{r},t), \tag{3.92}$$

where we have made use of the tridimensional version of expression (2.107). Equation (3.92) represents an instance of Huygens' principle: if the wave function $\psi(\mathbf{r},t)$ is known at a time t, it may be found at any later time t' by assuming that each point $\mathbf{r}$ at time t is a source of waves which propagate outward from $\mathbf{r}$. The strength of the wave amplitude arriving at point $\mathbf{r}'$ at time t' from the point $\mathbf{r}$ will be proportional to the original wave amplitude $\psi(\mathbf{r},t)$ and the constant of proportionality is given by $\imath G(\mathbf{r}',t';\mathbf{r},t)$. Morevoer, Eq. (3.92) is a consequence of the first-order character of the Schrödinger equation and of its linearity: the knowledge of $\psi(\mathbf{r},t)$ for all values of $\mathbf{r}$ and one particular t is enough to determine $\psi(\mathbf{r}',t')$ for all values of $\mathbf{r}'$ and any (subsequent or previous) time t', and the relation between the two wave functions is linear.

Green's functions are related to the resolvent of the Hamiltonian through the fact that the latter is the Fourier transform of the relative unitary operator, i.e.

$$\hat{R}_{\hat{H}}(\eta) = \frac{\imath}{\hbar}\int_0^{+\infty} d\tau e^{\frac{-\eta\tau}{\hbar}} e^{-\frac{\imath}{\hbar}\hat{H}\tau}, \tag{3.93}$$

where $\tau = t' - t$. If $\hat{O}$ is a linear operator in the Hilbert space $\mathcal{H}$, the *resolvent* $\hat{R}_{\hat{O}}(\eta)$ is the operator-valued function[14]

$$\hat{R}_{\hat{O}}(\eta) = (\hat{O} - \eta)^{-1}. \tag{3.94}$$

The function is defined for all complex values of η for which $(\hat{O} - \eta)^{-1}$ exists. Coming back to the case of the Hamiltonian, and making use of the projectors $\hat{P}_j$ on the j-th's eigenvalues of $\hat{H}$, we have

$$\hat{R}_{\hat{H}}(\eta)\hat{P}_j = \frac{\hat{P}_j}{E_j - \eta}. \tag{3.95}$$

This allows us to interpret the projectors P_j as the residues of the closed contour f_j in the complex plane η enclosing the point E_j located on the real axis of the complex plane (see Fig. 3.9), that is,

$$\hat{P}_j = \frac{1}{2\pi\imath}\oint_{f_j} d\eta \hat{R}_{\hat{H}}(\eta), \tag{3.96}$$

or, for the continuous part of the spectrum,

$$\hat{P}(\Delta j) = \frac{1}{2\pi\imath}\oint_{f(\Delta j)} d\eta \hat{R}_{\hat{H}}(\eta), \tag{3.97}$$

where the projectors $\hat{P}(\Delta j)$ project on a small interval around the continuous eigenvalue E_j. In the continuous case the function $\mathcal{R}$ may have a cut such that the parts above and

[14] For the problem of the spectrum of $\hat{O}$ and of the values of η, see [*Prugovečki* 1971, 475, 520–21] [*Taylor/Lay* 1958, 264–65].

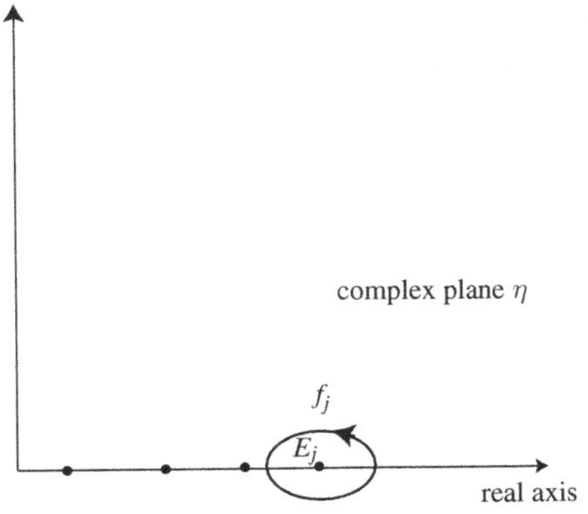

Figure 3.9 The projector P_j as the residue of the closed contour f_j in the complex plane *eta* enclosing the point E_j.

below it are not analytical continuations one of the other through the cut. In the case of a free particle ($\hat{H}_0 = \hat{\mathbf{p}}^2/2m = \hbar^2\mathbf{k}^2/2m$), the Hamiltonian has a pure continuous spectrum in the interval $[0, +\infty)$ and for this reason its resolvent

$$\hat{\mathrm{R}}_{H_0}(\eta) = \frac{1}{\hat{H}_0 - \eta} \tag{3.98}$$

is defined for all values of η that are not in the spectrum, and is a bounded operator (see Eq. (2.7)) defined on the entire Hilbert space whenever $\Re(\eta) < 0$ or $\Im(\eta) \neq 0$, i.e. when the argument of η is within the open interval $(0, 2\pi)$. Using Green's functions, we can then write the evolution of a free particle in space and time in the form[15]

$$\psi(\mathbf{r}', t') = \imath \int d\mathbf{r} \mathrm{G}_0(\mathbf{r}', t'; \mathbf{r}, t)\psi(\mathbf{r}, t), \tag{3.99}$$

for $t' > t$ and for all values of $0 < \eta < 2\pi$. G_0 is called the *free Green's function* and its explicit expression is

$$\mathrm{G}_0(\mathbf{r}', t'; \mathbf{r}, t) = -\imath \left[\frac{m}{2\pi\imath\hbar(t' - t)}\right]^{\frac{3}{2}} e^{\frac{\imath m|\mathbf{r}' - \mathbf{r}|^2}{2\hbar(t' - t)}}. \tag{3.100}$$

3.6 Different pictures

So far we have treated the operators associated with quantum-mechanical observables as time–independent quantities. Only the state vectors evolve according to the Schrödinger equation. For instance, the expectation value (see Eq. (2.78)) $\langle \psi \mid \hat{O} \mid \psi \rangle$ depends on time only because $\mid \psi \rangle$ is a function of time. This way of looking at the time evolution is called

15 Further details regarding the subject of this subsection can be found in [*Bjorken/Drell* 1964, 78–89] [*Prugovečki* 1971, 520–42].

the *Schrödinger picture*. However, it is not the only way of dealing with time evolution. In the present section we shall look at two additional pictures, the *Heisenberg* and the *Dirac* pictures. Needless to say, the physical quantities (probabilities, expectation values, etc.) will not depend on the picture chosen to represent time evolution of physical systems.

3.6.1 Heisenberg picture

In the Heisenberg picture the time dependence is completely transferred from the state vector to the observable. Let us consider the expectation value of an arbitrary observable $\hat{O}$ in the Schrödinger picture, which can be written as

$$ {}_{\rm S}\langle\psi(t)|\,\hat{O}^{\rm S}\,|\psi(t)\rangle_{\rm S}, \tag{3.101}$$

where the subscript S for the states and the superscript S for the observables denote the Schrödinger picture, and $\hat{O}^{\rm S}$ does not explicitly depend on time. Using Eq. (3.63), this expectation value may be written as

$$ {}_{\rm S}\langle\psi(t)|\,\hat{O}^{\rm S}\,|\psi(t)\rangle_{\rm S} = {}_{\rm S}\langle\psi(0)|\,\hat{U}_t^\dagger\hat{O}^{\rm S}\hat{U}_t\,|\psi(0)\rangle_{\rm S} = {}_{\rm H}\langle\psi\,|\,\hat{O}^{\rm H}(t)\,|\psi\rangle_{\rm H}, \tag{3.102}$$

where

$$ |\psi\rangle_{\rm H} = |\psi(0)\rangle_{\rm S} \tag{3.103a}$$

is the state vector in the Heisenberg picture and

$$ \hat{O}^{\rm H}(t) = \hat{U}_t^\dagger\hat{O}^{\rm S}\hat{U}_t \tag{3.103b}$$

is the time-dependent observable in the Heisenberg picture. It should also be emphasized that, according to Eqs. (3.103), the transformation from the Schrödinger picture to the Heisenberg picture is unitary and therefore leaves matrix elements and commutation relations invariant. From Eq. (3.103b) it follows that (see Prob. 3.14)

$$ \hat{H}^{\rm H} = \hat{H}^{\rm S} = \hat{H}. \tag{3.104}$$

To find the equation of motion for the observable, let us consider the time derivative of Eq. (3.102). We have

$$ \begin{aligned} &\frac{d}{dt}\left({}_{\rm S}\langle\psi(0)|\,\hat{U}_t^\dagger\hat{O}^{\rm S}\hat{U}_t\,|\psi(0)\rangle_{\rm S}\right) \\ &= {}_{\rm S}\langle\psi(0)|\left(\frac{\imath}{\hbar}\hat{H}\hat{U}_t^\dagger\hat{O}^{\rm S}\hat{U}_t - \frac{\imath}{\hbar}\hat{U}_t^\dagger\hat{O}^{\rm S}\hat{U}_t\hat{H} + \hat{U}_t^\dagger\frac{\partial\hat{O}^{\rm S}}{\partial t}\hat{U}_t\right)|\psi(0)\rangle_{\rm S} \\ &= {}_{\rm H}\langle\psi\,|\left(\frac{\imath}{\hbar}[\hat{H},\hat{O}^{\rm H}] + \frac{\partial}{\partial t}\hat{O}^{\rm H}(t)\right)|\psi\rangle_{\rm H}, \end{aligned} \tag{3.105}$$

where use has been made of Eqs. (3.103) and of the definition (2.86). The term $\partial \hat{O}^{\mathrm{S}}/\partial t$ will be different from zero only when $\hat{O}^{\mathrm{S}}$ explicitly depends on time. Since the expectation value of an observable cannot depend on the chosen picture, the rhs of Eq. (3.105) must be equal to

$$\frac{d}{dt}{}_{\mathrm{H}}\langle \psi \mid \hat{O}^{\mathrm{H}}(t) \mid \psi \rangle_{\mathrm{H}} \tag{3.106}$$

for any state vector $|\,\psi\rangle$. Therefore,

$$\imath\hbar\frac{d}{dt}\hat{O}^{\mathrm{H}}(t) = \imath\hbar\frac{\partial}{\partial t}\hat{O}^{\mathrm{H}}(t) + \left[\hat{O}^{\mathrm{H}}(t), \hat{H}\right]. \tag{3.107}$$

Equation (3.107) could also have been derived by direct differentiation of Eq. (3.103b) (see Prob. 3.15).

If the operator $\hat{O}$ does not depend explicitly on time in the Schrödinger picture,[16] then the first term of the rhs of Eq. (3.107) will be dropped, yielding

$$\imath\hbar\frac{d}{dt}\hat{O}^{\mathrm{H}}(t) = \left[\hat{O}^{\mathrm{H}}(t), \hat{H}\right]. \tag{3.108}$$

Equation (3.108) is known as the *Heisenberg equation* and is the counterpart of the Schrödinger equation in the Heisenberg picture. From Eq. (3.108) we also learn the important fact that an observable which does not explicitly depend on time and which commutes with the Hamiltonian is a *conserved quantity* or a *constant of motion*, as happens in the classical case (see Prob. 1.4).

Most importantly, Eq. (3.108) resembles very closely the classical canonical equations of motion (1.8). In fact, Eq. (3.108) can be obtained from Eqs. (1.8) with the substitution

$$\{\cdot,\cdot\cdot\} \longrightarrow \frac{1}{\imath\hbar}[\cdot,\cdot\cdot]. \tag{3.109}$$

This may be considered as a formal rule when passing from classical mechanics to quantum mechanics. For instance, if the Poisson bracket between a classical variable and the Hamiltonian is zero, this variable is a conserved quantity. For this reason, the Heisenberg evolution is formally similar to the classical time evolution. In fact, in classical mechanics there is no analogue of the Schrödinger evolution, or more precisely, this coincides with the "Heisenberg" evolution, since the state itself is just a collection of properties and therefore is itself an observable (see Sec. 1.1 and Subsec. 2.3.3).

Equations (3.107)–(3.108) may be also rewritten in terms of the expectation values as follows:

$$\imath\hbar\frac{d}{dt}\left\langle\hat{O}^{\mathrm{H}}(t)\right\rangle = \imath\hbar\left\langle\frac{\partial}{\partial t}\hat{O}^{\mathrm{H}}(t)\right\rangle + \left\langle\left[\hat{O}^{\mathrm{H}}(t), \hat{H}\right]\right\rangle, \tag{3.110}$$

where the analogue of Eq. (3.108) is obtained when the first term of the rhs of Eq. (3.110) is zero.

16 An explicitly time-dependent operator corresponds to a quantity which is classically time-dependent.

3.6.2 Dirac picture

The Dirac or interaction picture is very useful when the Hamiltonian can be split into a free part $\hat{H}_0$ and an interaction part $\hat{H}_\mathrm{I}$, i.e.

$$\hat{H} = \hat{H}_0 + \hat{H}_\mathrm{I}, \tag{3.111}$$

where, in general, both $[\hat{H}, \hat{H}_0]$ and $[\hat{H}, \hat{H}_\mathrm{I}]$ are different from zero and $\hat{H}_0$ does not explicitly depend on time.

In the interaction picture the time evolution is partly shared by both the state vector and the observable. In fact, we define

$$|\psi(t)\rangle_\mathrm{I} = \hat{U}^\dagger_{H_0,t} |\psi(t)\rangle_\mathrm{S}, \tag{3.112}$$

where $\hat{U}^\dagger_{H_0,t} = e^{\frac{\imath}{\hbar}\hat{H}_0 t}$. In order to establish the corresponding transformation for the observables, we need to write the expectation value of an arbitrary observable $\hat{O}$, as we have done in the case of the Heisenberg picture. This time we have

$${}_\mathrm{S}\langle\psi(t)|\,\hat{O}^\mathrm{S}\,|\psi(t)\rangle_\mathrm{S} = {}_\mathrm{I}\langle\psi(t)|\,\hat{U}^\dagger_{H_0,t}\hat{O}^\mathrm{S}\hat{U}_{H_0,t}\,|\psi(t)\rangle_\mathrm{I} = {}_\mathrm{I}\langle\psi(t)|\,\hat{O}^\mathrm{I}(t)\,|\psi(t)\rangle_\mathrm{I}, \tag{3.113}$$

so that the observable in the interaction picture is related to the corresponding observable in the Schrödinger picture by the unitary transformation

$$\hat{O}^\mathrm{I}(t) = \hat{U}^\dagger_{H_0,t}\hat{O}^\mathrm{S}\hat{U}_{H_0,t}. \tag{3.114}$$

The first consequence of Eq. (3.114) is that the free part of the Hamiltonian is invariant under the transformation to the interaction picture, i.e.

$$\hat{H}_0^\mathrm{I} = \hat{H}_0^\mathrm{S} = \hat{H}_0, \tag{3.115}$$

while

$$\hat{H}_0^\mathrm{H} = \hat{U}_t^\dagger \hat{H}_0 \hat{U}_t \tag{3.116}$$

is in general different from $\hat{H}_0^\mathrm{I}$. By differentiating Eq. (3.112) one obtains the evolution equation for the state in the Dirac picture (see Prob. 3.18)

$$\imath\hbar\frac{d}{dt}|\psi(t)\rangle_\mathrm{I} = \hat{H}_\mathrm{I}^\mathrm{I}(t)\,|\psi(t)\rangle_\mathrm{I}, \tag{3.117}$$

where

$$\hat{H}_\mathrm{I}^\mathrm{I}(t) = e^{\frac{\imath}{\hbar}\hat{H}_0 t}\hat{H}_\mathrm{I}e^{-\frac{\imath}{\hbar}\hat{H}_0 t}. \tag{3.118}$$

Similarly, differentiating Eq. (3.114), one obtains the equation of motion for the observable in the Dirac picture (see Prob. 3.19)

$$\imath\hbar\frac{d}{dt}\hat{O}^{\mathrm{I}}(t) = \imath\hbar\frac{\partial}{\partial t}\hat{O}^{\mathrm{I}}(t) + [\hat{O}^{\mathrm{I}}(t), \hat{H}_0], \tag{3.119}$$

where

$$\frac{\partial}{\partial t}\hat{O}^{\mathrm{I}}(t) = \hat{U}^{\dagger}_{H_0,t}\left(\frac{\partial}{\partial t}\hat{O}^{\mathrm{S}}\right)\hat{U}_{H_0,t}. \tag{3.120}$$

We should finally emphasize that the transformations to the Heisenberg or the Dirac pictures teach us that neither state vectors nor observables have a predominant role in the structure of the theory: it is possible to shift the time dependence from one to the other simply by applying a unitary transformation. As we have seen, the quantities which must remain invariant under these transformations are the matrix elements which represent probability amplitudes and therefore are the essential physical content of the theory.

3.7 Time derivatives and the Ehrenfest theorem

We have seen (Subsecs. 1.2.3 and 2.3.3) that true trajectories cannot be defined in quantum mechanics. This is so because a physical quantity which has a well-defined value at a certain time will not necessarily be determined at a subsequent time. As a consequence, we cannot define the time derivative of an observable in the way we are used to in classical mechanics. The most natural way to define the time derivative $\dot{\hat{O}}$ of a quantum-mechanical observable $\hat{O}$ in the Schrödinger picture is to assume that its expectation value $\left\langle\dot{\hat{O}}\right\rangle$ is equal to the time derivative of the expectation value of $\hat{O}$, i.e.

$${}_{\mathrm{S}}\langle\psi \mid \dot{\hat{O}}^{\mathrm{S}} \mid \psi\rangle_{\mathrm{S}} = \frac{d}{dt}\left({}_{\mathrm{S}}\langle\psi \mid \hat{O}^{\mathrm{S}} \mid \psi\rangle_{\mathrm{S}}\right). \tag{3.121}$$

Obviosuly, since we are dealing with this problem in the Schrödinger picture, the observable $\hat{O}$ does not evolve. The rhs of Eq. (3.121) may be easily computed by using Eq. (3.8) and its Hermitian conjugate, and gives

$$\begin{aligned}\frac{d}{dt}\left({}_{\mathrm{S}}\langle\psi \mid \hat{O}^{\mathrm{S}} \mid \psi\rangle_{\mathrm{S}}\right) &= {}_{\mathrm{S}}\langle\psi \mid \left(\frac{\imath}{\hbar}\hat{H}\hat{O}^{\mathrm{S}} - \frac{\imath}{\hbar}\hat{O}^{\mathrm{S}}\hat{H} + \frac{\partial}{\partial t}\hat{O}^{\mathrm{S}}\right) \mid \psi\rangle_{\mathrm{S}} \\ &= {}_{\mathrm{S}}\langle\psi \mid \left(\frac{\imath}{\hbar}[\hat{H}, \hat{O}^{\mathrm{S}}] + \frac{\partial}{\partial t}\hat{O}^{\mathrm{S}}\right) \mid \psi\rangle_{\mathrm{S}}.\end{aligned} \tag{3.122}$$

Since we have assumed that the lhs of Eq. (3.122) has to be equal to ${}_{\mathrm{S}}\langle\psi \mid \dot{\hat{O}}^{\mathrm{S}} \mid \psi\rangle_{\mathrm{S}}$ for any state vector $\mid \psi\rangle_{\mathrm{S}}$, we finally obtain

$$\imath\hbar\,\dot{\hat{O}}^{\mathrm{S}} = \imath\hbar\frac{\partial}{\partial t}\hat{O}^{\mathrm{S}} + [\hat{O}^{\mathrm{S}}, \hat{H}]. \tag{3.123}$$

Eq. (3.123) is very similar to Eq. (3.107) but has a rather different meaning: apart from the $\imath\hbar$ factor, the lhs of Eq. (3.107) represents the time derivative of the operator $\hat{O}$ in the

Heisenberg picture, whereas the lhs of Eq. (3.123) is the operator corresponding to the time derivative of the observable $\hat{O}$ in the Schrödinger picture.

Let us now consider the case $\hat{O}^{\mathrm{S}} = \hat{p}_x$ in Eq. (3.123). Then we have

$$\dot{\hat{p}}_x = \frac{1}{\imath\hbar}\left[\hat{p}_x, \hat{H}\right], \tag{3.124}$$

since $\hat{p}_x$ does not explicitly depend on time. Apart from the factor $\imath\hbar$, Eq. (3.124) is formally similar to the corresponding classical equation (1.8) if one replaces the classical Poisson brackets (1.9) with the quantum commutator (see Subsec. 2.1.5 and Eq. (3.109)). Moreover, if we apply the commutator $[\hat{p}_x, \hat{H}]$ to a generic wave function $\psi(x)$, we obtain

$$\left[\hat{p}_x, \hat{H}\right]\psi(x) = -\imath\hbar\left(\frac{\partial}{\partial x}\hat{H}\psi(x) - \hat{H}\frac{\partial}{\partial x}\psi(x)\right) = -\imath\hbar\frac{\partial \hat{H}}{\partial x}\psi(x), \tag{3.125}$$

where we have made use of Eq. (2.134) for the one-dimensional case. It follows that

$$\dot{\hat{p}}_x = -\frac{\partial \hat{H}}{\partial x}. \tag{3.126}$$

In a similar way, starting from

$$\dot{\hat{x}} = \frac{1}{\imath\hbar}\left[\hat{x}, \hat{H}\right], \tag{3.127}$$

we arrive at (see Prob. 3.20)

$$\dot{\hat{x}} = \frac{\partial \hat{H}}{\partial p_x}. \tag{3.128}$$

Both Eqs. (3.126) and (3.128) resemble the second Hamilton equation of motion (1.7), with the crucial difference that momentum and position are operators in quantum mechanics and numbers in classical mechanics. Eqs. (3.126) and (3.128) are the content of what is called the *Ehrenfest theorem.* The difference between quantum and classical mechanics is manifest by the following considerations. The analogy with classical mechanics should not be taken so extensively as to believe that the quantum–mechanical expectation values follow classical equations of motion. In fact, if this were the case, we should have

$$\left\langle\dot{\hat{p}}_x\right\rangle = -\frac{\partial}{\partial x}\hat{H}\left(\langle\hat{x}\rangle, \langle\hat{p}_x\rangle\right). \tag{3.129}$$

On the other hand, from Eq. (3.126) we have

$$\left\langle\dot{\hat{p}}_x\right\rangle = -\left\langle\frac{\partial}{\partial x}\hat{H}\left(\hat{x}, \hat{p}_x\right)\right\rangle, \tag{3.130}$$

which has not the same meaning of Eq. (3.129). Eqs. (3.129) and (3.130) would coincide only in the case when $\partial\hat{H}/\partial x$ is linear in $\hat{x}$ and $\hat{p}_x$, i.e. it is of the form $a\hat{x} + b\hat{p}_x + c$, where a, b and c are constants and for which the function of the expectation values is equal to the expectation value of the function. This condition is fulfilled by potential energies which are polynomials of at most second degree in the position $\hat{x}$, as it is the case for the free particle and the harmonic oscillator [see Secs. 3.4 and Sec. 4.4].

3.8 Energy–time uncertainty relation

Time in ordinary non-relativistic quantum mechanics – as in classical mechanics [see also Box 4.1: p. 152] – is essentially an external parameter, measured by classical clocks, by means of which we "label" the dynamics of a system. The Schrödinger equation in its current form only makes sense if we assume that space–time is "non-dynamical," i.e. it is not affected by the quantum–mechanical evolution of the system under study.[17] In other words, together with the three-dimensional ordinary space, time constitutes the "fixed" background which quantum mechanics is built on.

When considering the relationship between time and energy, we find a certain analogy with the position–momentum relationship. In fact, as momentum is the physical quantity that is conserved under space translation (momentum is the generator of the group of spatial translations) (see Subsec. 2.2.4), energy is the physical quantity that is conserved under time translations (and for this reason it is the generator of the group of time translations) (see Subsec. 3.5.3). It is then natural to expect that the position–momentum uncertainty relation has a counterpart in a time–energy uncertainty relation. As we shall see, it is indeed possible to write a sort of time–energy uncertainty relation, but its physical meaning has to be taken with extreme care.

Let us start with the simple example of the one-dimensional plane wave. The momentum of a plane wave is essentially a wave number (see Eq. (2.140)) and the position–momentum uncertainty relation basically describes the fact that one cannot localize a plane wave that intrinsically extends over the whole space. Analogously, energy is basically a frequency (see Eq. (1.25)) and therefore cannot be localized in time. As a consequence, the classical-wave Fourier relation $\Delta t \Delta \nu > (2\pi)^{-1}$ would directly translate into

$$\Delta E \Delta t > \hbar. \tag{3.131}$$

Superposing a large number of one-dimensional plane waves one obtains a wave packet of, say, width Δx and group velocity v_g (see Box 2.6: p. 80). As a consequence, the exact time at which the wave packet crosses a certain point is defined with an uncertainty

$$\Delta t \simeq \frac{\Delta x}{v_g}. \tag{3.132}$$

On the other hand, the wave packet has an energy uncertainty ΔE due to its spread in momentum space

$$\Delta E \simeq \frac{\partial E}{\partial p_x} \Delta p_x = v_g \Delta p_x. \tag{3.133}$$

[17] This is different from what happens in general relativity theory, which assumes that the evolution of the system and the structure of the space–time are self-consistently correlated and where the dynamical equations determine both the structure of space–time and how the system evolves.

The two previous equations yield

$$\Delta t \Delta E \simeq \Delta x \Delta p_x. \tag{3.134}$$

By using the momentum–position uncertainty relation, we derive

$$\Delta E \Delta t \geq \frac{\hbar}{2}, \tag{3.135}$$

which limits the product of the spread ΔE of the energy spectrum of the wave packet and the accuracy Δt of the prediction of the time of passage at a given point.

However, There are several features which make Eq. (3.135) profoundly different from Eq. (2.190):

Box 3.2 **Einstein's box**

At the sixth Solvay Conference in 1930, Einstein proposed a device consisting of a box with a hole in one of its sides and a shutter moved by means of a clock inside the box [Bohr 1949, 224–28]. If in its initial state the box contains a certain amount of radiation and the clock is set to open the shutter after a chosen short interval of time, it could be achieved that a single photon is released through the hole at a moment which is known as exactly as desired. Moreover, if we weigh the box before and after this event, we could measure the energy of the photon as exactly as we want, against the time–energy uncertainty relation.

Bohr's reply (see Fig. 3.10) was that any determination of the position of the balance's pointer is given with an accuracy Δx, which will involve an uncertainty Δp_x in the control of box' momentum according to Eq. (2.190). This uncertainty must be smaller than the total momentum which, during the whole interval δt of the balancing procedure, can be imparted by the gravitational field to a body with mass Δm, i.e.

$$\Delta p_x < \delta t \cdot g \cdot \Delta m, \tag{3.136}$$

where g is the gravity constant. The greater the accuracy of the reading x of the pointer, the longer must the balancing interval δt be if a given accuracy Δm of the weight is to be obtained. But according to the general relativity theory, when a clock is displaced in the direction of the gravitational force by an amount Δx, its rate will change in such a way that its reading in δt will differ by an amount Δt given by

$$\frac{\Delta t}{\delta t} = \frac{1}{c^2} g \Delta x. \tag{3.137}$$

By substituting the value of δt given by Eq. (3.137) into Eq. (3.136) we obtain

$$\Delta p_x < \frac{c^2 \Delta t \Delta m}{\Delta x}. \tag{3.138}$$

Finally, by applying Eq. (2.190) again with the equality sign, we obtain

$$\Delta t > \frac{\hbar}{c^2 \Delta m}. \tag{3.139}$$

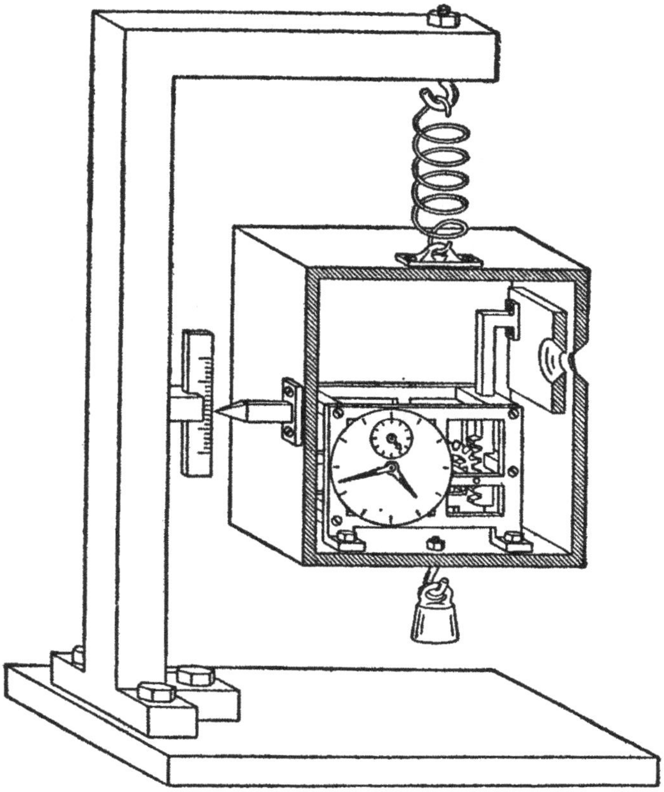

Figure 3.10 A graphical representation of the apparatus proposed in the Einstein–Bohr debate to test Eq. (3.135). Adapted from [Bohr 1949, 227]. See Box 3.2.

This, together with Einstein's formula

$$E = mc^2, \tag{3.140}$$

gives Eq. (3.131). We note that Bohr's argument – different to some of Bohr's formulations of complementarity [Bohr 1948] – is based upon a quantum-mechanical interpretation of the pointer, which is suitable because a device measuring a single photon must obey quantum laws (see Ch. 9).

- First, time is not an observable of the system in ordinary quantum mechanics. Therefore, we cannot introduce in a naïve way a time operator (see Sec. 3.9), and Eqs. (2.184) cannot be directly translated into a definition of Δt.
- Second, the position–momentum uncertainty relation expresses the fact that a valid state in quantum mechanics *cannot* display *simultaneously* certain values of Δx and Δp_x that violate Eq. (2.190). On the contrary, the energy of a system can be determined with arbitrary precision at any time.

- Finally, in Eq. (2.190) we consider two *simultaneous* measurements of position and momentum, whereas in Eq. (3.135) we consider the energy and the time of passage *at a given point*. In other words, the two uncertainty relations express two different and, in a certain sense, incompatible viewpoints.

Another way of looking at the problem of energy and time uncertainties is to derive the uncertainty relations by considering a generic observable $\hat{O}$ of the quantum system. According to Eq. (2.200), the uncertainty relation between the observable $\hat{O}$ and the energy is

$$(\Delta_\psi E) \cdot (\Delta_\psi O) \geq \frac{1}{2} \left| \left\langle \psi \left| \left[\hat{H}, \hat{O} \right] \right| \psi \right\rangle \right|. \tag{3.141}$$

On the other hand, making use of Eq. (3.108) and the reduced form of Eq. (3.110), the mean value of the commutator $\left[\hat{O}(t), \hat{H} \right]$ is related to the rate of change of the mean value of $\hat{O}$ by

$$\left\langle \psi \left| \left[\hat{O}(t), \hat{H} \right] \right| \psi \right\rangle = \imath \hbar \frac{d}{dt} \left\langle \psi \left| \hat{O}(t) \right| \psi \right\rangle, \tag{3.142}$$

where the evolution is in the Heisenberg picture but we have dropped the superscript H for the sake of notation. We may then combine Eqs. (3.141) and (3.142), obtaining

$$(\Delta_\psi E) \cdot (\Delta_\psi O) \geq \frac{\hbar}{2} \left| \imath \frac{d}{dt} \left\langle \psi \left| \hat{O}(t) \right| \psi \right\rangle \right|. \tag{3.143}$$

Dividing both sides of Eq. (3.143) by the absolute value of the rate of change of $\left\langle \psi \left| \hat{O}(t) \right| \psi \right\rangle$, we obtain

$$(\Delta_\psi E) \cdot \frac{(\Delta_\psi O)}{\left| \frac{d}{dt} \left\langle \psi \left| \hat{O}(t) \right| \psi \right\rangle \right|} \geq \frac{\hbar}{2}. \tag{3.144}$$

However, the time $\Delta_\psi t$ required for $\left\langle \psi \left| \hat{O} \right| \psi \right\rangle$ to change from its initial value (at a given time $t = t_0$) by a small positive amount $\Delta_\psi O$, neglecting higher-order terms in the Taylor expansion of $\hat{O}(t)$, is given by

$$\Delta_\psi t = \left. \frac{(\Delta_\psi O)}{\frac{d}{dt} \left\langle \psi \left| \hat{O}(t) \right| \psi \right\rangle} \right|_{t=t_0}, \tag{3.145}$$

where

$$\frac{d}{dt} \left\langle \psi \left| \hat{O}(t) \right| \psi \right\rangle > 0, \tag{3.146}$$

which leads to Eq. (3.135). In this case, the uncertainty relation between time and energy connects the energy uncertainty to a time interval that is characteristic of the system's rate of change.

An important application of Eq. (3.135) is the lifetime-width relation for unstable systems (radioactive nuclei, excited states of atoms, unstable elementary particles, etc.), i.e. systems which are not stationary and do not correspond to a well-defined value of the energy but rather to an energy spectrum with a certain spread ΔE, called the *level width*. The mean lifetime τ of the stable (or metastable) state here plays the role of the characteristic time considered above: One must wait (on average) for a time of order τ to observe an appreciable change in the properties of the system. As a consequence,

$$\tau \Delta E \simeq \hbar. \tag{3.147}$$

Sometimes, Eq. (3.135) is interpreted in the context of energy measurements in general. In this case, the accuracy ΔE of the energy measurement is connected with the time Δt required for the measurement itself.

Finally, an alternative approach to the problem is to consider the time as a proper dynamical observable of the system. In this case, we should provide an operatorial expression for time. Even though it is not straightforward to give a general formulation for a time operator, specific derivations are viable. In the next section we shall consider the so-called time-of-arrival operator.

3.9 Towards a time operator

Von Neumann[18] assumed that every observable can be represented by a self-adjoint operator (see Th. 2.1: p. 46). Though this must be correct in principle, the enterprise to build a valid operational representation of a given physical quantity may be very difficult in practice, as we shall see in this section. For instance, Wigner[19] showed that it is very difficult to find univocal quantum-mechanical counterparts to some simple classical expressions such as xp_x or $x^2 p_x^2$. This is obviously due to the non-commutability of position and momentum in quantum mechanics, which makes these expressions not self-adjoint operators.

As we have already mentioned, in quantum mechanics time can be considered from two points of view: as an external ordering parameter, for example representing the measuring time as indicated by an apparatus that is external to the measured system; and as an observable of the system itself, in particular as a variable which depends on the initial state of the system and on its dynamical evolution. In this section we are interested in the second interpretation.

A condition that one may reasonably impose on a time operator $\hat{t}$ is

$$\left\langle \psi(t_1) \left| \hat{t} \right| \psi(t_1) \right\rangle - \left\langle \psi(t_2) \left| \hat{t} \right| \psi(t_2) \right\rangle = t_1 - t_2, \tag{3.148}$$

for any $\mid \psi(t_1)\rangle$, $\mid \psi(t_2)\rangle$. However, Pauli[20] showed that it is impossible to find a time operator $\hat{t}$ such as to satisfy a commutation relation of the form $[\hat{t}, \hat{H}] = -\imath\hbar$, where time and energy are conjugate observables. In fact, this would conflict with the requirement that energy is bounded from below, i.e. the Hamiltonian operator does not possess a continuous spectrum from $-\infty$ to $+\infty$ (see Box 2.1: p. 47) – it must be so if we want a ground state of energy. Pauli's argument can be formulated as follows. Let $\left| \psi_E \right\rangle$ be an eigenstate of the Hamiltonian $\hat{H}$, such that

$$\hat{H} \left| \psi_E \right\rangle = E \left| \psi_E \right\rangle. \tag{3.149}$$

18 See [*von Neumann* 1932, 163–71] [*von Neumann* 1955, 324–25].

19 See [Wigner 1952].

20 See [*Pauli* 1980, 63]. See also [Paul 1962] [Engelmann/Fick 1959, Fick/Engelmann 1963a, Fick/Engelmann 1963b].

Then, we also have (see Prob. 3.21)

$$\hat{H}e^{\imath\alpha\hat{t}}\,|\,\psi_E\rangle = (E-\alpha\hbar)\,e^{\imath\alpha\hat{t}}\,|\,\psi_E\rangle, \tag{3.150}$$

where α is an arbitrary constant. As a consequence, also $e^{\imath\alpha\hat{t}}\,|\,\psi_E\rangle$ is an eigenstate of the energy with eigenvalue $E-\alpha\hbar$, and the spectrum of $\hat{H}$ cannot be bounded if $[\hat{t},\hat{H}] = -\imath\hbar$ must hold.

A possible solution of this problem is to consider specific formulations of a time operator. One of these is known as *time-of-arrival operator*.[21] In this case we consider the time-of-arrival of a particle at a detector in a fixed position X, a *trade-off* between an observable property *of* the system and an operational procedure *on* the system.

If we try to find a spectral decomposition of $\hat{t}$ according to Eq. (2.20), we encounter an immediate difficulty: for an arbitrary self-adjoint operator $\hat{O}$ we have

$$\int_{-\infty}^{+\infty} do\hat{P}(o) = \hat{I}, \tag{3.151}$$

while we have no reason for thinking that the same is valid for a "time" operator. In fact, it is not true that any state of a given system is certainly detected at some time, that is, we cannot impose as a property of the system that it *will* be detected at some time. Then, the spectral family $\hat{P}(t)$ is incomplete and we should say that $\hat{P}_{\hat{t}}$ only projects into the subspace $\mathcal{H}_{\mathrm{D}}$ – of the original Hilbert space $\mathcal{H}$ of the system – formed by the states detected at some time at position X of a given detector. If we try to define $\hat{P}(t)$ on the entire state space, we cannot distinguish between the initial state of the system (when it cannot be detected), i.e. when $\hat{P}(t=0)$, and the states in the space $\mathcal{H}_{\mathrm{D}'}$ that are never detected (indeed the former and the latter are all annihilated by $\hat{t}$).

Let us now consider a classical non-relativistic free particle in one dimension. The time-of-arrival of a particle with initial position x^0 and initial momentum p_x^0, detected at position X can be written

$$t(X) = \frac{m(X-x^0)}{p_x^0} \tag{3.152}$$

as a time–space inversion of the classical equation of motion

$$x(t;x^0,p_x^0) = \frac{p_x^0}{m}t + x^0. \tag{3.153}$$

Note that, except for the problem at $p_x^0 = 0$, the particle is always detected. In the Heisenberg picture for a quantum system we may write Eq. (3.152) as

$$\hat{t}(X) = \frac{m(X-\hat{x}^0)}{\hat{p}_x^0}, \tag{3.154}$$

which is of course problematic because $\hat{x}^0$ and $\hat{p}_x^0$ do not commute. In order to cure the problem, we try to construct a symmetric ordering for the operator in Eq. (3.154) in the following manner:

[21] See [Grot *et al.* 1996]. The idea of a time-of-arrival operator was originally developed by Allcock [Allcock 1969].

$$\hat{t}(X) = \frac{mX}{\hat{p}_x^0} - m\frac{1}{\sqrt{\hat{p}_x^0}}\hat{x}_0\frac{1}{\sqrt{\hat{p}_x^0}}, \tag{3.155}$$

or

$$\hat{t}(X) = -\imath\frac{m}{\hbar}\frac{1}{\sqrt{k_x}}\frac{d}{dk_x}\frac{1}{\sqrt{k_x}} + \frac{m}{\hbar}\frac{X}{k_x}, \tag{3.156}$$

where $\sqrt{k_x} = \imath\sqrt{|k_x|}$ for $k < 0$. Note that the 1-parameter family of operators $\hat{t}(X)$ can be generated unitarily via translations of the form

$$\hat{t}(X) = e^{\imath k_x X}\hat{t}(0)e^{-\imath k_x X}, \tag{3.157}$$

where $e^{-\imath k_x X}$ is the space counterpart of $\hat{U}_t$ (see Subsecs. 3.5.3 and 3.5.4).

Therefore, without loss of generality, it suffices to study the operator $\hat{t}(0)$ with the detector placed at the origin ($X = 0$). Hence from now on we drop the explicit X-dependence of $\hat{t}$ and write

$$\hat{t} = -\imath\frac{m}{\hbar}\frac{1}{\sqrt{k_x}}\frac{d}{dk_x}\frac{1}{\sqrt{k_x}}, \tag{3.158}$$

where the time-of-arrival operator should satisfy the condition

$$\hat{t}|t\rangle = t|t\rangle, \tag{3.159}$$

and the $|\, t\rangle$'s are the eigenkets of $\hat{t}$, so that, in the momentum representation, the eigenvalue equation for $\hat{t}$ becomes

$$\hat{t}\,\langle k_x \mid t\rangle = \left[-\imath\frac{m}{\hbar}\frac{1}{\sqrt{k_x}}\frac{d}{dk_x}\frac{1}{\sqrt{k_x}}\right]\langle k_x \mid t\rangle = t\,\langle k_x \mid t\rangle. \tag{3.160}$$

The biggest difficulty with such a time operator is that we have a singularity at the point $k_x = 0$. We may circumvent this singularity by means of a family of real bounded continuous odd functions $f_\epsilon(k)$ which approach $1/k_x$ pointwise, where ϵ is a small positive number. In this way we can overcome the problem of the singularity and construct a self-adjoint time-operator as a sequence of operators which are not themselves self-adjoint. We may choose

$$f_\epsilon(k_x) = \frac{1}{k_x} \text{ for } |k_x| > \epsilon, \tag{3.161a}$$

$$f_\epsilon(k_x) = \epsilon^{-2}k_x \text{ for } |k_x| < \epsilon. \tag{3.161b}$$

The "regulated" time-of-arrival operator becomes

$$\hat{t}_\epsilon = -\imath\frac{m}{\hbar}\sqrt{f_\epsilon(k_x)}\frac{d}{dk_x}\sqrt{f_\epsilon(k_x)}. \tag{3.162}$$

It is possible to show[22] that the following commutation relation between time and energy holds:

[22] See the original article for details.

$$\left[\hat{t}_\epsilon, \hat{H}\right] = -\imath\hbar(\hat{I} - g_\epsilon(k_x)), \tag{3.163}$$

where

$$g_\epsilon(k_x) = 1 - k_x f_\epsilon(k). \tag{3.164}$$

The function $g_\epsilon(k_x)$ vanishes for $|k_x| > \epsilon$, and in the small interval where it has support, it is bounded by 1, if we choose $f_\epsilon(k_x)$ as in Eqs. (3.161). For a particle in the state $|\psi\rangle$ the resulting energy–time uncertainty relation is

$$(\Delta t_\epsilon)^2 (\Delta E)^2 \geq \frac{\hbar^2}{4}\left(1 - \langle\psi| g_\epsilon(k_x)|\psi\rangle\right)^2, \tag{3.165}$$

which implies that, for sufficiently small ϵ and for all states with support away from the origin, we have $\Delta t_\epsilon \Delta E \geq \hbar/2$, in accordance with Eq. (3.135).

Summary

In this chapter we have developed the basic features of quantum dynamics. We may summarize the main results as follows:

- The quantum dynamical evolution equation is the *Schrödinger equation*, which is a first-order differential equation whose solution provides the state vector at any time t when the Hamiltonian $\hat{H}$ and the state vector at $t = 0$ are known.
- In order to solve the Schrödinger equation it is first necessary to find the *stationary states*, i.e. the eigenstates of the Hamiltonian operator, and then expand the initial state in terms of the stationary states.
- The Schrödinger equation is invariant under *Galilei transformations*.
- As a first example we have solved the quantum dynamics of a one-dimensional *particle in a box*, i.e. we have found the energy eigenstates and the corresponding eigenvalues.
- The evolution determined by the Schrödinger equation is *unitary*. This guarantees the reversibility of elementary quantum dynamics.
- Time evolution in quantum mechanics can be represented in *different pictures*. However, physical quantities, such as probabilities, expectation values, etc., will not depend on the chosen picture. If we keep the observables fixed and let the states evolve, we have the Schrödinger picture; if we keep the states fixed and let the observables evolve, we have the Heisenberg picture; and, finally, if we split the Hamiltonian into a free part and an interaction part, and therefore let both observables and states evolve, we obtain the Dirac picture. Transformations from one picture to the other are unitary.
- The *Ehrenfest theorem* shows that there is a formal analogy between the classical and quantum-mechanical equations of motion.

- The uncertainty relation between *energy* and *time* has been derived.
- A self-adjoint representation of *time* – as a dynamical variable that is intrinsic to the system – has been presented.

Problems

3.1 Derive Eq. (3.9) from Eq. (3.8).
(*Hint*: Multiply both sides of Eq. (3.8) by $\langle x \mid$ from the left and show that $\langle x \mid \hat{H} \mid \psi\rangle = \hat{H}\langle x \mid \psi\rangle$. For the last step write $\hat{H} = \frac{\hat{p}_x^2}{2m} + V(\hat{x})$ and expand $\mid \psi\rangle$ into the eigenvectors $\mid p_x\rangle$ of $\hat{p}_x$ (see Subsecs. 2.2.4–2.2.5).)

3.2 Prove that if $\mid \psi\rangle$ and $\left|\psi'\right\rangle$ are solutions of the same Schrödinger equation, also $c \mid \psi\rangle + c' \left|\psi'\right\rangle$ is a solution, where c and c' are arbitrary complex coefficients with $|c|^2 + \left|c'\right|^2 = 1$.

3.3 Find the stationary states for a free one-dimensional particle.

3.4 Write Eqs. (3.20) and (3.21) in terms of the wave function in the momentum representation.

3.5 Show that an initial normalized wave function will stay normalized under time evolution.
(*Hint:* Take advantage of Eq. (3.26).)

3.6 Prove that the wave function describing the ground state for a one-dimensional particle is even when the potential $V(x)$ is even.
(*Hint:* Take advantage of the continuity of the wave function and of properties (ii) and (iv) of the one-dimensional motion in Subsec. 3.2.3.)

3.7 Solve the time-dependent Schrödinger equation for a one-dimensional free particle whose state at time $t = 0$ is described by the wave function $\psi(x, 0) = \int dk\, c(k) e^{\imath kx}$, where $2\pi \int dk |c(k)|^2 = 1$.

3.8 Consider a traveling-wave solution $\psi(x, t) = Ce^{\frac{\imath}{\hbar}(p_x x - Et)}$ of the free-particle one-dimensional Schrödinger equation (3.11) and its transformed $\psi'(x', t')$ under Galilei transformations $x = x' + Vt$ and $t = t'$. Show that $\psi'(x', t')$ satisfies the corresponding Schrödinger equation for the primed variables. This result ensures the invariance of the Schrödinger equation under Galilei transformations.

3.9 Compute the normalization constant $\mathcal{N} = \sqrt{2/a}$ for the energy eigenfunctions of a particle in a box of dimension a (see Sec. 3.4).

3.10 Compute the uncertainty product for a particle in a box (see Sec. 3.4) when its wave function is a generic energy eigenfunction and verify that the position–momentum uncertainty relation is satisfied.
(*Hint:* Start from $\Delta p_x = \left(\left\langle \hat{p}_x^2\right\rangle - \left\langle \hat{p}_x\right\rangle^2\right)^{\frac{1}{2}}$ and $\Delta x = \left(\left\langle \hat{x}^2\right\rangle - \left\langle \hat{x}\right\rangle^2\right)^{\frac{1}{2}}$, and calculate explicitly the relevant mean values.)

3.11 Find the energy eigenfunctions and eigenvalues for a particle confined in a three-dimensional box, i.e. for a potential

$$V(x, y, z) = \begin{cases} 0 & 0 < x < a, 0 < y < b, 0 < z < c, \\ \infty & \text{otherwise.} \end{cases}$$

(*Hint:* Use the three-dimensional generalization of the stationary Schrödinger equation (3.41) and take advantage of the fact that the Hamiltonian may be written as $\hat{H} = \hat{H}_1(x) + \hat{H}_2(y) + \hat{H}_3(z)$.)

3.12 Given an initial superposition $|\psi\rangle = \sum_j c_j |\psi_j\rangle$ of energy eigenstates $|\psi_j\rangle$ (with $c_j \neq 0$ for at least two values of j) such that $\hat{H}|\psi_j\rangle = E_j|\psi_j\rangle$, show that under unitary time evolution $\hat{U}_t = e^{-\frac{\imath}{\hbar}\hat{H}t}$, $|\psi\rangle$ cannot evolve to a component $|\psi_k\rangle$.

3.13 Generalize the previous proof to any superposition and any unitary operator. In other words, prove that no unitary transformation $\hat{U}$ exists that can change any (arbitrary) superposition state $|\Psi\rangle = \sum_j c_j |\psi_j\rangle$ (with $c_j \neq 0$ for at least two values of j and $\langle\psi_k \mid \psi_l\rangle = \delta_{lk}$) into one of its components.

3.14 Prove Eq. (3.104).

3.15 Derive Eq. (3.107) by differentiating Eq. (3.103b).

3.16 Show that, if $\hat{O}$ and $\hat{O}'$ are two constants of motion (i.e. $[\hat{O}, \hat{H}] = [\hat{O}', \hat{H}] = 0$), then their commutator $[\hat{O}, \hat{O}']$ is also a constant of motion (see also Prob. 1.4).

3.17 Show that if $\hat{O}$ and $\hat{O}'$ are two observables such that $[\hat{O}^{\mathrm{S}}, (\hat{O}')^{\mathrm{S}}] = C$, then one also has that $[\hat{O}^{\mathrm{H}}, (\hat{O}')^{\mathrm{H}}]$ is equal to the same constant C.

3.18 Derive Eq. (3.117).

3.19 Derive Eq. (3.119).

3.20 Derive Eq. (3.128).
(*Hint*: Start from Eq. (3.127) and make use of the result of Prob. 2.27.)

3.21 Prove that, given the commutator $[\hat{t}, \hat{H}] = -\imath\hbar$, we have

$$\hat{H} e^{\imath\alpha\hat{t}} |\psi_E\rangle = (E - \alpha\hbar)\, e^{\imath\alpha\hat{t}} |\psi_E\rangle,$$

where $|\psi_E\rangle$ is an eigenstate of $\hat{H}$ with eigenvalue E.
(*Hint*: Use the result of Prob. 2.25.)

Further reading

Energy–time uncertainty relation

Bohr, Niels, Discussion with Einstein on epistemological problems in atomic physics, in *Albert Einstein. Philosopher-Scientist*, A. Schilpp (ed.), La Salle, IL: Open Court, 1949; 3rd edn. 1988, pp. 201–41.

Time operator

Grot, N., Rovelli, C., and Tate, R. S., Time-of-arrival in quantum mechanics. *Physical Review*, **A54** (1996), 4676–90.

4 Examples of quantum dynamics

In this chapter we shall discuss some elementary examples of quantum dynamics. In Sec. 4.1 we shall go back to the problem of a particle in a box, this time with finite potential wells. In Sec. 4.2 we shall analyze the effects of a potential barrier on a moving particle. In Sec. 4.3 we shall consider another quantum effect which has no analogue in the classical domain: a quantum particle can tunnel in a classically forbidden region. In Sec. 4.4 perhaps the most important dynamical typology (with a wide range of applications) is considered: the harmonic oscillator. Finally, in Sec. 4.5 several types of elementary fields are considered.

4.1 Finite potential wells

In Sec. 3.4 we have considered what is perhaps the simplest example of quantum dynamics, that is a free particle moving in a box with infinite potential walls. Consider the motion of, say, a one-dimensional particle in a rectangular potential well with finite steps. In Fig. 4.1 we show two of such potentials, symmetric in (a) and asymmetric in (b).

Let us consider the case pictured in Fig. 4.1(a) and indicate with V_0 the energy of the potential well. We may therefore distinguish three regions on the x-axis: region I ($x < 0$), where the potential energy is equal to V_0; region II ($0 \leq x \leq a$), where the particle is free; and region III ($x > a$), where the potential energy is again equal to V_0. From the discussion in Sec. 3.2 it is straightforward to conclude that the energy eigenvalues will have a lower bound, that is $E > V_{\rm min} = 0$. Moreover, for a particle's energy $E > V_0$ we shall have unbounded motion and therefore unbound states, corresponding to a continuous spectrum. For the discrete spectrum ($0 < E_n < V_0$), we may write the Schrödinger equation separately for the three regions. For region II, we shall have

$$\psi''(x) + k^2\psi(x) = 0, \tag{4.1a}$$

while, for regions I and III,

$$\psi''(x) - k'^2\psi(x) = 0, \tag{4.1b}$$

where

$$k = \frac{\sqrt{2mE}}{\hbar} \quad \text{and} \quad k' = \frac{\sqrt{2m(V_0 - E)}}{\hbar}. \tag{4.2}$$

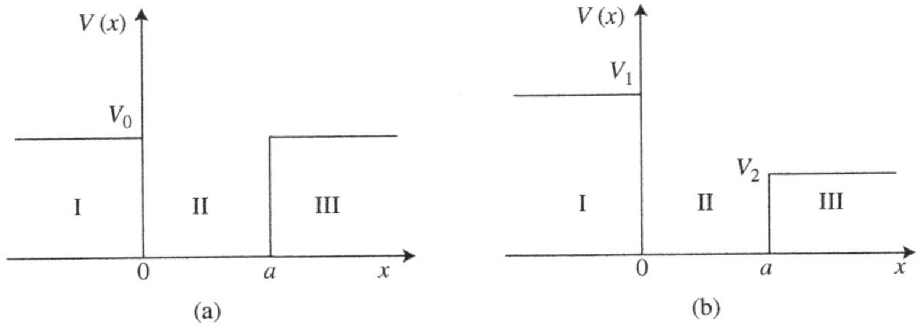

Figure 4.1 Schematic drawing of symmetric (a) and asymmetric (b) one-dimensional potential wells.

The general solution of Eqs. (4.1) will be of the form

$$\begin{cases} \psi_{\rm I}(x) = \mathcal{N}_{\rm I} e^{k'x} & x < 0, \\ \psi_{\rm II}(x) = \mathcal{N}_{\rm II} \sin(kx + \phi) & 0 \leq x \leq a, \\ \psi_{\rm III}(x) = \mathcal{N}_{\rm III} e^{-k'x} & x > a. \end{cases} \tag{4.3}$$

The sign of the exponents in $\psi_{\rm I}(x)$ and $\psi_{\rm III}(x)$ is dictated by the fact that ψ has to tend to zero when $x \to \pm\infty$. We now have to impose that $\psi(x)$ and its derivative $\psi'(x)$ be continuous and single-valued on the whole line. This gives four conditions for the continuity of ψ and ψ' at $x = 0$ and $x = a$, i.e.

$$\begin{cases} \psi_{\rm I}(0) = \psi_{\rm II}(0), \\ \psi'_{\rm I}(0) = \psi'_{\rm II}(0), \\ \psi_{\rm II}(a) = \psi_{\rm III}(a), \\ \psi'_{\rm II}(a) = \psi'_{\rm III}(a), \end{cases} \tag{4.4}$$

which, using Eqs. (4.3), translate into

$$\begin{cases} \mathcal{N}_{\rm I} = \mathcal{N}_{\rm II} \sin\phi, \\ \mathcal{N}_{\rm I} k' = \mathcal{N}_{\rm II} k \cos\phi, \\ \mathcal{N}_{\rm II} \sin(ka + \phi) = \mathcal{N}_{\rm III} e^{-k'a}, \\ \mathcal{N}_{\rm II} k \cos(ka + \phi) = -\mathcal{N}_{\rm III} k' e^{-k'a}. \end{cases} \tag{4.5}$$

These equations give the two conditions

$$\tan\phi = \frac{k}{k'} \quad \text{and} \quad \tan(ka + \phi) = -\frac{k}{k'}, \tag{4.6}$$

or

$$\phi = \arctan\frac{k}{k'} \quad \text{and} \quad \phi = -ka - \arctan\frac{k}{k'} + n\pi, \tag{4.7}$$

where arctan (k/k') with range in the interval $[-\pi/2, \pi/2]$ is the principal value of the multivalued inverse tangent function and n is an integer. In order to address this arbitrariness, we force ϕ to lie within the interval $(-\pi/2, +\pi/2)$. The requirements in Eqs. (4.7) are satisfied if and only if the rhss are equal. This can be achieved only for certain discrete values k_n of k, satisfying

$$n\pi - ka = 2\arctan\frac{k}{k'}, \tag{4.8}$$

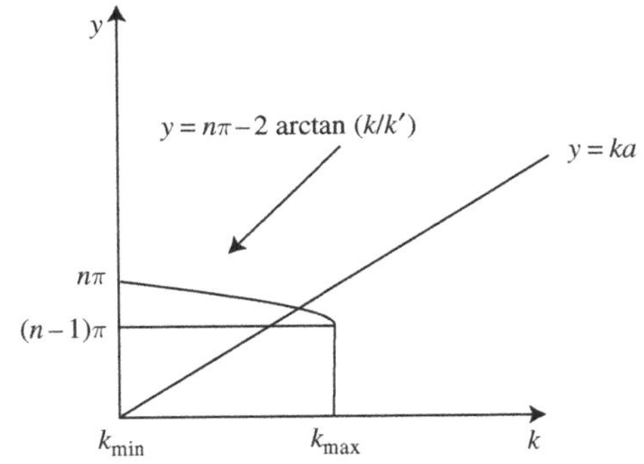

Figure 4.2 We can find a solution of Eq. (4.8) for a given value of n if and only if the line $y = ka$ crosses the curve $y = n\pi - 2\arctan(k/k')$. If the line $y = ka$ crosses several curves (each with a different value of n), then there are several discrete eigenvalues.

where the arctan function has values in the interval $[0, \pi/2]$. In order to find a solution, we observe that $\arctan(k/k') = 0$ when $k = k_{\min} = 0$ (i.e. $E = 0$), and $\arctan(k/k') = \pi/2$ when $k = k_{\max} = \sqrt{2mV_0}/\hbar$ (i.e. $E = V_0$). The situation is depicted in Fig. 4.2. The curve $y = n\pi - 2\arctan(k/k')$ and the line $y = ka$ will intersect once if and only if

$$k_{\max} a \geq (n-1)\pi, \tag{4.9}$$

that is

$$\frac{\sqrt{2mV_0}}{\hbar} a \geq (n-1)\pi. \tag{4.10}$$

The root $k_0 = 0$ of Eq. (4.8) is excluded because the corresponding eigenfunction vanishes identically for all values of x. The roots k_n with $n \geq 1$ determine the energy eigenvalues $E_n = \hbar^2 k_n^2/2m$. It is also clear that these eigenvalues will be arranged in increasing order of n, until the line $y = ka$ ceases to cross the above-mentioned curve. The number of energy eigenvalues is therefore finite. The normalization condition (see Eq. (2.108))

$$\int_{-\infty}^{+\infty} dx\, |\psi(x)|^2 = 1, \tag{4.11}$$

together with Eqs. (4.5), would allow us to determine the parameters $\mathcal{N}_{\mathrm{I}}$, $\mathcal{N}_{\mathrm{II}}$, and $\mathcal{N}_{\mathrm{III}}$ in Eqs. (4.3). We omit the straightforward algebra for the sake of space and only mention that Eq. (4.11) should be written as

$$\int_{-\infty}^{+\infty} dx\, |\psi(x)|^2 = \int_{-\infty}^{0} dx\, |\psi_{\mathrm{I}}(x)|^2 + \int_{0}^{a} dx\, |\psi_{\mathrm{II}}(x)|^2 + \int_{a}^{+\infty} dx\, |\psi_{\mathrm{III}}(x)|^2. \tag{4.12}$$

The first three eigenfunctions and the corresponding probability densities are shown in Fig. 4.3. We would like to stress that the present situation has a further point of departure from the classical analogue, with respect to the particle in a box (with infinite potential walls): besides energy quantization, the wave function has a support even in regions I and III. This means that there is non-zero probability of finding the quantum particle in these classically forbidden regions (see property (iii) of Subsec. 3.2.1). Here, we see an example

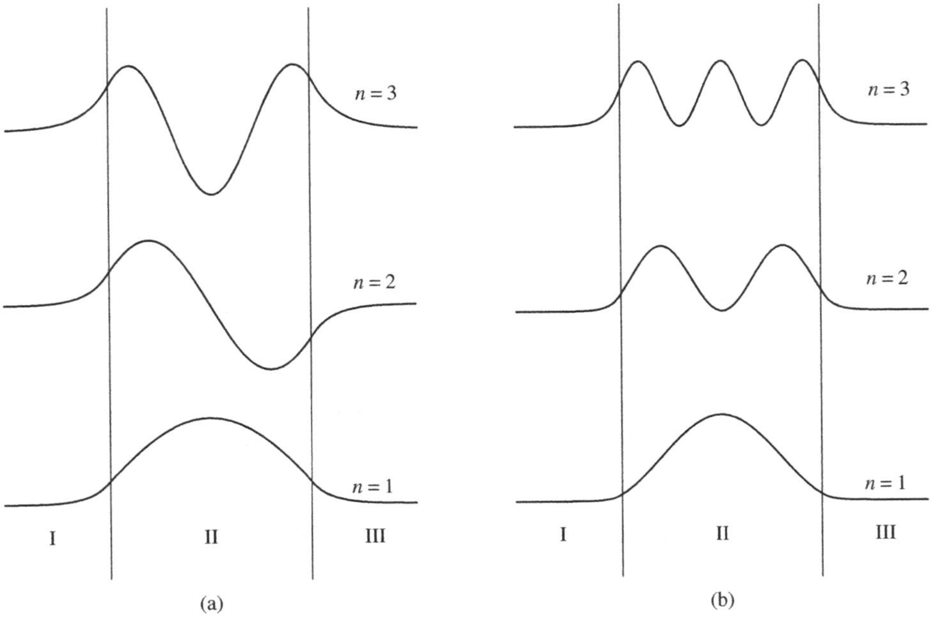

Figure 4.3 (a) Wave functions and (b) probability densities for the first three eigenfunctions for the symmetric finite-well potential.

of a genuine quantum phenomenon, namely tunnelling, which will be further discussed in Sec. 4.3. It should also be noted that, as anticipated in Subsec. 3.2.3, the number of nodes of the eigenfunction corresponding to the n-th eigenvalue E_n is equal to $n - 1$.

A final comment concerning the continuity of $\psi(x)$ and its spatial derivatives is in order here. The time-independent Schrödinger equation

$$-\frac{\hbar}{2m}\psi''(x) + V(x)\psi(x) = E\psi(x) \tag{4.13}$$

contains $\psi(x)$ and its second derivative. Moreover,

$$\psi'(x) - \psi'(b) = \int_b^x dx'\psi''(x'). \tag{4.14}$$

If the potential energy $V(x)$ is stepwise continuous, then from Eq. (4.13) also $\psi''(x)$ must be stepwise continuous and, from Eq. (4.14), $\psi'(x)$ turns out to be continuous (see also Fig. 4.4). However, this argument does not hold if the height of the potential step is infinite. This explains why, despite what we have done in the present section, in Sec. 3.4 we did not impose the continuity of $\psi'(x)$ at $x = 0$ and $x = a$. The discussion above can be summarized by saying that for a one-dimensional particle, if the potential energy $V(x)$ is stepwise continuous, with finite jumps, then also $\psi''(x)$ is stepwise continuous, whereas $\psi'(x)$ and $\psi(x)$ are continuous.

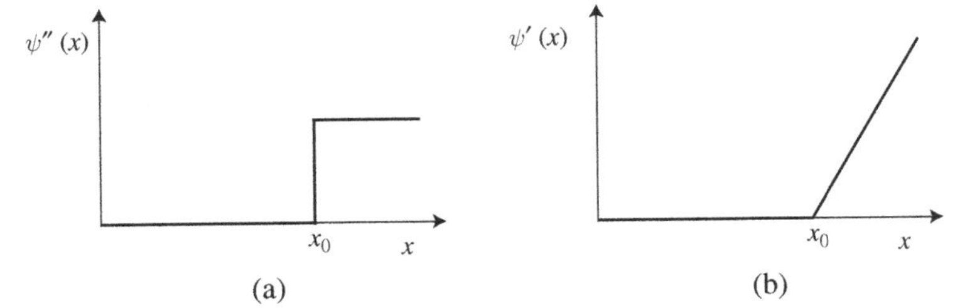

Figure 4.4 An enlargement of the first two derivatives of a wave function near the discontinuity $x = x_0$. If a wave function $\psi(x)$ has its second derivative $\psi''(x)$ stepwise continuous, its first derivative $\psi'(x)$ (and the wave function itself) must be continuous.

4.2 Potential barrier

Let us consider the generic case of an incoming one-dimensional particle (coming, say, from $-\infty$ and moving rightwards) which encounters some kind of *potential barrier*.

In this context, it is useful to introduce the concept of quantum current and current density. For this purpose, let us recall the three-dimensional time-dependent Schrödinger equation in the coordinate representation (Eq. (3.14))

$$\imath\hbar\frac{\partial}{\partial t}\psi(\mathbf{r},t) = \hat{H}\psi(\mathbf{r},t), \tag{4.15}$$

and its complex conjugate

$$-\imath\hbar\frac{\partial}{\partial t}\psi^*(\mathbf{r},t) = \hat{H}\psi^*(\mathbf{r},t), \tag{4.16}$$

and consider the expression

$$\imath\hbar\frac{\partial}{\partial t}\int_V d\mathrm{V}\rho(\mathbf{r},t) = \int_V d\mathrm{V}\left(\psi^*(\mathbf{r},t)\hat{H}\psi(\mathbf{r},t) - \hat{H}\psi^*(\mathbf{r},t)\psi(\mathbf{r},t)\right), \tag{4.17}$$

where $\rho(\mathbf{r},t) = \psi^*(\mathbf{r},t)\psi(\mathbf{r},t) = |\psi(\mathbf{r},t)|^2$ and $\rho(\mathbf{r},t)d\mathrm{V}$ is the probability of finding the particle in the infinitesimal volume $d\mathrm{V}$ (see also Eq. (2.108)), and where V is any finite region of the three-dimensional configuration space. To work out Eq. (4.17), we may take advantage of the fact that the Hamiltonian $\hat{H}$ is made of two parts, the kinetic energy and the potential energy. In the position representation the former is represented by the Laplacian Δ, whereas the latter is a multiplicative factor. This last part acts in the same way on $\psi(\mathbf{r},t)$ and $\psi^*(\mathbf{r},t)$ and disappears in the difference, so that

$$\begin{aligned}\imath\hbar\frac{\partial}{\partial t}\int_V d\mathrm{V}\rho(\mathbf{r},t) &= -\frac{\hbar^2}{2m}\int_V d\mathrm{V}\left[\psi^*(\mathbf{r},t)\Delta\psi(\mathbf{r},t) - \left(\Delta\psi^*(\mathbf{r},t)\right)\psi(\mathbf{r},t)\right]\\ &= -\frac{\hbar^2}{2m}\int_V d\mathrm{V}\nabla\cdot\left[\psi^*(\mathbf{r},t)\nabla\psi(\mathbf{r},t) - \left(\nabla\psi^*(\mathbf{r},t)\right)\psi(\mathbf{r},t)\right].\end{aligned} \tag{4.18}$$

Rearranging the terms, we arrive at

$$\int_V dV \left\{ \frac{\partial}{\partial t}\rho(\mathbf{r},t) + \frac{\hbar}{2\imath m}\nabla \cdot \left[\psi^*(\mathbf{r},t)\nabla\psi(\mathbf{r},t) - \left(\nabla\psi^*(\mathbf{r},t)\right)\psi(\mathbf{r},t)\right]\right\} = 0, \tag{4.19}$$

where the quantity

$$\mathbf{J} = \frac{\hbar}{2\imath m}\left[\psi^*(\mathbf{r},t)\nabla\psi(\mathbf{r},t) - \left(\nabla\psi^*(\mathbf{r},t)\right)\psi(\mathbf{r},t)\right] \tag{4.20}$$

is the *density of probability current*. Since Eq. (4.19) must hold for any finite integration region V, it follows that the integrand has to be zero, i.e.

$$\frac{\partial}{\partial t}\rho(\mathbf{r},t) + \nabla \cdot \mathbf{J} = 0. \tag{4.21}$$

The validity of this eqation could be directly checked by computing explicitly all the terms and verifing that they cancel. Equation (4.21) is the *quantum continuity equation* (see also Eq. (1.21)) and expresses the *local conservation of probability*. It is formally identical (and has a similar meaning) to the electrical current continuity equation, where ρ becomes the charge density and $\mathbf{J}$ the electrical current density.[1]

Two things should be noted in connection with Eq. (4.20). First, $\mathbf{J}$ is a real quantity since it may also be written as

$$\mathbf{J} = \Re\left(\psi^*(\mathbf{r},t)\frac{\hbar}{\imath m}\nabla\psi(\mathbf{r},t)\right). \tag{4.22}$$

Second, the quantity $(\hbar/\imath m)\nabla$ is nothing but the operator $\hat{\mathbf{v}} = \hat{\mathbf{p}}/m$, so that Eq. (4.22) is fully analogous to the definition $\mathbf{J} = \rho\mathbf{v}$ of the classical electrical current density.

4.2.1 Finite potential barrier

Let us now consider the case of a finite potential barrier, schematically depicted in Fig. 4.5. We may represent the incoming particle as a wave packet (see Box 2.6: p. 80), However, for the sake of simplicity we consider here a simple plane-wave component of the wave packet. For $E > V_0$ and at $-\infty$ and $+\infty$, the wave function has to be of the form $e^{\pm\imath kx}$ (see Sec. 4.1). It is interesting to apply the previously developed formalism of probability current density to this case. Crudely speaking, we have an incoming current that hits a potential barrier. The outcome of this dynamical process will be an outgoing current (moving towards $+\infty$) plus a reflected current (moving back to the left). This phenomenon is strictly quantum-mechanical. In fact, classically an incoming particle would be either transmitted or reflected by a potential barrier, depending on whether its energy is greater than V_0 or not. As we shall see, quantum-mechanically the particle is partially transmitted

[1] See [*Jackson* 1962, 2, 168–69].

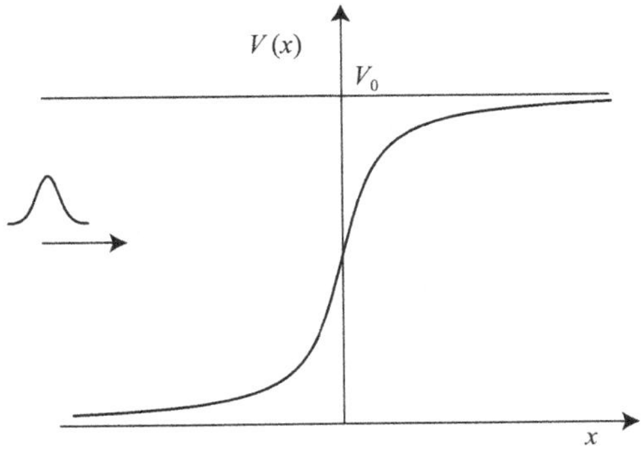

Figure 4.5 A wave packet entering from the left encounters a potential barrier with height V_0.

and partially reflected. Obviously, when considering a wave packet, this happens for any component wave of the packet. Let us take a generic wave function

$$\psi(x) = \mathcal{N}_+ e^{\imath kx}, \tag{4.23}$$

where $\mathcal{N}_+$ is some (complex) normalization factor, and compute the current density

$$J = \frac{\hbar}{2m\imath}\left[\imath k\psi(x)\psi^*(x) + \imath k\psi^*(x)\psi(x)\right] = \frac{\hbar k}{m}|\mathcal{N}_+|^2. \tag{4.24}$$

In particular, we may take

$$\psi(x \to +\infty) = \mathcal{N}_+ e^{\imath k_1 x}, \tag{4.25a}$$

$$\psi(x \to -\infty) = \mathcal{N}'_+ e^{\imath k_2 x} + \mathcal{N}_- e^{-\imath k_2 x}, \tag{4.25b}$$

where in the last equation the term $e^{\imath k_2 x}$ describes the particle moving to the right and $e^{-\imath k_2 x}$ describes the particle reflected at the potential barrier. $\mathcal{N}'_+$ and $\mathcal{N}_-$ are again two normalization factors, and without any loss of generality we may take $\mathcal{N}'_+ = 1$. Moreover (see Sec. 4.1), we also have

$$k_1 = \frac{1}{\hbar}\sqrt{2m(E - V_0)}, \quad k_2 = \frac{1}{\hbar}\sqrt{2mE}. \tag{4.26}$$

Given the definition of J, it is possible to define a transmission coefficient T^2 and a reflection coefficient R^2 that represent the probability of transmission and reflection, respectively[2]

$$\mathrm{T}^2 = \frac{J_\mathrm{T}}{J_\mathrm{I}}, \quad \mathrm{R}^2 = \frac{J_\mathrm{R}}{J_\mathrm{I}}, \tag{4.27}$$

where J_I is the incident current density, J_T is the transmitted current density and finally J_R is the reflected current density. Recalling Eq. (4.24), we have

[2] Usually, in the literature these are called T and R, respectively. Here, we adopt a slightly different convention for the sake of the uniformity with other contexts (see, for instance, Subsec. 2.3.4).

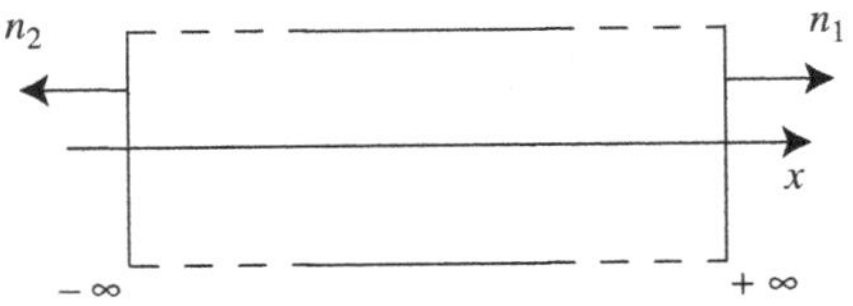

Figure 4.6 Schematic drawing of the closed surface to compute used the flux of **J**: n_1 and n_2 are two versors. Transmitted and reflected currents contribute positively to the flux, whereas the incident current contributes negatively.

$$J_{\rm I} = \frac{\hbar k_2}{m}, \quad J_{\rm T} = \frac{\hbar k_1}{m} |\mathcal{N}_+|^2, \quad J_{\rm R} = \frac{\hbar k_2}{m} |\mathcal{N}_-|^2, \tag{4.28}$$

and therefore

$$\mathrm{T}^2 = \frac{k_1}{k_2} |\mathcal{N}_+|^2, \quad \mathrm{R}^2 = |\mathcal{N}_-|^2. \tag{4.29}$$

Since the sum of the reflected and transmitted current densities should be equal to the incident current density, it is then clear that

$$\mathrm{T}^2 + \mathrm{R}^2 = 1 \quad \text{and} \quad |\mathcal{N}_-|^2 = 1 - \frac{k_1}{k_2} |\mathcal{N}_+|^2. \tag{4.30}$$

This result can also be derived from the continuity equation (4.21). In fact, we are considering a stationary situation, i.e. a situation where $\partial\rho/\partial t = 0$ and therefore $\nabla \cdot \mathbf{J} = 0$. For the Stokes theorem, this means that the flux of **J** through any closed surface must vanish. In particular, we may consider the infinite "surface" shown in Fig. 4.6 and compute the total flux of **J** through the surface. Then, we have

$$- k_2 + |\mathcal{N}_-|^2 k_2 + |\mathcal{N}_+|^2 k_1 = 0. \tag{4.31}$$

On dividing all terms by k_2, we obtain $\mathrm{T}^2 + \mathrm{R}^2 = 1$.

4.2.2 δ-shaped potential barrier

The particular cases of square barriers are considered in Probs. 4.3 and 4.4. A very special example is represented by a δ-function-like potential barrier (see Fig. 4.7), i.e.

$$V(x) = C\delta(x), \quad \text{with} \quad C > 0. \tag{4.32}$$

The peculiarity of this example lies in the fact that the barrier is both infitely high and infinitesimally wide. We may then divide the entire line into two different regions, I ($x < 0$) and II ($x > 0$). In both regions the potential energy $V(x)$ is zero, and therefore we may write the wave function of the particle as

$$\psi(x) = \begin{cases} \psi_{\rm I}(x) & x \leq 0, \\ \psi_{\rm II}(x) & x \geq 0, \end{cases} \tag{4.33}$$

with

$$\psi_{\rm I}(x) = e^{\imath kx} + A_1 e^{-\imath kx}, \tag{4.34a}$$
$$\psi_{\rm II}(x) = A_2 e^{\imath kx}, \tag{4.34b}$$

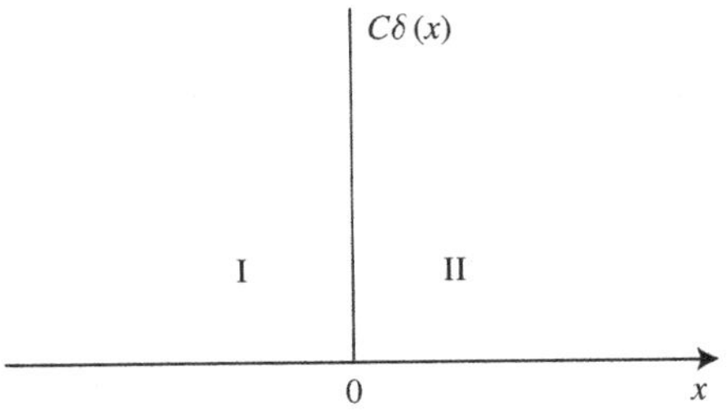

Figure 4.7 **Potential barrier given by a delta function $\delta(x)$ times a constant $C > 0$.**

where the energy of the incoming particle is given by

$$E = \frac{\hbar^2}{2m}k^2. \tag{4.35}$$

The Schrödinger equation takes the form

$$-\frac{\hbar^2}{2m}\psi''(x) + C\delta(x)\psi(x) = E\psi(x). \tag{4.36}$$

Rewriting the constant C as

$$C = \frac{\hbar^2}{2m}\eta, \tag{4.37}$$

Eq. (4.36) takes the simpler form

$$\psi''(x) = \left[\eta\delta(x) - k^2\right]\psi(x). \tag{4.38}$$

In order to find the first derivative of the wave function across the point $x = 0$, we take advantage of

$$\psi'_{\mathrm{II}}(0) - \psi'_{\mathrm{I}}(0) = \lim_{\epsilon\to 0^+}\int_{-\epsilon}^{\epsilon} dx\,\psi''(x).$$

Integrating Eq. (4.38) once gives

$$\psi'(x) = \int_{-\infty}^{x} dy\left[\eta\delta(y) - k^2\right]\psi(y). \tag{4.39}$$

However, we know that (see Eq. (2.23c))

$$\int_a^b dy\,\delta(y - y_0)\, f(y) = \begin{cases} f(y_0) & \text{if } y_0 \in (a,b), \\ 0 & \text{otherwise,} \end{cases} \tag{4.40}$$

and therefore

$$\psi'_{\mathrm{I}}(0) = -k^2\int_{-\infty}^{0} dy\,\psi(y), \tag{4.41a}$$

$$\psi'_{\mathrm{II}}(0) = -k^2\int_{-\infty}^{0} dy\,\psi(y) + \eta\psi(0). \tag{4.41b}$$

Equations (4.41) tell us that the δ-shaped potential induces a finite discontinuity at $x = 0$ in the first derivative of the wave function. As a consequence, the continuity equations for the wave function and its first derivative at $x = 0$ in this case read

$$\psi_{\text{II}}(0) = \psi_{\text{I}}(0), \tag{4.42a}$$

$$\psi'_{\text{II}}(0) - \psi'_{\text{I}}(0) = \eta\psi(0). \tag{4.42b}$$

Recalling Eqs. (4.34), Eqs. (4.42) translate into

$$\begin{cases} A_2 = 1 + A_1, \\ \imath k\,(A_2 + A_1 - 1) = \eta A_2, \end{cases} \tag{4.43}$$

that represents a linear system of two equations for the two unknowns A_1 and A_2, whose solution is given by

$$A_1 = \frac{\eta}{2\imath k - \eta} \quad \text{and} \quad A_2 = \frac{2\imath k}{2\imath k - \eta}. \tag{4.44}$$

Finally, it is possible to compute the transmission and reflection probabilities, that is,

$$\text{T}^2 = |A_2|^2 = \frac{4k^2}{4k^2 + \eta^2}, \tag{4.45a}$$

$$\text{R}^2 = |A_1|^2 = \frac{\eta^2}{4k^2 + \eta^2}. \tag{4.45b}$$

These probabilities teach us that, for any finite η, $0 < \text{T}^2 < 1$. For $\eta = 0$, we obviously have $\text{R} = 0$ and $\text{T} = 1$.

4.3 Tunneling

In Sec. 3.4 we have seen that the energy eigenfunctions of a particle in a box vanish outside the interval $(0, a)$. This is due to the fact that the potential walls at $x = 0$ and $x = a$ have infinite height and their width is not infinitesimal. In other words, as in classical mechanics, the particle is not able to penetrate the infinite potential barrier. However, the situation changes dramatically in the cases considered in Sec. 4.1 and in Subsec. 4.2.2. When the potential well is finite even for particle energies which are smaller than V_0 (discrete spectrum) (see Fig. 4.1(a)), or when the width of the infinite potential barrier is infinitesimal (see Fig. 4.7), there is a finite probability that the particle is found outside the classically allowed region. In the first case, the particle may be found outside the region II $(0 \leq x \leq a)$ (see Property (iii) of Subsec. 3.2.1). This is indicated by the fact that the wave function extends into the regions I $(x < 0)$ and III $(x > a)$, as seen in Eqs. (4.3) and in Fig. 4.3, even though they decrease exponentially outside region II. In the second case, the incoming particle is both partially trasmitted into region II and partially reflected back into region I. This phenomenon does not have a classical counterpart and is known as *tunneling* or *tunnel effect*. These expressions reflect the fact that a particle with energy $E < V_0$ has a finite chance to penetrate (to "tunnel") into the potential walls, entering the

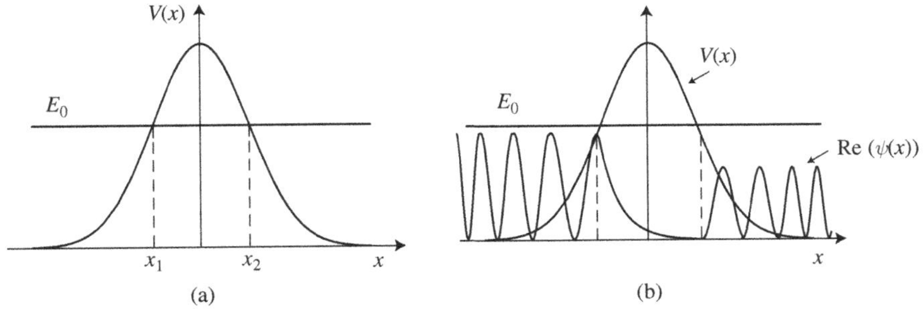

Figure 4.8 Classical turning points and quantum tunneling. (a) Classically, the motion is limited by the turning points x_1 and x_2, i.e. it may occur either in the region $x < x_1$ or in the region $x > x_2$. In other words, the classical particle coming from the left cannot penetrate the potential barrier beyond x_1. (b) In the quantum-mechanical case, the particle may penetrate the barrier beyond x_1 and even re-emerge beyond x_2. In the figure, the real part of the wave function is qualitatively shown.

classically forbidden regions. This phenomenon is symmetric to that considered in Subsec. 4.2.1, where a particle with $E > V_0$, which would be classically transmitted through the barrier, quantum-mechanically is partially reflected. Tunneling is a consequence of the very structure of quantum mechanics. In fact, in classical mechanics a particle cannot enter a forbidden region (it has to be reflected at the turning points) because otherwise its kinetic energy would become negative, which is clearly impossible.

In order to calculate the tunneling probability, one has to sum the contributions given by the integrals of the square modulus of the wave function in every classically forbidden region. In the example given in Sec. 4.1, for instance, for $E < V_0$ the tunneling probability would be given by

$$\wp_{\rm T} = \int_{-\infty}^{0} dx\, |\psi_{\rm I}(x)|^2 + \int_{a}^{+\infty} dx\, |\psi_{\rm III}(x)|^2 . \tag{4.46}$$

There is, however, a different situation where tunneling plays an important role. Let us consider the case depicted in Fig. 4.8(a): a particle with energy E_0 incoming from the left encounters a bell-shaped potential $V(x)$. Classically, the particle would be reflected at the turning point $x = x_1$ and could not enter the forbidden region $x > x_1$. Quantum-mechanically, instead, in the regions where $x \to \pm\infty$, the wave function must be a plane wave of the type $e^{\imath kx}$, while inside the classically forbidden region it will decrease exponentially. This situation is qualitatively shown in Fig. 4.8(b), which clearly displays the fact that a quantum particle can penetrate (with an exponentially decreasing probability) from the left into the right region.

A remarkable example of the tunneling phenomenon is represented by the emission of the α-particles by radioactive nuclei. α-particles are nuclei of helium (with two protons and two neutrons) and their potential energy inside the parent nucleus is schematically drawn in Fig. 4.9. This potential comes about as a result of the $1/r$ repulsive Coulomb potential (for large r) and the short-range nuclear attractive potential (for small r). Normally, α-particles have an energy E_0 smaller than the maximum of this potential and therefore, classically,

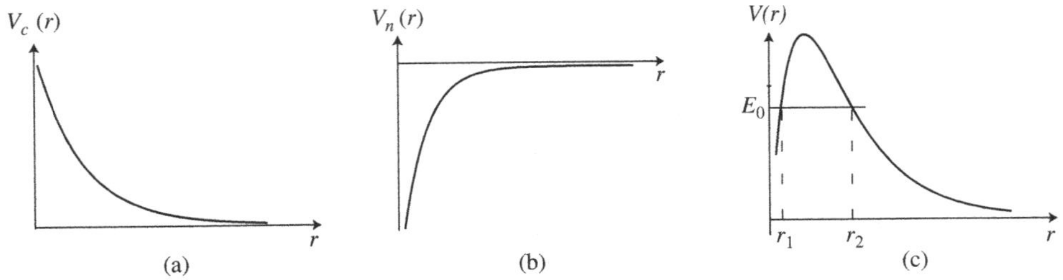

Figure 4.9 Tunneling of α-particles. (a) Repulsive Coulomb potential between the protons in the nucleus. (b) Short-range attractive nuclear (strong-force) potential. (c) The resulting potential for α-particles is given by the combination of (a) and (b). The origin represents the center of the nucleus and r the distance of the α-particles from this center. Denoting with E_0 the energy of the α-particle, r_1 and r_2 are the classical turning points. The potential $V(r)$ has a maximum for $r_1 < r < r_2$ and then decreases as $r \to \infty$.

could not be able to escape the region $0 < r < r_1$. As a matter of fact, they are emitted and enter the region $r > r_2$ with a positive kinetic energy in agreement with the genuinely quantum tunneling effect. Note that the tunnel effect may be very small and the mean life extraordinarily long.

Box 4.1 **Relativity and tunneling time**

In quantum mechanics there are in principle no limitations on the possible speed of quantum systems. This is a consequence of the fact that quantum mechanics is built on the background of classical mechanics (see Pr. 2.3: p. 72) and in the framework of Galilean relativity (see Sec. 3.3), for which there is no invariant velocity and hence no upper limit of speed. This has raised the question of whether there are any situations in which there is actually a violation of relativity. Tunneling seems to represent such a situation [Chiao/Steinberg 1997] because the phase velocity, i.e. the velocity $v_\phi = \omega/k$ (see Box 2.6: p. 80), at which the zero-crossing of the carrier wave would move, and the group velocity $d\omega/dk$ could be superluminal. In fact, this effect is not in contradiction with Einstein causality if we take into account the fact that the outgoing wave is always a fraction of the input wave. In other words, the tunneling probability is usually much smaller than one, so that the present effect cannot be used to actually transmit a signal at superluminal speed [Stenner *et al.* 2003].

Box 4.2 **Scanning tunneling microscopy**

In 1981, Gerd Binnig and Heinrich Rohrer invented a new type of imaging technique, the scanning tunneling microscope (STM). For this, they received the Nobel Prize for physics in 1986. It is nowadays a widespread surface science technique owing its popularity to the wide range of possible applications, and the ability to obtain a direct real-space image of conducting surfaces. Its most most important feature is the high spatial resolution of the order of 10^{-11} m, which allows users to scan and even to manipulate individual atoms (see Fig. 4.10).

This technique does not make use of any lenses, light, or electron sources. Rather, it is the tunneling effect which provides its physical foundation. In order to obtain an image of a surface, a small voltage is applied between a sharp metallic tip and the investigated surface, which are separated by a vacuum barrier. If the thickness of the potential barrier is of the order of a few atomic diameters, electrons are able to tunnel through it and a current will flow. This electric current depends exponentially on the barrier thickness. Hence, by scanning the tip over the surface at a constant current or barrier thickness, the record of the vertical tip motion will reflect the surface structure. The adoption of this technique opened the way to a large family of instruments generally known as "scanning probe microscopes."

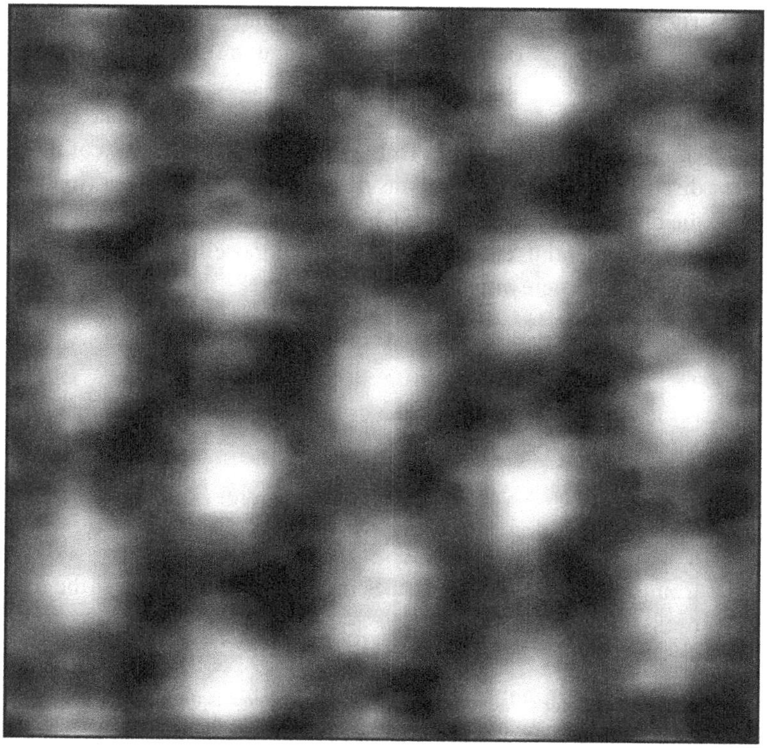

Figure 4.10 STM image of the regular arrangement of carbon atoms at the surface of graphite. The spatial extension of the region is 1 nm. Image adapted from the web page www.manep.ch/en/technological-challenges/spm.html.

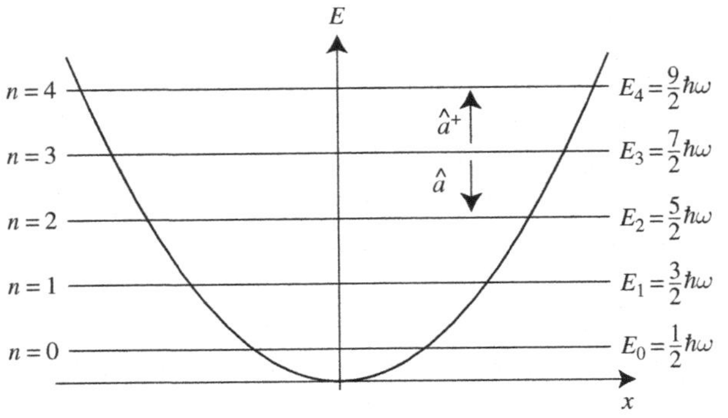

Figure 4.11 Potential and energy levels (see Eq. (4.61)) of the harmonic oscillator. The action of the creation and annihilation operators (see Subsec. 4.4.2) is illustrated by the arrows.

4.4 Harmonic oscillator

Up to now we have solved some simple dynamical problems in quantum mechanics: The free particle, the particle in a box, and that of a one-dimensional particle subjected to several types of potential barriers and wells. The next simplest problem is the *harmonic oscillator*. In the one-dimensional case, this is a particle subjected to small linear oscillations and, as we shall see, will serve as a guidance tool for the solution of more complicated problems.[3] In fact, we shall find applications of the solutions and the methods discussed here to the problems of quantum measurement (in Ch. 9) and of quantization of the electromagnetic field (in Ch. 13).

The potential energy of a harmonic oscillator is given by $V(x) = \frac{1}{2}m\omega^2 x^2$ (see Fig. 4.11), where m is the mass and

$$\omega = 2\pi\nu \tag{4.47}$$

is the angular frequency of the oscillator. Therefore, the quantum Hamiltonian we have to solve has the form

$$\hat{H} = \frac{\hat{p}_x^2}{2m} + \frac{1}{2}m\omega^2\hat{x}^2 = -\frac{\hbar^2}{2m}\frac{\partial^2}{\partial x^2} + \frac{1}{2}m\omega^2 x^2, \tag{4.48}$$

from which we derive the Schrödinger equation for the one-dimensional harmonic oscillator

[3] The harmonic oscillator has always been considered as the simplest non-trivial problem and the basic model for any mechanical paradigm. We remember that, at the very beginning of classical physics, Galileo Galilei found (1583) a law that described the isochronous character of small oscillations of a simple pendulum:

$$\tau = 2\pi\sqrt{\frac{l}{g}},$$

where τ is the period of the small oscillation, l is the length of the pendulum, and g is the acceleration due to gravity.

$$\psi''(x) + \frac{2m}{\hbar^2}\left(E - \frac{1}{2}m\omega^2x^2\right)\psi(x) = 0. \tag{4.49}$$

From a qualitative point of view, we may already say that the probability of finding the particle at $x \to \pm\infty$ must tend to zero, since $\lim_{x\to\pm\infty} V(x) = \infty$ (see property (iii) of Subsec. 3.2.1). The harmonic oscillator spectrum will then be similar to that of a particle in a box (see Sec. 3.4 and Fig. 3.6) and we expect that also in this case the spectrum will be discrete with no degenerate levels (see property (iii) of Subsec. 3.2.2 and property (i) of Subsec. 3.2.3).

Given the importance of this system and the fact that it represents an ideal example of how quantum mechanics works in practice, in the following subsections we shall solve the linear oscillator according to two different methods.

4.4.1 Heisenberg's solution

First, we shall discuss the solution given by Werner Heisenberg in 1925 in the context of matrix mechanics [Heisenberg 1925] (see also Subsec. 1.5.7). We start from Eq. (4.48) and work in the Heisenberg picture (see Subsec. 3.6.1), choosing a basis $\{|n\rangle\}$ in which $\hat{H}$ is diagonal (an eigenbasis of $\hat{H}$):

$$\hat{H}\,|n\rangle = E_n\,|n\rangle, \tag{4.50}$$

where the energy eigenvalue E_n corresponds to the eigenvector $|n\rangle$, i.e.

$$E_n = H_{nn} = \left\langle n\left|\hat{H}\right|n\right\rangle, \tag{4.51}$$

since, as we know, $\langle m \mid n\rangle = \delta_{nm}$ for every energy eigenvectors $|m\rangle$, $|n\rangle$. In order to solve the equations of motion, we preliminarily compute the matrix elements $\dot{O}_{nm}$ of the time derivative of a generic operator $\hat{O}$ given its matrix elements O_{nm}. If $\hat{O}$ does not explicitly depend on time, from Eq. (3.108)

$$\frac{d}{dt}\hat{O}(t) = \frac{1}{\imath\hbar}[\hat{O}(t), \hat{H}], \tag{4.52}$$

we have

$$\begin{aligned}
\dot{O}_{nm} &= \left\langle n\left|\dot{\hat{O}}\right|m\right\rangle = \left\langle n\left|\frac{1}{\imath\hbar}[\hat{O}(t), \hat{H}]\right|m\right\rangle = \frac{1}{\imath\hbar}\left\langle n\left|\left(\hat{O}\hat{H} - \hat{H}\hat{O}\right)\right|m\right\rangle \\
&= \frac{1}{\imath\hbar}(E_m - E_n)\left\langle n\left|\hat{O}\right|m\right\rangle = \imath\frac{E_n - E_m}{\hbar}O_{nm} \\
&= \imath\omega_{nm}O_{nm},
\end{aligned} \tag{4.53}$$

where we have made use of Eq. (4.50) and where the ω_{nm} are Bohr's quantized angular frequencies, given by (see Eq. (1.76))

$$\omega_{nm} = \frac{E_n - E_m}{\hbar}. \tag{4.54}$$

Our first aim is now to find the energy levels of the harmonic oscillator. To reach our goal, let us consider the quantum counterpart of the Hamilton equations (3.126) and (3.128), which, as we have seen, hold at an operatorial level, i.e.

$$\dot{\hat{x}} = \frac{\partial \hat{H}}{\partial p} = \frac{\hat{p}_x}{m} \quad \text{and} \quad \dot{\hat{p}}_x = -\frac{\partial \hat{H}}{\partial x} = -m\omega^2 \hat{x}. \tag{4.55}$$

From Eqs. (4.55) we obtain

$$\ddot{\hat{x}} + \omega^2 \hat{x} = 0. \tag{4.56}$$

Heisenberg suggested that this relation must hold for every matrix element, that is

$$\left(\ddot{\hat{x}} + \omega^2 \hat{x}\right)_{nk} = \ddot{\hat{x}}_{nk} + \omega^2 \hat{x}_{nk} = 0, \tag{4.57}$$

where

$$\begin{aligned} x_{nk} = \langle n \left| \hat{x} \right| k \rangle &= \int dx \, \langle n \mid x \rangle \, x \, \langle x \mid k \rangle \\ &= \int dx \psi_n^*(x) x \psi_k(x). \end{aligned} \tag{4.58}$$

The function $\psi_j(x) = \langle x \mid j \rangle$ is the eigenfunction corresponding to the eigenvalue E_j. In order to obtain the matrix elements of $\ddot{\hat{x}}$ we should differentiate Eq. (4.53) once again with respect to time, which yields

$$\ddot{x}_{nm} = -\omega_{nm}^2 x_{nm}. \tag{4.59}$$

We may then rewrite Eq. (4.57) as

$$\left(\omega^2 - \omega_{nk}^2\right) x_{nk} = 0. \tag{4.60}$$

It is clear that the operator $\hat{x}$ cannot have all the matrix elements equal to zero. From Eq. (4.60), x_{nk} can be different from zero only when the quantity $\omega^2 - \omega_{nk}^2$ is equal to zero, i.e. when $\omega_{nk} = \pm\omega$, or, using Eq. (4.54), when

$$E_n - E_k = \pm\hbar\omega. \tag{4.61}$$

Eq. (4.61), together with the fact that the energy levels of a one-dimensional harmonic oscillator are not degenerate, teaches us that the energy levels E_n are equally spaced (see Fig. 4.11). Equation (4.61) also allows us to order the levels in terms of growing energy values, i.e. $E_{j+1} - E_j = \hbar\omega$.

Since $\hat{H}$ is real, without loss of generality we may choose a real set of eigenfunctions $\psi_n(x) = \langle x \mid n \rangle = \psi_n^*(x)$ (see Eq. (4.48) and property (iv) of Subsec. 3.2.2). Moreover, $\hat{H}$ is a positive definite operator, i.e.

$$H_{nn} = \left\langle n \left| \hat{H} \right| n \right\rangle = \int dx \psi_n^*(x) \hat{H} \psi_n(x) = \int dx \psi_n(x) \hat{H} \psi_n(x) > 0. \tag{4.62}$$

This can be shown by the following argument. Since the potential energy $V(x) > 0$, to prove Eq. (4.62) we only have to ascertain that $1/2m \left\langle n \left| \hat{p}_x^2 \right| n \right\rangle > 0$. We then have

$$\left\langle n \left| \hat{p}_x^2 \right| n \right\rangle = \left[\langle n \mid \hat{p}_x \right] \left[\hat{p}_x \mid n \rangle \right] = \left[\langle n \mid \hat{p}_x^\dagger \right] \left[\hat{p}_x \mid n \rangle \right] > 0, \tag{4.63}$$

where we have made use of the fact that $\hat{p}_x$ is Hermitian. The fact that $\hat{H}$ is positive definite tells us that there is a minimum value for E, which can be taken for $n = 0$ (see Fig. 4.11).

Going back to Eq. (4.60), we may conclude that $x_{nk} \neq 0$ only if $k = n - 1$ or $k = n + 1$. Loosely speaking, only neighboring levels "talk" to each other. Moreover, we observe that x_{nk} (given by Eq. (4.58)) is real, because we have chosen a real eigenbasis and $\hat{x}$ is a real operator. On the other hand, $\hat{x}$ is also Hermitian, and, as a consequence, it must be symmetric too, i.e. $x_{nk} = x_{kn}$. We are now in a position to calculate all the non-zero matrix elements of $\hat{x}$, i.e. $x_{n,n-1}$ and $x_{n,n+1}$. For this purpose we may exploit the Heisenberg commutation relation (2.173a) to arrive at

$$\hat{x}\hat{\dot{x}} - \hat{\dot{x}}\hat{x} = \imath\frac{\hbar}{m}\hat{I}, \tag{4.64}$$

or

$$\sum_l (x_{nl}\dot{x}_{lk} - \dot{x}_{nl}x_{lk}) = \imath\frac{\hbar}{m}\delta_{nk}, \tag{4.65}$$

which implies

$$\sum_l (x_{nl}\dot{x}_{ln} - \dot{x}_{nl}x_{ln}) = \imath\frac{\hbar}{m}. \tag{4.66}$$

Using Eq. (4.53), we may write Eq. (4.66) as

$$\sum_l (x_{nl}\omega_{ln}x_{ln} - \omega_{nl}x_{nl}x_{ln}) = \frac{\hbar}{m}, \tag{4.67}$$

where ω_{nm} is antisymmetric on its two indices since (from Eq. (4.54)) $\omega_{nm} = -\omega_{mn}$. Hence, $2\sum_l x_{nl}^2\omega_{ln} = \hbar/m$. Since l can only take the values $l = n - 1$ or $l = n + 1$, we finally obtain

$$x_{n,n-1}^2\omega_{n-1,n} + x_{n,n+1}^2\omega_{n+1,n} = \frac{\hbar}{2m}. \tag{4.68}$$

Furthermore, $\omega_{n+1,n} = \omega$ and $\omega_{n-1,n} = -\omega$, so that we may start with $n = 0$ and construct the matrix element

$$x_{01} = \sqrt{\frac{\hbar}{2m\omega}}. \tag{4.69}$$

From the knowledge of x_{01} and proceeding recursively for growing values of n, we find the general relation (see Prob. 4.6)

$$x_{n,n+1} = \sqrt{\frac{(n+1)\hbar}{2m\omega}}, \tag{4.70}$$

which permits to find all the desired matrix elements of the position operator. The knowledge of the matrix elements of $\hat{x}$ allows to find the matrix elements of all the other relevant operators (e.g. $\hat{p}_x = m\hat{\dot{x}}$, $\hat{H}$, etc.). In particular,

$$\begin{aligned} E_n = H_{nn} &= \frac{1}{2}m\left(\dot{x}^2\right)_{nn} + \frac{1}{2}m\omega^2\left(x^2\right)_{nn} \\ &= \frac{1}{2}m\sum_l \dot{x}_{nl}\dot{x}_{ln} + \frac{1}{2}m\omega^2\sum_l x_{nl}x_{ln} \\ &= -\frac{1}{2}m\sum_l \omega_{nl}\omega_{ln}x_{nl}x_{ln} + \frac{1}{2}m\omega^2\sum_l x_{nl}x_{ln}, \end{aligned} \tag{4.71}$$

where we have used $\dot{x}_{nl} = \imath\omega_{nl}x_{nl}$ (see Eq. (4.53)). Since x_{nl} is symmetric and ω_{nl} antisymmetric, we have

$$\begin{aligned} E_n &= \frac{1}{2}m\sum_l \left(\omega_{nl}^2 x_{nl}^2 + \omega^2 x_{nl}^2\right) \\ &= \frac{1}{2}m\left(\omega^2 x_{n,n+1}^2 + \omega^2 x_{n,n-1}^2 + \omega^2 x_{n,n+1}^2 + \omega^2 x_{n,n-1}^2\right) \\ &= m\omega^2\left(x_{n,n+1}^2 + x_{n,n-1}^2\right) \\ &= m\omega^2\left(\frac{(n+1)\hbar}{2m\omega} + \frac{n\hbar}{2m\omega}\right) = \left(n+\frac{1}{2}\right)\hbar\omega, \end{aligned} \tag{4.72}$$

where we have made use of Eqs. (4.70) and (4.54). Thus we have found the energy eigenvalues of the harmonic oscillator. We immediately see that they have the desired properties: they are equally spaced (the energy difference between neighboring levels being $\hbar\omega$) and labelled according to growing values of energy. The minimum energy eigenvalue is attained for $n = 0$, and we have $E_0 = (1/2)\hbar\omega$, which is called the *zero-point energy* (see Fig. 4.11). From the discussion above, it appears clear that the quantity $\hbar\omega$ may be interpreted as an energy quantum for a harmonic oscillator of frequency ω (see also Eq. (1.22)): the oscillator may "jump" from a level j to a level k only when an energy $|j-k|\hbar\omega$ is either absorbed ($j < k$) or emitted ($j > k$).

Now that we have found the energy eigenvalues, the harmonic oscillator problem is solved. In fact, the knowledge of the matrix elements $x_{nm}(t)$ and $\dot{x}_{nm}(t)$ allows us to calculate any physical quantity at any time. In the next subsection, however, we shall see a different type of solution of the harmonic oscillator problem, in which, in addition to the eigenvalues, we shall also determine the eigenfunctions.

4.4.2 Algebraic solution

It is very interesting and instructive to realize that the harmonic oscillator problem can be solved by exploiting the Heisenberg commutation relations only. To see how this is possible, let us introduce the operators

$$\hat{a} = \sqrt{\frac{m}{2\hbar\omega}}\left(\omega\hat{x} + \imath\frac{\hat{p}_x}{m}\right), \tag{4.73a}$$

$$\hat{a}^\dagger = \sqrt{\frac{m}{2\hbar\omega}}\left(\omega\hat{x} - \imath\frac{\hat{p}_x}{m}\right), \tag{4.73b}$$

which are called *annihilation* (lowering) and *creation* (raising) operators, respectively. It should be emphasized that these operators are not Hermitian – one is the Hermitian conjugate or adjoint (see p. 46) of the other – and therefore are not observables (see Th. 2.1: p. 46). They possess the important property (see Prob. 4.9)

$$\left[\hat{a}, \hat{a}^\dagger\right] = \hat{I}. \tag{4.74}$$

The reason for their names lies in the way they act on the energy eigenstates. In particular, we have

$$a^{\dagger}_{nk} = \left\langle n \left| \hat{a}^{\dagger} \right| k \right\rangle = \sqrt{\frac{m}{2\hbar\omega}} \left[\omega x_{nk} - \imath \left(\imath \omega_{nk} x_{nk} \right) \right] = \sqrt{\frac{m}{2\hbar\omega}} \left(\omega + \omega_{nk} \right) x_{nk}. \tag{4.75}$$

We already know that x_{nk} is different from zero only when $k = n \pm 1$. However, for $k = n + 1$, $\omega_{n,n+1} = -\omega$ and therefore $a^{\dagger}_{n,n+1} = 0$. The only non-zero matrix element of $\hat{a}^{\dagger}$ is then $a^{\dagger}_{n,n-1}$, for which we have (see Eq. (4.70))

$$\omega_{n,n-1} = \omega \quad \text{and} \quad x_{n,n-1} = \sqrt{\frac{n\hbar}{2m\omega}}. \tag{4.76}$$

It follows that

$$a^{\dagger}_{nk} = \delta_{k,n-1}\sqrt{n}. \tag{4.77a}$$

Similarly, we have

$$a_{nk} = \delta_{k,n+1}\sqrt{n+1}. \tag{4.77b}$$

Let us consider the product operator

$$\begin{aligned}
\hat{a}^{\dagger}\hat{a} &= \frac{m}{2\hbar\omega}\left(\omega\hat{x} - \imath\frac{\hat{p}_x}{m}\right)\left(\omega\hat{x} + \imath\frac{\hat{p}_x}{m}\right) \\
&= \frac{m}{2\hbar\omega}\left(\omega^2\hat{x}^2 + \frac{\imath\omega}{m}\hat{x}\hat{p}_x - \frac{\imath\omega}{m}\hat{p}_x\hat{x} + \frac{\hat{p}_x^2}{m^2}\right) \\
&= \frac{m}{2\hbar\omega}\left(\omega^2\hat{x}^2 + \frac{\imath\omega}{m}\left[\hat{x}, \hat{p}_x\right] + \frac{\hat{p}_x^2}{m^2}\right) \\
&= \frac{1}{\hbar\omega}\left(\frac{1}{2}m\omega^2\hat{x}^2 + \frac{\hat{p}_x^2}{2m}\right) - \frac{1}{2},
\end{aligned} \tag{4.78}$$

where we have made use of the commutation relation (2.173a). The term in brackets in Eq. (4.78) is just the harmonic-oscillator Hamiltonian (4.48). Therefore, we can write

$$\hat{H} = \hbar\omega\left(\hat{a}^{\dagger}\hat{a} + \frac{1}{2}\right), \tag{4.79}$$

or, in terms of matrix elements,

$$H_{nn} = \hbar\omega\left[\left(a^{\dagger}a\right)_{nn} + \frac{1}{2}\right], \tag{4.80}$$

where

$$\left(a^{\dagger}a\right)_{nn} = \sum_k a^{\dagger}_{nk}a_{kn} = a^{\dagger}_{n,n-1}a_{n-1,n} = n. \tag{4.81}$$

Therefore, in the energy representation the operator $\hat{N} = \hat{a}^{\dagger}\hat{a}$ is diagonal and its n-th diagonal term is just equal to n. This is the reason why it is called the *number* operator, i.e.

$$\hat{N}\left|n\right\rangle = n\left|n\right\rangle, \tag{4.82}$$

which may be considered as the eigenvalue equation of the number operator. On the other hand, we have

$$\hat{a}\,|n\rangle = \sum_m |m\rangle \langle m|\hat{a}|n\rangle = \sum_m a_{mn}\,|m\rangle . \tag{4.83}$$

In the last summation the only term which contributes to the sum is the term for which $m = n - 1$, i.e.

$$\hat{a}\,|n\rangle = a_{n-1,n}\,|n-1\rangle = \sqrt{n}\,|n-1\rangle . \tag{4.84}$$

A similar expression may be found for $\hat{a}^\dagger\,|n\rangle$, so that we have

$$\hat{a}\,|n\rangle = \sqrt{n}\,|n-1\rangle, \quad \hat{a}^\dagger\,|n\rangle = \sqrt{n+1}\,|n+1\rangle . \tag{4.85}$$

Equations (4.85) explain why $\hat{a}$ and $\hat{a}^\dagger$ are called annihilation (lowering) and creation (raising) operators, respectively. We may then interpret the set of the harmonic oscillators eigenvectors as a ladder which we may climb through $\hat{a}^\dagger$ and descend through $\hat{a}$. In matricial form, the explicit expressions for the number, annihilation, and creation operators are given by

$$\hat{N} = \begin{bmatrix} 0 & 0 & 0 & 0 & \cdots \\ 0 & 1 & 0 & 0 & \cdots \\ 0 & 0 & 2 & 0 & \cdots \\ 0 & 0 & 0 & 3 & \cdots \\ \cdots & \cdots & \cdots & \cdots & \cdots \end{bmatrix}, \tag{4.86a}$$

$$\hat{a} = \begin{bmatrix} 0 & \sqrt{1} & 0 & 0 & \cdots \\ 0 & 0 & \sqrt{2} & 0 & \cdots \\ 0 & 0 & 0 & \sqrt{3} & \cdots \\ 0 & 0 & 0 & \cdots & \cdots \\ \cdots & \cdots & \cdots & \cdots & \cdots \end{bmatrix}, \tag{4.86b}$$

$$\hat{a}^\dagger = \begin{bmatrix} 0 & 0 & 0 & 0 & \cdots \\ \sqrt{1} & 0 & 0 & 0 & \cdots \\ 0 & \sqrt{2} & 0 & 0 & \cdots \\ 0 & 0 & \sqrt{3} & 0 & \cdots \\ \cdots & \cdots & \cdots & \cdots & \cdots \end{bmatrix}, \tag{4.86c}$$

where the only non-zero elements of the annihilation and creation operators are located immediately above and below the principal diagonal, respectively. The minimum allowed value for n is zero (see p. 156) and therefore we must have

$$\hat{a}\,|0\rangle = 0, \quad \text{or} \quad \left(\omega\hat{x} + \iota\,\dot{\hat{x}}\right)|0\rangle = 0. \tag{4.87}$$

In the coordinate representation, where $\psi_n(x) = \langle x \mid n\rangle$, this translates into

$$\left(\omega\hat{x} + \frac{\iota\,\hat{p}_x}{m}\right)\psi_0(x) = 0. \tag{4.88}$$

Using the differential form of the operator $\hat{p}_x$ in the x-representation (see Subsec. 2.2.4), we obtain

$$\frac{\partial}{\partial x}\psi_0(x) = -\frac{m\omega}{\hbar}x\psi_0(x), \tag{4.89}$$

whose solution is

$$\psi_0(x) = \mathcal{N}e^{-\frac{m\omega}{2\hbar}x^2}, \tag{4.90}$$

where $\mathcal{N}$ is a normalization constant which can be determined by imposing the condition (see Prob. 4.14) $\int dx|\psi_0(x)|^2 = 1$. Finally, we obtain (see Prob. 4.15)

$$\psi_0(x) = \left(\frac{m\omega}{\pi\hbar}\right)^{\frac{1}{4}} e^{-\frac{m\omega}{2\hbar}x^2}. \tag{4.91}$$

This is the lowest harmonic oscillator energy eigenfunction. However, the energy pertaining to $\psi_0(x)$ is different from zero, since we have taken as zero of the energy scale the minimum value of the potential energy. In fact, from Eq. (4.79) we obtain

$$E_0 = \frac{1}{2}\hbar\omega, \tag{4.92}$$

the zero-point energy (see p. 158). Equation (4.91) tells us that the wave function for the ground state of the harmonic oscillator is a pure Gaussian for which (see Subsec. 2.3.1 and also Prob. 4.20) the uncertainty relation is "saturated," i.e. we have $\Delta\hat{p}_x\Delta\hat{x} = \hbar/2$. This is a peculiar feature of Gaussian wave functions in general, and of the ground state of the harmonic oscillator in particular, and is the reason why such a ground state is called a *coherent state*.[4] Coherent states have exactly the property in some contexts of minimizing the uncertainty product and will be treated in much greater detail in Ch. 13.

Now, we have to determine all the other eigenfunctions for $n > 0$. The simplest way to do it is by reiterating application of operator $\hat{a}^\dagger$ (following the second Eq. (4.85)). We start from $\psi_0(x)$ and obtain $\psi_1(x)$ as

$$\psi_1(x) = \hat{a}^\dagger\psi_0(x). \tag{4.93}$$

In general, we may find $\psi_n(x)$ if we know $\psi_{n-1}(x)$ thanks to the relation

$$\psi_n(x) = \frac{1}{\sqrt{n}}\hat{a}^\dagger\psi_{n-1}(x) = \sqrt{\frac{m}{2\hbar n\omega}}\left(\omega x - \frac{\hbar}{m}\frac{\partial}{\partial x}\right)\psi_{n-1}(x), \tag{4.94}$$

where we have made use of Eq. (4.73b) and of the second Eq. (4.85). It is convenient to introduce the new variable

$$\xi = \sqrt{\frac{m\omega}{\hbar}}x, \tag{4.95}$$

[4] The origin of this name lies in the fact that the harmonic oscillator dynamics (see next subsection) preserves the shape and the widths of the coherent-state wave function in position and in momentum representations. In other words, an initial coherent state remains coherent at any subsequent time. For an explicit evidence of this statement see Prob. 13.23.

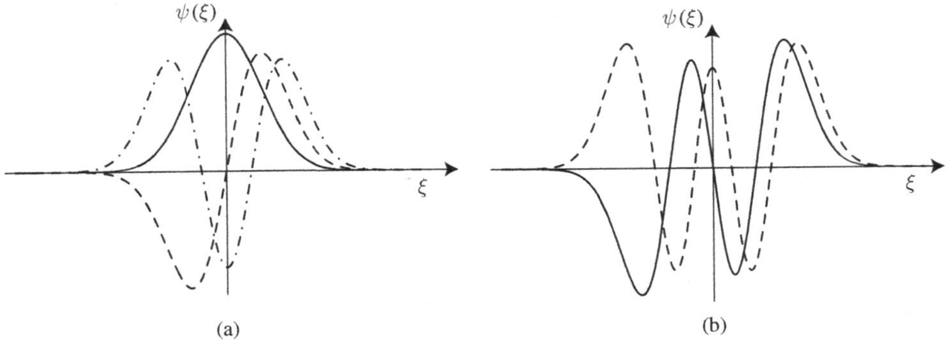

Figure 4.12 Eigenfunctions for the one-dimensional harmonic oscillator. (a) Eigenfunctions for $n = 0$ (solid line), $n = 1$ (dashed line), and $n = 2$ (dot–dashed line). (b) Eigenfunctions for $n = 3$ (solid line) and $n = 4$ (dashed line).

so that we have

$$\psi_n(\xi) = \frac{1}{\sqrt{2n}}\left(\xi - \frac{\partial}{\partial \xi}\right)\psi_{n-1}(\xi) = -\frac{1}{\sqrt{2n}} e^{\frac{\xi^2}{2}} \frac{\partial}{\partial \xi}\left[e^{-\frac{\xi^2}{2}}\psi_{n-1}(\xi)\right]. \tag{4.96}$$

It is easy to verify that the solution of Eq. (4.96) is given by (see Fig. 4.12)

$$\psi_n(\xi) = \frac{\pi^{-\frac{1}{4}} e^{-\frac{\xi^2}{2}}}{2^{\frac{n}{2}}\sqrt{n!}} \mathrm{H}_n(\xi), \tag{4.97}$$

where $\mathcal{N}$ is the normalization constant (see again Prob. 4.14) and H_n is the n-th Hermite polynomial (see Prob. 4.16), defined by

$$\mathrm{H}_n(\zeta) = (-1)^n e^{\zeta^2} \frac{d^n}{d\zeta^n} e^{-\zeta^2}. \tag{4.98}$$

As $\psi_n(\xi)$ are eigenfunctions of an Hermitian operator, they also satisfy the orthonormality and completeness conditions,[5] i.e.

$$\int_{-\infty}^{+\infty} d\xi\, \psi_n^*(\xi)\psi_k(\xi) = \delta_{nk}, \tag{4.99a}$$

$$\sum_{n=0}^{\infty} \psi_n^*(\xi)\psi_n(\xi') = \delta(\xi - \xi'). \tag{4.99b}$$

It is most interesting to note that the $\psi_n(\xi)$, as the Hermite polynomials (see Prob. 4.16), are either even (when n is even) or odd (when n is odd). This happens because the Hamiltonian is invariant under parity transformation $x \mapsto -x$. In fact, given Eq. (4.48), we have

$$\hat{H}\psi_n(x) = E_n\psi_n(x) \quad \text{and} \quad \hat{H}\psi_n(-x) = E_n\psi_n(-x). \tag{4.100}$$

[5] See [*Gradstein/Ryshik* 1981] [*Abramowitz/Stegun* 1964].

Box 4.3 **Example of harmonic oscillator's dynamics**

To settle the ideas, let us consider a specific example: a one-dimensional harmonic oscillator of mass m and angular frequency ω can be found – at time $t_0 = 0$ – to have the energy $E_1 = (3/2)\hbar\omega$ with probability $\wp_1 = 3/4$ or the energy $E_2 = (5/2)\hbar\omega$ with probability $\wp_2 = 1/4$. We also know that, initially, the mean value of the position is zero while the mean value of the momentum is positive. We will find the state vector $|\psi(t)\rangle$ from which all "future" information about the system can be extracted. We then have

$$|\psi(0)\rangle = c_1^{(0)}|1\rangle + c_2^{(0)}|2\rangle, \tag{4.101}$$

where $|c_1^{(0)}|^2 = \wp_1$ and $|c_2^{(0)}|^2 = \wp_2$. We may write $c_1^{(0)} = \varrho_1 e^{\imath\theta_1}$ and $c_2^{(0)} = \varrho_2 e^{\imath\theta_2}$, where $\varrho_1 = \sqrt{3}/2$ and $\varrho_2 = 1/2$. Let us now compute the mean value of $\hat{x}$ in the state $|\psi(0)\rangle$:

$$\begin{aligned}\langle\hat{x}\rangle_0 &= \langle\psi(0)|\hat{x}|\psi(0)\rangle \\ &= \varrho_1^2 x_{11} + \varrho_1\varrho_2 e^{\imath(\theta_2-\theta_1)}x_{12} + \varrho_1\varrho_2 e^{\imath(\theta_1-\theta_2)}x_{21} + \varrho_2^2 x_{22}.\end{aligned} \tag{4.102}$$

Recalling Eq. (4.70) and $x_{11} = x_{22} = 0$, we obtain

$$\begin{aligned}\langle\hat{x}\rangle_0 &= \varrho_1\varrho_2 e^{\imath(\theta_2-\theta_1)}\sqrt{\frac{\hbar}{m\omega}} + \varrho_1\varrho_2 e^{\imath(\theta_1-\theta_2)}\sqrt{\frac{\hbar}{m\omega}} \\ &= 2\varrho_1\varrho_2\sqrt{\frac{\hbar}{m\omega}}\cos(\theta_2-\theta_1).\end{aligned} \tag{4.103}$$

Since $\langle\hat{x}\rangle_0 = 0$, Eq. (4.103) implies that $\cos(\theta_2-\theta_1) = 0$. We compute now the mean value of $\hat{p}_x$ in $|\psi(0)\rangle$:

$$\begin{aligned}\langle\hat{p}_x\rangle_0 &= \langle\psi(0)|\hat{p}_x|\psi(0)\rangle \\ &= m\varrho_1^2\dot{x}_{11} + m\varrho_1\varrho_2 e^{\imath(\theta_2-\theta_1)}\dot{x}_{12} + m\varrho_1\varrho_2 e^{\imath(\theta_1-\theta_2)}\dot{x}_{21} + m\varrho_2^2\dot{x}_{22}.\end{aligned} \tag{4.104}$$

Recalling again Eq. (4.70) and using $\dot{x}_{n,n+1} = -\imath\omega x_{n,n+1}$ (see Eq. (4.53)), we obtain

$$\begin{aligned}\langle\hat{p}_x\rangle_0 &= -\imath\omega m\varrho_1\varrho_2 x_{12}\left(e^{\imath(\theta_2-\theta_1)} - e^{\imath(\theta_1-\theta_2)}\right) \\ &= 2\varrho_1\varrho_2\sqrt{\hbar\omega m}\sin(\theta_2-\theta_1).\end{aligned} \tag{4.105}$$

Since $\langle\hat{p}_x\rangle_0 > 0$, Eq. (4.105) implies that $\sin(\theta_2-\theta_1) > 0$. The only way to combine $\cos(\theta_2-\theta_1) = 0$ and $\sin(\theta_2-\theta_1) > 0$ is that $\theta_2 - \theta_1 = \pi/2$ or $\theta_2 = \theta_1 + \pi/2$. Collecting all these results, the initial state vector is completely defined by

$$|\psi(0)\rangle = \frac{e^{\imath\theta_1}}{2}\left(\sqrt{3}|1\rangle + \imath|2\rangle\right), \tag{4.106}$$

and therefore

$$|\psi(t)\rangle = \frac{e^{\imath\theta_1}}{2}\left(\sqrt{3}e^{-\frac{3}{2}\imath\omega t}|1\rangle + \imath e^{-\frac{5}{2}\imath\omega t}|2\rangle\right). \tag{4.107}$$

It should be noted that the phase factor $e^{\imath\theta_1}$ in Eqs. (4.106)–(4.107) is a global phase factor and is irrelevant. Equation (4.107) determines completely the state of the system at time t and from it all the relevant physical quantities may be calculated. In particular, we have (see Prob. 4.18)

$$\langle\hat{x}\rangle_t = \frac{\sqrt{3}}{2}\sqrt{\frac{\hbar}{m\omega}}\sin\omega t, \quad \langle\hat{p}_x\rangle_t = \frac{\sqrt{3}}{2}\sqrt{m\hbar\omega}\cos\omega t. \tag{4.108}$$

This means that

$$\psi_n(x) = C\psi_n(-x), \tag{4.109}$$

where C is a constant. Since the $\psi_n(x)$ are normalized, then $|C| = 1$ and, due to the reality of $\psi_n(x)$, $C = \pm 1$ (see property (iv) of Subsec. 3.2.3). Then, the eigenfunctions must be either even or odd. It is also interesting to recognize that the number of zeros of these eigenfunctions grows with n (see again Prob. 4.16). In particular, $\psi_n(x)$ has n zeros.

The same happens with the eigenfunctions for a particle in a box (see Sec. 3.4 and, in particular, Fig. 3.7), where $\psi_n(x)$ has $n-1$ zeros in the interval $(0, a)$. This is a general property of the one-dimensional Schrödinger equation: the n-th excited eigenfunction vanishes n times (see property (ii) of Subsec. 3.2.3).

4.4.3 Dynamics

We have seen in Subsec. 3.1.3 that the solution of the time-independent Schrödinger equation is a necessary and fundamental step towards the determination of the dynamics of a quantum system: given an initial state vector $|\,\psi(0)\rangle$ at $t_0 = 0$, we may easily find the time-evolved state vector $|\,\psi(t)\rangle$. Since in the case of the harmonic oscillator the energy spectrum is discrete, the general procedure consists of expanding $|\,\psi(0)\rangle$ into a (finite or infinite) sum of harmonic-oscillator energy eigenvectors (see Eqs. (3.20), (4.50), and (4.82)), i.e.

$$|\,\psi(0)\rangle = \sum_n c_n^{(0)}\,|\,n\rangle, \tag{4.110}$$

with complex $c_n^{(0)} = \langle n \mid \psi(0)\rangle$. Once that the Schrödinger equation has been solved, this is often a straightforward step. Then, the time-evolved state vector is simply given by (see Eq. (3.21))

$$|\,\psi(t)\rangle = e^{-\frac{\imath}{\hbar}\hat{H}t}\,|\,\psi(0)\rangle = \sum_n c_n^{(t)}\,|\,n\rangle, \tag{4.111}$$

where $c_n^{(t)} = e^{-\frac{\imath}{\hbar}E_n t}c_n^{(0)}$. For a specific example of dynamics see Box 4.3.

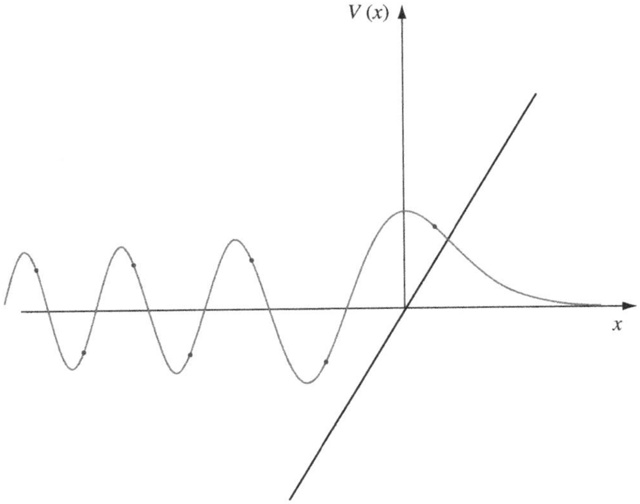

Figure 4.13 Potential energy corresponding to a particle in a uniform field. The slope of the linear potential is determined by the strength of the field (*mg* in the gravitational case and eE in the electric one). The curve shows the wave function proportional to $\mathrm{A}(-\varsigma)$.

4.5 Quantum particles in simple fields

So far we have discussed a number of "simple" quantum problems, i.e. of Hamiltonians whose Schrödinger equation can be solved. In the present section, we shall deal with the dynamics of a quantum particle subjected to an external field. For the sake of simplicity, we shall consider three elementary examples: the uniform field, the triangular well, and the static electromagnetic field. More sophisticated cases will be examined in Subsec. 6.2.2 and in Part III.

4.5.1 Uniform field

Let us consider the motion of a one-dimensional particle subjected to a constant force F, as for example a mass m in a constant gravitational field ($F = -mg$, where g is the gravity acceleration), or a charge e in a uniform electric field ($F = -\mathrm{eE}$, E being the electric field). The potential energy (see Fig. 4.13) will then be given by

$$V(x) = -Fx + C, \tag{4.112}$$

where the constant C may be taken equal to zero by imposing the condition $V(0) = 0$. From the examination in Sec. 3.2, it is evident that the motion is infinite for $x \to -\infty$ (and finite for $x \to +\infty$), and the spectrum is continuous and covers all energy eigenvalues from $-\infty$ to $+\infty$. The Schrödinger equation reads

$$\frac{\hbar^2}{2m}\frac{\partial^2}{\partial x^2}\psi(x) + (E + xF)\,\psi(x) = 0. \tag{4.113}$$

It is useful to introduce the change of variable[6]

$$\zeta(x) = \left(x + \frac{E}{F}\right)\left(\frac{2m|F|}{\hbar^2}\right)^{\frac{1}{3}}, \tag{4.114}$$

in terms of which the Schrödinger equation simply reads as

$$\psi''(\zeta) + \mathrm{sgn}(F)\zeta\psi(\zeta) = 0, \tag{4.115}$$

where the dependence on energy is disappeared. This equation can be solved by using the Laplace method, and the solution, for an arbitrary energy value and apart from a normalization constant, is given by

$$\psi(\zeta) \propto \mathrm{A}(-\zeta), \tag{4.116}$$

where $\mathrm{A}(\zeta)$ is the Airy function[7]

$$\mathrm{A}(\zeta) = \frac{1}{\pi}\int_0^{+\infty} du \cos\left(u\zeta + \frac{u^3}{3}\right). \tag{4.117}$$

As expected from the discussion above, and as it is shown in Fig. 4.13, the $\psi(\zeta)$ function oscillates for (large) negative values of x and exponentially tends to zero for $x \to +\infty$.

The normalization constant may be obtained by imposing on the eigenfunctions $\psi(\zeta)$ the normalization condition (see Subsec. 2.2.2)

$$\int_{-\infty}^{+\infty} dx \psi[\zeta(x)]\psi\left[\zeta'(x)\right] = \delta(E' - E), \tag{4.118}$$

where

$$\zeta(x) = \left(x + \frac{E'}{F}\right)\left(\frac{2m|F|}{\hbar^2}\right)^{\frac{1}{3}}. \tag{4.119}$$

4.5.2 Triangular well

An interesting variation of the uniform-field case is the so-called *triangular-well* potential. Here, besides the constant force for $x > 0$ that pushes toward negative x's, we have an infinite wall at $x = 0$ (see Fig. 4.14). Its classical counterpart is the simple problem of a non dissipative ball of mass m bouncing – under the effect of a constant gravitational field **g** – on a completely elastic and infinitely heavy wall. Alternatively, it may be viewed as a particle of charge e in a constant electric field **E** with an infinite barrier at $x = 0$. In the quantum case, the motion is bounded and the spectrum is discrete for an arbitrary value of the energy. As we have learnt in Sec. 3.4, the wave function will be identically zero for $x \leq 0$. For $x > 0$, the Schrödinger equation reads

[6] See [*Landau/Lifshitz* 1976b, Sec. 24].

[7] In this case, given the extension of the derivation, which is out of the scope of this book, we limit our exposition to the solution and for its derivation we refer the reader to [*Landau/Lifshitz* 1976b]. Note, in particular, that the Airy function may be expressed in terms of a Bessel function of order 1/3. See also [*Gradstein/Ryshik* 1981].

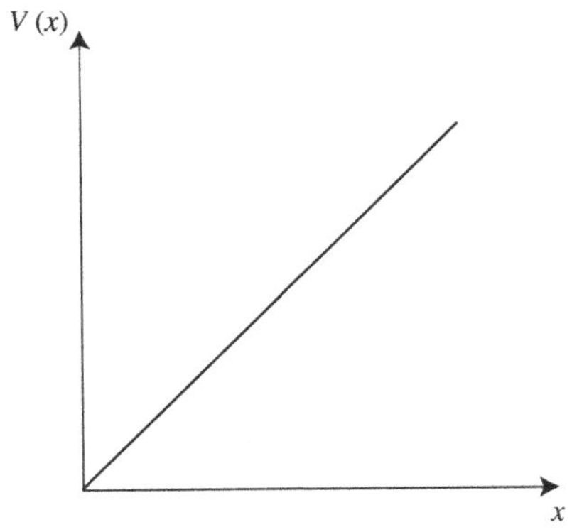

Figure 4.14 Schematic drawing of a triangular well constituted by a potential *V(x)* and an infinite wall at $x = 0$.

$$-\left(\frac{\hbar^2}{2m}\frac{\partial^2}{\partial x^2} + xF\right)\psi(x) = E_n\psi(x), \qquad (4.120)$$

where F is negative and, in the two examples we have discussed, is equal to $-mg$ or $-\mathrm{eE}$. As in the previous subsection, the change of variable

$$\zeta = \left(x + \frac{E_n}{F}\right)\left(\frac{2m|F|}{\hbar^2}\right)^{\frac{1}{3}} \qquad (4.121)$$

leads to

$$\psi''(\zeta) - \zeta\psi(\zeta) = 0, \qquad (4.122)$$

whose solution

$$\psi(\zeta) = \mathcal{N}\mathrm{A}(\zeta) \qquad (4.123)$$

is again given by the Airy function, and $\mathcal{N}$ is a normalization constant. In this case, however, since $\psi(x)$ needs to vanish for $x = 0$, the discrete eigenvalues must satisfy the condition

$$E_n = -\left(\frac{F^2\hbar^2}{2m}\right)^{\frac{1}{3}}\mathrm{a}_n, \qquad (4.124)$$

where the a_n are the zeroes of the Airy function, approximately given by[8]

$$\mathrm{a}_n = -\left[\frac{3\pi}{2}\left(n - \frac{1}{4}\right)\right]^{\frac{2}{3}}, \quad n = 1, 2, 3, \ldots. \qquad (4.125)$$

Accordingly, the energy eigenvalues may be written as

$$E_n \simeq \left(\frac{F^2\hbar^2}{2m}\right)^{\frac{1}{3}}\left[\frac{3\pi}{2}\left(n - \frac{1}{4}\right)\right]^{\frac{2}{3}}, \quad n = 1, 2, 3, \ldots. \qquad (4.126)$$

[8] See [*Gradstein/Ryshik* 1981].

Finally, it is interesting to note that a similar model, where the infinite wall is assumed to oscillate around $x = 0$, has been introduced by Enrico Fermi in 1949 in order to explain the acceleration of cosmic rays with the presence of varying magnetic fields, and is known as the *Fermi accelerator* model.[9]

4.5.3 Static electromagnetic field

In this subsection we shall derive the Hamiltonian which describes the interaction of a particle with charge e with a static electromagnetic field. In Subsec. 6.3.4 we shall study the motion of a quantum particle subject to a homogenous magnetic field. A more complete account of the dynamics of the electromagnetic field and its interaction with atoms will be given in Chs. 11 (Sec. 11.3) and 13 (see in particular Secs. 13.1 and 13.7).

In the non-relativistic limit, the classical equations of motion for a charged particle in an electromagnetic field are given in terms of the Lorentz force[10]

$$\frac{d\mathbf{p}}{dt} = \mathrm{e}\left(\mathbf{E} + \frac{\mathbf{p}}{m} \times \mathbf{B}\right), \tag{4.127}$$

where

$$\mathbf{E} = -\nabla \mathrm{U} - \frac{\partial}{\partial t}\mathbf{A} \quad \text{and} \quad \mathbf{B} = \nabla \times \mathbf{A} \tag{4.128}$$

are the electric and magnetic fields, respectively, U is the scalar potential, and $\mathbf{A}$ is the vector potential.[11] It is interesting to notice that, whereas in the classical case the vector and scalar potentials have no direct physical significance and represent instead mathematical tools, in quantum mechanics they manifest themselves in typical physical effects (see also Sec. 13.8).

Equation (4.127) can be derived from the classical Lagrangian

$$\mathrm{L} = \frac{1}{2}m\dot{r}^2 + \mathrm{e}\,(\dot{\mathbf{r}} \cdot \mathbf{A} - \mathrm{U}). \tag{4.129}$$

To this purpose, we can make use of the Lagrange equations (1.15) (see Prob. 4.22).

Let us now derive the classical expression for the Hamiltonian. In order to obtain this result we can use the formula (see Eq. (1.6))

$$H = \sum_j \dot{q}_j \frac{\partial \mathrm{L}}{\partial \dot{q}_j} - \mathrm{L} = \dot{\mathbf{r}} \cdot \frac{\partial \mathrm{L}}{\partial \dot{\mathbf{r}}} - \mathrm{L}, \tag{4.130}$$

which yields

$$\begin{aligned} H &= \frac{\mathbf{p}}{m} \cdot (\mathbf{p} + \mathrm{e}\mathbf{A}) - \frac{\mathbf{p}^2}{2m} - \frac{\mathrm{e}}{m}\mathbf{p} \cdot \mathbf{A} + \mathrm{eU} \\ &= \frac{1}{2m}\mathbf{p}^2 + \mathrm{eU} \\ &= \frac{1}{2m}(\mathbf{P} - \mathrm{e}\mathbf{A})^2 + \mathrm{eU}, \end{aligned} \tag{4.131}$$

[9] See [Fermi 1949] [Saif *et al.* 1998].

[10] Here and throughout this book we make use of the SI system of units.

[11] The central issue of the Gauge transformations will be discussed in Ch. 13.

where

$$\mathbf{P} = \frac{\partial L}{\partial \dot{\mathbf{r}}} = \mathbf{p} + e\mathbf{A} \tag{4.132}$$

is the generalized momentum conjugate to **r**. As usual, in order to obtain the quantum expression for the Hamiltonian, we impose the canonical commutation relation $[\hat{r}_j, \hat{P}_k] = \imath\hbar\delta_{jk}$ (see Eq. (2.174)) and replace the generalized momentum $\hat{\mathbf{P}}$ by the operator $-\imath\hbar\nabla$, so that

$$\hat{H} = \frac{1}{2m}(-\imath\hbar\nabla - e\mathbf{A})^2 + eU. \tag{4.133}$$

Then, the Schrödinger equation for a charged particle in a static electromagnetic field may be written as

$$\imath\hbar\frac{\partial}{\partial t}\psi(\mathbf{r},t) = \frac{1}{2m}\left[-\imath\hbar\nabla - e\mathbf{A}(\mathbf{r},t)\right]^2 \psi(\mathbf{r},t) + eU(\mathbf{r},t)\psi(\mathbf{r},t), \tag{4.134}$$

which can also be reformulated as

$$\begin{aligned}\imath\hbar\frac{\partial}{\partial t}\psi(\mathbf{r},t) &= \left[\frac{\hat{\mathbf{P}}^2}{2m} - \frac{e}{2m}\left(\hat{\mathbf{P}}\cdot\mathbf{A} + \mathbf{A}\cdot\hat{\mathbf{P}}\right) + \frac{e^2}{2m}\mathbf{A}^2 + eU\right]\psi(\mathbf{r},t) \\ &= \left[-\frac{\hbar^2}{2m}\Delta + \frac{e\imath\hbar}{2m}\nabla\cdot\mathbf{A} + \frac{e\imath\hbar}{m}\mathbf{A}\cdot\nabla + \frac{e^2}{2m}\mathbf{A}^2 + eU\right]\psi(\mathbf{r},t).\end{aligned} \tag{4.135}$$

In the last expression we have taken into account the fact that $\hat{\mathbf{P}}$ and **A** do not commute, and their (scalar) commutator is given by (see Prob. 2.26)

$$\left[\hat{\mathbf{P}}, \mathbf{A}\right] = \hat{\mathbf{P}}\cdot\mathbf{A} - \mathbf{A}\cdot\hat{\mathbf{P}} = -\imath\hbar\nabla\cdot\mathbf{A}. \tag{4.136}$$

Summary

In this chapter we have applied the methods previously presented to some simple examples of quantum dynamics:

- We have discussed the behavior of a one-dimensional particle in a rectangular *potential well with finite steps*. This has introduced us to a novel quantum effect: the presence of the particle in regions that are classically forbidden.
- We have examined the quantum motion in the presence of a *potential barrier*, where we have introduced transmission and reflection probabilities and a quantum continuity

equation. Even when the potential barrier is higher than the energy of the incoming particle, this is able to pass the classical turning point.

- The examples above have led us to consider the genuinely quantum phenomenon of *tunneling*: a quantum particle can penetrate (i.e. "tunnel") into regions that are classically forbidden. This effect may be seen as a consequence of the uncertainty relation: when the particle is very localized, the momentum and energy uncertainty (spread) allows it to overcome a potential barrier that is higher than its kinetic energy.
- We have analyzed the (one-dimensional) *harmonic oscillator* that may be considered as a guidance tool for the solution of more complex problems. In particular, we have introduced two different methods of solving this problem: the Heisenberg and the algebraic methods. We have found that the spectrum of the harmonic oscillator is discrete and that energy levels are equally spaced. With the help of a specific example, we have also presented the dynamics of a quantum particle subject to a harmonic-oscillator potential.
- Finally, we have introduced some examples of a particle in simple fields. In particular, we have introduced the *uniform field*, the *triangular-well potential*, and the *static electromagnetic field*.

Problems

4.1 Find the energy levels and eigenfunctions for a particle in a rectangular potential, i.e. for the three-dimensional motion in a potential $V(\mathbf{r})$ which is zero for $0 < x < a$, $0 < y < b$, $0 < z < c$ and infinity outside this region.
(*Hint*: In the Schrödinger equation it is possible to write $E = E_x + E_y + E_z$ and, correspondingly, $\psi(\mathbf{r}) = \psi(x)\psi(y)\psi(z)$. In other words, the three-dimensional problem is separable into three one-dimensional problems.)

4.2 Find the energy eigenstates and eigenvalues of the discrete spectrum for the problem of a one-dimensional particle in an asymmetric potential well (see Fig 4.1(b)) and check that the solution reduces to that obtained in Sec. 4.1 when $V_1 = V_2 = V_0$.
(*Hint*: Proceed as in Sec. 4.1 and define $k = \sqrt{2mE}/\hbar$, $k_1 = \sqrt{2m(V_1 - E)}/\hbar$, and $k_2 = \sqrt{2m(V_2 - E)}/\hbar$.)

4.3 Consider a one-dimensional particle of energy E which moves from left to right and encounters the potential barrier depicted in Fig. 4.15 with $V_0 < E$.

(a) Find the coefficients $\mathcal{N}_+$ and $\mathcal{N}_-$ of the wave function

$$\psi(x > 0) = \psi_>(x) = \mathcal{N}_+ e^{\imath k_1 x},$$
$$\psi(x < 0) = \psi_<(x) = e^{\imath k_2 x} + \mathcal{N}_- e^{-\imath k_2 x}.$$

(b) Given that the ratio $E/V_0 = 16/15$, what is the value of the ratio $\mathrm{T}^2/\mathrm{R}^2$ between the transmission and the reflection probabilities?

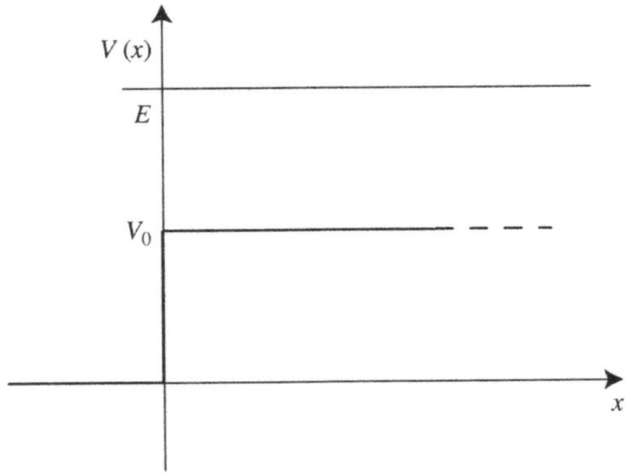

Figure 4.15 A quantum particle with energy E encounters a potential step of height $V_0 < E$.

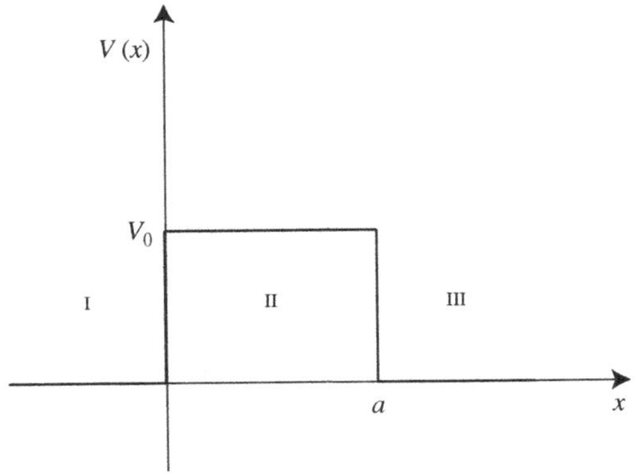

Figure 4.16 The rectangular potential barrier with finite width a. The potential profile determines the regions I, II, and III.

4.4 Consider the rectangular potential barrier with finite width a as in Fig. 4.16. Calculate the transmission coefficient of a particle moving from left to right.

4.5 Obtain result (4.53) in the Schrödinger picture.
(*Hint*: Observe that an eigenstate $|n\rangle$ of the energy evolves as $|n(t)\rangle = e^{-\imath \frac{E_n t}{\hbar}} |n\rangle$.)

4.6 Prove Eq. (4.70).

4.7 Prove that the mean value of the position operator in any energy eigenstate of the harmonic oscillator ($\hat{H} = \frac{\hat{p}_x^2}{2m} + \frac{1}{2} m\omega^2 \hat{x}^2$) is equal to zero.

4.8 Prove that for a general one-dimensional potential and the energy eigenstate $|n\rangle$ of the discrete spectrum, one has $\langle n | \hat{p}_x | n \rangle = 0$.

4.9 Prove the commutation relation (4.74).

4.10 Compute the product operator $\hat{a}\hat{a}^\dagger$ (see Eqs. (4.73)) in terms of $\hat{x}$ and $\hat{p}_x$. Then, write the harmonic-oscillator Hamiltonian in terms of $\hat{a}\hat{a}^\dagger$ and in terms of both $\hat{a}\hat{a}^\dagger$ and $\hat{a}^\dagger\hat{a}$.

4.11 Prove the commutation relations $[\hat{a}, \hat{N}] = \hat{a}$ and $[\hat{a}^\dagger, \hat{N}] = -\hat{a}^\dagger$.

4.12 Prove that $[\hat{a}, (\hat{a}^\dagger)^2] = 2\hat{a}^\dagger$ and $[\hat{a}^2, \hat{a}^\dagger] = 2\hat{a}$ and that we have, in general,

$$[\hat{a}, (\hat{a}^\dagger)^n] = n(\hat{a}^\dagger)^{n-1} \quad \text{and} \quad [\hat{a}^n, \hat{a}^\dagger] = n\hat{a}^{n-1}. \tag{4.137}$$

4.13 Prove the relation $| n \rangle = \frac{\left(\hat{a}^\dagger\right)^n}{\sqrt{n!}} | 0 \rangle$, where $|n\rangle$ is the energy eigenstate of the harmonic oscillator with energy eigenvalue $E_n = (n + 1/2)\hbar\omega$.

4.14 Verify that the ground-state wave function of the harmonic oscillator is normalized.

4.15 Check that the ground-state wave function of the harmonic oscillator $\psi_0(x)$ (Eq. (4.91)) satisfies the time-independent Schrödinger equation for the harmonic oscillator, i.e.

$$-\frac{\hbar^2}{2m}\frac{\partial^2}{\partial x^2}\psi(x) + \left(\frac{1}{2}m\omega^2 x^2 - E\right)\psi(x) = 0. \tag{4.138}$$

4.16 The Hermite polynomials are defined by the relation

$$\mathrm{H}_n(\zeta) = (-1)^n e^{\zeta^2} \frac{d^n}{d\zeta^n} e^{-\zeta^2}$$

for any integer $n \geq 0$.

(i) Write the first six Hermite polynomials.

(ii) Verify that $\mathrm{H}_n(\zeta)$ is a polynomial of degree n in ζ which has n zeros and that is even (odd) when n is even (odd).

(iii) Verify that Hermite polynomials satisfy the recursion relations

$$\frac{d}{d\zeta}\mathrm{H}_n(\zeta) = 2n\mathrm{H}_{n-1}(\zeta),$$
$$2\zeta\mathrm{H}_n(\zeta) = \mathrm{H}_{n+1}(\zeta) + 2n\mathrm{H}_{n-1}(\zeta),$$
$$\left(2\zeta - \frac{d}{d\zeta}\right)\mathrm{H}_n(\zeta) = \mathrm{H}_{n+1}(\zeta),$$

where the third equation is a consequence of the first two. On the other hand, the third equation and one of the first two imply together the other one.

(iv) Verify that the Hermite polynomials satisfy the differential equation

$$\left(\frac{d^2}{d\zeta^2} - 2\zeta\frac{d}{d\zeta} + 2n\right)\mathrm{H}_n(\zeta) = 0.$$

4.17 Solve in a direct way the Schrödinger equation (4.49) for the one-dimensional harmonic oscillator.
(*Hint*: First, change variable to $\xi = x\sqrt{\frac{m\omega}{\hbar}}$. Second, make the ansatz

$$\psi(\xi) = e^{-\frac{\xi^2}{2}}\varphi(\xi).$$

Finally, use property (iv) of Prob. 4.16 to solve the resulting differential equation.)

4.18 Compute $\langle \hat{x} \rangle_t$ and $\langle \hat{p}_x \rangle_t$ in Eqs. (4.108).

4.19 Make a comparison between the mean values of $\hat{x}^2$ and $\hat{p}_x^2$ for a harmonic oscillator in the classical and in the quantum case. Comment the results in the light of Ehrenfest theorem (see Sec. 3.7).

4.20 Use the results of Prob. 4.19 to prove that the uncertainty relation between position and momentum for a harmonic oscillator in the state $|n\rangle$ may be written as

$$\Delta x \Delta p_x = \left(n + \frac{1}{2}\right)\hbar.$$

4.21 Show that any initial quantum state evolving under the harmonic oscillator Hamiltonian with frequency ω acquires, after a period $\tau = 2\pi/\omega$, a phase equal to $-\pi$.

4.22 Derive Eq. (4.127) from the Lagrangian (4.129) by making use of the Lagrange equations (1.15).
(*Hint*: Notice that the total differential

$$\frac{d\mathbf{A}}{dt} = \frac{\partial \mathbf{A}}{\partial t} + (\mathbf{v} \cdot \boldsymbol{\nabla})\, \boldsymbol{A} \tag{4.139}$$

has two parts, the variation $(\partial \mathbf{A}/\partial t)dt$ of the vector potential as a function of time at a given point, and the variation obtained by moving from a location in space to another one at an infinitesimal distance $d\mathbf{r}$, given by $(\mathbf{v} \cdot \boldsymbol{\nabla})\, \boldsymbol{A}$.)

5 Density matrix

The density matrix is a very useful tool that enlarges the concept of state from a vector in a Hilbert space to a true operator from which all properties of a quantum system can be extracted. This generalization is particularly relevant because it allows compound systems to be treated in a natural way, especially when dealing with specific properties of one subsystem. For this reason, as we shall see in Secs. 9.4 and 14.2, it plays a central role in the dynamics of open systems (non-unitary evolutions).

In this chapter we shall first deal with the formalism of the density matrix (Sec. 5.1). Then, we shall apply this formalism to calculations for expectation values and outcomes of a measurement (Sec. 5.2) and present the time evolution of the density matrix for different pictures (Sec. 5.3). In Sec. 5.4 we shall discuss the relationship between classical and quantum statistics. In Sec. 5.5 the fundamental concept of entanglement will be examined. In addition, we shall learn how to trace out a subsystem and briefly consider the Schmidt decomposition. Finally, in Sec. 5.6, we shall give a useful representation of pure and mixed states.

5.1 Basic formalism

From Sec. 1.3 and Subsec. 2.3.3 we know that the state vector $|\psi\rangle$ contains the maximal information about a quantum system. Associated with any state vector $|\psi\rangle$, it is always possible to define the projector $\hat{P}_\psi = |\psi\rangle\langle\psi|$ (see Eq. (1.28)). We now introduce the *density matrix* or *density operator* of the system, denoted by the symbol $\hat{\rho}$. This operator is a generalization of the projector $\hat{P}_\psi$. Here we limit ourselves to the discrete case. Let us first discuss the case where the system admit a wave-function description, i.e. where $\hat{\rho} = \hat{P}_\psi$. Then, for a normalized $|\psi\rangle$, $\hat{\rho}$ possesses the important properties (see also Eq. (1.41b) and Prob. 5.1)

$$\hat{\rho}^2 = \hat{\rho}, \tag{5.1a}$$

$$\mathrm{Tr}(\hat{\rho}) = 1, \tag{5.1b}$$

where we recall that $\mathrm{Tr}(\hat{O}) = \sum_j \langle b_j | \hat{O} | b_j \rangle$ is the trace of the operator $\hat{O}$ and $\{b_j\}$ an orthonormal basis (see Eqs. (2.56)–(2.57)) on the Hilbert space of the system. We remark that, if a density operator $\hat{\rho}$ is not normalized, i.e. $\mathrm{Tr}(\hat{\rho}) \neq 1$, but finite, we can always normalize it as follows (see also Eq. (2.111)):

$$\hat{\rho}' = \frac{\hat{\rho}}{\mathrm{Tr}(\hat{\rho})}. \tag{5.2}$$

Obviously, there are also cases in which the density matrix is not normalizable (see Subsec. 2.2.2 and Prob. 5.2). We also note that, if the density operator can be expressed as a single projector, we shall have $\mathrm{Tr}(\hat{\rho}^2) = 1$.

Let us consider the example of the polarization state

$$|\psi\rangle = c_h |h\rangle + c_v |v\rangle \tag{5.3}$$

of a photon, where (as in Sec. 1.3, Box 2.2, and Subsec. 2.1.3) $|h\rangle$ and $|v\rangle$ are states of horizontal and vertical polarization, and $|c_h|^2 + |c_v|^2 = 1$. The density matrix corresponding to such a state can be written as

$$\begin{aligned}\hat{\rho} &= |\psi\rangle\langle\psi| \\ &= |c_h|^2 |h\rangle\langle h| + |c_v|^2 |v\rangle\langle v| + c_h c_v^* |h\rangle\langle v| + c_h^* c_v |v\rangle\langle h|.\end{aligned} \tag{5.4}$$

Equation (5.4) may be easily generalized to a form that is valid for any density matrix. If $\{|j\rangle\}$ is an orthonormal basis on the Hilbert space of the system, we may exploit the resolution of identity $\sum_j |j\rangle\langle j| = \hat{I}$ ($\sum_k |k\rangle\langle k| = \hat{I}$) to obtain (see also Eq. (2.17))

$$\begin{aligned}\hat{\rho} &= \sum_j \sum_k |j\rangle\langle j|\hat{\rho}|k\rangle\langle k| \\ &= \sum_{j,k} \rho_{jk} |j\rangle\langle k|,\end{aligned} \tag{5.5}$$

where $\langle j | k\rangle = \delta_{jk}$, and $\rho_{jk} = \langle j |\hat{\rho}| k\rangle$ are the matrix elements of $\hat{\rho}$.

In matrix form, where the eigenstates of the two-dimensional polarization Hilbert space are defined as (see Eqs. (2.11))

$$|h\rangle = \begin{pmatrix} 1 \\ 0 \end{pmatrix}, \quad |v\rangle = \begin{pmatrix} 0 \\ 1 \end{pmatrix}, \tag{5.6}$$

Equation (5.4) may be cast in the form

$$\hat{\rho} = \begin{pmatrix} c_h \\ c_v \end{pmatrix} \begin{pmatrix} c_h^* & c_v^* \end{pmatrix} = \begin{bmatrix} |c_h|^2 & c_h c_v^* \\ c_h^* c_v & |c_v|^2 \end{bmatrix}. \tag{5.7}$$

Now, Eqs. (5.4) and (5.7) describe an ensemble of (say, N) photons which are all in the same state $|\psi\rangle$. In other words, any subensemble of the original ensemble will be described by the same density matrix $\hat{\rho}$. For reasons which are evident from Eq. (5.7), the first and second terms in Eq. (5.4) are called the *diagonal* terms, whereas the third and fourth terms in Eq. (5.4) are called the *off-diagonal* terms of the density matrix $\hat{\rho}$. Note that the diagonal elements of $\hat{\rho}$ are the square moduli of the coefficients of the basis vectors $|h\rangle$, $|v\rangle$, and, since we know that we must have $|c_h|^2 + |c_v|^2 = 1$, we see here the reason of property (5.1b): this property expresses the conservation of norm or of probability.

We may also consider the case of a classical statistical ensemble of N photons in which a fraction $N_h/N = |c_h|^2$ of the photons is in the state $|h\rangle$ of horizontal polarization and a fraction $N_v/N = |c_v|^2$ (with $N_h + N_v = 1$) is in the state $|v\rangle$ of vertical polarization. Such an ensemble will be necessarily described by a density operator of the type

$$\hat{\tilde{\rho}} = |c_h|^2 \, |h\rangle \langle h| + |c_v|^2 \, |v\rangle \langle v| = \begin{bmatrix} |c_h|^2 & 0 \\ 0 & |c_v|^2 \end{bmatrix}, \tag{5.8}$$

where the off-diagonal terms are not present. Equation (5.8) may be rewritten as

$$\hat{\tilde{\rho}} = |c_h|^2 \hat{P}_h + |c_v|^2 \hat{P}_v \tag{5.9}$$

where, as usual, $\hat{P}_h = |h\rangle \langle h|$ and $\hat{P}_v = |v\rangle \langle v|$. Density operators of the form (5.8) do not satisfy Eq. (5.1a) if c_h and c_v are both different from zero, even though they still satisfy Eq. (5.1b). On the contrary, we have (see also Prob. 5.3)

$$\mathrm{Tr}(\hat{\tilde{\rho}}^2) < \mathrm{Tr}(\hat{\tilde{\rho}}) = 1. \tag{5.10}$$

This implies that $\hat{\tilde{\rho}}^2 \neq \hat{\tilde{\rho}}$. We may summarize the situation we have presented so far by dividing all possible states (represented by density matrices) into two major classes: *pure states* $\hat{\rho}$, which satisfy Eqs. (5.1), and *mixtures* $\hat{\tilde{\rho}}$, which satisfy Eq. (5.10). It is also clear that the state of a "classical object"[1] will always belong to the second class. For instance, the statistical mixture describing the state of a classical dice before the outcome of the throw is read can be written as $\hat{\tilde{\rho}} = (1/6) \sum_{j=1}^{6} |j\rangle \langle j|$.

In the example above, as long as one measures the polarization in the basis $\{|h\rangle, |v\rangle\}$, it is not possible to detect a difference between $\hat{\rho}$ and $\hat{\tilde{\rho}}$: The probabilities $\wp(h)$ and $\wp(v)$ of detecting horizontal and vertical polarization, respectively, are in both cases equal to $|c_h|^2$ and $|c_v|^2$. However, the differences between $\hat{\rho}$ and $\hat{\tilde{\rho}}$ come about when one decides to measure the polarization in a rotated basis, for example in the basis $\{|\nearrow\rangle, |\nwarrow\rangle\}$, where (see Subsec. 2.1.2 and Eq. (1.79))

$$|\nearrow\rangle = \frac{1}{\sqrt{2}}(|h\rangle + |v\rangle), \tag{5.11a}$$

$$|\nwarrow\rangle = \frac{1}{\sqrt{2}}(|h\rangle - |v\rangle). \tag{5.11b}$$

From Eqs. (5.11) one obtains

$$|h\rangle = \frac{1}{\sqrt{2}}(|\nearrow\rangle + |\nwarrow\rangle), \tag{5.12a}$$

$$|v\rangle = \frac{1}{\sqrt{2}}(|\nearrow\rangle - |\nwarrow\rangle), \tag{5.12b}$$

[1] By a "classical object" we mean here a macroscopic object that follows the laws of classical mechanics.

which, when substituted in Eq. (5.3), yield the state vector in the basis rotated by 45°

$$|\psi\rangle = \frac{c_h + c_v}{\sqrt{2}} |\nearrow\rangle + \frac{c_h - c_v}{\sqrt{2}} |\nwarrow\rangle \tag{5.13}$$

corresponding to a density matrix

$$\hat{\rho}_{\nwarrow\nearrow} = \frac{1}{2}\begin{bmatrix} |c_h + c_v|^2 & (c_h + c_v)(c_h^* - c_v^*) \\ (c_h^* + c_v^*)(c_h - c_v) & |c_h - c_v|^2 \end{bmatrix}. \tag{5.14}$$

On the other hand, the density matrix $\hat{\tilde{\rho}}$, according to the transformation (5.12), changes into

$$\begin{aligned} \hat{\tilde{\rho}}_{\nwarrow\nearrow} &= |c_h|^2 \left[\frac{1}{\sqrt{2}}(|\nearrow\rangle + |\nwarrow\rangle)\right]\left[\frac{1}{\sqrt{2}}(\langle\nearrow| + \langle\nwarrow|)\right] \\ &\quad + |c_v|^2 \left[\frac{1}{\sqrt{2}}(|\nearrow\rangle - |\nwarrow\rangle)\right]\left[\frac{1}{\sqrt{2}}(\langle\nearrow| - \langle\nwarrow|)\right] \\ &= \frac{|c_h|^2 + |c_v|^2}{2} |\nearrow\rangle\langle\nearrow| + \frac{|c_h|^2 + |c_v|^2}{2} |\nwarrow\rangle\langle\nwarrow| \\ &\quad + \frac{|c_h|^2 - |c_v|^2}{2} (|\nearrow\rangle\langle\nwarrow| + |\nwarrow\rangle\langle\nearrow|), \end{aligned} \tag{5.15}$$

so that the probabilities of detecting polarization along the 45° and 135° directions, respectively, are given by

$$\wp'(45°) = \wp'(135°) = \frac{|c_h|^2 + |c_v|^2}{2}. \tag{5.16}$$

These values strongly differ from the values (see Prob. 5.5)

$$\wp(45°) = \frac{|c_h + c_v|^2}{2}, \tag{5.17a}$$

$$\wp(135°) = \frac{|c_h - c_v|^2}{2}, \tag{5.17b}$$

which are obtained from the density matrix (5.14). For this reason, the off-diagonal terms of a density matrix (see e.g. Eq. (5.7)) are often called *quantum coherences* because they are responsible for the interference effects typical of quantum mechanics that are absent in classical dynamics.

5.2 Expectation values and measurement outcomes

The concept of expectation value of an observable (see Subsec. 2.1.3) allows us to give a general definition of the density matrix, that is,

$$\left\langle \hat{O} \right\rangle_{\hat{\rho}} = \mathrm{Tr}\left(\hat{\rho}\hat{O}\right). \tag{5.18}$$

In the case of a pure state, Eq. (5.18) becomes (see Eq. (2.78))

$$\left\langle \hat{O} \right\rangle_{\psi} = \sum_{j} \left\langle o_j \left| \hat{\rho} \hat{O} \right| o_j \right\rangle = \sum_{j} \langle o_j \mid \psi \rangle \langle \psi \mid \hat{O} \mid o_j \rangle. \tag{5.19}$$

However, when the system is in a mixed state, the rhs of Eq. (5.19) loses its meaning and we take Eq. (5.18) as the *definition* of the density operator $\hat{\rho}$.[2] In fact, if the density operator is defined by

$$\hat{\tilde{\rho}} = \sum_{j} w_j \mid \psi_j \rangle \langle \psi_j \mid, \tag{5.20}$$

and if $\{\mid \psi_j \rangle\}$ is an orthonormal basis on the Hilbert space, then Eq. (5.18) becomes

$$\left\langle \hat{O} \right\rangle_{\hat{\tilde{\rho}}} = \mathrm{Tr}\left(\hat{\tilde{\rho}} \hat{O} \right) = \sum_{j} w_j \left\langle \psi_j \left| \hat{O} \right| \psi_j \right\rangle. \tag{5.21}$$

In Subsec. 1.3.2 we have defined – on a rather mathematical level – the scalar product $\langle \varphi \mid \psi \rangle$ of two state vectors $\mid \psi \rangle$ and $\mid \varphi \rangle$. However, we know that scalar products give probability amplitudes (see Sec. 1.4) and we have seen that these probability amplitudes are coefficients of some basis vectors (see Subsecs. 2.1.1–2.1.3 and Sec. 2.2). Then, from a physical point of view, $\langle \varphi \mid \psi \rangle$ tells us how much the two vectors $\mid \psi \rangle$ and $\mid \varphi \rangle$ overlap. When $\mid \psi \rangle$ and $\mid \varphi \rangle$ are orthogonal, $\langle \varphi \mid \psi \rangle = 0$, while, when $\mid \psi \rangle = \mid \varphi \rangle$, $\langle \varphi \mid \psi \rangle = 1$. If $\mid \varphi \rangle$ is an eigenvector of an observable $\hat{O}$ such that

$$\hat{O} \mid \varphi \rangle = o \mid \varphi \rangle, \tag{5.22}$$

then $\mid \langle \varphi \mid \psi \rangle \mid^2$ gives the probability to obtain the outcome o when measuring the observable $\hat{O}$ on the state $\mid \psi \rangle$. In terms of the density matrix, this probability may be rewritten as

$$\mid \langle \varphi \mid \psi \rangle \mid^2 = \langle \varphi \mid \psi \rangle \langle \psi \mid \varphi \rangle = \langle \varphi \mid \hat{\rho} \mid \varphi \rangle, \tag{5.23}$$

or

$$\mid \langle \varphi \mid \psi \rangle \mid^2 = \mathrm{Tr}\left(\hat{\rho} \hat{P}_{\varphi} \right), \tag{5.24}$$

where $\hat{P}_{\varphi} = \mid \varphi \rangle \langle \varphi \mid$ is the projector associated with the state $\mid \varphi \rangle$.

Finally, since any density matrix can be written in the form

$$\hat{\rho} = \sum_{j} w_j \hat{P}_j, \tag{5.25}$$

where the $\hat{P}_j$'s are projectors, it is always possible to find an eigenbasis of $\hat{\rho}$. Since from Eq. (5.23) it is evident that $\left\langle \varphi \left| \hat{\rho} \right| \varphi \right\rangle \geq 0$, the weights w_j must be non-negative, and we

[2] The general validity of formula (5.18) for calculating quantum probabilities for systems of dimensions > 2 was proved, on a rather abstract mathematical level, in [Gleason 1957]. On this point see also Sec. 16.4.

also have $\sum_j w_j = 1$. Since the weights w_k are non-negative, the density operator must be Hermitian and positive.[3]

If $\{|b_k\rangle\}$ is an orthonormal basis of eigenvectors of $\hat{\rho}$, we can write

$$\hat{\rho}\,|b_k\rangle = r_k\,|b_k\rangle. \tag{5.26}$$

We also see here a further evidence that the state (in this case the density operator) of a quantum system cannot be determined through a single measurement: by the measurement of a certain observable, we only obtain one of its eigenvalues and *not* its probability distribution in the initial state.

Let us consider this problem from a slightly different point of view. The density matrix in Eq. (5.26) is either a pure state or a mixture. Since $\{|b_k\rangle\}$ is an eigenbasis of $\hat{\rho}$, if $\hat{\rho}$ is a pure state, r_k may be either 1 or 0, i.e. $r_k = 1$ for a single value $k = j$, when $|b_j\rangle$ is the state vector such that $\hat{\rho} = |b_j\rangle\langle b_j|$, and 0 in all other cases (see Prob. 5.6). However, measuring the projector $\hat{P}_j = |b_j\rangle\langle b_j|$ (as any other projector) does not determine the state, since it corresponds to a test through which we ask the system *if* it is in the state $|b_j\rangle$ (or if at least it has a non-zero overlap with the state $|b_j\rangle$). If $\hat{\rho}$ is a mixture, the measurement of a projector will again give only partial information about the system's state. Stated in different terms, we may ask a quantum system *if* it is in a given state among an orthogonal set (and the system will answer "yes" or "not"), but we cannot ask in *what* state the system is in, which would imply the ability to discriminate between non-orthogonal states. We shall deal with the fundamental problem of state measurement in quantum mechanics in Ch. 15.

5.3 Time evolution and density matrix

We wish now to deal with the time evolution of the density operator. Since for a state $|\psi(t)\rangle$ the density matrix can be written as $\hat{\rho}(t) = |\psi(t)\rangle\,\langle\psi(t)|$, it is also clear that, for a pure state under unitary evolution, we have

$$\dot{\hat{\rho}} = \frac{\hat{H}}{\imath\hbar}\,|\psi(t)\rangle\,\langle\psi(t)| + |\psi(t)\rangle\,\langle\psi(t)|\,\frac{\hat{H}}{-\imath\hbar} = \frac{\imath}{\hbar}\left(\hat{\rho}\hat{H} - \hat{H}\hat{\rho}\right) = \frac{\imath}{\hbar}[\hat{\rho},\hat{H}], \tag{5.27}$$

or

$$\imath\hbar\dot{\hat{\rho}} = [\hat{H},\hat{\rho}], \tag{5.28}$$

[3] This statement is certainly true in principle. In practice, however, this Hermitian operator may be arbitrarily complex (as some combination of other operators) and needs not coincide with any simple, directly accessible, physical quantity.

which is formally similar to Eq. (3.107) but with inverted sign. This expression is known as the *von Neumann equation*. It tells us that pure states evolve into pure states under Schrödinger evolution. As we shall see (in Chs. 9 and 14), Eq. (5.28) needs to be corrected in presence of dissipative processes, when the time evolution is no longer unitary.

It is useful to note that, as for the state vector, the time evolution of the density matrix can be represented both in the Schrödinger and Heisenberg pictures. In fact, since $| \psi(t)\rangle_S = \hat{U}_t \, | \, \psi(0)\rangle_S$, we have

$$\hat{\rho}_S(t) = \hat{U}_t \, | \, \psi\rangle_H \, \langle \psi \, |_H \, \hat{U}_t^\dagger = \hat{U}_t \hat{\rho}_H \hat{U}_t^\dagger, \tag{5.29a}$$

or

$$\hat{\rho}_H = \hat{U}_t^\dagger \hat{\rho}_S \hat{U}_t = \hat{\rho}_S(0). \tag{5.29b}$$

It should be noted, however, that, contrarily to usual observables, the density operator in the Heisenberg picture does not depend on time.

Finally, in the Dirac picture, we have from Eq. (3.112)

$$\hat{\rho}_I(t) = \hat{U}_{H_0,t}^\dagger \hat{\rho}_S(t) \hat{U}_{H_0,t}. \tag{5.30}$$

5.4 Statistical properties of quantum mechanics

In the two previous sections, the introduction of the density operator has allowed us to point out two important statistical features of quantum mechanics:

(i) From the knowledge of the density operator it is possible to compute the mean value of any observable; and
(ii) the density operator itself evolves according to the von Neumann equation.

Here we want to emphasize the remarkable analogy between classical and quantum statistics, i.e. between the classical phase space density ρ (see Sec. 1.1) and the quantum density operator $\hat{\rho}$. In fact, Eq. (5.18) is the quantum counterpart of the classical equation (1.19), and Eq. (5.28) is the quantum analogue of the classical Liouville equation (see Eq. (1.21)). The central point that we can learn from this parallelism is that one may turn the fundamental equations of classical mechanics into those of quantum theory by replacing variables by operators, Poisson brackets by commutators up to the multiplicative factor $\imath\hbar$ according to the rule (3.109), and integrating over the phase space by the trace. This is another formulation of the correspondence principle (see p. 72) and is very useful when trying to generalize the classical statistical mechanics in the quantum context. A complete derivation, which is outside the scope of this book,[4] shows that the thermodynamic functions of a quantum system are derived from the partition function

[4] See [*Huang* 1963, 171–91] [*Kittel* 1958].

$$Z(\beta) = \mathrm{Tr}\left(\hat{\rho}\right) = \mathrm{Tr}\left(e^{-\beta\hat{H}}\right), \tag{5.31}$$

as in classical mechanics, where $\beta = (k_B T)^{-1}$, T the temperature of the system, and k_B is the Boltzmann constant. In particular, for the energy and the entropy we have

$$E = \left\langle \hat{H} \right\rangle = -\frac{\partial}{\partial\beta} \ln Z, \tag{5.32a}$$

$$S = k_B \left(\ln Z - \beta \frac{\partial}{\partial\beta} \ln Z \right), \tag{5.32b}$$

respectively. From the correspondence principle, it is possible to obtain a more general expression for the entropy of a system, which is proportional to the mean value of the operator $\ln \hat{\rho}$, i.e.

$$S_{\mathrm{VN}} = -k_B \mathrm{Tr}\left(\hat{\rho} \ln \hat{\rho}\right), \tag{5.33}$$

which is known as the *von Neumann entropy* and is the quantum analogue of the Boltzmann (see Eq. (1.60)) or Shannon entropy.[5]

Given the above definition of entropy, the equilibrium quantum state can be determined by maximizing the entropy for a given energy expectation value. The result describes the state of a quantum system in thermodynamic equilibrium at temperature T, that is,[6]

$$\hat{\rho} = \mathcal{N} e^{-\frac{\hat{H}}{k_B T}}, \tag{5.34}$$

$\mathcal{N}$ being a normalization constant such that $\mathrm{Tr}(\hat{\rho}) = 1$.

5.5 Compound systems

As in classical mechanics, also quantum systems may be part of a larger closed system. As we shall see in Chs. 9 and 13, this has important consequences in quantum dynamics that makes it profoundly different from classical dynamics. In classical physics, the knowledge of the properties of composite systems implies the determination of the properties of the subsystems. In quantum mechanics, however, there may be cases in which even though the wave function of the composite system is known, the subsystems do not admit a description in terms of an independent wave function, such that the compound state be a mere product of the states of the subsystems. In such cases, where the total wave function is not factorizable, the concept of density matrix is crucial for a proper description of these subsystems.

5.5.1 Entanglement

When one deals with systems that consist of two or more subsystems, quantum mechanics has some extremely puzzling features if observed from a classical point of view. Let us

[5] This will be the subject of Sec. 17.1.
[6] See [*Gardiner* 1991, 36–37].

illustrate this with the help of an example. We consider the polarization state of a pair of photons. The Hilbert space $\mathcal{H}$ of the total system will be the direct product $\mathcal{H}_1 \otimes \mathcal{H}_2$ of the two-dimensional Hilbert spaces of the two particles, and $\{| h\rangle_1, | v\rangle_1\}$ and $\{| h\rangle_2, | v\rangle_2\}$ the two basis for photon 1 and photon 2, respectively. A complete basis of the direct product Hilbert space $\mathcal{H}_1 \otimes \mathcal{H}_2$ for the total system of the two photons will then be given by the direct product basis

$$\{| h\rangle_1 \otimes | h\rangle_2, | h\rangle_1 \otimes | v\rangle_2, | v\rangle_1 \otimes | h\rangle_2, | v\rangle_1 \otimes | v\rangle_2\}, \tag{5.35}$$

which is obtained by performing the direct product of each basis vector of $\mathcal{H}_1$ and each basis vector of $\mathcal{H}_2$. For instance, the polarization state of the two photons could be described by $| h\rangle_1 \otimes | h\rangle_2$, which would mean that both photons are horizontally polarized, as well as by any other of the four combinations above. However, as we know, a state vector can consist of an arbitrary superposition of the basis vectors. In particular, we may think of formally writing the state vector of the two photons as a symmetric superposition of the second and third element of the basis (5.35), i.e.

$$| \Psi\rangle_{12} = \frac{1}{\sqrt{2}}(| h\rangle_1 \otimes | v\rangle_2 + | v\rangle_1 \otimes | h\rangle_2), \tag{5.36}$$

where $1/\sqrt{2}$ is an appropriate normalization factor. This is a perfectly legitimate state vector in quantum mechanics: since $| h\rangle_1 \otimes | v\rangle_2$ and $| v\rangle_1 \otimes | h\rangle_2$ are two possible states for the polarization of two photons, according to the superposition principle (p. 18) a linear combination of them is also an allowed state. How can we interpret such a state? It is clearly a superposition state of a particular type. In each of the terms in the rhs of Eq. (5.36) the two photons are in orthogonal states. For example, if we measure the polarization of photon 1 and find it to be horizontal, we know with certainty that the polarization of photon 2 will be vertical, and *vice versa*. However, it is impossible to assign a certain state either to photon 1 or to photon 2. In other words, we cannot write the state $| \Psi\rangle_{12}$ of Eq. (5.36) in the *factorized* form

$$| \Psi\rangle_{12} = | \psi\rangle_1 \otimes | \varphi\rangle_2, \tag{5.37}$$

which can be considered as a definition of a state that is not entangled. This would be the case, for instance, when we had the two-photon system in the state $| h\rangle_1 \otimes | h\rangle_2$, or in the state (see Prob. 5.7)

$$| \Psi\rangle_{12} = \frac{1}{2}(| h\rangle_1 \otimes | h\rangle_2 + | h\rangle_1 \otimes | v\rangle_2 + | v\rangle_1 \otimes | h\rangle_2 + | v\rangle_1 \otimes | v\rangle_2). \tag{5.38}$$

Generalizing the above argument in the context of the density-matrix formalism, we can say that any state $\hat{\rho}$ is *separable* if it can be written as

$$\hat{\rho} = \sum_j w_j \hat{\rho}_1^{(j)} \otimes \hat{\rho}_2^{(j)}, \tag{5.39}$$

where $w_j \geq 0$ and $\sum_j w_j = 1$. Equation (5.39) reflects the fact that a separable state can be prepared by two distant observers who receive instructions from a common source. On the contrary, we say that the photons in a state of the type (5.36) are *entangled*. This means that the two photons (through their polarization degree of freedom) constitute an inseparable whole and cannot be treated as separate systems.

Summarizing the above discussion, we can state the following definition:

Definition 5.1 (Entanglement) *Two systems $\mathcal{S}_1$ and $\mathcal{S}_2$ are said to be entangled with respect to a certain degree of freedom if their total state $\mid\Psi\rangle_{12}$, relative to that degree of freedom,* cannot *be written in a factorized form as a product $\mid\psi\rangle_1 \otimes \mid\varphi\rangle_2$.*

In other words, an entangled state is a state of a composite system whose subsystems are not probabilistically independent. While the superposition character of a state vector is basis-dependent, it should be understood that Def. 5.1 has to hold in *any* basis, i.e. a truly entangled state cannot be factorized in any basis (see also Subsec. 6.4.3). It should also be stressed that, while entanglement and non-factorizability of the state vector are equivalent concepts, entanglement implies superposition, i.e. a superposition state is not necessarily entangled. As we have seen, it is a natural consequence of the linearity of quantum mechanics.[7] In Ch. 16 we shall examine the extreme consequences of entanglement and show how it can be physically generated, while in Ch. 17 we shall introduce a measure of entanglement.

5.5.2 Partial trace

The concept of density matrix is particularly useful when one deals with systems which consist of two or more subsystems. In the previous section we have learnt that it is not possible to assign proper state vectors to entangled subsystems. However, using the concept of density matrix, it is possible to partly overcome this difficulty. Let us go back to the example of the polarization-entangled state of a pair of photons (see Eq. (5.36)). The density matrix associated with such a (pure) state can be written as

$$\hat{\rho}_{12} = \mid\Psi\rangle_{12}\,\langle\Psi\mid = \frac{1}{2}\left(\mid h\rangle_1\,\langle h\mid\otimes\mid v\rangle_2\,\langle v\mid + \mid v\rangle_1\,\langle v\mid\otimes\mid h\rangle_2\,\langle h\mid\right.$$
$$\left. +\ \mid h\rangle_1\,\langle v\mid\otimes\mid v\rangle_2\,\langle h\mid + \mid v\rangle_1\,\langle h\mid\otimes\mid h\rangle_2\,\langle v\mid\right). \quad (5.40)$$

In Eq. (5.40) and in the following repeated indices are dropped. Is it possible to write the density matrix for, say, photon 1 irrespectively of photon 2? To answer this question, we may average $\hat{\rho}_{12}$ over all possible states of polarization of photon 2, that is, we may perform the *trace* of the density matrix (5.40) with respect to photon 2:

$$\hat{\varrho}_1 = \mathrm{Tr}_2\left(\hat{\rho}_{12}\right) = \sum_{j=h,v} {}_2\langle j\mid\hat{\rho}_{12}\mid j\rangle_2 = \frac{1}{2}\left(\mid h\rangle_1\,\langle h\mid + \mid v\rangle_1\,\langle v\mid\right). \quad (5.41a)$$

This expression is called the *reduced* density matrix of photon 1 and represents the maximum information which is available about photon 1 alone, irrespective of the polarization of photon 2. In a similar way,

[7] The linearity of quantum mechanics comes from the superposition principle (see p. 18) and manifests itself in the linear nature of the Schrödinger equation.

$$\hat{\varrho}_2 = \mathrm{Tr}_1\left(\hat{\rho}_{12}\right) = \sum_{j=h,v} {}_1\langle j \mid \hat{\rho}_{12} \mid j\rangle_1 = \frac{1}{2}\left(\mid h\rangle_2 \, \langle h \mid + \mid v\rangle_2 \, \langle v \mid\right). \tag{5.41b}$$

It is evident that Eqs. (5.41) are both mixed density matrices and describe ensembles of photons, half of which have horizontal polarization and the other half vertical polarization. Equations (5.41) express the fact that, when measuring the polarization of one of the two photons, there is 50% probability of obtaining either of the possible results. As we shall see in Ch. 9, this formalism is particularly useful when dealing with the measurement problem.

The mixed character of $\hat{\varrho}_1$ and $\hat{\varrho}_2$ is precisely due to the entangled character of $\hat{\rho}_{12}$. Conversely, it should be noted that a partial trace performed over a non-entangled composite state of two subsystems gives rise to a pure state density matrix (see Probs. 5.9 and 5.10).

Let us now consider a more complicated example. Let the compound state of two two-level subsystems be described by the ket

$$\begin{aligned} \mid \Psi\rangle &= c_{00} \mid 0\rangle_1 \mid 0\rangle_2 + c_{01} \mid 0\rangle_1 \mid 1\rangle_2 + c_{10} \mid 1\rangle_1 \mid 0\rangle_2 + c_{11} \mid 1\rangle_1 \mid 1\rangle_2 && (5.42) \\ &= \sum_{j,k=0}^{1} c_{jk} \mid j\rangle \otimes \mid k\rangle . && (5.43) \end{aligned}$$

For the sake of notational simplicity, the direct product sign $\otimes$ will be omitted here and henceforth except where confusion may arise. The corresponding density operator may be written as

$$\hat{\rho} = \mid \Psi\rangle \langle \Psi \mid = \sum_{j,k,l,m}^{0,1} c_{jk} c^*_{lm} \mid j \rangle \langle l \mid \otimes \mid k \rangle \langle m \mid, \tag{5.44}$$

which in matrix form reads

$$\hat{\rho}_{12} = \left[\begin{array}{cc|cc} |c_{00}|^2 & c_{00}c^*_{01} & c_{00}c^*_{10} & c_{00}c^*_{11} \\ c^*_{00}c_{01} & |c_{01}|^2 & c_{01}c^*_{10} & c_{01}c^*_{11} \\ \hline c^*_{00}c_{10} & c^*_{01}c_{10} & |c_{10}|^2 & c_{10}c^*_{11} \\ c^*_{00}c_{11} & c^*_{01}c_{11} & c^*_{10}c_{11} & |c_{11}|^2 \end{array}\right]. \tag{5.45}$$

The reduced density matrix $\hat{\varrho}_1$ obtained from $\hat{\rho}_{12}$ by tracing out the second subsystem, is

$$\begin{aligned} \mathrm{Tr}_2\left(\hat{\rho}_{12}\right) &= \sum_n \left\langle n \left| \hat{\rho}_{12} \right| n \right\rangle \\ &= \sum_{j,k,l,m,n} c_{jk} c^*_{lm} \mid j \rangle \langle l \mid \otimes \langle n \mid k\rangle \langle m \mid n\rangle \\ &= \sum_{j,k,l,m,n} c_{jk} c^*_{lm} \mid j \rangle \langle l \mid \delta_{nk} \delta_{mn} \\ &= \sum_{j,l,n} c_{jn} c^*_{ln} \mid j \rangle \langle l \mid , \end{aligned} \tag{5.46}$$

which in matrix form reads (see also Prob. 5.11)

$$\hat{\varrho}_1 = \left[\begin{array}{cc} |c_{00}|^2 + |c_{01}|^2 & c_{00}c^*_{10} + c_{01}c^*_{11} \\ c^*_{00}c_{10} + c^*_{01}c_{11} & |c_{10}|^2 + |c_{11}|^2 \end{array}\right]. \tag{5.47}$$

From Eq. (5.47) it is clear that the four elements of $\hat{\varrho}_1$ are simply given by the sum of the diagonal elements of each of the four blocks in which $\hat{\rho}_{12}$ can be cast according to Eq. (5.45).

5.5.3 Schmidt decomposition

Let us consider a composite system made of two finite subsystems with the same dimension N, in the state

$$\mid\Psi\rangle = \sum_{j,k=1}^{N} C_{jk} \left| a_j \right\rangle \mid b_k\rangle, \tag{5.48}$$

where $\langle a_k \mid a_j \rangle = \langle b_k \mid b_j \rangle = \delta_{jk}$. It is interesting to note that it is always possible[8] to convert the double sum into a single sum, i.e.

$$\mid\Psi\rangle = \sum_{n=1}^{N} c_n \mid v_n\rangle \mid w_n\rangle \tag{5.49}$$

by means of the unitary transformations

$$\mid v_n\rangle = \sum_j U_{nj} \left| a_j \right\rangle, \quad \mid w_n\rangle = \sum_k U'_{nk} \mid b_k\rangle, \tag{5.50}$$

where

$$C_{jk} = \sum_n U^{\mathrm{T}}_{nj} D_{nn} U'_{nk}, \tag{5.51}$$

$U^{\mathrm{T}}_{nj} = U_{jn}$, and $c_n = D_{nn}$. This is called *Schmidt decomposition*, and the number of non-zero coefficients c_n's is called the *Schmidt number*. The c_n^2 are the singular values of the matrix $\hat{C}$, i.e. the non-vanishing eigenvalues of the Hermitian matrices $\hat{C}\hat{C}^\dagger$ and $\hat{C}^\dagger\hat{C}$, whose set of eigenvectors are $\{\mid v_n\rangle\}$ and $\{\mid w_n\rangle\}$, respectively (see Prob. 5.12). If $\{\left| a_j \right\rangle\}$ and $\{\mid b_k\rangle\}$ are two orthonormal bases for the two distinct Hilbert spaces, then, for the unitarity of the transformations $\hat{U}$ and $\hat{U}'$, $\{\mid v_n\rangle\}$ and $\{\mid w_n\rangle\}$ are also two othornormal bases for the respective spaces. It is interesting to observe[9] that a (pure) state $\mid\Psi\rangle$ of a composite system is a product state if and only if (iff) it has a Schmidt number 1. Moreover, it is a product state iff the corresponding reduced density matrices of the two subsystems are pure (see Prob. 5.13).

It is possible to extend the above proof to the case of two systems with different dimensions and state the following theorem:

Theorem 5.1 (Schmidt decomposition) *For any pure state $\mid\Psi\rangle_{12}$ of a composite system $\mathcal{S}_{12}$ it is always possible to find orthonormal sets of vectors $\{\mid j\rangle_1\}$ for system $\mathcal{S}_1$ and $\{\mid j\rangle_2\}$ for system $\mathcal{S}_2$ such that*

[8] See [Schmidt 1907] [Peres 1995].
[9] See [*Nielsen/Chuang* 2000, 110].

$$| \Psi \rangle_{12} = \sum_j c_j \, | j \rangle_1 \, | j \rangle_2 , \quad (5.52)$$

where c_j are non-negative real numbers with $\sum_j c_j^2 = 1$.

If, $\{|a_j\rangle\}$ and $\{|b_k\rangle\}$ are two orthonormal bases for two distinct Hilbert spaces with different dimensions, then $\{| j\rangle_1\}$ and $\{| j\rangle_2\}$ are are two, possibly incomplete, sets of orthonormal vectors for these two spaces. It should be stressed that Schmidt decomposition must *not* be understood as a way of *reducing* the total Hilbert space. As a matter of fact, the involved transformations are not universal but rather depend on the coefficients considered and, therefore, are state-dependent (see also Subsec. 3.5.3).

In the previous subsection we have said that a partial trace of a pure density matrix may yield a mixture. Here we show how, starting with a mixed state, we may write down a compound pure state of which the "initial" mixed state is a reduced state. Suppose, for instance, that we have a system $\mathcal{S}$ in a mixed state

$$\hat{\rho}_{\mathcal{S}} = \sum_n w_n \, | a_n \rangle \langle a_n |, \quad (5.53)$$

where $w_n \geq 0$ (and $w_n \neq 0$ for $k \geq 2$ values of n), and $\{| a_n\rangle\}$ is a basis for the Hilbert space $\mathcal{H}_{\mathcal{S}}$ of $\mathcal{S}$. Now it is always possible to find a larger system $\mathcal{S} + \mathcal{S}'$ (provided that $\mathcal{S}'$ has at least k independent vectors, with $k \geq 2$) such that the state vector of the total system $\mathcal{S} + \mathcal{S}'$ is of the form (5.49), i.e.

$$| \Psi \rangle_{\mathcal{S}+\mathcal{S}'} = \sum_n \sqrt{w_n} \, | a_n \rangle \, | b_n \rangle , \quad (5.54)$$

where $\{| b_n\rangle\}$ is a basis for the Hilbert space $\mathcal{H}_{\mathcal{S}'}$ of $\mathcal{S}'$. It can be easily shown that, by tracing out $\mathcal{S}'$,

$$\begin{aligned} \mathrm{Tr}_{\mathcal{S}'} \left(| \Psi \rangle_{\mathcal{S}+\mathcal{S}'} \langle \Psi | \right) &= \sum_{mn} \sqrt{w_m w_n} \, | a_m \rangle \langle a_n | \, \mathrm{Tr} \left(| b_m \rangle \langle b_n | \right) \\ &= \sum_{mn} \sqrt{w_m w_n} \, | a_m \rangle \langle a_n | \, \delta_{mn} \\ &= \hat{\rho}_{\mathcal{S}}, \end{aligned} \quad (5.55)$$

we exactly recover the state (5.53). In other words, we have first considered the Hilbert space $\mathcal{H}_{\mathcal{S}}$ of a system $\mathcal{S}$ in a mixed state. Then, we have considered a second system $\mathcal{S}'$ with a Hilbert space $\mathcal{H}_{\mathcal{S}'}$ and built a compound system of these two in a larger Hilbert space $\mathcal{H}_{\mathcal{S}+\mathcal{S}'} = \mathcal{H}_{\mathcal{S}} \otimes \mathcal{H}_{\mathcal{S}'}$, which is described by a pure state. Finally, we have traced out the system $\mathcal{S}'$. This shows that a mixed state of a certain system can always be obtained as a partial trace of a pure state of a larger system.

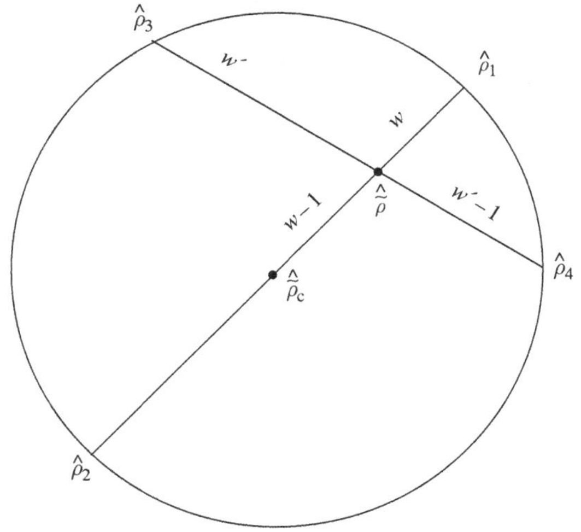

Figure 5.1 Representation of pure and mixed states on a sphere.

5.6 Pure- and mixed-state representation

It is always possible to write a density matrix as a linear combination of the density matrices corresponding to pure states. In particular, for a two-level system we can write a density operator $\hat{\rho}$ in the convex form (see also Eq. (5.25))

$$\hat{\rho} = w\hat{\rho}_1 + (1 - w)\hat{\rho}_2, \tag{5.56}$$

where $\hat{\rho}_1$ and $\hat{\rho}_2$ are two appropriate and different pure states and $0 \leq w \leq 1$ is some weight. If the state described by Eq. (5.56) is pure, $w = 1$ or $w = 0$. Otherwise, it represents a mixture. Hence, in the space of states the set of pure states, which may be represented by projection operators, is not convex. It should then be clear, as we have said in Sec. 5.1, that the formalism of density operators is a generalization of the formalism of projectors and state vectors. In the case of pure states, this representation perfectly corresponds to the Poincaré sphere representation introduced in Subsec. 1.3.3. Indeed, we may build a sphere of the density operators and see that any state $| j \rangle$ on the Poincaré sphere corresponds to a projector $\hat{P}_j = | j \rangle \langle j |$ on the sphere of density operators. We may represent any pure state ($w = 0$ or $w = 1$) as a point on the surface of a sphere and any mixture ($0 < w < 1$) as a point lying in the interior of the sphere.[10] Supposing that Eq. (5.56) represents a mixture $\hat{\hat{\rho}}$, $\hat{\rho}_1$ and $\hat{\rho}_2$ are two pure states in which $\hat{\rho}$ is decomposed. In Fig. 5.1 they are taken to be orthogonal (they are opposite points on a diameter passing through $\hat{\hat{\rho}}$). However, the decomposition of a mixture is not unique. In fact, we can take any other line

[10] See [*Poincaré* 1902, 89–90].

(not necessarily a diameter) passing through $\hat{\hat{\rho}}$ – in Fig. 5.1 the decomposition in terms of $\hat{\rho}_3$ and $\hat{\rho}_4$ is shown. It is clear that in this case $\hat{\rho}_3$ and $\hat{\rho}_4$ are not orthogonal.

Since Tr $(\hat{\rho}^2) = 1$ for a pure state and Tr $(\hat{\rho}^2) < 1$ for a mixed state (see Sec. 5.1), a good measure of the *purity* of a state is represented by

$$\mathrm{M} = \mathrm{Tr}\left(\hat{\rho}^2\right). \tag{5.57}$$

It is clear that states inside the sphere of density operators will have $\mathrm{M} < 1$, while on the surface $\mathrm{M} = 1$. Let us consider the center $\hat{\hat{\rho}}_{\mathrm{c}}$ of the sphere. As any other mixed state, this state has an infinite number of possible representations. If we take the vertical diameter, we have

$$\hat{\hat{\rho}}_{\mathrm{c}} = \frac{1}{2}\left(|\uparrow\rangle\langle\uparrow| + |\downarrow\rangle\langle\downarrow|\right), \tag{5.58}$$

where $|\uparrow\rangle\langle\uparrow|$ and $|\downarrow\rangle\langle\downarrow|$ represent here the north and south poles of the sphere, respectively. Therefore, the matrix form of $\hat{\hat{\rho}}_{\mathrm{c}}$ in the basis $\{|\uparrow\rangle, |\downarrow\rangle\}$ will be given by

$$\hat{\hat{\rho}}_{\mathrm{c}} = \frac{1}{2}\begin{bmatrix} 1 & 0 \\ 0 & 1 \end{bmatrix} = \frac{1}{2}\hat{I}, \tag{5.59}$$

and

$$\hat{\hat{\rho}}_{\mathrm{c}}^2 = \frac{1}{4}\begin{bmatrix} 1 & 0 \\ 0 & 1 \end{bmatrix} = \frac{1}{4}\hat{I}, \tag{5.60}$$

which yields $\mathrm{M} = 1/2$, and this corresponds to the minimum value of Tr $(\hat{\rho}^2)$ for a bidimensional system.

Going to higher Hilbert-space dimensions, it is straightforward to generalize the previous result so as to obtain

$$\mathrm{M}_c^{(n)} = \frac{1}{n}, \tag{5.61}$$

where n is the dimension of the space, and we can see that the purity of a completely mixed and equally weighted n-dimensional density matrix (of a center) is simply the inverse of the number of possible levels (see Prob. 5.14). It is also clear that for $n \to \infty$, $\mathrm{M}_c^{(n)} \to 0$.

Summary

In this chapter we have developed the basic features of the density matrix. We may summarize the main results as follows:

- The concept of *density matrix* has led us to classify quantum-mechanical states into *pure states*, which can be represented as projectors, and *mixed* states.
- Quantum-mechanical *statistics* present some specific features.
- Multiparticle systems can present the feature of *entanglement*, which describes a situation where two or more (not necessarily interacting) systems, for what concerns their physical properties, are not separable but have to be considered as a whole.

Problems

5.1 Prove Eqs. (5.1) for pure states.

5.2 Show that the density matrix corresponding to a non–normalizable wave function (see Eq. (2.117)) is not normalizable, i.e. $\mathrm{Tr}\,(\hat{\rho}) = \infty$.

5.3 Show that the mixed density matrix $\hat{\tilde{\rho}}$ of Eq. (5.8) satisfies Eq. (5.10) if both c_h and c_v are different from zero.

5.4 Prove that pure states evolve into pure states under unitary time evolution. (*Hint*: For the property (5.1b) use the cyclic property of the trace (see Box 3.1)).

5.5 The fact that the probabilities (5.16) differ from probabilities (5.17) suggests an important conclusion about the relation between density matrices (5.14) and (5.15).

5.6 Consider the state $|\,\psi\rangle = (1/\sqrt{2})\,(|\,v\rangle + |\,h\rangle\,)$ on the two-dimensional polarization Hilbert space. Write the corresponding density matrix $\hat{\rho} = |\,\psi\rangle\,\langle\psi\,|$ and show that it can be written as a projector, and find the eigenvalues and eigenvectors of $\hat{\rho}$. What are the main differences between this density matrix and

$$\hat{\rho}' = \begin{bmatrix} \frac{1}{2} & 0 \\ 0 & \frac{1}{2} \end{bmatrix}?$$

5.7 Prove that the state of Eq. (5.38) may be written as a direct product $|\,\Psi\rangle_{12} = |\,\psi\rangle_1 \otimes |\varphi\rangle_2$.

5.8 Prove that the state of the form (5.36), that is,

$$|\,\Psi\rangle_{12} = c_{hv}\,|\,h\rangle_1\,|\,v\rangle_2 + c_{vh}\,|\,v\rangle_1\,|\,h\rangle_2,$$

where $|c_{hv}|^2 + |c_{vh}|^2 = 1$ and both coefficients are non-zero, cannot be written as a product state $|\,\Psi'\rangle_{12} = |\,\psi\rangle_1 \otimes |\varphi\rangle_2$, with

$$|\,\psi\rangle_1 = c_v\,|\,v\rangle_1 + c_h\,|\,h\rangle_1,$$
$$|\varphi\rangle_2 = c_v'\,|\,v\rangle_2 + c_h'\,|\,h\rangle_2.$$

5.9 Consider again the state vector $|\,\Psi\rangle_{12}$ of Eq. (5.38) describing the polarization of two photons. Show that the reduced density matrices $\hat{\varrho}_1 = \mathrm{Tr}_2\,(\hat{\rho}_{12})$, and $\hat{\varrho}_2 = \mathrm{Tr}_1\,(\hat{\rho}_{12})$ describe pure states, where $\hat{\rho}_{12} = |\,\Psi\rangle_{12}\,\langle\Psi\,|$.

5.10 Generalize the result of Prob. 5.9 and prove that the reduced density matrix obtained starting from a non-entangled pure state describes a pure state.

5.11 Compute the reduced density matrix $\hat{\varrho}_2$ of the matrix (5.45).

5.12 Consider two systems, 1 and 2, with bases $\{|\,0\rangle, |\,1\rangle\,\}_1$ and $\{|\,0\rangle, |\,1\rangle\,\}_2$, respectively, where

$$|\,0\rangle = \begin{pmatrix} 1 \\ 0 \end{pmatrix}, \quad |\,1\rangle = \begin{pmatrix} 0 \\ 1 \end{pmatrix}.$$

Let us assume that the compound system is in the entangled state

$$\begin{aligned} |\Psi\rangle &= \frac{1}{\sqrt{2}}(|0\rangle_1 |1\rangle_2 + |1\rangle_1 |0\rangle_2) \\ &= \sum_{j,k=0}^{1} C_{jk} |j\rangle_1 |k\rangle_2 . \end{aligned}$$

Apply the Schmidt decomposition in order to

- find the eigenvalues of the matrices $\hat{C}\hat{C}^\dagger$ and $\hat{C}^\dagger\hat{C}$;
- choose the states $|v_n\rangle$ and $|w_n\rangle$ of Eq. (5.49);
- determine the unitary transformations $\hat{U}$ and $\hat{U}'$ of Eqs. (5.50);
- verify that the relation (5.51) holds.

5.13 Prove that

- a (pure) state $|\Psi\rangle$ of a composite system is a product state if and only if (iff) it has a Schmidt number 1,
- it is a product state iff the corresponding reduced density matrices of the two subsystems are pure.

(*Hint*: The first part is straightforward. For the second part, take advantage of the solutions of Prob. 5.10.)

5.14 Prove Eq. (5.61).

Further reading

Density matrix

Fano, Ugo, Description of states in quantum mechanics by density matrix and operator techniques. *Review of Modern Physics*, **29** (1957), 74–93.

Entanglement

Clifton, R., Butterfield, J., and Halvorson, H. (eds.), *Quantum Entanglement: Selected Papers*, Oxford: Oxford University Press, 2004.

PART II

MORE ADVANCED TOPICS

6 Angular momentum and spin

Up to now we have laid the foundations and the basic principles of quantum mechanics and have considered some examples of quantum dynamics. In this chapter we go back to the problem of the definition of quantum observables and the search for their eigenvalues and eigenfunctions. In particular, we discuss here angular momentum (in Sec. 6.1), some special examples (in Sec. 6.2), and spin (in Sec. 6.3). In Sec. 6.4 we discuss some examples of composition of angular momenta pertaining to different particles and total angular momentum. Finally, in Sec. 6.5 we analyze the uncertainty relations between angular momentum and angle, and present an operational representation of the angle observable.

6.1 Orbital angular momentum

When moving from one-dimensional problems to higher dimensions, rotations come into play. These have taken a central scientific role since the beginning of human culture, when the most ancient civilizations began a systematic observation of celestial bodies and their recurrent motion. However, also in this case the passage from classical to quantum physics introduces additional questions that are the subject of this section.

6.1.1 General features

Classically, the orbital angular momentum of a point particle is defined as the vector product

$$\mathbf{L} = \mathbf{r} \times \mathbf{p}, \tag{6.1}$$

where $\mathbf{r}$ is the position and $\mathbf{p}$ the linear momentum with respect to a certain Cartesian reference frame $Oxyz$ (see Fig. 6.1). It should be noted that the angular momentum is an *axial vector* (or *pseudovector*), i.e. a quantity that transforms like a vector under a proper rotation, but gains an additional sign flip under an improper rotation (a transformation that can be expressed as an inversion followed by a proper rotation). The conceptual opposite of a pseudovector is a (true) vector or a *polar vector*. Further examples of axial vectors are represented by magnetic field and torque. In general, quantities resulting from vector products are axial vectors.

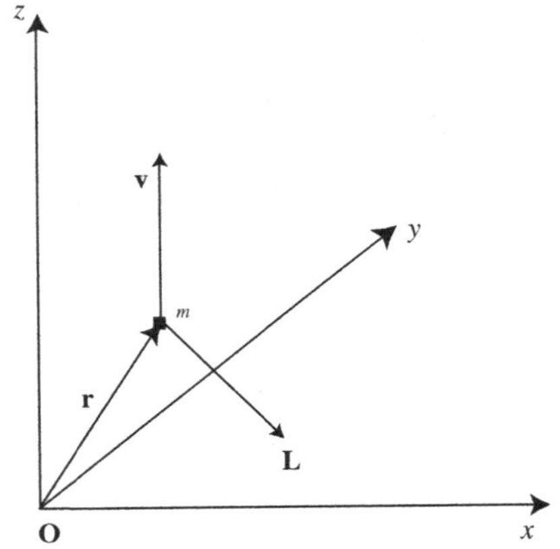

Figure 6.1 A classical particle with mass m, velocity $\mathbf{v}$, and position $\mathbf{r}$ with respect to a certain reference frame $Oxyz$ has an angular momentum $\mathbf{L} = \mathbf{r} \times \mathbf{p} = m\mathbf{r} \times \mathbf{v}$.

Making use of the correspondence principle (see p. 72), we can define the orbital angular momentum in the same way also in quantum mechanics, but by taking into account the commutation relations (2.173). Then, in the coordinate representation, we have the following three equations for the orbital angular momentum (see Eq. (2.134))

$$\hat{L}_x = \hat{y}\hat{p}_z - \hat{z}\hat{p}_y = -\imath\hbar\left(y\frac{\partial}{\partial z} - z\frac{\partial}{\partial y}\right), \tag{6.2a}$$

$$\hat{L}_y = \hat{z}\hat{p}_x - \hat{x}\hat{p}_z = -\imath\hbar\left(z\frac{\partial}{\partial x} - x\frac{\partial}{\partial z}\right), \tag{6.2b}$$

$$\hat{L}_z = \hat{x}\hat{p}_y - \hat{y}\hat{p}_x = -\imath\hbar\left(x\frac{\partial}{\partial y} - y\frac{\partial}{\partial x}\right), \tag{6.2c}$$

or, in more compact form,

$$\hat{L}_j = \epsilon_{jkn}\hat{r}_k\hat{p}_n, \tag{6.3}$$

where j, k, n are indices which may assume the values $1, 2, 3$, $\hat{r}_1 = \hat{x}$, $\hat{r}_2 = \hat{y}$, and $\hat{r}_3 = \hat{z}$, and similarly $\hat{L}_1 = \hat{L}_x$, $\hat{L}_2 = \hat{L}_y$, and $\hat{L}_3 = \hat{L}_z$. The symbol ϵ refers to the Levi–Civita tensor, which is antisymmetric (it changes sign by exchange of two indices), is zero when two indices are equal, and $\epsilon_{123} = 1$ (see Fig. 6.2). In Eq. (6.3) and in the following of this chapter, a summation over repeated indices is understood.

The first important aspect to note in Eqs. (6.2) and (6.3) is that different components of the angular momentum do not commute. In fact, the commutation relations for the first two components can be calculated as follows:

$$\begin{aligned}\left[\hat{L}_x, \hat{L}_y\right] &= \left[\hat{y}\hat{p}_z - \hat{z}\hat{p}_y, \hat{z}\hat{p}_x - \hat{x}\hat{p}_z\right] \\ &= \left[\hat{y}\hat{p}_z, \hat{z}\hat{p}_x\right] + \left[\hat{z}\hat{p}_y, \hat{x}\hat{p}_z\right] \\ &= \hat{y}\left[\hat{p}_z, \hat{z}\right]\hat{p}_x + \hat{p}_y\left[\hat{z}, \hat{p}_z\right]\hat{x} \\ &= \imath\hbar\left(\hat{x}\hat{p}_y - \hat{y}\hat{p}_x\right) \\ &= \imath\hbar\hat{L}_z,\end{aligned} \tag{6.4}$$

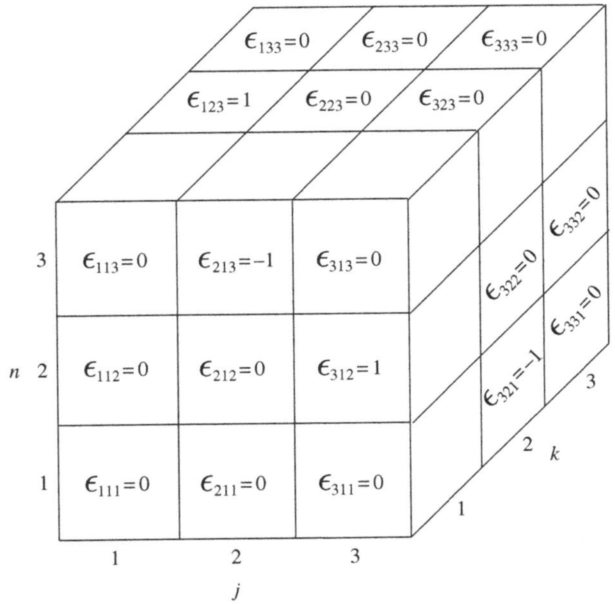

Figure 6.2 Pictorial representation of the Levi-Civita tensor. A cube is segmented into 27 boxes, according to the values (1, 2, or 3) of the indices *j, k, n*. Each box then represents one component of the tensor. Only the values of the visible boxes are shown. Of the 27 components of the Levi-Civita tensor only six (those for which the indices are different from each other) are different from zero: they can have the values +1 or -1 depending on whether the sequence of the indices can be obtained from the sequence 123 with an even or an odd number of permutations, respectively.

where we have made use of property (2.95), of the commutation relation $\left[\hat{z}, \hat{p}_z\right] = \imath\hbar$ [see Eq. (2.173b)], and of Eq. (6.2c). Proceeding in a similar manner, we can calculate the other commutation relations (see Prob. 6.1). The corresponding results are

$$\left[\hat{L}_x, \hat{L}_y\right] = \imath\hbar\hat{L}_z, \quad \left[\hat{L}_y, \hat{L}_z\right] = \imath\hbar\hat{L}_x, \quad \left[\hat{L}_z, \hat{L}_x\right] = \imath\hbar\hat{L}_y, \tag{6.5}$$

which can be written in the more compact form

$$\left[\hat{L}_j, \hat{L}_k\right] = \imath\hbar\epsilon_{jkn}\hat{L}_n. \tag{6.6}$$

In other words, the three components of the angular momentum are not simultaneously measurable (see Cor. 2.1: p. 67). As we shall see below, this is due to the fact that $\hat{\mathbf{L}}$ generates the group of rotations. A second important feature of the angular momentum is that any of its components commutes with its square (see Prob. 6.2), i.e.

$$\left[\hat{L}_j, \hat{\mathbf{L}}^2\right] = 0, \tag{6.7}$$

with $j = x, y, z$. This result tells us that it is possible to measure simultaneously $\hat{\mathbf{L}}^2$ and one (and only one) of the components $\hat{L}_j$. This is again a consequence of the fact that $\hat{\mathbf{L}}$ is the generator of the rotation group ($\hat{\mathbf{L}}^2$ is a scalar for the rotations). We shall investigate the relation between rotation invariance and properties of the angular momentum in the next subsection.

Finally, in the context of quantum–classical physics correspondence, where the Poisson brackets are replaced by commutators (see Sec. 3.7), notice that, classically, two components of $\mathbf{L}$ cannot simultaneously be canonical momenta. On the contrary, the modulus of $\mathbf{L}$ and any of the L_j components can simultaneously be canonical momenta. This shows once again that quantum commutation relations have a certain analogy in classical mechanics.

6.1.2 Rotations and angular momentum

We have already seen how it is possible to introduce the linear momentum $\hat{\mathbf{p}}$ as the invariant for space translations (or, equivalently, the generator of spatial translations) (see Subsec. 2.2.4 and also Subsec. 8.2.1) and the total energy (the Hamiltonian $\hat{H}$) as the invariant for time translations (or, equivalently, the generator of time translations) (see Subsecs. 3.1.1 and 3.5.3).

Let us now discuss the two consequences drawn at the end of the previous subsection with more details. We consider now how a generic ket $|\,\psi\rangle$ is modified by the action of the unitary rotation operator (see Subsec. 3.5.4), that is,

$$\hat{U}_\theta = e^{\frac{\imath}{\hbar}\theta \mathbf{n}\cdot\hat{\mathbf{L}}} = e^{\imath\theta\mathbf{n}\cdot\hat{\mathbf{l}}}, \tag{6.8}$$

which describes a rotation by an angle θ about the direction given by the versor $\mathbf{n}$, and where

$$\hat{\mathbf{l}} = \frac{\hat{\mathbf{L}}}{\hbar} = \hat{\mathbf{r}} \times \hat{\mathbf{k}}, \tag{6.9}$$

$\hat{\mathbf{k}} = \hat{\mathbf{p}}/\hbar = -\imath\nabla$ being the propagation vector (see Eq. (2.140)). Without loss of generality we may consider infinitesimal rotations, i.e. rotations by an angle $\delta\theta \to 0$. In such a case, Taylor-expanding the unitary operator (6.8) to the first order, we have

$$\hat{U}_{\delta\theta} \simeq 1 + \imath\delta\theta\mathbf{n}\cdot\hat{\mathbf{l}} = 1 + \imath\delta\theta\hat{R}, \tag{6.10}$$

where $\hat{R} = \mathbf{n}\cdot\hat{\mathbf{l}}$ is called the *generator of the rotation* about the direction given by the vector $\mathbf{n}$. The unitary operator (6.8) induces, on a generic operator $\hat{O}$, the transformation (see Eq. (3.68))

$$\begin{aligned}\hat{O} \mapsto \hat{O}' &= \hat{U}_{\delta\theta}\hat{O}\hat{U}^\dagger_{\delta\theta} = \left(1 + \imath\delta\theta\hat{R}\right)\hat{O}\left(1 - \imath\delta\theta\hat{R}\right)\\ &= \hat{O} + \delta\hat{O},\end{aligned} \tag{6.11}$$

where, to the first order in $\delta\theta$,

$$\delta\hat{O} \simeq \imath\delta\theta\left[\hat{R}, \hat{O}\right]. \tag{6.12}$$

As a consequence, if $\hat{O}$ is a scalar for rotations, it has to commute with the generator $\hat{R}$ of the rotation, which is the projection of $\hat{\mathbf{L}}$ along the direction $\mathbf{n}$. In such a case, $\delta\hat{O} = 0$. This then is the reason why any component of $\hat{\mathbf{L}}$ commutes with $\hat{\mathbf{L}}^2$, $\hat{R}$, and $\hat{\mathbf{p}}^2$, all of which are scalars for the rotations.

Let us write the explicit form of the generator of rotations. In the general three-dimensional case (see Eq. (2.38) for the two-dimensional case), the matrix which describes a rotation about the axes x, y, z (by the Euler angles β, ϕ, and θ, respectively) is[1]

$$\hat{\mathbf{R}}(\beta,\phi,\theta) = \begin{bmatrix} \cos\beta\cos\phi\cos\theta - \sin\beta\sin\theta & \sin\beta\cos\phi\cos\theta + \cos\beta\sin\theta & -\sin\phi\cos\theta \\ -\cos\beta\cos\phi\sin\theta - \sin\beta\cos\theta & -\sin\beta\cos\phi\sin\theta + \cos\beta\cos\theta & \sin\phi\sin\theta \\ \cos\beta\sin\phi & \sin\beta\sin\phi & \cos\phi \end{bmatrix}. \tag{6.13}$$

Let us now consider a specific rotation by an angle θ about the z-axis. The rotation matrix is (see Prob. 6.3)

$$\hat{\mathbf{R}}(\theta) = \begin{bmatrix} \cos\theta & \sin\theta & 0 \\ -\sin\theta & \cos\theta & 0 \\ 0 & 0 & 1 \end{bmatrix}. \tag{6.14}$$

Then, any vector $\mathbf{v} \equiv (v_x, v_y, v_z)$ is transformed by the rotation (6.14) of an angle $\delta\theta$ into a vector $\mathbf{v}' \equiv (v'_x, v'_y, v'_z)$ such that ($\mathbf{v}' = \mathbf{v} - \delta\mathbf{v}$)

$$v'_x = \cos\delta\theta\, v_x + \sin\delta\theta\, v_y, \tag{6.15a}$$
$$v'_y = -\sin\delta\theta\, v_x + \cos\delta\theta\, v_y, \tag{6.15b}$$
$$v'_z = v_z. \tag{6.15c}$$

Since $v'_x \simeq v_x + \delta\theta v_y$ and $v'_y \simeq -\delta\theta v_x + v_y$, we obtain

$$\delta v_x = -\delta\theta v_y, \quad \delta v_y = \delta\theta v_x, \quad \delta v_z = 0. \tag{6.16}$$

Comparing Eq. (6.16) with Eq. (6.12), we arrive at

$$\left[\hat{l}_z, v_x\right] = \imath v_y, \quad \left[\hat{l}_z, v_y\right] = -\imath v_x, \quad \left[\hat{l}_z, v_z\right] = 0. \tag{6.17}$$

Since this is true of any vector, we have e.g. (see Prob. 6.4)

$$\left[\hat{L}_z, \hat{x}\right] = \imath\hbar\hat{y} \quad \text{and} \quad \left[\hat{L}_z, \hat{p}_x\right] = \imath\hbar\hat{p}_y. \tag{6.18}$$

[1] We follow for the Euler angles the $z - y - z$-convention [*Byron/Fuller* 1969–70, 11].

6.1.3 Eigenvalues of the angular momentum

We wish now to find the eigenvalues and eigenfunctions of $\hat{\mathbf{l}}$. For this purpose, we have to choose a complete set (see comments to Cor. 2.1: p. 67) of angular momentum observables. This is not trivial in the present case since, as we have seen, in contrast to the case of position and momentum, different components of the angular momentum do not commute with each other. We have then to choose a pair of commuting observables we wish to jointly diagonalize. Let us select e.g. $\hat{\mathbf{l}}^2$ and $\hat{l}_z$ (see Eq. (6.7)). First, we note that we must have $\hat{\mathbf{l}}^2 - \hat{l}_z^2 = \hat{l}_x^2 + \hat{l}_y^2 \geq 0$ (see Prob. 6.6). Let us choose for our Hilbert space basis a set of states $|l, m_l\rangle$ that have definite values of $\hat{\mathbf{l}}^2$ and $\hat{l}_z$. In this context, the ket $|l, m_l\rangle$ is an eigenket of both $\hat{\mathbf{l}}^2$ and $\hat{l}_z$, where m_l is the eigenvalue of $\hat{l}_z$, and, as we shall see below, it is also connected to the eigenvalue of $\hat{\mathbf{l}}^2$. Then, we have

$$\left\langle l, m_l \left| \hat{\mathbf{l}}^2 - \hat{l}_z^2 \right| l, m_l \right\rangle = \left\langle l, m_l \left| \hat{l}_x^2 + \hat{l}_y^2 \right| l, m_l \right\rangle \geq 0. \tag{6.19}$$

This means that the eigenvalues of $\hat{l}_z^2$ cannot exceed the eigenvalues of $\hat{\mathbf{l}}^2$, i.e. once the eigenvalue of $\hat{\mathbf{l}}^2$ is fixed, $\hat{l}_z^2$ is bounded[2]. Let us indicate by l the maximal eigenvalue of $\hat{l}_z$, that is

$$-l \leq m_l \leq l, \tag{6.20}$$

where l is called the *azimuthal quantum number*, while m_l, representing the eigenvalue of $\hat{l}_z$, is called the *magnetic quantum number*, so that we can write

$$\hat{l}_z |l, m_l\rangle = m_l |l, m_l\rangle . \tag{6.21}$$

In order to investigate the algebra of angular momentum[3] it is convenient to introduce raising and lowering operators (see also Subsec. 4.4.2)

$$\hat{l}_\pm = \hat{l}_x \pm \imath \hat{l}_y, \tag{6.22}$$

which satisfy

$$\hat{l}_- = \hat{l}_+^\dagger \tag{6.23}$$

and the commutation relations (see Prob. 6.7)

$$\left[\hat{l}_z, \hat{l}_\pm\right] = \pm\hat{l}_\pm, \left[\hat{l}_+, \hat{l}_-\right] = 2\hat{l}_z, \left[\hat{\mathbf{l}}^2, \hat{l}_\pm\right] = 0. \tag{6.24}$$

Let us now consider the action of $\hat{l}_z$ on the state vector $\hat{l}_+ |l, m_l\rangle$

$$\begin{aligned} \hat{l}_z\hat{l}_+ |l, m_l\rangle &= \left(\hat{l}_+\hat{l}_z + \left[\hat{l}_z, \hat{l}_+\right]\right) |l, m_l\rangle \\ &= m\hat{l}_+ |l, m_l\rangle + \hat{l}_+ |l, m_l\rangle \\ &= (m_l + 1)\hat{l}_+ |l, m_l\rangle . \end{aligned} \tag{6.25}$$

[2] Of course, this holds true also in the classical case, where the length of the projection of a vector along a certain direction cannot exceed the length of the original vector.

[3] On the abstract concept of algebra see also Subsecs. 8.4.3 and 8.4.4.

Table 6.1 Ordering of the "ascending" and "descending" angular momentum eigenstates and corresponding eigenvalues of $\hat{l}_z$

"Ascending" eigenstates	"Descending" eigenstates	Eigenvalue of $\hat{l}_z$
$\hat{l}_+^{2l}\|l,-l\rangle \propto \|l,l\rangle$	$\|l,l\rangle$	l
$\hat{l}_+^{2l-1}\|l,-l\rangle \propto \|l,l-1\rangle$	$\hat{l}_-\|l,l\rangle \propto \|l,l-1\rangle$	$l-1$
...	$\hat{l}_-^2\|l,l\rangle \propto \hat{l}_-\|l,l-1\rangle$ $\propto \|l,l-2\rangle$	$l-2$
...	...	...
$\hat{l}_+^2\|l,-l\rangle \propto \hat{l}_+\|l,-l+1\rangle$ $\propto \|l,-l+2\rangle$	...	$-l+2$
$\hat{l}_+\|l,-l\rangle \propto \|l,-l+1\rangle$	$\hat{l}_-^{2l-1}\|l,l\rangle \propto \|l,-l+1\rangle$	$-l+1$
$\|l,-l\rangle$	$\hat{l}_-^{2l}\|l,l\rangle \propto \|l,-l\rangle$	$-l$

Therefore, $\hat{l}_+ \left|l, m_l\right\rangle$ is an eigenvector of $\hat{l}_z$ with eigenvalue $m+1$. Then,

$$\hat{l}_+ \left|l, m_l\right\rangle \propto \left|l, m_l + 1\right\rangle . \tag{6.26a}$$

Proceeding in the same way for the state vector $\hat{l}_- \left|l, m_l\right\rangle$, we obtain

$$\hat{l}_- \left|l, m_l\right\rangle \propto \left|l, m_l - 1\right\rangle . \tag{6.26b}$$

Now, recalling that l is the maximal eigenvalue of $\hat{l}_z$ and that $-l$ is the minimal eigenvalue of $\hat{l}_z$, we also must have that $\hat{l}_+ \left|l, l\right\rangle = 0$ and $\hat{l}_- \left|l, -l\right\rangle = 0$. It is then possible to order the angular momentum eigenstates in descending order of the eigenvalue of $\hat{l}_z$, as in Tab. 6.1. Then, for any value of l, we have $2l+1$ possible states and, since $2l+1$ must be an integer, l should consequently be either integer or half-integer.

In order to find the eigenvalues of $\hat{\mathbf{l}}^2$, consider that

$$\begin{aligned}\hat{l}_-\hat{l}_+ &= \left(\hat{l}_x - \imath\hat{l}_y\right)\left(\hat{l}_x + \imath\hat{l}_y\right) = \hat{l}_x^2 + \hat{l}_y^2 + \imath\left[\hat{l}_x, \hat{l}_y\right] \\ &= \hat{\mathbf{l}}^2 - \hat{l}_z^2 - \hat{l}_z,\end{aligned} \tag{6.27}$$

from which it follows that

$$0 = \hat{l}_-\hat{l}_+ \left|l, l\right\rangle = \left(\hat{\mathbf{l}}^2 - \hat{l}_z^2 - \hat{l}_z\right)\left|l, l\right\rangle , \tag{6.28}$$

which implies

$$\left[\hat{\mathbf{l}}^2 - l(l+1)\right]\left|l, l\right\rangle = 0, \tag{6.29}$$

or

$$\hat{\mathbf{l}}^2 \left|l, l\right\rangle = l(l+1)\left|l, l\right\rangle , \tag{6.30}$$

which means that the eigenvalue of $\hat{\mathbf{l}}^2$ is $l(l+1)$. It is straightforward to generalize Eq. (6.30) to obtain (see Prob. 6.9)

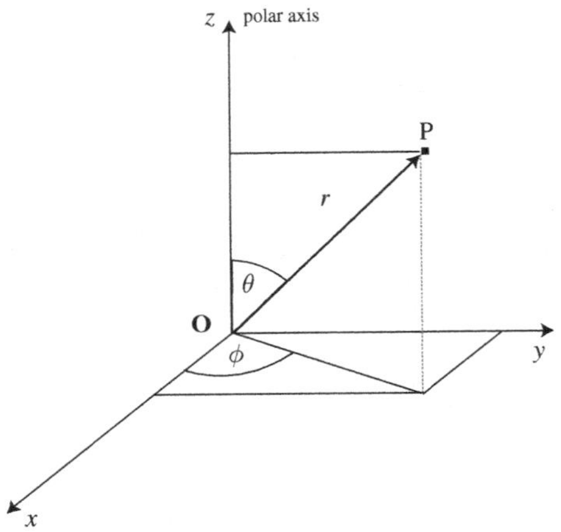

Figure 6.3 Relationship between rectangular (Cartesian) coordinates (x, y, z) and spherical coordinates (r, ϕ, θ) of a point P in the three-dimensional space, where r is the distance from the origin, ϕ is the azimuthal angle (from the x-axis, with $0 \le \phi < 2\pi$), and θ the polar angle (from the z-axis, with $0 \le \theta \le \pi$).

$$\hat{\mathbf{l}}^2 \left| l, m_l \right\rangle = l(l+1) \left| l, m_l \right\rangle . \tag{6.31}$$

This is a peculiarity of quantum mechanics, in that the eigenvalue of the square of $\hat{\mathbf{l}}$ is not the square of the eigenvalue of $\hat{\mathbf{l}}$. As we see from Eq. (6.27), it is a direct consequence of the fact that the angular momentum's components (in particular $\hat{l}_x$ and $\hat{l}_y$) do not commute with each other. Indeed, if they did commute, the last term in brackets in Eq. (6.28) would vanish and the eigenvalue of $\hat{\mathbf{l}}^2$ would be equal to l^2.

6.1.4 Eigenfunctions of the angular momentum

We have just built an abstract algebra of the angular momentum operator[4]. on the basis of commutation relations, i.e. the general features of the angular momentum's eigenvalues and eigenkets. For this algebra only integer or half-integer values of l are suitable. However, there are also further constraints which do not depend on this algebra. In particular, we must impose that the wave functions be probability amplitudes and therefore they must be single-valued. Due to the rotational invariance, it is better to use the spherical coordinates (see Fig. 6.3)

$$x = r \sin\theta \cos\phi, \quad y = r \sin\theta \sin\phi, \quad z = r \cos\theta, \tag{6.32}$$

[4] See again Subsec. 8.4.4.

whose inverse transformations are given by

$$r = \sqrt{x^2 + y^2 + z^2}, \quad \phi = \arctan\frac{y}{x}, \quad \theta = \arctan\frac{\sqrt{x^2 + y^2}}{z}. \tag{6.33}$$

Then, we have (see Eq. (6.2c))

$$\begin{aligned} -\imath\hbar\frac{\partial}{\partial\phi} &= -\imath\hbar\left(\frac{\partial x}{\partial\phi}\frac{\partial}{\partial x} + \frac{\partial y}{\partial\phi}\frac{\partial}{\partial y}\right) \\ &= -\imath\hbar\left(-\hat{y}\frac{\partial}{\partial x} + \hat{x}\frac{\partial}{\partial y}\right) = \hat{L}_z. \end{aligned} \tag{6.34}$$

This is a consequence of the fact that $\hat{L}_z$ generates the rotations about the z axis, i.e. the translations in ϕ. Let us denote by $\psi(r,\theta,\phi)$ the angular momentum eigenfunctions. This means that $\psi(r,\theta,\phi)$ must be simultaneously eigenfunctions of $\hat{l}_z$ and $\hat{\mathrm{l}}^2$. In particular, we have

$$\hat{l}_z\psi(r,\theta,\phi) = -\imath\frac{\partial}{\partial\phi}\psi(r,\theta,\phi) = m_l\psi(r,\theta,\phi). \tag{6.35}$$

Therefore, the eigenfunction of $\hat{l}_z$ corresponding to the eigenvalue m_l must be proportional to $e^{\imath m_l\phi}$ and we may write

$$\psi(r,\theta,\phi) = f(r,\theta)\frac{e^{\imath m_l\phi}}{\sqrt{2\pi}}. \tag{6.36}$$

If we make a rotation of 2π about the z-axis ($\phi \mapsto \phi + 2\pi$), we will obviously return to the initial position. Then, we must also have the same value of the wave function $\psi(r,\theta,\phi)$, and, as a consequence, m_l is an integer. Given that l is the maximum value of m_l, l also must be an integer (and so also $-l$). Moreover, in a superposition of the wave functions $\psi = fe^{\imath m_l\phi}$ and $\psi' = f'e^{\imath m_l'\phi}$, only relative phases are relevant (see Subsec. 2.1.3) so that in the expression

$$fe^{\imath m_l\phi} + f'e^{\imath m_l'\phi} = e^{\imath m_l\phi}\left(f + f'e^{\imath(m_l'-m_l)\phi}\right) \tag{6.37}$$

only the difference $m_l' - m_l$ is relevant, and therefore must be an integer – also the value $l = 0$ must exist.[5] Then, all the values of l must be integers. As we shall see in Sec. 6.3, this is not true for the spin angular momentum, because it is an internal degree of freedom that does not depend on the coordinates.

Let us now introduce the notation

$$\mathrm{F}_m(\phi) = \frac{e^{\imath m_l\phi}}{\sqrt{2\pi}}. \tag{6.38}$$

[5] In this case we have that ψ is a function of r only, and this represents a situation of spherical symmetry, as we shall see in the next section.

Then (see Eq. (2.145)),

$$\int_0^{2\pi} d\phi \mathrm{F}_m^*(\phi)\mathrm{F}_{m'}(\phi) = \frac{1}{2\pi}\int_0^{2\pi} d\phi e^{\imath(m_l'-m_l)\phi} = \delta_{m_l m_l'}, \tag{6.39}$$

i.e. the functions $\mathrm{F}_m(\phi)$ are orthonormal. We wish now to find the matrix elements of $\hat{l}_\pm$. Let us first calculate the mean value of the operator $\hat{l}_-\hat{l}_+$. We have

$$\left\langle l, m_l \left| \hat{l}_-\hat{l}_+ \right| l, m_l \right\rangle = \sum_{j=-l}^{l} \left\langle l, m_l \left| \hat{l}_- \right| l, j \right\rangle \left\langle l, j \left| \hat{l}_+ \right| l, m_l \right\rangle. \tag{6.40}$$

Since

$$\hat{l}_- \left| l, j \right\rangle \propto \left| l, j-1 \right\rangle \quad \text{and} \quad \hat{l}_+ \left| l, m_l \right\rangle \propto \left| l, m_l + 1 \right\rangle, \tag{6.41}$$

and the eigenvectors of $\hat{l}_z$ with different values of m_l must be orthonormal, the only term of the sum (6.40) that is different from zero is for $j = m_l + 1$, that is

$$\left\langle l, m_l \left| \hat{l}_-\hat{l}_+ \right| l, m_l \right\rangle = \left\langle l, m_l \left| \hat{l}_- \right| l, m_l + 1 \right\rangle \left\langle l, m_l + 1 \left| \hat{l}_+ \right| l, m_l \right\rangle. \tag{6.42}$$

Taking into account Eq. (6.27), we also have

$$\left\langle l, m_l \left| \hat{l}_-\hat{l}_+ \right| l, m_l \right\rangle = \left\langle l, m_l \left| \hat{\mathbf{l}}^2 - \hat{l}_z(\hat{l}_z + 1) \right| l, m_l \right\rangle. \tag{6.43}$$

Since $\hat{l}_- = \hat{l}_+^\dagger$, from Eqs. (6.21), (6.31), and (6.42)–(6.43), it follows that

$$\left| \left\langle l, m_l + 1 \left| \hat{l}_+ \right| l, m_l \right\rangle \right|^2 = l(l+1) - m_l(m_l + 1), \tag{6.44}$$

which in turn implies

$$\left\langle l, m_l + 1 \left| \hat{l}_+ \right| l, m_l \right\rangle = \sqrt{l(l+1) - m_l(m_l + 1)}. \tag{6.45}$$

Then, it easily follows that

$$\hat{l}_+ \left| l, m_l \right\rangle = \sqrt{l(l+1) - m_l(m_l + 1)} \left| l, m_l + 1 \right\rangle. \tag{6.46a}$$

By using the substitution $\hat{l}_+ = \hat{l}_-^\dagger$ in Eq. (6.42), we also obtain

$$\hat{l}_- \left| l, m_l \right\rangle = \sqrt{l(l+1) - m_l(m_l - 1)} \left| l, m_l - 1 \right\rangle. \tag{6.46b}$$

From the previous results we also have that

$$\hat{l}_-\hat{l}_+ \left| l, m_l \right\rangle = (l - m_l)(l + m_l + 1) \left| l, m_l \right\rangle, \tag{6.47a}$$

$$\hat{l}_+\hat{l}_- \left| l, m_l \right\rangle = (l + m_l)(l - m_l + 1) \left| l, m_l \right\rangle. \tag{6.47b}$$

In order to make some concrete examples, in the following we shall analyze the two simplest cases $l = 0$ and $l = 1$. The case $l = 0$ is trivial: we have only one (spherically symmetric) eigenstate, $|0, 0\rangle$, i.e.

$$\hat{\mathbf{l}}^2 \,|0, 0\rangle = 0. \tag{6.48}$$

For $l = 1$, we have $m_l = -1, 0, +1$, and therefore the relevant operators are represented by 3×3 matrices. In particular, it is easy to derive (see Prob. 6.10)

$$\hat{l}_z = \begin{bmatrix} 1 & 0 & 0 \\ 0 & 0 & 0 \\ 0 & 0 & -1 \end{bmatrix}, \quad \hat{l}_+ = \begin{bmatrix} 0 & \sqrt{2} & 0 \\ 0 & 0 & \sqrt{2} \\ 0 & 0 & 0 \end{bmatrix}, \quad \hat{l}_- = \begin{bmatrix} 0 & 0 & 0 \\ \sqrt{2} & 0 & 0 \\ 0 & \sqrt{2} & 0 \end{bmatrix}, \tag{6.49}$$

where we have defined the basis vectors as

$$|1, 1\rangle = \begin{pmatrix} 1 \\ 0 \\ 0 \end{pmatrix}, \quad |1, 0\rangle = \begin{pmatrix} 0 \\ 1 \\ 0 \end{pmatrix}, \quad |1, -1\rangle = \begin{pmatrix} 0 \\ 0 \\ 1 \end{pmatrix}. \tag{6.50}$$

The matrices corresponding to $\hat{l}_x$ and $\hat{l}_y$ can be calculated in terms of $\hat{l}_+$ and $\hat{l}_-$. The matrix $\hat{l}_x$ is given by

$$\hat{l}_x = \frac{\hat{l}_+ + \hat{l}_-}{2} = \frac{1}{2}\begin{bmatrix} 0 & \sqrt{2} & 0 \\ \sqrt{2} & 0 & \sqrt{2} \\ 0 & \sqrt{2} & 0 \end{bmatrix}, \tag{6.51a}$$

whereas $\hat{l}_y$ is given by

$$\hat{l}_y = \frac{\hat{l}_+ - \hat{l}_-}{2\imath} = \frac{1}{2\imath}\begin{bmatrix} 0 & \sqrt{2} & 0 \\ -\sqrt{2} & 0 & \sqrt{2} \\ 0 & -\sqrt{2} & 0 \end{bmatrix}. \tag{6.51b}$$

As we have already done for $\hat{l}_z = -\imath \partial/\partial\phi$ (see Eq. (6.35)), it is useful to express $\hat{l}_+$ and $\hat{l}_-$ in spherical coordinates. We start from the definitions of $\hat{l}_x$ and $\hat{l}_y$ (see Eqs. (6.2a) and (6.2b)), and we have then to express the partial derivatives with respect to x, y, and z in terms of the spherical coordinates (6.33) (see Prob. 6.11). We find

$$\begin{aligned} \frac{\partial}{\partial x} &= \frac{\partial r}{\partial x}\frac{\partial}{\partial r} + \frac{\partial \phi}{\partial x}\frac{\partial}{\partial \phi} + \frac{\partial \theta}{\partial x}\frac{\partial}{\partial \theta} \\ &= \sin\theta\cos\phi\frac{\partial}{\partial r} - \frac{1}{r}\frac{\sin\phi}{\sin\theta}\frac{\partial}{\partial\phi} + \frac{1}{r}\cos\theta\cos\phi\frac{\partial}{\partial\theta}, \end{aligned} \tag{6.52a}$$

$$\begin{aligned} \frac{\partial}{\partial y} &= \frac{\partial r}{\partial y}\frac{\partial}{\partial r} + \frac{\partial \phi}{\partial y}\frac{\partial}{\partial \phi} + \frac{\partial \theta}{\partial y}\frac{\partial}{\partial \theta} \\ &= \sin\theta\sin\phi\frac{\partial}{\partial r} + \frac{1}{r}\frac{\cos\phi}{\sin\theta}\frac{\partial}{\partial\phi} + \frac{1}{r}\cos\theta\sin\phi\frac{\partial}{\partial\theta}, \end{aligned} \tag{6.52b}$$

$$\begin{aligned} \frac{\partial}{\partial z} &= \frac{\partial r}{\partial z}\frac{\partial}{\partial r} + \frac{\partial \phi}{\partial z}\frac{\partial}{\partial \phi} + \frac{\partial \theta}{\partial z}\frac{\partial}{\partial \theta} \\ &= \cos\theta\frac{\partial}{\partial r} - \frac{1}{r}\sin\theta\frac{\partial}{\partial\theta}. \end{aligned} \tag{6.52c}$$

Back-substituting these expressions into Eqs. (6.2), we obtain

$$\hat{l}_x = \imath \left(\sin\phi \frac{\partial}{\partial\theta} + \cos\phi \cot\theta \frac{\partial}{\partial\phi} \right), \tag{6.53a}$$

$$\hat{l}_y = -\imath \left(\cos\phi \frac{\partial}{\partial\theta} - \sin\phi \cot\theta \frac{\partial}{\partial\phi} \right), \tag{6.53b}$$

from which we can easily build $\hat{l}_{\pm}$, i.e.

$$\hat{l}_+ = e^{\imath\phi} \left(\frac{\partial}{\partial\theta} + \imath \cot\theta \frac{\partial}{\partial\phi} \right), \tag{6.54a}$$

$$\hat{l}_- = e^{-\imath\phi} \left(\frac{\partial}{\partial\theta} - \imath \cot\theta \frac{\partial}{\partial\phi} \right). \tag{6.54b}$$

Similarly, for $\hat{\mathbf{l}}^2$ we have from Eqs. (6.34) and (6.53)

$$\begin{aligned} \hat{\mathbf{l}}^2 &= \hat{l}_x^2 + \hat{l}_y^2 + \hat{l}_z^2 \\ &= -\frac{1}{\sin\theta}\frac{\partial}{\partial\theta}\left(\sin\theta\frac{\partial}{\partial\theta}\right) - \frac{1}{\sin^2\theta}\frac{\partial^2}{\partial\phi^2}, \end{aligned} \tag{6.55}$$

from which we see that $\hat{\mathbf{l}}^2$ is essentially the angular part of the Laplacian in spherical coordinates (see Prob. 6.12). Therefore,

$$\hat{\mathbf{l}}^2 \psi(r,\theta,\phi) = l(l+1)\psi(r,\theta,\phi) \tag{6.56}$$

is similar to the Laplace equation and its solutions were already found by Laplace. In other words, we may assume that the total eigenfunctions $\psi(r,\phi,\theta)$ are a product of two functions, one depending on r alone, and the other on ϕ and θ, i.e.

$$\psi(r,\theta,\phi) = \mathrm{R}(r)Y(\phi,\theta). \tag{6.57}$$

Given the eigenvalue equations (6.21) and (6.31), the functions $Y(\phi,\theta)$ must depend on the angular momentum quantum numbers l and m. We may therefore label such functions by these indices, so that, since both $\hat{\mathbf{l}}^2$ and $\hat{l}_z$ do not depend on r (see Eqs. (6.34) and (6.55)), we have

$$\hat{\mathbf{l}}^2 Y_{lm}(\phi,\theta) = l(l+1)Y_{lm}(\phi,\theta), \tag{6.58a}$$

$$\hat{l}_z Y_{lm}(\phi,\theta) = m_l Y_{lm}(\phi,\theta). \tag{6.58b}$$

These functions have to be normalized according to

$$\int d\Omega Y^*_{l'm'}(\phi,\theta)Y_{lm}(\phi,\theta) = \delta_{ll'}\delta_{mm'}, \tag{6.59}$$

where $d\Omega = \sin\theta d\theta d\phi$ is the element of solid angle. The functions $Y_{lm}(\phi,\theta)$ that verify these conditions are the so-called *spherical harmonics*.[6] Since we already know the eigenfunctions of $\hat{l}_z$ (see Eq. (6.36)), by making use of the definition (6.38) we have

[6] See [*Byron/Fuller* 1969–70, 253–61].

$$Y_{lm}(\phi, \theta) = \mathrm{F}_m(\phi)\Theta_{lm}(\theta) = \frac{e^{\imath m_l \phi}}{\sqrt{2\pi}}\Theta_{lm}(\theta), \tag{6.60}$$

and (see Tab. 6.1)

- $\hat{l}_+ Y_{ll}(\phi, \theta) = 0$;
- $\hat{l}_- Y_{lm}(\phi, \theta) \propto Y_{l,m-1}(\phi, \theta)$.

From the first condition, we have (see Eq. (6.54a))

$$\left(\frac{\partial}{\partial \theta} + \imath \cot\theta \frac{\partial}{\partial \phi}\right) e^{\imath l \phi}\Theta_{ll}(\theta) = 0, \tag{6.61}$$

that is

$$e^{\imath l \phi}\left(\frac{\partial}{\partial \theta} - l \cot\theta\right)\Theta_{ll}(\theta) = 0, \tag{6.62}$$

which implies

$$\frac{\partial}{\partial \theta}\Theta_{ll}(\theta) = l \cot\theta \Theta_{ll}(\theta), \tag{6.63}$$

whose solution is $\Theta_{ll}(\theta) = \mathcal{N}(\sin\theta)^l$ (see Prob. 6.13). Now, the constant $\mathcal{N}$ can be obtained from the normalization condition (see Prob. 6.14), so that we finally have[7]

$$\Theta_{ll}(\theta) = \frac{(-\imath)^l}{2^l l!}\sqrt{\frac{(2l+1)!}{2}}(\sin\theta)^l. \tag{6.64}$$

In general, the functions $\Theta_{lm}(\theta)$ are solutions of the Legendre's associated equation,[8] i.e. Legendre associated polynomials. The most general form of the Y_{lm} functions is therefore

$$Y_{lm}(\phi, \theta) = \frac{e^{\imath m_l \phi}}{\sqrt{2\pi}}(-\imath)^l\sqrt{\frac{(2l+1)(l+m_l)!}{2(l-m_l)!}}\frac{1}{2^l l!}\frac{1}{(\sin\theta)^{m_l}}\frac{d^{l-m_l}}{d\cos\theta^{l-m_l}}(\sin\theta)^{2l}, \tag{6.65}$$

where the expressions

$$\frac{d^{l-m_l}}{d\cos\theta^{l-m_l}}(\sin\theta)^{2l} \tag{6.66}$$

are (up to a constant factor) the Legendre polynomials (as functions of $\sin\theta$ and $\cos\theta$). We explicitly calculate the first ones. For $l = m_l = 0$, we have

$$Y_{0,0}(\phi, \theta) = \frac{1}{\sqrt{4\pi}}. \tag{6.67}$$

7 The factor $(-\imath)^l$ in Eq. (6.64) corresponds to a phase choice that has been made for convenience.

8 See [*Byron/Fuller* 1969–70, 260].

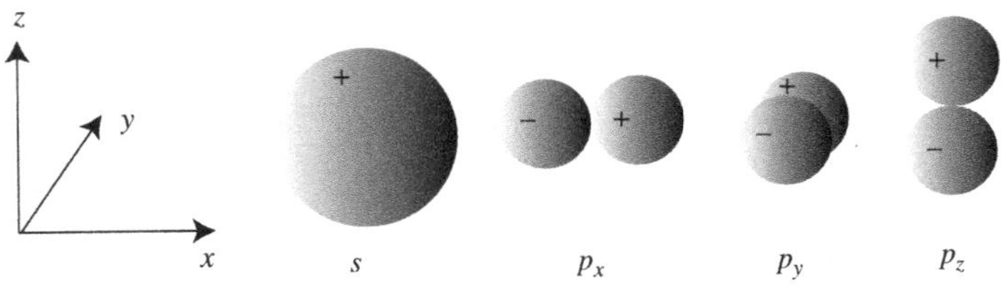

Figure 6.4 Pictorial representation of s- and p-states. s represents a spherically symmetric state and the circle corresponds to a contour of constant probability density (the grey scale is used here only for pictorial purposes). For p_x, p_y, p_z a contour of constant probability density will be a pair of closed surfaces. Note that in Eqs. (6.67), (6.70), and (6.71) we have not considered the radial part (see Eq. (6.57)).

For $l = 1$, we have

$$Y_{1,0}(\phi,\theta) = \imath\sqrt{\frac{3}{4\pi}}\cos\theta, \tag{6.68a}$$

$$Y_{1,\pm 1}(\phi,\theta) = \pm\imath\sqrt{\frac{3}{8\pi}}e^{\pm\imath\phi}\sin\theta, \tag{6.68b}$$

and, for $l = 2$,

$$Y_{2,0}(\phi,\theta) = -\sqrt{\frac{5}{16\pi}}\left(3\cos^2\theta - 1\right), \tag{6.69a}$$

$$Y_{2,\pm 1}(\phi,\theta) = \pm\sqrt{\frac{15}{8\pi}}e^{\pm\imath\phi}\sin\theta\cos\theta, \tag{6.69b}$$

$$Y_{2,\pm 2}(\phi,\theta) = \pm\sqrt{\frac{15}{32\pi}}e^{\pm 2\imath\phi}\sin^2\theta. \tag{6.69c}$$

States for $l = 0$ are called s-states.[9] The symbol s stands for *sharp*: at the very beginning of atomic spectroscopy, this term was used to indicate the spectral lines that did not show the presence of further sub-lines. As we have said, they are spherically symmetric (see Fig. 6.4 and also 11.6). For states of $l = 1$, it may be convenient to write the spherical harmonics (6.68a) as $Y_{1,0}(\phi,\theta) = \imath\sqrt{3/4\pi}\,z/r$. For the (6.68b), it may be convenient to take the combination $(Y_{1,1} \pm Y_{1,-1})/\sqrt{2}$, so that we have the three states

$$\sqrt{\frac{3}{4\pi}}\frac{x}{r}, \quad \sqrt{\frac{3}{4\pi}}\frac{y}{r}, \quad \sqrt{\frac{3}{4\pi}}\frac{z}{r}, \tag{6.70}$$

which are called p-states (p standing here for *principal*). The first two p-states are states with zero angular momentum around x- and y-axes, respectively. They are analogous to the three components of a polar vector: they are zero in the central plane perpendicular to the axes, positive on one side of the plane, and negative on the other. States with $l = 2$ are called d-states (d stands for *diffusive*) and the (not normalized) five cubic harmonics can be chosen to have symmetry of the type

[9] See [*Harrison* 2000, 34–36].

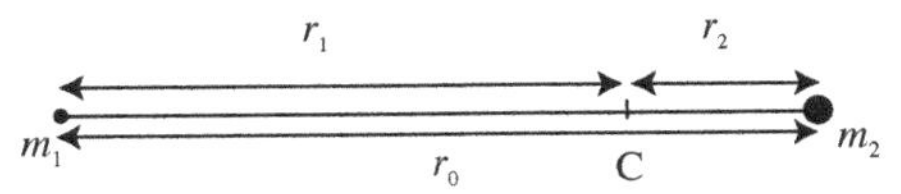

Figure 6.5 Rigid rotator model. Two point-like masses m_1 and m_2 are connected by a rigid light bar of length r_0. C is the center of mass of the system, and r_1 and r_2 are the distances of the two masses from the center of mass.

$$\frac{xy}{r^2}, \quad \frac{yz}{r^2}, \quad \frac{zx}{r^2}, \quad \frac{3z^2 - r^2}{r^2}, \quad \frac{x^2 - y^2}{r^2}. \tag{6.71}$$

corresponding to $d_{xy}, d_{yz}, d_{xz}, d_{z^2}$, and $d_{x^2-y^2}$, respectively.

In general, the correspondence between the name of the wave and the angular momentum is given by

$$\begin{array}{ccccccc} \text{waves} & s & p & d & f & g & h \\ l & 0 & 1 & 2 & 3 & 4 & 5 \end{array}. \tag{6.72}$$

6.2 Special examples

In this section, we shall provide some first applications of the previous formalism: the rigid rotator (Subsec. 6.2.1), the central potential (Subsec. 6.2.2), the particle in a constant magnetic field (Subsec. 6.2.3), and the harmonic oscillator in several dimensions (Subsec. 6.2.4).

6.2.1 Rigid rotator

Let us consider the motion of two point-like masses m_1 and m_2 rigidly connected by an infinitely thin massless bar (see Fig. 6.5). The center of mass of the rotator is located at distances r_1 and r_2 from masses m_1 and m_2, respectively, where

$$r_1 = \frac{m_2}{m_1 + m_2} r_0, \quad r_2 = \frac{m_1}{m_1 + m_2} r_0. \tag{6.73}$$

As any rigid object in the three-dimensional physical space, the rotator has six degrees of freedom: the three translational degrees of freedom of the center-of-mass motion and the three rotational degrees of freedom about the center of mass, represented, e.g., by the three Euler angles (see Subsec. 6.1.2). If, for the time being, we forget about the "trivial" center-of-mass translations, we are left with the remaining degrees of freedom described by the dynamical variables and their conjugate angular momenta.

The classical energy of a rigid rotator is given by its usual kinetic component

$$T = \frac{1}{2} \mathrm{I} \omega^2, \tag{6.74}$$

where I is the moment of inertia of the system about the rotation axis and ω is the angular frequency (see Eq. (4.47)). Since the angular momentum of the system is given by

$$\mathbf{L} = \mathrm{I}\boldsymbol{\omega}, \tag{6.75}$$

Eq. (6.74) may be rewritten as

$$T = \frac{\mathbf{L}^2}{2\mathrm{I}}. \tag{6.76}$$

The moment of inertia of the rotator about the rotation axis, i.e.

$$\mathrm{I} = m_1 r_1^2 + m_2 r_2^2, \tag{6.77}$$

may be also expressed as

$$\mathrm{I} = m r_0^2, \tag{6.78}$$

where

$$m = \frac{m_1 m_2}{m_1 + m_2} \tag{6.79}$$

is the reduced mass. The quantum counterpart of Eq. (6.76) directly gives the Hamiltonian

$$\hat{H} = \frac{\hat{\mathbf{L}}^2}{2\mathrm{I}}, \tag{6.80}$$

whose eigenfunctions are given by the spherical harmonics (6.65) and whose eigenvalues are equal to (see Eq. (6.58a) and Fig. 6.6)

$$E_l = \frac{\hbar^2}{2\mathrm{I}} l(l+1). \tag{6.81}$$

We stress the fact that these eigenvalues should be referred to as the total angular momentum (which also includes the spin). Since, however, we are not considering the spin here, we limit ourselves to the orbital momentum component, developing a full formalism in other parts of the book (see Sec. 12.3). The transitions between adjacent levels involve emission (absorbtion) of radiation quanta of frequency (see Eqs. (1.75) and (1.76), and also Prob. 6.15)

$$\nu_{l,l+1} = \frac{\Delta E_{l,l+1}}{h} = 2\mathrm{B}(l+1), \tag{6.82}$$

where

$$\mathrm{B} = \frac{h}{8\pi^2 \mathrm{I}} \tag{6.83}$$

is a rotational constant.

6.2.2 Central potential

In this subsection we wish to establish the most general features of the motion of a quantum particle subjected to a central potential, which is strictly connected with the matter we have dealt with in the previous section. As we shall see below, in this case the angular part of the energy eigenfunctions coincides with the spherical harmonics. Moreover, due to the

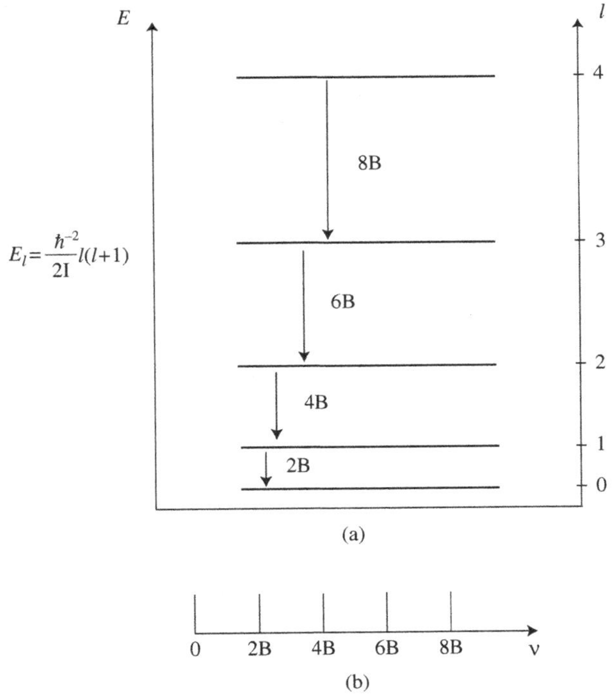

Figure 6.6 (a) Energy levels for a rigid rotator as a function of the angular-momentum eigenvalue l. The arrows show the transitions between adjacent levels. (b) The frequency spectrum of the rotator is a set of equidistant lines. The rotational constant B is defined in Eq. (6.83).

central symmetry of the problem, and the consequent rotational invariance, changing the direction of the orbital angular momentum does not change the energy of the particle.

Let us consider the general Hamiltonian

$$\hat{H} = -\frac{\hbar^2}{2m}\Delta + V(r), \tag{6.84}$$

where the potential $V(r)$ does not depend on θ or ϕ. Since the Hamiltonian $\hat{H}$ is spherically symmetric, it commutes with each of the components of the angular momentum, and, as a consequence, also with $\hat{\mathbf{l}}^2$. Moreover, $\hat{l}_z$ commutes with $\hat{\mathbf{l}}^2$ (see Eq. (6.7)), so that $\hat{H}$, $\hat{\mathbf{l}}^2$ and $\hat{l}_z$ represent a complete set of simultaneously measurable observables (see Subsecs. 2.1.5 and 2.2.7; see also Prob. 6.16). In other words, simultaneous eigenfunctions of $\hat{\mathbf{l}}^2$ and $\hat{l}_z$ are in this case also eigenfunctions of the Hamiltonian. We may then write (as in Eq. (6.57)) the eigenfunctions of the Hamiltonian as

$$\psi(\mathbf{r}) = \mathrm{R}(r)Y_{lm}(\phi, \theta), \tag{6.85}$$

where $Y_{lm}(\phi, \theta)$ are just the spherical harmonics (6.65), so that the problem reduces to that of finding the radial wave function $\mathrm{R}(r)$ $(0 \leq r < \infty)$. Recalling Eqs. (6.84), (6.255) (see Prob. 6.12), and (6.58a), the Schrödinger equation

$$\hat{H}\psi(\mathbf{r}) = E\psi(\mathbf{r}) \tag{6.86}$$

may be rewritten as

$$\left[-\frac{\hbar^2}{2mr^2}\frac{\partial}{\partial r}\left(r^2\frac{\partial}{\partial r}\right)+\frac{\hbar^2 l(l+1)}{2mr^2}+V(r)-E\right]\mathrm{R}(r)=0. \tag{6.87}$$

If we introduce the operator corresponding to the radial component of the momentum

$$\hat{p}_r=-\imath\hbar\frac{1}{r}\frac{\partial}{\partial r}r=-\imath\hbar\left(\frac{\partial}{\partial r}+\frac{1}{r}\right), \tag{6.88}$$

the Hamiltonian $\hat{H}$ may be rewritten in the form

$$\hat{H}=\frac{1}{2m}\left(\hat{p}_r^2+\frac{\hbar^2\hat{\mathbf{l}}^2}{r^2}\right)+V(r), \tag{6.89}$$

which is identical to the classical Hamiltonian in the spherical coordinates.[10]

We now introduce the substitution $\mathrm{R}(r)=\xi(r)/r$, so that we have

$$\begin{aligned}\frac{1}{r^2}\frac{\partial}{\partial r}\left(r^2\frac{\partial}{\partial r}\right)\frac{\xi}{r}&=\frac{1}{r^2}\frac{\partial}{\partial r}\left(r\xi'-\xi\right)\\&=\frac{\xi''}{r},\end{aligned} \tag{6.90}$$

where

$$\xi'(r)=\frac{d}{dr}\xi(r)=\frac{\partial}{\partial r}\xi(r). \tag{6.91}$$

Therefore, Eq. (6.87) reduces finally to

$$-\frac{\hbar^2}{2m}\xi''+\left[\frac{\hbar^2 l(l+1)}{2mr^2}+V(r)-E\right]\xi=0, \tag{6.92}$$

which is identical to the one-dimensional Schrödinger equation (see, e.g., Eq. (3.10)) apart from the additive term

$$V_c(r)=\frac{\hbar^2 l(l+1)}{2mr^2}, \tag{6.93}$$

which plays the role of a "centrifugal barrier." Using the normalization condition of the spherical harmonics (6.59), the three-dimensional normalization condition for the wave functions (see Eq. (2.108))

$$\int d^3r\psi^*(\mathbf{r})\psi(\mathbf{r})=1 \tag{6.94}$$

becomes

$$\int_0^\infty dr\mathrm{R}^*(r)\mathrm{R}(r)r^2=1, \tag{6.95}$$

[10] See [*Goldstein* 1950, Secs. 9.4 and 9.7].

or also

$$\int_0^\infty dr \xi^*(r)\xi(r) = 1. \tag{6.96}$$

If $l = 0$, the centrifugal-barrier term vanishes. In this case we say that the system is in an s-wave (see Fig. 6.4). Therefore, for s-waves, the central-potential Schrödinger equation reduces to a one-dimensional problem with the same potential – with the extra condition that $\xi = 0$ at the origin (otherwise $\mathrm{R}(r)$ would diverge). Clearly, it is possible to reformulate the problem in terms of a new variable x $(-\infty < x < +\infty)$ by means of the potential transformation

$$V(x) = \begin{cases} V(r) + \hbar^2 l(l+1)(2mr^2)^{-1} & \text{for } x = r > 0, \\ \infty & \text{for } x < 0, \end{cases} \tag{6.97}$$

that is, for $x < 0$, we have an infinite potential barrier. For this reason, the wave function must vanish for $x \leq 0$ – see property (iii) of the Schrödinger equation in Subsec. 3.2.1. Neglecting the zero at the origin, the ground-state wave function $\xi(r)$ has no zeros (see property (iv) of Subsec. 3.2.1), the first excited state just one, and so on (see property (ii) of Subsec. 3.2.3). The number n of non-trivial zeros of the wave function is called the *radial quantum number*. Therefore, a complete three-dimensional eigenfunction may be labelled by the three quantum numbers n, l, m_l.

Let us now consider the case in which

$$\lim_{r \to 0} r^2 V(r) = 0, \tag{6.98}$$

Then, for small r, the centrifugal term dominates, and the behavior of the wave function is ruled by

$$\xi(r \to 0) \simeq r^a. \tag{6.99}$$

In fact, taking the limit of Eq. (6.92) for $r \to 0$, we obtain

$$l(l+1) - a(a-1) = 0, \tag{6.100}$$

which is satisfied for $a = l + 1$ and for $a = -l$. The latter, however, is not a good solution because the wave function would not vanish at the origin. Therefore, we have

$$\xi(r \to 0) \simeq r^{l+1}, \quad \mathrm{R}(r \to 0) \simeq r^l. \tag{6.101}$$

6.2.3 Particle in constant magnetic field

Let us consider the three-dimensional motion of a (spinless) quantum particle of mass m and charge e embedded in a constant magnetic field **B** directed along the z-axis.[11] In the

[11] For a treatment of this problem, taking into account the spin degree of freedom, see Subsec. 6.3.4.

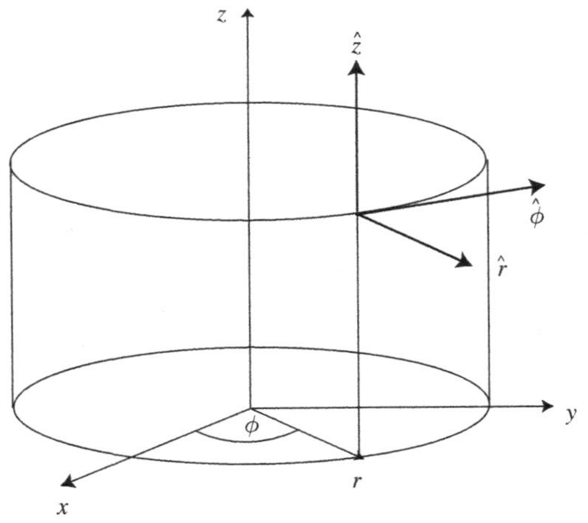

Figure 6.7 **Change of variables from the Cartesian reference frame (x,y,z) to the cylindrical coordinates (r,ϕ,z).**

absence of an electric field ($\mathbf{E} = 0$), from Eqs. (4.128) we may assume that the scalar potential vanishes ($\mathrm{U} = 0$) and that the vector potential is given by[12]

$$\mathbf{A} = -\frac{B}{2}\hat{y}\boldsymbol{\imath} + \frac{B}{2}\hat{x}\boldsymbol{\jmath} + 0\boldsymbol{k}, \tag{6.102}$$

where B is the intensity of the magnetic field. Recalling Eq. (4.133), the total Hamiltonian of the particle can be written in the form

$$\begin{aligned}\hat{H} &= \frac{1}{2m}\left[\left(\hat{p}_x + \frac{eB}{2}\hat{y}\right)\boldsymbol{\imath} + \left(\hat{p}_y - \frac{eB}{2}\hat{x}\right)\boldsymbol{\jmath} + \hat{p}_z\boldsymbol{k}\right]^2 \\ &= \frac{\hat{p}_z^2}{2m} + \frac{1}{2m}\left(\hat{p}_x^2 + \hat{p}_y^2\right) - \frac{e\hbar B}{2m}\hat{l}_z + \frac{e^2B^2}{8m}\left(\hat{x}^2 + \hat{y}^2\right).\end{aligned} \tag{6.103}$$

The observables $\hat{p}_z$, $\hat{l}_z$, and

$$\hat{H}_r = \frac{1}{2m}\left(\hat{p}_x^2 + \hat{p}_y^2\right) - \frac{e\hbar B}{2m}\hat{l}_z + \frac{e^2B^2}{8m}\left(\hat{x}^2 + \hat{y}^2\right), \tag{6.104}$$

which is the "planar" part of the Hamiltonian, constitute a complete set of commuting constants of motion (see Subsec. 3.6.1 and Prob. 6.17). In order to find the energy levels of the quantum particle in the homogeneous magnetic field, we must solve the Schrödinger equation

$$\hat{H}\psi(r,\phi,z) = E\psi(r,\phi,z), \tag{6.105}$$

where we have moved to cylindrical coordinates r, ϕ, z (see Fig. 6.7), such that

$$x = r\cos\phi, \quad y = r\sin\phi, \quad z = z. \tag{6.106}$$

[12] Notice that the expression (6.102) of the vector potential satisfies the Coulomb-gauge condition $\nabla \cdot \mathbf{A} = 0$ [see Eq. (13.7)].

Given the presence in Eq. (6.103) of the term

$$\frac{\hat{p}_z^2}{2m}, \tag{6.107}$$

which is proportional to $\partial^2/\partial z^2$, and of the term

$$-\frac{e\hbar B}{2m}\hat{l}_z, \tag{6.108}$$

which is proportional to $\partial/\partial\phi$, it is natural to make the ansatz

$$\psi(r,\phi,z) = e^{\frac{\imath}{\hbar}p_z z}e^{\imath m_l\phi}\varphi(r), \tag{6.109}$$

which allows the problem to be separated. If we replace Eq. (6.109) into Eq. (6.105), then we are left with a simple harmonic oscillator equation (in the variable r) for the function $\varphi(r)$, with frequency $\omega = eB/m$. It turns out that the constant of motion $\hat{p}_z$ may assume any value from $-\infty$ to $+\infty$, whereas $\hat{l}_z$ may take the values $m_l = 0, \pm 1, \pm 2, \ldots$. Furthermore, the third constant of motion $\hat{H}_r$ has harmonic-oscillator-type eigenvalues

$$(2n+1)\frac{e\hbar B}{2m}. \tag{6.110}$$

In conclusion, the total energy eigenvalues of the particle are given by

$$E_n = \frac{\hbar^2 k_z^2}{2m} + (2n+1)\frac{e\hbar B}{2m}. \tag{6.111}$$

6.2.4 Harmonic oscillator in several dimensions

It is possible to consider a particle subjected to a harmonic-oscillator potential (see Sec. 4.4) in k-dimensions, with $k > 1$. When the frequencies of the harmonic oscillator are the same in all directions, the problem is isotropic. In this case, the total Hamiltonian is

$$\hat{H} = \sum_{j=1}^{k} \hat{H}_j, \tag{6.112}$$

$\hat{H}_j$ being the Hamiltonian of the j-th dimension of the harmonic oscillator

$$\hat{H}_j = \frac{\hat{p}_j^2}{2m} + \frac{1}{2}m\omega^2\hat{r}_j^2, \tag{6.113}$$

where $\hat{r}_j$ is the j-th component of the position observable $\hat{\mathbf{r}}$. Therefore, this problem is separable and the total Hilbert space is the direct sum of the Hilbert spaces of the k one-dimensional harmonic oscillators. In the following, we shall consider the three-dimensional case and let the reader solve the two-dimensional one (see Prob. 6.19).[13]

[13] For a general treatment of the k-dimensional isotropic harmonic oscillator see [*Messiah* 1958, 451–54].

When $k = 3$, we have

$$\hat{H} = \frac{\hat{\mathbf{p}}^2}{2m} + \frac{1}{2m}\omega^2\hat{\mathbf{r}}^2$$
$$= \frac{1}{2m}\left[\left(\hat{p}_x^2 + m^2\omega^2\hat{x}^2\right) + \left(\hat{p}_y^2 + m^2\omega^2\hat{y}^2\right) + \left(\hat{p}_z^2 + m^2\omega^2\hat{z}^2\right)\right]. \quad (6.114)$$

A complete orthonormal set of eigenvectors of this Hamiltonian will be given by

$$\left|n_x, n_y, n_z\right\rangle = \left|n_x\right\rangle \otimes \left|n_y\right\rangle \otimes \left|n_z\right\rangle, \quad (6.115)$$

where each of the n_x, n_y, n_z is a positive integer or zero. In fact, we have (see Eqs. (4.50) and (4.72))

$$\hat{H}_x\left|n_x\right\rangle = \left(n_x + \frac{1}{2}\right)\hbar\omega\left|n_x\right\rangle, \quad (6.116a)$$

$$\hat{H}_y\left|n_y\right\rangle = \left(n_y + \frac{1}{2}\right)\hbar\omega\left|n_y\right\rangle, \quad (6.116b)$$

$$\hat{H}_z\left|n_z\right\rangle = \left(n_z + \frac{1}{2}\right)\hbar\omega\left|n_z\right\rangle, \quad (6.116c)$$

so that

$$\hat{H}\left|n_x, n_y, n_z\right\rangle = \left(n + \frac{3}{2}\right)\hbar\omega\left|n_x, n_y, n_z\right\rangle. \quad (6.117)$$

where

$$n = n_x + n_y + n_z. \quad (6.118)$$

In other words, the total energy eigenvalue

$$E_n = \left(n + \frac{3}{2}\right)\hbar\omega \quad (6.119)$$

depends only on the sum of the component eigenvalues. This means that for any given value of n there are several possible combinations of values for the three components n_x, n_y, n_z. In the general case, this number is given by

$$d = \frac{(n + k - 1)\,!}{n!\,(k - 1)\,!}, \quad (6.120)$$

i.e. the number of possible ways of casting n indistinguishable objects into k boxes (see also Eq. (7.22)), which may also be seen as the degree of degeneracy of the eigenvalue E_n (see Th. 2.2: p. 47, and Subsec. 3.1.4). In the case $k = 3$ this number reduces to $(n + 2)(n + 1)/2$.

We shall denote the vacuum (ground) state by

$$\left|\mathbf{0}\right\rangle = \left|0, 0, 0\right\rangle, \quad (6.121)$$

as the state where there are no energy quanta (see p. 158). The energy eigenvalues, the eigenvectors, and the total number of energy quanta for the ground state and the first few excited states are reported in Tab. 6.2. Figure 6.8 shows a schematic drawing of the populated energy levels as a function of the energy and angular momentum eigenvalues.

Table 6.2 Eigenvalues, number of energy quanta, eigenvectors, degree of degeneracy, and orbital quantum number for the three-dimensional harmonic oscillator

E	n	Eigenvectors	d	l
$\frac{3}{2}\hbar\omega$	0	$\lvert\mathbf{0}\rangle = \lvert 0,0,0\rangle$	1	0
$\frac{5}{2}\hbar\omega$	1	$\lvert 0,0,1\rangle, \lvert 0,1,0\rangle, \lvert 1,0,0\rangle$	3	1
$\frac{7}{2}\hbar\omega$	2	$\lvert 0,1,1\rangle, \lvert 1,0,1\rangle, \lvert 1,1,0\rangle,$ $\lvert 0,0,2\rangle, \lvert 0,2,0\rangle, \lvert 2,0,0\rangle$	6	0, 2
$\frac{9}{2}\hbar\omega$	3	$\lvert 1,1,1\rangle, \lvert 2,1,0\rangle, \lvert 2,0,1\rangle, \lvert 0,2,1\rangle, \lvert 1,2,0\rangle,$ $\lvert 0,1,2\rangle, \lvert 1,0,2\rangle, \lvert 3,0,0\rangle, \lvert 0,3,0\rangle, \lvert 0,0,3\rangle$	10	1, 3
...	...	...	...	...

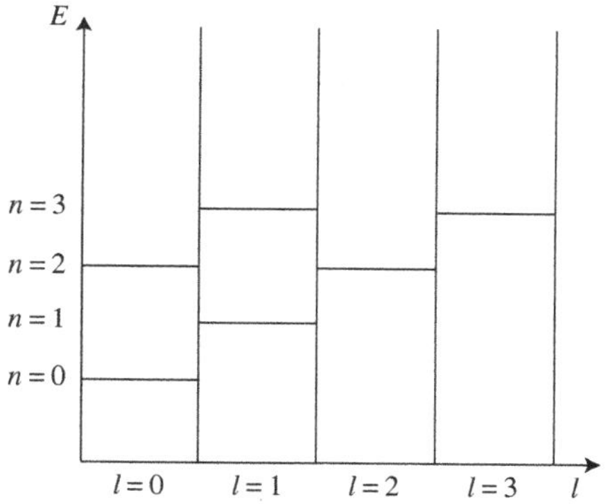

Figure 6.8 A schematic representation of the first few energy-levels of the three-dimensional harmonic oscillator with respect to *n* and *l*.

As in the one-dimensional case (see Eqs. (4.73)), it is convenient at this point to introduce the annihilation operators

$$\hat{a}_x = \sqrt{\frac{m}{2\hbar\omega}}\left(\omega\hat{x} + \imath\frac{\hat{p}_x}{m}\right), \tag{6.122a}$$

$$\hat{a}_y = \sqrt{\frac{m}{2\hbar\omega}}\left(\omega\hat{y} + \imath\frac{\hat{p}_y}{m}\right), \tag{6.122b}$$

$$\hat{a}_z = \sqrt{\frac{m}{2\hbar\omega}}\left(\omega\hat{z} + \imath\frac{\hat{p}_z}{m}\right), \tag{6.122c}$$

and the corresponding creation operators $\hat{a}_x^\dagger, \hat{a}_y^\dagger, \hat{a}_z^\dagger$, which satisfy the commutation relations (see also Eq. (4.74))

$$\left[\hat{a}_j, \hat{a}_l^\dagger\right] = \delta_{jl}, \tag{6.123a}$$

$$\left[\hat{a}_j, \hat{a}_l\right] = \left[\hat{a}_j^\dagger, \hat{a}_l^\dagger\right] = 0. \tag{6.123b}$$

Using these definitions, we may straightforwardly write

$$\hat{a}_x \left|\mathbf{0}\right\rangle = \hat{a}_y \left|\mathbf{0}\right\rangle = \hat{a}_z \left|\mathbf{0}\right\rangle = 0. \tag{6.124}$$

Moreover (see also Prob. 4.13),

$$\left|n_x, n_y, n_z\right\rangle = \frac{\left(\hat{a}_x^\dagger\right)^{n_x} \left(\hat{a}_y^\dagger\right)^{n_y} \left(\hat{a}_z^\dagger\right)^{n_z}}{\sqrt{n_x! n_y! n_z!}} \left|\mathbf{0}\right\rangle . \tag{6.125}$$

A complete set of commuting observables is obviously represented by the operators

$$\left\{\hat{N}_x = \hat{a}_x^\dagger \hat{a}_x, \;\; \hat{N}_y = \hat{a}_y^\dagger \hat{a}_y, \;\; \hat{N}_z = \hat{a}_z^\dagger \hat{a}_z\right\}, \tag{6.126}$$

each of which may be interpreted as the number of quanta in the *mode* x, y, and z, respectively. The corresponding basis is exactly $\{\left|n_x, n_y, n_z\right\rangle\}$. In a similar way, we may define the observable corresponding to the total number of quanta as

$$\hat{N} = \hat{N}_x + \hat{N}_y + \hat{N}_z, \tag{6.127}$$

in terms of which the total Hamiltonian (6.114) may be written as

$$\hat{H} = \left(\hat{N} + \frac{3}{2}\right)\hbar\omega. \tag{6.128}$$

In order to further investigate the degeneracy of the energy eigenvalues, it is convenient to introduce the angular momentum $\hat{\mathbf{L}} = \hat{\mathbf{r}} \times \hat{\mathbf{p}}$. Just as $\{\hat{N}_x, \hat{N}_y, \hat{N}_z\}$, also

$$\left\{\hat{H}, \hat{L}^2, \hat{L}_z\right\} \tag{6.129}$$

represents a complete set of commuting observables (see Subsec. 6.2.2). Their eigenvectors are labelled by the corresponding quantum numbers n, l, m_l, whose eigenvalues are given by

$$\left(n + \frac{3}{2}\right)\hbar\omega, \quad l\,(l+1)\,\hbar^2, \quad m_l\hbar. \tag{6.130}$$

The kets $\left|n, l, m_l\right\rangle$ constitute a complete set of eigenvectors of the Hamiltonian and therefore are orthonormal as well: they may obtained from $\left|n_x, n_y, n_z\right\rangle$ by a unitary transformation, i.e. a change of basis (see Subsec. 2.1.2).

Let us now make some general considerations about the degeneracy of the energy eigenvalues (see again Tab. 6.2, especially the last column, and Subsec. 3.1.4). In order to have energy degeneracy, there must be a conserved quantity in addition to energy. Indeed, if there are degeneracies, the Hamiltonian, restricted to the space of the degenerate eigenvectors, will be a multiple of the identity, and therefore any operator in that space would commute with it. A very interesting case is when the conserved quantity exists classically. In systems with a central potential, the angular momentum is conserved and states with non-zero angular momentum form a multiplet that are degenerate in energy. That is, a state with angular momentum l belongs to a multiplet where there are $2l + 1$ states. More complex cases are possible if there are additional conserved observables. For a two-dimensional rotation, for instance, when there is an additional constant of motion, the classical orbits of the finite motion must be closed (e.g. ellipses, as in the Kepler problem), as opposed to open orbits in absence of conservation.

In the case of the quantum harmonic oscillator, we may explain the degeneracy of the energy eigenvalues with the fact that, beyond the conservation of the angular momentum and the energy, we also have conservation of the quantity $\hat{H}_j - \hat{H}_k$, where j and k may take the values x, y, z and $j \neq k$. We expect that this operator does not change the energy of the system. For instance, the wave functions corresponding to the three vectors

$$|2,0,0\rangle, \quad |0,2,0\rangle, \quad |0,0,2\rangle \tag{6.131}$$

which apart from a common factor proportional to $e^{-\omega r^2/2\hbar}$, are equal to the Hermite polynomial of degree two in x, y, z, respectively (see Eq. (4.97)). In other words, we have

$$\langle r \mid (|2,0,0\rangle + |0,2,0\rangle + |0,0,2\rangle) = \langle r \mid \eta\rangle \propto \left(x^2 + y^2 + z^2 + C\right) e^{-\omega r^2/2\hbar}, \tag{6.132}$$

where C is a constant, and $|r\rangle = |x\rangle\ |y\rangle\ |z\rangle$. This wave function is spherically symmetric, and therefore the state vector $|\eta\rangle$ has $l = 0$: it is the first radial excitation in an s-wave (see Subsec. 6.1.4). Moreover,

$$\frac{1}{\hbar\omega}\left(\hat{H}_x - \hat{H}_y\right)|\eta\rangle = 2\,(|2,0,0\rangle - |0,2,0\rangle), \tag{6.133}$$

which corresponds to the angular momentum $l = 2$. In other terms, the observable $\hat{H}_x - \hat{H}_y$ (and similar combinations) is a tensor and allows a transformation from the eigenvectors of a given l to the eigenvectors corresponding to $l = \pm 2$.

Another example is given by the classical treatment of the Coulomb potential, where the additional conserved quantity is represented by the so-called Runge–Lenz vector[14]

$$\mathbf{A} = \frac{1}{m}\mathbf{p} \times \mathbf{L} - \mathrm{e}^2 \frac{\mathbf{r}}{|r|}. \tag{6.134}$$

The detailed computation shows that in this case the existence of this conserved quantity implies that the first radially excited s-state is degenerate with the first $l = 1$ state.

6.3 Spin

6.3.1 Spin as an intrinsic and quantum-mechanical degree of freedom

We have already encountered (in Subsec. 1.5.6) the spin as a new quantization feature of quantum systems. The existence of an intrinsic magnetic moment for the electron was first proposed by Uhlenbeck and Goudsmit, because the spectrum of the hydrogen atom presents levels which, for $l > 0$, are split into doublets (see Fig. 6.9). They postulated that, for the electron, this intrinsic momentum be equal to $\hbar/2$. This hypothesis is able to explain the experimental results previously obtained by Stern and Gerlach: they had used a

[14] See [*Landau/Lifshitz* 1976a, §15].

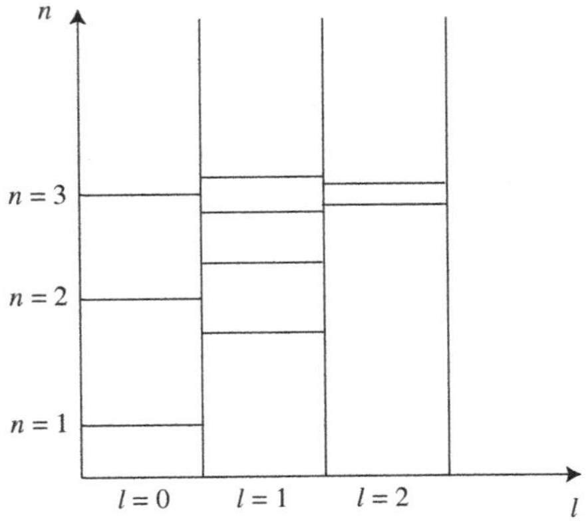

Figure 6.9 A schematic representation of the first few levels in the spectrum of a hydrogen atom, where n denotes the principal quantum number (see also Sec. 11.2). The levels corresponding to $l > 0$ are split in doublets.

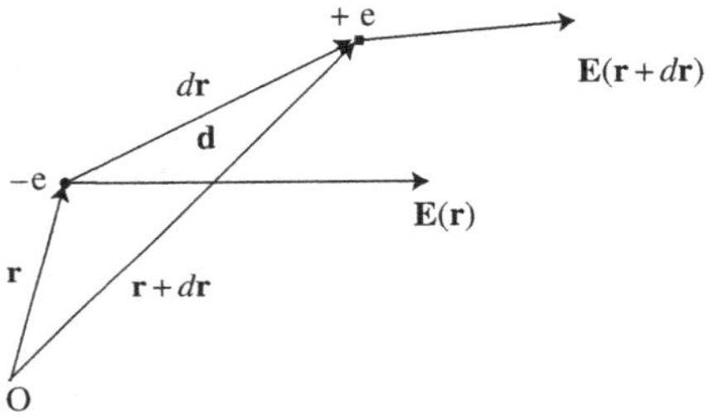

Figure 6.10 An electric dipole with charges +e and −e in an electric field gradient. The distance between the opposite charges is denoted by $d\mathbf{r}$, and $\mathbf{d} = e d\mathbf{r}$ is the electric dipole moment.

beam of silver atoms in a magnetic field to show that the atoms behave as magnetic dipoles subjected to the magnetic field gradient in the same way electric dipoles do when subjected to an electric field gradient.[15] In the latter case (see Fig. 6.10), if $d\mathbf{r}$ is small compared to the variation length of the electric field, the resulting classical force acting on the dipole would be given by

$$\mathbf{F}_{\mathrm{e}} = \mathrm{e}\,[\mathbf{E}(\mathbf{r} + d\mathbf{r}) - \mathbf{E}(\mathbf{r})] = (\mathbf{d} \cdot \nabla)\,\mathbf{E}(\mathbf{r}), \tag{6.135a}$$

where $\mathbf{E}(\mathbf{r})$ is the electric field at position $\mathbf{r}$, $-\mathrm{e}$ is the electron charge, and $\mathbf{d} = e d\mathbf{r}$ is the electric dipole moment. In the magnetic case, we have, similarly,

$$\mathbf{F}_m = (\boldsymbol{\mu} \cdot \nabla)\,\mathbf{B}, \tag{6.135b}$$

where $\mathbf{B}$ is the magnetic field and $\boldsymbol{\mu}$ is the magnetic dipole moment. If the gradient is along the z-axis, Eq. (6.135b) reduces to $\mathbf{F}_m = \mu_z \partial_z \mathbf{B}$.

[15] For the classical treatment see [*Jackson* 1962, 37–38, 61].

The fact that the spin gives rise to an intrinsic magnetic moment is an indication that we should consider spin as an intrinsic angular momentum. When we developed the angular momentum algebra (see Subsec. 6.1.3), we observed that the value l of the orbital angular momentum could in principle assume half–integer values (see Tab. 6.1 and comments). This possibility is ruled out for the orbital angular momentum because – for the wave function to be single-valued – it can only assume integer values (see Eqs. (6.36)–(6.37) and comments). However, this possibility is allowed for the spin, since it is an internal variable which has nothing to do with position. Moreover, the experiments tell us that the electron spin indeed assumes values which are half-integer multiples of $\hbar$. As a matter of fact, the two spots found on the screen by Stern and Gerlach (see Fig. 1.18) correspond to the two possible values of the spin, $\hbar/2$ and $-\hbar/2$. It is also possible to perform similar experiments for nuclei, but, given that the mass of the proton is about 2000 times that of the electron, its contribution to the magnetic moment is irrelevant relative to that of the electron – indeed we have

$$\mu_p \simeq \frac{1}{1000}\mu_e, \tag{6.136}$$

where μ_p and μ_e are the magnetic momenta of proton and electron, respectively. If we want to introduce the spin as a new quantum observable, we have to define it as an operator (see Subsec. 2.1.1). Since it is an intrinsic angular momentum, the spin operator

$$\hat{\mathbf{S}} = \hbar\hat{\mathbf{s}}, \tag{6.137}$$

where $\hat{\mathbf{s}}$ is a vector whose components are matrices that will be determined later, will obey the general properties of the angular momentum algebra (see Subsec. 6.1.1), i.e.

$$\left[\hat{s}_j, \hat{s}_k\right] = \iota\epsilon_{jkn}\hat{s}_n, \tag{6.138a}$$

$$\left[\hat{s}_j, \hat{l}_k\right] = 0, \quad \left[\hat{s}_j, \hat{r}_k\right] = 0, \quad \left[\hat{s}_j, \hat{p}_k\right] = 0. \tag{6.138b}$$

It is important to emphasize that in the classical limit – i.e. when $\hbar \to 0$ (see Pr. 2.3: p. 72) – the orbital angular momentum maps onto its classical counterpart, whereas for the spin we have $s \to 0$ and, as expected, it has no classical analogue.

As in the case of the orbital angular momentum – where a complete set of operators is given by $\hat{\mathbf{l}}^2$ and one of the projections of $\hat{\mathbf{l}}$, e.g. $\hat{l}_z$ – also in this context, besides the spin value $s = 1/2, 1, 3/2, \ldots$, we have to include into our description the eigenvalues of the projection of the spin along a certain direction. For convenience, we again choose the z orientation. This means that we have to take into account a new quantum number – determined by the projection of the spin along the z direction – which we call m_s, the *spin magnetic quantum number*. For example, in the case $s = 1/2$, m_s may take the values $\pm 1/2$. Generalizing, a spin-matrix is of rank $(2s + 1)$, i.e. (analogously to the $2l + 1$ values of m_l) there are always $2s + 1$ possible values of m_s for a particle of spin number s.

Therefore, in general terms, the description of the state of a particle should account both for the probability amplitudes of the different positions in space (in terms of the three continuous variables x, y, z), and also for the probability of the different spin values. As a consequence, the wave function has to account for the spin degree of freedom. To this end, the "usual" wave function has to be multiplied by a spin part, which, in the case of spin 1/2, has two components in order to account for the two possible values of the spin. The resulting function is called *spinor* and will be indicated by

$$\psi(\mathbf{r}, s) = \begin{pmatrix} \psi_\uparrow(\mathbf{r}) \\ \psi_\downarrow(\mathbf{r}) \end{pmatrix}, \tag{6.139}$$

where $\uparrow$ denotes the spin value $+1/2$ and $\downarrow$ the spin value $-1/2$. In other words, $\psi_\uparrow(\mathbf{r})$ is the eigenfunction of the spin projection along z with eigenvalue $+1/2$ and $\psi_\downarrow(\mathbf{r})$ the eigenfunction corresponding to the eigenvalue $-1/2$. The spinor represents an ensemble of coordinate functions which correspond to different values of $\hat{s}_z$. Therefore,

$$\int_\Omega d\mathrm{V}|\psi_{s_z}(\mathbf{r})|^2 \tag{6.140}$$

represents the probability that the particle be found in a volume Ω with a particular value of the spin, depending upon whether $m_s = +1/2$ or $m_s = -1/2$. The probability that the particle be in a volume element $d\mathrm{V}$ with any spin value is in turn

$$d\mathrm{V} \sum_{s_z=-\frac{1}{2}}^{+\frac{1}{2}} |\psi_{s_z}(\mathbf{r})|^2. \tag{6.141}$$

The normalization condition (2.108) will necessarily translate into

$$\int d\mathrm{V} \sum_{s_z=-\frac{1}{2}}^{+\frac{1}{2}} |\psi_{s_z}(\mathbf{r})|^2 = 1, \tag{6.142}$$

where the volume integration has to be understood over the whole space.

6.3.2 Spin matrices

Derivation for the bidimensional case

Consider again the case of the electron, for which the spin has experimentally been found to be 1/2. Since we have two possible values, and consequently the spinor wave functions have two components, the spin-1/2 matrices must be 2×2 matrices. Denoting again the two basis spin wave-functions by $\psi_\uparrow(\mathbf{r})$ and $\psi_\downarrow(\mathbf{r})$, the total spinor wave-function can be expressed as

$$\psi(\mathbf{r}, s) = \begin{pmatrix} \psi_\uparrow(\mathbf{r}) \\ \psi_\downarrow(\mathbf{r}) \end{pmatrix} = \psi_\uparrow(\mathbf{r}) \begin{pmatrix} 1 \\ 0 \end{pmatrix} + \psi_\downarrow(\mathbf{r}) \begin{pmatrix} 0 \\ 1 \end{pmatrix}. \tag{6.143}$$

Therefore,

$$\left\{ \begin{pmatrix} 1 \\ 0 \end{pmatrix}, \begin{pmatrix} 0 \\ 1 \end{pmatrix} \right\} \tag{6.144}$$

represents a spin eigenbasis in the case $s = 1/2$. In Dirac notation,

$$|\uparrow\rangle_z = \begin{pmatrix} 1 \\ 0 \end{pmatrix}, \quad |\downarrow\rangle_z = \begin{pmatrix} 0 \\ 1 \end{pmatrix}, \tag{6.145}$$

and

$$\psi(\mathbf{r}, s) = \langle \mathbf{r} \mid \psi \rangle = \langle \mathbf{r} \mid \psi_\uparrow \rangle |\uparrow\rangle_z + \langle \mathbf{r} \mid \psi_\downarrow \rangle |\downarrow\rangle_z . \tag{6.146}$$

Since we know that

$$\hat{s}_z |\uparrow\rangle_z = \frac{1}{2}\begin{pmatrix} 1 \\ 0 \end{pmatrix}, \quad \hat{s}_z |\downarrow\rangle_z = -\frac{1}{2}\begin{pmatrix} 0 \\ 1 \end{pmatrix}, \tag{6.147}$$

then it follows that

$$\hat{s}_z = \frac{1}{2}\begin{bmatrix} 1 & 0 \\ 0 & -1 \end{bmatrix}. \tag{6.148}$$

Let us now define raising and lowering spin operators $\hat{s}_+$ and $\hat{s}_-$ (see Eq. (6.22)), respectively. Then, we have

$$\hat{s}_\pm = \hat{s}_x \pm \imath \hat{s}_y. \tag{6.149}$$

Proceeding as in the case of the orbital angular momentum, we have

$$\left[\hat{s}_z, \hat{s}_\pm\right] = \pm \hat{s}_\pm, \tag{6.150}$$

where we have made use of Eq. (6.138a). From this result, it follows that

$$\hat{s}_z \hat{s}_+ |\downarrow\rangle = \frac{1}{2}\hat{s}_+ |\downarrow\rangle, \tag{6.151a}$$

where we have made use of the basis (6.144), and which in turn means that the ket $\hat{s}_+ |\downarrow\rangle$ is the eigenvector of $\hat{s}_z$ with eigenvalue $+1/2$, i.e.

$$\hat{s}_+ |\downarrow\rangle = |\uparrow\rangle. \tag{6.151b}$$

Similarly,

$$\hat{s}_z \hat{s}_- |\uparrow\rangle = -\frac{1}{2}\hat{s}_- |\uparrow\rangle, \tag{6.151c}$$

from which it follows that

$$\hat{s}_- |\uparrow\rangle = |\downarrow\rangle. \tag{6.151d}$$

On the other hand, for obvious reasons, we also have

$$\hat{s}_+ |\uparrow\rangle = 0, \quad \hat{s}_- |\downarrow\rangle = 0, \tag{6.151e}$$

because one cannot further raise the vector $|\uparrow\rangle$, nor further lower the vector $|\downarrow\rangle$. Then, we can deduce the explicit expression for the raising and lowering spin operators, which are given by

$$\hat{s}_+ = \begin{bmatrix} 0 & 1 \\ 0 & 0 \end{bmatrix} \quad \text{and} \quad \hat{s}_- = \begin{bmatrix} 0 & 0 \\ 1 & 0 \end{bmatrix}, \tag{6.152}$$

so that, since

$$\hat{s}_x = \frac{\hat{s}_+ + \hat{s}_-}{2} \quad \text{and} \quad \hat{s}_y = \frac{\hat{s}_+ - \hat{s}_-}{2\imath}, \tag{6.153}$$

we can explicitly derive the two expressions for the x and y components, which, together with $\hat{s}_z$, give the spin–1/2 operators as

$$\hat{s}_x = \begin{bmatrix} 0 & \frac{1}{2} \\ \frac{1}{2} & 0 \end{bmatrix} = \frac{1}{2}\hat{\sigma}_x, \quad \hat{s}_y = \begin{bmatrix} 0 & -\frac{\imath}{2} \\ \frac{\imath}{2} & 0 \end{bmatrix} = \frac{1}{2}\hat{\sigma}_y, \quad \hat{s}_z = \begin{bmatrix} \frac{1}{2} & 0 \\ 0 & -\frac{1}{2} \end{bmatrix} = \frac{1}{2}\hat{\sigma}_z. \tag{6.154}$$

The zero-trace matrices $\hat{\sigma}_x$, $\hat{\sigma}_y$, and $\hat{\sigma}_z$ are known as *Pauli spin matrices*.[16] It is easy to derive their commutation relations (see Prob. 6.20)

$$\left[\hat{\sigma}_j, \hat{\sigma}_k\right] = 2\imath\epsilon_{jkn}\hat{\sigma}_n. \tag{6.155}$$

Moreover, it is straightforward to verify that

$$\hat{\sigma}_x^2 = \hat{\sigma}_y^2 = \hat{\sigma}_z^2 = \hat{I}. \tag{6.156}$$

Another important property of the Pauli matrices is that they anticommute (see Prob. 6.21), i.e.

$$\left[\hat{\sigma}_j, \hat{\sigma}_k\right]_+ = 2\hat{I}\delta_{jk}. \tag{6.157}$$

Finally, it is not difficult to prove (see Prob. 6.22) that the eigenvectors and eigenvalues of $\hat{\sigma}_x$ and $\hat{\sigma}_y$ are given by the following equations:

$$\hat{\sigma}_x \mid\uparrow\rangle_x = \mid\uparrow\rangle_x, \quad \hat{\sigma}_x \mid\downarrow\rangle_x = - \mid\downarrow\rangle_x, \tag{6.158a}$$

$$\hat{\sigma}_y \mid\uparrow\rangle_y = \mid\uparrow\rangle_y, \quad \hat{\sigma}_y \mid\downarrow\rangle_y = - \mid\downarrow\rangle_y, \tag{6.158b}$$

where

$$\mid\uparrow\rangle_x = \frac{1}{\sqrt{2}}\left(\mid\uparrow\rangle_z + \mid\downarrow\rangle_z\right), \quad \mid\downarrow\rangle_x = \frac{1}{\sqrt{2}}\left(\mid\uparrow\rangle_z - \mid\downarrow\rangle_z\right), \tag{6.159a}$$

$$\mid\uparrow\rangle_y = \frac{1}{\sqrt{2}}\left(\mid\uparrow\rangle_z + \imath\mid\downarrow\rangle_z\right), \quad \mid\downarrow\rangle_y = \frac{1}{\sqrt{2}}\left(\mid\uparrow\rangle_z - \imath\mid\downarrow\rangle_z\right). \tag{6.159b}$$

It is worth emphasizing that Pauli matrices are simultaneously Hermitian and unitary (see Prob. 6.24). It is very interesting to note that the Pauli matrices, together with the 2×2 identity, form an operatorial basis for spin-1/2 Hilbert space: any operator in the spinors space can be written as a combination of $\hat{\sigma}_x$, $\hat{\sigma}_y$, $\hat{\sigma}_z$, and $\hat{I}$ as

[16] See [Pauli 1927, 614].

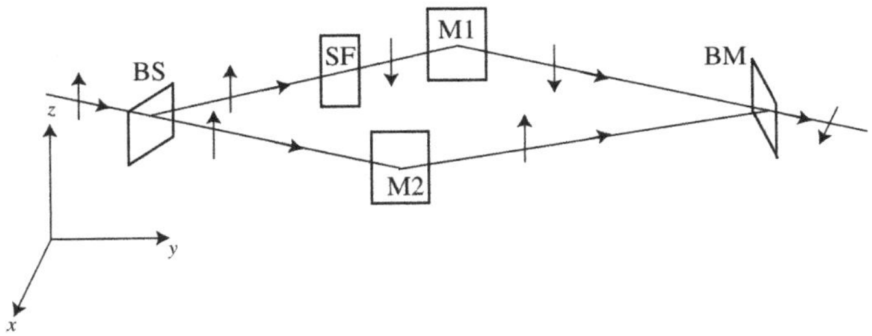

Figure 6.11 Schematic setup for the realization of spin superposition. The arrows indicate incoming spin-up particles along the z direction and outgoing particles with spin typically in the xy-plane (here in the x direction). The beam, incoming from the left, is initially split into two beams. Then, a spin flipper (SF) changes the spin up into spin down. Finally, after the two mirrors, the two beams are recombined by the beam merger (BM). In the actual experimental setup [Summhammer *et al.* 1983] this is realized through single-crystal neutron interferometry.

$$\hat{O} = \alpha \hat{I} + \boldsymbol{\beta} \cdot \hat{\boldsymbol{\sigma}} = \begin{bmatrix} \alpha + \beta_z & \beta_x - \imath \beta_y \\ \beta_x + \imath \beta_y & \alpha - \beta_z \end{bmatrix}, \tag{6.160}$$

where $\hat{\boldsymbol{\sigma}} = (\hat{\sigma}_x, \hat{\sigma}_y, \hat{\sigma}_z)$ and $\hat{O}$ is Hermitian if and only if α and $\boldsymbol{\beta}$ are real. Finally, the Pauli matrices represent the lowest-dimensional realization of infinitesimal rotations in three-dimensional space (see also Sec. 17.5).

An ideal experiment

Let us now consider the experiment shown in Fig. 6.11. The initial state of the particle may be described by (see Eqs. (6.143) and (6.145))

$$\psi(\mathbf{r}, s) = \psi_0(x, y, z) \,|\uparrow\rangle_z \,. \tag{6.161}$$

After the initial beam splitter the state becomes

$$\psi'(\mathbf{r}, s) = \psi_1(x, y, z) \,|\uparrow\rangle_z + \psi_2(x, y, z) \,|\uparrow\rangle_z \,, \tag{6.162}$$

where the transmission and reflection coefficients T and R (see Subsecs. 2.3.4 and 3.5.2) are absorbed into the amplitudes ψ_1 and ψ_2. And, after the spin flipper in path 1, changes into

$$\psi''(\mathbf{r}, s) = \psi_1(x, y, z) \,|\downarrow\rangle_z + \psi_2(x, y, z) \,|\uparrow\rangle_z \,. \tag{6.163}$$

The effect of the beam merger is to recombine the two sub-beams so that we finally obtain

$$\psi'''(\mathbf{r}, s) = \psi_3(x, y, z) \left(|\downarrow\rangle_z + |\uparrow\rangle_z\right) = \psi_3(x, y, z)\sqrt{2}\,|\uparrow\rangle_x \,. \tag{6.164}$$

The aforementioned experiment bears an interesting relationship to the interferometry experiment introduced in Sec. 1.2 and developed further throughout the book (see also

Subsec. 2.1.1 and 3.5.2). As in that experiment, in fact, we have here a beam (neutrons rather than photons) that is essentially subjected to three operations: splitting, phase shifting (spin flipping), and beam merging. These two experiments are so linked to each other that they effectively may be understood as two different realizations of the same ideal experiment.

Other dimensions

Up to now we have considered the case $s = 1/2$. It is clear that the wave function for spin zero ($s = 0$) has only one component ($m_s = 0$), therefore $\hat{\sigma}_z \psi = 0$. This is just the case where spin is absent.

For $s = 1$ we have three possible values $m_s = -1, 0, +1$, and the spin matrices are the following (see Prob. 6.26):

$$\hat{s}_x = \frac{1}{\sqrt{2}} \begin{bmatrix} 0 & 1 & 0 \\ 1 & 0 & 1 \\ 0 & 1 & 0 \end{bmatrix}, \quad \hat{s}_y = \frac{1}{\sqrt{2}} \begin{bmatrix} 0 & -\imath & 0 \\ \imath & 0 & -\imath \\ 0 & \imath & 0 \end{bmatrix}, \quad \hat{s}_z = \frac{1}{\sqrt{2}} \begin{bmatrix} 1 & 0 & 0 \\ 0 & 0 & 0 \\ 0 & 0 & -1 \end{bmatrix}. \tag{6.165}$$

6.3.3 Spin and magnetic field

Let us go back to Eq. (6.103), describing the Hamiltonian of a particle of charge e and mass m in a constant magnetic field directed along the z-axis. For an electron in an atom, the last term is negligibly small since $\hat{x}^2 + \hat{y}^2$ is of the order of 10^{-20} m^2. The third term, instead, is proportional to $\mathbf{B} \cdot \hat{\mathbf{l}}$, and, from the correspondence principle (see p. 72), describes the interaction between the magnetic field $\mathbf{B}$ and the magnetic moment

$$\hat{\boldsymbol{\mu}}_l = \frac{e\hbar}{2m} \hat{\mathbf{l}} \tag{6.166}$$

of a (spinless) charged massive particle having orbital angular momentum $\hat{\mathbf{l}}$.[17] Such a magnetic moment is simply proportional to the orbital angular momentum. The quantity

$$\mu_B = \frac{e\hbar}{2m} \tag{6.167}$$

is called the *Bohr magneton.*

The argument above, together with the discussion of Subsec. 6.3.1, suggests that to each angular momentum corresponds a magnetic moment. Moreover, their ratio is called the *gyromagnetic ratio*. This is indeed the central theory of the Stern and Gerlach experiment as

[17] The classical interaction (potential) energy between a magnetic moment $\boldsymbol{\mu}$ and a magnetic field $\mathbf{B}$ is given by $E = -\boldsymbol{\mu} \cdot \mathbf{B}$ [*Jackson* 1962, 186]. Traditionally, one used to distinguish between the magnetic field $\mathbf{H}$ and the magnetic induction $\mathbf{B} = \mu_0 \mathbf{H}$, where μ_0 is the magnetic permeability. Modern approaches consider $\mathbf{B}$ as the basic quantity, sometimes called the magnetic field. We follow this convention.

well as Uhlenbeck's hypothesis about the electron's spin. In the presence of spin, therefore, we are forced to assume a further interaction term in the Hamiltonian, of the type

$$\hat{H}_s = -\hat{\boldsymbol{\mu}}_s \cdot \hat{\mathbf{B}}, \tag{6.168}$$

where

$$\hat{\boldsymbol{\mu}}_s = g\frac{e\hbar}{2m}\hat{\mathbf{s}} \tag{6.169}$$

is the spin magnetic moment. The factor g depends on the specific particle under consideration. For example, in the case of the electron, experiments tell us that $g_e = -2.0023193 \simeq -2$. Similar experiments may also be carried out for nucleons, yielding $g_p = 5.5856947$ for protons and $g_n = -3.8260855$ for neutrons. Eq. (6.169) is the explanation of Eq. (6.136). We finally comment that the values of the g-factors above can be predicted only with the help of the relativistic extension of quantum mechanics.[18]

6.3.4 Quantum motion in a homogeneous magnetic field

Let us consider the motion of a quantum particle subject to a homogeneous magnetic field. In Subsec. 4.5.3 we have already derived the Hamiltonian (4.133) and the Schrödinger equation (4.135) for a spinless particle in a static electromagnetic field, and in Subsec. 6.2.3 we have already considered a constant magentic field but without taking into account the spin degree of freedom. In the present case, we have to add the magnetic term (6.168) and neglect the electrostatic term eU in Eq. (4.135). Moreover, instead of condition (6.102), it is convenient to choose the vector potential of the homogeneous field in the form

$$A_x = -By, \quad A_y = A_z = 0, \tag{6.170}$$

and

$$\mathbf{B} = B\boldsymbol{k}. \tag{6.171}$$

The Hamiltonian has then the form (see Eqs. (6.169) and (6.103))

$$\hat{H} = \frac{1}{2m}\left(\hat{p}_x + eBy\right)^2 + \frac{\hat{p}_y^2}{2m} + \frac{\hat{p}_z^2}{2m} - \tilde{\mu}\hat{s}_z B, \tag{6.172}$$

where

$$\tilde{\mu} = g\frac{e\hbar}{2m}. \tag{6.173}$$

The spin operator $\hat{s}_z$ does commute with the Hamiltonian (since the other components of the spin are absent). This means that the projection of the spin onto the z-axis is conserved, and therefore we can substitute to $\hat{s}_z$ its eigenvalue s_z. As a consequence the spinorial character of the wave function becomes immaterial, and we can limit ourselves to the consideration of its dependence on the spatial coordinates. The Schrödinger equation may then be written as

[18] See [*Mandl/Shaw* 1984, 200–203, 231–34].

$$\left[\frac{1}{2m}\left(\hat{p}_x + eBy\right)^2 + \frac{\hat{p}_y^2}{2m} + \frac{\hat{p}_z^2}{2m} - \tilde{\mu}s_z B\right]\psi(\mathbf{r}) = E\psi(\mathbf{r}). \tag{6.174}$$

Since the Hamiltonian does not contain the operators $\hat{x}$ and $\hat{z}$, it also commutes with the components $\hat{p}_x$ and $\hat{p}_z$ of the momentum, that is, these components are also conserved (see Eq. (3.108) and comments). It is then natural to make the ansatz (see Eq. (2.146))

$$\psi(\mathbf{r}) = e^{\frac{i}{\hbar}(p_x x + p_z z)}\varphi(y), \tag{6.175}$$

where p_x and p_z may take on any value from $-\infty$ to $+\infty$. Moreover, the z component of the generalized momentum (4.132) $\hat{\mathrm{P}}_z$ coincides with the corresponding component of the ordinary momentum $\hat{p}_z$, because $A_z = 0$. As a consequence, it is often said that the motion along the direction of the field is not quantized. By substituting Eq. (6.175) into Eq. (6.174), we obtain a Schrödinger equation for $\varphi(y)$ (see Prob. 6.27)

$$\varphi''(y) + \frac{2m}{\hbar^2}\left[\left(E + \tilde{\mu}s_z B - \frac{p_z^2}{2m}\right) - \frac{m}{2}\omega_B^2\left(y - y_0\right)^2\right]\varphi(y) = 0, \tag{6.176}$$

where

$$\omega_B = \frac{eB}{m} \quad \text{and} \quad y_0 = -\frac{p_x}{eB}. \tag{6.177}$$

Equation (6.176) is the Schrödinger equation (4.49) for a harmonic oscillator with frequency ω_B. This means that the quantity

$$E + \tilde{\mu}s_z B - \frac{p_z^2}{2m} \tag{6.178}$$

that plays the role of the energy of the oscillator in Eq. (6.176) can only assume the discrete values $(n + 1/2)\hbar\omega_B$, with $n = 0, 1, 2, \ldots$. Therefore, we are finally able to write

$$E = \left(n + \frac{1}{2}\right)\hbar\omega_B - \tilde{\mu}s_z B + \frac{p_z^2}{2m}. \tag{6.179}$$

The first term represents the discrete energy eigenvalues corresponding to the motion onto the plane orthogonal to the field, which are called *Landau levels* (see also Probs. 6.28 and 6.29).

It should be noted that the energy levels are degenerate with continuous multiplicity, because Eq. (6.179) does not contain the continuous variable p_x. In the case of the electron, there is an extra degeneration, since the energy levels are identical for states with quantum numbers n and $s_z = 1/2$, as well as $n + 1$ and $s_z = -1/2$.

6.4 Composition of angular momenta and total angular momentum

In classical mechanics, the composition of angular momenta of different systems simply reduces to the sum of the corresponding vectors. In quantum mechanics, instead, given

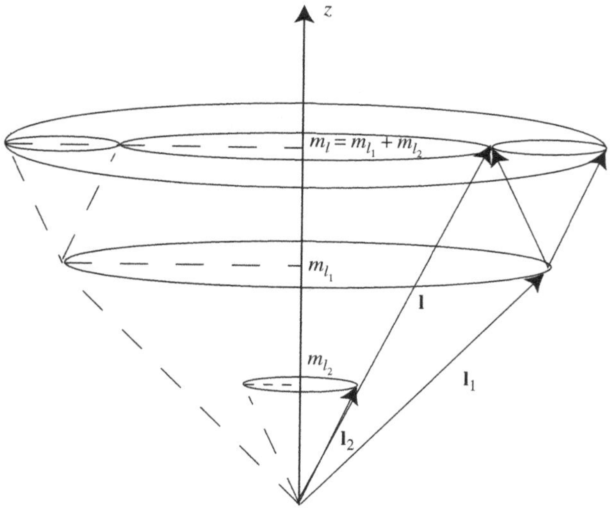

Figure 6.12 Landé semiclassical vectorial model for angular momentum. Given two angular momenta $\hat{\mathbf{l}}_1$ and $\hat{\mathbf{l}}_2$, with projections along the z-axis equal to, respectively, m_{l_1} and m_{l_2}, the resulting angular momentum gives rise to a projection along the z-axis equal to $m_l = m_{l_1} + m_{l_2}$. See also [*White* 1934, 154–55].

the particular features of the angular momentum algebra and the existence of the spin as an intrinsic angular momentum, we have to find a rule that will allow us to compose different angular momenta in agreement with the algebra itself (see Fig. 6.12). In other words, we want to answer the following question: given two particles, each endowed with its own angular momentum, what can we say about the angular momentum of the system? In particular, our objective will be to find eigenvalues and eigenvectors of the resulting angular momentum given the eigenvalues and eigenvectors of the angular momenta of the subsystems. In order to reach our result, we first consider two independent orbital angular momenta (in Subsec. 6.4.1), then we use these results to establish the general properties of the total angular momentum (orbital + spin) of a particle (in Subsec. 6.4.2). In Subsec. 6.4.3 we consider the composition of two spin angular momenta, whereas in Subsec. 6.4.4 we conclude by considering the change of basis in the context of the angular momentum algebra.

6.4.1 Independent orbital angular momenta

Let us suppose we have two particles ($\mathcal{S}_1$ and $\mathcal{S}_2$) that have independent angular momenta, i.e. $\left[\hat{\mathbf{l}}_1, \hat{\mathbf{l}}_2\right] = 0$. The resulting orbital angular momentum is $\hat{\mathbf{l}} = \hat{\mathbf{l}}_1 + \hat{\mathbf{l}}_2$, so that we have

$$\left[\hat{l}_j, \hat{l}_k\right] = \left[\hat{l}_{1_j} + \hat{l}_{2_j}, \hat{l}_{1_k} + \hat{l}_{2_k}\right] = \imath\epsilon_{jkn}\hat{l}_n, \tag{6.180}$$

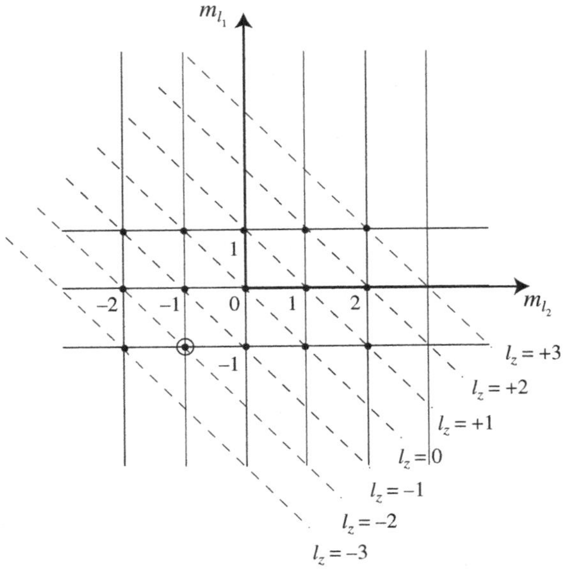

Figure 6.13 Graphical representation of the distribution of eigenvalues of the z component of the angular momenta of two independent particles (see text). Diagonal lines are lines of constant $\hat{l}_z$ (the z component of the resulting angular momentum). For instance, the circled dot represents the state $|1,-1\rangle_1 |2,-1\rangle_2$.

since

$$\left[\hat{l}_{1_j},\ \hat{l}_{1_k}\right] = \imath\epsilon_{jkn}\hat{l}_{1_n}, \quad \left[\hat{l}_{2_j}, \hat{l}_{2_k}\right] = \imath\epsilon_{jkn}\hat{l}_{2_n}, \tag{6.181}$$

where the indices j, k, and n label any of the orientations x, y, and z. It is clear that we can diagonalize the angular momenta of the two particles simultaneously. In other words, we have

$$\left(\hat{\mathbf{l}}\right)^2 \left|l, m_l\right\rangle = l\,(l+1)\left|l, m_l\right\rangle, \tag{6.182}$$

$$\hat{l}_{1,2_z}\left|l, m_{l_{1,2}}\right\rangle = m_{l_{1,2}}\left|l, m_{l_{1,2}}\right\rangle. \tag{6.183}$$

Therefore, an obvious basis for the product space would be given by

$$\left\{\left|l_1, m_{l_1}; l_2, m_{l_2}\right\rangle\right\} = \left\{\left|l_1, m_{l_1}\right\rangle \otimes \left|l_2, m_{l_2}\right\rangle\right\}. \tag{6.184}$$

For example, in the case $l_1 = 1$ and $l_2 = 2$ we would have 15 possible basis kets. However, if we are interested in the resulting angular momentum, a complete set of operators is represented by $\hat{\mathbf{l}}, \hat{l}_z, \hat{\mathbf{l}}_1^2, \hat{\mathbf{l}}_2^2$ ($\hat{l}_z = \hat{l}_{1,2_z} = \hat{l}_{1_z} + \hat{l}_{2_z}$), while $\hat{l}_{1_z}, \hat{l}_{2_z}$ are not compatible with this basis in particular with $\hat{\mathbf{l}}$ (see Prob. 6.30). Moreover, if the system Hamiltonian has to be invariant under rotation, it must commute with $\hat{\mathbf{l}}^2$ and $\hat{l}_z$, but it does not necessarily commute with $\hat{\mathbf{l}}_1^2$ and $\hat{\mathbf{l}}_2^2$.

Suppose again that $l_1 = 1$ and $l_2 = 2$. Then, l can only assume the values $l = 1, 2, 3$. That this is the case can be understood with the help of Fig. 6.13, which represents the 15 possible combinations for m_{l_1} and m_{l_2}. The fact that $\max(l_z) = 3$ tells us that there must exist a value $l = 3$, which actually is a multiplet with seven states. If we move to the next diagonal line ($l_z = 2$), we see that there are two instances that give rise to this situation: one is the value $m_l = 2$ coming from the previous multiplet and the other is the value $m_l = 2$ coming from a new multiplet ($l = 2$) with five states. Moving on to the next diagonal line ($l_z = 1$), we have now three instances with the value $m_l = 1$: one coming from the multiplet with $l = 3$, the second coming from the multiplet $l = 2$, and the third coming from a new multiplet ($l = 1$) with three states. Proceeding in this way, we realize that there are no other new multiplets, so that we can arrange the 15 states into three different multiplets: $l = 3$ (seven states), $l = 2$ (five states), and $l = 1$ (three states). In general, for any fixed m_l there are several possible combinations of m_{l_1} and m_{l_2}:

$$\left(\hat{l}_{1_z} + \hat{l}_{2_z}\right) \left| l_1, m_{l_1} \right\rangle \otimes \left| l_2, m_{l_2} \right\rangle = \left(m_{l_1} + m_{l_2}\right) \left| l_1, m_{l_1} \right\rangle \otimes \left| l_2, m_{l_2} \right\rangle. \tag{6.185}$$

The results shown in Fig. 6.13, for $l_2 > l_1$ can be generalized as follows:[19]

$$n(l_z) = \begin{cases} 0 & \text{if} \quad |l_z| > l_1 + l_2 \\ l_1 + l_2 + 1 - |l_z| & \text{if} \quad l_1 + l_2 \geq |l_z| \geq |l_1 - l_2| \\ 2l_1 + 1 & \text{if} \quad |l_1 - l_2| \geq |l_z| \geq 0 \end{cases}, \tag{6.186}$$

where $n(l_z)$ is the number of points situated on the diagonal $l_z = m_{l_1} + m_{l_2}$.

To sum up, we can state the following theorem:

Theorem 6.1 (Addition of angular momenta) *When summing the angular momenta of moduli l_1 and l_2, we obtain angular momenta l such that*

$$|l_1 - l_2| \leq l \leq l_1 + l_2, \tag{6.187}$$

all of them interspaced by a unity. To any value l correspond $2l + 1$ eigenvectors $\left| l, m_l \right\rangle$ of the resulting angular momentum.

6.4.2 Total angular momentum

As we have seen, besides the orbital angular momentum, in quantum mechanics we also have an intrinsic and genuinely quantum spin angular momentum. Therefore, we can speak of a total angular momentum $\hat{\mathbf{J}} = \hbar \hat{\boldsymbol{\jmath}}$ which is the sum of the orbital angular momentum $\hat{\mathbf{L}}$ and of the spin angular momentum $\hat{\mathbf{S}}$

$$\hat{\mathbf{J}} = \hat{\mathbf{L}} + \hat{\mathbf{S}}. \tag{6.188}$$

[19] See [*Messiah* 1958, 557].

$\hat{\mathbf{J}}$ is the generator of rotations and has the same commutation relations as $\hat{\mathbf{S}}$ and $\hat{\mathbf{L}}$, that are

$$\left[\hat{J}_k, \hat{J}_n\right] = \imath\hbar\epsilon_{knr}\hat{J}_r. \tag{6.189}$$

To make some examples, the explicit matrices of the total angular momentum when $j = 1/2$, $l = 0$, and $s = j$, are given by

$$\hat{J}_x = \frac{1}{2}\hbar\begin{bmatrix} 0 & 1 \\ 1 & 0 \end{bmatrix}, \quad \hat{J}_y = \frac{1}{2}\hbar\begin{bmatrix} 0 & -\imath \\ \imath & 0 \end{bmatrix},$$
$$\hat{J}_z = \frac{1}{2}\hbar\begin{bmatrix} 1 & 0 \\ 0 & -1 \end{bmatrix}, \quad \hat{\mathbf{J}}^2 = \frac{3}{4}\hbar^2\begin{bmatrix} 1 & 0 \\ 0 & 1 \end{bmatrix}, \tag{6.190}$$

which, apart from a constant factor, are the Pauli matrices. When $j = 1$, $l = 0$, and $s = j$, we have

$$\hat{J}_x = \frac{\hbar}{\sqrt{2}}\begin{bmatrix} 0 & 1 & 0 \\ 1 & 0 & 1 \\ 0 & 1 & 0 \end{bmatrix}, \quad \hat{J}_y = \frac{\hbar}{\sqrt{2}}\begin{bmatrix} 0 & -\imath & 0 \\ \imath & 0 & -\imath \\ 0 & \imath & 0 \end{bmatrix},$$
$$\hat{J}_z = \hbar\begin{bmatrix} 1 & 0 & 0 \\ 0 & 0 & 0 \\ 0 & 0 & -1 \end{bmatrix}, \quad \hat{\mathbf{J}}^2 = 2\hbar^2\begin{bmatrix} 1 & 0 & 0 \\ 0 & 1 & 0 \\ 0 & 0 & 1 \end{bmatrix}. \tag{6.191}$$

When $l = 1$ and $s = 1/2$, we have two possible cases:

- $j = l - s = 1/2$, which reduces to the first case considered above;
- and $j = l + s = 3/2$, for which we have

$$\hat{J}_x = \frac{1}{2}\hbar\begin{bmatrix} 0 & \sqrt{3} & 0 & 0 \\ \sqrt{3} & 0 & 2 & 0 \\ 0 & 2 & 0 & \sqrt{3} \\ 0 & 0 & \sqrt{3} & 0 \end{bmatrix}, \quad \hat{J}_y = \frac{1}{2}\hbar\begin{bmatrix} 0 & -\imath\sqrt{3} & 0 & 0 \\ \imath\sqrt{3} & 0 & -2\imath & 0 \\ 0 & -2\imath & 0 & -\imath\sqrt{3} \\ 0 & 0 & \imath\sqrt{3} & 0 \end{bmatrix},$$
$$\hat{J}_z = \frac{1}{2}\hbar\begin{bmatrix} 3 & 0 & 0 & 0 \\ 0 & 1 & 0 & 0 \\ 0 & 0 & -1 & 0 \\ 0 & 0 & 0 & -3 \end{bmatrix}, \quad \hat{\mathbf{J}}^2 = \frac{15}{4}\hbar^2\begin{bmatrix} 1 & 0 & 0 & 0 \\ 0 & 1 & 0 & 0 \\ 0 & 0 & 1 & 0 \\ 0 & 0 & 0 & 1 \end{bmatrix}. \tag{6.192}$$

6.4.3 Singlet and triplet states

Following Th. 6.1, if we add the angular momenta of two spin-1/2 particles, we may obtain for the total system spin number s either 1 or 0. In the former case ($s = 1$), we shall have a multiplet with three possible values of m_s ($m_s = -1, 0, +1$), i.e. a triplet, whereas in the latter case ($s = 0$), we shall obtain a singlet (a state with antiparallel spins and total spin zero). Taking as quantum numbers the total spin s and its z component, we may write the triplet states as (see also Prob. 6.31)

$$\begin{aligned} \mid s=1, m_s=1\rangle_{12_z} &= \left| s_1=\frac{1}{2}, m_{s_1}=\frac{1}{2}\right\rangle_1 \left| s_2=\frac{1}{2}, m_{s_2}=\frac{1}{2}\right\rangle_2 \\ &= \mid\uparrow\rangle_{1_z} \mid\uparrow\rangle_{2_z}, \end{aligned} \tag{6.193a}$$

$$\mid s=1, m_s=-1\rangle_{12_z} = \left|\frac{1}{2}, -\frac{1}{2}\right\rangle_1 \left|\frac{1}{2}, -\frac{1}{2}\right\rangle_2 = \mid\downarrow\rangle_{1_z} \mid\downarrow\rangle_{2_z}, \tag{6.193b}$$

$$\begin{aligned} \mid s=1, m_s=0\rangle_{12_z} &= \frac{1}{\sqrt{2}}\left(\left|\frac{1}{2}, \frac{1}{2}\right\rangle_1 \left|\frac{1}{2}, -\frac{1}{2}\right\rangle_2 + \left|\frac{1}{2}, -\frac{1}{2}\right\rangle_1 \left|\frac{1}{2}, \frac{1}{2}\right\rangle_2\right) \\ &= \frac{1}{\sqrt{2}}\left(\mid\uparrow\rangle_{1_z} \mid\downarrow\rangle_{2_z} + \mid\downarrow\rangle_{1_z} \mid\uparrow\rangle_{2_z}\right), \end{aligned} \tag{6.193c}$$

and the singlet state as

$$\begin{aligned} \mid s=0, m_s=0\rangle_{12_z} &= \frac{1}{\sqrt{2}}\left(\left|\frac{1}{2}, \frac{1}{2}\right\rangle_1 \left|\frac{1}{2}, -\frac{1}{2}\right\rangle_2 - \left|\frac{1}{2}, -\frac{1}{2}\right\rangle_1 \left|\frac{1}{2}, \frac{1}{2}\right\rangle_2\right) \\ &= \frac{1}{\sqrt{2}}\left(\mid\uparrow\rangle_{1_z} \mid\downarrow\rangle_{2_z} - \mid\downarrow\rangle_{1_z} \mid\uparrow\rangle_{2_z}\right). \end{aligned} \tag{6.194}$$

The triplet states (6.193) are symmetric (i.e. they do not change by interchanging the two particles), whereas the *singlet state* (6.194) is completely antisymmetric (see also Ch. 7).

As it is shown in Prob. 6.32, the singlet state (6.194) assumes the same form when written in the other basis $\{\mid\uparrow\rangle_x, \mid\downarrow\rangle_x\}$ or $\{\mid\uparrow\rangle_y, \mid\downarrow\rangle_y\}$ for particles 1 and 2. Explicitly, we have

$$\mid 0, 0\rangle_{12_x} = \frac{e^{\imath\pi}}{\sqrt{2}}\left(\mid\uparrow\rangle_{1_x} \mid\downarrow\rangle_{2_x} - \mid\downarrow\rangle_{1_x} \mid\updownarrow\rangle_{2_x}\right), \tag{6.195a}$$

$$\mid 0, 0\rangle_{12_y} = \frac{e^{\imath\pi/2}}{\sqrt{2}}\left(\mid\uparrow\rangle_{1_y} \mid\downarrow\rangle_{2_y} - \mid\downarrow\rangle_{1_y} \mid\updownarrow\rangle_{2_y}\right). \tag{6.195b}$$

Equations (6.195) tell us that the singlet state, just because it corresponds to a zero-total angular momentum state, is invariant under rotations (see Subsec. 6.1.2).

Another very important property of the singlet state is worthy of mention here. Let us consider an ideal experiment in which we decide to measure the spin projection of the two particles along a certain direction (say the z-direction) on the state (6.194). Using the notation

$$\mid\Psi_0\rangle = \mid s=0, \quad m_s=0\rangle_{12_z}, \tag{6.196}$$

and considering the eigenvalue equation

$$\hat{\sigma}_{k_z} \mid\Psi_0\rangle = s_{k_z} \mid\Psi_0\rangle, \tag{6.197}$$

we immediately see that if the result of the measurement gives the eigenvalue $s_{1_z} = 1/2$ for particle 1, then we must also have the eigenvalue $s_{2_z} = -1/2$ for particle 2. Conversely, if the result of the measurement gives the eigenvalue $s_{1_z} = -1/2$ for particle 1, then we must have $s_{2_z} = 1/2$ for particle 2. In other words, the singlet state is an entangled state (see Subsec. 5.5.1) and cannot be factorized. It is therefore interesting to note that this property holds true also in both the x- and y-basis. As we shall see in Ch. 16, this type of state plays a crucial role in the understanding of entanglement and correlations in quantum mechanics.

6.4.4 Clebsch–Gordan coefficients

Suppose again that we have two particles with angular momenta $\hat{\boldsymbol{j}}_1$ and $\hat{\boldsymbol{j}}_2$. Obviously, we have $\left[\hat{\boldsymbol{j}}_1, \hat{\boldsymbol{j}}_2\right] = 0$, where $\hat{\boldsymbol{j}} = \hat{\boldsymbol{j}}_1 + \hat{\boldsymbol{j}}_2$ is the total angular momentum, with $\hat{j}_z = \hat{j}_{1_z} + \hat{j}_{2_z}$ and $\hat{\boldsymbol{j}}^2 = \hat{\boldsymbol{j}}_1^2 + \hat{\boldsymbol{j}}_2^2 + 2\hat{\boldsymbol{j}}_1 \cdot \hat{\boldsymbol{j}}_2$. We have already seen (see Subsec. 6.4.1) that we may define two different bases for the total Hilbert space, given by the tensor product of the Hilbert spaces of the two subsystems. We may choose:

- either the product eigenvectors of $\hat{\boldsymbol{j}}_1, \hat{j}_{1_z}$ and $\hat{\boldsymbol{j}}_2, \hat{j}_{2_z}$, i.e.

$$\left| j_1, m_{j_1} \right\rangle \otimes \left| j_2, m_{j_2} \right\rangle = \left| j_1, j_2, m_{j_1}, m_{j_2} \right\rangle, \tag{6.198}$$

- or, since $\left[\hat{\boldsymbol{j}}^2, \hat{\boldsymbol{j}}_1^2\right] = \left[\hat{\boldsymbol{j}}^2, \hat{\boldsymbol{j}}_2^2\right] = 0$, the eigenvectors

$$\left| j_1, j_2; j, m_j \right\rangle, \tag{6.199}$$

which characterize the total system.

In other words, both sets of operators form a complete set of commuting observables. The unitary transformation from one basis to the other is given by (see Subsec. 2.1.2)

$$\left| j_1, j_2; j, m_j \right\rangle = \sum_{m_{j_1}, m_{j_2}} \left| j_1, j_2, m_{j_1}, m_{j_2} \right\rangle \left\langle j_1, j_2, m_{j_1}, m_{j_2} \,\middle|\, j_1, j_2; j, m_j \right\rangle, \tag{6.200}$$

where the coefficients of this expansion

$$\left\langle j_1, j_2, m_{j_1}, m_{j_2} \mid j_1, j_2; j, m_j \right\rangle \tag{6.201}$$

are called the *Clebsch–Gordan coefficients*. A more compact notation is

$$\langle j_1, j_2, m_1, m_2 \mid j, m \rangle, \tag{6.202}$$

where in the present subsection $m_1 = m_{j_1}$, $m_2 = m_{j_2}$, and $m = m_j$. These amplitudes have a purely geometrical character and only depend on the angular momenta and their orientation and not upon the physical nature of the dynamical variables of particles 1 and 2 from which the angular momenta are constructed.

Many properties of the Clebsch–Gordan coefficients follow from their definition. In order for $\langle j_1, j_2, m_1, m_2 \mid j, m \rangle$ to be different from zero, we must have $m_1 + m_2 = m$ and $|j_1 - j_2| \leq j \leq j_1 + j_2$ (see Th. 6.1). Since they are coefficients of a unitary transformation, they must obey the orthonormality relations

$$\sum_{m_1, m_2} \langle j_1, j_2, m_1, m_2 \mid j, m \rangle \left\langle j_1, j_2, m_1, m_2 \mid j', m' \right\rangle = \delta_{jj'} \delta_{mm'}, \tag{6.203a}$$

$$\sum_{j, m} \langle j_1, j_2, m_1, m_2 \mid j, m \rangle \, \langle j_1, j_2, m_{1'}, m_{2'} \mid j, m \rangle = \delta_{m_1 m_{1'}} \delta_{m_2 m_{2'}}. \tag{6.203b}$$

Table 6.3 Values of j and m and the corresponding number of possible states in the case of the addition of angular momenta $j_1 = j_2 = 1$

j	m	Number of possible states
2	$-2 \leq m \leq +2$	5
1	$-1 \leq m \leq +1$	3
0	0	1
Total number of states		9

In the following we establish a general procedure that can be exploited in order to determine the Clebsch–Gordan coefficients. The successive example will help to make the matter clear.

In the simplest case ($j = j_1 + j_2$ and $m = j$) we can determine the Clebsch–Gordan coefficients directly. We first notice that

$$| j_1, j_2, j_1 + j_2, j_1 + j_2 \rangle = | j_1, j_2, j_1, j_2 \rangle , \tag{6.204}$$

because the double sum in Eq. (6.200) actually reduces to a single instance. Then, by repeated application of $\hat{j}_- = \hat{j}_{1-} + \hat{j}_{2-}$ to both sides of Eq. (6.204), we build all the $| j_1, j_2; j = j_1 + j_2, m\rangle$ (with $m = -j, -j+1, \ldots, j-1, j$) corresponding to $j = j_1 + j_2$. We successively build all the vectors of the series $j = j_1 + j_2 - 1$, that is, all the kets $| j_1, j_2; j = j_1 + j_2 - 1, m\rangle$ beginning with that corresponding to $m = j$, unambiguously defined by the conditions of reality and positivity, and by its property of being orthogonal to $| j_1, j_2, j_1 + j_2, j_1 + j_2 - 1\rangle$. We finally form all other kets $| j_1, j_2; j, m\rangle$, where $|j_1 - j_2| \leq j \leq j_1 + j_2 - 2$ and $-j \leq m \leq +j$, by repeated application of $\hat{j}_-$. Proceeding in this way, it is possible to determine all the basis vectors and the corresponding Clebsch–Gordan coefficients.

Example

In order to understand how the above procedure works in practice, let us consider the case $j_1 = j_2 = 1$, for which $0 \leq j \leq 2$. In this case the possible states are summarized in Tab. 6.3.

Case $j = 2$

- $j = 2, m = 2$. In this case, making use of Eq. (6.202), Eq. (6.200) simply reduces to

$$| 1, 1; 2, 2\rangle = \langle 1, 1, 1, 1 \mid 2, 2\rangle \, | 1, 1, 1, 1\rangle , \tag{6.205}$$

since the only state that can contribute to $| 1, 1; 2, 2\rangle$ is $| 1, 1, 1, 1\rangle$. The corresponding Clebsch–Gordan coefficient is then given by

$$\langle 1, 1, 1, 1 \mid 2, 2\rangle = 1. \tag{6.206}$$

- $j = 2, m = -2$. The solution here is

$$|\,1,1;2,-2\rangle = \langle 1,1,-1,-1\,|\,2,-2\rangle\,|\,1,1,-1,-1\rangle\,. \tag{6.207}$$

As before, the Clebsch–Gordan coefficient may be written as

$$\langle 1,1,-1,-1\,|\,2,-2\rangle = 1. \tag{6.208}$$

In order to find the other coefficients we take advantage of the properties of the raising and lowering angular momentum operators (see Eqs. (6.46))

$$\begin{aligned}\hat{\jmath}_{\pm}\,|\,j_1,j_2;j,m\rangle &= \sqrt{(j\pm m+1)\,(j\mp m)}\,|\,j_1,j_2;j,m\pm 1\rangle \\ &= \sum_{m_1,m_2}\langle j_1,j_2,m_1,m_2\,|\,j,m\rangle\,|\,j_1,j_2,m_1,m_2\rangle\,,\end{aligned} \tag{6.209a}$$

and

$$\begin{aligned}(\hat{\jmath}_{1\pm}+\hat{\jmath}_{2\pm})\,|\,j_1,j_2,m_1,m_2\rangle &= \sqrt{(j_1\pm m_1+1)\,(j_1\mp m_1)}\,|\,j_1 j_2,m_1\pm 1,m_2\rangle \\ &+\sqrt{(j_2\pm m_2+1)\,(j_2\mp m_2)}\,|\,j_1,j_2,m_1,m_2\pm 1\rangle\,.\end{aligned} \tag{6.209b}$$

- $j = 2, m = 1$. We must then have $-1 \le m_1 \le +1$, $-1 \le m_2 \le +1$, and, therefore, either $(m_1, m_2) = (1,0)$ or $(m_1, m_2) = (0, 1)$, which implies

$$|\,1,1;2,1\rangle = \langle 1,1,1,0\,|\,2,1\rangle\,|\,1,1,1,0\rangle + \langle 1,1,0,1\,|\,2,1\rangle\,|\,1,1,0,1\rangle\,. \tag{6.210}$$

In order to find the coefficients, we apply $\hat{\jmath}_-$ to both sides of (see Eqs. (6.205)– (6.206))

$$|\,1,1;2,2\rangle = |\,1,1,1,1\rangle\,, \tag{6.211}$$

and write

$$2\,|\,1,1;2,1\rangle = \sqrt{2}\,|\,1,1,0,1\rangle + \sqrt{2}\,|\,1,1,1,0\rangle\,, \tag{6.212}$$

from which it follows that

$$\langle 1,1,1,0\,|\,2,1\rangle = \frac{1}{\sqrt{2}}, \quad \langle 1,1,0,1\,|\,2,1\rangle = \frac{1}{\sqrt{2}}. \tag{6.213}$$

- $j = 2, m = 0$. In this case, the possible values of the pair m_1, m_2 are $(+1,-1)$, $(0,0)$, and $(-1,+1)$. Therefore, we have

$$\begin{aligned}|\,1,1;2,0\rangle &= \langle 1,1,1,-1\,|\,2,0\rangle\,|\,1,1,1,-1\rangle \\ &+\langle 1,1,0,0\,|\,2,0\rangle\,|\,1,1,0,0\rangle + \langle 1,1,-1,1\,|\,2,0\rangle\,|\,1,1,-1,1\rangle.\end{aligned} \tag{6.214}$$

Applying $\hat{\jmath}_-$ to both sides of Eq. (6.212), we obtain

$$\begin{aligned}\sqrt{6}\,|\,1,1;2,0\rangle &= \frac{1}{\sqrt{2}}\Big(\sqrt{2}\,|\,1,1,0,0\rangle + \sqrt{2}\,|\,1,1,1,-1\rangle \\ &+\sqrt{2}\,|\,1,1,-1,1\rangle + \sqrt{2}\,|\,1,1,0,0\rangle\Big).\end{aligned} \tag{6.215}$$

Therefore, the corresponding Clebsch–Gordan coefficients are

$$\langle 1,1,0,0 \mid 2,0\rangle = \sqrt{\frac{2}{3}}, \tag{6.216a}$$

$$\langle 1,1,1,-1 \mid 2,0\rangle = \frac{1}{\sqrt{6}}, \tag{6.216b}$$

$$\langle 1,1,-1,1 \mid 2,0\rangle = \frac{1}{\sqrt{6}}. \tag{6.216c}$$

- $j = 2, m = -1$. In this case, the possible values of m_1, m_2 are $(0,-1)$ and $(-1,0)$. We further apply $\hat{\jmath}_-$ to both sides of Eq. (6.215) and obtain

$$\begin{aligned}\sqrt{6}\,|\,1,1;2,-1\rangle &= \frac{1}{\sqrt{6}}\Big[\sqrt{2}\,|\,1,1,0,-1\rangle + \sqrt{2}\,|\,1,1,-1,0\rangle \\ &\quad + 2\left(\sqrt{2}\,|\,1,1,-1,0\rangle + \sqrt{2}\,|\,1,1,0,-1\rangle\right)\Big],\end{aligned} \tag{6.217}$$

from which it follows that

$$|\,1,1;2,-1\rangle = \frac{1}{\sqrt{2}}\,|\,1,1,0,-1\rangle + \frac{1}{\sqrt{2}}\,|\,1,1,-1,0\rangle\,. \tag{6.218}$$

In conclusion, the Clebsch–Gordan coefficients are

$$\langle 1,1,0,-1 \mid 2,-1\rangle = \frac{1}{\sqrt{2}}, \tag{6.219a}$$

$$\langle 1,1,-1,0 \mid 2,-1\rangle = \frac{1}{\sqrt{2}}. \tag{6.219b}$$

Case $j = 1$

- $j = 1, m = -1$. Again, the possible values of m_1, m_2 are $(0,-1)$ and $(-1,0)$.

$$\begin{aligned}|\,1,1;1,-1\rangle &= \langle 1,1,0,-1 \mid 1,-1\rangle\,|\,1,1,0,-1\rangle \\ &\quad + \langle 1,1,-1,0 \mid 1,-1\rangle\,|\,1,1,-1,0\rangle \\ &= \alpha\,|\,1,1,0,-1\rangle + \beta\,|\,1,1,-1,0\rangle\,,\end{aligned} \tag{6.220}$$

with $\alpha, \beta \in \Re$. This time we cannot take advantage of $\hat{\jmath}_-$, and in order to find the coefficients we may use the orthonormalization conditions

$$\langle 1,1;1,-1 \mid 1,1;1,-1\rangle = 1, \quad \langle 1,1;2,-1 \mid 1,1;1,-1\rangle = 0. \tag{6.221}$$

The conditions (6.221), together with Eq. (6.218) imply that

$$\alpha^2 + \beta^2 = 1, \quad \frac{1}{\sqrt{2}}\alpha + \frac{1}{\sqrt{2}}\beta = 0, \tag{6.222}$$

since

$$\left[\langle 1,1,0,-1\,|\,\frac{1}{\sqrt{2}} + \langle 1,1,-1,0\,|\,\frac{1}{\sqrt{2}}\right][\alpha\,|\,1,1,0,-1\rangle + \beta\,|\,1,1,-1,0\rangle\,] = 0. \tag{6.223}$$

From (6.222), we have $\alpha = -\beta$ and therefore

$$\langle 1,1,0,-1 \mid 1,-1\rangle = \frac{1}{\sqrt{2}}, \tag{6.224a}$$

$$\langle 1,1,-1,0 \mid 1,-1\rangle = -\frac{1}{\sqrt{2}}. \tag{6.224b}$$

In order to find the other coefficients for the remaining values of m, we make use of the formulae (6.209) and repeatedly apply $\hat{j}_+$ to both sides of Eq. (6.220), which can now be rewritten as

$$|\,1,1;1,-1\rangle = \frac{1}{\sqrt{2}}\,|\,1,1,0,-1\rangle - \frac{1}{\sqrt{2}}\,|\,1,1,-1,0\rangle\,. \tag{6.225}$$

- $j=1, m=0$. The possible values of m_1, m_2 are $(-1,1)$, $(0,0)$, and $(1,-1)$. First, we follow the above procedure:

$$\sqrt{2}\,|\,1,1;1,0\rangle = \frac{1}{\sqrt{2}}\Big[\sqrt{2}\,|\,1,1,1,-1\rangle + \sqrt{2}\,|\,1,1,0,0\rangle - \Big(\sqrt{2}\,|\,1,1,0,0\rangle + \sqrt{2}\,|\,1,1,-1,1\rangle\Big)\Big]. \tag{6.226}$$

Then,

$$|\,1,1;1,0\rangle = \frac{1}{\sqrt{2}}\,[|\,1,1,1,-1\rangle - |\,1,1.-1,1\rangle\,]. \tag{6.227}$$

Therefore, Clebsch–Gordan coefficients are equal to

$$\langle 1,1,1,-1 \mid 1,0\rangle = \frac{1}{\sqrt{2}}, \tag{6.228a}$$

$$\langle 1,1,0,0 \mid 1,0\rangle = 0, \tag{6.228b}$$

$$\langle 1,1,-1,1 \mid 1,0\rangle = -\frac{1}{\sqrt{2}}. \tag{6.228c}$$

- $j=1, m=1$. Here the possible values of m_1, m_2 are $(1,0)$ and $(0,1)$. We further apply the raising angular momentum operator to $|\,1,1;1,0\rangle$ and obtain

$$\sqrt{2}\,|\,1,1;1,1\rangle = \frac{1}{\sqrt{2}}\Big(\sqrt{2}\,|\,1,11,0\rangle - \sqrt{2}\,|\,1,1,0,1\rangle\Big), \tag{6.229}$$

from which the Clebsch–Gordan coefficients may be derived:

$$\langle 1,1,1,0 \mid 1,1\rangle = \frac{1}{\sqrt{2}}, \tag{6.230a}$$

$$\langle 1,1,0,1 \mid 1,1\rangle = -\frac{1}{\sqrt{2}}. \tag{6.230b}$$

Case $j = 0$

- $j=0, m=0$. The possible values of m_1, m_2 are $(1,-1)$, $(0,0)$, and $(-1,1)$:

$$|\,1,1;0,0\rangle = \langle 1,1,1,-1 \mid 0,0\rangle\,|\,1,1,1,-1\rangle + \langle 1,1,0,0 \mid 0,0\rangle\,|\,1,1,0,0\rangle + \langle 1,1,-1,1 \mid 0,0\rangle\,|\,1,1,-1,1\rangle\,. \tag{6.231}$$

As in the case $j = 1, m = -1$, we have to resort to the orthonormalization conditions

$$\langle 1,1;0,0 \mid 1,1;0,0\rangle = 1, \quad \langle 1,1;1,0 \mid 1,1;0,0\rangle = 0, \quad \langle 1,1;2,0 \mid 1,1;0,0\rangle = 0. \tag{6.232}$$

Denoting $\langle 1,1,1,-1 \mid 0,0\rangle = \alpha$, $\langle 1,1;0,0 \mid 0,0\rangle = \beta$, and $\langle 1,1,-1,1 \mid 0,0\rangle = \gamma$, all real, from the orthonormalization conditions and Eqs. (6.215) and (6.227), we have the following system of three equations:

$$\begin{cases} \alpha^2 + \beta^2 + \gamma^2 = 1, \\ \frac{\alpha}{\sqrt{2}} - \frac{\gamma}{\sqrt{2}} = 0, \\ \frac{\alpha}{\sqrt{6}} + 2\frac{\beta}{\sqrt{6}} + \frac{\gamma}{\sqrt{6}} = 0. \end{cases} \tag{6.233}$$

Then, the Clebsch–Gordan coefficients are

$$\langle 1,1,1,-1 \mid 0,0\rangle = \frac{1}{\sqrt{3}}, \tag{6.234a}$$

$$\langle 1,1,0,0 \mid 0,0\rangle = -\frac{1}{\sqrt{3}}, \tag{6.234b}$$

$$\langle 1,1,-1,1 \mid 0,0\rangle = \frac{1}{\sqrt{3}}. \tag{6.234c}$$

Generalization

It is possible to write a general formula for the Clebsch–Gordan coefficients:[20]

$$\langle j_1, j_2, m_1, m_2 \mid j, m\rangle = (-1)^{j_1 - j_2 + m}\sqrt{2j+1}\begin{pmatrix} j_1 & j_2 & j \\ m_1 & m_2 & -m \end{pmatrix}, \tag{6.235}$$

where

$$\begin{aligned} \begin{pmatrix} j_1 & j_2 & j_3 \\ m_1 & m_2 & m_3 \end{pmatrix} &= \sqrt{\frac{(j_1+j_2-j_3)!\,(j_1-j_2+j_3)!\,(-j_1+j_2+j_3)!}{(j_1+j_2+j_3+1)!}} \\ &\times \sqrt{(j_1+m_1)!\,(j_1-m_1)!\,(j_2+m_2)!\,(j_2-m_2)!\,(j_3+m_3)!\,(j_3-m_3)!} \\ &\times \sum_{z\in N} \frac{(-1)^{z+j_1-j_2-m_3}}{z!\,(j_1+j_2-j_3-z)!\,(j_1-m_1-z)!\,(j_2+m_2-z)!} \\ &\times \frac{1}{(j_3-j_2+m_1+z)!\,(j_3-j_1-m_2+z)!} \end{aligned} \tag{6.236}$$

are the so-called $3j$-symbols, originally calculated by Wigner. In principle, the sum in Eq. (6.236) is infinite. However, since j_1, j_2 and j_3 are finite numbers, the sum is truncated due to the fact that the factorial of a negative number in the term $(j_1 + j_2 - j_3 - z)!$ is infinite.

It is useful to cast the Clebsch–Gordan coefficients into tables. We give in the following such tables for three cases. When $j_1 = j_2 = 1/2$, we have the results shown in Tab. 6.4 (the empty spaces can be considered as filled by zeros). When $j_1 = 1$ and $j_2 = 1/2$, the results are shown in Tab. 6.5. Finally, for $j_1 = j_2 = 1$ (the example explicitly examined above), the results are summarized in Tab. 6.6.

[20] See [*Edmonds* 1957, 42–50], and in particular formulae (3.6.11) and (3.7.3).

Table 6.4 Clebsch–Gordan coefficients resulting from the composition of angular momenta $j_1 = j_2 = 1/2$

	$j=1$ $m=1$	$j=1$ $m=0$	$j=0$ $m=0$	$j=1$ $m=-1$
$m_1=\frac{1}{2}, m_2=\frac{1}{2}$	1			
$m_1=\frac{1}{2}, m_2=-\frac{1}{2}$		$\frac{1}{\sqrt{2}}$	$\frac{1}{\sqrt{2}}$	
$m_1=-\frac{1}{2}, m_2=\frac{1}{2}$		$\frac{1}{\sqrt{2}}$	$-\frac{1}{\sqrt{2}}$	
$m_1=-\frac{1}{2}, m_2=-\frac{1}{2}$				1

Table 6.5 Clebsch–Gordan coefficients resulting from the composition of angular momenta $j_1 = 1$ and $j_2 = 1/2$

	$j=\frac{3}{2}$ $m=\frac{3}{2}$	$j=\frac{3}{2}$ $m=\frac{1}{2}$	$j=\frac{1}{2}$ $m=\frac{1}{2}$	$j=\frac{3}{2}$ $m=-\frac{1}{2}$	$j=\frac{1}{2}$ $m=-\frac{1}{2}$	$j=\frac{3}{2}$ $m=-\frac{3}{2}$
$m_1=1, m_2=\frac{1}{2}$	1					
$m_1=1, m_2=-\frac{1}{2}$		$\frac{1}{\sqrt{3}}$	$\sqrt{\frac{2}{3}}$			
$m_1=0, m_2=\frac{1}{2}$		$\sqrt{\frac{2}{3}}$	$-\frac{1}{\sqrt{3}}$			
$m_1=0, m_2=-\frac{1}{2}$				$\sqrt{\frac{2}{3}}$	$\frac{1}{\sqrt{3}}$	
$m_1=-1, m_2=\frac{1}{2}$				$\frac{1}{\sqrt{3}}$	$-\sqrt{\frac{2}{3}}$	
$m_1=-1, m_2=-\frac{1}{2}$						1

Table 6.6 Clebsch–Gordan coefficients resulting from the composition of angular momenta $j_1 = j_2 = 1$

	2 2	2 1	1 1	2 0	1 0	0 0	2 −1	1 −1	2 −2
1,1	1								
1,0		$\frac{1}{\sqrt{2}}$	$\frac{1}{\sqrt{2}}$						
0,1		$\frac{1}{\sqrt{2}}$	$-\frac{1}{\sqrt{2}}$						
1,−1				$\frac{1}{\sqrt{6}}$	$\frac{1}{\sqrt{2}}$	$\frac{1}{\sqrt{3}}$			
0,0				$\sqrt{\frac{2}{3}}$	0	$-\frac{1}{\sqrt{3}}$			
−1,1				$\frac{1}{\sqrt{6}}$	$-\frac{1}{\sqrt{2}}$	$\frac{1}{\sqrt{3}}$			
0,−1							$\frac{1}{\sqrt{2}}$	$\frac{1}{\sqrt{2}}$	
−1,0							$\frac{1}{\sqrt{2}}$	$-\frac{1}{\sqrt{2}}$	
−1,−1									1

6.5 Angular momentum and angle

We have found (Eq. (6.34)) that the z component of the orbital angular momentum can be expressed as the differential operator $\hat{L}_z = -\imath\hbar\partial/\partial\phi$, ϕ being the azimuthal angle. The latter relation is formally similar to (see Eq. (2.134))

$$\hat{p}_x = -\imath\hbar\frac{\partial}{\partial x}. \tag{6.237}$$

Taking into account that in classical mechanics angular momentum and angle are conjugate variables, one would then be tempted to consider $\hat{L}_z$ and ϕ as quantum-mechanical conjugate observables, just as we do for $\hat{x}$ and $\hat{p}_x$. In this respect, one would like to write an uncertainty relation similar to Eq. (2.190), i.e.

$$\Delta\phi\Delta L_z \geq \frac{\hbar}{2}. \tag{6.238}$$

However, a close inspection of Eq. (6.238) shows that it cannot be correct. In fact, the maximum value of $\Delta\phi$ is certainly smaller than π, if one considers the interval $-\pi < \phi \leq \pi$ (see Prob. 6.33), so that, when ΔL_z decreases, Eq. (6.238) cannot be satisfied. The reason why this happens is that the operator corresponding to $-\imath\hbar\partial/\partial\phi$ is not self-adjoint (see Subsec. 2.1.1).

Let us consider the function

$$f(\eta) = \int_{-\pi}^{+\pi} d\phi\psi^*(\phi+\eta)\phi^2\psi(\phi+\eta). \tag{6.239}$$

One sees that $(\Delta\phi)^2$ is the minimum value of $f(\eta)$ (see Prob. 6.34). Using the Schwartz inequality as in the derivation of the uncertainty relation between position and momentum (see Subsec. 2.3.2), one obtains

$$(\Delta L_z)^2 f(\eta) \geq \frac{1}{4}\hbar^2[1 - 2\pi\psi^*(\eta+\pi)\psi(\eta+\pi)]^2. \tag{6.240}$$

From this it follows that[21]

$$\Delta L_z \frac{\Delta\phi}{1 - 3(\Delta\phi)^2/\pi^2} \geq 0.15\hbar. \tag{6.241}$$

The factor 0.15 is due to the particular derivation, but with a more general one it can be replaced by 1/2. Notice that here the angle is not represented by an operator, but it is not always trivial to build a Hermitian operator representing a quantum observable (see also Sec. 3.9). In the following, we will show how it is possible to define a Hermitian angle operator.[22]

As recalled above (see Eq. (6.34)), $\hat{L}_z$ can be represented as a differential operator on the function space $L^2([0, 2\pi), d\phi)$, i.e.

[21] See [Judge 1963].

[22] See [Lévy-Leblond 1976] [Carruthers/Nieto 1968] [*Busch et al.* 1995, 73–74].

$$\hat{L}_z \mathrm{F}'(\phi) = -\imath\hbar \frac{\partial}{\partial\phi}\mathrm{F}'(\phi). \tag{6.242}$$

The spectrum of $\hat{L}_z$ is the set $\mathbb{Z}$ of integers, and the eigenfunctions $\mathrm{F}_m(\phi)$ corresponding to the eigenvectors $|m_l\rangle$ of $\hat{L}_z$ are given by (see Eqs. (6.38), and (2.135) and comments)

$$\mathrm{F}_m(\phi) = \langle \phi \mid m_l \rangle = e^{\imath m_l \phi}, \quad m_l \in \mathbb{Z}, \tag{6.243}$$

with $\mathrm{F}'_m(\phi) = \sqrt{2\pi}\mathrm{F}_m(\phi)$, and where the eigenvectors $|\phi\rangle$ of the angle operator are

$$|\phi\rangle = \frac{1}{\sqrt{2\pi}} \sum_{m_l=-\infty}^{+\infty} \langle m_l \mid \phi \rangle \, |m_l\rangle = (2\pi)^{-\frac{1}{2}} \sum_{m_l=-\infty}^{+\infty} e^{-\imath m_l \phi}|m_l\rangle. \tag{6.244}$$

Let us introduce an additive unitary shift operator

$$\hat{U}_k|m_l\rangle = |m_l + k\rangle, \tag{6.245}$$

where (see Th. 3.1: p. 123, and Eq. (8.30))

$$\hat{U}_k = e^{\imath k\hat{\phi}}. \tag{6.246}$$

Given the spectral decomposition

$$\begin{aligned} \hat{I} &= \int d\phi \, |\phi\rangle \, \langle\phi| = (2\pi)^{-1} \sum_{k,m_l} \int d\phi \, |m_l\rangle \, \langle m_l \mid \phi\rangle \, \langle \phi \mid k\rangle \, \langle k| \\ &= (2\pi)^{-1} \sum_{k,m_l} \int d\phi e^{\imath(k-m_l)\phi} |m_l\rangle\langle k|, \end{aligned} \tag{6.247}$$

we can define the following self-adjoint angle operator (see Eq. (2.20)):

$$\hat{\phi} = \int_0^{2\pi} d\phi\,\phi \; |\phi\rangle \, \langle\phi| = \pi\hat{I} + \sum_{m_l \neq k} \frac{1}{\imath(k - m_l)} |m_l\rangle\langle k|, \tag{6.248}$$

where ϕ are its eigenvalues.

Concerning the commutation relations, we notice that, since the range of ϕ is finite, the angle observable, in the ϕ representation, should be taken in quantum mechanics as the multiplication, not by ϕ (that goes from $-\infty$ to $+\infty$), but rather by[23]

$$Y(\phi) = \phi - 2\pi \sum_{n=0}^{\infty} \Xi[\phi - (2n+1)\pi] + 2\pi \sum_{n=0}^{\infty} \Xi[-\phi - (2n+1)\pi], \tag{6.249}$$

where Ξ is the Heaviside step function, given by

$$\Xi(x) = \begin{cases} 1 & \text{if} \quad x \geq 0 \\ 0 & \text{if} \quad x < 0 \end{cases}. \tag{6.250}$$

[23] See [Judge/Lewis 1963].

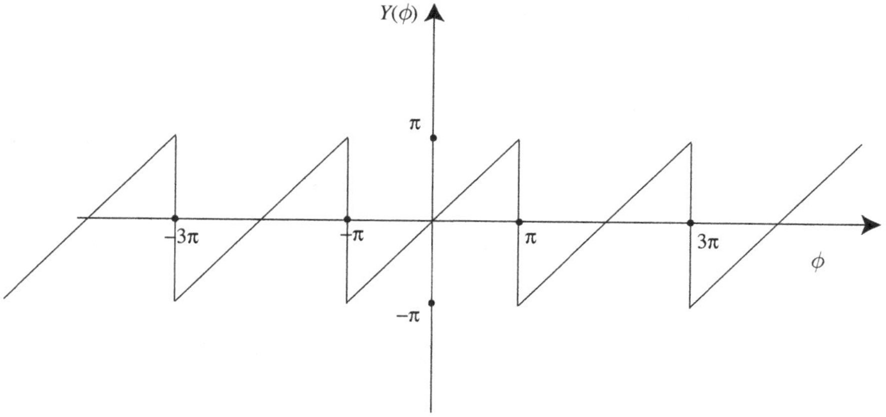

Figure 6.14 **The step function $Y(\phi)$ representing the angle observable (see Eq. (6.249)) in the ϕ representation and in the range $[-\pi, \pi]$.**

In other words (see Fig. 6.14), $Y(\phi)$ is just $\phi(\text{mod } 2\pi)$ in the range $[-\pi, \pi]$. From Eq. (6.249),

$$\left[\hat{L}_z, Y(\phi)\right] = -\imath\hbar\left[1 - 2\pi\sum\delta\{\phi - (2n+1)\pi\}\right] \tag{6.251}$$

follows, which represents the commutation relation between $\hat{L}_z$ and $Y(\phi)$ in the $Y(\phi)$ representation. We emphasize that Eq. (6.251) is simply given by $\imath\hbar\{L_z, Y(\phi)\}$, where $\{,\}$ are the classical Poisson brackets (see Sec. 3.7). In terms of the commutator (6.251), by calculating the expectation value of the commutator on the state $\psi(\phi) = \langle\phi \mid \psi\rangle$, which is normalized in a 2π interval, and making use of the generalized uncertainty relation (2.200), it is possible to derive a more satisfying uncertainty relation between angular momentum and angle, that is,[24]

$$\Delta_{\psi(\phi)}L_z\Delta_{\psi(\phi)}\phi \geq \frac{\hbar}{2}\left[1 - 2\wp(\pi)\right], \tag{6.252}$$

where $\wp(\pi) = |\psi(\pi)|^2$ is the probability density for finding the system at the angle π.

Summary

In this chapter we have discussed the total angular momentum, which is the sum of the orbital angular momentum and the spin.

- We have first introduced the *orbital angular momentum* following its classical definition and proposed an abstract algebra based on the commutation relations. Here, the main

[24] See [Pegg *et al.* 2005] for a review.

feature is that different components of the angular momentum *do not commute* with each other, whereas the square of the angular momentum does with each of these components. Then, we have shown that the angular momentum is the generator of the *group of rotations*, and calculated its eigenvalues and eigenfunctions.

- Taking advantage of the angular-momentum formalism, we have considered a couple of special potentials: the *central potential* and the *constant magnetic field*.
- We have analyzed the *harmonic-oscillator* problem in *several dimensions*, where the *degeneracy* of energy eigenvalues plays an important role.
- We have also introduced the *spin* degree of freedom, i.e. the intrinsic magnetic moment of a particle. This physical quantity has no classical analogue.
- The *composition* of different angular momenta has been discussed. The explicit formalism for spin triplet and singlet states has also been presented. In particular, the *singlet* state is an example of *entangled* state that has a wide range of applications.
- We have derived the explicit form of the *Clebsch–Gordan coefficients*. They represent probability amplitudes that allow us to move from the representation of the eigenvectors of the component angular momenta to the representation of the eigenvectors of the angular momentum resulting from their composition.
- Finally, we have considered the problem of defining an *uncertainty relation* between angular momentum and angle, and have presented an operational representation of the *angle observable*.

Problems

6.1 Calculate the commutation relations between the y and z components and between the z and x components of the orbital angular momentum (see Eq. (6.6)).

6.2 Prove Eq. (6.7).
(*Hint*: Write $\hat{\mathbf{L}}^2$ as $\hat{L}_x^2 + \hat{L}_y^2 + \hat{L}_z^2$ and make use of Eq. (6.6).)

6.3 Write the matrices which describe rotations of an angle θ about the axes x and y.

6.4 Taking advantage of the explicit expression of $\hat{L}_z$, derive Eqs. (6.18).

6.5 Derive the commutation relations $\left[\hat{l}_j, \hat{l}_k\right] = \imath\epsilon_{jkn}\hat{l}_n$.

6.6 Prove that $\hat{\mathbf{l}}^2 - \hat{l}_z^2 = \hat{l}_x^2 + \hat{l}_y^2 \geq 0$.

6.7 Prove the relevant commmutation relations pertaining to the angular momentum raising and lowering operator $\hat{l}_\pm$, which are summarized in Eq. (6.24).

6.8 Derive the commutation relations

$$\left[\hat{l}_x, \hat{l}_\pm\right] = \mp\hat{l}_z, \quad \left[\hat{l}_y, \hat{l}_\pm\right] = -\imath\hat{l}_z. \tag{6.253}$$

6.9 Prove the eigenvalue equation (6.31): $\hat{\mathbf{l}}^2 \left| l, m_l \right\rangle = l(l+1) \left| l, m_l \right\rangle$.

6.10 Derive the matrices (6.49).
(*Hint*: Make use of Eqs. (6.21), (6.46a), and (6.46b).)

6.11 Compute the Jacobian matrix

$$\frac{\partial(r,\phi,\theta)}{\partial(x,y,z)} = \begin{bmatrix} \frac{\partial r}{\partial x} & \frac{\partial r}{\partial y} & \frac{\partial r}{\partial z} \\ \frac{\partial \phi}{\partial x} & \frac{\partial \phi}{\partial y} & \frac{\partial \phi}{\partial z} \\ \frac{\partial \theta}{\partial x} & \frac{\partial \theta}{\partial y} & \frac{\partial \theta}{\partial z} \end{bmatrix}. \tag{6.254}$$

6.12 Compute the Laplacian in spherical coordinates and show that

$$\nabla^2 = \frac{1}{r^2}\frac{\partial}{\partial r}\left(r^2 \frac{\partial}{\partial r}\right) - \frac{\hat{\mathbf{l}}^2}{r^2}. \tag{6.255}$$

6.13 Verify that $\Theta_{ll} = \mathcal{N}(\sin\theta)^l$ is a solution of the differential equation (6.63).

6.14 Find the normalization constant $\mathcal{N}$ and derive Eq. (6.64).

6.15 Prove Eq. (6.82).

6.16 Prove that a spherically symmetric Hamiltonian commutes with each component of the angular momentum and with $\hat{\mathbf{L}}^2$.

6.17 Prove that $\hat{l}_z$ commutes with $\hat{H}_r$ given by Eq. (6.104).

6.18 In the case $l = 1$, compute explictly the matrices $\hat{l}_x\hat{l}_y$ and $\hat{l}_y\hat{l}_x$ and verify that $\left[\hat{l}_x, \hat{l}_y\right] = \imath\hat{l}_z$.

6.19 Following the guide of Subsec. 6.2.4 solve the two-dimensional harmonic oscillator.

6.20 Derive the commutation relations for the Pauli spin matrices.

6.21 Prove Eq. (6.157).

6.22 Calculate the eigenkets and eigenvalues of $\hat{\sigma}_x$ and $\hat{\sigma}_y$ in the basis $\{|\uparrow\rangle_z, |\downarrow\rangle_z\}$ in order to prove Eqs. (6.158) and (6.159).

6.23 Calculate the Pauli matrices as sums of projectors, by using the basis $\{|\uparrow\rangle_z, |\downarrow\rangle_z\}$.

6.24 Prove that the Pauli matrices are Hermitian and unitary.

6.25 A known example of the fact that any operator in the spinor's space can be written as a combination of $\hat{\sigma}_x$, $\hat{\sigma}_y$, $\hat{\sigma}_z$, and $\hat{I}$, is represented by the operator

$$(\hat{\sigma}\cdot\mathbf{f})(\hat{\sigma}\cdot\mathbf{f}') = (\mathbf{f}\cdot\mathbf{f}')\hat{I} + \imath\hat{\sigma}\cdot(\mathbf{f}\times\mathbf{f}'),$$

where $\mathbf{f}, \mathbf{f}'$ are vectors. Prove this result.

6.26 Deduce the matrices (6.165).
(*Hint*: Follow the general lines of the derivation of Pauli spin matrices.)

6.27 Derive Eq. (6.176).

6.28 Find the Landau levels for the electron.

6.29 Find the eigenfunction $\varphi_n(y)$ corresponding to the energy levels (6.179).

6.30 Prove, for the case discussed in Subsec. 6.4.1, first that $\hat{\mathbf{l}}, \hat{l}_z, \hat{\mathbf{l}}_1^2$, and $\hat{\mathbf{l}}_2^2$ commute with each other, and then that $\hat{l}_{1_z}, \hat{l}_{2_z}$ are not compatible with $\hat{\mathbf{l}}$.

6.31 Compute the total spin and its component along z for the states (6.193) and (6.194).

6.32 Take advantage of Eqs. (6.159) and perform a change of basis into Eq. (6.194) in order to derive Eqs. (6.195).

6.33 Prove that the maximum uncertainty value of $\Delta\phi$ is $\pi/\sqrt{3}$.
(*Hint*: Assume that the uniform distribution of angles ϕ between 0 and 2π maximizes the angle uncertainty.)

6.34 Prove that the minimum value of $f(\eta)$ in Eq. (6.239) is equal to $(\Delta\phi)^2$.

6.35 Apply Robertson's formula (2.200) in order to calculate the uncertainty relations between different components of the total angular momentum.

Further reading

Edmonds, A. R., *Angular Momentum in Quantum Mechanics*, Princeton: Princeton University Press, 1957, 1960, 1985.

Wigner, Eugene P., *Group Theory*, New York: Academic, 1959.

7 Identical particles

In classical physics, the question whether there are subsets of identical particles is not particularly interesting. After all, whether the particles are identical or not, we can always track each of them at any time. As we shall see, in quantum mechanics identical particles are indistinguishable because we cannot track trajectories, and this fact induces some important symmetry properties on the total wave function describing the system of particles under consideration. Generally speaking, a symmetry is an effect of an invariance, that is, the system does not change if one effects certain transformations (see Ch. 8). In this case, the system is not modified under exchange of two identical particles. Moreover, such a symmetry yields a most interesting connection between spin and statistics.

After a short introduction to the significance of statistics in quantum mechanics (in Sec. 7.1), we define the permutation operator and discuss the symmetry properties of the wave function of N identical particles (Sec. 7.2). In Sec. 7.3 we show that quantum particles, differently from classical particles (that obey the Maxwell–Boltzmann statistics), may be subject to two different statistics, namely the Fermi–Dirac statistics for fermions and the Bose–Einstein statistics for bosons. In Sec. 7.4 we treat the so-called exchange interaction and, finally, in Sec. 7.5 we consider some applications of the two statistics.

7.1 Statistics and quantum mechanics

Identical particles are those particles whose Hamiltonian is symmetric under exchange of one particle with the other. Of course, they must bear identical intrinsic properties (mass, charge, spin, etc.). In classical mechanics, given the forces acting on a system, once we have assigned $\mathbf{r}$ and $\mathbf{p}$ to all particles at a certain time, their trajectories are determined (see Sec. 1.1 and Subsec. 2.3.3) and therefore we are able to "follow" each particle at any future time: the particles are distinguishable, at least in principle. Instead, in quantum mechanics there is no trajectory (see Subsec. 2.3.3 again) and for this reason we are not able to track identical particles separately: in this case we are forced to conclude that the particles are *indistinguishable*. This fact has also the consequence that several "classical" configurations may correspond to the same quantum state. Therefore, in this respect, the number of possible *independent* states is smaller in quantum mechanics than in classical mechanics. Hence, statistics has in quantum mechanics a fundamental theoretical importance.

In order to put this matter on a more concrete ground, let us consider the following example (see Fig. 7.1), inspired by an experiment performed by Mandel and co-workers

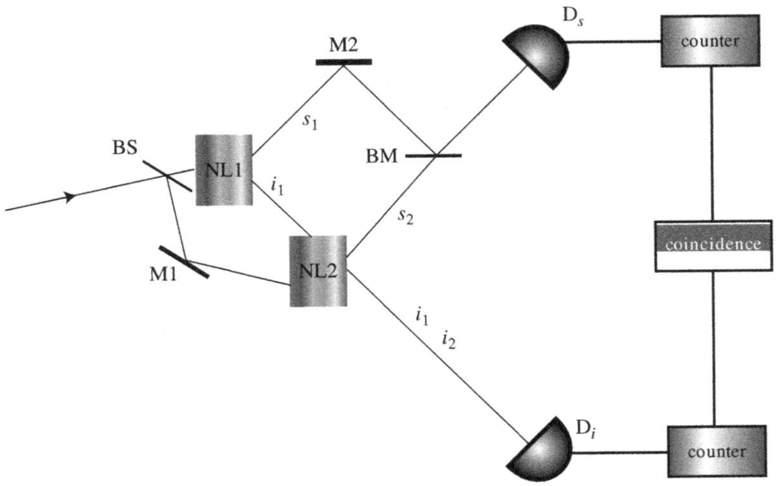

Figure 7.1 Interferometric example of indistinguishability. An initial (pump) beam is split by a beam splitter (BS) and the resulting beams enter two non-linear crystals of $LiIO_3$ (NL1 and NL2). From NL1 a signal photon (s_1) and an idler photon (i_1) emerge: the i-photon passes through NL2 and will be aligned with the second idler (i_2), which is emitted by NL2 together with the second signal photon (s_2). The two s-photons are combined by the beam merger (BM) and the outgoing beam falls on detector D_s, whereas the two idler photons fall on detector D_i. The two detectors are connected to a coincidence counter.

(see also Subsec. 9.5.1). We make use here of a quantum-optical phenomenon, called *parametric down-conversion*: a photon passes through a non-linear crystal,[1] which absorbs the original photon and correspondingly emits two photons, conventionally called *idler photon* (i-photon) and *signal photon* (s-photon). It is evident that momentum and energy must be conserved in this process, and this is also the reason for the name "down-conversion": a photon of frequency ν – and energy $h\nu$ (see Subsec. 1.2.1) – is converted into two photons of smaller frequencies ν_i and ν_s (and energies $h\nu_i$ and $h\nu_s$), respectively, with $\nu = \nu_i + \nu_s$. It is interesting to contrast parametric down-conversion with beam splitting: in parametric down-conversion two photons leave a non-linear crystal when one enters, whilst in beam splitting the same number of photons goes in and out, which is why it is a linear and unitary transformation (see Subsec. 3.5.2). The two i-photons represented in Fig. 7.1 fall on the same detector in the same state. For this reason, a coincidence counting cannot discriminate between the case in which the s-photon and i_1 are detected and the case in which the s-photon and i_2 are detected. This results into interference fringes (obtained from the various displacements of the beam merger BM) that are absent when the idler photon i_1 is blocked by a filter placed between NL1 and NL2. Such an interference phenomenon is only understandable if one admits that the two photons i_1 and i_2 are physically indistinguishable.

[1] The word *non-linear* simply means here that the output state cannot be described as the action of a linear operator on the input state.

7.2 Wave function and symmetry

Suppose we define the wave function of two identical particles as $\Psi(\xi_1, \xi_2) = \Psi(\mathbf{r}_1, \mathbf{r}_2, s_{1_z}, s_{2_z})$, where $\xi_j = (\mathbf{r}_j, s_{j_z})$, $j = 1, 2$, and s_{j_z} describes the j–th's z component of the spin degree of freedom. Let us introduce the permutation operator $\hat{U}_P$ that, in the case of two particles, acts as follows:

$$\hat{U}_P \Psi(\xi_1, \xi_2) = \Psi(\xi_2, \xi_1), \tag{7.1}$$

where all coordinates of the particles are exchanged. However, as we have already emphasized, the wave functions $\Psi(\xi_1, \xi_2)$ and $\Psi(\xi_2, \xi_1)$ cannot be in general different. Therefore (see Subsec. 2.1.3), they may vary at most for a global phase factor $e^{\imath\phi}$, and we must have

$$\Psi(\xi_2, \xi_1) = e^{\imath\phi} \Psi(\xi_1, \xi_2). \tag{7.2}$$

If we exchange the two particles once more, we must have

$$\hat{U}_P^2 \Psi(\xi_1, \xi_2) = \Psi(\xi_1, \xi_2), \tag{7.3}$$

and as a consequence $\hat{U}_P^2 = \hat{I}$.[2] According to Eqs. (7.1)–(7.2) we have

$$\hat{U}_P^2 \Psi(\xi_1, \xi_2) = e^{2\imath\phi} \Psi(\xi_1, \xi_2), \tag{7.4}$$

so that, taking into account Eq. (7.3), we may argue that $e^{2\imath\phi} = 1$ and $e^{\imath\phi} = \pm 1$.[3] From the previous equations, we can conclude that

$$\Psi(\xi_2, \xi_1) = \pm \Psi(\xi_1, \xi_2). \tag{7.5}$$

In other words, the wave functions of a system of two identical particles are either symmetric or antisymmetric with respect to the exchange of the particles. It is also clear that the possible wave functions of a certain two-particle system must be either all symmetric or all antisymmetric. In fact, if we consider the superposition of two wave functions, one symmetric (ψ_S) and the other antisymmetric (ψ_A), i.e.

$$\Psi = \alpha\psi_S + \beta\psi_A, \tag{7.6}$$

the resulting wave function has no definite parity relatively to the permutation operation, because

$$\hat{U}_P (\alpha\psi_S + \beta\psi_A) = \alpha\psi_S - \beta\psi_A, \tag{7.7}$$

which is neither symmetric nor antisymmetric. This can be generalized to N particles as follows:

[2] It should be noted that this is strictly true only up to a phase factor, that is, after two successive permutations, the system may not return exactly to the same wave function. We shall discuss this point more extensively in Sec. 13.8. We notice here that $\hat{U}_P$ is not a representation of the group of permutations.

[3] States with $e^{\imath\phi} \neq \pm 1$ can be realized in two dimensions: they are relevant for condensed matter physics and are called anyons.

Principle 7.1 (Symmetrization) *The dynamical states of a system of N identical particles are necessarily either all symmetrical or all antisymmetric with respect to the exchange of any two particles.*

Since the Hamiltonian is symmetric relative to two-particle exchange, this symmetry must be conserved in time and, as a consequence, for any time evolution we have

$$\hat{U}_P \left| \psi_{S/A}(t) \right\rangle = \hat{U}_P \hat{U}_{t-t_0} \left| \psi_{S/A}(t_0) \right\rangle = \hat{U}_{t-t_0} \hat{U}_P \left| \psi_{S/A}(t_0) \right\rangle, \tag{7.8}$$

and therefore

$$\left[\hat{U}_P, \hat{U}_{t-t_0} \right] = 0. \tag{7.9}$$

The property $\hat{U}_P^2 = \hat{I}$ is not general, but depends on the fact that we have chosen the example of two particles: two successive exchanges of two particles do not change the initial situation (state). In a general N-particles framework, the permutation operator is still unitary. In fact, let us consider an orthonormal basis for a system of N identical particles. It is clear that a permutation acting on a basis vector will transform it into a different vector in the same basis. Therefore, permutations must represent a bijective correspondence between vectors on the original orthonormal basis. This in turn means that the permutation operators are unitary, i.e.

$$\hat{U}_P \hat{U}_P^\dagger = \hat{U}_P^\dagger \hat{U}_P = \hat{I}. \tag{7.10}$$

Permutations have also the group property[4] that the action of two successive permutations $\hat{U}'_P$ and $\hat{U}''_P$ correspond to a single permutation $\hat{U}_P = \hat{U}''_P \hat{U}'_P$.

Consider now the cases of the permutation of two and three particles among N particles. In the case of two-particle permutation among N particles, we can write

$$\hat{U}_P^{12} \left| \mathbf{r}_\alpha^{(1)} \mathbf{r}_\beta^{(2)} \cdots \mathbf{r}_\omega^{(N)} \right\rangle = \left| \mathbf{r}_\beta^{(1)} \mathbf{r}_\alpha^{(2)} \cdots \mathbf{r}_\omega^{(N)} \right\rangle, \tag{7.11}$$

where, for the sake of simplicity, we have chosen the eigenkets

$$\left| \mathbf{r}_\alpha^{(1)} \mathbf{r}_\beta^{(2)} \cdots \mathbf{r}_\omega^{(N)} \right\rangle = \left| \mathbf{r}_\alpha^{(1)} \right\rangle \otimes \left| \mathbf{r}_\beta^{(2)} \right\rangle \otimes \cdots \otimes \left| \mathbf{r}_\omega^{(N)} \right\rangle \tag{7.12}$$

of the position operators $\hat{\mathbf{r}}^{(1)}, \hat{\mathbf{r}}^{(2)}, \ldots, \hat{\mathbf{r}}^{(N)}$.

In the case of the permutation of three particles, for example the cyclic permutation

$$\hat{U}_P^{123} \left| \mathbf{r}_\alpha^{(1)} \mathbf{r}_\beta^{(2)} \mathbf{r}_\gamma^{(3)} \right\rangle = \left| \mathbf{r}_\gamma^{(1)} \mathbf{r}_\alpha^{(2)} \mathbf{r}_\beta^{(3)} \right\rangle, \tag{7.13}$$

we have

$$\begin{aligned} \hat{U}_P^{123} \left| \Psi \right\rangle &= \sum_{\alpha\beta\gamma} \hat{U}_P^{123} \left| \mathbf{r}_\alpha^{(1)} \mathbf{r}_\beta^{(2)} \mathbf{r}_\gamma^{(3)} \right\rangle \left\langle \mathbf{r}_\alpha^{(1)} \mathbf{r}_\beta^{(2)} \mathbf{r}_\gamma^{(3)} \mid \Psi \right\rangle \\ &= \sum_{\alpha\beta\gamma} \left| \mathbf{r}_\gamma^{(1)} \mathbf{r}_\alpha^{(2)} \mathbf{r}_\beta^{(3)} \right\rangle \left\langle \mathbf{r}_\alpha^{(1)} \mathbf{r}_\beta^{(2)} \mathbf{r}_\gamma^{(3)} \mid \Psi \right\rangle, \end{aligned} \tag{7.14}$$

[4] See Sec. 8.4.

where $| \Psi \rangle$ is a generic three-particle state vector. Renaming the summation indices, we obtain

$$\hat{U}_P^{123} = \sum_{\alpha\beta\gamma} \left| \mathbf{r}_\alpha^{(1)} \mathbf{r}_\beta^{(2)} \mathbf{r}_\gamma^{(3)} \right\rangle \left\langle \mathbf{r}_\beta^{(1)} \mathbf{r}_\gamma^{(2)} \mathbf{r}_\alpha^{(3)} \right| . \tag{7.15}$$

7.3 Spin and statistics

7.3.1 The two statistics

What we have said up to now is true of every type of particle. However, following the above distinction between symmetric and antisymmetric wave functions, particles can be cast into two main groups according to their spin:

- *Bosons*, so called after the name of the Indian physicist Satyendranath Bose.[5] Their wave function is symmetric and they have integer spin.
- *Fermions*, so called after the name of the Italian physicist Enrico Fermi.[6] Their wave function is antisymmetric and they have half-integer spin.

The connection between spin and statistics is put forward by relativistic arguments and will not be proved here.[7] Notice, however, that Fermi and Bose formulated their theories in terms of the occupation number, i.e. the number of particles which may occupy the same energetic level. The reformulation of this problem in terms of antisymmetric and symmetric wave functions is due to Dirac.[8]

Examples of bosons are represented by photons, π mesons, gluons, and W and Z particles, and examples of fermions are represented by electrons, protons, neutrons, quarks, and leptons (see Tab. 7.1). On the elementary-particle level, we may say that matter is made up of fermions while forces are intermediated by bosons.

The previous properties can be summarized by the following theorem:

Theorem 7.1 (Spin and statistics) *Particles with integer spin follow Bose–Einstein statistics (symmetric wave function), while particles with half-integer spin follow Fermi–Dirac statistics (antisymmetric wave function).*

Compound particles can be either bosons or fermions depending on the number of constituting fermions. Following the law of composition of angular momenta (see Eq. (6.187)),

[5] See [Bose 1924]. For a history of the problem see [*Mehra/Rechenberg* 1982–2001, I, 554–78].
[6] See [Fermi 1926a, Fermi 1926b].
[7] See [*Streater/Wightman* 1978].
[8] See [Dirac 1926b].

Table 7.1 The table of elementary particles, fermions (subdivided in leptons and quarks) and bosons. Fermions are the elementary constituents of matter. There are three generations of fermions: the electron and its neutrino, and the quarks down and up; the muon and its neutrino, and the quarks strange and charmed; the τ lepton and its neutrino, and the quarks bottom and top. Protons and neutrons, for reasons explained below, are also fermions. The bosons represent the carrier of forces and are represented by the photons for electromagnetic force, by W and Z bosons for weak force, and by the (eight) gluons for strong force (called "color" force, i.e. bounds between quarks). The electron, muon, and τ leptons have all charge -1. All neutrinos have charge 0. Quarks u, c, t have charge 2/3 while quarks d, s, b have charge $-1/3$. Finally, all bosons have charge 0. Antiparticles, when different from their respective particle, have the same masses and spin, but the charge is opposite in sign

Elementary particles		
Fermions		Bosons
Leptons	Quarks	
e, ν_e	u, d	γ
μ, ν_μ	c, s	Z^0, W^-, W^+
τ, ν_τ	t, b	gluons

if this number is even, the resulting particle is a boson, if it is odd, the resulting particle is a fermion. For instance, neutrons and protons are fermions because both are composed of three quarks, whereas the hydrogen atom is composed of two fermions, i.e. an electron plus a proton, which makes a boson. Similarly, nuclei may also be bosons or fermions depending on the number of neutrons and protons.

Box 7.1 **Rasetti's discovery**

In this context, a remarkable example is the historical development of the nuclear theory of the atom. At the end of 1920s, in order to explain the atomic and mass numbers of the different atoms, the hypothesis was that some electrons could also be present in the nucleus. According to this hypothesis, for example, the ^{14}N nucleus would be made of 14 protons and 7 electrons, i.e. 21 fermions. However, Franco Rasetti (1901–2001), by observing the Raman spectrum of $^{14}N_2$ molecules, established that it was a boson. The problem was solved in 1932 when James Chadwick (1891–1974) discovered the neutron: the ^{14}N nucleus was then found to be made of 7 protons and 7 neutrons, i.e. 14 fermions. Incidentally, we note that the hypothetical presence of electrons in the nucleus would suffer serious drawbacks on the basis of the uncertainty relation: in order to confine an electron inside a region of the order of the typical dimension of the nucleus (10^{-15} m), the electron itself would need to possess a huge kinetic energy and this in turn would mean a huge binding potential energy, which is difficult to account for.

As an example of application of the two statistics, let us consider two identical particles in a box with infinite walls (see Sec. 3.4). If the two populated energy levels are E_{n_1} and E_{n_2} ($n_1 \neq n_2$), in the case of bosons the total wave function is

$$\Psi_S(\xi_1, \xi_2) = \frac{1}{\sqrt{2}} \left[\psi_{n_1}(\xi_1)\psi_{n_2}(\xi_2) + \psi_{n_1}(\xi_2)\psi_{n_2}(\xi_1) \right], \tag{7.16a}$$

while in the case of fermions is given by

$$\Psi_A(\xi_1, \xi_2) = \frac{1}{\sqrt{2}} \left[\psi_{n_1}(\xi_1)\psi_{n_2}(\xi_2) - \psi_{n_1}(\xi_2)\psi_{n_2}(\xi_1) \right], \tag{7.16b}$$

where $1/\sqrt{2}$ is a normalization factor.

7.3.2 Fermions

Fermions[9] obey the Fermi–Dirac statistics (antisymmetry of the wave function and half-integer spin). Therefore, the wave function of N fermions changes sign each time two particles are interchanged. On the other hand, the permutation of N particles may always be decomposed into a certain number of pairwise exchanges. In general, therefore, we shall have

$$\Psi(\xi_1, \xi_2, \ldots, \xi_N) = (-1)^l \Psi(\xi_{l_1}, \xi_{l_2}, \ldots, \xi_{l_N}), \tag{7.17}$$

where $\xi_{l_1}, \xi_{l_2}, \ldots, \xi_{l_N}$ is a permutation of $\xi_1, \xi_2, \ldots, \xi_N$ and l is the rank of such permutation.

In the example of Eq. (7.16b), we see that, putting $n_1 = n_2$ we would obtain an identical zero. This result is an expression of the *Pauli exclusion principle*,[10] which states that

Principle 7.2 (Pauli exclusion principle) *Two fermions cannot occupy the same individual state.*

This means, for example, that two electrons with the same spin cannot occupy the same energy level of an atom. Hence, suppose that we have k possible states in which we want to distribute N fermions ($N \leq k$). Then, the number ζ_A of possible ways of distributing N fermions on k states is given by the so-called combination without repetition, i.e. by the ratio between the number of possible permutations[11] $k!/(k-N)!$ and the number $N!$ of the possible permutations of the N particles. In other words, ζ_A would coincide with the number of possible permutations only if the symmetrization principle did not apply: the combinations without repetition are those permutations that are considered different only

9 See [*Huang* 1963, 241–76] for a treatment of the Fermi gas.

10 See [Pauli 1925]. For a historical reconstruction of this point see [*Mehra/Rechenberg* 1982–2001, I, 422–45].

11 Permutations are distinguished if we either have a different numerical distribution or a different order of the elements.

Table 7.2 The four possible distributions of three fermions in four states. Dots represent fermions while the symbol ∅ represents unoccupied states

Distribution	State 1	State 2	State 3	State 4
I	·	·	·	∅
II	·	∅	·	·
III	·	·	∅	·
IV	·	·	·	∅

if the numerical distribution is different, that is, the number of ways one can choose N among k objects. In formulae we have

$$\zeta_A = \frac{k!}{N!(k-N)!} = \begin{pmatrix} k \\ N \end{pmatrix}. \tag{7.18}$$

For example, suppose that we wish to cast three fermions in four states. Then, we have four possible arrangements described in Tab. 7.2.

There is a simple way to represent antisymmetric wave functions in the case of non-interacting fermions. For instance, a simple inspection of Eq. (7.16b) shows that it can be rewritten in the form

$$\Psi(\xi_1, \xi_2) = \frac{1}{\sqrt{2}} \begin{vmatrix} \psi_{n_1}(\xi_1) & \psi_{n_1}(\xi_2) \\ \psi_{n_2}(\xi_1) & \psi_{n_2}(\xi_2) \end{vmatrix}. \tag{7.19}$$

This representation can be generalized to N particles:

$$\begin{aligned} \Psi(\xi_1, \xi_2, \ldots, \xi_N) &= \frac{1}{\sqrt{N!}} \sum_{\text{perm.}} (-1)^l \psi_{l_{n_1}}(\xi_1) \ldots \psi_{l_{n_N}}(\xi_N) \\ &= \frac{1}{\sqrt{N!}} \begin{vmatrix} \psi_{n_1}(\xi_1) & \psi_{n_1}(\xi_2) & \ldots & \psi_{n_1}(\xi_N) \\ \psi_{n_2}(\xi_1) & \psi_{n_2}(\xi_2) & \ldots & \psi_{n_2}(\xi_N) \\ \ldots & \ldots & \ldots & \ldots \\ \psi_{n_N}(\xi_1) & \psi_{n_N}(\xi_2) & \ldots & \psi_{n_N}(\xi_N) \end{vmatrix}, \end{aligned} \tag{7.20}$$

where the determinant is called the *Slater determinant*: an exchange of two particles means an exchange of two columns in the determinant, which accounts for the resulting change of sign. It is easy to see that, if we have $n_j = n_k$ for certain $j \neq k$, i.e. if two particles are in the same state, then two rows are equal and, as a consequence, the Slater determinant is zero. This is a mathematical representation of Pauli exclusion principle.

Notice that the table of elements can be explained by the Pauli exclusion principle, since, by adding an electron to a given element, we obtain the next element in Mendeleev's periodic table.

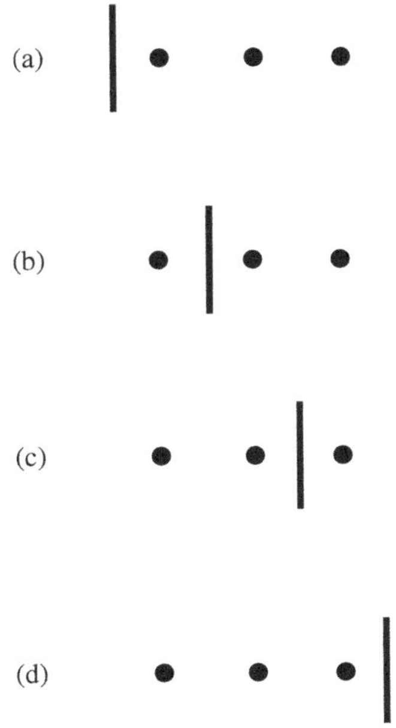

Figure 7.2 Example illustrating the method used for counting the number of possible configurations of $N = 3$ bosons in $k = 2$ states. In this case we have only one partition (the separation between the two states) and the desired number is obtained by moving this wall from left to right (from (a) to (d)). The resulting configurations are summarized in Tab. 7.3.

7.3.3 Bosons

Bosons[12] obey the Bose–Einstein statistics (symmetry of the wave function and integer spin). In this case, Eq. (7.17) translates into

$$\psi(\xi_1, \xi_2, \ldots, \xi_N) = \psi(\xi_{l_1}, \xi_{l_2}, \ldots, \xi_{l_N}), \tag{7.21}$$

so that, in general, the symmetric wave function of a system of N bosons will be given by the sum of all possible terms of the type (7.21). Suppose again we wish to cast N particles into k possible states. Unlike fermions, bosons do not undergo the constraint imposed by the Pauli exclusion principle (see Pr. 7.2). For this reason, any number of bosons can occupy any state, and in this case the number of possible configurations is given by the so-called combinations with repetition.[13] The number of configurations can be obtained by considering the following argument. The total number N of particles can be partitioned in different ways among the k available states. In each distinct configuration, it is as if we would insert $k - 1$ ideal walls (the separations between the states) in between the particles. As a consequence, the number of possible configurations is given by the number of ways in which we can choose $k - 1$ elements among $N + k - 1$ (see Fig. 7.2).

[12] See [*Huang* 1963, 278–302] for a treatment of the Bose gas.
[13] In this case, obviously, N is not limited by k.

Table 7.3 The four possible configurations of three bosons in two different states. Dots represent bosons while the symbol ∅ represents unoccupied states

Configuration	State 1	State 2
I	...	∅
II	∅	...
III	.	..
IV	..	.

In conclusion, the number ζ_S of possible ways of distributing N bosons among k states is given by the combinations without repetition of $k-1$ objects among $N+k-1$ (see also the solution of Prob. 1.10 and Eq. (6.120)), that is,

$$\zeta_S = \binom{k+N-1}{k-1}. \tag{7.22}$$

In conclusion, in the case of both bosons and fermions the individuality of the particles is of no relevance: only the number of particles that occupy a certain state is important.

7.4 Exchange interaction

The *exchange interaction* is another important effect of quantum statistics and more generally of the symmetry properties of the wave function. Consider a Hamiltonian which is independent of the spin and two identical bosons with spin zero. The wave function of this system can be written as $\Psi(\mathbf{r}_1, \mathbf{r}_2)$. Suppose that the particles are subject to a central force (see Subsec. 11.2.1) so that Ψ can be decomposed into a radial part and a spherical-harmonics part (see Subsec. 6.1.4), i.e.

$$\Psi(\mathbf{r}_1, \mathbf{r}_2) = R(|\mathbf{r}_1 - \mathbf{r}_2|) Y_{lm}(\theta, \phi). \tag{7.23}$$

Exchanging the two particles is equivalent to inverting the vector $\mathbf{r}_1 - \mathbf{r}_2$ (i.e. changing its sign). With such an inversion, the spherical harmonics undergo the transformation

$$Y_{lm}(\theta, \phi) \to Y_{lm}(\pi - \theta, \phi + \pi) = (-1)^l Y_{lm}(\theta, \phi). \tag{7.24}$$

However, we know that the wave function must be symmetric under bosonic exchange and therefore only even l's are admitted. This is a constraint that was not present in the Hamiltonian and that we must therefore add from the outside on the basis of statical arguments.

In the case of fermions of spin-1/2, and with the spin-independent Hamiltonian, the complete wave function (taking into account both the spatial and spin degree of freedom) will be given by

$$\Phi(\mathbf{r}_1, \mathbf{r}_2; s_1, s_2) = \Psi(\mathbf{r}_1, \mathbf{r}_2)\varsigma(s_1, s_2) = R(|\mathbf{r}_1 - \mathbf{r}_2|)Y_{lm}(\theta, \phi)\varsigma(s_1, s_2). \tag{7.25}$$

Since the total wave function must be antisymmetric with respect to a fermionic exchange, we have two possible cases: either spin symmetric and position antisymmetric (l's odd) or spin antisymmetric and position symmetric (l's even).

In conclusion, what we call exchange interaction are the constraints imposed by the statistics on the possible dynamical states of bosons and fermions. Due to its nature, the exchange interaction plays an important role in the physics of chemical bonds.

An interesting consequence of both Fermi–Dirac and Bose–Einstein statistics is the different behavior of bosons and fermions in a Coulomb potential $V(r) \sim 1/r$. The energy of N bosons turns out to be proportional to $N^{4/3}$ so that, when $N \to \infty$, the energy density E/N grows indefinitely. In the case of N fermions, however, their energy is proportional to N, so that, for $N \to \infty$, the energy density E/N remains constant.

7.5 Two recent applications

7.5.1 Bose-Einstein condensate

A macroscopic Bose gas, i.e. a system made of a very large number of bosons, behaves at room temperature following the laws of classical statistical mechanics, i.e. the Boltzmann statistics. In fact, the uncertainty relation establishes that to any boson a wave packet (see Box 2.6) may be assigned, which has a dimension of the order of the thermal wavelength λ_T, given by (see Prob. 7.6)

$$\lambda_T = \sqrt{\frac{h^2}{2\pi m k_B T}}, \tag{7.26}$$

where, as usual, T is the temperature, m the mass of the boson, and k_B the Boltzmann constant. At room temperature, this wavelength is much smaller than the mean distance between the bosons, and the gas is typically disordered. However, when a dilute gas is cooled, the wavelength increases until a critical point where it is of the order of the mean distance between the bosons. When the temperature falls below near 100 μK, all the bosons occupy the same state (a phase transition that can be compared to the transition from incoherent light to a coherent laser). It is interesting to note that this is a major example of a macroscopic system that follows quantum-mechanical predictions and shows that quantum effects are not necessarily lost on a macroscopic scale.

Bose condensates may be realized in different forms. We briefly report here some realizations. In 1995 the realization of a Bose–Einstein condensate was obtained in two laboratories in the USA, one at the JILA (Joint Institute for Laboratory Astrophysics) in Boulder, Colorado, by the group of Carl E. Wieman and Eric A. Cornell,[14] and the other at the MIT (Massachusetts Institute of Technology) in Boston, by the team of Wolfgang

[14] [Anderson *et al.* 1995].

Ketterle.[15] This cooling can be obtained by a laser cooling technique (in which the bosons are slowed down) followed by an "evaporation" technique: the bosons are magnetically isolated from the environment and, by using a radio frequency field in resonance with the energy difference between "up" and "down" atoms, the atoms with greater energy are lost, so that the temperature decreases. The achievement of Bose–Einstein condensation is one of the greatest successes of quantum statistics: in fact, this result had been already predicted by Einstein. Also the degeneracy of a Fermi–Bose mixture has been obtained[16] following the original proposal by Bose.

A stunning experiment was done recently,[17] showing that, when two independent Bose–Einstein condensates are prepared in two traps, with a spatial separation of a fraction of millimeter, and a coherent light pulse is stored in one of the two, a wave packet travels through this condensate, successively reaching and traversing also the second one. Once this happens, it is possible to regenerate the initial light pulse, which would not have been possible without the presence of the second condensate. The only way to understand this result is that the two condensates are quantum indistinguishable systems, and that the ground state function (before or after the pulse is gone) would have a component in both traps simultaneously.

7.5.2 Quantum dot

A striking example of the validity of the Pauli exclusion principle is represented by the so-called *quantum dots* or artificial atoms.[18] It is possible to confine a cloud of electrons to a bidimensional surface by squeezing it between two layers of insulating material. Successively, one must generate a field able to electrostatically attract the electrons onto a given point. We obtain then a "zero"-dimensional cloud of electrons (in the sense that the electrons are not free to move along any of the three Cartesian space directions). This cloud may vary from few electrons to some hundreds. While in the natural atom the potential well is hyperbolic (i.e. $V(\mathbf{r}) \sim 1/|\mathbf{r}|$) in the case of the quantum dot the form of the potential well is elliptic (see Fig. 7.3). Moreover, while the dimension of the natural atom is of the order of 0.1 nanometers, that of the artificial atom is of the order of 10–100 nanometers. In the quantum dot, the electrons are naturally placed in atomic levels following the Pauli exclusion principle even if there is no nucleus. It is quite possible that there will be many applications of quantum dots for electronic microcircuitry. Indeed, they can be understood as wells where electrons may be stocked. Furthermore, electronic lines bind several quantum dots: when an electron is needed from one well to another, a manipulation of the potentials or a laser may extract it from a given quantum dot.

15 See [Townsend *et al.* 1997]. The 2001 Nobel prize in physics was awarded to Eric A. Cornell, Wolfgang Ketterle and Carl E. Wieman for the achievement of Bose–Einstein condensation in dilute gases of alkali atoms, and for early fundamental studies of the properties of the condensates.

16 See [Vichi *et al.* 1998].

17 See [Ginsberg *et al.* 2007] [Fleischhauer 2007].

18 See [Kastner 1993].

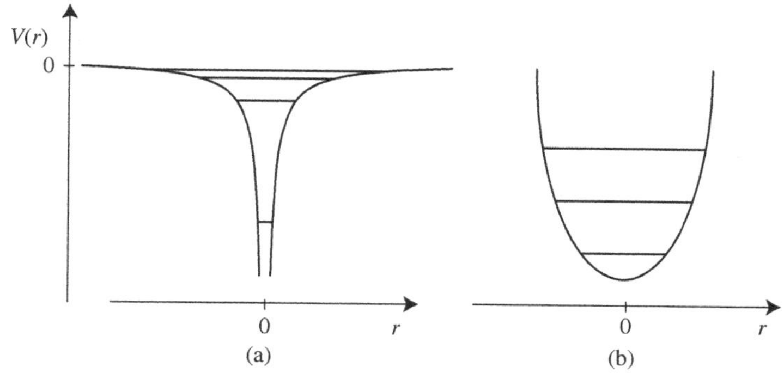

Figure 7.3 Potential wells of a natural atom (a) and of a quantum dot (b). While the potential well of a natural atom is hyperbolic, the potential well of a quantum dot is either elliptic or parabolic.

Summary

- In this chapter we have introduced the unitary permutation operator: any permutation can be decomposed in a product of pairwise exchanges. Together with the fact that identical particles are *indistinguishable* in quantum mechanics, this has led us to the definition of the two fundamental statistics of quantum mechanics.
- The *Fermi–Dirac* statistics, for particles with half-integer spin and antisymmetric wave function under permutation, is characterized by the fact that two fermions cannot share the same state (Pauli exclusion principle).
- Bosons, particles with integer spin, are subject to the *Bose–Einstein statistics*, i.e. have symmetric wave function under permutation (by interchanging two bosons we obtain exactly the same state). Here any number of particles may occupy the same state.
- The *exchange interaction* is a truly quantum effect, due to the additional constraints imposed by statistics on the particles' interaction.
- Finally, we have discussed some major examples illustrating the quantum behavior of fermions and bosons: The experimental realization of *Bose–Einstein condensation* obtained by laser-cooling technics in dilute gases of alkaline atoms, and the confinement of electrons in artificial atoms that respect the Pauli exclusion principle, known as *quantum dots*.

Problems

7.1 Consider Eqs. (7.14)–(7.15). Write the analogue of Eq. (7.14) in terms of the wave function

$$\Psi(\mathbf{r}_\alpha^{(1)}\mathbf{r}_\beta^{(2)}\mathbf{r}_\gamma^{(3)}) = \left\langle \mathbf{r}_\alpha^{(1)}\mathbf{r}_\beta^{(2)}\mathbf{r}_\gamma^{(3)} \mid \Psi \right\rangle. \tag{7.27}$$

7.2 Write the explicit form of the permutation operator for four particles among N particles.

7.3 Calculate the number of possible configurations of five identical fermions distributed among four states.

7.4 Prove that, when there are two possible states ($k = 2$), Eq. (7.22) reduces to $\zeta_S = N + 1$.

7.5 The nitrogen nucleus is a boson or a fermion? Why?

7.6 Justify Eq. (7.26) on the basis of arguments connected with the uncertainty relation.

Further reading

Huang, Kerson, *Statistical Mechanics*, New York: Wiley, 1963, 1987.

Kittel, Charles, *Elementary Statistical Physics*, 1958; New York: Dover.

Landau, Lev D. and Lifshitz, E. M., *Statistical Physics. Part I–II* (Engl. trans.), vols. V and IX of *The Course of Theoretical Physics*, Oxford: Pergamon, 1976.

Streater, R. F. and Wightman, A. S., *PCT, Spin and Statistics, and All That*, 2nd edn. 1978; Princeton: Princeton University Press, 2000.

8 Symmetries and conservation laws

A *symmetry* is an equivalence of different physical situations, and is therefore strictly associated with the concept of *invariance*. One of the main aims of physics is the search in nature for symmetries and, therefore, for observables that are invariant under certain classes of transformation. In this context, it becomes particularly interesting to look for the conditions under which a certain symmetry is eventually broken. We have already discussed some symmetries of quantum mechanics: time translations (Ch. 3), space translations (Ch. 2) and rotations (Ch. 6). Another example of invariance is represented by the indistinguishability of identical particles, which we discussed in the previous chapter.

Symmetries are induced by physical transformations that generate some invariant properties (see Sec. 8.1). In Sec. 8.2, the *continuous* transformations induced by rotations and space translations are analyzed, whereas in Sec. 8.3, space reflection and time reversal (*discrete* symmetries) are discussed. As we shall see in the next chapter, time reversibility is lost in the measurement process. The rigorous mathematical framework of a theory of transformations and of their invariance is called group theory (the subject of Sec. 8.4).

8.1 Quantum transformations and symmetries

8.1.1 Active and passive transformations

We have seen that time evolution may be represented either in the Schrödinger picture (state vectors evolve and observables are kept constant) or in the Heisenberg picture (states vectors remain constant and observables evolve) (see Subsec. 3.6.1 and Sec. 5.3). Similarly, *all* quantum transformations may be classified in transformations of the state vector (*active* transformations) and transformations of the observables (*passive* transformations). Any active transformation $\hat{\mathcal{U}}$ on an arbitrary ket $|\,\psi\rangle$ may be written as $\hat{\mathcal{U}}\,|\,\psi\rangle = |\psi^{'}\rangle$. In other words, in the case of active transformations we consider the physical situation in the same reference frame by moving the vector representing the state.

A passive transformation may be understood as a change of the coordinate axes since it corresponds to a "view" of the (original) state in a new reference frame (from the "point of view" of a new observable). In other words, it is as if we had changed the basis of the system and therefore rotated the axes (see Subsec. 2.1.2). The relationship between active and passive transformations may be seen in Fig. 8.1. An example of passive and active transformations is given in Box 8.1.

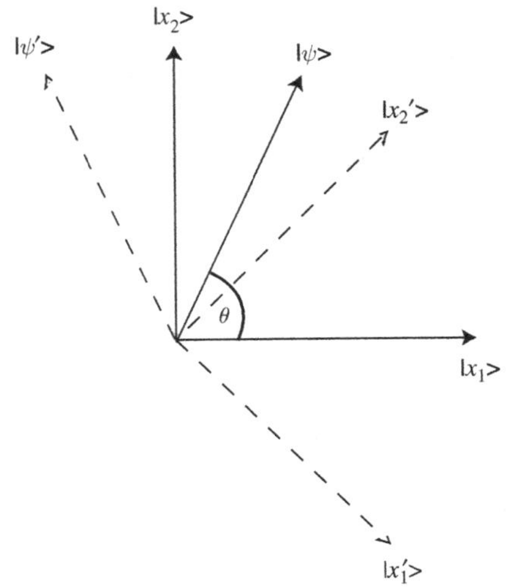

Figure 8.1 Passive and active transformations. A transformation may be considered from two viewpoints, as an active transformation, where the state vector $|\psi\rangle$ is changed into the state vector $\left|\psi'\right\rangle$; and as a passive transformation, where the reference frame is changed from the basis vectors $|x_1\rangle, |x_2\rangle$ (where $|\psi\rangle = \cos\theta\,|x_1\rangle + \sin\theta\,|x_2\rangle$ and θ is the angle between the directions of $|\psi\rangle$ and $|x_1\rangle$) to the vectors $\left|x_1'\right\rangle, \left|x_2'\right\rangle$ (where $|\psi\rangle$ in the new basis is equivalent to $\left|\psi'\right\rangle$ in the old basis).

Let us now face the problem of quantum symmetries on the most general grounds. Suppose we have a system which has been prepared in some way in the state $|\psi\rangle$. Consider a certain transformation that changes the state $|\psi\rangle$ into

$$\left|\psi'\right\rangle = \hat{\mathcal{U}}\,|\psi\rangle. \tag{8.1}$$

Similarly, for a state $|\varphi\rangle$, we shall have

$$\left|\varphi'\right\rangle = \hat{\mathcal{U}}\,|\varphi\rangle. \tag{8.2}$$

Box 8.1 **Example of passive and active transformations**

Let us consider the example of beam splitting (see Subsec. 3.5.2). We may write the initial state as

$$|\psi\rangle = |1\rangle. \tag{8.3}$$

A beam splitter induces the active transformation

$$\begin{aligned} |\psi\rangle \mapsto \hat{U}_{BS}\,|\psi\rangle &= \left|\psi'\right\rangle \\ &= \frac{1}{\sqrt{2}}(|1\rangle + |2\rangle). \end{aligned} \tag{8.4}$$

The previous transformation induces a change of basis (passive transformation) which is the inverse with respect to Eq. (8.4), i.e.

$$\left|1'\right\rangle = \frac{1}{\sqrt{2}}(|1\rangle - |2\rangle), \quad \left|2'\right\rangle = \frac{1}{\sqrt{2}}(|1\rangle + |2\rangle). \tag{8.5}$$

This means that it is equivalent to consider that $|\psi\rangle$ has been transformed into $\left|\psi'\right\rangle$ or that the state has remained $|\psi\rangle$ and we have performed the change of basis (8.5) (see Prob. 8.1). In fact, let us consider the observable of path distinguishability (see Subsec. 2.3.4)

$$\hat{\mathcal{P}} = |1\rangle\langle 1| - |2\rangle\langle 2| = \begin{bmatrix} 1 & 0 \\ 0 & -1 \end{bmatrix}. \tag{8.6}$$

We have (see Eq. (8.13) and Prob. 8.2)

$$\left\langle \psi' \left| \hat{\mathcal{P}} \right| \psi' \right\rangle = \left\langle \psi \left| \hat{\mathcal{P}}' \right| \psi \right\rangle = 0, \tag{8.7}$$

where

$$\hat{\mathcal{P}}' = \left|1'\right\rangle\left\langle 1'\right| - \left|2'\right\rangle\left\langle 2'\right|, \tag{8.8}$$

where $|1'\rangle$ and $|2'\rangle$ are given by Eq. (8.5). The result (8.7) can be understood if we consider that $|\psi\rangle$ and $|\psi'\rangle$ are states of maximal interference relative to the bases of the observables $\hat{\mathcal{P}}'$ (whose eigenstates are $|1'\rangle$ and $|2'\rangle$) and $\hat{\mathcal{P}}$ (whose eigenstates are $|1\rangle$ and $|2\rangle$), respectively. In other words, $|\psi\rangle$ and $|\psi'\rangle$ may be considered as eigenkets of two visibility-of-interference-fringes observables $\hat{\mathcal{V}}$ and $\hat{\mathcal{V}}'$, respectively (see again Subsec. 2.3.4).

The scalar product $\langle \psi \mid \varphi \rangle$ gives the probability amplitude that a measurement performed on $|\psi\rangle$ give $|\varphi\rangle$ as outcome (see Sec. 5.2). As we know, what one can actually measure is the square modulus $|\langle \psi \mid \varphi \rangle|^2$ of the probability amplitude, i.e. the probability. It is clear that if the system must be invariant under the transformation $\hat{\mathcal{U}}$, this probability must be the same independently of whether we compute (measure) it on the states $|\psi\rangle$ and $|\varphi\rangle$ or on $\left|\psi'\right\rangle$ and $\left|\varphi'\right\rangle$. We may therefore conclude that the system has a symmetry whenever

$$\left|\left\langle \psi' \mid \varphi' \right\rangle\right|^2 = |\langle \psi \mid \varphi \rangle|^2. \tag{8.9}$$

This condition imposes severe constraints on the transformation operator $\hat{\mathcal{U}}$. In particular, we have only two possibilities:

$$\left\langle \psi' \mid \varphi' \right\rangle = \left\langle \psi \left| \hat{\mathcal{U}}^\dagger \hat{\mathcal{U}} \right| \varphi \right\rangle = \langle \psi \mid \varphi \rangle, \tag{8.10a}$$

$$\left\langle \psi' \mid \varphi' \right\rangle = \left\langle \psi \left| \hat{\mathcal{U}}^\dagger \hat{\mathcal{U}} \right| \varphi \right\rangle = \langle \varphi \mid \psi \rangle. \tag{8.10b}$$

In the first case, $\hat{\mathcal{U}}$ must be a *unitary* operator (see the first property in Subsec. 3.5.1) whereas in the case of Eq. (8.10b) it must be represented by an *anti unitary* operator. An antiunitary transformation $\tilde{\hat{U}}$ is defined by the following properties:[1]

$$\left\langle \tilde{\hat{U}}\varphi \mid \tilde{\hat{U}}\psi \right\rangle = \langle \psi \mid \varphi \rangle, \tag{8.11a}$$

$$\tilde{\hat{U}}\left[\mid \psi\rangle + \mid \varphi\rangle\right] = \tilde{\hat{U}} \mid \psi\rangle + \tilde{\hat{U}} \mid \varphi\rangle, \tag{8.11b}$$

$$\tilde{\hat{U}} c \mid \psi\rangle = c^* \tilde{\hat{U}} \mid \psi\rangle. \tag{8.11c}$$

Summarizing, all quantum symmetries are ruled by *Wigner's theorem*:

Theorem 8.1 (Wigner) *A symmetry, i.e. a mapping that transforms the vectors* $|\psi\rangle$ *and* $|\varphi\rangle$ *into* $\hat{\mathcal{U}}|\psi\rangle$ *and* $\hat{\mathcal{U}}|\varphi\rangle$*, respectively, and that preserves the square modulus of the inner product*

$$|\langle \hat{\mathcal{U}}\psi | \hat{\mathcal{U}}\varphi\rangle|^2 = |\langle \psi | \varphi\rangle|^2, \tag{8.12}$$

is either linear and unitary or antilinear and antiunitary.

In other words, if between the vectors of the Hilbert space $\mathcal{H}$ there exists a one-to-one correspondence $\hat{\mathcal{U}}$ which is defined up to an arbitrary constant and which conserves the square modulus of the scalar product, then the arbitrary phases can be chosen so as to make $\hat{\mathcal{U}}$ either unitary and linear or antiunitary and antilinear. Note that the canonical commutation relations between observables are conserved in the transformation if $\hat{\mathcal{U}}$ is linear and are replaced by their complex conjugate if $\hat{\mathcal{U}}$ is antilinear (see Prob. 8.5).

Continuous symmetries must contain the identity transformation as a limiting case, and therefore are represented by unitary transformations. On the other hand, *discrete* symmetries do not need to be unitary, and condition (8.10b) is allowed.

In order to determine the action of a passive continuous transformation on a generic observable $\hat{O}$, let us consider the mean value of $\hat{O}$ with respect to the transformed state $|\psi'\rangle$ (see also Subsec. 3.5.1). This mean value must be equal to the mean value of the transformed observable $\hat{O}'$ with respect to the original state vector $|\psi\rangle$, i.e.

$$\left\langle \psi' \left| \hat{O} \right| \psi' \right\rangle = \left\langle \psi \left| \hat{O}' \right| \psi \right\rangle. \tag{8.13}$$

Our aim here is to determine the relationship between $\hat{O}$ and $\hat{O}'$ in terms of the unitary transformation $\hat{U}$. We have

$$\left\langle \psi' \left| \hat{O} \right| \psi' \right\rangle = \left\langle \psi \left| \hat{U}^\dagger \hat{O} \hat{U} \right| \psi \right\rangle = \left\langle \psi \left| \hat{O}' \right| \psi \right\rangle, \tag{8.14}$$

from which it follows that

[1] If an operator satisfies the properties (8.11b) and (8.11c) is called *antilinear*.

$$\hat{O}' = \hat{U}^\dagger \hat{O} \hat{U} = \hat{U}^{-1} \hat{O} \left(\hat{U}^{-1}\right)^\dagger . \tag{8.15}$$

In other words, an active unitary transformation $\hat{U}$ on a state vector $|\psi\rangle$ is completely equivalent to (and therefore indistinguishable from) a passive transformation made on the observables in which the reference frame is rotated according to $\hat{U}^{-1}$.

8.1.2 Noether's theorem

If a system has a symmetry under a certain transformation (for example, under rotations or translations), its Hamiltonian must be invariant under the same transformation, and this property is conserved in time (see for instance Subsec. 6.2.3). In this case, the symmetry is called *exact* or conserved. Under this circumstance, the Hamiltonian must commute with the infinitesimal generator of this symmetry, i.e. the generator must be a constant of motion (see Subsec. 6.1.2). Therefore, to any symmetry of a system (i.e. to any invariance of the Hamiltonian with respect to a certain transformation) corresponds a constant of motion. This fact directly follows from the Heisenberg equation of motion (3.108)

$$\imath\hbar \frac{d}{dt}\hat{O} = \left[\hat{O}, \hat{H}\right] \tag{8.16}$$

for an observable $\hat{O}(t)$ (where for the sake of notation we have dropped the superscript H) that does not explicitly depend on time. Using Stone's theorem (see p. 123), any unitary and continuous transformation may be written as $\hat{U}(\alpha) = e^{\imath\alpha\hat{G}}$, where $\hat{G} = \hat{G}^\dagger$ is a Hermitian operator and represents the generator of the transformation, and α is a continuous parameter. For an infinitesimal transformation, to the first order in $\delta\alpha$, $\hat{U}$ is given by

$$\hat{U}(\delta\alpha) = 1 + \imath\hat{G}\delta\alpha + O(\delta\alpha^2). \tag{8.17}$$

If an operator $\hat{\xi}$ is transformed into $\hat{\xi} + \delta\hat{\xi}$ by the transformation (8.17), we have, again to the first order,

$$\begin{aligned} \hat{\xi} + \delta\hat{\xi} = \hat{U}\hat{\xi}\hat{U}^\dagger &= \left(1 + \imath\delta\alpha\hat{G}\right)\hat{\xi}\left(1 - \imath\delta\alpha\hat{G}\right) \\ &\simeq \hat{\xi} + \imath\delta\alpha\left(\hat{G}\hat{\xi} - \hat{\xi}\hat{G}\right), \end{aligned} \tag{8.18}$$

and

$$\delta\hat{\xi} = \imath\delta\alpha\left[\hat{G}, \hat{\xi}\right]. \tag{8.19}$$

As we have said above, if our system is invariant under this transformation, also the Hamiltonian must be invariant. This means that, replacing $\hat{\xi}$ by $\hat{H}$, $\delta\hat{H}$ must be zero and therefore $\left[\hat{G}, \hat{H}\right] = 0$. In conclusion, according to Heisenberg equation, $\hat{G}$ is a constant of motion (see Prob. 8.4). This result is also known as *Noether's theorem*, due to Emmy Noether, and may be formulated as follows:

Theorem 8.2 (Noether) *Every parameter of a continuous and exact symmetry has a corresponding constant of motion given by the observable that represents the generator of the corresponding transformation.*

8.2 Continuous symmetries

If the transformation $\mathcal{U}$ is continuous (which implies a continuous symmetry), it must also be unitary, because it must include transformations that are arbitrarily close to the identity, which is unitary (see Prob. 8.6). On the other hand, discrete symmetries may be either unitary or antiunitary.

It is interesting to note that the transformations in space, time and of angle we have considered in previous chapters and that are induced by dynamical observables – by momentum, i.e. translational symmetry (see Subsec. 2.2.4), by energy, i.e. time symmetry (see Sec. 3.5), and by angular momentum, i.e. rotational symmetry (see Subsec. 6.1.2) – are linear and therefore are also continuous and unitary (see also Subsec. 3.5.4). Among the continuous symmetries, the roto-translations (rotations combined with translations) are said to constitute the Euclidean group (see Sec. 8.4).

8.2.1 Space translations and momentum transformations

Consider a three-dimensional particle with position $\hat{\mathbf{r}}$, momentum $\hat{\mathbf{p}}$, and spin $\hat{\mathbf{s}}$. If we apply the most simple space translation, i.e. a one-dimensional displacement by a distance a along the x-axis, the nine fundamental variables (the components of $\hat{\mathbf{r}}$, $\hat{\mathbf{p}}$, and $\hat{\mathbf{s}}$) are invariant, with the exception of x, which goes over into $x - a$: if we have the eigenvalue equation (2.122), that is,

$$\hat{x} \,|\, x\rangle = x \,|\, x\rangle \tag{8.20}$$

for an arbitrary eigenstate $|\, x\rangle$, and denote by $\hat{U}_x(a)$ the unitary transformation such that

$$\hat{U}_x(a)\hat{x}\hat{U}_x^\dagger(a) \,|\, x\rangle = (x - a) \,|\, x\rangle, \tag{8.21}$$

we may infer

$$\hat{U}_x(a)\hat{x}\hat{U}_x^\dagger(a) = \hat{x} - a, \tag{8.22}$$

The Hermitian generator $\hat{G}_x$ therefore obeys the commutation relations

$$[\hat{G}_x, \hat{x}] = -\imath, \quad [\hat{G}_x, \hat{y}] = \ldots = [\hat{G}_x, \hat{p}_x] = \ldots = [\hat{G}_x, \hat{s}_z] = 0, \tag{8.23}$$

which imply (see Eq. (2.173a))

$$\hat{G}_x = \frac{\hat{p}_x}{\hbar} + k_0, \quad (8.24)$$

where k_0 is an arbitrary constant that can be put equal to zero. The finite-transformation operator $\hat{U}_x(a)$ can be then chosen to be (see Subsec. 2.2.4)

$$\hat{U}_x(a) = e^{-\frac{\imath}{\hbar}\hat{p}_x a}. \quad (8.25)$$

As a consequence, the three-dimensional translation transformation can be therefore written as

$$\begin{aligned}\hat{U}_{\mathbf{r}}(\mathbf{a}) &= \hat{U}(a_x, a_y, a_z) \\ &= e^{-\frac{\imath}{\hbar}\hat{p}_x a_x} e^{-\frac{\imath}{\hbar}\hat{p}_y a_y} e^{-\frac{\imath}{\hbar}\hat{p}_z a_z} \\ &= e^{-\frac{\imath}{\hbar}\hat{\mathbf{p}}\cdot\mathbf{a}}. \end{aligned} \quad (8.26)$$

In a similar way, it can be shown that three-dimensional momentum translations by a vectorial amount $\mathbf{v}$ are described by the unitary transformation: $\hat{U}_{\mathbf{p}}(\mathbf{v}) = e^{\frac{\imath}{\hbar}\mathbf{v}\cdot\hat{\mathbf{r}}}$ or the expression

$$\hat{U}_{\mathbf{p}}(\mathbf{v})\hat{\mathbf{p}}\hat{U}^{\dagger}_{\mathbf{p}}(\mathbf{v}) = \hat{\mathbf{p}} - \mathbf{v} \quad (8.27)$$

Both space and momentum translation transformations form a group that is isomorphic to the respective translation group, i.e.

$$\hat{U}(\xi)\hat{U}(\eta) = \hat{U}(\eta)\hat{U}(\xi) = \hat{U}(\xi + \eta), \quad (8.28)$$

where ξ and η are two parameters.

8.2.2 Time translations

The case of time translations is formally similar to that of spatial translations, where the momentum is replaced by the Hamiltonian $\hat{H}$ and the distance a is replaced by the time t. This kind of transformation has been extensively treated in Subsecs. 3.1.1 and 3.5.3.

8.2.3 Rotations

A rotation $\mathbf{R}$ of an angle ϕ about an axis $\mathbf{n} = (n_x, n_y, n_z)$ in a Cartesian frame is given by (see Subsec. 6.1.2)

$$\hat{\mathbf{R}} = \hat{I} - \imath\phi\sum_k n_k \hat{s}_k, \quad (8.29)$$

where $\hat{s}_k, k = x, y, z$ are the spin matrices for three dimensions (see Eqs. (6.165)).

The corresponding unitary transformation is given by

$$\hat{U}_{\mathbf{R}}(\phi) = e^{-\frac{i}{\hbar}\phi\mathbf{n}\cdot\hat{\mathbf{J}}}, \tag{8.30}$$

where $\hat{\mathbf{J}}$ is the total angular momentum.

8.3 Discrete symmetries

As we have seen, discrete symmetries may be either unitary or antiunitary. Discrete symmetries also give rise to constants of motion. This time, however, these will not be represented by the infinitesimal generators (which do not exist!) but directly by the transformation operators. In the following we shall briefly discuss three major examples of these transformations.

8.3.1 Space reflection or parity

The group of reflections with respect to a point is constituted by the reflection operator $\hat{U}_{\mathcal{R}}$ and the identity operator $\hat{I}$. In fact, obviously, $\hat{U}_{\mathcal{R}}^2 = \hat{I}$. In the transformation induced by $\hat{\mathcal{R}}$, the polar vectors $\mathbf{r}$ and $\mathbf{p}$ change sign while the axial vectors $\mathbf{r} \times \mathbf{p}$ and $\mathbf{s}$ are invariant. Since $\mathbf{r}$ and $\mathbf{p}$ simultaneously change sign, the transformation conserves the commutation relations of the orbital variables. It also conserves those involving the spin, that is,

$$\hat{U}_{\mathcal{R}}\hat{\mathbf{r}}\hat{U}_{\mathcal{R}}^{\dagger} = -\hat{\mathbf{r}}, \quad \hat{U}_{\mathcal{R}}\hat{\mathbf{p}}\hat{U}_{\mathcal{R}}^{\dagger} = -\hat{\mathbf{p}}, \quad \hat{U}_{\mathcal{R}}\hat{\mathbf{L}}\hat{U}_{\mathcal{R}}^{\dagger} = \hat{\mathbf{L}}, \quad \hat{U}_{\mathcal{R}}\hat{\mathbf{s}}\hat{U}_{\mathcal{R}}^{\dagger} = \hat{\mathbf{s}}. \tag{8.31}$$

Therefore, the reflection operator will act on a wave function as

$$\hat{U}_{\mathcal{R}}\psi(\mathbf{r}) = \psi(-\mathbf{r}). \tag{8.32}$$

We take $\hat{U}_{\mathcal{R}}$ is taken to be unitary.[2]

8.3.2 Time reversal

Time reversal is the transformation $t \to -t$, i.e. it changes the direction of time. Unlike space reflections, time reversal is an antiunitary mapping. In fact, consider the unitary evolution

$$|\,\psi(t_2)\rangle = \hat{U}_{t_2-t_1}\,|\,\psi(t_1)\rangle\,. \tag{8.33}$$

[2] Abstractly, both space reflection and time reversal could be taken as well as unitary or antiunitary transformations. The fact that space reflection is assumed to be unitary and time reversal to be antiunitary stems from relativistic considerations [*Bjorken/Drell* 1964, 24–25, 71–75] [*Bjorken/Drell* 1965, 118–23].

If the dynamical properties of the system are invariant under time reversal, then there is a mapping

$$|\psi(t)\rangle \mapsto \hat{\tilde{U}}_{\mathcal{T}} |\psi(t)\rangle = \left|\psi^{\mathcal{T}}(t)\right\rangle \tag{8.34}$$

such that the same unitary operator which transforms $|\psi(t_1)\rangle$ into $|\psi(t_2)\rangle$, also transforms $\left|\psi^{\mathcal{T}}(t_2)\right\rangle$ into $\left|\psi^{\mathcal{T}}(t_1)\right\rangle$

$$\left|\psi^{\mathcal{T}}(t_1)\right\rangle = \hat{U}_{t_2-t_1}\left|\psi^{\mathcal{T}}(t_2)\right\rangle. \tag{8.35}$$

For the properties of the scalar product and of both unitary and antiunitary (see Eq. (8.11a)) transformations, we have

$$\left\langle\psi(t_2) \mid \psi^{\mathcal{T}}(t_1)\right\rangle = \left\langle\hat{U}_{t_1-t_2}\psi(t_2) \mid \hat{U}_{t_1-t_2}\psi^{\mathcal{T}}(t_1)\right\rangle. \tag{8.36}$$

Moreover, using Eqs. (8.33)–(8.35), we may write Eq. (8.36) as

$$\left\langle\psi(t_2) \mid \psi^{\mathcal{T}}(t_1)\right\rangle = \left\langle\psi(t_1) \mid \psi^{\mathcal{T}}(t_2)\right\rangle, \tag{8.37}$$

i.e. time reversal is an antiunitary transformation.

The reversibility of the solutions of the Schrödinger equation with respect to time is due to the invariance of the Hamiltonian when $\mathbf{p}$ is changed into $-\mathbf{p}$ (see also Subsec. 2.2.6). We have time reversal invariance whenever $\mathbf{r}$ and $\mathbf{p}$ transform into $\mathbf{r}$ and $-\mathbf{p}$, respectively. In accordance with the fact that $\hat{\tilde{U}}_{\mathcal{T}}$ is antiunitary, it changes the sign of the canonical commutation relations. Since $\mathbf{L} = \mathbf{r} \times \mathbf{p}$, it also changes the sign of the orbital angular momentum and of the spin. Therefore, we also have $\hat{\tilde{U}}_{\mathcal{T}}\hat{\mathbf{J}}\hat{\tilde{U}}_{\mathcal{T}}^{\dagger} = -\hat{\mathbf{J}}$, and consequently the time reversal operator anticommutes with the generator of rotations.

8.3.3 Charge conjugation

Charge conjugation is a symmetry that interchanges particles and antiparticles. For example, electrons and positrons, protons and antiprotons, quarks and antiquarks. It is unitary and antisymmetric. This symmetry plays a major role in quantum field theory (i.e. the relativistic extension of quantum mechanics) and will not be discussed here in detail. We limit ourselves to note that a fundamental theorem of quantum field theory (the so-called *CPT theorem*) establishes that, even though charge conjugation, parity, and time reversal may be not individually conserved, their product is universally conserved.

8.4 A brief introduction to group theory

8.4.1 Definition of groups

A very important idea in modern mathematics is that it is possible and convenient to study quite general algebraic structures that are characterized by a small number of properties, without specifying the system one is considering. In the past, the object of study was always represented by concrete mathematical structures (e.g. real numbers, complex numbers, matrices). In this modern abstract approach it is possible to derive in one shot the properties of all structures that share certain general features, even though these structures may be quite diverse. Of course, the properties that are peculiar to a given structure cannot be derived in this way.

Although the number of characterizing properties is small, a certain number of theorems can be derived. This approach has the advantage that many concrete and quite different realizations belong to the same algebraic structure, so that one avoids the repetition of the same proof in different contexts and, at the same time, it becomes possible to grasp the origin of the various properties.[3]

A very common algebraic structure is the *group*. A set of elements forms a group if, for any elements a and b pertaining to the group, there is a binary operation, i.e. a function with two arguments, such that the result, say c, of this operation is also an element of the set, and this operation satisfies the following properties:[4]

- Associativity: $a * (b * c) = (a * b) * c$ or $a + (b + c) = (a + b) + c$, where a, b, and c are elements of the group.
- Existence of a neutral element (denoted by the identity 1 if we use the multiplicative notation, by 0 if we use the additive notation) such that $a * 1 = 1 * a = a$ ($a + 0 = 0 + a = a$).
- Existence of the inverse (opposite): For each element a there is an inverse a^{-1} (in the additive notation the opposite $-a$) such that $a * a^{-1} = a^{-1} * a = 1$ ($a - a = 0$),

Sometimes one uses shortcuts in the notation: $a * b^{-1}$ is written as a/b. In the additive notation $a + (-b)$ is written as $a - b$.[5]

The group is defined by giving the set of elements and the binary operation, satisfying the previous three properties. In the case of a group having a finite number of elements N, i.e. a finite group, the binary operation can be defined by giving the multiplication (addition) table, i.e. by specifying for each pair of elements (a and b) the value of the product (sum). In this case the table contains N^2 elements. For $N > 1$, not all of the N^{N^2} different multiplication tables are consistent with the definition of a group (see Prob. 8.7).

The simplest example of group contains only one element, the neutral element. Here the multiplication (addition) table is trivial: $1 * 1 = 1$ ($0 + 0 = 0$). Other examples of groups are as follows (see Prob. 8.8):

1. The N complex N-th roots of the identity, where the group operation is the standard multiplication. The group is denoted by $\mathbb{Z}_N$.
2. The integer numbers, where the group operation is the standard addition. The group is denoted by $\mathbb{Z}$.
3. The real numbers (denoted by $\mathbb{R}$), or the rational numbers (denoted by $\mathbb{Q}$), where the group operation is the standard addition.

[3] A similar philosophy is behind generic programming with templates in C++ [*Plauger et al.* 2000].

[4] The symbol used for this binary operation may be represented – according to the circumstances and to the tradition – by "$*$" or "+," multiplicative and additive notation, respectively.

[5] The first minus is the unary minus, the second one is the binary minus, which are two different concepts, as should be well known to programmers.

4. The positive real numbers, (denoted by $\mathbb{R}^+$) or the positive rational numbers (denoted by $\mathbb{Q}^+$), where the group operation is the standard multiplication.
5. The $n \times n$ matrices with real elements, with non-zero determinant, where the group operation is the standard matrix multiplication.

The real numbers, with the standard multiplication, do not form a group because the number 0 does not have an inverse.

The fact that both the usual arithmetic addition and multiplication are the operations associated with well-known groups explains why the additive and multiplicative notations are used, although the additive notation is typically used when the group is commutative (or *Abelian*), i.e. when the operation satisfies the fourth property:

- Commutativity: $a * b = b * a$ $(a + b = b + a)$.

8.4.2 Some important concepts

It is very useful to introduce the concept of *isomorphism* of two groups: we say that two groups A (with binary operation $*$) and B (with binary operation $\cdot$) are isomorphic if there exist two invertible functions, $f(a)$ (where $a \in A$) and $g(b)$ (where $b \in B$) that bring the elements of A in B ($f(a) \in B$) and the elements of B in A ($g(b) \in A$), respectively, such that $f(g(b)) = b$ and $g(f(a)) = a$. In other words, there must be a biunivocal correspondence among the elements of A and B that preserves the group operation

$$f(a_1) \cdot f(a_2) = f(a_1 * a_2). \tag{8.38}$$

A noticeable example of isomorphism is provided by the groups 3. and 4. of the previous subsection, where the functions f and g are respectively exp and ln. The reader can readily check the correctness of this statement using the relation

$$\exp(a_1) * \exp(a_2) = \exp(a_1 + a_2). \tag{8.39}$$

Two isomorphic groups may be considered as different instances of the same mathematical object: the same abstract elements are named in a different way in each instance.

Another very important concept is the *representation of a group*: a set R is a representation of the group G if the binary operation $g \circ r$ (g and r being respectively elements of G and of R) indicating the application of a group element to a representation element is defined and satisfies the associative property

$$(a * b) \circ r = a * (b \circ r). \tag{8.40}$$

We must also have that $1 \circ r = r$. Obviously, a group is a representation of itself. For example, N-dimensional vectors form a representation of the $N \times N$ dimensional matrices.

If R_1 and R_2 are two representations of the group G, we can define their sum $R = R_1 \oplus R_2$ i.e. all the pairs of elements such that the first element is in R_1 and the second element is in R_2, with

$$g \circ (r_1, r_2) = (g \circ r_1, g \circ r_2). \tag{8.41}$$

The representations that cannot be written as the sum of two representations, are called *irreducible*.

If only the neutral element of the group has the property that $g \circ r = r$ for all r, we say that the representation is *faithful*. If we know how a group acts on a faithful representation, we can trace back the group operation. Therefore, a group may be defined by the way it acts on a faithful representation.

For example, the representation of the group of the $N \times N$-dimensional matrices (M_N, where the group operation is the standard matrix multiplication) given by the set of N-dimensional vectors is a faithful representation. Let us consider another representation of the group M_N, i.e. the set of real numbers $\mathbb{R}$, where the action of an element m of M_N on an element r of $\mathbb{R}$ is given by

$$m \circ r = \det(m) r. \tag{8.42}$$

For $N > 1$, the representation is not faithful: all matrices m with determinant equal to 1 satisfy the relation $m \circ r = \det(m) r = r$.

Sometimes, the representation of a group may have a dimension much higher than the group itself. If we consider the group of rotations on a plane, the set of all the two-dimensional geometric objects form a representation of the group, which is reducible. Even irreducible representations may have a dimension much higher than that of the group they represent. For example, we can consider the group $O(3)$ of rotations in three dimensions, composed by all three-dimensional orthogonal matrices, i.e. real matrices such that

$$\sum_{b=1}^{3} R^{\mathrm{T}}_{ab} R_{bc} = \delta_{ac}. \tag{8.43}$$

In this case, the spin degree of freedom of the wave function of a spin-l particle provides a representation of the group $O(3)$ (see Subsec. 6.3.2).

We could also consider tensors with l indices, where the rotations act on each index, e.g. in the case $l = 3$ we have

$$(R * t)_{abc} = \sum_{d,e,f=1}^{3} R_{ad} R_{be} R_{cf} t_{def}. \tag{8.44}$$

For $l = 1$ we recover the usual vectors. For $l > 1$ the tensors form a representation of the rotation group that is not reducible. Reducible representations are provided for example by symmetric tensors, i.e. tensors that are invariant under permutations of the indices and that also satisfy the properties of being traceless, i.e. for $l = 3$

$$\sum_{a=1}^{3} t_{aab} = \sum_{a=1}^{3} t_{aba} = \sum_{a=1}^{3} t_{baa} = 0. \tag{8.45}$$

It may be interesting to note that these tensorial representations are isomorphic to the representations constructed with the spin degrees of freedom of a spin-l particle. However, a proof of this fact would take us very far from the main subject of this book.

As we have seen, the group is a very general concept that permeates the whole mathematics. Advanced group theory implies the introduction of many classifications and many extra concepts. We mention here only a few, mainly to let the reader become acquainted with the usual terminology:

- A group is finite if it contains a finite number of elements.
- Given two groups F and G, we define their direct product as the group whose elements are the pairs of elements of F and G such that
$$\{f_1, g_1\} \otimes \{f_2, g_2\} = \{f_1 * f_2, g_1 \cdot g_2\}. \tag{8.46}$$
- Given two groups R and T, such that R is a representation of T, we define a semi-direct product as the group whose elements are the pairs of elements of R and T such that
$$\{r_1, t_1\} \circ \{r_2, t_2\} = \{r_1 * r_2, t_2 \cdot (r_2 * t_1)\} \tag{8.47}$$
In this case we have to check by an explicit computation that we have actually constructed a group.

  A well-known example is the case where the group R is that of the rotations in a d-dimensional space and the group T that of the translations in a d-dimensional space. Both groups admit as faithful representation the set of d-dimensional vectors. By applying first a translation and then a rotation to a vector we can define an element of the roto-translation group. A roto-translation applied to a vector v gives
$$r * (t + v) = \sum_{b=1}^{d} r_{ab}(t_b + v_b). \tag{8.48}$$
It is evident that translations can be represented as vectors, and therefore the latter are a representation of the rotations. We have defined a roto-translation as a translation followed by a rotation on d-dimensional vectors. The roto-translation group is the semi-direct product of the rotations and translations. It is not a direct product: indeed if we apply two roto-translations to a generic vector, we find out that
$$\{r_1, t_1\} * \{r_2, t_2\} = \{r_1 * r_2, t_2 + r_2 * t_1\}, \tag{8.49}$$
which is the formula for the semi-direct product where we have used additive notation for translations (the translations constitute a commutative group).
- A group that is not the direct product of two groups is simple.
- A group that is not the semi-direct product of two groups is semi-simple.

In the case of *semigroups* only a small part of the requirements for having a group are satisfied: the identity may not exist and, if it exists, the elements may not have an inverse. Non-negative real (or integer) numbers form a semigroup with respect to the addition.

The prototype of a semigroup is provided by the positive real numbers, where the group operation is the addition. This semigroup is isomorphic to the semigroup of real numbers greater than 1, where the group operation is the multiplication. This example may be generalized: all operators of the form (see Subsec. 3.5.4)
$$e^{-\hat{O}t} \tag{8.50}$$
acting on an Hilbert space form a semigroup in the space of bounded operators (see Box 2.1) for positive t, if the spectrum of $\hat{O}$ is bounded from below.

Conversely, any semigroup of operators that depend only on one parameter ($S(t)$) which is defined for $t \geq 0$, which is isomorphic to the previously introduced semigroup, i.e. $S(t) * S(u) = S(t + u)$, can be written in the form (8.50) if $S(0) = 1$.

8.4.3 Rings, fields, vector spaces, and algebras

Rings are also very common algebraic structures, but they are more complex than groups. There are two operations: one is addition, denoted by "+," that forms a commutative group, the other is multiplication, denoted by "∗," that nearly forms a group – the only requirement missing is that some elements (among them the null element of the additive group, usually denoted by a zero) do not have an inverse. The two operations must also satisfy the distributive law

$$a * (b + c) = a * b + a * c. \tag{8.51}$$

If we add the requirement that the only element which does not have an inverse is the neutral element of the addition, we obtain a *field*. According to the recent notation, the multiplication must be commutative. Non-commutative fields are called *skew fields* (or *divisions rings*). In any case, the set of $n \times n$ matrices does not form a field (for $n > 1$), because all matrices with zero determinant are not invertible.

The most familiar example of a field is the set $\mathbb{Q}$ of the rational number. If we introduce a topology over the rationals and take the topological closure, we obtain the real numbers if we use the usual topology (if a different topology is used, different results are obtained, e.g. the p-adic numbers). Another example of a field is provided by the complex numbers $\mathbb{C}$.

A commutative group V is called a *vector space* over a field S if we can define on it the multiplication by the elements of the field S (the elements of S are usually called the scalars). Therefore, in order to define a vector space, we need to consider both the commutative group V and the field S. The following conditions must be satisfied:[6]

- $s * (v_1 + v_2) = s * v_1 + s * v_2$;
- $(s_1 + s_2) * v = s_1 * v + s_2 * v$;
- if 1 is the identity of the field, $1 * v = v$.

Real and complex Hilbert spaces are vector spaces over the fields of real and complex numbers, respectively. The previous definition of vector space is quite general and covers both the finite- and infinite-dimensional cases.

In the same way that commutative groups are promoted to vector spaces by introducing the extra multiplication by elements of an external field, the same operation promote rings to algebras.

The space of all operators over an Hilbert space is the standard example of an algebra (over the real or over the complex numbers, depending if the Hilbert space is real or complex, respectively).

[6] In the following relations s stands for scalars, i.e. elements of the field, and v stands for vectors, i.e. elements of the commutative group.

8.4.4 Finite-dimensional Lie groups and algebras

The general theory of Lie groups is rather complex; here we shall not present a precise definition of a Lie group but shall only give some examples and some definitions, mainly for the finite-dimensional case, i.e. when the group is a finite-dimensional manifold. In fact, generally Lie groups contain an infinite number of elements. They are also topological spaces in the sense that there is a topology and we can define the concept of limit and neighborhood.

Examples of Lie groups are as follows:

- Real numbers.
- The real numbers in the interval [0, 1], where the sum is performed modulo one.
- The rotations in a finite-dimensional space of dimension d, usually denoted by $O(d)$.
- Generalized rotations in a finite-dimensional space of dimension d that preserve a metric $g_{a,b}$, i.e.

$$\sum_{a',b'=1}^{d} R^{\mathrm{T}}_{aa'} g_{a'b'} R_{b'b} = g_{ab}, \tag{8.52}$$

which for $g_{ab} = \delta_{ab}$ and $d = 3$ reduces to Eq. (8.43).

It is particularly interesting to introduce the case of the simple metric

$$g_{ab} = \delta_{ab} f_a, \tag{8.53}$$

where $f_a = 1$ for $a \le d_1$ and $f_a = -1$ otherwise. This group is called the indefinite orthogonal group $O(d_1, d_2)$, where $d_1 + d_2 = d$. The group $O(3, 1)$ is particularly interesting in physics and especially in relativity: it is the so-called Lorentz group, i.e. the group of those transformations that leave invariant the Minkowsky metric.
- Any subgroup (that is not a finite group) of the previously defined groups (this definition covers many interesting cases).

Let us consider connected Lie groups, that is, those groups that are topologically connected (i.e. there is a continuous path connecting any two arbitrary elements of the groups). In connected Lie groups all elements can be reached by a finite number of products of elements belonging to the neighborhood of the origin, so that the structure of the group near the origin is crucial.

In the case of a finite-dimensional Lie group, a small neighborhood of the identity is topologically isomorphic to a neighborhood of the origin in an n-dimensional space: in other words all elements sufficiently near to the identity can be written in a unique way as $g(\mathbf{v})$ where $\mathbf{v}$ is a n-dimensional vector and $g(\mathbf{0}) = 1$. We say that n is the dimension of the group.

Let us consider a simple example, the orthogonal matrices in a three-dimensional space, usually denoted by $O(3)$, i.e. the group of rotations in three dimensions. The determinant of an orthogonal matrix can be ± 1 and matrices with determinant equal to -1 cannot be transformed continuously to matrices with determinant equal to $+1$, so the group $O(3)$ must have at least two connected components (it actually has two connected components).

If a rotation r is near to the identity, we can write it in an unique way as

$$r = e^m \quad \text{or} \quad m = \log(r), \tag{8.54}$$

if m is a matrix that is near to the origin. The matrix m is antisymmetric. The set M of matrices m belongs to an algebra (the Lie algebra of the group): a matrix m of M can be written as

$$m = \sum_{i=1}^{3} R_i v_i, \tag{8.55}$$

where the R are the generators of the group; the generators can be chosen as $\hat{L}_x,\ \hat{L}_y,\ \hat{L}_z$ (see Subsec. 6.1.2 and Sec. 8.2).

The reader should notice that, for matrices far from the identity, we cannot continue to associate to a group element an element of the algebra in an unique and continuous way. For example, two-dimensional rotations can be characterized by the rotation angle θ: the Lie algebra is just given by the real, but the group is given by the angles in the interval $[-\pi, \pi]$. Moreover there are different groups that have the same algebras (e.g. the first two examples of Lie groups).

Similar constructions can be done for all Lie groups. In the general case the Lie algebra is non-Abelian. One can prove that

$$[T_i, T_k] = \sum_l C_{ikl} T_l, \tag{8.56}$$

where the structure constants (C) contain important information on the nature of the Lie algebra and the T_j are generators of generic transformations.

It should be clear that the same group may be represented in many ways: for example the three-dimensional rotation group may act on the space of a traceless symmetric tensor (a five-dimensional space) and in this case the elements of the group are represented as five-dimensional matrices. The explicit form of the generators changes, but the structure constants remain the same.

All finite-dimensional Lie algebra may be classified and the list is relatively simple. The classification of Lie groups is more complex, since the Lie algebras give information only on the component connected to the origin; to specify the Lie group we need more information, e.g. the number of connected components and about the group transformation from one component to another component.

There are quadratic forms in the generators (at least one) that commute with all elements of the algebra: they are called Casimir operators and they play an important role in the classification of irreducible representation. Indeed, for each irreducible representation R we have that

$$C * r = Ar, \tag{8.57}$$

where the number A does not depend on the element r. For example in the case of the $O(3)$ group the only Casimir is

$$C = \hat{l}_x^2 + \hat{l}_y^2 + \hat{l}_z^2 \tag{8.58}$$

and the values of A are $l(l+1)$, l being the spin (angular momentum) of the representation.

Lie groups are very important in physics as far as continuous symmetries are associated with Lie groups. Moreover, in the Hamiltonian formalism, the Noether theorem (see Subsec. 8.1.2) tells us that there are conserved quantities associated with the corresponding generators of the Lie algebra. As we have seen, this classical Noether theorem can be extended to the quantum case. In both classical and in quantum mechanics, energy conservation is related to the time invariance of the physical laws in the same way that momentum conservation is related to translational invariance.

Summary

In this chapter we have discussed quantum-mechanical symmetries and groups:

- First, we have seen that the transformations induced by the Schrödinger picture correspond to an *active transformation* (i.e. to the rotation of the state vector in a given representation), whereas the transformations induced by the Heisenberg picture can be seen as a *passive transformation* (a rotation of the reference frame by keeping the state vector fixed).
- Since the probability of measurement outcomes must be conserved under a symmetry transformation, there are two possible types of such transformations, unitary and antiunitary (*Wigner's theorem*). Continuous transformations have to be unitary, whereas discrete ones may be either unitary or antiunitary.
- We have examined the relationship between symmetries and conservation laws and seen that to any symmetry of the system corresponds a constant of motion (*Noether's theorem*).
- We have considered *rotations* and *space–time translations* as instances of continuous transformation.
- We have considered *space reflection* and *time reversal* as examples of discrete transformation: space reflection is unitary whereas time reflection is antiunitary.
- Finally, a short summary of *group theory* has been given: in quantum mechanics symmetries may be mathematically formulated in terms of groups.

Problems

8.1 Consider the example presented in Box 8.1. Show that the state $| 1\rangle$, expressed in the basis $\left\{\left| 1'\right\rangle, \left| 2'\right\rangle\right\}$ and given by (8.5), may be written as

$$| 1\rangle = \frac{1}{\sqrt{2}}\left(\left| 1'\right\rangle + \left| 2'\right\rangle\right),$$

i.e. with the same expansion coefficients as $\left|\psi'\right\rangle$ in the basis $\{| 1\rangle, | 2\rangle\}$.

8.2 Prove Eq. (8.7).

8.3 We have said (in Subsec. 3.5.4) that unitary transformations may be represented as rotations on the Poincaré sphere. Making use of the expansion presented in Subsec. 1.3.3, show what rotation has to be done when passing from the which-path representation to the visibility-of-interference representation in the example considered in Box 8.1.

8.4 Show that a transformation $\hat{\mathcal{U}}$ gives rise to an exact symmetry only if $\hat{\mathcal{U}}$ is a constant of motion.

8.5 Given a canonical commutation relation

$$\left[\hat{O}, \hat{O}'\right] = \imath\hbar,$$

prove that it is invariant under unitary transformation of observables $\hat{O}$ and $\hat{O}'$.

8.6 Prove that a continuous symmetry must be necessarily represented by a unitary transformation.

8.7 Justify why, given a group with N elements, there are N^{N^2} multiplication (addition) tables, and prove that not all of them can be associated with a group.

8.8 Verify that all examples described as 1, 2, 3, 4, and 5 in Subsec. 8.4.1 satisfy the group properties.

Further reading

Symmetries

Eisberg, R. and Resnick, R., *Quantum Physics of Atoms, Molecules, Solids, Nuclei, and Particles*, Chichester: Wiley, 1974; 2nd edn. 1985.

Weyl, Hermann, *Symmetry*, Princeton: Princeton University Press, 1952, 1980, 1982, 1989.

Group theory

Fuchs, J. and Schweigert, C., *Symmetries, Lie Algebras and Representations*, Cambridge: Cambridge University Press, 1997.

Greiner, W. and Müller, B., *Quantum Mechanics Symmetries*, Tun: H. Deutsch, 1984; Berlin: Springer, 1989, 1994.

Weyl, Hermann, *The Theory of Groups and Quantum Mechanics*, Dover Publications, 1950, Engl. trans. of *Gruppentheorie und Quantenmechanik*, 1928, 1931.

Wigner, Eugene P., *Group Theory*, New York: Academic, 1959.

9 The measurement problem in quantum mechanics

In most textbooks, measurement does not receive the full attention it deserves and sometimes is even not treated at all, apart a brief and cryptic mention of the "reduction of the wave packet." However, in the last decades, the situation has profoundly changed and it is time to consider measurement a fundamental part of quantum mechanics, even, to a certain extent, an important generalization of the traditional theory (see also Chs. 14–15).

This chapter consists of three major parts. In the first block (Secs. 9.1–9.4) we develop the main physical features of the measurement process: the heart of the argument is here represented by the aspects related to the interpretation. In the second part (Secs. 9.5–9.8) we discuss several special (and partly interdependent) topics of measurement: the heart here is represented by experimental aspects. In the third part (Secs. 9.9–9.12) we deal with the measurement process on a more formal plane, making use, in particular, of the generalization represented by the concepts of *effect* and *positive operator valued measure* (POVM).

As we have said, the measurement problem is one of the most fundamental issues in the conceptual structure of quantum mechanics (as described in Sec. 9.1) and has a long history that will be examined in Sec. 9.2. In this context, the existence of apparently paradoxical quantum states comes about: the so-called Schrödinger cat states (see Sec. 9.3). Among the several approaches that have been proposed, one possible way out of this paradox is provided by the decoherence approach (see Sec. 9.4).

In Sec. 9.5 we show that measurement consists of two aspects: the destruction of the interference and the acquisition of information, and it is shown that only the latter is irreversible. In Sec. 9.6 we return on interaction-free measurements, while in Sec. 9.7 a puzzling feature of the theory is shown: the apparent possibility of manipulating the past (delayed choice). Another "strange" feature of the theory is discussed in Sec. 9.8: the quantum Zeno effect.

In Sec. 9.9 the subject of conditional measurements and postselection is presented. A very interesting and relatively new development is represented by the possibility of performing unsharp measurements (the subject of Sec. 9.10). In Sec. 9.11 the quantum non-demolition measurement is presented. Finally, some important statistical issues of quantum measurement are treated in Sec. 9.12.

9.1 Statement of the problem

When measuring, we expect the result of our measurement to be a determined value (see Subsecs. 2.1.1, 2.1.3, and 3.5.3). This is what actually happens when an experimenter records the value of an observable. On the other hand (see Pr. 2.2: p. 57), we know that a possible result of a measurement is an eigenvalue of the measured observable $\hat{O}$, and we also know that this eigenvalue, say o_k, is strictly associated with the corresponding eigenvector, say $|o_k\rangle$, since to obtain a given eigenvalue as outcome means that the system is in the state represented by the corresponding eigenvector.

In this context, it is important to distinguish between preparation and measurement. In a *preparation* we submit the system to a number of constraints that *select* a certain output state. For instance, the filter P1 introduced in Subsec. 1.3.1 represents a preparation of the incoming photons in a vertical polarization state. In such a process, a number of the incoming systems is discarded (those not satisfying the selection requirement). A *measurement*, on the other hand, is a process through which we *ask* the system about the value of a certain observable (the measurement outcome). Obviously, the answer of the system is *not* controllable by the experimenter. In other words, a preparation is determinative, while a measurement is interrogative. As we shall see below, the preparation of the object system on which measurement is performed is, in general, a preliminary step of measurement.

Suppose now that the state before measuring is (or has been prepared in) a superposition relative to $\hat{O}$, i.e.

$$|\psi\rangle = \sum_j c_j |o_j\rangle. \tag{9.1}$$

Successively, assume that, as a result of the measurement, we obtain the component $|o_k\rangle$ of the initial state: now, there is no way to obtain this result from any superposition state (that is, a superposition that contains also other components but $|o_k\rangle$) by means of a unitary evolution (see Probs. 3.12 and 3.13), though, as we know (see Sec. 3.5), time evolution in quantum mechanics is unitary. This is the essence of the measurement problem: how can we reconcile the ordinary (unitary) quantum-mechanical time evolution with the experimental evidence of measurement? A conflict of this type (between a general theoretical framework and some experimental evidences) is very common in science – see, for example, Planck's problem (in Subsec. 1.5.1) – and it is the very source of scientific research because it pushes us to seek for other and more general hypotheses beyond the existing theories.[1]

Let us now reformulate the problem in terms of the density matrix formalism. The density matrix corresponding to the initial pure state (9.1) can be written as

$$\hat{\rho} = |\psi\rangle\langle\psi| = \sum_j |c_j|^2 |o_j\rangle\langle o_j| + \sum_{j\neq k} c_j c_k^* |o_j\rangle\langle o_k|. \tag{9.2}$$

The two sums on the rhs of Eq. (9.2) represent the diagonal and off-diagonal (coherent) parts, respectively (see Sec. 5.1). It is clear that, in order to obtain a determined outcome

[1] This is the essence of abduction: see footnote 8, p. 573.

(i.e. an eigenvalue of the measured observable corresponding to a given eigenstate), the latter sum has to vanish after the measurement: in this case, we must obtain a classical statistical mixture. This means that not all forms of mixtures are adequate in order to describe a measurement. The necessary condition for obtaining a determined result when measuring is that the apparatus $\mathcal{A}$ that performs the measurement on the object system $\mathcal{S}$ can extract information from $\mathcal{S}$ (it can tell what the value is of the observable we are measuring). This is in turn possible only if the state of the object system is a mixture of eigenstates of the measured observable. An example is represented by the statistical mixture describing the state of a classical dice before the outcome of the throw is read. Summarizing, after the measurement we must obtain a mixture of the type

$$\hat{\tilde{\rho}} = \sum_j |c_j|^2 |o_j\rangle\langle o_j|. \tag{9.3}$$

Again, also in this formalism, there is no unitary evolution that can account for the transformation from the state before the measurement, represented by Eq. (9.2), to the state after the measurement, represented by Eq. (9.3) (see Prob. 9.1). We can rewrite the change from Eq. (9.2) to Eq. (9.3) in the form of the requirement

$$\hat{\rho} \leadsto \sum_j \hat{P}_j \hat{\rho} \hat{P}_j \tag{9.4}$$

for the different eigenvalues o_j of the measured observable $\hat{O}$, and where $\hat{P}_j = |o_j\rangle\langle o_j|$ are the projectors on the different eigenstates of $\hat{O}$. This expression is known as *Lüders mixture*.[2] It is a mathematical formulation of what is commonly known as *projection postulate*. We recall that the trace of the state does not change with this operation and therefore normalization is preserved.

Let us consider the example of a two-level system, and suppose, for the sake of simplicity, that

$$|\psi\rangle = c_\downarrow | \downarrow\rangle + c_\uparrow | \uparrow\rangle, \tag{9.5}$$

where $| \downarrow\rangle$ and $| \uparrow\rangle$ are the eigenkets of the measured observable $\hat{O}$ and form an orthonormal basis on the two-dimensional Hilbert space $\mathcal{H}_\mathcal{S}$ of the system. Then, it is straightforward to obtain

$$\begin{aligned}
\sum_{j=\downarrow,\uparrow} \hat{P}_j \hat{\rho} \hat{P}_j &= | \downarrow\rangle\langle\downarrow | \left(c_\downarrow | \downarrow\rangle + c_\uparrow | \uparrow\rangle\right) \left(\langle\downarrow | c_\downarrow^* + \langle\uparrow | c_\uparrow^*\right) | \downarrow\rangle\langle\downarrow | \\
&\quad + | \uparrow\rangle\langle\uparrow | \left(c_\downarrow | \downarrow\rangle + c_\uparrow | \uparrow\rangle\right) \left(\langle\downarrow | c_\downarrow^* + \langle\uparrow | c_\uparrow^*\right) | \uparrow\rangle\langle\uparrow | \\
&= |c_\downarrow|^2 | \downarrow\rangle\langle\downarrow | + |c_\uparrow|^2 | \uparrow\rangle\langle\uparrow | = |c_\downarrow|^2 \hat{P}_\downarrow + |c_\uparrow|^2 \hat{P}_\uparrow \\
&= \hat{\tilde{\rho}}_{\uparrow,\downarrow},
\end{aligned} \tag{9.6}$$

2 See [Lüders 1951].

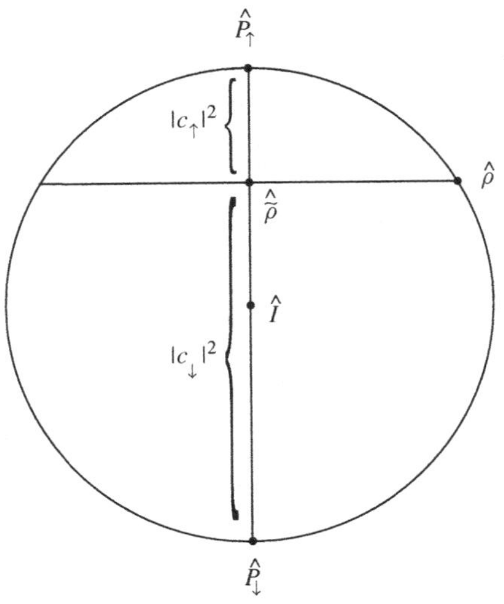

Figure 9.1 Geometric representation of a measurement on the sphere of density matrices. The eigenstates of the measured observable, described by the two projectors $\hat{P}_\downarrow$ and $\hat{P}_\uparrow$, are here represented as the south and north poles, respectively. Suppose that the input (prepared) state $\hat{\rho}$ is represented by an arbitrary point on the surface. Then, it suffices to draw the orthogonal line from this point to the unitary diameter passing through the points $\hat{P}_\downarrow$ and $\hat{P}_\uparrow$. This will individuate the mixture $\hat{\tilde{\rho}}$ and divide this diameter into two parts, whose lengths correspond to $|c_\downarrow|^2$ and $|c_\uparrow|^2$, respectively.

where we have taken into account that, due to orthonormality,

$$\langle \downarrow \mid \uparrow \rangle = \langle \uparrow \mid \downarrow \rangle = 0 \quad \text{and} \quad \langle \downarrow \mid \downarrow \rangle = \langle \uparrow \mid \uparrow \rangle = 1. \tag{9.7}$$

We can represent this situation on the sphere of the density matrices (see Fig. 9.1 and Sec. 5.6).

In order to examine more exactly what happens when measuring, we need to introduce the concept of *apparatus* in the context of quantum measurement. Strictly speaking, the apparatus $\mathcal{A}$ can be divided into a *little measuring device*, also called a *meter*, which interacts directly with $\mathcal{S}$, an *amplifier*, and a *pointer* on a graduated scale. The role of the meter is that of establishing a correlation between the system and the apparatus, while the role of the amplifier is that of transforming a "microscopic input" (which is a consequence of the interaction between the system and the meter) into a macroscopic output, i.e. the result that can be read by an observer. In the most simple case, the possible outcomes may be represented as ticks on a reading scale (see Subsec. 2.1.1). In the discrete case, we can understand the reading scale as a partition of the space of the possible outcome values. The *pointer* is some device that can move along this reading scale by associating a given number to a certain outcome result.

Let us now write a quantum-mechanical definition of the apparatus $\mathcal{A}$ in terms of an arbitrary eigenbasis $\{|a_j\rangle\}$ of the pointer observable $\hat{O}_\mathcal{A}$ in its Hilbert space $\mathcal{H}_\mathcal{A}$, i.e. of the observable describing the position of a pointer on a discrete scale (any eigenvalue a_j associated with an eigenket $|a_j\rangle$ can be understood as a value on this scale). Then, we shall write the generic state of the apparatus as

$$|\mathcal{A}\rangle = \sum_j c_{a_j} |a_j\rangle . \qquad (9.8)$$

In order to distinguish the pointer observable from the measured observable of the object system $\mathcal{S}$, let us write $\hat{O}_\mathcal{S}$ for the latter. If the apparatus functions properly, we expect that, having obtained an eigenvalue o_k of $\hat{O}_\mathcal{S}$, the apparatus $\mathcal{A}$ will register a corresponding value a_k. In other words, the apparatus and the object system must be coupled in such a way that there is a one-to-one correspondence between values o_k of the observable $\hat{O}_\mathcal{S}$ and the values a_k that $\mathcal{A}$ registers. One of the most important points to be emphasized here is that the interaction between $\mathcal{S}$ and $\mathcal{A}$ should (and can) be described as a quantum-mechanical interaction. If the interaction Hamiltonian dominates during the interaction time τ over the free terms of the Hamiltonian, the unitary operator describing the evolution during the interaction time may be written as

$$\hat{U}_\tau^{\mathcal{SA}} = e^{-\frac{\imath}{\hbar}\int_0^\tau dt \hat{H}_{\mathcal{SA}}(t)}, \qquad (9.9)$$

where $\hat{H}_{\mathcal{SA}}$ is the system-apparatus interaction Hamiltonian. A simple form of this Hamiltonian can be expressed as

$$\hat{H}_{\mathcal{SA}}(t) = \varepsilon_{\mathcal{SA}}(t)\hat{O}_\mathcal{A} \otimes \hat{O}_\mathcal{S}, \qquad (9.10)$$

where $\varepsilon_{\mathcal{SA}}$ is a coupling function. Suppose that we start with an initial state at $t_0 = 0$ when apparatus and system are uncoupled, i.e.

$$|\Psi_{\mathcal{SA}}(0)\rangle = |\psi(0)\rangle \otimes |\mathcal{A}(0)\rangle . \qquad (9.11)$$

We have[3]

$$\hat{O}_\mathcal{S}\hat{O}_\mathcal{A} |\Psi_{\mathcal{SA}}(0)\rangle = \sum_j c_j \hat{O}_\mathcal{S} |o_j\rangle \hat{O}_\mathcal{A} |a_0\rangle , \qquad (9.12)$$

because the states $|o_j\rangle$ and $|a_0\rangle$ belong to different Hilbert spaces and are initially uncorrelated, and where $|a_0\rangle = |\mathcal{A}(0)\rangle$ is the initial state (at rest) of the pointer observable, which, for the sake of simplicity, we have assumed to be an eigenstate of the pointer. Then, it is clear that, at the end of the interaction between $\mathcal{S}$ and $\mathcal{A}$, we would like to have

$$\begin{aligned} |\Psi_{\mathcal{SA}}(\tau)\rangle &\equiv \hat{U}_\tau^{\mathcal{SA}} |\psi(0)\rangle |a_0\rangle \\ &= \sum_j c_j e^{\imath\phi_j} |o_j\rangle |a_j\rangle, \end{aligned} \qquad (9.13)$$

where $e^{\imath\phi_j}$ is a phase factor. In fact, state (9.13) displays a perfect correlation between system and apparatus states.

In order to exemplify how such a correlation may be achieved, we can choose a simple model: we consider here only the meter $\mathcal{M}$ instead of the whole apparatus $\mathcal{A}$, and assume that both the system $\mathcal{S}$ and the meter $\mathcal{M}$ are represented by two-level systems. Then, we consider only the interaction between $\mathcal{S}$ and $\mathcal{M}$. In particular, we choose the operator $\hat{\sigma}_z^\mathcal{S}$ for the observable of $\mathcal{S}$ and $\hat{\sigma}_x^\mathcal{M}$ for the observable of the meter (see the matrices (6.154)).

[3] In the following, for the sake of notation, we shall generally omit the direct product ($\otimes$) sign between the states (or operators) belonging to different Hilbert states.

The system $\mathcal{S}$ is initially prepared in a superposition of the two eigenstates of $\hat{\sigma}_z^{\mathcal{S}}$, which are the spin-up and the spin-down state, respectively, given by

$$|\uparrow\rangle_{\mathcal{S}} = \begin{pmatrix} 1 \\ 0 \end{pmatrix} \qquad |\downarrow\rangle_{\mathcal{S}} = \begin{pmatrix} 0 \\ 1 \end{pmatrix}. \tag{9.14}$$

The meter is initially in the z spin-down state, $|\downarrow\rangle_{\mathcal{M}}$, so that if, after the interaction, the system is in $|\downarrow\rangle_{\mathcal{S}}$, the state of $\mathcal{M}$ remains unchanged. Otherwise, it will become $|\uparrow\rangle_{\mathcal{M}}$. The interaction Hamiltonian (9.10) can then be explicitly written as

$$\hat{H}_{\mathcal{SM}} = \varepsilon_{\mathcal{SM}} \left(1 + \hat{\sigma}_z^{\mathcal{S}}\right) \hat{\sigma}_x^{\mathcal{M}}. \tag{9.15}$$

The first step we need to make in order to calculate the action of the unitary operator $\hat{U}_\tau^{\mathcal{SM}}$ on the initial state $|\Psi(0)\rangle_{\mathcal{SM}}$, is to diagonalize the matrix $\hat{H}_{\mathcal{SM}}$. The operator $\hat{\sigma}_z^{\mathcal{S}}$ is already diagonal with respect to the basis states (9.14). In fact, we have (see Prob. 9.2)

$$\hat{\sigma}_z^{\mathcal{S}} |\uparrow\rangle_{\mathcal{S}} = |\uparrow\rangle_{\mathcal{S}} \quad \text{and} \quad \hat{\sigma}_z^{\mathcal{S}} |\downarrow\rangle_{\mathcal{S}} = -|\downarrow\rangle_{\mathcal{S}}. \tag{9.16}$$

On the other hand, the two eigenvalues of $\hat{\sigma}_x^{\mathcal{M}}$ are ± 1 and the eigenkets are given by

$$|\uparrow\rangle_x^{\mathcal{M}} = \frac{1}{\sqrt{2}} \begin{pmatrix} 1 \\ 1 \end{pmatrix} \quad \text{and} \quad |\downarrow\rangle_x^{\mathcal{M}} = \frac{1}{\sqrt{2}} \begin{pmatrix} 1 \\ -1 \end{pmatrix}, \tag{9.17}$$

respectively. In terms of the z spin-up and spin-down states, these eigenkets are

$$|\uparrow\rangle_x^{\mathcal{M}} = \frac{1}{\sqrt{2}} (|\uparrow\rangle_{\mathcal{M}} + |\downarrow\rangle_{\mathcal{M}}), \quad |\downarrow\rangle_x^{\mathcal{M}} = \frac{1}{\sqrt{2}} (|\uparrow\rangle_{\mathcal{M}} - |\downarrow\rangle_{\mathcal{M}}). \tag{9.18}$$

Now, in order to find the time evolution of the quantum state of the compound system, we only need to write its initial state in terms of the eigenkets (9.14) and (9.17) of $\hat{H}_{\mathcal{SM}}$

$$\begin{aligned} |\Psi(\tau)\rangle_{\mathcal{SM}} &= \hat{U}_\tau^{\mathcal{SM}} |\Psi(0)\rangle_{\mathcal{SM}} \\ &= e^{-\frac{\imath}{\hbar}\tau\varepsilon_{\mathcal{SM}}(1+\hat{\sigma}_z^{\mathcal{S}})\hat{\sigma}_x^{\mathcal{M}}} \left[\left(c_\uparrow |\uparrow\rangle_{\mathcal{S}} + c_\downarrow |\downarrow\rangle_{\mathcal{S}}\right) |\downarrow\rangle_{\mathcal{M}} \right]. \end{aligned} \tag{9.19}$$

As we know from Subsec. 2.2.6, the action of the Hamiltonian onto its eigenkets simply returns the corresponding eigenvalues. Therefore, since

$$|\downarrow\rangle_{\mathcal{M}} = \frac{1}{\sqrt{2}} \left(|\uparrow\rangle_x^{\mathcal{M}} - |\downarrow\rangle_x^{\mathcal{M}} \right), \tag{9.20}$$

we have

$$\begin{aligned} |\Psi(\tau)\rangle_{\mathcal{SM}} = \frac{1}{\sqrt{2}} \Big(& c_\uparrow e^{-\frac{2\imath}{\hbar}\tau\varepsilon_{\mathcal{SM}}} |\uparrow\rangle_{\mathcal{S}} |\uparrow\rangle_x^{\mathcal{M}} + c_\downarrow |\downarrow\rangle_{\mathcal{S}} |\uparrow\rangle_x^{\mathcal{M}} \\ & - c_\uparrow e^{+\frac{2\imath}{\hbar}\tau\varepsilon_{\mathcal{SM}}} |\uparrow\rangle_{\mathcal{S}} |\downarrow\rangle_x^{\mathcal{M}} - c_\downarrow |\downarrow\rangle_{\mathcal{S}} |\downarrow\rangle_x^{\mathcal{M}} \Big), \end{aligned} \tag{9.21}$$

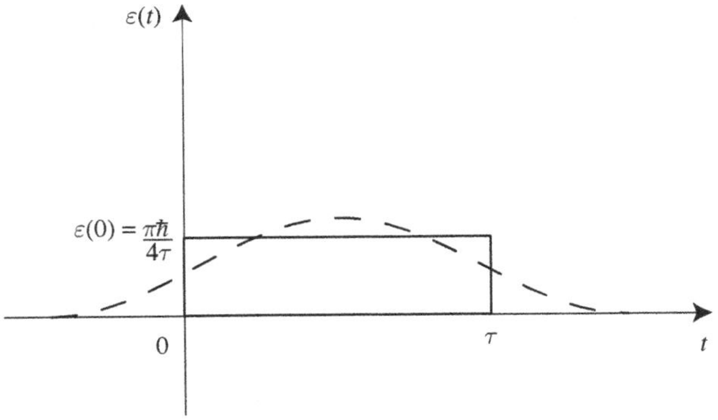

Figure 9.2 Two ways of tuning the coupling function $\varepsilon_{\mathcal{SM}}(t)$ for entangling the system and the meter: in the solid line, the interaction is tuned on at $t = 0$ and off at $t = \tau$, i.e. $\varepsilon_{\mathcal{SM}}(t) = \varepsilon_{\mathcal{SM}}(0)$ $[\Xi(t) - \Xi(\tau)]$, where $\Xi(x)$ is the Heaviside step function (see Eq. (6.250)). Alternatively, one might imagine switching the interaction on and off in a smooth manner, as in the dashed line, in order to reach an equivalent result.

where the exponential $e^{-\frac{\imath}{\hbar}\tau\varepsilon_{\mathcal{SM}}(1+\hat{\sigma}_z^{\mathcal{S}})\hat{\sigma}_x^{\mathcal{M}}}$ is 1 when the eigenvalue of $\hat{\sigma}_z^{\mathcal{S}}$ is -1, i.e. for the state $|\downarrow\rangle_{\mathcal{S}}$. Substituting expressions (9.18) into Eq. (9.21), we obtain

$$\begin{aligned} |\Psi(\tau)\rangle_{\mathcal{SM}} = \frac{1}{\sqrt{2}}\Bigg(& -c_\uparrow \frac{2\imath}{\sqrt{2}} \sin\frac{2\tau\varepsilon_{\mathcal{SM}}}{\hbar} |\uparrow\rangle_{\mathcal{S}} |\uparrow\rangle_{\mathcal{M}} + \sqrt{2} c_\downarrow |\downarrow\rangle_{\mathcal{S}} |\downarrow\rangle_{\mathcal{M}} \\ & + c_\uparrow \frac{2\imath}{\sqrt{2}} \cos\frac{2\tau\varepsilon_{\mathcal{SM}}}{\hbar} |\uparrow\rangle_{\mathcal{S}} |\downarrow\rangle_{\mathcal{M}} \Bigg). \end{aligned} \tag{9.22}$$

Choosing now

$$\frac{2\tau\varepsilon_{\mathcal{SM}}}{\hbar} = \frac{\pi}{2}, \quad \text{or} \quad \tau = \frac{\pi\hbar}{4\varepsilon_{\mathcal{SM}}}, \tag{9.23}$$

i.e. by fine-tuning the interaction time, we finally obtain

$$|\Psi(\tau)\rangle_{\mathcal{SM}} = +c_\downarrow |\downarrow\rangle_{\mathcal{S}} |\downarrow\rangle_{\mathcal{M}} - \imath c_\uparrow |\uparrow\rangle_{\mathcal{S}} |\uparrow\rangle_{\mathcal{M}}, \tag{9.24}$$

which is the required coupling between the system and the meter. In other words, we are assuming that the interaction is switched on at $t = 0$ and off at $t = \tau$. Alternatively, one might envisage a situation where the interaction is smoothly turned on and off (see Fig. 9.2). This formalism can be easily translated in the density-matrix formalism above (see Prob. 9.4).

Generalizing again, we see, in conclusion, that the stunning result we have obtained is that, if the initial state of $\mathcal{S}$ is prepared as a superposition relatively to the measured observable $\hat{O}_{\mathcal{S}}$, then the resulting total state of $\mathcal{S} + \mathcal{A}$ is entangled (see Subsec. 5.5.1). The "intermediate" state (9.24) between the initial state of the form (9.19) and the final state after the measurement, which we express here as the mixture

$$\hat{\rho}_{\mathcal{SM}} = |c_\downarrow|^2 |\downarrow\rangle_{\mathcal{S}} \langle\downarrow| \otimes |\downarrow\rangle_{\mathcal{M}} \langle\downarrow| + |c_\uparrow|^2 |\uparrow\rangle_{\mathcal{S}} \langle\uparrow| \otimes |\uparrow\rangle_{\mathcal{M}} \langle\uparrow|, \tag{9.25}$$

may be considered a first step of the measurement process, called *premeasurement*, where the correlation between system and apparatus is provided (see also the general form (9.13)). We should keep distinct the preparation, where some selection of a given state occurs, and the premeasurement, which is not selective as far as we expect that the interrogative selection is provided by the successive step of measurement. In other words, due to premeasurement, the superposition state of $\mathcal{S}$ seems to "affect" the apparatus, too: the price we have to pay in order to fulfil the requirement of the one-to-one correlation is a joint state that does not seem to instantiate a determined result, and where even the apparatus does not behave in a classical manner, as we normally expect from an apparatus. This situation should be explained. Before we begin to discuss this problem, it is convenient to take a look at the different proposals that have been made in order to overcome this difficulty. As a matter of fact, one can learn a lot from a critical examination of the history of these proposals.

9.2 A brief history of the problem

9.2.1 Projection postulate

Bohr and the supporters of the Copenhagen interpretation (see Subsec. 1.5.7) gave no clue on how to frame the measurement problem in terms of the quantum-mechanical formalism. This is due to the fact that Bohr thought that quantum theory can only be accounted for in terms of classical, macroscopic experience: Bohr[4] argued that the account of all evidence in quantum mechanics must be expressed in classical terms, since all terms, by which we perceive and experience, have an unambiguous meaning only in the frame of macroscopic ordinary experience, the only one that can unambiguously be called "experience." When we translate this approach to the terminology of the measurement theory, Bohr's requirement can be expressed as postulating the necessity of a classical apparatus.

However, Bohr's position appears untenable if taken literally. Indeed, it can be shown that the pointer cannot be classical. Assume, on the contrary, that the pointer observable $\hat{O}_{\mathcal{A}}$ is classical. Then, in the unitary measurement coupling $\hat{U}_\tau^{\mathcal{SA}} = e^{\imath \hat{H}_{\mathcal{SA}}\tau}$, $\hat{H}_{\mathcal{SA}}$ commutes with $\hat{O}_{\mathcal{A}}$. In fact, any quantum observable has to commute with any classical variable (c-number). Therefore, $\hat{O}_{\mathcal{A}}$ also commutes with $\hat{U}_\tau^{\mathcal{SA}}$ and it must follow that the probability distribution of $\hat{O}_{\mathcal{A}}$ is completely independent from the measured observable. In other words, the apparatus remains uncoupled from the object system. In fact, we have

$$\begin{aligned}\left\langle \hat{O}_{\mathcal{A}} \right\rangle_\tau &= \langle \Psi_{\mathcal{SA}}(\tau)| \hat{O}_{\mathcal{A}} |\Psi_{\mathcal{SA}}(\tau)\rangle \\ &= \langle \Psi_{\mathcal{SA}}(0)| \left(\hat{U}_\tau^{\mathcal{SA}}\right)^\dagger \hat{O}_{\mathcal{A}} \hat{U}_\tau^{\mathcal{SA}} |\Psi_{\mathcal{SA}}(0)\rangle\end{aligned}$$

4 See [Bohr 1948, 327] [Bohr 1949, 209–210].

$$= \langle \Psi_{\mathcal{SA}}(0)| \left(\hat{U}_{\tau}^{\mathcal{SA}}\right)^{\dagger} \hat{U}_{\tau}^{\mathcal{SA}} \hat{O}_{\mathcal{A}} |\Psi_{\mathcal{SA}}(0)\rangle$$
$$= \langle \Psi_{\mathcal{SA}}(0)| \hat{O}_{\mathcal{A}} |\Psi_{\mathcal{SA}}(0)\rangle. \tag{9.26}$$

Such a result is incompatible with measurement unless $\hat{O}_{\mathcal{S}}$ is trivial (i.e. constant). Therefore, we have to include the apparatus (or some part of it) in the quantum description of the measurement.

Another fundamental issue in this context is whether the state of the system changes during the interaction with the apparatus. For this reason, in 1927 Pauli distinguished two types of measurement: of first and of second kind.[5] We have a measurement of the first kind when the probability distribution of the (unknown) values of the measured observable does not change with the measurement. Otherwise, i.e. when the measurement process changes the probability distribution of the measured observable, we have a measurement of the second kind. The necessary and sufficient condition of a measurement of the first kind of an observable $\hat{O}_{\mathcal{S}}$, where a value o_j represented by $\hat{P}_j$ is obtained, is that $\hat{P}_j$ commutes with the state $\hat{\rho}_{\mathcal{S}}$ in which the system is before measuring.

Von Neumann was the first to try to formalize the problem of measurement [*von Neumann* 1932]. He assumed that there are basically two types of evolution in quantum mechanics. The first is the usual unitary evolution, while the other is represented by measurement, which presents the following features: it is a discontinuous, non-causal, instantaneous, non-unitary, and irreversible change of state. Von Neumann called this abrupt change *reduction of the wave packet* (a passage from a superposition to one of its components). Then, von Neumann postulated that, if the observable $\hat{O}_{\mathcal{S}}$ is measured on a system $\mathcal{S}$ in an arbitrary state $|\psi\rangle$, then the latter is projected after the measurement onto one of the basis vectors $|o_j\rangle$ of the representation in which $\hat{O}_{\mathcal{S}}$ is diagonal, i.e. in an eigenstate of $\hat{O}_{\mathcal{S}}$ for which the probability is $|\langle o_j|\psi\rangle|^2$. This has been known as the *projection postulate*, and is formally expressed by the requirement (9.4). Since it seemed to him that there was no possibility of accounting for this abrupt change in terms of the quantum formalism, von Neumann introduced the observer's consciousness in order to justify it, and said that "the measurement or the related process of the subjective perception is a new entity relative to the physical environment and is not reducible to the latter. Indeed, subjective perception leads us to the intellectual inner life of the individual, which is extra-observational by its very nature" [*von Neumann* 1955, 418]. It seems rather strange that a physical measurement is here considered to be equivalent to a subjective perception that von Neumann conceived as an extra-observational phenomenon. It is true that von Neumann defended a form of psycho-physical parallelism, but it seems not to be completely compatible with this explanation.[6] In fact, there is a fundamental ambiguity in von Neumann's formulation, and there are three possible ways of resolving the conflict between his theory of measurement and psychophysical parallelism. If the only reality is the physical one (or if psychological reality reflects or can be led to reflect physical reality), then one

[5] See [*Pauli* 1980, 75].

[6] The psycho-physical parallelism is the hypothesis that the mind and the physical world are parallel processes without causal relationship between each other. Generally, it is believed that the mind represents the physical reality. Von Neumann also attributed psychophysical parallelism to Bohr, which is surely a mistake.

should account for the reduction in terms of a form of illusion in the observer's mind. Only by this price, can one maintain the parallelism. Another possibility is that the reduction is real, and then no different reality can exist, if one does not wish to run into contradiction. Therefore, the mind should be somehow capable of acting on the physical world. In this case no parallelism is tenable. Finally, the reduction "happens" in the mind and there is no relationship between "mental" and "material" realities.[7]

9.2.2 The statistical interpretations

It should be mentioned that Einstein was one of the strongest opponents to the quantum theory in its mature form (see also Sec. 16.1), i.e. as a theory based on the superposition and quantization principles (see Subsecs. 1.2.3 and 2.1.1). His interpretation can be synthesized as follows [Einstein 1949, 671–72]: the wave function (or the state vector) provides a description of certain statistical properties of an ensemble of systems, but is not a complete description of an individual system, a description which, according to him, should be in principle possible. In other words, Einstein believed in the possibility of building a more fundamental and deterministic theory that could stand in the same relationship to quantum mechanics as classical mechanics stands to statistical mechanics, and could therefore show that the probabilistic features of quantum theory (see Sec. 1.4 and Subsec. 3.1.2) are not fundamental. Following this interpretation, the problem of measurement does not exist at all since the apparent abrupt change from a superposition to an eigenstate of the measured observable amounts to the selection of a single system among a statistical ensemble of systems (see Sec. 5.1). Later, proponents of a statistical interpretation of quantum mechanics[8] admitted that no sound reduction of quantum mechanics to a deterministic theory was possible. We have already seen that in the last 20 years experiments on individual quantum systems have become accessible (see the experiment of Aspect *et al.* in Subsec. 1.2.3). This is clearly a confutation of the statistical interpretation. On the other hand, we shall see (in Sec. 16.3) that also the reduction of quantum mechanics to a classically (local) deterministic theory is not possible.

Two further approaches have originated from the statistical interpretation. One is the thermodynamic approach.[9] The central idea of Daneri *et al.* is to understand measurement as a type of "ergodic amplification" when a macroscopic apparatus is considered. But Jauch, Wigner, and Yanase,[10] inspired by the work of Renninger,[11] showed that, in the case of interaction-free measurements (see Subsec. 1.2.4 and also Sec. 9.6), no amplification occurs. So that, even if this model can be applied in specific situations, it does not work as a general account of measurement.

[7] On these problems see [Tarozzi 1996]. Later developments of von Neumann's position (the so-called "Wigner's friend paradox") can be found in [Wigner 1961], where the same ambiguity remains.
[8] See for example [*Blokhintsev* 1965, Blokhintsev 1976, Ballentine 1970].
[9] See [van Hove 1955, van Hove 1957, van Hove 1959, Daneri *et al.* 1962].
[10] See [Jauch *et al.* 1967].
[11] See [Renninger 1960].

The other statistical approach is the many world interpretation (MWI), originally due to Everett[12] and further developed by DeWitt.[13] The fundamental idea is that, during a measurement, the wave function is never reduced but any component is instantiated in a different world (or is relative to a different observer), so that the "multiversal" wave function remains a superposition. Everett's approach has strongly contributed to pointing out the necessity to write the state vector representing the apparatus as a function of the state vector representing the object system. Though this approach has found wide interest, it suffers from problem of basis degeneracy.[14] In fact, suppose that we have the transition

$$|\psi\rangle|\mathcal{A}(0)\rangle \mapsto \sum_j c_j|o_j\rangle|a_j\rangle, \tag{9.27}$$

where one could say, following the MWI, that we have "measured" the observable $\hat{O}_\mathcal{S}$ whose spectral decomposition is $\hat{O}_\mathcal{S} = \sum_j o_j|o_j\rangle\langle o_j|$. However, we can also choose another basis (see Subsec. 2.1.2) and write

$$\begin{aligned}\sum_j c_j|o_j\rangle|a_j\rangle &= \sum_k |a'_k\rangle \sum_j c_j \left\langle a'_k \mid a_j\right\rangle \left|o_j\right\rangle \\ &= \sum_k |a'_k\rangle|o'_k\rangle,\end{aligned} \tag{9.28}$$

where we have

$$|a_j\rangle = \sum_k \left\langle a'_k \mid a_j\right\rangle |a'_k\rangle, \tag{9.29}$$

while

$$|o'_k\rangle = \sum_j c_j \left\langle a'_k \mid a_j\right\rangle \left|o_j\right\rangle \tag{9.30}$$

are the *relative* states of the system with respect to the apparatus' states. The conclusion we may draw from Eq. (9.28) is that $\mathcal{A}$ contains not only information about the observable $\hat{O}_\mathcal{S}$ but also about other observables $\hat{O}'_\mathcal{S} = \sum_k o'_k|o'_k\rangle\langle o'_k|$, even if in general $\left[\hat{O}_\mathcal{S}, \hat{O}'_\mathcal{S}\right] \neq 0$ (see Cor. 2.1: p. 67). Hence, in order to respect the uncertainty relations, we are forced to admit that, when measured, the system is in a state whose components are diagonalized only *relatively to one observable*, and we are forced to understand with measurement the process by which we obtain a determined result. On the other hand, the MWI allows the possibility that the same system can be measured relatively to two non-commuting observables. In other words, we have a problem shift: the MWI tried to answer von Neumann's problem by supposing that *any value* is somehow instantiated (in a different world or for a different mind), but in so doing it proposed a situation in which also *any observable* is in principle measured. This leads eventually to an ambiguity in the very concept of measurement by mixing the concepts of measurement and premeasurement.

[12] See [Everett 1957].
[13] See [DeWitt 1970].
[14] As shown in [Zurek 1981, 1516].

A further development of Everett's (and in part of Wigner's) approach is provided by Lockwood.[15] The main idea is that any component of a superposition state of the object system is correlated not only with a possible apparatus' state but also with a component of the observer's mind, in reality a multimind (called Mind). Therefore, when we observe a particular result, our mind is only perspectively connected with a special "Everett branch," but, in turn, our perspective is only a component of the Mind. In other words, in addition to a temporal dimension, the Mind has a "coherent dimension" where all its possible states are displayed. In the same way that we cannot have an access to our future states, our mind cannot have access to the other Mind's components that are correlated with different components of the apparatus and the object system.

An important problem of all the "multibranch" interpretations is that it is very difficult to explain the fact that in general different components of a superposition have different probabilities to be observed, given that all components are realized (in different worlds or in different components of the Mind): as a consequence, all states should be equiprobable. One might object that the probability distributions are true only in a given world (or for a given mind), and so are different among different worlds or Mind's components. However, in general these probability distributions are assumed to be ad hoc.

Another very interesting approach is called *decoherence* and consists of the loss of coherence between the components of a superposition in terms of the action of an external environment on the coupling between system and apparatus. Due to the importance of this approach, we shall discuss it separately (in Sec. 9.4).

A position somehow related to the MWI and to decoherence is that of the consistent and decohering histories. Taking inspiration from Griffiths,[16] Gell-Man and Hartle developed a model of decohering histories.[17] A *history* is a particular sequence $\{\hat{P}\} = \{\hat{P}_1(t_1), \hat{P}_2(t_2), \ldots \hat{P}_n(t_n)\}$ of events (represented by projectors), which are in general alternative to at least another sequence $\{\hat{P}'\} = \{\hat{P}'_1(t_1), \hat{P}'_2(t_2), \ldots\}$.

A complete *fine-grained* history is specified by giving the values of a complete set of operators at all times. One history is a *coarse-graining* of another one if the set $\{\hat{P}\}$ of the first consists of elements that are sums of the elements of the set $\{\hat{P}'\}$. The completely coarse grained history is the unitary operator. The reciprocal relationships of coarse and fine graining histories constitute only a partial ordering of sets of alternative histories. For more details see Box 9.1.

Box 9.1 **Decohering histories**

Let us now discuss in a little bit more detail the decohering histories approach. The formalism is based on an application of Feynman's path integrals – we shall present this formalism in Sec. 10.8. We introduce this concept here only to the minimal extent necessary for understanding the decohering history approach. We specify the amplitude for a completely

[15] See [Lockwood 1996a].

[16] See [Griffiths 1984]. See also [Omnès 1988, Omnès 1989, Omnès 1990].

[17] See [Gell-Mann/Hartle 1990, Gell-Mann/Hartle 1993].

fine-grained history in a particular basis of generalized one-dimensional coordinates $\hat{x}_j(t)$, say all fundamental field variables at all points in space. This amplitude is proportional to (see Eq. (10.181))

$$\vartheta[\hat{x}_j(t)] \propto e^{\frac{\imath}{\hbar}S[\hat{x}_j(t)]}, \tag{9.31}$$

where S is the action.

We suppose that at least some pairs of histories, which lead to a common final event, do not present interference terms, i.e. they are not in superposition (see Fig. 9.3). We express this formally by introducing a decoherence functional $\mathcal{D}$. $\mathcal{D}$ is a complex functional on any pair of histories in a given set of alternative histories (for example the set of all histories with a common final state) and, for two histories, can have the form

$$\mathcal{D}[\hat{x}'_j(t), \hat{x}_j(t)] = \delta\left(\hat{x}'_j(t_f) - \hat{x}_j(t_f)\right) \exp\left[\frac{\imath(S[\hat{x}'_j(t)] - S[\hat{x}_j(t)])}{\hbar}\right] \hat{\rho}^U_i(\hat{x}'_j(t_j), \hat{x}_j(t_j)), \tag{9.32}$$

where $\hat{\rho}^U_i$ is the initial density matrix of the universe in the $\hat{x}_j$ representation. The decoherence functional for coarse-grained histories is obtained from Eq. (9.32) according to the superposition principle by summing over all terms which are not specified by the coarse graining (see Fig. 9.3)

$$\mathcal{D}\left([\Delta^{(2)}], [\Delta^{(1)}]\right) = \int_{[\Delta^2]} \delta\hat{x}' \int_{[\Delta^1]} \delta\hat{x}\delta\left(\hat{x}'_j(t_f) - \hat{x}_j(t_f)\right)$$
$$\times \exp\left[\frac{\imath(S[\hat{x}'_j(t)] - S[\hat{x}_j(t)])}{\hbar}\right] \hat{\rho}^U_i(\hat{x}'_j(t_j), \hat{x}_j(t_j)), \tag{9.33}$$

where the integrations are performed over all possible trajectories that are allowed in the interval range $\left(\Delta^{(1)} \text{ or } \Delta^{(2)}\right)$ of a given history. This approach is very interesting, but its weakness seems to be the fact that one applies the concept of decoherence to the whole universe, whereas, as we shall see, it is physically reasonable to treat it in terms of the action of a large environment on a small object system.

9.2.3 *Ad-hoc* approaches

Other approaches, for example that followed by Ghirardi, Rimini, and Weber,[18] have tried to correct the formalism of quantum mechanics by introducing additional parameters modifying the Schrödinger equation in order to account for the localization (or determination) when macroscopic bodies are involved. In 1982 Barchielli, Lanz, and Prosperi[19] had already proposed a dynamical equation of the form

$$\frac{d}{dt}\hat{\rho}(t) = -\frac{\imath}{\hbar}\left([\hat{H}, \hat{\rho}(t)] - \frac{\eta}{4}[\hat{x}, [\hat{x}, \hat{\rho}(t)]]\right), \tag{9.34}$$

[18] See [Ghirardi *et al.* 1986].

[19] See [Barchielli *et al.* 1982] .

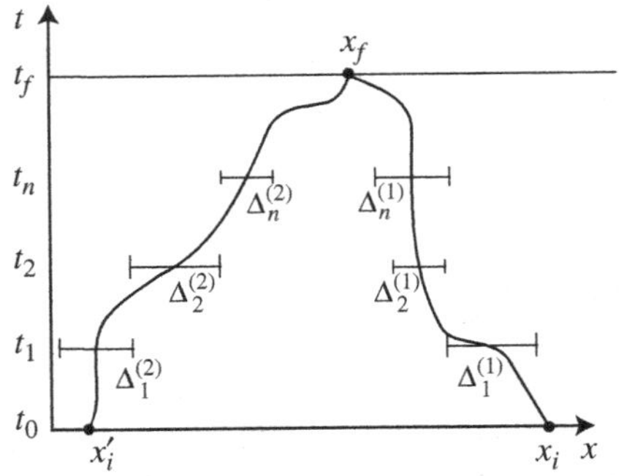

Figure 9.3 **Decohering histories. Here two histories, $\Delta^{(1)}$ and $\Delta^{(2)}$ are shown, leading to a common final event $\hat{P}_{x_f}$. We have considered the set of ranges $\left\{\Delta_k^{(1)}\right\}$ of values of $\hat{x}_j$ at times t_k, $k = 1, \ldots, n$. A set of alternatives at any time t_k consists of ranges $\Delta_k^{(1)}, \Delta_k^{(2)}, \Delta_k^{(3)}, \ldots, \Delta_k^{\mu}$, which exhaust the possible values of x_j as μ ranges of all integers. An *individual* history is specified by a particular $\Delta_k^{(\mu)}$ at particular times t_k, $k = 1, \ldots, n$. We write $[\Delta^{(\mu)}] = (\Delta_1^{(\mu)}, \ldots, \Delta_n^{(\mu)})$ to indicate a particular history.**

which can be considered a type of master equation – we shall discuss of master equations in Sec. 14.2: we limit ourselves here to the main points that can explain the nature of this approach. Equation (9.34) is similar to the Schrödinger or von Neumann equation (5.28) plus an additional term on the rhs. This term could account for the "localization" of a density operator, whose evolution would be otherwise unitary: it is composed of a localization parameter η times a double commutator. Ghirardi, Rimini, and Weber considered, in particular, an N-particle compound system and rewrote Eq. (9.34) as

$$\frac{d}{dt}\hat{\rho} = -\frac{\imath}{\hbar}[\hat{H}, \hat{\rho}] - \sum_{j=1}^{N} \eta_j \left(\hat{\rho} - \mathcal{T}_j[\hat{\rho}_j]\right), \tag{9.35}$$

where $\hat{\rho}, \hat{\rho}_j$ are the density operators for the total system and for the j-th particle, respectively, η_j is the frequency of the process undergone by the j-th element, and the transformation $\mathcal{T}_j$ acts as follows:

$$\mathcal{T}_j[\hat{\rho}_j] = \sqrt{\frac{\zeta}{\pi}} \int_{-\infty}^{+\infty} dx\, e^{-\frac{\zeta}{2}(\hat{x}_j - x)^2} \hat{\rho} e^{-\frac{\zeta}{2}(\hat{x}_j - x)^2}, \tag{9.36}$$

where ζ is a localization constant. The authors suggested attributing the following values to the introduced constants:

$$\eta \simeq 10^{-16}\,\text{sec}^{-1}, \tag{9.37a}$$

$$\frac{1}{\zeta} \simeq 10^{-5}\,\text{cm}. \tag{9.37b}$$

Condition (9.37a) means that the localization happens spontaneously every 10^8–10^9 years for a single particle. A detailed computation shows that the localization time-rate is shorter the larger is the system. On the other hand, the value of $1/\zeta$ is much larger than the length of a de Broglie wave typical for a particle. This means that both constants are very important only for many-particle systems or for systems with wavelength larger than 10^{-5} cm. Therefore, they leave almost unaffected the microscopic world of quantum mechanics, but allow a description of how macro-objects may be localized and "reduced."

In conclusion, we wish to point out that the weakest point of these approaches is that they have an ad hoc nature since they introduce quantities that are not physically observed.

9.2.4 Conclusion

This short review does not cover all proposals that have been made in order to solve the measurement problem before 1980.[20] The important point is that up to the beginning of the 1980s there was still no clear solution. This can account for the diffusion of Feynman's dictum that nobody understands quantum theory but everybody uses it. As a picture of the situation, it seems appropriate. As a methodology, it is certainly not adequate. It is clear that, facing problems of knowledge, one should in the first instance find a hypothesis that can work. However, once such a hypothesis has been found, the scientific enterprise consists in finding out why or at least how it works. It is clear that we shall find, in the best case, a hypothesis that will eventually be accepted because it works. But, in posing that question, we have enlarged the field of our enquiry, and this represents a true progress of knowledge. As a matter of fact, in the 1980s something new began. However, before we discuss what is believed by a large part of the scientific community to be the solution of the measurement problem – the decoherence approach – it is necessary first to understand a major issue of measurement and, in general, of the relationship between microscopic and macroscopic physics. This problem was originally formulated by Schrödinger and is the subject of the next section.

9.3 Schrödinger cats

9.3.1 Historical introduction

Schrödinger was among the first physicists to understand the conceptual difficulties posed by quantum mechanics. In 1935, in a series of papers that later became milestones in the development of this theory [Schrödinger 1935], he discussed, in particular, some far-reaching consequences of the superposition principle and of entanglement. He considered some "absurd" macroscopical situations that could arise if quantum mechanics was to be taken as a universally applicable theory. This is expressed paradigmatically by the so-called Schrödinger's cat paradox. The original *Gedankenexperiment* is the following. A cat is in a box together with a very small quantity of radioactive material (see Fig. 9.4) – the decay probability is, say, one atom per hour. The decayed atom would activate a Geiger counter

[20] The reader interested in other approaches and also in a more detailed investigation of the problem is referred to [*Auletta* 2000], and references therein.

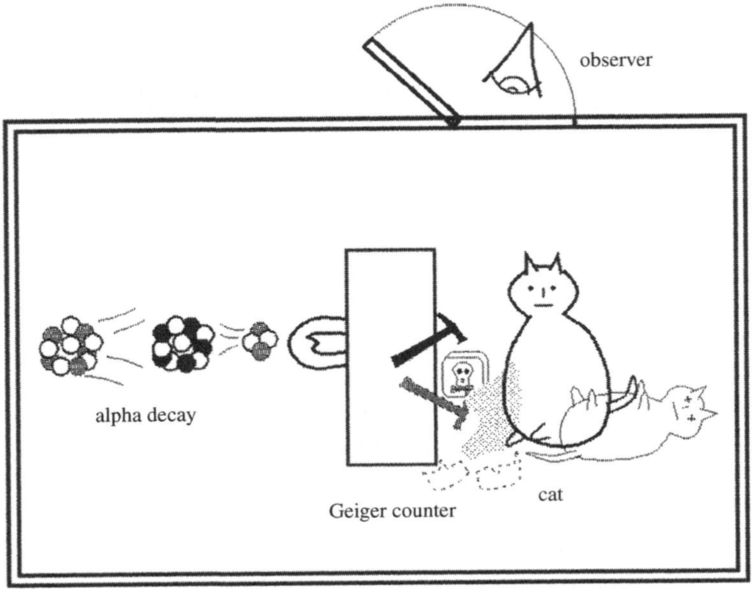

Figure 9.4 **Pictorial representation of the Schrödinger's cat thought-experiment. Adapted from www.lassp.cornell.edu/ardlouis/dissipative/cat.gif.**

which is connected through a relay to a hammer that may break an ampulla, thus releasing some poison. Now, according to the probabilistic character of the radioactive decay (see Sec. 1.4), after some time the wave function describing the system should represent a superposition of *living cat* and *dead cat* (see also Eq. (9.13)). But this seems impossible, since nobody has ever observed such a situation at the macroscopic level. On the other hand, we may assume that a measurement, in our case the opening of the box to see the cat, will univocally determine the state of the cat (*either* alive *or* dead), thus eliminating the superposition – according to the projection postulate. If so, we ascribe an enormous and perhaps unjustified power to the observation, that is, the power to realize *ex novo* a very special physical situation: the so-called wave function reduction or collapse.

From a formal viewpoint, let us write as $| d \rangle_a$ and $| u \rangle_a$ the eigenstates corresponding to the decayed and undecayed atoms, respectively. Similarly, $| \smile \rangle_c$ and $| \frown \rangle_c$ shall represent a living and a dead cat, respectively. Then, after some time, say half an hour, the state of the system "atom + cat" shall be described by the entangled superposition state

$$| \Psi \rangle = \frac{1}{\sqrt{2}} (| u \rangle_a | \smile \rangle_c + | d \rangle_a | \frown \rangle_c). \tag{9.38}$$

It should be noted that the state (9.38) is quite paradoxical, and, in particular, two important facts have to be noted: first, due to the correlation between the system and cat states, the superposition character of the atomic state has been mapped onto the compound system. Second, it is an entangled state between a microscopic system (the atom) a macroscopic one (the cat). As a consequence, a measurement of the state of either system would immediately transfer the information onto the other (see also Ch. 16 and Sec. 9.1).

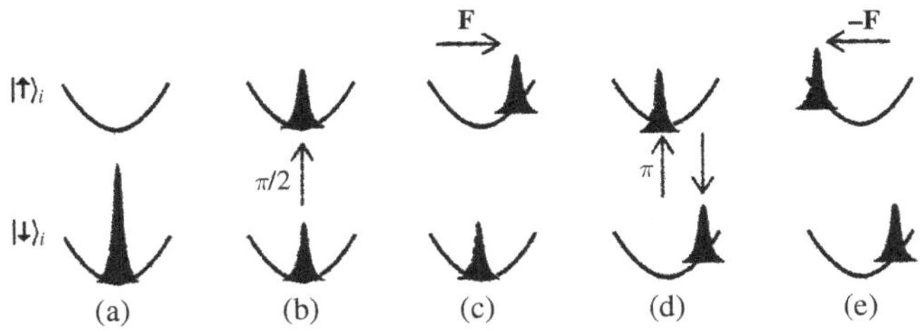

Figure 9.5 Experimental realization of a Schrödinger cat with a trapped ion. The figure shows the evolution of the position–space atomic wave packet entangled with the internal (ground and excited) states $|\downarrow\rangle$ and $|\uparrow\rangle$ during the creation of a Schrödinger cat state. The wave packets are snapshots in time, taken when the atom is in the harmonic trap potential (represented by the parabolas). The area of the wave packets is proportional to the probability of finding the atom in the given internal state. (a) The initial wave packet corresponds to the quantum ground state of motion after laser-cooling. (b) The wave packet is split after a $\pi/2$ laser pulse (see also Subsec. 14.5.2). (c) The $|\uparrow\rangle$ wave packet is excited to a coherent state (see p. 161 and also Subsec. 13.4.2) by the force **F** provided by a displacement laser beam. Note that **F** acts only on $|\uparrow\rangle$, hence entangling the internal and the motional degrees of freedom. (d) The $|\uparrow\rangle$ and $|\downarrow\rangle$ wave packets are interchanged following a π-pulse. (e) The $|\uparrow\rangle$ wave packet is excited to a coherent state by the displacement-beam force −**F**, which is out of phase with respect to the force in (c). The state in (e) corresponds to a Schrödinger cat state. Adapted from [Monroe *et al.* 1996, 1133].

9.3.2 Experimental realizations

It is evident that the existence (or non-existence) of Schrödinger cat states changes dramatically the reciprocal relationship between the macroscopic and the microscopic world. In particular, the difficulty lies in the answer to the following question: where is the border between the two worlds? To answer this question, several investigations and experiments have been carried out, exploring the intermediate region between the two limits, that of the so-called *mesoscopic* phenomena. For example, a Schrödinger cat-like state of the electromagnetic field involving an average number of 3.3 photons has been realized in Paris by the group of S. Haroche.[21] A similar state, but involving the motion of the center of mass of a beryllium ion, had been previously observed in the experiment performed by D. Wineland and co-workers[22] (see Fig. 9.5). In this case, states of the form

$$|\Psi\rangle = \frac{1}{\sqrt{2}}(|\uparrow\rangle\,|\alpha\rangle + |\downarrow\rangle\,|-\alpha\rangle) \tag{9.39}$$

are produced, where $|\uparrow\rangle$ and $|\downarrow\rangle$ are the internal degrees of freedom, and $|\alpha\rangle$ and $|-\alpha\rangle$ (when $|\alpha|$ is large) are macroscopically distinguishable coherent states of the center-of-mass motion of the ion (see also Box 14.1). In this situation, the internal degree of freedom plays the role of the microscopic system, whereas the center-of-mass motion plays the role of the macroscopic one. It is interesting to note that these two different degrees of freedom, though pertaining to the same physical system (the ion), are written in a form

21 See [Brune *et al.* 1996]. We shall return on this experiment in Subsec. 14.5.2.

22 See [Monroe *et al.* 1996]. By further refining this technique, Wineland's team has also produced a six-atom Schrödinger cat state [Leibfried *et al.* 2005].

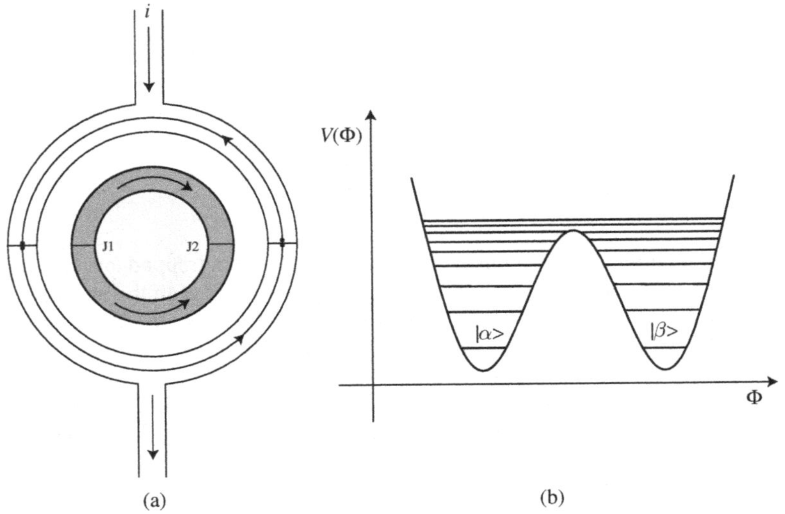

Figure 9.6 (a) The inner (grey) ring is a SQUID with two Josephson junctions J1 and J2. The flux can be detected using the outer ring, which is another SQUID working as a magnetometer when injected with a bias current *i*. (b) Schematic representation of the double well potential of the SQUID as a function of the flux in the symmetric case.

that is completely analogous to an entanglement of two different subsystems – compare Eq. (9.39) with Eq. (9.38). As a matter of fact, they pertain to two different Hilbert spaces, and this justifies the above formalism.

Another realization of conceptually similar experiments involves Superconducting QUantum Interference Devices, or SQUIDs: these are extremely sensitive magnetic flux-to-voltage transducers.[23] The SQUID is among the most sensitive detectors that can be used in biomagnetic applications (magnetoencephalography) and magnetocardiography. Other practical applications are geomagnetism, nondestructive testing, radio frequency amplification and the measurement of fundamental constants. It is not our aim here to enter into the details of the theory of superconductivity: we shall rather limit ourselves to show how this device can be used to produce Schrödinger cats.

The basic element of a SQUID is a superconducting ring[24] containing one (radio-frequency SQUID) or more (direct-current SQUID) Josephson junctions[25] (see Fig. 9.6(a)). A Josephson junction consists essentially of two superconductors weakly coupled through a thin non-superconducting material. The SQUID has unique electrical and magnetic properties: when a small magnetic field is applied to the superconducting loop, a persistent current is induced. Such a current may flow clockwise or counterclockwise, in order to either reduce or enhance the applied flux, thus approaching an integer number of superconducting flux quanta $\Phi_0 = h/2e = 2.067833636 \times 10^{-15}$ Wb. The dynamics of

[23] On this device see [*Tinkham* 1996].
[24] See [Leggett 1989].
[25] See [Josephson 1962].

the SQUID is driven by the difference between the flux Φ that goes through the loop and the external flux Φ_{ext} applied to the loop, and can be described by the one-dimensional potential energy

$$V(\Phi) = C\left(\frac{\Phi - \Phi_{ext}}{\Phi_0}\right)^2 + C' \cos\left(2\pi\frac{\Phi}{\Phi_0}\right), \tag{9.40}$$

where C and C' are constants which depend on the physical parameters of the device (inductance of the ring, capacitance, and critical current of the junction). When $\Phi_{ext} = \Phi_0/2$, the potential is symmetric and is schematically depicted in Fig. 9.6(b) together with the first few energy levels. The lower levels are confined in either the left or the right well and represent the "classical" states, i.e. states for which the current flows *either* clockwise *or* counterclockwise. Quantum-mechanically, however, the system may tunnel from the left to the right well or *vice versa* through the central barrier (see Sec. 4.3). Moreover, states which are superpositions of left and right states, that is, states of the type

$$|\psi\rangle = \frac{1}{\sqrt{2}}(|\alpha\rangle \pm |\beta\rangle), \tag{9.41}$$

can be viewed as Schrödinger cat states. We stress here that Eq. (9.41) is slightly different from Eq. (9.38) in that it represents a superposition relative to a single degree of freedom and not an entangled state. As we have said, truly speaking, Schrödinger cat states are given by entangled states in which one of the degrees of freedom is macroscopic and the two involved components are macroscopically distinguishable, as in Eqs. (9.38) and (9.39). However, even states of the form (9.41) can be conceptually considered as belonging to the Schrödinger cat family, provideed that the components be macroscopically distinguishable.

States of the kind of Eq. (9.41) have been created and detected by two experimental groups using slightly different setups.[26] In these experiments, the SQUID dynamics is driven by the motion of approximately 10^9 electron pairs. Moreover, the states $|\alpha\rangle$ and $|\beta\rangle$ in the superposition (9.41) differ by an amount of magnetic flux which corresponds to a current of a few μA that can be then considered a macroscopic quantity. Both these points make the SQUID one of the ideal candidates for probing macroscopic quantum effects at the border of classical and quantum physics.[27]

A very interesting representation is provided in Fig. 9.7, where the Wigner function – a quasi-probability distribution corresponding to a density matrix of a system – is shown (see also Subsec. 13.5.4).

9.3.3 Schrödinger cats and the macroscopic world

We do not observe Schrödinger cats in our everyday experience due to the fact that the effects of Schrödinger cat states are difficult to detect, but it is also true that these effects

[26] See [Friedman *et al.* 2000] [van der Wal *et al.* 2000].

[27] There also are further techniques for producing Schrödinger cats, for instance by taking advantage of photon-number states [Ourjoumtsev *et al.* 2007]. The advantage of this technique is that is that it allows for the generation of arbitrarily large squeezed Schrödinger cats [see also Subsec. 13.4.3].

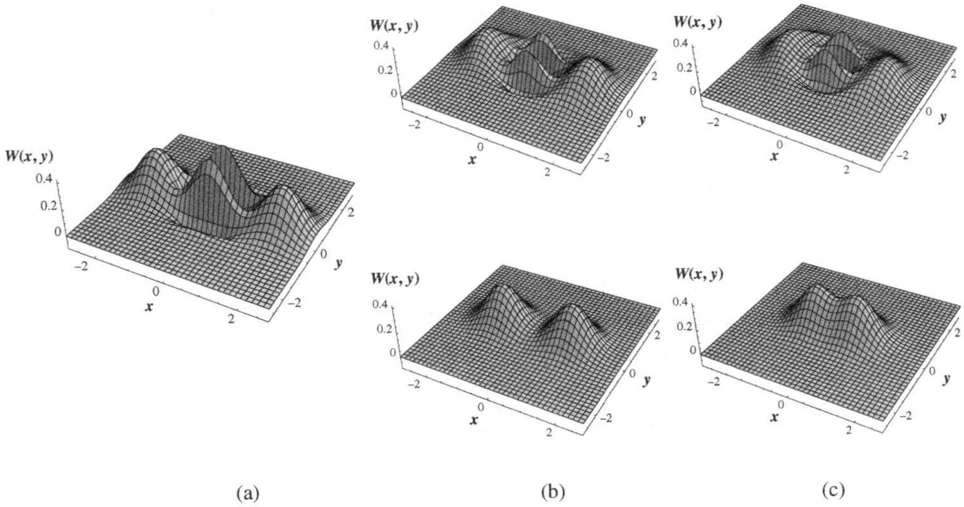

Figure 9.7 (a) Wigner function of an initial Schrödinger cat state of the form $|\psi\rangle = \mathcal{N}_-(|\alpha\rangle - |-\alpha\rangle)$, $\mathcal{N}_-$ being a normalization constant and $|\alpha|^2 = 3.3$. The two central bumps represent the interference terms of the corresponding density matrix. The state under consideration is fragile and sensitive to dissipation, and needs care to be preserved (here represented by an appropriate feedback action). (b) Wigner function of (top) the same state after 13 feedback cycles and (bottom) after one relaxation time in the absence of feedback. In the last case the off-diagonal terms tend to disappear. (c) Wigner function of (top) the same state after 25 feedback cycles and (bottom) after two relaxation times in the absence of feedback. Summing up, in absence of feedback, the Wigner function becomes quickly positive definite and takes a classical aspect, corresponding, in (b), bottom, to a mixed density operator, and, in (c), bottom, assuming almost the form describing a single measurement outcome, while, in the presence of feedback ((b) and (c), top), the quantum aspects of the state remain well visible. Adapted from [Fortunato *et al.* 1999b].

have been detected under special circumstances, as we have seen. This forces us to consider such states, rather than as an absurd consequence of quantum-mechanical formalism, as observable features of mesoscopic systems at present level of technological possibilities. This opens the way to the observation of similar effects in the macroscopic domain. In this context, one should consider that quantum mechanics is already used successfully for the interpretation of a number of macroscopic coherent phenomena (superconductivity, Bose–Einstein condensation (see Subsec. 7.5.1), superfluidity, etc.). More recently, entanglement has been shown to exist in the insulating magnetic salt $LiHo_xY_{1-x}F_4$.[28]

Therefore, one of the lessons we learn from these experiments is that we are able to progressively move the border between micro and macro. This is in contrast to Bohr's idea that there is a sharp border line between the two worlds.[29] This situation is very similar to the one we encountered in Subsec. 2.3.4 in the context of wave-particle duality: in both cases there is a smooth transition between two extreme behaviors.

[28] See [Ghosh *et al.* 2003] [Vedral 2003].

[29] An evidence of this is the role ascribed from Bohr to the measuring apparatus (see Subsec. 9.2.1).

9.4 Decoherence

In the 1970s and 1980s there was growing evidence that, when a quantum system is coupled to a large reservoir, *dephasing* arises, i.e. the system loses coherence: the off-diagonal terms of the density matrix (9.2) tend quickly to zero. The dephasing models gave the hint in considering the environment as a possible source of solution of the measurement problem. In fact, *environment* can be considered as the largest reservoir at our disposal,[30] and can be defined as a large system which is permanently coupled to any microsystem and is not controllable by the observer. In other words, we are assuming that any quantum system is truly an *open system* and that part of the initial information contained in the system is lost in the environment, so that we are not able to extract it. For a more general and detailed account of the interaction between a system and its environment, see Sec. 14.2.

9.4.1 Zurek's model

Instead of considering the act of measurement as a mere interaction between the apparatus $\mathcal{A}$ and the object system $\mathcal{S}$, Zurek explicitly introduced[31] the environment $\mathcal{E}$ as a third player that is always present when measuring, assuming therefore that quantum systems are essentially open to the environment. The concepts of environment and open quantum systems had been known for many years,[32] but this was the first time these had been applied to the measurement process. This is to a certain extent the opposite of what happens in classical mechanics where it is assumed that physical systems can be considered, at least in principle, as isolated. Zurek's fundamental idea is that, when considering the measurement process, we first write the state vector for $\mathcal{S}+\mathcal{A}+\mathcal{E}$, $|\Psi_{\mathcal{SAE}}\rangle$ and let $\mathcal{S}$, $\mathcal{E}$, and $\mathcal{E}$ interact. Then, we perform a partial trace and obtain the reduced density matrix of $\mathcal{S}+\mathcal{A}$ only. In this way, while the total system $\mathcal{S}+\mathcal{A}+\mathcal{E}$ still remains entangled and is subjected to a unitary evolution according to the Schrödinger equation, the reduced system $\mathcal{S}+\mathcal{A}$ is a mixture relatively to the "point of view" of the apparatus (see Secs. 5.5 and 9.1).

The measurement process can be schematically divided into two distinct steps followed by the partial trace which completes the process. The initial state (at time $t=0$) is for the sake of simplicity a factorized state of $\mathcal{S}$, $\mathcal{E}$, and $\mathcal{A}$. Then, at time $t=t_1$, due to the interaction between $\mathcal{S}$ and $\mathcal{A}$, these become entangled (we have a premeasurement). In the time interval $t_1 \le t \le t_2$, also the environment entangles with $\mathcal{S}$ and $\mathcal{A}$. At time t_2 the interaction is switched off. Formally,

30 Zeh was the first to introduce this important idea [Zeh 1970].

31 See [Zurek 1981, Zurek 1982].

32 Due especially to Louisell's work. For a summary see [*Louisell* 1973].

$$|\Psi_{\mathcal{SAE}}(0)\rangle = |\psi\rangle|a_0\rangle|\mathcal{E}(0)\rangle \tag{9.42a}$$

$$\mapsto |\Psi_{\mathcal{SAE}}(t = t_1)\rangle = \left[\sum_j c_j(|o_j\rangle|a_j\rangle)\right]|\mathcal{E}(t_1)\rangle \tag{9.42b}$$

$$\mapsto |\Psi_{\mathcal{SAE}}(t \geq t_2)\rangle = \sum_j c_j|o_j\rangle|a_j\rangle|e_j\rangle, \tag{9.42c}$$

where $\{|e_j\rangle\}$ is some basis for the environment. Transformation (9.42b) has provided the connection between $\mathcal{A}$ and $\mathcal{S}$ and is described by the unitary operator (9.9). This is a premeasurement because it does not yet represent the desired result. Instead, transformation (9.42c) shows the action of the environment, and can be described by another unitary operator which couples $\mathcal{E}$ to $\mathcal{S}+\mathcal{A}$, i.e.

$$\hat{U}_t^{\mathcal{SA},\mathcal{E}} = e^{-\frac{i}{\hbar}t\hat{H}_{\mathcal{SA},\mathcal{E}}} \tag{9.43}$$

for $t_1 < t < t_2$.

Now we can trace out the environment by writing the corresponding reduced density matrix (see Subsec. 5.5.2):

$$\begin{aligned}\hat{\varrho}_{\mathcal{SA}} &= \mathrm{Tr}_{\mathcal{E}}\left(\hat{\rho}_{\mathcal{SAE}}\right) \\ &= \mathrm{Tr}_{\mathcal{E}}\left[|\Psi_{\mathcal{SAE}}(\tau)\rangle\langle\Psi_{\mathcal{SAE}}(\tau)|\right] \\ &= \sum_j |c_j|^2|a_j\rangle\langle a_j| \otimes |o_j\rangle\langle o_j|.\end{aligned} \tag{9.44}$$

Equation (9.44) has been obtained under the simplified hypothesis (9.42c). The physical meaning of this tracing out is then the following: the information contained in the off-diagonal elements of the coupled system $\mathcal{S}+\mathcal{A}$ is not destroyed but *downloaded in the environment* (see also Sec. 17.3). However, from the point of view of the apparatus, it is inaccessible and is for this reason completely lost (it would be accessible only under the hypothesis that we can exactly reconstruct the state of the whole system $\mathcal{S}+\mathcal{A}+\mathcal{E}$). Since the quantum coherences are lost (see the end of Sec. 5.1), this interpretation of measurement is known as *decoherence*. Strictly speaking, the correlation between $\mathcal{S}$, $\mathcal{E}$, and $\mathcal{A}$ may be not perfect. As a matter of fact, the environment is a huge complex of systems that can be more or less interrelated and in which random fluctuations are also present. Then, a perfect correlation between $\mathcal{S}$ and $\mathcal{E}$ can be taken as an ideal limit. As a result, one would obtain a reduced density matrix in which the off-diagonal terms are suppressed but not completely eliminated. In other words, during the measurement, they tend very quickly to zero. Hence, we never obtain a perfect diagonalization of the density matrix describing the object system relatively to the measured observable[33] (see also Sec. 9.10).

[33] Zurek, in his original articles, invokes environment-induced superselection rules, which imply an abrupt change, but this requirement seems to be not necessary.

This phenomenon helps us to better understand the mechanism of the transition from the quantum to the classical (see Subsec. 9.3.3 and also Sec. 10.5). The most important point to be focussed on is the physical action of the environment on $\mathcal{S}$ and $\mathcal{A}$. The environment makes all information about the premeasured system unavailable with only one exception: the pointer of the apparatus will contain the information about the observables which commute with the interaction Hamiltonian $\hat{H}_{\mathcal{SA}}$. In such a case these particular observables will not be disturbed.

The fact that only observables commuting with the interaction Hamiltonian $\hat{H}_{\mathcal{SA}}$ are actually measured, is an important consequence of the uniqueness of triorthogonal decomposition. In fact, the relevant point here is not the obvious circumstance that the off-diagonal terms of the density matrix vanish in some basis,[34] but that the basis in which an approximate diagonalization occurs does not depend on the initial conditions (see also Subsec. 3.5.3). As we have seen (in Subsec. 9.2.2), a total system composed of two subsystems can be decomposed into different basis corresponding to different observables (basis degeneracy). This is, however, not true for triorthogonal decomposition.[35] In fact, for three or more subsystems, there is only one possible basis that turns out to be the basis with respect to which the measured observable $\hat{O}_{\mathcal{S}}$ is diagonal. In order to prove such statement, let us assume the following lemma:[36]

Lemma 9.1 (Elby–Bub) *Let $\{|a_j\rangle\}$ and $\{|b_j\rangle\}$ be linearly independent sets of vectors, respectively in the Hilbert spaces $\mathcal{H}_1, \mathcal{H}_2$ for two generic systems $\mathcal{S}_1, \mathcal{S}_2$. Let $\{|b'_j\rangle\}$ be a linearly independent set of vectors that differs non-trivially[37] from $\{|b_j\rangle\}$. If $|\Psi\rangle = \sum_j c_j|a_j\rangle \otimes |b_j\rangle$, then $|\Psi\rangle = \sum_j c'_j|a'_j\rangle \otimes |b'_j\rangle$ only if at least one of the $\{|a'_j\rangle\}$ vectors is a linear combination of (at least two) $\{|a_j\rangle\}$ vectors.*

We use now Lemma 9.1 to prove *per contradictionem* the uniqueness of triorthogonal decomposition.

Proof

Take a vector $|\Psi\rangle = \sum_j c_j|a_j\rangle \otimes |b_j\rangle \otimes |e_j\rangle$, where $\{|a_j\rangle\}, \{|b_j\rangle\}, \{|e_j\rangle\}$ are orthogonal sets of vectors respectively in the Hilbert spaces $\mathcal{H}_1, \mathcal{H}_2, \mathcal{H}_3$ for three generic systems $\mathcal{S}_1, \mathcal{S}_2, \mathcal{S}_3$, respectively. Then, we claim that, even if some of the $|c_j|$'s are equal, no alternative orthogonal sets $\{|a'_j\rangle\}, \{|b'_j\rangle\}, \{|e'_j\rangle\}$ exist such that $|\Psi\rangle = \sum_j c'_j|a'_j\rangle \otimes |b'_j\rangle \otimes |e'_j\rangle$, unless each alternative set of vectors differs only trivially from the set it replaces.

[34] Since the density matrix is an Hermitian operator, there is always a basis in which its off-diagonal terms are exactly zero (see Prob. 9.3).

[35] As proved in [Elby/Bub 1994].

[36] See the original article [Elby/Bub 1994] for the proof of the lemma.

[37] "Trivially different" qualifies states that differ only for a multiplicative phase factor (see also the end of Subsec. 2.1.3).

Assume, without loss of generality, that $\{|e'_j\rangle\}$ differs not trivially from $\{|e'_j\rangle\}$, and let us write $|\Psi\rangle = \sum_j c_j |w_j\rangle \otimes |e_j\rangle$, where $|w_j\rangle = |a_j\rangle \otimes |b_j\rangle$. Now, suppose that $|\Psi\rangle = \sum_j c'_j |w'_j\rangle \otimes |e'_j\rangle$, where $|w'_j\rangle = |a'_j\rangle \otimes |b'_j\rangle$. Then, it is clear that we cannot rewrite the factorized state $|a'_j\rangle \otimes |b'_j\rangle$ as an entangled state (see Subsec. 5.5.1).

But, according to lemma 9.1, since $|\Psi\rangle = \sum_j c_j |w_j\rangle \otimes |e_j\rangle$ and since $\{|e_j\rangle\}$ differs not trivially from $\{|e'_j\rangle\}$, then we have $|\Psi\rangle = \sum_j c'_j |w'_j\rangle \otimes |e'_j\rangle$ only if $|w'_k\rangle = \sum_j c_j^{(k)} |w_j\rangle$, where at least two of the $c_j^{(k)}$'s are non-zero. But since $|w_j\rangle = |a_j\rangle \otimes |b_j\rangle$, it follows that $|w'_k\rangle$ is an entangled state, i.e. $|w'_k\rangle = \sum_j c_j^{(k)} |a_j\rangle \otimes |b_j\rangle$, which is in contradiction with the fact that $|w'_k\rangle$ was not entangled.

Q.E.D

The uniqueness of triorthogonal decomposition is a very important point. In fact, while the tracing out is only relative to the system and the apparatus, the uniqueness of the triorthogonal decomposition introduces an objective character in the measurement theory that can account for irreversibility. However, the transition from the quantum to the classical world cannot have an unphysical sharp border line. This requirement is satisfied by the fact that, when measuring, we have no complete elimination of the off-diagonal terms of a density matrix.[38] Then, as already shown in the Schrödinger cat experiments, there are good reasons to suppose that there is a tiny tail of quantum "noise" also in our macroscopic world.

9.4.2 Scully, Shea, and McCullen's model

One of the earliest attempts to explain the transition from a pure state density matrix as in Eq. (9.2) to a reduced density matrix (9.3) induced by a measurement process is due to Scully, Shea, and McCullen.[39] Their model, which we are going to illustrate below, even though simple and perhaps not completely realistic, had the historical merit of giving a new direction and new emphasis toward the solution of the measurement problem, a field dominated from 1932 up to the end of the 1970s by the *standard* von Neumann's interpretation (see Subsec. 9.2.1). Scully and co-workers showed that a reduction to a mixed state can occur by making a partial trace on the degrees of freedom of the apparatus (represented in their model by an atom) and that we obtain a mixture even if we do not actively extract any information. In other words, "reduction" occurs with the sole presence of the detector and without necessarily reading the measurement outcome. This is in strong contrast to von Neumann's hypothesis that requires the presence of an observer and the occurrence of the act of observation.

Consider the arrangement shown in Fig. 9.8. We begin with a fermionic wave packet (see Box 2.6: p. 80) in the state $|\varphi\rangle$ of spin–1/2 molecules in a superposition state of

[38] A series of experiments lead by H. Rauch shows that, even when there is no interference in the configuration space, interference effects are not lost in the momentum space [Rauch 2000].

[39] See [Scully *et al.* 1978].

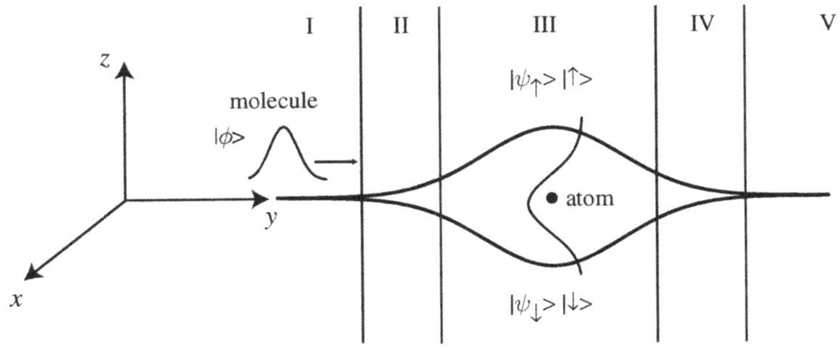

Figure 9.8 Schematic representation of the experiment proposed by Scully and co-workers. A spin-1/2 molecule in the state $|\varphi\rangle$ moves from the left to the right along the $+y$ direction. The initial spin superposition is split in the region II into components $|\psi_\uparrow\rangle\,|\uparrow\rangle$ and $|\psi_\downarrow\rangle\,|\downarrow\rangle$, and then merged again in the region IV. A (not well-localized) atom is placed in the region III so that, with the passage of the molecule, it can undergo a transition from the initial ground state to the excited state.

spin-up and spin-down along the z direction. This wave packet moves in the $+y$ direction (region I), and is then passed (region II) through a magnet with field $\mathbf{B}$ and gradient $\partial_z \mathbf{B}$ in the z direction (see Sec. 6.3). As a result, the beam is split (region III) into a spatial component with spin up ($|\psi_\uparrow\rangle$) and another one with spin down ($|\psi_\downarrow\rangle$). The two partial beams are then merged (region IV) by applying an opposite magnetic gradient. In region V the state will be the same as in absence of magnetic field, except for the inevitable spreading of the wave packet.

The initial state of the molecule (spin plus spatial degree of freedom) is given by

$$|\varphi(0)\rangle = \frac{1}{\sqrt{2}}\left(|\psi_\uparrow(0)\rangle\,|\uparrow\rangle_z + |\psi_\downarrow(0)\rangle\,|\downarrow\rangle_z\right), \tag{9.45}$$

where $|\psi_\uparrow(0)\rangle = |\psi_\downarrow(0)\rangle = |\psi(0)\rangle$ represents the initial (spatial) wave packet, and

$$|\uparrow\rangle_z = \begin{pmatrix} 1 \\ 0 \end{pmatrix}, \quad |\downarrow\rangle_z = \begin{pmatrix} 0 \\ 1 \end{pmatrix} \tag{9.46}$$

denote the eigenstates of the spin along the z direction.

We imagine two alternatives:

- In the described experiment the final density matrix $\hat{\rho}_m$ of the molecule, within the limits of a perfect overlap between $|\psi_\uparrow\rangle$ and $|\psi_\downarrow\rangle$, will be (see Prob. 9.5)

$$\hat{\rho}_m(t) = |\varphi(t)\rangle\,\langle\varphi(t)| = |\psi(t)\,\rangle\,\langle\psi(t)|\begin{bmatrix} 1 & 1 \\ 1 & 1 \end{bmatrix}, \tag{9.47}$$

  where $|\psi(t)\rangle = |\psi_\uparrow(t)\rangle = |\psi_\downarrow(t)\rangle$.
- Let us now insert a meter represented by a two-level atom placed along the path of the molecules. Such an atom may be found in the ground state $|g\rangle_a$ or in the (metastable) excited state $|e\rangle_a$. With the introduction of the meter, our Hilbert space enlarges in such a way that, leaving aside for the time being the spatial degree of freedom, the new basis states will be given by the direct product of the meter states times the spin states of the molecule

$$|e\rangle_a \otimes |\uparrow\rangle_z = \begin{pmatrix} 1 \\ 0 \\ 0 \\ 0 \end{pmatrix}, \quad |e\rangle_a \otimes |\downarrow\rangle_z = \begin{pmatrix} 0 \\ 1 \\ 0 \\ 0 \end{pmatrix},$$

$$|g\rangle_a \otimes |\uparrow\rangle_z = \begin{pmatrix} 0 \\ 0 \\ 1 \\ 0 \end{pmatrix}, \quad |g\rangle_a \otimes |\downarrow\rangle_z = \begin{pmatrix} 0 \\ 0 \\ 0 \\ 1 \end{pmatrix}. \tag{9.48}$$

Let us suppose that the atom and the molecule interact via a certain interaction Hamiltonian. Skipping the details of this interaction,[40] it suffices here to imagine that this interaction occurs only when the molecule is in contact with the atom and that entangles the atomic and the molecular states. In particular, it is assumed that the meter is affected by, but does not influence, the state of the system. Moreover, such an interaction allows for transitions $|g\rangle_a \rightsquigarrow |e\rangle_a$ and *vice versa.*

Starting with a ground-state atom, the total state (molecular spatial degree of freedom + molecular spin degree of freedom + atomic excitation) after region II but before interaction with the atom will be given by

$$\begin{aligned} |\Psi(0)\rangle\rangle &= |g\rangle_a \otimes \left(|\psi_\uparrow(0)\rangle\rangle\, |\uparrow\rangle_z + |\psi_\downarrow(0)\rangle\rangle\, |\downarrow\rangle_z\right) \\ &= \frac{1}{\sqrt{2}} \begin{pmatrix} 0 \\ 0 \\ |\psi_\uparrow(0)\rangle \\ |\psi_\downarrow(0)\rangle \end{pmatrix}, \end{aligned} \tag{9.49}$$

where we have taken again $t = 0$ for the sake of simplicity. At later times, due to the atom–molecule interaction, there will be a finite probability that the meter be in its excited state, i.e.

$$\begin{aligned} |\Psi(t)\rangle\rangle &= \frac{1}{\sqrt{2}} \Big(\alpha\, |\psi_\uparrow(t)\rangle\, |e\rangle_a |\uparrow\rangle_z + \beta\, |\psi_\downarrow(t)\rangle\, |e\rangle_a |\downarrow\rangle_z \\ &\quad + \gamma\, |\psi_\uparrow(t)\rangle\, |g\rangle_a |\uparrow\rangle_z + \delta\, |\psi_\downarrow(t)\rangle\, |g\rangle_a |\downarrow\rangle_z\Big) \\ &= \frac{1}{\sqrt{2}} \left[|\psi_\uparrow(t)\rangle \left(\alpha\, |e\rangle_a + \gamma\, |g\rangle_a\right) |\uparrow\rangle_z + |\psi_\downarrow(t)\rangle \left(\beta\, |e\rangle_a + \delta\, |g\rangle_a\right) |\downarrow\rangle_z\right] \\ &= \frac{1}{\sqrt{2}} \begin{pmatrix} \alpha\, |\psi_\uparrow(t)\rangle \\ \beta\, |\psi_\downarrow(t)\rangle \\ \gamma\, |\psi_\uparrow(t)\rangle \\ \delta\, |\psi_\downarrow(t)\rangle \end{pmatrix}. \end{aligned} \tag{9.50}$$

Due to the conservation of the probability, we must have $|\alpha|^2 + |\gamma|^2 = 1$ and $|\beta|^2 + |\delta|^2 = 1$. As we see from Eq. (9.50), when α and β are both non-vanishing, the meter

[40] For a complete treatment see the original article [Scully *et al.* 1978].

is triggered by both paths of the apparatus. The final density matrix of the compound system, constructed on the basis of Eq. (9.50) will be

$$\hat{\rho}_{am} = \begin{bmatrix} |\alpha|^2 \left|\psi_\uparrow\right\rangle\left\langle\psi_\uparrow\right| & \alpha\beta^* \left|\psi_\uparrow\right\rangle\left\langle\psi_\downarrow\right| & \alpha\gamma^* \left|\psi_\uparrow\right\rangle\left\langle\psi_\uparrow\right| & \alpha\delta^* \left|\psi_\uparrow\right\rangle\left\langle\psi_\downarrow\right| \\ \alpha^*\beta \left|\psi_\downarrow\right\rangle\left\langle\psi_\uparrow\right| & |\beta|^2 \left|\psi_\downarrow\right\rangle\left\langle\psi_\downarrow\right| & \gamma^*\beta \left|\psi_\downarrow\right\rangle\left\langle\psi_\uparrow\right| & \delta^*\beta \left|\psi_\downarrow\right\rangle\left\langle\psi_\downarrow\right| \\ \alpha^*\gamma \left|\psi_\uparrow\right\rangle\left\langle\psi_\uparrow\right| & \beta^*\gamma \left|\psi_\uparrow\right\rangle\left\langle\psi_\downarrow\right| & |\gamma|^2 \left|\psi_\uparrow\right\rangle\left\langle\psi_\uparrow\right| & \delta^*\gamma \left|\psi_\uparrow\right\rangle\left\langle\psi_\downarrow\right| \\ \alpha^*\delta \left|\psi_\downarrow\right\rangle\left\langle\psi_\uparrow\right| & \beta^*\delta \left|\psi_\downarrow\right\rangle\left\langle\psi_\downarrow\right| & \gamma^*\delta \left|\psi_\downarrow\right\rangle\left\langle\psi_\downarrow\right| & |\delta|^2 \left|\psi_\downarrow\right\rangle\left\langle\psi_\downarrow\right| \end{bmatrix}, \tag{9.51}$$

where, for the sake of simplicity, we have not indicated the time dependence. When $\alpha = \beta = 0$ ($\gamma = \delta = 1$), we have a free evolution of the system and no interaction with the meter. A "good" meter will jump to the excited state when the molecule is in the upper path (spin-up) and not when the molecule is in the lower path (spin-down). Therefore, in order to have a perfect correlation, we should have $\alpha \simeq \delta \simeq 1$ and $\beta \simeq \gamma \simeq 0$.

Suppose that, after the region V, the molecule beam passes through a Stern–Gerlach magnet, with the field gradient oriented in the x direction:

$$\mathbf{B} = \hat{\imath} B_x(x, y). \tag{9.52}$$

This device will split the beam into two packets moving along the directions $+x$ and $-x$.

In absence of the detector (Eq. (9.47)), the probability densities for the beam to be deflected in the $+x$ or $-x$ direction are given by

$$\begin{aligned} \wp_{+x}(\mathbf{r}) &= \langle \mathbf{r}|_x \langle \uparrow | \hat{\rho}_m | \uparrow \rangle_x |\mathbf{r}\rangle = \frac{1}{2}(1 \quad 1)\begin{bmatrix} 1 & 1 \\ 1 & 1 \end{bmatrix}\begin{pmatrix} 1 \\ 1 \end{pmatrix} |\psi(\mathbf{r})|^2 \\ &= 2|\psi(\mathbf{r})|^2, \end{aligned} \tag{9.53a}$$

$$\begin{aligned} \wp_{-x}(\mathbf{r}) &= \langle \mathbf{r}|_x \langle \downarrow | \hat{\rho}_m | \downarrow \rangle_x |\mathbf{r}\rangle = \frac{1}{2}(1 \quad -1)\begin{bmatrix} 1 & 1 \\ 1 & 1 \end{bmatrix}\begin{pmatrix} 1 \\ -1 \end{pmatrix} |\psi(\mathbf{r})|^2 \\ &= 0, \end{aligned} \tag{9.53b}$$

where (see Eqs. (6.159a))

$$|\uparrow\rangle_x = \frac{1}{\sqrt{2}}\left(|\uparrow\rangle_z + |\downarrow\rangle_z\right), \quad |\downarrow\rangle_x = \frac{1}{\sqrt{2}}\left(|\uparrow\rangle_z - |\downarrow\rangle_z\right), \tag{9.54}$$

and

$$\int d\mathbf{r} |\psi(\mathbf{r})|^2 = \frac{1}{2}. \tag{9.55}$$

Hence no particles would be found in the $-x$ direction if a detector was to be placed. This is the effect of a constructive interference in the $+x$ region and of a destructive interference in the $-x$ region (see Subsecs. 1.2.2 and 2.3.4).

Now we ask the same question when the "atomic detector" is present (see Eq. (9.51)). Assuming that we are not able to "read" the detector, we look for probabilities $\wp_{\pm x}(\mathbf{r})$ irrespective of the state of the detector. These may be calculated into two steps. First, by performing a partial trace on the atom we derive the reduced density matrix of the molecule alone, i.e.

$$\hat{\tilde{\rho}}_m = \mathrm{Tr}_a\left(\hat{\rho}_{am}\right) = {}_a\!\left\langle e \left| \hat{\rho}_{am} \right| e \right\rangle_a + {}_a\!\left\langle g \left| \hat{\rho}_{am} \right| g \right\rangle_a. \tag{9.56}$$

Second, we compute the probabilities

$$\wp_{+x}(\mathbf{r}) = \langle \mathbf{r} |_x \langle \uparrow | \hat{\tilde{\rho}}_m | \uparrow \rangle_x | \mathbf{r} \rangle, \qquad (9.57a)$$

$$\wp_{-x}(\mathbf{r}) = \langle \mathbf{r} |_x \langle \downarrow | \hat{\tilde{\rho}}_m | \downarrow \rangle_x | \mathbf{r} \rangle, \qquad (9.57b)$$

where, by choosing suitable values of the parameters α, β, γ, and δ, the reduced density matrix of the molecule (see Eqs. (5.41)) takes the form

$$\hat{\tilde{\rho}}_m = \begin{bmatrix} 1 & 0 \\ 0 & 1 \end{bmatrix} | \psi \rangle \langle \psi |. \qquad (9.58)$$

From Eq. (9.58) we now calculate the probabilities

$$\wp_{+x}(\mathbf{r}) = \tfrac{1}{2}(1 \quad 1) \begin{bmatrix} 1 & 0 \\ 0 & 1 \end{bmatrix} \begin{pmatrix} 1 \\ 1 \end{pmatrix} | \psi(\mathbf{r}) |^2 = | \psi(\mathbf{r}) |^2, \qquad (9.59a)$$

$$\wp_{-x}(\mathbf{r}) = \tfrac{1}{2}(1 \quad -1) \begin{bmatrix} 1 & 0 \\ 0 & 1 \end{bmatrix} \begin{pmatrix} 1 \\ -1 \end{pmatrix} | \psi(\mathbf{r}) |^2 = | \psi(\mathbf{r}) |^2. \qquad (9.59b)$$

Hence, in the presence of a "detector" we have a 50% probability of finding the particle in the $+x$ direction and a 50% probability of finding the particle in the $-x$ direction. This means that the sole presence of the atom, together with the ignorance of its final state, formally expressed by the partial trace (9.58), suffices to destroy the constructive and destructive interference that had led to probabilities (9.53).

This is very strong evidence of the fact that in quantum mechanics not only is it important which operation is actually performed, but also which operation may be in principle performed. This is in contrast with a generalized thermodynamical solution to the measurement problem (see Subsec. 9.2.2). We shall discuss this point more extensively below (in Sec. 9.5).

9.4.3 Cini's model

The important work by Cini[41] has the merit of presenting another interesting model that shows how the action of large systems can cause a transition from a pure state to a mixture onto a microscopic object. The main idea here is that the apparatus $\mathcal{A}$ is constituted of a large number of particles. For this reason, no explicit use of the environment is made. In this sense this model is independent from Zurek's one, but has the advantage of showing, in a simple way, that after the interaction the system–apparatus density matrix $\hat{\rho}_{\mathcal{AS}}$ is indistinguishable from a mixture formed by the different states representing the possible outcomes of the measurement, each one consisting of the particle in a given state plus the apparatus in the corresponding macroscopically definite state. As we shall see, it is the macroscopic distance between the different states of the measuring device that leads to this equivalence, justified by the very small magnitude of the interference terms.

---

[41] See [Cini 1983]. See also [Cini *et al.* 1979].

Cini's aim is to make a specific model of a certain kind of counters, i.e. devices which record the presence or the passage of a particle. We assume the object system $\mathcal{S}$ to be a two-level quantum system in the initial state

$$|\mathcal{S}\rangle = c_+ | \uparrow\rangle + c_- | \downarrow\rangle, \tag{9.60}$$

where $\{| \uparrow\rangle, | \downarrow\rangle\}$ is a basis in the two-dimensional Hilbert space of $\mathcal{S}$. The apparatus $\mathcal{A}$ is composed of N bosons, each with the same (symmetric) spatial wave function and an internal degree of freedom represented by two possible states: the ground state $|a_0\rangle$ and the excited state $|a_1\rangle$. The system–apparatus interaction Hamiltonian is assumed to be

$$\hat{H}_{\mathcal{SA}}(\hat{a}) = \frac{1}{2}\varepsilon'_{\mathcal{SA}}(1+\hat{\sigma}_z)\left(\hat{a}_0^\dagger \hat{a}_1 + \hat{a}_0 \hat{a}_1^\dagger\right), \tag{9.61}$$

where $\hat{\sigma}_z$ is the third Pauli spin matrix (see Eq. (6.154)), $\varepsilon'_{\mathcal{SA}}$ is the coupling constant, and $\hat{a}_0^\dagger, \hat{a}_0, \hat{a}_1^\dagger, \hat{a}_1$ are creation and annihilation operators (see Subsec. 4.4.2 and also Sec. 13.1) on $|a_0\rangle, |a_1\rangle$, respectively, satisfying

$$[\hat{a}_0^\dagger, \hat{a}_1^\dagger] = [\hat{a}_0, \hat{a}_1] = 0. \tag{9.62}$$

It should be noted that this choice of the interaction Hamiltonian implies that only one of the two independent states of $\mathcal{S}$ ($| \uparrow\rangle$) is capable of interacting with the counter's particles, the other one ($| \downarrow\rangle$) being isolated and, therefore, stationary. This is what Cini means by "polarized counter," namely a counter that selects between the different values of the measured observable.

A given state of the counter will be defined through the number n of particles in the ground state $|a_0\rangle$. Hence, $N - n$ will be the number of particles in the excited state $|a_1\rangle$. A generic state of the apparatus is defined by (see Prob. 4.13)

$$|n\rangle = |n, N-n\rangle = \frac{1}{\sqrt{n!}}\frac{1}{\sqrt{(N-n)!}}\left(\hat{a}_0^\dagger\right)^n \left(\hat{a}_1^\dagger\right)^{N-n} |0\rangle, \tag{9.63}$$

where $|0\rangle$ is the vacuum state of the apparatus, i.e. the state with no particle. When the initial state of $\mathcal{A}$ is the neutral state $n = N$, the time necessary to excite the first particle will be of the order (see Prob. 9.6)

$$\tau_0 \simeq \frac{\hbar}{\varepsilon'_{\mathcal{SA}}\sqrt{N}}. \tag{9.64}$$

However, from a physical viewpoint, τ_0 should be approximately independent of the number of particles, leading us to redefine the coupling constant as

$$\varepsilon'_{\mathcal{SA}} = \frac{\varepsilon_{\mathcal{SA}}}{\sqrt{N}}. \tag{9.65}$$

Starting with an uncorrelated initial state of $\mathcal{S} + \mathcal{A}$ defined by

$$|\Psi_{\mathcal{SA}}(t_0)\rangle = |\mathcal{S}\rangle \otimes |N\rangle, \tag{9.66}$$

where $|\mathcal{S}\rangle$ is given by Eq. (9.60), we have

$$|\Psi_{\mathcal{SA}}(t)\rangle = e^{-\frac{i}{\hbar}\hat{H}_{\mathcal{SA}}t}|\Psi_{\mathcal{SA}}(t_0)\rangle. \tag{9.67}$$

Cini shows that one can rewrite this equation as (see Prob. 9.7)

$$|\Psi_{\mathcal{SA}}(t)\rangle = c_+ | \uparrow\rangle \left[\sum_{n=0}^{N} \hat{a}_n(t)|n\rangle \right] + c_- | \downarrow\rangle |N\rangle, \tag{9.68}$$

where

$$\hat{a}_n(t) = \frac{\imath^{N-n}\sqrt{N!}}{\sqrt{n!}\sqrt{(N-n)!}} \left[\cos\left(\frac{\varepsilon_{\mathcal{SA}} t}{\hbar\sqrt{N}} \right) \right]^n \left[\sin\left(\frac{\varepsilon_{\mathcal{SA}} t}{\hbar\sqrt{N}} \right) \right]^{N-n}. \tag{9.69}$$

Considering the superposition $\sum_{n=0}^{N} \hat{a}_n(t)|n\rangle$ in Eq. (9.68), the probability $\wp_n$ of finding n particles in the ground state at time t is given by

$$\wp_n(t) = \left(\begin{array}{c} N \\ n \end{array} \right) p(t)^n p'(t)^{N-n}, \tag{9.70}$$

where

$$p(t) = \cos^2\left(\frac{\varepsilon_{\mathcal{SA}} t}{\hbar\sqrt{N}} \right) \quad \text{and} \quad p'(t) = 1 - p(t) = \sin^2 \frac{\varepsilon_{\mathcal{SA}} t}{\hbar\sqrt{N}}. \tag{9.71}$$

Equation (9.70) represents a binomial distribution, and, for large N, is sharply peaked around its maximum. Defining

$$\langle n(t)\rangle = Np(t), \tag{9.72}$$

we can immediately obtain

$$\wp_{\langle n\rangle} = \frac{N!}{\langle n\rangle!\,(N - \langle n\rangle)!} \left(\frac{\langle n\rangle}{N} \right)^{\langle n\rangle} \left(\frac{N - \langle n\rangle}{N} \right)^{N-\langle n\rangle}, \tag{9.73}$$

which yields (see Prob. 9.8)

$$\wp_{\langle n\rangle} \simeq 1. \tag{9.74}$$

This shows that the probability of finding $n \neq \langle n(t)\rangle$ is small. In conclusion, in the limit of very large values of N, one can approximately write Eq. (9.68) in the form

$$|\Psi_{\mathcal{SA}}(t)\rangle = c_+ \,|\, \langle n(t)\rangle\rangle \,|\uparrow\rangle + c_- |N\rangle| \downarrow\rangle. \tag{9.75}$$

This simple model raises several important remarks:

- The two components of Eq. (9.75) pertaining to the apparatus are macroscopically distinguishable [see Sec. 9.3]. This means that a suitable interaction has realized a one-to-one correlation between the states of the microsystem and those of the measuring device. That is just what one would expect from an ideal instrument. It should be emphasized that this result has been obtained precisely due to the large number of particles constituting the apparatus, in contrast to the examples of Sec. 9.1 or Subsec. 9.4.2, where the meter is a microsystem.
- Equation (9.75) does not correspond to a mixed state but rather to an entangled system–apparatus state. However, the macroscopic "distance" between the apparatus' components makes the resulting interference (off-diagonal) terms negligibly small.

In this respect, even though a reduction process has not occurred, a sort of collapse comes out from the ordinary laws of quantum mechanics, without adding hypotheses ad hoc (see for instance Subsec. 9.2.3).

- The correlation between the microsystem and the apparatus occurs at times of the order of $\hbar/\varepsilon_{SA}$ that, again in the limit of a very large N, are much smaller than the time required for a complete excitation of the counter, of the order of $\hbar\sqrt{N}/\varepsilon_{SA}$.

9.4.4 Decoherence time

The problem of the time scale at which decoherence phenomena occur is of central importance. Indeed, the Zurek's approach relies on the fact that the decoherence time is inversely proportional to the number of involved particles. This problem has been explicitly formalized by Zurek, but already Caldeira and Leggett, though in the context of a specific model, had proposed the introduction of such a parameter.[42] In the case of a spatial superposition of two Gaussian wave functions, Zurek proposed the definition of a decoherence time parameter τ_d as

$$\tau_d \simeq \gamma^{-1}\left(\frac{\lambda_T}{\Delta x}\right)^2, \tag{9.76}$$

where γ is the relaxation rate, i.e. the coupling constant of the system–environment interaction,[43] Δx is the separation between the two Gaussian peaks, and λ_T is the thermal de Broglie wavelength (see Eq. (7.26))

$$\lambda_T = \frac{\hbar}{\sqrt{2mk_B T}}. \tag{9.77}$$

This model assumes high temperatures (the environment is therefore often represented by a heath bath). As a matter of fact, the result (9.76) has been explicitly derived for a particle moving in one dimension, subject to a harmonic oscillator potential, linearly interacting with an environment described by a large number of harmonic oscillators (a *reservoir* or a *bath*).[44]

It is interesting to emphasize that in such a model the "macroscopic" size of the (initial) superposition state is given by the separation parameter Δx: the more the Gaussian peaks are far apart (the more macroscopic the superposition state is), the smaller the decoherence time. As a consequence, "microscopic" superposition states are little affected by decoherence, whereas "macroscopic" ones decohere on a very fast time scale.

[42] See [Zurek 1986] [Caldeira/Leggett 1985].
[43] As we shall see in Sec. 14.2.
[44] See [Paz *et al.* 1993, 490–94] and also [Brune *et al.* 1992, 5205].

9.4.5 Final considerations about decoherence

At the end of this section it is fair to ask ourselves the following question: is decoherence a *good* solution of the measurement problem? We must immediately say that decoherence is *a* solution that does not present contradictions with experience. We cannot exclude that there will be other more general and satisfying solutions. More than this: we cannot state that all the other approaches that have been presented in Sec. 9.2 have to be discarded. However, the decoherence solution has certainly some advantages with respect to other proposals that have been advanced. First, it is a solution of the measurement problem that does not introduce elements which are external to the domain of quantum mechanics. It is worth emphasizing here that it is a fundamental methodological principle to try to account for phenomena in the frame of the discipline where they are studied (this is not the case, for example, for von Neumann's solution). Second, up to now it has proved to be in good agreement with experimental results, in particular concerning the decoherence time (see for instance Subsec. 14.5.2).

From a philosophical viewpoint, it is clear that, in quantum mechanics, a measurement is ultimately an act of decision, i.e. we choose – by introducing a particular coupling between the system and the apparatus (see Eq. (9.10)) – to measure a particular observable (see Subsec. 2.3.3) and the environment "helps" us in diagonalizing or quasi-diagonalizing the density matrix of the system relatively to it. This operational feature of the theory gives the quantum theory of measurement a deeper meaning than to the role assigned to the measurement process in classical mechanics. According to the latter, in fact, measurement is considered as a registration of already given data.

This act of decision does not necessarily imply a form of subjectivism. Spontaneous "measurements" are also thought to exist in nature: it suffices that at least two systems that are open to the environment interact in order to spontaneously obtain a determination relatively to an observable.[45] In other words, measurement is only a particular case of a larger class of dynamical interactions (as we shall see in Sec. 14.2).

9.5 Reversibility/irreversibility

In Sec. 9.1 we have distinguished between premeasurement and measurement. Moreover, in the previous section, we have seen some example of decoherence in which a microscopic meter interacts with an object system and a tracing out is performed (as in Subsec. 9.4.2), or an object system interacts with a macroscopic apparatus (and different possible outcomes become distinguishable) even if no explicit tracing out is performed (as in Subsec. 9.4.3). It is therefore convenient to split also the strict measurement act into two steps, and, therefore, apart from the initial preparation of the object system, to distinguish between:

- premeasurement, which consists in entangling an object system with an apparatus that now we in general assume to be a macroscopic device;

[45] See [Joos/Zeh 1985].

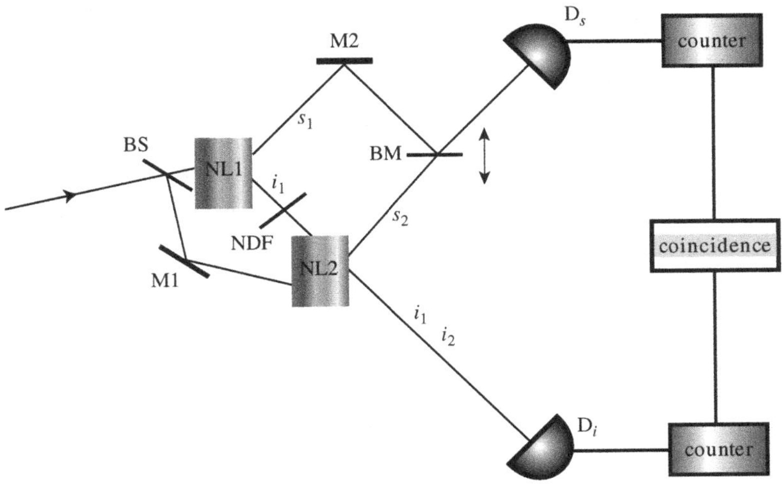

Figure 9.9 Schematic representation of Mandel's experiment. An initial photon beam is split by a beam splitter (BS) and the resulting beams, after being deflected by the mirror M1, travel toward two non-linear crystals of $LiIO_3$ (NL1 and NL2). From NL1 a signal photon (s_1) and an idler photon (i_1) emerge: the i-photon passes through NL2 and will be aligned with the second idler (i_2), which is emitted by NL2 together with the second signal photon (s_2). The two s-photons are combined by the mirror M2 and the beam merger (BM), and the outgoing beam falls on detector D_s, whereas the two idler photons fall on detector D_i. BM may be vertically displaced. A neutral density filter (NDF) is inserted between NL1 and NL2. In the case where the transmittivity of the NDF is 100%, when examining the coincidences we cannot distinguish between the two idler and signal photon pairs from either NL1 or NL2: this is the so-called fourth-order interference (see Sec. 13.6). When the transmittivity of the NDF is 0, then i_1 is blocked and a coincidence can only result from the signal and idler photon pair emitted by NL2. Here there is no ambiguity and no interference. For values of transmittivity between 0 and 1 we have intermediate possibilities (see also Subsec. 2.3.4).

- the first measurement step consisting in washing out of interference, which is in general obtained by some tracing out; and
- the actual acquiring (and eventual storing) of information, which is obtained through some form of detection.

The first step of measurement is reversible and requires no actual interaction between the system and the apparatus, as we shall see now, while the information acquiring is necessarily irreversible. We present now some interesting experimental contexts that can throw light on this distinction.

9.5.1 Mandel, Wang and Zou's experiment

This distinction can be seen at work by the following experiment, performed by Wang, Zou, and Mandel.[46] Two non-linear crystals give rise to two spontaneous parametric down-conversions (see Fig. 9.9; see also Sec. 7.1). When the transmittivity $|T|^2$ of the neutral

[46] See [Wang *et al.* 1991b].

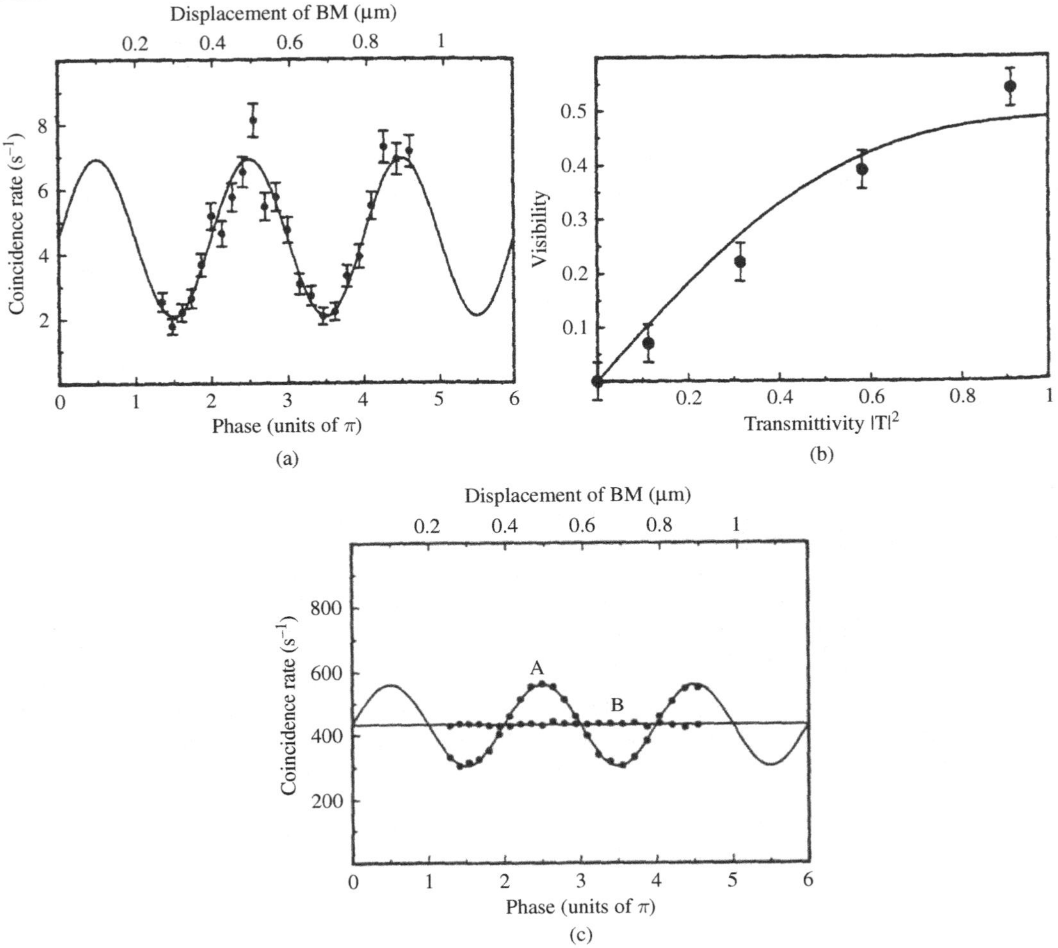

Figure 9.10 (a) The results of measuring the coincidence rate of detectors D_s and D_i for various displacements of BM are here shown. The solid line represents the best-fitting sinusoidal function of the expected periodicity. Error bars show the statistical uncertainty. (b) Measured visibility $\mathcal{V}$ of the interference (see Subsec. 2.3.4) registered by the coincidence counting rate of detectors D_s and D_i for various filter transmittivity $|T|^2$. (c) Measured signal photon counting rate of the sole detector D_s as a function of BM displacement. Curve A: filter transmittivity $|T|^2 = 0.91$; curve B: $|T|^2 = 0$. Adapted from [Wang *et al.* 1991b].

density filter is 100%, one cannot distinguish between pairs $|s_1\rangle, |i_1\rangle$ and $|s_2\rangle, |i_2\rangle$: this reflects itself into the presence of interference fringes in the coincidence rate of detection at D_i and D_s, obtained for different vertical displacements of BM (see Fig. 9.10(a)). On the contrary, when $|T|^2 = 0$, no interference is visible (see Fig. 9.10(b)). It is particularly interesting to note that the counting rate registered by D_s alone suffices to exhibit interference (second order interference) when $|T|^2 > 0$ (see Fig. 9.10(c)), because D_s cannot distinguish if the s-photon comes from NL1 or from NL2. Here, it is less obvious why blocking i_1 washes out the interference: since the down-conversions in both NL1 and NL2 are spontaneous, detector D_s should not be able to distinguish whether the s-photon comes from NL1 or NL2. Why, then, do the results show no interference for $|T|^2 = 0$? It is clear that this phenomenon cannot be caused by some disturbance of the signal photons. The sole *in principle* distinguishability of the two "paths" giving rise to the interference (signal photon coming from NL1 *or* from NL2) suffices to destroy the latter (see also Sec. 9.6). As a matter of fact, if i_1 is blocked, we *could* use the information from detector D_i to establish whether the s-photon comes from NL1 (no coincidence) or from NL2 (coincidence). In this respect, it looks like the state vector describes not only what is *actually* known but also what is *in principle* knowable (see Subsec. 2.3.3) – we shall come back to the meaning of the state in quantum mechanics in Ch. 15. In conclusion, the state vector has "registered" the destruction of the interference even if nobody has actually and irreversibly acquired this information.

9.5.2 Quantum eraser

In the following conceptual experiment we present a variant of the complementarity between wave-like and corpuscular behaviors (see also Subsec. 1.2.4 and 2.3.4). This experiment was first proposed by Scully, Englert, and Walther.[47] Since then, the so-called *quantum eraser* has been experimentally realized, even though with different setups.[48]

In the original proposal, we have an atomic beam which goes through two slits of wall I (see Fig. 9.11); behind this wall there is a further series of slits which are used as collimators to define two atomic beams that reach the narrow slits of wall II where the interference originates. Between wall I and wall II and after the collimators the atomic beams are orthogonally intersected by an intense source of light, for instance by a LASER (Light Amplification by Stimulated Emission of Radiation) beam,[49] which brings the internal state $|\iota\rangle$ of the two-level atoms from an unexcited (ground) state $|g\rangle$ into an excited state $|e\rangle$. Thereafter, each of the two beams passes through a microcavity.[50] Finally, they fall on a screen. The atomic source is adjusted in such a way that there is at most one atom at a time in the apparatus.

In the interference region, the wave function describing the center-of-mass motion of the atoms is the superposition of the two terms referring to slit 1 and slit 2 ($\mathbf{r}$ indicates the center-of-mass coordinate), so that the total (center-of-mass plus internal) wave function is given by

$$\Psi(\mathbf{r}) = \frac{1}{\sqrt{2}} [\psi_1(\mathbf{r}) + \psi_2(\mathbf{r})] \, |\iota\rangle, \tag{9.78}$$

47 See [Scully *et al.* 1991].
48 See e.g. [Herzog *et al.* 1995] .
49 See Box 13.1: p. 494.
50 See Sec. 13.7.

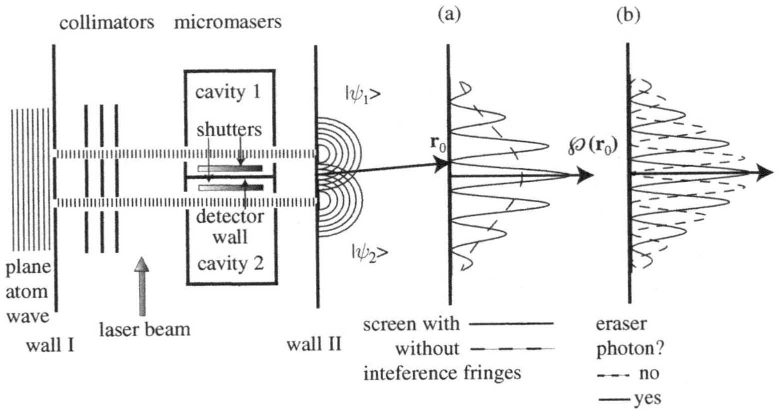

Figure 9.11 Scully, Englert, and Walther's proposed experiment. A set of slits (after the wall I) collimates two atomic beams that pass through the narrow slits (wall II) where the interference pattern originates. This setup is supplemented by two high-quality microcavities and a laser beam to provide which-path information. (a) Quantum erasure configuration in which electro-optic shutters separate microwave photons in the two cavities from the thin-film semiconductor (the central detector wall) which absorbs microwave photons and acts as a photodetector. In the absence of the laser beam, there is no possibility of obtaining which-path information and we have interference on the screen (solid line). Instead, introducing the laser beam, we may acquire which-path information and the interference is destroyed (dashed line). (b) The shutters are open. The probability density of the particles on the screen depends upon whether a photocount is observed in the detector wall ("yes") or not ("no"), demonstrating that the correlation between the event on the screen and the eraser photocount is necessary to retrieve the interference pattern.

and the probability density of particles falling on the screen at the point $\mathbf{r} = \mathbf{r}_0$ will be given by

$$\wp(\mathbf{r}_0) = \frac{1}{2}\left[|\psi_1(\mathbf{r}_0)|^2 + |\psi_2(\mathbf{r}_0)|^2 + \psi_1^*(\mathbf{r}_0)\psi_2(\mathbf{r}_0) + \psi_2^*(\mathbf{r}_0)\psi_1(\mathbf{r}_0)\right]\langle\iota|\iota\rangle. \tag{9.79}$$

The cavity frequency is tuned in resonance with the energy difference between the excited and ground states of the atoms. The velocity of the atoms may be selected in such a way that, after being prepared in an excited state by the laser beam, on passing through either one of the cavities each atom will emit a microwave photon (which stays in the cavity) and leave which-path information. After the atom has passed through the cavity it is again in force-free space and its momentum keeps the initial value.

Passing through the cavities and making the transition from $|e\rangle$ to $|g\rangle$, the state of the global system (atomic beam plus cavity) is given by

$$\Psi(\mathbf{r}) = \frac{1}{\sqrt{2}}\left[\psi_1(\mathbf{r})|1_1 0_2\rangle + \psi_2(\mathbf{r})|0_1 1_2\rangle\right]|g\rangle, \tag{9.80}$$

where $|1_1 0_2\rangle$ denotes the state in which there is one photon in cavity 1 and none in cavity 2. In contrast to Eq. (9.79), the probability density on the screen is now

Box 9.2 **Recoil-free *which path* detectors?**

The assumption that *which path* detectors are recoil-free has been questioned by Storey and co-workers [Storey *et al.* 1994]. They pointed out that the which-path determination would not be possible without a double momentum transfer between detector and photon (when a photon is emitted and then reabsorbed from the opposite direction), whose magnitude is in the limits of the uncertainty relation between position and momentum. However, Scully and co-workers [Englert *et al.* 1995] showed the correctness of their results that have also been experimentally confirmed (see Subsec. 2.3.4 and also Sec. 9.7). Therefore, there is no significant change in the spatial wave function of the atoms. It is only the correlation between the center-of-mass wave function and the photon's degrees of freedom in the cavities that is responsible for the loss of interference.

$$\wp(\mathbf{r}_0) = \frac{1}{2}\big[|\psi_1(\mathbf{r}_0)|^2 + |\psi_2(\mathbf{r}_0)|^2 + \psi_1^*(\mathbf{r}_0)\psi_2(\mathbf{r}_0)\langle 1_1 0_2|0_1 1_2\rangle + \psi_2^*(\mathbf{r}_0)\psi_1(\mathbf{r}_0)\langle 0_1 1_2|1_1 0_2\rangle\big]\langle g|g\rangle. \tag{9.81}$$

Since the two cavity state vectors $|1_1 0_2\rangle$, $|0_1 1_2\rangle$ are orthogonal to each other, the interference terms vanish in Eq. (9.81) and diffraction fringes are washed out, so that Eq. (9.81) reduces to

$$\wp(\mathbf{r}_0) = \frac{1}{2}\left[|\psi_1(\mathbf{r}_0)|^2 + |\psi_2(\mathbf{r}_0)|^2\right]. \tag{9.82}$$

It should be noted that the *which path* detectors are recoil-free (see Box 9.2).

Let us now separate the detectors in the cavity by a shutter–detector combination, so that, when the shutters are closed, the photons are forced to remain either in the upper or in the lower cavity. However, if the shutters are opened, light will be allowed to interact with the photodetector wall and in this way the radiation will be absorbed and the memory of the passage erased (such an operation is called *quantum erasure*). After the erasure, will we again obtain the interference fringes which were eliminated before? The answer is yes, so that interference effects can be restored by manipulating the *which path* detectors long after the atoms have passed and before reaching the final (detection) wall.

This result can be formally expressed as follows. Let us include the photodetector walls into the description. These are initially in the ground state $|g\rangle_D$, so that Eq. (9.80) modifies to

$$\Psi(\mathbf{r}) = \frac{1}{\sqrt{2}}\left[\psi_1(\mathbf{r})|1_1 0_2\rangle + \psi_2(\mathbf{r})|0_1 1_2\rangle\right]|g\rangle_A|g\rangle_D. \tag{9.83}$$

After absorbing the photon, the photodetector passes to an excited state $|e\rangle_D$. If we introduce symmetric and antisymmetric atomic states

$$\psi_\pm(\mathbf{r}) = \frac{1}{\sqrt{2}}\left[\psi_1(\mathbf{r}) \pm \psi_2(\mathbf{r})\right], \tag{9.84}$$

together with symmetric and antisymmetric states of the radiation fields contained in the cavities

$$|\pm\rangle = \frac{1}{\sqrt{2}}\left[|1_1 0_2\rangle \pm |0_1 1_2\rangle\right], \quad (9.85)$$

we can rewrite Eq. (9.83) as

$$\Psi(\mathbf{r}) = \frac{1}{\sqrt{2}}\left[\psi_+(\mathbf{r})|+\rangle + \psi_-(\mathbf{r})|-\rangle\right]|g\rangle_{\mathrm{A}}|g\rangle_{\mathrm{D}}. \quad (9.86)$$

The action of the quantum eraser on the system is to change Eq. (9.86) into

$$\Psi'(\mathbf{r}) = \frac{1}{\sqrt{2}}\left[\psi_+(\mathbf{r})|0_1 0_2\rangle|e\rangle_{\mathrm{D}} + \psi_-(\mathbf{r})|-\rangle|g\rangle_{\mathrm{D}}\right]|g\rangle_{\mathrm{A}}. \quad (9.87)$$

The reason is that the interaction Hamiltonian between radiation and photodetectors only depends on symmetric combinations of radiation variables so that the antisymmetric state remains unchanged.

Now, as long as the final state of the photodetector is unknown, the atomic probability density at the screen is

$$\begin{aligned}\wp(\mathbf{r}_0) &= \mathrm{Tr}_{A,F,D}\left[\Psi'^*(\mathbf{r}_0)\Psi'(\mathbf{r}_0)\right] \quad (9.88)\\ &= \frac{1}{2}\left[\psi_+^*(\mathbf{r}_0)\psi_+(\mathbf{r}_0) + \psi_-^*(\mathbf{r}_0)\psi_-(\mathbf{r}_0)\right] = \frac{1}{2}\left[\psi_1^*(\mathbf{r}_0)\psi_1(\mathbf{r}_0) + \psi_2^*(\mathbf{r}_0)\psi_2(\mathbf{r}_0)\right],\end{aligned}$$

where the trace has been performed on the atomic, field, and detector degrees of freedom. Clearly, Eq. (9.88) does not show any interference terms. However, if we compute the probability density for finding both the photodetector excited and the atom at $\mathbf{r}_0$ on the screen, we have

$$\begin{aligned}\wp_{e_{\mathrm{D}}}(\mathbf{r}_0) &= \mathrm{Tr}_{A,F}\left[|e\rangle_{\mathrm{D}}\langle e|\,\Psi'^*(\mathbf{r}_0)\Psi'(\mathbf{r}_0)\right]\\ &= |\psi_+(\mathbf{r}_0)|^2 = \frac{1}{2}\left[|\psi_1(\mathbf{r}_0)|^2 + |\psi_2(\mathbf{r}_0)|^2\right] + \Re\left[\psi_1^*(\mathbf{r}_0)\psi_2(\mathbf{r}_0)\right], \quad (9.89)\end{aligned}$$

which exhibits the same interference term as Eq. (9.79). In similar way, the probability of finding both the photodetector in the ground state and the atom at $\mathbf{r}_0$ on the screen is

$$\begin{aligned}\wp_{g_{\mathrm{D}}}(\mathbf{r}_0) &= \mathrm{Tr}_{A,F}\left[|g\rangle_{\mathrm{D}}\langle g|\,\Psi'^*(\mathbf{r}_0)\Psi'(\mathbf{r}_0)\right]\\ &= |\psi_-(\mathbf{r}_0)|^2 = \frac{1}{2}\left[|\psi_1(\mathbf{r}_0)|^2 + |\psi_2(\mathbf{r}_0)|^2\right] - \Re\left[\psi_1^*(\mathbf{r}_0)\psi_2(\mathbf{r}_0)\right], \quad (9.90)\end{aligned}$$

giving rise to the dashed antifringes indicated in Fig. 9.11(b).

The discussion of the above conceptual experiment naturally leads us to ask the following questions:

- Do we want to know whether we registered a "slit 1" atom or a "slit 2" atom.
- Are we interested in ascertaining one of the two situations, i.e. having the microwave-photon sensor either excited ($|e\rangle_{\mathrm{D}}$) or not ($|g\rangle_{\mathrm{D}}$)?

We cannot answer both questions at the same time. In other words, we *either* know the *atom path* without using the eraser and hence without knowing anything about the photodetector and its state (first alternative), *or* we desire to know the latter (second alternative) and we must reproduce the *interference* by losing all information about the particles' path. In the former case the absence of interference (Eqs. (9.82) and (9.88)) is due to the sum of probabilities (9.89) and (9.90). It is important not to forget that the photon is a quantum object: when the shutters are opened, the two cavities become a single larger one. Now, the photon's wave is a combination of the two partial waves, such that either the two waves reinforce each other (constructive interference) and the photosensory detects the photon, or they mutually extinguish each other, with the consequence that the photosensor detects no photon (destructive interference). In other words, there is only a 50% probability of detecting the photon.

From this conceptual experiment, there are two lessons to be drawn: (1) the washing out of the interference is reversible if we do not actually acquire and/or store information; (2) when measuring, one cannot consider the object system as separate from the context of the experiment being performed.[51]

The above examination, in particular the discussion presented in Box 9.2, supports the statement that the complementarity principle is genuinely a fundamental principle of quantum mechanics and not a mere consequence of the uncertainty relation (see also Subsec. 2.3.4).

9.6 Interaction-free measurement

We have already briefly discussed an example of interaction-free measurement (in Subsec. 1.2.4). Let us now deal with this matter more extensively. M. Renninger was the first to propose an experiment in which one can measure some properties of a system without interacting with it.[52] He called this type of measurements *negative-result measurements*, even though today the term *interaction-free measurements* is more commonly used.

A more recent experimental proposal is based on the use of a Mach–Zehnder interferometer[53] (see Fig. 9.12). The beam splitters BS1 and BS2 are symmetric, and, after BS2, if the phase difference ϕ is zero, detector D3 detects all photons (due to constructive interference) while D4 is the dark-output (due to destructive interference). At any time, there is at most one photon in the interferometer. We are testing whether an object O is located along the upper path 2. Let us suppose that O is actually present in path 2. Then, there are three possible outcomes of the experiment:

- no detector clicks;
- D3 clicks;
- D4 clicks.

In the first case (with probability 1/2) the photon has been absorbed by the object. In the second case (with probability 1/4) the photon reaches D_3 (it could have also reached it if the object were not present, with probability 1). The third case may occur with probability 1/4.

51 See [Bohr 1935b, Bohr 1948, Bohr 1949].

52 See [Renninger 1960]. We have already quoted his pioneering article in showing the impossibility of generalizing a thermodynamic model of measurement (see Subsec. 9.2.2).

53 See [Elitzur/Vaidman 1993, 988–91] .

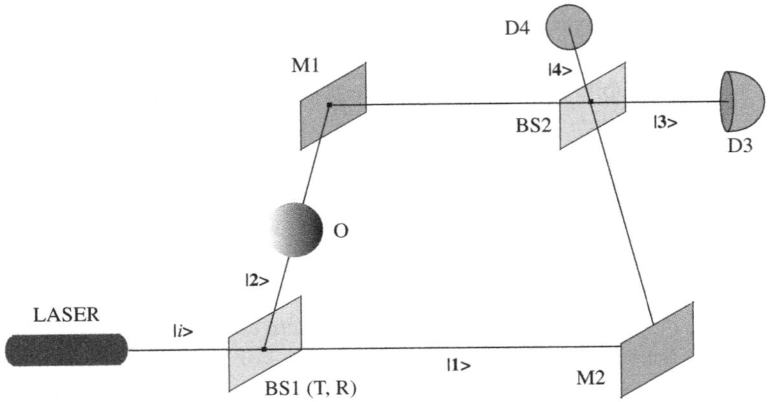

Figure 9.12 **An interaction-free measurement setup proposed by Elitzur and Vaidman. If the relative phase $\phi = 0$, detector D4 clicks only if one of the arms is blocked by the object O.**

If one performs the test and detector D4 clicks, one may be informed that the object is located in the upper path without having interacted with it. In order to investigate the quantum-mechanical roots of this conclusions, let the evolution of the system *in absence of the object* be described as follows (see also Subsec. 2.3.4)

$$\begin{aligned} |i\rangle &\overset{\text{BS1}}{\mapsto} \frac{1}{\sqrt{2}}(|1\rangle + |2\rangle) \overset{\text{M1,M2}}{\mapsto} \frac{1}{\sqrt{2}}(|1\rangle + |2\rangle) \\ &\overset{\text{BS2}}{\mapsto} \frac{1}{2}(|3\rangle - |4\rangle) + \frac{1}{2}(|3\rangle + |4\rangle) = |3\rangle, \end{aligned} \tag{9.91}$$

where $|i\rangle$ is the initial state (an incoming photon entering the interferometer from the left), $|1\rangle, \ldots, |4\rangle$ are the state vectors of the photon associated with the corresponding paths as indicated in Fig. 9.12. Then, the photon leaves the interferometer moving to the right, and it is detected by D3. If *the object O is present* the evolution is described by

$$\begin{aligned} |i\rangle &\overset{\text{BS1}}{\mapsto} \frac{1}{\sqrt{2}}(|1\rangle + |2\rangle) \overset{\text{O}}{\mapsto} \frac{1}{\sqrt{2}}(|1\rangle + |a\rangle) \\ &\overset{\text{M1,M2}}{\mapsto} \frac{1}{\sqrt{2}}(|1\rangle + |a\rangle) \overset{\text{BS2}}{\mapsto} \frac{1}{2}(|3\rangle - |4\rangle) + \frac{1}{\sqrt{2}}|a\rangle, \end{aligned} \tag{9.92}$$

where $|a\rangle$ represents the state of the photon when it is absorbed by the object. From Eq. (9.92) we have the three possible outcomes described before: absorption by the object, detection by D3, or detection by D4, with probabilities of 1/2, 1/4, and 1/4, respectively. As we have said, with probability 1/4, we can detect the presence of an object without interacting with it.

One may ask whether it is possible to enhance the probability to test the presence of some object without interacting with it. The answer is yes, since one can envisage[54] an apparatus consisting of a series of N interferometers, with N large (see Fig. 9.13(a)). The reflectivity $|\mathrm{R}|^2$ of each of the N BSs is chosen to be $\cos^2(\pi/2N)$ and the relative phases between corresponding paths in the upper and lower halves to be zero. The result is that the

54 See [Kwiat *et al.* 1995a].

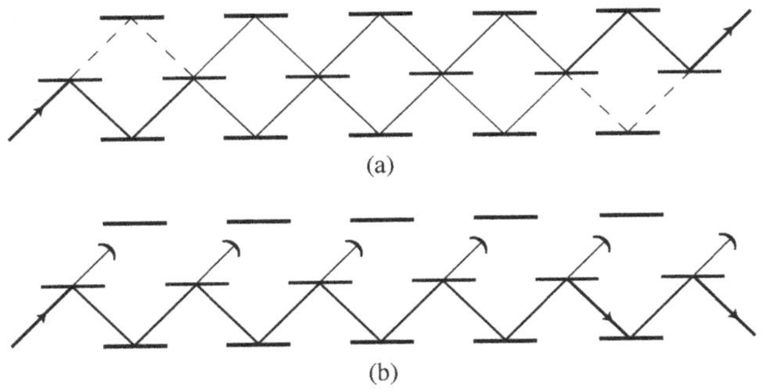

Figure 9.13 (a) The principle of coherently repeated interrogation. (a) A single photon incident from the lower left gradually transfers to the upper right half of the system. After N stages, where N depends on the BS reflectivities, the photon will certainly exit via the upper port of the last BS. (b) Introduction of detectors prevents the interference. At each stage the state is projected back into the bottom half of the system if the respective detector does not fire. After all stages there is a good chance that the photon now exits via the lower port of the last BS, indicating the presence of the detectors.

amplitude of the photon undergoes a gradual transfer from the lower to the upper halves of the interferometers.

After all N stages, the photon will certainly leave by the upper exit (see Prob. 9.9). Now, we insert a series of detectors (which together represent here the "obstacle") in the upper half of the apparatus which prevent the interference (see Fig. 9.13.(b)). At each beam splitter, there is a small chance that the photon takes the upper path and triggers a detector, and a large probability $\cos^2(\pi/2N) \simeq 1 - \pi^2/4N^2$ that it continues to travel on the lower path. The non-firing of each detector projects the state onto the lower half. The probability that the photon will be found in the lower exit after N stages is then

$$\wp = \left[\cos^2\left(\frac{\pi}{2N}\right)\right]^N \simeq \left[1 - \left(\frac{\pi}{2N}\right)^2\right]^N. \tag{9.93}$$

This represents the probability of "success," i.e. the probability that we are able to detect the presence of the obstacle without interacting with it. The probability that the photon will be absorbed by the object is very low and is given by $1 - \wp$. To summarize, there are three possible outcomes: the photon leaves the apparatus by the upper path, and we may infer with certainty that the object is absent; the photon leaves the apparatus in the lower part, and we may infer with certainty that the object is present; finally, no photon leaves the apparatus, and this means that the photon has been absorbed by the object. However, as we have seen, the probability of the third outcome may be made arbitrarily small as N grows. Hence, it is clear that, as the number of stages becomes very large, we may approach an efficiency (probability of success) arbitrarily close to 1 (see Fig. 9.14).

The expanded version shown in Fig. 9.13 may be "compressed" into a single device by making use of two identical "cavities" weakly coupled by a highly reflective beam splitter (see Fig. 9.15). At time $t = 0$ a photon is inserted in the left cavity (for timing purposes, it is important that the length of the photon wave be shorter than the cavity length for the duration of the experiment). For a beam splitter of reflectivity $|\mathrm{R}|^2 = \cos^2(\pi/2N)$ and in

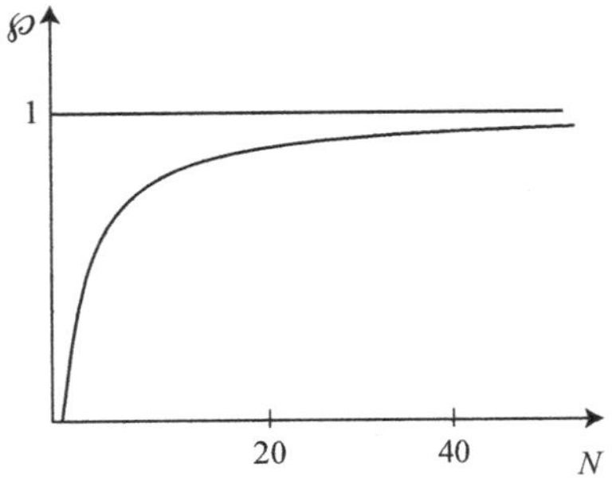

Figure 9.14 **Probability of success in repeated interaction-free measurements as a function of the number *N* of stages: the efficiency rapidly tends to 1 as *N* grows.**

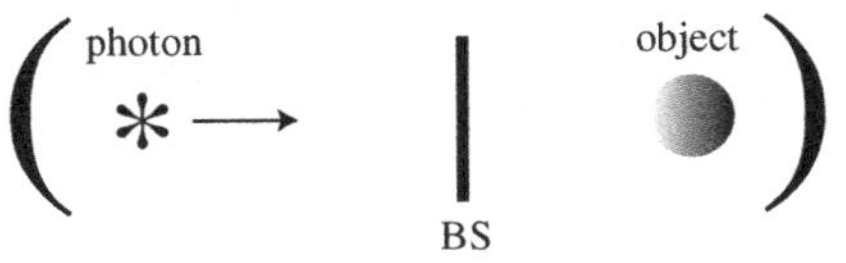

Figure 9.15 **Interaction-free measurement with two cavities. *N* represents here the number of cycles and will depend on the reflectivity of the coupling beam splitter.**

absence of any absorber the photon will certainly be located in the right cavity at time $t_N = N \times$ (round-trip time), due to interference effects. For this reason, if a detector is inserted in the left cavity at time t_N, it would not fire. However, if there is an object in the right cavity, the photon wave function is at each cycle projected back onto the left cavity, giving rise to a (close to 1) probability of detecting the photon in the left side.

It is interesting to comment on some interpretational issues pertaining to interaction-free measurement. The interaction-free measurement is due to an instantaneous "reduction" of the superposition of both paths in the interferometer due to the simple presence of an obstacle. As a consequence of this reduction, the photon is now localized. Hence, interaction-free measurements pertain to the same class of experiments as those with "reductions" with the sole presence of the detector (see Subsec. 9.4.2 and Mandel and co-workers's experiment reported in Subsec. 9.5.1). In the latter case, for instance, it is the mere possibility of obtaining which-path information that destroys the interference while no actual measurements need to be made. Stated in other terms, it is the non-local character of a single photon that allows us to infer the presence of an object even though a posteriori we may conclude that the photon has taken a given path.

There is another possible interpretation of the above results. This interpretation was anticipated by de Broglie[55] and successively developed by Vigier, Selleri, Tarozzi, Croca, and their co-workers.[56] The main idea is that both wave and particles have full physical

[55] See [*de Broglie* 1956]. It is not our intention here to discuss de Broglie's interpretation of quantum mechanics, an issue that goes far beyond the scope of this book – however, see also Subsec. 16.3.2. We shall limit ourselves to the examination of a specific consequence that, in the scientific community, has raised much attention in the 1980s–1990s.

[56] See [Croca 1987] [Garuccio *et al.* 1982] [Hardy 1992] [Selleri 1969, Selleri 1982] [Tarozzi 1985].

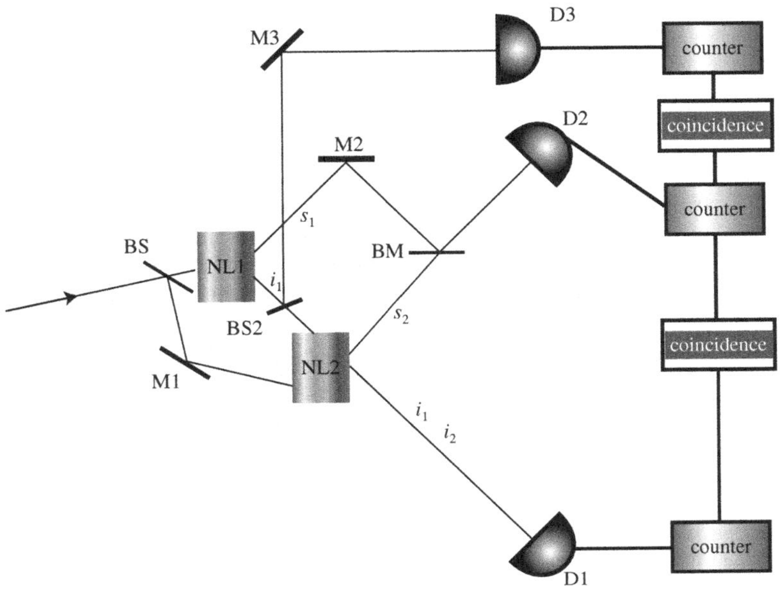

Figure 9.16 Schematic representation of Mandel's experiment on empty waves. The essential setup is the same as in Fig. 9.9. The main difference is that, in the place of a neutral density filter, we have now a second beam splitter (BS2), whose output beams travel toward the second non-linear crystal (NL2) and toward the third detector (D3). Now, if detector D3 clicks, from the point of view of the empty wave theory an empty wave (i_1) should fall on D1 and still induce coherence between s_1 and s_2. Experimental results showed no coherence in this case, which supports the interpretation of quantum waves as probability amplitudes.

reality. While almost all energy-momentum is associated with the particle, a small fraction is smeared on the wave. However, even if we cannot measure physical quantities of a wave without particle (the so-called *empty wave*), we could notwithstanding detect some fundamental effects of the wave on the probability distribution. In particular, interference effects could be sensibly different if an empty wave is present or not. The empty wave hypothesis could account for the superposition produced in an interferometer in this way: the particle is always localized in a pathway but in the other pathway an empty wave travels such that at the second beam splitter its interference effects with the photon manifest themselves. Returning to the interaction-free measurement, this could then be explained by assuming that an obstacle prevents the empty wave from traveling and for this reason there are no longer interference effects at the second beam splitter. This would be clearly a classical explanation. Such hypothesis was tested by Mandel and co-workers, who showed that the hypothesis of an empty wave is inconsistent with experimental results. The apparatus used by Mandel and co-workers is a modification of that presented in Sec. 9.5 and is shown in Fig. 9.16.[57]

[57] See [Zou *et al.* 1992]. In this experimental setup it is supposed that a non-null empty wave may go through the second non-linear crystal. In another, more complicated, experimental set up [Wang *et al.* 1991a] there is no necessity of this assumption.

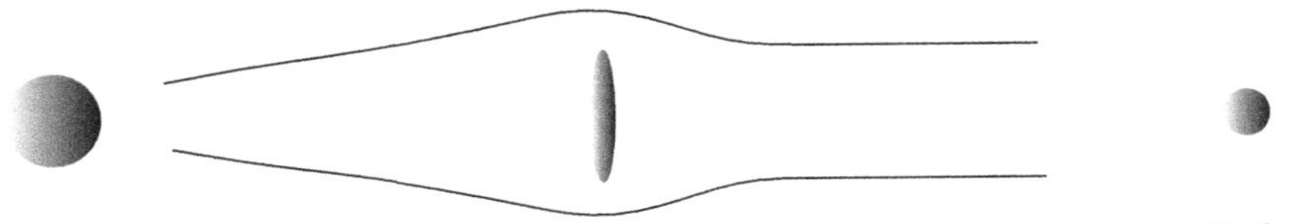

Figure 9.17 Depiction of Wheeler's experiment. A far away quasar sends a light beam to us. Before arriving to the Earth, the light encounters on its route a lenticular galaxy. In this case the light can take two different paths. However, whether it actually takes the two paths or if it follows a single path seems to be decided by the measurement performed on Earth millions of years later.

The technical consequences of interaction-free measurements can be very far-reaching: more recently, this effect has been used to obtain images of objects without illuminating them (interaction-free imaging). In order to obtain such a result, it suffices to work with beams rather than with single photons and to associate to each photon a "pixel" of the image.[58]

9.7 Delayed-choice experiments

In what follows we discuss an important issue of quantum theory, proposed for the first time by Wheeler in 1978. This issue can be better illustrated using a cosmological example [Wheeler 1983, 190–99]. Let us consider a quasar (QUAsi-StellAR radio source), whose light, before reaching the Earth, encounters a lenticular galaxy. As is well known, such a galaxy can act as a gravitational lens, such that the light can take two simultaneous paths (see Fig. 9.17). The problem is the following: we can choose here on Earth either to observe an interference phenomenon (wave-like behavior), hence merging the light from both paths and detecting the outgoing beams, or to detect the light on a determinate path (corpuscular behavior). At a first sight, it is as if we could decide here and now a certain (wave-like or corpuscular) behavior about an event which seems *to have already happened* millions of years ago (see also Sec. 2.4). How can we solve this problem?

If we want to solve the problem without introducing retrocausation,[59] the only reasonable possibility is that the assumption "the event has already occurred" is false. Even classically, we can only speak of past events if we are able somehow to interact with their present effects. In other terms, what we call "past" (the light of million years ago) is actually only a present thing, i.e. the light, *not* of million years ago, but the light which we detect now, that is, the light that, in order to arrive to us, has traveled all the way (and all the time) to us. Therefore, by performing the quoted experiment, we are not at all acting on the past, but only on the *effects which past events have transmitted to our present*. Summarizing, we cannot deal with the past without interacting with its effects on the present, and, whenever we do that, we are also free to choose the form to receive them, i.e. to interrogate them.

[58] See [Kwiat *et al.* 1996] [A. White *et al.* 1998].
[59] See [Wheeler 1978, 41] [Wheeler 1983, 183–84].

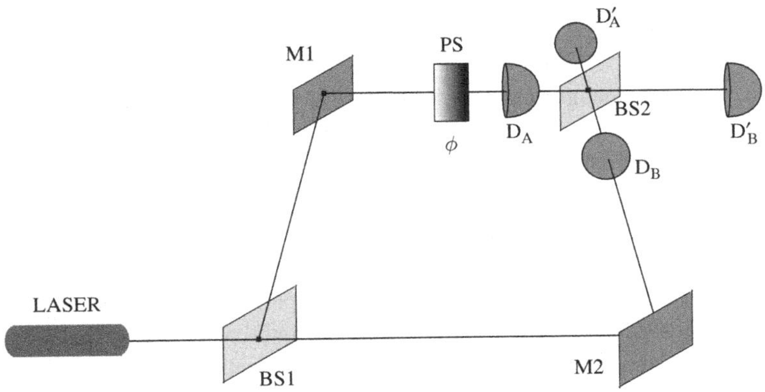

Figure 9.18 **Interferometry experiment for testing delayed choice. The set up is essentially a Mach–Zehnder interferometer in which the two detectors may be switched from positions D_A, D_B (which-path detectors) to positions D'_A, D'_B (interference detectors). This decision will be taken after the photon has already passed through BS1.**

From a quantum-mechanical perspective, we can push ourselves even further, and say that, in the specific example of the quasar, *no event* has occurred at all. In fact, we are allowed to say that an event has occurred only when there is a physical trace such as a detection (see Sec. 9.5). The delayed choice shows on a chronological scale the fundamental quantum result that we cannot speak (not even on the level of the definition) of a phenomenon without interacting with it in some way. Using Wheeler's words, we can state that "no elementary phenomenon is a phenomenon until it is a registered (observed) phenomenon."

Walther and co-workers [Hellmuth *et al.* 1986] have realized an interferometry experiment in order to verify the delayed-choice framework following, in a slightly different version, Wheeler's original proposal (see Fig. 9.18; see also Fig. 2.11). In a Mach–Zehnder like set up, the detectors may be switched from positions D_A, D_B to positions D'_A, D'_B and *vice versa*. This may be done after the beam has already passed the first beam splitter. In the arrangement D_A, D_B we detect the path of the photon, whereas in the arrangement D'_A, D'_B we detect the interference (see Subsec. 2.3.4). Comparing the results obtained in the second case with an "ordinary" interferometry experiment, no differences have been detected: we are totally free to perform delayed-choice experiments without altering quantum predictions.

The above discussion on delayed-choice experiments teaches us a general lesson: There are time intervals – in the case shown in Fig. 9.18, the time interval in which the photon travels from BS1 to the detectors – where we cannot assume that an event happened, whereas, after this interval, an event may have occurred – the photon has been registered. We are then forced to admit that there must also be a reality *before* an event has been registered, since events can only come out of some form of reality. Therefore, reality cannot be made only of a collections of events. This suggests to interpret reality as a dynamical interplay between local events and non-local behaviors (or correlations) that are described

Box 9.3 Complementarity

In the paper published in 1928, which founded the Copenhagen interpretation (see Subsecs. 1.2.4 and 1.5.7), Bohr thought of the complementarity principle in terms of a sharp yes/no alternative between wave-like and corpuscular behaviors and interpreted this complementarity as between the quantum "superposed" dynamics – ruled by the Schrödinger equation (see Ch. 3) – and localizations in space and time induced by measurement. We have already said that complementarity cannot be seen as a sharp yes/no alternative (see Subsec. 2.3.4). As we have seen in the present chapter, there is no sharp boundary between measurement and other quantum dynamical processes, so that also the latter point of Bohr's interpretation is not fully correct. This reformulation of the complementarity principle allows also a reconciliation between the apparently antithetic aspects of the theory, i.e. the unitary time-evolutions and the acquisition of information during a measurement, as it was supposed by von Neumann (see Subsec. 9.2.1).

by the wave function of a (multiparticle) system[60] (see Chs. 15 and 16). Local events and non-local behaviors can be taken to be complementary (see Box 9.3). Moreover, quantum systems are necessarily open to the environment, i.e. they can never be completely isolated. In this respect, measurement is a special case of a more general class of interactions (see Secs. 9.4–9.5, and also Ch. 14).

Finally, note that "quantum eraser" experiments (see Subsec. 9.5.2) have some relationship with the "delayed choice." In fact, it has been proposed and successively realized[61] to consider the quantum eraser as a stronger form of delayed choice to the extent that one can postpone the decision to read or to erase some information (typically a "which-path" information) until after detection. Only, the quantum eraser has more to do with the reversible behavior of the (quantum) detector, and in this sense it is partly related to a different aspect of the theory.

9.8 Quantum Zeno effect

In this chapter we have already analyzed several puzzling features of quantum measurement. Another "strange" effect, which a measurement may give rise to in quantum mechanics, is the so-called *quantum Zeno effect*,[62] which was introduced by Misra and

[60] See [Auletta 2003] [Auletta 2006].

[61] See [Herzog *et al.* 1995] . For the realization see [Kim *et al.* 2000].

[62] This name has been taken from the famous Greek philosopher Zeno of Elea (490 BC–425 BC), who proposed several "paradoxes" about motion aiming to show that any apparently moving object should in reality be "frozen" in its position. For instance, an arrow can never reach its target because the distance the arrow should cover can be divided into infinite space intervals (for example progressively dividing by two the initial interval). Since each small segment requires a finite time, it should take an infinite time for the arrow to reach the target. Of course, Zeno supposed the infinite divisibility of space but not of time. Moreover, ancient philosophers did not know the concept of infinitesimal quantities.

Sudarshan[63] even if some antecedents had already been developed earlier. In a few words, the Zeno effect is the inhibition of spontaneous or induced transitions between quantum states by frequent measurements, so that a system remains in its initial state throughout a given time interval.[64]

It should be emphasized that inhibition of spontaneous transitions – such as the decay of an unstable state, originally proposed by Misra and Sudarshan – are very difficult to realize experimentally. In order to illustrate the main features of the quantum Zeno effect, we start with a simple model:[65] we suppose that a certain observable $\hat{O}$ of a system $\mathcal{S}$ is being monitored continuously and that the initial state of $\mathcal{S}$ is the n-th eigenstate $|\psi_n\rangle$ of $\hat{O}$. We may treat the continuous measurements as a limiting case of discrete measurements separated by small time intervals τ. As we know from Ch. 3, between two measurements (in the interval τ) the evolution of $\mathcal{S}$ is governed by the Schrödinger equation (see also Sec. 14.4). This evolution will cause $\mathcal{S}$ – with some probability – to make a transition from the eigenstate $|\psi_n\rangle$ to some other one. The first interval begins at $t = 0$, and for sufficiently small times τ we may expand the Schrödinger equation into a power series, obtaining, just before the first measurement,

$$|\psi(\tau)\rangle \simeq \left[1 + \frac{\hat{H}\tau}{\imath\hbar} + \frac{1}{2}\left(\frac{\hat{H}\tau}{\imath\hbar}\right)^2 + \cdots\right] |\psi_n\rangle, \tag{9.94}$$

where $\hat{H}$ is the Hamiltonian of the system $\mathcal{S}$. The probability that $\mathcal{S}$ will still be in the n-th eigenstate (that no transition occurred) is clearly given by

$$\begin{aligned}\wp_{nn} &= |\langle\psi_n \mid \psi(\tau)\rangle|^2 \simeq \left|1 + \frac{\tau}{\imath\hbar}\left\langle\psi_n\left|\hat{H}\right|\psi_n\right\rangle - \frac{\tau^2}{2\hbar^2}\left\langle\psi_n\left|\hat{H}^2\right|\psi_n\right\rangle\right|^2 \\ &\simeq 1 - \frac{\tau^2}{\hbar^2}\left[\left\langle\psi_n\left|\hat{H}^2\right|\psi_n\right\rangle - \left\langle\psi_n\left|\hat{H}\right|\psi_n\right\rangle^2\right],\end{aligned} \tag{9.95}$$

up to second order in τ. Since the quantity in square brackets in the rhs of Eq. (9.95) is the variance of the energy in the initial state $|\psi_n\rangle$ (see Subsec. 2.3.1), we can rewrite it in the form

$$\wp_{nn} = 1 - \frac{\tau^2}{\hbar^2}(\Delta E_n)^2. \tag{9.96}$$

We may repeat the same procedure k times, each step consisting of a free evolution during the time-interval τ plus an instantaneous measurement of $\hat{O}$: the probability that, at the end, $\mathcal{S}$ is still in the initial eigenstate is the k-th power of the expression (9.96):

$$\wp_{nn}(k\tau) = \wp_{nn}^k(\tau) = \left[1 - \frac{\tau^2}{\hbar^2}(\Delta E_n)^2\right]^k, \tag{9.97}$$

[63] See [Misra/Sudarshan 1977].

[64] The so-called *watchdog effect* is a particular case of the quantum Zeno effect. It is the suppression of the response of a quantum object when the energy of the system is monitored continuously.

[65] See [*Braginsky/Khalili* 1992, 95–104].

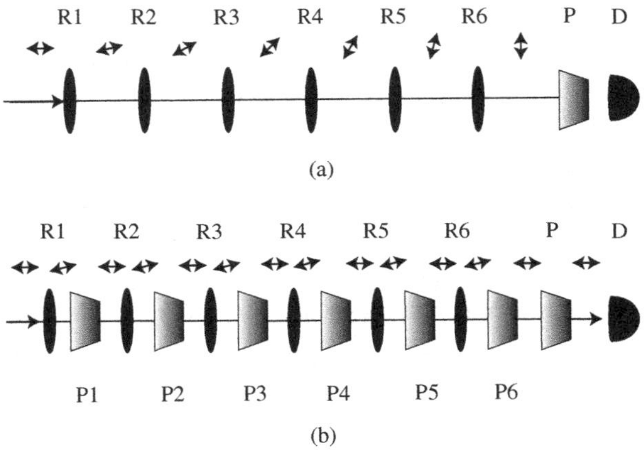

Figure 9.19 Simple optical version of the Zeno effect. (a) A series of polarization rotators (R1, R2, . . . , R6) is used to rotate the polarization of the input photon from horizontal to vertical. All light is then absorbed by the horizontal polarization filter P and nothing is observed in the final detector D. (b) When a series of horizontal polarizers (P1, P2, . . . , P6) is interspersed between the rotators, the light is projected back into a state of horizontal polarization at every stage, resulting in the detection of light by the final detector D. If the number of stages is equal to five, more than 50% of the input light will be transmitted. In particular, for the case shown ($N = 6$), the chance of transmission is very nearly twice the chance of absorption.

where $k\tau$ is the total time for the k steps. The continuous limit for $\tau \to 0$ (but keeping the total measurement time $k\tau$ fixed) is given by[66]

$$\lim_{\tau \to 0} \wp_{nn}(k\tau) \to \exp\left[-\frac{k\tau^2}{\hbar^2}(\Delta E_n)^2\right], \tag{9.98}$$

for small τ. Finally, in the limit of vanishing τ, the probability that no transition occurs becomes

$$\wp_{nn}(k\tau) = 1. \tag{9.99}$$

In other words, $\mathcal{S}$ is "frozen" in the initial state $|\psi_n\rangle$. In conclusion, when any observable with discrete spectrum is continuously monitored with infinite accuracy, the system is "forced" to remain in the initial state.

More recently,[67] a combination of the interaction–free device (see Sec. 9.6) with that of a Zeno experiment has been proposed. If a photon with horizontal polarization enters the device shown in Fig. 9.19(a) from the left, its polarization is gradually rotated to vertical by the sequence of polarization rotators. As a consequence no light can pass the horizontal polarizer P and be detected. However, if we insert a series of horizontal polarizers in between the rotators, the state will be "back projected" each time to the initial state of horizontal polarization and, with a certain probability, will be detected by the final detector.

[66] Since $\lim_{n\to\infty}\left(1+\frac{x}{n}\right)^n = e^x$.

[67] See [Kwiat *et al.* 1995b] and also [Kwiat *et al.* 1996].

9.9 Conditional measurements or postselection

As we have seen (eq. (5.18)), we may express expectation values of an observable $\hat{O}$ in terms of a trace of the form $\mathrm{Tr}\left(\hat{\rho}\hat{O}\right)$, where $\hat{\rho}$ represents the state of a physical system. The probability of obtaining the event described by the projector $\hat{P}_j$ given the state $\hat{\rho}$ can be written as

$$\wp\left(\hat{P}_j|\hat{\rho}\right) = \mathrm{Tr}\left(\hat{\rho}\hat{P}_j\right). \tag{9.100}$$

Suppose now that, given the state $\hat{\rho}$, the quantum event (detection) $\hat{P}_j$ actually occurs. Now it is easy to show (see Prob. 9.10) that, if $\mathrm{Tr}\left(\hat{\rho}\hat{P}_j\right) \neq 0$ (i.e. if the probability of the event $\hat{P}_j$ is non-zero),

$$\hat{\rho}_j = \frac{\hat{P}_j\hat{\rho}\hat{P}_j}{\mathrm{Tr}\left(\hat{\rho}\hat{P}_j\right)} \tag{9.101}$$

is again a density operator and represents, according to von Neumann, the conditional state given that the event $\hat{P}_j$ has occurred (see also Eq. (5.2)). Then, the conditional probability of another event $\hat{P}_k$ once $\hat{P}_j$ has occurred is given by

$$\wp(\hat{P}_k|\hat{\rho}_j) = \mathrm{Tr}\left(\hat{\rho}_j\hat{P}_k\right) = \frac{\mathrm{Tr}\left(\hat{\rho}\hat{P}_j\hat{P}_k\hat{P}_j\right)}{\mathrm{Tr}\left(\hat{\rho}\hat{P}_j\right)}, \tag{9.102}$$

and is known as the *von Neumann formula*.

It is often useful to take advantage of projective measurements (measurements that satisfy the projection postulate) to "guide" the evolution of a quantum system. This procedure is known as *conditional measurement*,[68] and will be illustrated in the following. Consider a system $\mathcal{S}$ in an initial state $|\psi_{\mathcal{S}}(0)\rangle$ interacting with a second system $\mathcal{S}'$ in an initial state $\left|\psi_{\mathcal{S}'}(0)\right\rangle$. After their interaction for a time τ, the combined (entangled) state of the total system will be given by

$$\left|\Psi_{\mathcal{S}\mathcal{S}'}(\tau)\right\rangle = e^{-\frac{i}{\hbar}\tau\hat{H}}\,|\psi_{\mathcal{S}}(0)\rangle \otimes \left|\psi_{\mathcal{S}'}(0)\right\rangle, \tag{9.103}$$

where $\hat{H}$ is the Hamiltonian of the total system. Suppose that we now make a measurement on an observable $\hat{O}$ of the system $\mathcal{S}'$ which gives the result o. It follows that the conditional state of $\mathcal{S}$ after the measurement will be given by

$$\left|\psi^{\mathrm{C}}_{\mathcal{S}}(\tau)\right\rangle = \frac{1}{\left\|\left\langle o \mid \Psi_{\mathcal{S}\mathcal{S}'}(\tau)\right\rangle\right\|}\left\langle o \mid \Psi_{\mathcal{S}\mathcal{S}'}(\tau)\right\rangle, \tag{9.104}$$

where $\| \langle o|\Psi_{\mathcal{S}\mathcal{S}'}(\tau)\rangle \|^2 = \wp(o)$ is precisely the probability that the measurement on $\mathcal{S}'$ gives the result o. It is then clear that, if this probability is different from zero, by carefully "choosing" the result o of the measurement, we may *select* a predetermined final state of $\mathcal{S}$. Of course, we cannot *decide* a priori the result of the measurement, but we know that there

68 This approach has been suggested by [Sherman/Kurizki 1992]. See also the "quantum state engineering" proposed by [Vogel *et al.* 1993].

Box 9.4 Example of postselection

In order to better understand how postselection works, we further illustrate it by means of a simple example. Consider a two-level system $\mathcal{S}$ with basis states $\{|a\rangle, |b\rangle\}$ and another two-level system $\mathcal{S}'$ with basis states $\{|c\rangle, |d\rangle\}$. Suppose also that the initial state of $\mathcal{S}$ is $|\psi_{\mathcal{S}}(0)\rangle = |a\rangle$ and that the initial state of $\mathcal{S}'$ is $|\psi_{\mathcal{S}'}(0)\rangle = |c\rangle$. If the total Hamiltonian $\hat{H}$ is such that the entangled combined state at time τ is

$$\left|\Psi_{\mathcal{S}\mathcal{S}'}(\tau)\right\rangle = \frac{1}{\sqrt{2}}\left(|a\rangle \otimes |c\rangle + |b\rangle \otimes |d\rangle\right), \tag{9.105}$$

then, when performing a measurement on $\mathcal{S}'$, the probability of obtaining the state $|d\rangle$ is 1/2 and the postselected state $|b\rangle$ is given by

$$\left|\psi_{\mathcal{S}}^{\mathrm{C}}(\tau)\right\rangle = \frac{1}{\left\|\left\langle d \mid \Psi_{\mathcal{S}\mathcal{S}'}(\tau)\right\rangle\right\|}\left\langle d \mid \Psi_{\mathcal{S}\mathcal{S}'}(\tau)\right\rangle = |b\rangle. \tag{9.106}$$

In other words, in this way we have guided the evolution of $\mathcal{S}$ from the initial state $|a\rangle$ to the final state $|b\rangle$ with probability 1/2, thanks to the selective measurement made on the ancillary system $\mathcal{S}'$.

is a certain finite probability that the desired outcome *will* be obtained. Then, if we want to guide the evolution of the system $\mathcal{S}$ from $|\psi_{\mathcal{S}}(0)\rangle$ to a certain final state $\left|\psi_{\mathcal{S}}^{\mathrm{C}}(\tau)\right\rangle$, we may repeat the procedure above over and over, until we obtain the measurement that will give the desired result. This will happen with probability $\wp(o)$ and therefore the procedure above will have to be repeated – on average – a number of times of the order of $(\wp(o))^{-1}$ before the desired goal is reached. Each run for which the measurement gives a result different from the one that is desired has to be discarded. Therefore, the price one has to pay for using this scheme in order to guide the evolution of $\mathcal{S}$ is the probability of success, which by necessity is less than the unity. This procedure is also known as postselection (because we select a posteriori the result of the measurement) or *selective measurement* and has to be contrasted to the *non-selective* measurement, where, even though a measurement is equally performed on $\mathcal{S}'$, the postselection is not made. From the discussion above, it appears natural to consider postselection as a sort of preparation (see Subsec. 1.3.1 and conclusions of Sec. 9.1).

Finally, we note that the generalization of Eq. (9.104) to non-pure states is simply given by

$$\hat{\rho}_{\mathcal{S}}^{\mathrm{C}}(\tau) = \frac{1}{\wp}\mathrm{Tr}_{\mathcal{S}'}\left[\hat{\rho}_{\mathcal{S}\mathcal{S}'}(\tau)\,|o\rangle\,\langle o|\right], \tag{9.107}$$

where $\hat{\rho}_{\mathcal{S}\mathcal{S}'}(\tau)$ is the evolved density operator of the compound system, $\hat{\rho}_{\mathcal{S}}^{\mathrm{C}}(\tau)$ is the density operator of $\mathcal{S}$ alone after the conditional measurement, and

$$\wp = \mathrm{Tr}_{\mathcal{S}}\left\{\mathrm{Tr}_{\mathcal{S}'}\left[\hat{\rho}_{\mathcal{S}\mathcal{S}'}(\tau)\,|o\rangle\,\langle o|\right]\right\} \tag{9.108}$$

is the success probability of the conditional measurement.

Box 9.5 Example of operation

In order to explain the concept of operation in simple terms, let us consider the example of the polarization filter introduced in Subsec. 1.3.1 and further developed in Subsec. 2.1.5. Suppose we have a photon described by an initial state vector

$$|\psi_i\rangle = \cos\theta\,|h\rangle + \sin\theta\,|v\rangle, \tag{9.109}$$

where $|h\rangle$ and $|v\rangle$ are the usual horizontal and vertical polarization states, respectively. The density operator corresponding to the state $|\psi_i\rangle$ is

$$\hat{\rho}_i = \cos^2\theta\,|h\rangle\langle h| + \sin^2\theta\,|v\rangle\langle v| + \sin\theta\cos\theta\,(|h\rangle\langle v| + |v\rangle\langle h|). \tag{9.110}$$

The photon impinges on a vertical polarization filter. The test $\mathcal{T}$ in this case is represented by

$$\mathcal{T}(\hat{\rho}_i) = |v\rangle\langle v|\,\hat{\rho}_i\,|v\rangle\langle v| = \hat{P}_v\hat{\rho}_i\hat{P}_v. \tag{9.111}$$

In fact, there are two possible outcomes of this "experiment": either the photon is absorbed (i.e. it does not pass the test) or it emerges from the filter (i.e. it passes the test) in the state $\hat{\rho}_f = |v\rangle\langle v|$ – if we are interested only in the final states that pass the test, we have performed a selective operation. This result also follows from direct application of Eq (9.114) (see Prob. 9.11).

9.10 Positive operator valued measure

The traditional (von Neumann's) formal treatment of measurement makes use of projectors (and of the projection postulate) (see Secs. 9.1–9.2). In other words, any observable that can be represented by an operator on the Hilbert space of the system under consideration is expanded in terms of *complete* (see Eq. (1.41a)) and *orthogonal* operators (see Eq. (1.41b)), i.e. of projectors. Measurements represented by a set of projectors are therefore called *orthogonal*. However, as we have seen, measurements defined by projectors generate a discontinuous change in the evolution of the state. If a measurement did not perturb the system to be measured, it could in principle be repeated over and over in order to improve at each subsequent measurement the knowledge of the state of the system (i.e. the information that can be extracted from the system). However, repeated orthogonal measurements do not improve our knowledge of the measured observable, simply because subsequent orthogonal measurements can only deterministically repeat the result of the first (see Eq. (1.40)).

These arguments suggest that it would be suitable to define measurements not directly associated with traditional observables. In this new framework, measurements associated with a projector could be considered as an ideal limiting case of a more general class of measurements for which the orthogonality condition is abandoned. We then introduce (1) a wider class of operations than is allowed by the traditional von Neumann theory

(Subsec. 9.10.1), (2) a new class of non-orthogonal operators that replace projectors (Subsec. 9.10.2), and (3) a new kind of observables which can be expanded in terms of these non-orthogonal operators rather than in terms of projectors (Subsec. 9.10.3).

9.10.1 Operations

In order to generalize our definition of measurement, it is convenient to introduce the concept of operation, first from a mathematical standpoint. It represents a useful generalization of the class of quantum transformations. An *operation* is defined as follows:[69]

Definition 9.1 (Operation) *An operation $\mathcal{T}$ is a positive linear mapping from a state space into another (possibly itself), which satisfies the following requirement:*

$$0 \leq \mathrm{Tr}(\mathcal{T}\hat{\rho}) \leq \mathrm{Tr}(\hat{\rho}) \tag{9.112}$$

and the norm $\| \mathcal{T} \|$ is defined as $\| \mathcal{T} \| = \mathrm{Sup}\left\{\mathrm{Tr}\left[\mathcal{T}(\hat{\rho})\right]\right\}$ for each $\hat{\rho} \in \mathcal{H}$, where $\mathcal{H}$ is the Hilbert state space.

We symbolize the operation $\mathcal{T}$ on $\hat{\rho}$ by $\mathcal{T}(\hat{\rho})$. We may think of $\mathcal{T}$ as a "test" which is undergone by the system in the state $\hat{\rho}$. Then, the probability of the transmission of a state $\hat{\rho}$ by an operation $\mathcal{T}$ (i.e. the probability that the system state passes the test) is

$$\mathrm{Tr}\left[\mathcal{T}\left(\hat{\rho}\right)\right]. \tag{9.113}$$

In this context, the output or final state, upon transmission, is taken to be (see Sec. 9.9)

$$\hat{\rho}_f = \frac{\mathcal{T}\left(\hat{\rho}_i\right)}{\mathrm{Tr}\left[\mathcal{T}\left(\hat{\rho}_i\right)\right]}, \tag{9.114}$$

where $\hat{\rho}_i$ is the initial state.

Following the discussion in Sec. 9.9, we may distinguish[70] between non-selective and selective operations (preparations): when we are interested in the transformations performed by an apparatus, we speak of *non-selective operations* (depending on the equivalence class of the initial state only). In other words, by keeping the apparatus fixed and by varying the initial state, we obtain all possible state transformations induced by the apparatus. We speak, instead, of *selective operations*, if we are interested in a given subset of the possible final states, satisfying a certain requirement. For this reason, a selective operation may be considered a *preparation*.

[69] See [*Davies* 1976, 17–18].
[70] See [*Kraus* 1983, 13–17, 39–40, 71].

Then, we are able to give an operational definition of *state*: since different preparation procedures may be statistically equivalent (they yield the same statistics for all possible measurements), the state corresponds to an *equivalence class of preparations*. In other words, two states are said to be equivalent if they reproduce the same statistics and therefore cannot be experimentally distinguished (see Subsec. 9.10.3). Given the nature of quantum theory, and especially the non-commutativity and the uncertainty principle, this is the most suitable definition of a state. On the other hand, it is also possible to give an operational definition of *observable*: since different experimental coupling procedures may select the same observable, the observable is an *equivalence class of premeasurements*.

However, according to the discussion of Subsec. 8.1.1, there must exist some sort of equivalence between the application of an operation to the system's state or to a given observable of the system. This equivalence is formally expressed by the following theorem:[71]

Theorem 9.1 (Kraus) *For an operation $\mathcal{T}$ there exist operators $\hat{\vartheta}_k$ (where k is an integer in a finite or infinite set K) on the Hilbert space satisfying*

$$\sum_{k \in K} \hat{\vartheta}_k^\dagger \hat{\vartheta}_k = \hat{I}, \tag{9.115}$$

such that, for a given system's observable $\hat{O}$ and an arbitrary state $\hat{\rho}$, the operations $\mathcal{T}$ and $\mathcal{T}^*$ are given by (see also Eq. (9.4))

$$\mathcal{T}\left(\hat{\rho}\right) = \sum_{k \in K} \hat{\vartheta}_k \hat{\rho} \hat{\vartheta}_k^\dagger, \tag{9.116a}$$

$$\mathcal{T}^*\left(\hat{O}\right) = \sum_{k \in K} \hat{\vartheta}_k^\dagger \hat{O} \hat{\vartheta}_k, \tag{9.116b}$$

respectively, and where (see Prob. 9.12)

$$\mathrm{Tr}\left[\mathcal{T}\left(\hat{\rho}\right)\hat{O}\right] = \mathrm{Tr}\left[\hat{\rho}\mathcal{T}^*\left(\hat{O}\right)\right]. \tag{9.117}$$

The operation described by Th. 9.1 is not necessarily unitary, i.e. in general it is not possible to write down the previous formulae with only a single operator ϑ, and can be used to describe active and passive transformations (see again Subsec. 8.1.1). Indeed, Eqs. (9.116) describe the same dynamics from two different perspectives: the operation (9.116a) performed by $\mathcal{T}$ corresponds to the generalization of Schrödinger picture, while the operation (9.116b) performed by $\mathcal{T}^*$ corresponds to the generalization of the Heisenberg picture. In the next subsection we shall see how the operators $\hat{\vartheta}_k^\dagger$ and $\hat{\vartheta}_k$ can be interpreted. Incidentally, we note that the action of the operators $\hat{\vartheta}_k^\dagger$ and $\hat{\vartheta}_k$ in Eqs. (9.116) shows a formal similarity with the action of the "environment" operators in the master equation formalism (see Sec. 14.2). As we shall see this is not fortuitous.

[71] See [*Kraus* 1983, 42].

9.10.2 System, apparatus, and the back action

As we have said, we may understand the traditional (von Neumann) exact measurement as a limiting case of a wider class of measurements: we renounce the condition of orthogonality, which characterizes projectors, but not that of completeness.[72] To this purpose, we make use of a new type of operator called *effect*.[73] Effects describe the most general form of measurement, and, in particular, they provide a definition of a measurement apparatus that performs a *realistic* – and, as a consequence, necessarily approximate – measurement of a certain observable. Limiting ourselves for the time being to the one-dimensional case, let us consider the conditional probability $\wp(x_m|x_S)$ that the measuring device registers the measurement outcome x_m when the observable $\hat{x}$ is actually in the eigenstate $|x_S\rangle$ with value x_S, that is, when our apparatus has not precisely registered the exact value of the measured observable. If $\hat{\rho}_i$ represents the initial state of the system, we express the probability of obtaining a result x_m by means of these conditional probabilities as

$$\wp(x_m) = \int_{-\infty}^{+\infty} dx_S \wp(x_m|x_S)\wp(x_S), \tag{9.118}$$

where

$$\wp(x_S) = \langle x_S|\hat{\rho}_i|x_S\rangle \tag{9.119}$$

is the initial state's a priori probability distribution for the values of $\hat{x}$. Equation (9.118) is a form of Bayes' rule (see also Eq. (9.170)). Probability (9.118) can be rewritten, in a form similar to the standard Eq. (5.18), as[74]

$$\wp(x_m) = \mathrm{Tr}\left[\hat{E}(x_m)\hat{\rho}_i\right], \tag{9.120}$$

where the effects $\hat{E}(x_m)$ are given by

$$\hat{E}(x_m) = \int_{-\infty}^{+\infty} dx_S \wp(x_m|x_S)|x_S\rangle\langle x_S|, \tag{9.121}$$

which expresses the relationship between projectors $\hat{P}(x_S) = |x_S\rangle\langle x_S|$ and effects $\hat{E}(x_m)$ and can be therefore considered a clarifying expression and a first approximation to the meaning of effects. They are considered here as weighted averages of projectors, with weights given by the probabilities $\wp(x_m|x_S)$ that describe the approximate behavior of the measuring apparatus. For the sake of simplicity, in this context we limit ourselves to consider the probabilities $\wp(x_m|x_S)$ as the apparatus' properties. We shall return to this

72 Due to the necessity of a proper normalization of the output state, *any* measurement must satisfy the completeness condition (see Eq. (9.114)).

73 See [*Braginsky/Khalili* 1992, 33–37].

74 See [Busch 2003].

point in Subsec. 9.11.1, where we shall explore the quantum nature of these probabilities, and in Sec. 14.3, where a generalization of this formalism will be provided.

Now, by analogy with Eq. (9.101), we write the density operator describing the output state of $\mathcal{S}$ after the whole measurement process in the form

$$\hat{\rho}_f(x_m) = \frac{1}{\wp(x_m)}\hat{\vartheta}(x_m)\hat{\rho}_i\hat{\vartheta}^\dagger(x_m), \tag{9.122}$$

where the operator $\hat{\vartheta}(x_m)$ (whose form is to be determined) can be called an amplitude operator and completely describes the whole measurement process. From Eq. (9.120) and the normalization condition $\mathrm{Tr}[\hat{\rho}_f(x_m)] = 1$, it follows that (see Prob. 9.13)

$$\hat{\vartheta}^\dagger(x_m)\hat{\vartheta}(x_m) = \hat{E}(x_m). \tag{9.123}$$

Using the polar decomposition of an operator,[75] $\hat{\vartheta}(x_m)$ can be represented as

$$\hat{\vartheta}(x_m) = \hat{U}_t(x_m)\hat{E}^{\frac{1}{2}}(x_m), \tag{9.124}$$

where $\hat{U}_t(x_m)$ is a unitary evolution operator whose form depends on the measurement and where

$$\hat{E}^{\frac{1}{2}}(x_m) = \int_{-\infty}^{+\infty} dx_{\mathcal{S}} \left[\wp(x_m|x_{\mathcal{S}})\right]^{\frac{1}{2}} |x_{\mathcal{S}}\rangle\langle x_{\mathcal{S}}| \tag{9.125}$$

is uniquely determined by the apparatus' conditional probability $\wp(x_m|x_{\mathcal{S}})$ and commutes with $\hat{x}$. Therefore, when we consider a single output state, according to Eq. (9.116a) we can write the whole measurement process as

$$\begin{aligned}\hat{\rho}_f(x_m) &= \frac{1}{\wp(x_m)}\hat{\vartheta}(x_m)\hat{\rho}_i\hat{\vartheta}^\dagger(x_m)\\ &= \frac{1}{\wp(x_m)}\hat{U}_t(x_m)\hat{E}^{\frac{1}{2}}(x_m)\hat{\rho}_i\hat{E}^{\frac{1}{2}}(x_m)\hat{U}_t^\dagger(x_m).\end{aligned} \tag{9.126}$$

Therefore, measurement can be formally viewed as a two-component process consisting in the unitary transformation

$$\hat{\rho}_f(x_m) = \hat{U}_t(x_m)\hat{\rho}'\hat{U}_t^\dagger(x_m), \tag{9.127a}$$

which is a premeasurement, and the reduction

$$\hat{\rho}'(x_m) = \frac{1}{\wp(x_m)}\hat{E}^{\frac{1}{2}}(x_m)\hat{\rho}_i(x_m)\hat{E}^{\frac{1}{2}}(x_m). \tag{9.127b}$$

[75] It is so called in analogy with the decomposition of a complex number λ as $\lambda = e^{\imath\phi}|\lambda|$, where $|e^{\imath\phi}| = 1$ [*Taylor/Lay* 1958, 379].

The matrix $\hat{\rho}'$ is a density matrix again determined by the measurement process. Thus, as we have illustrated in Sec. 9.5, there are two aspects of the measurement: the unitary transformation, which is reversible; and information acquiring, which is irreversible. For the time being, this decomposition of the measurement process is purely formal. In the next section we shall learn more about the physical mechanism that generates such a situation. The step (9.127b) leaves the measured value of the observable unchanged, that is, we have (see Prob. 9.14)

$$\int_{-\infty}^{+\infty} dx_m \wp(x_m) \left\langle x_{\mathcal{S}} \left| \hat{\rho}'(x_m) \right| x_{\mathcal{S}} \right\rangle = \left\langle x_{\mathcal{S}} \left| \hat{\rho}_i \right| x_{\mathcal{S}} \right\rangle, \tag{9.128}$$

since the operator $\hat{E}^{\frac{1}{2}}(x_m)$ commutes with $\hat{x}$. In general, a measurement will cause a perturbation of the state of the object system induced by the apparatus, which is called *back action*.[76] The back action is caused only by the unitary evolution of the form (9.127a), while the reduction is completely determined by transition probabilities of the form $\wp(x_m|x_{\mathcal{S}})$, which is the informational content which we can extract from a measurement. In other words, the unitary transformation does not change the entropy of $\mathcal{S}$ and therefore cannot produce any new information. However, as results from Eqs. (9.120)–(9.121), the strength of the perturbation depends on the information which we *can obtain* in a measurement process, i.e. from $\wp(x_m|x_{\mathcal{S}})$. As we shall see with further details in the following, this formalism provides the definitive solution of the measurement problem that we have discussed from the beginning of the present chapter.

Summarizing the above discussion, an effect is to a certain extent an analog of a projector but with important differences. In fact, effects respect the completeness condition (see Eq. (1.41a)), i.e. for a set of effects $\{\hat{E}_j\}$, we have

$$\sum_j \hat{E}_j = \hat{I}. \tag{9.129}$$

However, while projectors have only 0 and 1 as eigenvalues and, as a consequence, for any set of projectors $\{\hat{P}_n\}$, $\hat{P}_j\hat{P}_k = \delta_{jk}\hat{P}_k$ (see Subsec. 1.3.2), effects in the set $\{\hat{E}_n\}$ may be *unsharp*, that is, we may have $\hat{E}_j\hat{E}_k \neq \delta_{jk}\hat{E}_k$.[77] Also effects which represent combinations of projectors pertaining to the same set (i.e. which project on one of the vectors of a certain orthonormal basis on the Hilbert space of the system) commute, so that, for instance, we have

$$\left[\hat{E}_j, \hat{E}_k\right] = 0. \tag{9.130}$$

Moreover, we can say that projectors are orthogonal extremes of the convex set of effects.

[76] A recent experimental evidence of quantum back action has been provided, even at a macroscopic level, as an effect of a superconducting transistor (the measuring device) on a nanomechanical resonator [Naik *et al.* 2006]. See also [Roukes 2006].

[77] See [*Busch et al.* 1991, 10] .

9.10.3 Unsharp observables

In the previous subsection we have made an important but still rather specific step towards a generalization of the concept of observables. Indeed, up to now we have considered a set of effects that commute. However, effects are also a very important tool for widening the concept of measurement in order to comprehend the joint measurement of non-commuting observables. We can distinguish between *sharp properties* (the values of an observable), which are defined by projectors, and *unsharp properties*, which are defined by effects. We may also distinguish between *sharp observables*, defined by projector valued measures (PVMs), and *unsharp observables* defined by positive operator valued measures (POVMs): PVMs may be considered as a combination of projectors, whereas POVMs may be seen as combinations of effects that do not necessarily commute. In other words, PVMs are resolutions of the identity in terms of orthogonal projectors, whereas POVMs are resolutions of the identity in terms of non-orthogonal and in general non-commuting effects.

Unsharpness must not be confused with subjective ignorance: it rather expresses the objective uncertainty intrinsic in the joint measurement of non-commuting (incompatible) observable. For example, in the case of the Mach–Zehnder interferometer described in Subsec. 2.3.4, it is possible to construct intermediate situations between a perfect determination of the path of a particle and perfect visibility of the interference pattern. This intermediate situation is an unsharp feature.

One of the main goals of POVMs is that they allow a coherent quantum-mechanical description of joint measurement of non-commuting observables. While PVMs just tell us that it is not possible to perform ideal joint measurements of non-commuting observables, POVMs allow us to proceed a step further: it is possible to perform *non-ideal* measurements of incompatible observables – which justifies our formulation of complementarity principle in terms of a smooth relation.

Let us consider a simple example.[78] Consider the experimental arrangement shown in Fig. 9.20. While BS1 and BS2 are symmetric, the beam splitter BS3 has a transmission parameter $\sqrt{\eta}$ (see Prob. 9.15). We may distinguish between three possible cases:

- $\eta = 1$. In this case, the outcome probabilities are given by (Prob. 9.16)

$$\wp_k = \wp(D_k) = \mathrm{Tr}\left[\hat{P}_k^{\mathcal{V}} \hat{\rho}\right] = \left\langle \hat{P}_k^{\mathcal{V}} \right\rangle_{\psi}, \quad \text{for} \quad k = 1, 2 \tag{9.131}$$

and

$$\wp_3 = \wp(D_3) = 0, \tag{9.132}$$

where

$$\hat{P}_2^{\mathcal{V}} = \begin{bmatrix} \cos^2 \frac{\phi}{2} & -\frac{\imath}{2} \sin\phi \\ \frac{\imath}{2} \sin\phi & \sin^2 \frac{\phi}{2} \end{bmatrix}, \quad \hat{P}_1^{\mathcal{V}} = \hat{I} - \hat{P}_2^{\mathcal{V}}, \tag{9.133}$$

and

$$| \psi \rangle = | 1 \rangle \tag{9.134}$$

[78] See [de Muynck *et al.* 1991] [Martens/De Muynck 1990].

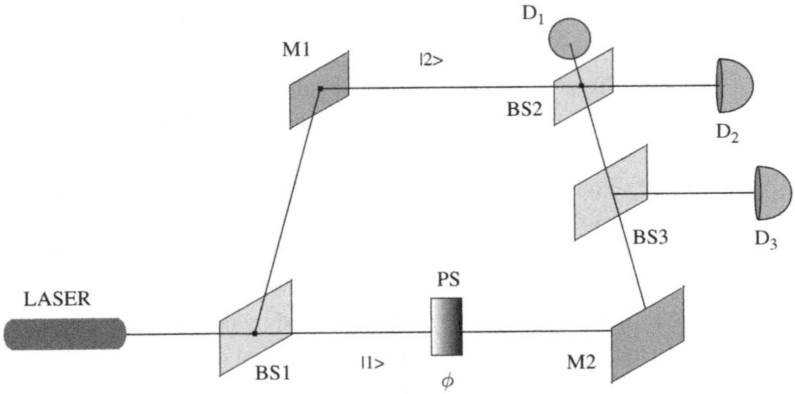

Figure 9.20 Example of POVMs by means of an interferometry experiment. A single-photon state enters from the left a Mach–Zehnder interferometer with phase shift ϕ. In the right arm of the interferometer an additional beam splitter BS3 is inserted, with transmission parameter $\sqrt{\eta}$. Three detectors are placed at three outputs of the interferometer.

is the initial state of the photon, whose corresponding density matrix is $\hat{\rho} = |\psi\rangle\langle\psi|$. In this case, the projectors (9.133) project into the eigenstates of the interference or superposition observable, which can then be written as

$$\hat{\mathcal{V}} = \hat{P}_1^{\mathcal{V}} - \hat{P}_2^{\mathcal{V}}. \tag{9.135}$$

- $\eta = 0$. In this case we can clearly distinguish an upper and lower path, where we have (see Prob. 9.17)

$$\wp_u = \wp_1 + \wp_2 = \left\langle \hat{P}_u^{\mathcal{P}} \right\rangle_\psi, \quad \wp_d = \wp_3 = \left\langle \hat{P}_d^{\mathcal{P}} \right\rangle_\psi, \tag{9.136}$$

where

$$\hat{P}_d^{\mathcal{P}} = \frac{1}{2}\begin{bmatrix} 1 & 1 \\ 1 & 1 \end{bmatrix}, \quad \hat{P}_u^{\mathcal{P}} = \hat{I} - \hat{P}_d^{\mathcal{P}}. \tag{9.137}$$

Then, these projectors represent a resolution of the path observable

$$\hat{\mathcal{P}} = \hat{P}_u^{\mathcal{P}} - \hat{P}_d^{\mathcal{P}}. \tag{9.138}$$

Here, we obviously have a typical classical probability according to a projection-like (or von Neumann's) reduction.

- In all cases where $0 < \eta < 1$ we have a POVM. In this case the probabilities are given by (see Prob. 9.18)

$$\wp_j = \mathrm{Tr}\left[\hat{E}_j \hat{\rho}\right] = \left\langle \hat{E}_j \right\rangle_\psi, \quad \text{for } j = 1, 2, 3, \tag{9.139}$$

where (see Prob. 9.19)

$$\hat{E}_1 = \frac{1}{2}\left[\hat{P}_d^{\mathcal{P}} + \eta \hat{P}_u^{\mathcal{P}} - \sqrt{\eta}\left(\hat{P}_2^{\mathcal{V}} - \hat{P}_1^{\mathcal{V}}\right)\right], \tag{9.140a}$$

$$\hat{E}_2 = \frac{1}{2}\left[\hat{P}_d^{\mathcal{P}} + \eta \hat{P}_u^{\mathcal{P}} + \sqrt{\eta}\left(\hat{P}_2^{\mathcal{V}} - \hat{P}_1^{\mathcal{V}}\right)\right], \tag{9.140b}$$

$$\hat{E}_3 = (1 - \eta)\, \hat{P}_u^{\mathcal{P}}. \tag{9.140c}$$

Therefore, the POVM observable may be written as

$$\hat{O}_{\rm POV} = o_1 \hat{E}_1 + o_2 \hat{E}_2 + o_3 \hat{E}_3. \tag{9.141}$$

As we shall see in Sec. 15.4, due to the possibility of jointly measuring non-commuting observables, it is possible to establish an important connection between POVM and quantum-state reconstruction.

9.11 Quantum non-demolition measurements

A quantum non-demolition (QND) *measurement* is a measurement in which an apparatus extracts information only on the observable to be measured and transfers the whole back action onto the canonical conjugate observable. In other words, the observable to be measured remains unperturbed, while the canonically conjugate one is perturbed precisely to the minimal extent allowed by the uncertainty relations. In order to examine the properties of a QND measurement, we need to introduce first the concept of indirect measurement.

9.11.1 Indirect measurement

We can use the distinction between the two aspects of measurement (introduced in Subsec. 9.10.2) to present the concept of *indirect measurement*, which we contrast to the standard *direct measurement*.[79] The latter form of measurement is an interaction between a quantum system $\mathcal{S}$ and a macroscopic apparatus $\mathcal{A}$, while the indirect measurement is characterized by two separate steps: first, $\mathcal{S}$ interacts with another quantum system $\mathcal{S}_{\rm P}$, the quantum probe, whose initial state has been accurately prepared on purpose in advance. This is an intermediate system with which the object system $\mathcal{S}$ interacts, and from which the apparatus $\mathcal{A}$ extracts information about $\mathcal{S}$. During the first step there is no reduction at all, and the evolution is completely unitary, resulting in a correlation between $\mathcal{S}$ and $\mathcal{S}_{\rm P}$. In other words, the system and the probe become entangled (see Sec. 9.1). The second step consists of a direct measurement of some chosen observable of $\mathcal{S}_{\rm P}$: the state of the probe (and therefore, due to the entanglement, also of the object system) is reduced and the information acquired. As we shall see below, the indirect measurement is a necessary but not sufficient element of a quantum non-demolition (QND) measurement.

[79] See [*Braginsky/Khalili* 1992, 40–49].

Let us now introduce two conditions referring to the two steps defined above:

- The reduction, i.e. the second step of measurement should begin only when the unitary evolution, i.e. the first step has already finished.
- The second step should not contribute significantly to the total error of the measurement.

If these conditions are satisfied, we can infer the magnitudes of the error in the measurement and therefore of the perturbation (back action) of $\mathcal{S}$ from an analysis of the first step only, i.e. of the unitary evolution, because the only source of error is due to the intrinsic uncertainties of the initial state of $\mathcal{S}_{\mathrm{P}}$.

We can describe the indirect measurement in a formal way as follows. Let the *first step* be represented by the transformation

$$\hat{\rho}_i^{\mathcal{S}}\hat{\rho}^{\mathcal{S}_{\mathrm{P}}} \mapsto \hat{U}_t\hat{\rho}_i^{\mathcal{S}}\hat{\rho}^{\mathcal{S}_{\mathrm{P}}}\hat{U}_t^\dagger, \tag{9.142}$$

where $\hat{\rho}_i^{\mathcal{S}}\hat{\rho}^{\mathcal{S}_{\mathrm{P}}}$ is the total density matrix of the system $\mathcal{S}+\mathcal{S}_{\mathrm{P}}$, and $\hat{U}_t$ is the coupling unitary-evolution operator. The corresponding state of $\mathcal{S}_{\mathrm{P}}$ alone, after the interaction, is given by the reduced density matrix (see Subsec. 5.5.2)

$$\hat{\tilde{\rho}}^{\mathcal{S}_{\mathrm{P}}} = \mathrm{Tr}_{\mathcal{S}}\left(\hat{U}_t\hat{\rho}_i^{\mathcal{S}}\hat{\rho}^{\mathcal{S}_{\mathrm{P}}}\hat{U}_t^\dagger\right). \tag{9.143}$$

Suppose that we want to measure the observable $\hat{x}$ on $\mathcal{S}$. Thanks to the entanglement, it is possible to achieve a one-to-one correspondence between the observable $\hat{x}$ of $\mathcal{S}$ and a carefully chosen observable of $\mathcal{S}_{\mathrm{P}}$, say $\hat{p}_x$. We can then perform a "direct" measurement of $\hat{p}_x$ on $\mathcal{S}_{\mathrm{P}}$. Since this measurement contributes negligibly to the experiment's overall error, we can idealize it as arbitrarily accurate. Then, we can infer from the value of $\hat{p}_x$ on $\mathcal{S}_{\mathrm{P}}$ the value x_m of the observable $\hat{x}$ on $\mathcal{S}$. Because of the one-to-one correspondence we can use $\hat{x}$ as a substitute for $\hat{p}_x$ and hence use x_m not only as the inferred value of $\hat{x}$ but also as the result of a measurement on $\mathcal{S}_{\mathrm{P}}$ *itself*, that is, the associated eigenstate of the probe can be denoted by $|x_m\rangle$. Just before the second step of the measurement, the probability distribution of the measured value x_m will be simply given by

$$\wp(x_m) = \mathrm{Tr}_{\mathcal{S}_{\mathrm{P}}}\left[\hat{P}_{x_m}\hat{\tilde{\rho}}^{\mathcal{S}_{\mathrm{P}}}\right] = \mathrm{Tr}_{\mathcal{S}_{\mathrm{P}}}\left[|x_m\rangle\langle x_m|\mathrm{Tr}_{\mathcal{S}}(\hat{U}_t\hat{\rho}_i^{\mathcal{S}}\hat{\rho}^{\mathcal{S}_{\mathrm{P}}}\hat{U}_t^\dagger)\right], \tag{9.144}$$

which, using the linearity and the cyclic property of the trace (see Box 3.1), can be rewritten as (see Eq. (9.120))

$$\wp(x_m) = \mathrm{Tr}_{\mathcal{S}}\left[\hat{E}(x_m)\hat{\rho}_i^{\mathcal{S}}\right], \tag{9.145}$$

where

$$\hat{E}(x_m) = \mathrm{Tr}_{\mathcal{S}_{\mathrm{P}}}\left[\hat{U}_t^\dagger|x_m\rangle\langle x_m|\hat{U}_t\hat{\rho}^{\mathcal{S}_{\mathrm{P}}}\right]. \tag{9.146}$$

The back action of the entire two-step measurement on $\mathcal{S}$ is embodied in the final state of the object system, which we have supposed to have the form (9.122), as we shall prove in the following. In fact, the above considerations imply that such a normalized final state be

$$\hat{\rho}_f^{\mathcal{S}}(x_m) = \frac{1}{\wp(x_m)}\langle x_m|\hat{U}_t\hat{\rho}_i^{\mathcal{S}}\hat{\rho}^{\mathcal{S}_{\mathrm{P}}}\hat{U}_t^\dagger|x_m\rangle. \tag{9.147}$$

If we express the initial state of $\mathcal{S}_P$ as (see also Eq. (5.25))

$$\hat{\rho}^{\mathcal{S}_P} = \sum_k w_k |\psi_k\rangle\langle\psi_k|, \tag{9.148}$$

where $|\psi_k\rangle\langle\psi_k|$ are some projectors on the probe's Hilbert space, and, by substituting this expression into Eq. (9.147), we obtain

$$\hat{\rho}_f^{\mathcal{S}}(x_m) = \frac{1}{\wp(x_m)} \sum_k w_k \hat{\vartheta}_k(x_m)\hat{\rho}_i^{\mathcal{S}}\hat{\vartheta}_k^\dagger(x_m), \tag{9.149}$$

where

$$\hat{\vartheta}_k(x_m) = \langle x_m|\hat{U}_t|\psi_k\rangle. \tag{9.150}$$

Differently from Eq. (3.87), Eq. (9.150) does not contain a probability amplitude because it is a unitary operator that represents the coupling of the probe and the system, whereas the ket and the bra belong to the probe's Hilbert space only. As a result, it represents an amplitude *operator*. As we have said, this amplitude operator describes all steps of the measurement of a given observable: premeasurement ($|\psi_k\rangle$), i.e. preparation of the initial state of the probe; unitary evolution of the probe together with the object system ($\hat{U}_t$); and reduction of the probe ($\langle x_m|$), and gives therefore the "probability amplitude" for the quantum probe to have evolved from the initial state $|\psi_k\rangle$ to the final state $|x_m\rangle$, if the interaction with $\mathcal{S}$ is given by $\hat{U}_t$. Eq. (9.149) shows that the final state of the object system is a mixture of states of the type (9.122) with weighting factors w_k. Therefore, the final state (9.149) has additional uncertainties relative to the state (9.122), which derive from the fact that we allowed $\mathcal{S}_P$ to begin in the mixed state (9.148) and not in a pure one. These additional uncertainties are of classical origin as they are relative to the uncertainties in the initial state of the probe.

9.11.2 QND measurement

The conclusions drawn from Subsecs. 9.10.2 and 9.11.1 allow us to discuss the QND measurement. The central ingredient that makes the QND procedure[80] realizable is just the two-step measurement process described previously. It is then clear that some features of the indirect measurement also characterize the QND measurement. In general, we may say that, in a QND measurement, the system $\mathcal{S}$ interacts only with a probe $\mathcal{S}_P$, and the interaction between $\mathcal{S}$ and $\mathcal{S}_P$ is such that $\mathcal{S}_P$ is influenced only by one observable, or a set of observables, that are not affected by the back action of $\mathcal{S}_P$ on $\mathcal{S}$. More precisely, the system's observables which influence the probe must all commute with each other – i.e. they should belong to the same complete set of observables (see end of Subsec. 2.1.5).

[80] See [*Braginsky/Khalili* 1992, 55–67].

Moreover, a QND measurement can be performed only on observables that are conserved during the object's free evolution, i.e. on constants of motion (see Subsec. 3.6.1): in the absence of external forces, the observable is conserved both during the measurement (because the back action is transferred to the canonically conjugate observable only) and during the unitary evolution between consecutive measurements (because it is an integral of motion).[81]

The above considerations imply that a QND measurement does not add any perturbation to the observable to be measured, so that the uncertainty of the measured observable after the measurement is only a consequence of the a priori uncertainty of its value.

Then, the observable $\hat{O}_{\mathrm{ND}}$ associated with a QND measurement must satisfy the following two requirements:[82]

- At any time it must commute with itself at a different time,

$$\left[\hat{O}_{\mathrm{ND}}(t), \hat{O}_{\mathrm{ND}}(t')\right] = 0, \quad t' \neq t. \tag{9.151a}$$

- It must commute with the time-displacement unitary operator $\hat{U}_t$

$$\left(\hat{U}_t^{\dagger} \hat{O}_{\mathrm{ND}} \hat{U}_t - \hat{O}_{\mathrm{ND}}\right) |\psi\rangle = 0, \tag{9.151b}$$

where $|\psi\rangle$ is the probe's initial state and the expression within the brackets is the Heisenberg-picture change in $\hat{O}_{\mathrm{ND}}$ produced by the interaction between $\mathcal{S}$ and $\mathcal{S}_{\mathrm{P}}$. Equation (9.151b) represents a necessary and sufficient condition of a QND observable.[83]

We can therefore say that a QND measurement is characterized by the *repeatability*, so that the first measurement – which determines the values for all subsequent QND ones – is a preparation of $\mathcal{S}$ in the desired state, and the others are the determination of the value. As a consequence, a QND measurement is a measurement of the first kind – but, obviously, not necessarily *vice versa*. Formally,

$$\hat{O}_{\mathrm{ND}}(t_k) = f_k\left[\hat{O}_{\mathrm{ND}}(t_0)\right], \tag{9.152}$$

[81] One of the most interesting applications of QND-like measurement schemes is represented by the detection of gravitational waves. See [Braginsky *et al.* 1980, 751–52] [Unruh 1979] [Thorne *et al.* 1978]. Moreover, the first quantum-optical experimental verifications of the validity of this measurement scheme were realized in the mid 1980s.

[82] See also [Caves *et al.* 1980, 364].

[83] Equivalently, we could express the first condition by stating that a QND observable commutes with all its time derivatives.

where t_k is some arbitrary time after the initial t_0 (the time of the first measurement), and f_k is some real-valued function. Note that condition (9.152) implies condition (9.151a) (see Prob. 2.12). The generalization of Eq. (9.152) to continuous measurements is given by

$$\hat{O}_{\rm ND}(t) = f\left[\hat{O}_{\rm ND}(t_0); t, t_0\right], \tag{9.153}$$

which defines a *continuous* QND observable, while, if an observable satisfies Eq. (9.152) only at selected discrete times, it is called a *stroboscopic* QND observable. Examples of the last one are the position and the momentum of a harmonic oscillator (due to the periodicity of the evolution). The simplest way to satisfy Eq. (9.153) is to choose an observable which is conserved in the absence of interactions.

On the other hand, condition (9.151b) is certainly satisfied if the expression between brackets vanishes, i.e. if $\hat{O}_{\rm ND}$ returns to its initial value after the measurement.[84] In this case, we may simply write the necessary and sufficient condition for a QND measurement as

$$\left[\hat{O}_{\rm ND}, \hat{U}_t\right] |\psi\rangle = 0. \tag{9.154}$$

The previous condition requires the form of $\hat{U}_t$ to be known. Due to the practical difficulty in determining $\hat{U}_t$, the evolution operator of the coupled system, one can alternatively choose a necessary but not sufficient criterion for a QND observable: that the measured observable be an integral of motion for the total system $\mathcal{S} + \mathcal{S}_{\rm P}$, i.e. that it remains constant during the interaction,

$$\imath\hbar \frac{\partial \hat{O}_{\rm ND}}{\partial t} + \left[\hat{O}_{\rm ND}, \hat{H}\right] = 0, \tag{9.155}$$

where $\hat{H}$ is the total Hamiltonian of the compound system (see also Eq. (3.107)). Equation (9.155), in the case that $\hat{O}_{\rm ND}$ has no explicit time-dependence, reduces to the condition

$$\left[\hat{O}_{\rm ND}, \hat{H}_{\mathcal{S}_{\rm P}\mathcal{S}}\right] = 0, \tag{9.156}$$

where $\hat{H}_{\mathcal{S}_{\rm P}\mathcal{S}}$ is the interaction Hamiltonian between $\mathcal{S}$ and $\mathcal{S}_{\rm P}$. Equation (9.156) is a more severe condition than (9.154), because, if it is satisfied, the measurement is a QND measurement for any time interval of the interaction between the systems and for any probe's initial state. If the measured observable is an integral of the free motion of $\mathcal{S}$, then it also satisfies

$$\imath\hbar \frac{\partial \hat{O}_{\rm ND}}{\partial t} + \left[\hat{O}_{\rm ND}, \hat{H}_{\mathcal{S}}\right] = 0. \tag{9.157}$$

From this equation and the obvious $\left[\hat{O}_{\rm ND}, \hat{H}_{\mathcal{S}_{\rm P}}\right] = 0$ we obtain Eq. (9.156).

84 Rigorously speaking, Eq. (9.151b) is satisfied also if the initial state of the probe is an eigenstate with null eigenvalue of the operator given by the difference in brackets. However, this possibility has unpractical consequences and will not be discussed here.

9.11.3 No measurement without a measurement

We have seen that a QND measurement does not add any perturbation to the measured observable. This means that the standard deviation of the probability distribution of the measured observable is not altered by a QND measurement. One might think that the entire probability distribution of the measured observable is not altered by a QND measurement. If this were the case, then repeated QND measurements of the same observable could increase the amount of information which we can extract from the system. We shall see (in Subsec. 15.2.2) that, due to unitarity, repeated (even QND) measurements on a single system *cannot* increase the observers' knowledge. In this subsection we show that, in agreement with the general theory of quantum measurement, if a QND measurement performed on a system $\mathcal{S}$ does not alter the probability density of the measured observable, then the measurement process does not provide *any* information about the measured observable itself .[85]

Let us start from the definition of the amplitude operator given by Eq. (9.150)

$$\hat{\vartheta}(x_m, \hat{x}_{\mathcal{S}}) = \langle x_m | \hat{U}_t(\hat{x}_m, \hat{x}_{\mathcal{S}}) | \psi_m \rangle, \tag{9.158}$$

where $\hat{x}_m$ is the measured observable of the probe (meter), x_m is its measured value, $\hat{x}_{\mathcal{S}}$ is the (QND) observable of the system $\mathcal{S}$, and $| \psi_m \rangle$ is the initial state of the meter. In Eq. (9.158) we have explicitly introduced the dependence of the amplitude operator, which completely describes the measurement, on $\hat{x}_{\mathcal{S}}$ through the unitary operator $\hat{U}_t$.

The QND condition for a back-action-evading measurement then means that $\hat{x}_{\mathcal{S}}$ and $\hat{\vartheta}$ share the same eigenstates:

$$\hat{\vartheta}(\hat{x}_{\mathcal{S}}, x_m) \,|\, x_{\mathcal{S}} \rangle = \vartheta(x_m, x_{\mathcal{S}}) \,|\, x_{\mathcal{S}} \rangle, \tag{9.159a}$$

$$\hat{\vartheta}^{\dagger}(\hat{x}_{\mathcal{S}}, x_m) \,|\, x_{\mathcal{S}} \rangle = \vartheta^{*}(x_{\mathcal{S}}, x_m) \,|\, x_{\mathcal{S}} \rangle . \tag{9.159b}$$

After a measurement on the meter which gives the result x_m, the system is therefore described by the density matrix (9.122)

$$\hat{\rho}_f(x_m) = \frac{1}{\wp(x_m)} \hat{\vartheta}(x_{\mathcal{S}}, x_m) \hat{\rho}_i \hat{\vartheta}^{\dagger}(x_{\mathcal{S}}, x_m), \tag{9.160}$$

where $\wp(x_m)$ is given by Eqs. (9.118) and (9.120) and may be rewritten as

$$\begin{aligned} \wp(x_m) &= \mathrm{Tr}_{\mathcal{S}} \left[\hat{\vartheta}(\hat{x}_{\mathcal{S}}, x_m) \hat{\rho}_i \hat{\vartheta}^{\dagger}(\hat{x}_{\mathcal{S}}, x_m) \right] \\ &= \int dx_{\mathcal{S}} \left\langle x_{\mathcal{S}} \left| \hat{\vartheta}(\hat{x}_{\mathcal{S}}, x_m) \hat{\rho}_i \hat{\vartheta}^{\dagger}(\hat{x}_{\mathcal{S}}, x_m) \right| x_{\mathcal{S}} \right\rangle. \end{aligned} \tag{9.161}$$

Now, the probability density of the measured observable after the measurement is given by

$$\begin{aligned} \wp_f(x_{\mathcal{S}}) &= \left\langle x_{\mathcal{S}} \left| \hat{\rho}_f(x_m) \right| x_{\mathcal{S}} \right\rangle \\ &= \frac{1}{\wp(x_m)} \left\langle x_{\mathcal{S}} \left| \hat{\vartheta}(\hat{x}_{\mathcal{S}}, x_m) \hat{\rho}_i \hat{\vartheta}^{\dagger}(\hat{x}_{\mathcal{S}}, x_m) \right| x_{\mathcal{S}} \right\rangle. \end{aligned} \tag{9.162}$$

85 This subsection has been suggested by section 7 of [Fortunato *et al.* 1999a]. See also [Alter/Yamamoto 1998].

Applying the QND conditions (9.159) we obtain

$$\wp_f(x_S) = \frac{1}{\wp(x_m)} |\vartheta(x_S, x_m)|^2 \wp(x_S), \quad (9.163)$$

where $\wp(x_S)$ is the initial state's a priori probability distribution of $\hat{x}_S$, given by Eq. (9.119).

If we require that the probability density (9.163) does not change due to the measurement process, i.e. $\wp_f(x_S) = \wp(x_S)$, then

$$|\vartheta(x_S, x_m)|^2 = \wp(x_m) \quad (9.164)$$

must hold. However, $\wp(x_m)$ is not a function of x_S (the eigenvalues of the measured observable) and therefore also the eigenvalues $\vartheta(x_S, x_m)$ of $\hat{\vartheta}(\hat{x}_S, x_m)$ are independent of x_S. Since the operator $\hat{\vartheta}$ must completely describe the measurement process, if its eigenvalues are independent of the eigenvalues of $\hat{x}_S$, the measurement obviously gives no information about $\hat{x}_S$, unless the initial state is an eigenstate of the measured observable.

9.12 Decision and estimation theory

9.12.1 Classical decision theory

First, we expose some elementary notions of the classical decision theory.[86] We call a strategy a *decision procedure* aiming to choose among several hypotheses. It is supposed that only one of these hypotheses is true. If such a strategy involves a random element, then we may define a probability $\wp_{\mathrm{H}_j}(\mathbf{D})$ that the hypothesis H_j $(j = 1, 2, \ldots, N)$ is chosen among the N alternative hypotheses when the set $\mathbf{D} = \{\mathrm{D}_1, \mathrm{D}_2, \ldots, \mathrm{D}_n\}$ of data is observed. The hypothesis H_j here means that the observed system is supposed to be in a certain state j. The quantity $\wp_k(\mathbf{D}) = \wp(\mathbf{D}|\mathrm{H}_k)$ represents the probability density function that the particular set $\mathbf{D}$ of data is observed when the system is actually in state k (which is precisely the state that verifies the hypothesis H_k), and is also known as the *likelihood function*. The hypothesis H_k is true with a priori probability $\wp_k^{\mathrm{A}}$. Then, the probability that the hypothesis H_j is chosen, when the hypothesis H_k is true (thus allowing for the possibility of making an error if j is different from k), is given by

$$\wp(\mathrm{H}_j|\mathrm{H}_k) = \int_{\mathbb{R}^n} d\mathbf{D}\, \wp_{\mathrm{H}_j}(\mathbf{D})\, \wp_k(\mathbf{D}), \quad (9.165)$$

where $\mathbb{R}^n$ is the n-dimensional space of the data. Equation (9.165) embodies an interesting refinement of the classical measurement theory (as it was presented in Sec. 1.1 when dealing with the postulate of the reduction to zero of the measurement error), in that it explicitly allows the treatment of errors (here represented by wrong decisions). In fact, if the hypothesis H_j is false, then also the consequent decision will be possibly wrong: making a wrong decision has a cost. Accordingly, we define C_{jk} as the cost incurred by

86 See [*Helstrom* 1976, 8–10].

choosing hypothesis H_j when the hypothesis H_k is true – in other words, the numbers C_{jk} assign relative weights to the different possible errors and correct decisions. Of course, the decision procedure has to be repeated many times, and the aim of statistical decision theory is precisely to minimize the *average* cost of the procedure. As H_k is true with a priori probability $\wp_k^A$, the average cost of the strategy is

$$\langle C \rangle = \sum_{j=1}^{N} \sum_{k=1}^{N} \wp_k^A C_{jk} \wp\left(H_j | H_k\right). \tag{9.166}$$

Defining for each (potentially chosen) H_j a risk function as follows:

$$R_j(\mathbf{D}) = \sum_{k=1}^{N} \wp_k^A C_{jk} \wp_k(\mathbf{D}), \tag{9.167}$$

and taking into account the expression (9.165), we may write the average cost as

$$\langle C \rangle = \int_{\mathbb{R}^n} d\,\mathbf{D} \sum_{j=1}^{N} R_j(\mathbf{D}) \wp_{H_j}(\mathbf{D}). \tag{9.168}$$

The central problem of decision theory is the following: we seek the N functions $\wp_{H_j}(\mathbf{D})$ that satisfy general probability conditions and make the average cost as small as possible. The solution to this minimization problem (see Prob. 9.21) is given by the posterior risks $R_j^P(\mathbf{D})$ of the hypothesis in view of the data observed, which are proportional to the risk functions $R_j(\mathbf{D})$, i.e.

$$R_j^P(\mathbf{D}) = \sum_{i=1}^{N} C_{ji} \wp^P(H_i | \mathbf{D}) = \frac{R_j(\mathbf{D})}{\wp(\mathbf{D})}, \tag{9.169}$$

where

$$\wp^P(H_i | \mathbf{D}) = \wp_i^A \frac{\wp_i(\mathbf{D})}{\wp(\mathbf{D})} \tag{9.170}$$

is the a posteriori probability of hypothesis H_i (given the set $\mathbf{D}$ of observed data), while

$$\wp(\mathbf{D}) = \sum_{k=1}^{N} \wp_k^A \wp_k(\mathbf{D}) \tag{9.171}$$

is the overall joint probability density function of the data. Equation (9.170) is Bayes' rule, while the formula (9.171) is the formula for the total probability.[87] We stress that the expression (9.170) gives the probability that the hypothesis H_i is *true* given the data $\mathbf{D}$, whereas the expression $\wp_{H_j}(\mathbf{D})$ represents the probability that the hypothesis H_j is *chosen* when the data $\mathbf{D}$ are observed.

It can be shown that the best strategy is the one that chooses the hypothesis for which the posterior risk is minimal. This choice does not necessarily coincide with the choice of

[87] See [*Gnedenko* 1969, 56–58].

the hypothesis for which the posterior probability (9.170) is greatest, as was first proposed by Bayes.[88] This last strategy comes from the choice of costs, either

$$C_{ij} = 1 \quad \text{for} \quad i \neq j, \qquad C_{ii} = 0\,; \tag{9.172a}$$

or

$$C_{ij} = 0 \quad \text{for} \quad i \neq j, \qquad C_{ii} = -1. \tag{9.172b}$$

We can arrive at similar results by considering the continuous case.[89]

In order to give a feeling about what kind of problems the classical decision theory presented above is able to handle, consider the following example. In a radar system, the time delay between the transmission of a pulse and the reception of the corresponding echo is proportional to the distance to the target. In order to determine it, a radar receiver must estimate the time-of-arrival τ of the signal $s(t-\tau)$ from the voltage $v(t)$ appearing at the terminals of its antenna during some observation intervals $(0, t')$. Because of the presence of the noise, the arrival time τ cannot be estimated with perfect accuracy. The question is then how to design a receiver to estimate τ with minimal error. The different information-bearing parameters of a signal (its amplitude, frequency, and so on) are to be estimated as accurately as possible by the receiver.

As we shall see in the next subsection, the formalism developed here is susceptible to a natural quantum generalization.

9.12.2 Quantum decision theory

In the quantum case,[90] the hypothesis H_j is represented by the assumption that the system is in the state $\hat{\rho}_j$.[91] We may associate to each hypothesis H_j an effect $\hat{E}_{\mathrm{H}_j}$, which we may call the *detection operator*, i.e. a non-negative definite Hermitian operator which satisfies the condition $\sum_j \hat{E}_{\mathrm{H}_j} = \hat{I}$ (see Eq. (9.129)), i.e. the quantum analogue of the classical completeness of the set of the hypotheses. The set of effects is determined by the conditional probability that one chooses the hypothesis H_j when H_k is true, i.e. (see Eqs. (9.100) and (9.120))

$$\wp\left(\mathrm{H}_j|\mathrm{H}_k\right) = \mathrm{Tr}\left(\hat{\rho}_k \hat{E}_{\mathrm{H}_j}\right), \tag{9.173}$$

which is the quantum analogue of Eq. (9.165). In general, as we have seen in Sec. 9.10, the detection operators $\hat{E}_{\mathrm{H}_j}$ are not represented by projectors. For example, a strategy that merely guesses a hypothesis out of the N available ones, choosing an arbitrary one with probability N^{-1}, without performing an actual measurement, is described by the POVM

$$\hat{E}_{\mathrm{H}_j} = N^{-1}\hat{I}, \quad j = 1, 2, \ldots, N. \tag{9.174}$$

88 See [Bayes 1763].

89 See [*Helstrom* 1976, 25–31].

90 See [*Helstrom* 1976, 90–100].

91 We recall the substantial difference between the classical and the quantum state (see Subsecs. 2.3.3 and 3.1.2).

The system is in the state $\hat{\rho}_k$ with an a priori probability $\wp_k^{\mathrm{A}}$. Then, the average cost of the decision strategy is given by (see Eq. (9.166))

$$\langle \mathcal{C} \rangle = \sum_{j=1}^{N} \sum_{k=1}^{N} \wp_k^{\mathrm{A}} C_{jk} \mathrm{Tr} \left(\hat{\rho}_k \hat{E}_{\mathrm{H}_j} \right) = \mathrm{Tr} \left(\sum_{j=1}^{N} \hat{\mathcal{R}}_j \hat{E}_{\mathrm{H}_j} \right), \tag{9.175}$$

where the Hermitian detection risk operator $\hat{\mathcal{R}}_j$ is defined by (see Eq. (9.167))

$$\hat{\mathcal{R}}_j = \sum_{k=1}^{N} \wp_k^{\mathrm{A}} C_{jk} \hat{\rho}_k. \tag{9.176}$$

Let us define a Hermitian operator $\hat{\mathrm{L}}$ – which is the quantum analogue of the function L (see the solution to the Prob. 9.21) – as

$$\hat{\mathrm{L}} = \sum_{j=1}^{N} \hat{E}_{\mathrm{H}_j} \hat{\mathcal{R}}_j = \sum_{j=1}^{N} \hat{\mathcal{R}}_j \hat{E}_{\mathrm{H}_j}, \tag{9.177}$$

since $\hat{\mathrm{L}}$, $\hat{E}_{\mathrm{H}_j}$, and $\hat{\mathcal{R}}_j$ are all Hermitian operators. Then, it can be proved that the requirement of a cost-minimizing strategy is fulfilled by the conditions

$$\left(\hat{\mathcal{R}}_j - \hat{\mathrm{L}} \right) \hat{E}_{\mathrm{H}_j} = 0, \tag{9.178a}$$

$$\hat{\mathcal{R}}_j - \hat{\mathrm{L}} \geq 0 \tag{9.178b}$$

for $j = 1, \ldots, N$. It can be seen that the role of $\hat{\mathrm{L}}$ is that of a Lagrangian multiplier, and we call it the *Lagrange operator*. Hence, from Eqs. (9.175) and (9.177), and taking into account the conditions (9.178), the minimum Bayes cost is

$$\langle \mathcal{C} \rangle_{\mathrm{min}} = \mathrm{Tr} \left(\hat{\mathrm{L}} \right). \tag{9.179}$$

Moreover, the difference between the cost incurred by using another POVM $\{\hat{E}'_{\mathrm{H}_j}\}$ (see Eq. (9.175)) and the cost arising from the optimal strategy (Eq. (9.179)) is given by

$$\langle \mathcal{C} \rangle - \langle \mathcal{C} \rangle_{\mathrm{min}} = \mathrm{Tr} \left[\sum_{j=1}^{N} \left(\hat{\mathcal{R}}_j - \hat{\mathrm{L}} \right) \hat{E}'_{\mathrm{H}_j} \right], \tag{9.180}$$

since the set $\{\hat{E}'_{\mathrm{H}_j}\}$ satisfies the completeness conditions. Equation (9.180), given the validity of Eq. (9.178b), implies

$$\langle \mathcal{C} \rangle - \langle \mathcal{C} \rangle_{\mathrm{min}} \geq 0. \tag{9.181}$$

The optimum POVMs form a convex set (see Eq. (5.56)), i.e. if $\{\hat{E}'_{\mathrm{H}_j}\}$ and $\{\hat{E}''_{\mathrm{H}_j}\}$ are two optimum POVMs, then all POVMs of the form

$$\hat{E}_{\mathrm{H}_j} = w \hat{E}'_{\mathrm{H}_j} + (1 - w) \hat{E}''_{\mathrm{H}_j}, \quad \forall \hat{E}_{\mathrm{H}_j} \text{ and for } 0 < w < 1, \tag{9.182}$$

are also optimal. Indeed, the corresponding Lagrange operator is

$$\hat{\mathrm{L}}_w = w \hat{\mathrm{L}}' + (1 - w) \hat{\mathrm{L}}'', \tag{9.183}$$

by means of which, for any j, we may rewrite Eq. (9.178b) in the form

$$\hat{\mathcal{R}}_j - \hat{\mathrm{L}}_w = w(\hat{\mathcal{R}}_j - \hat{\mathrm{L}}') + (w-1)(\hat{\mathcal{R}}_j - \hat{\mathrm{L}}'') \geq 0. \tag{9.184}$$

The Lagrange operator can be eliminated from the optimization Eqs. (9.178) by writing them in the form[92]

$$\hat{E}_{\mathrm{H}_j}(\hat{\mathcal{R}}_k - \hat{\mathcal{R}}_j)\hat{E}_{\mathrm{H}_k} = 0 \tag{9.185}$$

for all pairs j, k.

We now wish to analyze specifically the case of pure states. Let us consider a system in a pure state $|\psi_k\rangle$ under each hypothesis H_k ($k = 1, 2, \ldots, N$). In other words we are assuming $\hat{\rho}_k = |\psi_k\rangle\langle\psi_k|$. The optimum POVM can be confined in an N-dimensional subspace $\mathcal{H}_N$ (spanned by the N vectors $|\psi_k\rangle$) of the Hilbert space $\mathcal{H}_\mathcal{S}$ of the system. We proceed in two steps: first, we verify that each component outside $\mathcal{H}_N$ does not contribute to the decision probability $\wp(\mathrm{H}_j|\mathrm{H}_k)$; and second, we try to find a solution to the optimization equation (9.185). In order to verify the first statement, let $\hat{P}_N$ project arbitrary vectors belonging to $\mathcal{H}_\mathcal{S}$ onto $\mathcal{H}_N$ and write the POVM according to the identity

$$\hat{E}_{\mathrm{H}_j} = \hat{P}_N\hat{E}_{\mathrm{H}_j}\hat{P}_N + \left(\hat{I} - \hat{P}_N\right)\hat{E}_{\mathrm{H}_j}\hat{P}_N + \hat{P}_N\hat{E}_{\mathrm{H}_j}\left(\hat{I} - \hat{P}_N\right) + \left(\hat{I} - \hat{P}_N\right)\hat{E}_{\mathrm{H}_j}\left(\hat{I} - \hat{P}_N\right). \tag{9.186}$$

Combining Eq. (9.173) with Eq. (9.186), we obtain

$$\begin{aligned}\wp(\mathrm{H}_j|\mathrm{H}_k) &= \mathrm{Tr}\left(\hat{\rho}_k\hat{E}_{\mathrm{H}_j}\right) = \left\langle\psi_k\left|\hat{E}_{\mathrm{H}_j}\right|\psi_k\right\rangle \\ &= \mathrm{Tr}\left(\hat{\rho}_k\hat{E}'_{\mathrm{H}_j}\right),\end{aligned} \tag{9.187}$$

where

$$\hat{E}'_{\mathrm{H}_j} = \hat{P}_N\hat{E}_{\mathrm{H}_j}\hat{P}_N, \tag{9.188}$$

since $(\hat{I} - \hat{P}_N)|\psi_k\rangle = 0$. As a consequence, the optimum POVM can be taken as the generalized resolution of the identity

$$\sum_{j=1}^{N}\hat{E}'_{\mathrm{H}_j} + \left(\hat{I} - \hat{P}_N\right) = \hat{I}, \tag{9.189}$$

and, since the term $(\hat{I} - \hat{P}_N)$ has no effect on the subspace $\mathcal{H}_N$, we can drop it and confine ourselves to subspace $\mathcal{H}_N$.

Now we pass to the second step of our procedure and try to find a solution of the optimization equation (9.185) by means of the projectors

$$\hat{E}_{\mathrm{H}_j} = \hat{P}_{\mathrm{H}_j} = |\varphi_j\rangle\langle\varphi_j|, \tag{9.190}$$

which project onto N orthonormal vectors $|\varphi_j\rangle$ spanning $\mathcal{H}_N$, called *measurement states*. Then, Eq. (9.185) (where the effects may be replaced by projectors) will be satisfied for all pairs j, k, such that

$$\left\langle\varphi_k\left|\left(\hat{\mathcal{R}}_j - \hat{\mathcal{R}}_k\right)\right|\varphi_j\right\rangle = 0, \tag{9.191a}$$

[92] See derivation in [Holevo 1973a].

or (see Eq. (9.176))

$$\sum_{i=1}^{N} \wp_i^{\mathrm{A}} \left(C_{ji} - C_{ki}\right) \langle\varphi_k|\psi_i\rangle\langle\psi_i|\varphi_j\rangle = 0, \tag{9.191b}$$

which provides $N(N-1)/2$ equations for the N^2 unknown amplitudes

$$\vartheta_{ij} = \langle\varphi_i \mid \psi_j\rangle. \tag{9.192}$$

These are the components of the N state vectors $|\psi_j\rangle$ along the axes $|\varphi_i\rangle$ of $\mathcal{H}_N$. The inequality (9.178b) imposes the requirement that the matrices $\hat{M}^{(i)}$, whose elements are given by

$$M_{mn}^{(i)} = \sum_{j=1}^{N} \wp_j^{\mathrm{A}} \left(C_{ij} - C_{mj}\right) \vartheta_{mj}\vartheta_{nj}^*, \tag{9.193}$$

be non-negative definite so that the Bayes cost can be at a true minimum. The special cost function (9.172b)

$$C_{ij} = -\delta_{ij} \tag{9.194}$$

corresponds to minimizing the average probability of error $\wp^e = \langle\mathcal{C}\rangle + 1$. Then, Eq. (9.191b) reduces to the equality

$$\wp_m^{\mathrm{A}}\vartheta_{km}\vartheta_{mm}^* = \wp_k^{\mathrm{A}}\vartheta_{kk}\vartheta_{mk}^*, \tag{9.195}$$

and the minimum attainable error probability is then given by

$$\wp_{\min}^e = 1 - \sum_{j=1}^{N} \wp_j^{\mathrm{A}}|\vartheta_{jj}|^2. \tag{9.196}$$

For this cost function and for N linearly independent states $|\psi_j\rangle$ the optimum POVM is indeed a PVM. However, if the states are not linearly independent, the N vectors $|\psi_j\rangle$ span a subspace of $\mathcal{H}_\mathcal{S}$ of dimension smaller than N and this subspace cannot accommodate N different orthogonal projectors of type (9.190). As a consequence, the optimum POVM cannot be a PVM.

9.12.3 Estimate of the wave function

Quantum estimation theory deals with the estimation of a number N of parameters $(\zeta_1, \zeta_2, \ldots, \zeta_n) = \boldsymbol{\zeta}$ of a certain density operator $\hat{\rho}(\boldsymbol{\zeta})$.[93] The observational strategy for estimating the parameters $\{\zeta_k\}$ is a POVM. The resulting estimates $\{\zeta_k'\}$ are random variables, and the probability that they lie in a value region Υ of the parameter space $\mathcal{Z}$ is given by (see Eq. (9.173))

$$\wp(\boldsymbol{\zeta}' \in \Upsilon|\boldsymbol{\zeta}) = \mathrm{Tr}\left[\hat{\rho}(\boldsymbol{\zeta})\hat{E}(\Upsilon)\right], \tag{9.197}$$

93 See [*Helstrom* 1976, 235–43]. See also [*Holevo* 1982, 106, 169–74].

where we assume that the effects for finite regions can be formed as integrals of infinitesimal operators, i.e.

$$\hat{E}(\Upsilon) = \int_{\Upsilon} d\boldsymbol{\zeta}' \hat{E}(\boldsymbol{\zeta}'). \tag{9.198}$$

The joint conditional probability density function $\wp(\boldsymbol{\zeta}'|\boldsymbol{\zeta})$ is then given by

$$\wp(\boldsymbol{\zeta}'|\boldsymbol{\zeta}) = \mathrm{Tr}\left[\hat{\rho}(\boldsymbol{\zeta})\hat{E}(\boldsymbol{\zeta}')\right]. \tag{9.199}$$

The main aim of quantum estimation theory is therefore to find the best POVM $\hat{E}(\Upsilon)$ suitable for estimating the desired parameters of a certain density operator. This goal can be attained by carrying over the same calculations as in Subsec. 9.12.2 to the continuum case (the discrete data are replaced by continuous random variables). As we have said, the optimum estimator will in general be a POVM. The maximum likelihood estimator can be found by solving the optimization equations (9.178), which here translate into

$$\left[\hat{\mathcal{R}}(\boldsymbol{\zeta}') - \hat{\mathrm{L}}(\boldsymbol{\zeta}')\right] d\hat{E}(\boldsymbol{\zeta}') = 0, \tag{9.200a}$$

$$\hat{\mathcal{R}}(\boldsymbol{\zeta}') - \hat{\mathrm{L}}(\boldsymbol{\zeta}') \geq 0, \tag{9.200b}$$

where the Hermitian risk operator is given by

$$\hat{\mathcal{R}}(\boldsymbol{\zeta}') = \int_{\mathcal{Z}} d\boldsymbol{\zeta}\, \wp^{\mathrm{A}}(\boldsymbol{\zeta}) C(\boldsymbol{\zeta}', \boldsymbol{\zeta}) \hat{\rho}(\boldsymbol{\zeta}), \tag{9.201}$$

and the Lagrange operator is

$$\hat{\mathrm{L}}(\boldsymbol{\zeta}') = \int_{\mathcal{Z}} d\hat{E}(\boldsymbol{\zeta}')\hat{\mathcal{R}}(\boldsymbol{\zeta}'). \tag{9.202}$$

In terms of $\hat{\mathcal{R}}(\boldsymbol{\zeta}')$ the average cost is simply (see Eq. (9.179))

$$\langle C \rangle = \mathrm{Tr}\left(\int_{\mathcal{Z}} d\hat{E}(\boldsymbol{\zeta}')\hat{\mathcal{R}}(\boldsymbol{\zeta}')\right) = \mathrm{Tr}\left[\hat{\mathrm{L}}(\boldsymbol{\zeta}')\right]. \tag{9.203}$$

A specific problem of great interest is the estimate of the wave function (see also Ch. 15). A wave function can be estimated, provided that the corresponding state vector $|\psi\rangle$ is known to lie in a Hilbert space of finite dimensions, and a residual uncertainty is accepted. Let us select an orthonormal basis $\{|b_k\rangle\}$ and consider an estimate of the complex vector $\mathbf{c} = (c_1, c_2, \ldots, c_n)$ that defines the "true" state vector through the relation

$$|\psi\rangle = \sum_{k=1}^{n} c_k |b_k\rangle. \tag{9.204}$$

The n complex numbers $c_k = c_{k_x} + \imath c_{k_y}$ are parameters of the density operator $\hat{\rho} = |\psi\rangle\langle\psi|$. Because the vector $|\psi\rangle$ must have unit length, the point $\mathbf{c} = (c_{1_x}, c_{1_y}, \ldots, c_{n_x}, c_{n_y})$ must lie on a $(2n-1)$-dimensional hypersphere $\mathbf{S}_{2n}$ of radius 1 (for a two-dimensional system the Poincaré sphere is a possible representation). However, only $2n-2$ of the coordinates are relevant. For instance, in the bidimensional case, we have two coefficients (c_1 and c_2), and, since $|c_1|^2 = 1 - |c_2|^2$ and a global phase factor is irrelevant, we may set, say, c_1 as real in such a way that it suffices the modulus and the phase of c_2 to determine the density operator. Suppose now that we want to estimate $|\psi\rangle$ by means of a state vector

$$\left|\psi'\right\rangle = \sum_{k=1}^{n} c_k' \,|b_k\rangle \,. \tag{9.205}$$

Nothing being known in advance about the state of the system, we are forced to assume that the point $\mathbf{c}$ may be anywhere on the hypersurface, and we may assign to our estimating point $\mathbf{c}'$ an a priori probability density function $\wp^{\mathrm{A}}(\mathbf{c}')$ that must satisfy

$$\int d\mathbf{c}' \wp^{\mathrm{A}}(\mathbf{c}') = 1. \tag{9.206}$$

It follows that this probability density must be a constant equal to the inverse of the area A_{2n} of the hypersurface, i.e.

$$\wp^{\mathrm{A}}(\mathbf{c}') = \frac{1}{2}\Gamma(n)\pi^{-n}, \tag{9.207}$$

where Γ is the Euler Gamma function.[94] The maximum likelihood estimator is the POVM

$$\begin{aligned} d\hat{E}(\mathbf{c}') &= nA_{2n}^{-1}|\psi'\rangle\langle\psi'|d\mathbf{S} \\ &= nA_{2n}^{-1}\sum_{k=1}^{n}\sum_{j=1}^{n} c_k'(c_j')^{*}|b_k\rangle\langle b_j|d\mathbf{S}, \end{aligned} \tag{9.208}$$

where $d\mathbf{S}$ is the area element of the sphere, which satisfies the requirement (see Prob. 9.24)

$$\int_{\mathbf{S}_{2n}} d\hat{E}(\mathbf{c}') = \hat{I}. \tag{9.209}$$

The Lagrange operator defined by Eqs. (9.177) and (9.202) is

$$\hat{\mathrm{L}} = nA_{2n}^{-2}\int_{\mathbf{S}_{2n}} d\,\mathbf{S}|\psi'\rangle\langle\psi'| = A_{2n}^{-1}\hat{I}, \tag{9.210}$$

so that

$$\hat{\mathcal{R}}_j(\mathbf{c}) - \hat{\mathrm{L}} = A_{2n}^{-1}\left[\hat{I} - \hat{\rho}\right]. \tag{9.211}$$

The operator on the lhs of Eq. (9.211) turns out to be non-negative definite as required by Eqs. (9.178b) and (9.200b) (see Prob. 9.25). The joint conditional probability density function of the estimate $\mathbf{c}'$ is therefore

$$\wp(\mathbf{c}'|\mathbf{c}) = nA_{2n}^{-2}|\langle\psi'|\psi\rangle|^2, \tag{9.212}$$

and is independent of the basis vectors $\{|b_k\rangle\}$. It is also independent of the phase of the estimate $|\psi'\rangle$: such a phase is not statistically significant in agreement with the fact that the phase of a state vector is physically meaningless (see end of Subsec. 2.1.3). We can therefore consider the ket (9.205) to be an estimate of the state vector $|\psi\rangle$. The absolute value $\eta = |\langle\psi'|\psi\rangle|$ measures how close the estimated $|\psi'\rangle$ lies to the true $|\psi\rangle$ (it is equal to 1 if

[94] See [*Gradstein/Ryshik* 1981, 8.3].

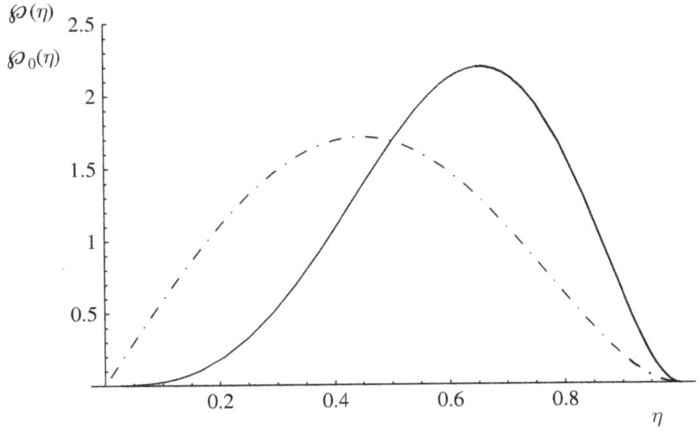

Figure 9.21 Plot of the probability density functions $\wp(\eta)$ (solid line) and $\wp_0(\eta)$ (dashed line) (Eqs. (9.213) and (9.214)) for $n = 4$: The peak of $\wp(\eta)$ is always closer to 1 than the corresponding peak of $\wp_0(\eta)$.

the estimate is exact). The relevance of this procedure can be appreciated by considering the following argument. The conditional probability density function of η is given by[95]

$$\wp(\eta) = 2n(n-1)\eta^3(1-\eta^2)^{n-2}. \tag{9.213}$$

If one had chosen the point $\mathbf{c}'$ at random on the unit hypersphere $\mathbf{S}_{2n}$, the probability density function of η would be equal to

$$\wp_0(\eta) = 2(n-1)\eta(1-\eta^2)^{n-2}. \tag{9.214}$$

As it is shown in Fig. 9.21, the peak of the probability (9.213) is closer to 1 than the probability calculated by choosing $\mathbf{c}'$ at random.[96]

Summary

- In this chapter we have discussed one of the most important and puzzling aspects of quantum mechanics: measurement. First, we have shown what the problem consists of in the fact that the measurement process appears to be ruled by a *non-unitary and non-reversible dynamics*, in contrast to the "ordinary" quantum dynamics. Different positions and interpretations have been historically developed in order to solve this problem.
- A *Gedankenexperiment* proposed by Schrödinger in 1935 has become an experimental reality in recent years: the so-called *Schrödinger cat*. It is the possibility of building

[95] See [*Helstrom* 1976, 292–93].

[96] By a similar procedure it is possible to provide an estimate of the position, the angle of rotation, and the time-of-arrival (see [*Helstrom* 1976, 239–80].

mesoscopic (and perhaps in the future also macroscopic) objects that shows the fundamental quantum feature of entanglement. This implies that there is no sharp break between the micro- and macroworlds, as was initially proposed by the supporters of the Copenhagen interpretation.

- In the *decoherence approach* to the measurement problem, instead of considering only the interaction between a measurement apparatus and an object system, one also takes into account the action of the *environment*, which, by "absorbing" the off-diagonal elements of the density matrix describing the object system, allows the diagonalization of the matrix in the eigenbasis of the measured observable – via the formal mechanism of the partial trace. Decoherence presupposes no sharp break between the measurement process and the unitary dynamics of quantum theory. It is an interpretation that is consistent with the theory, that does not need ad hoc assumptions and is also able to make predictions that have been experimentally confirmed (e.g. the decoherence time).
- Through some interesting experiments (for instance using the so-called quantum eraser) it has been shown that the measurement act consists of two parts: the washing out of interference and the *acquisition of information*. Only the latter is irreversible. The conclusion is that the state vector describes not only what is *actually* known but what is *in principle* knowable.
- We have formulated in more rigorous terms the *interaction-free measurement*, previously treated in Ch. 1.
- *Delayed-choice experiments* seem to imply that one may retroact on the past. In fact, it has been shown that we never have to deal with the past but only with the present effects of past events. According to quantum mechanics, reality consists not only of events but also of non-local interdependencies and that there is a complementary dynamical relationship between these two features.
- Another surprising feature is represented by the quantum *Zeno effect*, i.e. the possibility of freezing the state of a system by continuously monitoring one of its observables with a discrete spectrum.
- Thanks to its nature, the measurement of quantum systems may also be exploited as a tool to "guide" their evolution: selective or *conditional measurements* have been contrasted to non-selective ones.
- An important and relatively recent development is represent by *POVMs* that allow the possibility to measure *unsharp* observables. While PVMs are combinations of projectors, POVMs are combinations of *effects* – resolutions of identity that satisfy the completeness condition but are non-orthogonal and in general do not commute with each other.
- *Quantum non-demolition* measurements allow the possibility of transfering the perturbation due to measurement to an observable that is canonically conjugate to the one we want to measure.
- Finally, we have dealt with *decision and estimation theory*, whose aim – given a certain set of experimental data – is to find the lowest-cost strategy able to choose among several hypotheses and to guess an unknown parameter. In the quantum case, the best strategy is in general provided by a POVM.

Problems

9.1 Prove that, starting with the pure state $\hat{\rho}$, after an unitary evolution, we again obtain a pure state.

9.2 Diagonalize the interaction Hamiltonian (9.15).

9.3 Prove that there always exists a basis in which the off-diagonal terms of a density matrix are exactly zero.

9.4 Formalize the example of Sec. 9.1 in density-matrix terms.

9.5 Derive explicitly Eq. (9.47).

9.6 Compute the matrix element $\langle\uparrow|\left\langle N\left|\hat{H}_{\mathcal{SA}}\right|N-1\right\rangle|\uparrow\rangle$, where $\hat{H}_{\mathcal{SA}}$ is given by Eq. (9.61) in order to derive Eq. (9.64).

9.7 Making use of the definitions

$$\hat{b}_0 = \frac{1}{\sqrt{2}}\left(\hat{a}_1 + \hat{a}_0\right) \quad \text{and} \quad \hat{b}_1 = \frac{1}{\sqrt{2}}\left(\hat{a}_1 - \hat{a}_0\right), \tag{9.215}$$

(i) diagonalize the Hamiltonian (9.61);
(ii) derive Eq. (9.68).

9.8 Making use of the Stirling approximation (see the solution of Prob. 1.10), derive the result (9.74).

9.9 By making use of the general transformations induced by the beam splitters, prove the result shown in Fig. 9.13(a) with $|\mathrm{R}|^2 = \cos^2(\pi/2N)$.

9.10 Prove that, if $\hat{\rho}$ is an initial state and $\mathrm{Tr}\left(\hat{\rho}\hat{P}_j\right) \neq 0$, also

$$\hat{\rho}_j = \frac{\hat{P}_j\hat{\rho}\hat{P}_j}{\mathrm{Tr}\left(\hat{\rho}\hat{P}_j\right)}$$

is a density operator.

9.11 Use Eq. (9.114) to calculate the final state $\hat{\rho}_f$ in the polarization example of Box 9.5.

9.12 Prove Eq. (9.117).

9.13 Derive Eq. (9.123).

9.14 Prove Eq. (9.128).

9.15 Consider the arrangement of Fig. 9.20. Given the input single-photon state

$$|1\rangle = \begin{pmatrix} 1 \\ 0 \end{pmatrix},$$

compute the output state (just before any detection takes place) and determine the detection probabilities at the three detectors D1, D2, and D3, for arbitrary values of the phase shift ϕ and of the transmission parameter $\sqrt{\eta}$ of BS3. In particular, consider the limiting cases $\eta = 0, 1$.

9.16 Calculate the probabilities (9.131), that is when $\eta = 1$, by making use of the mean value of the projectors occurring therein, and show that they are identical to the results obtained by solving the previous problem.

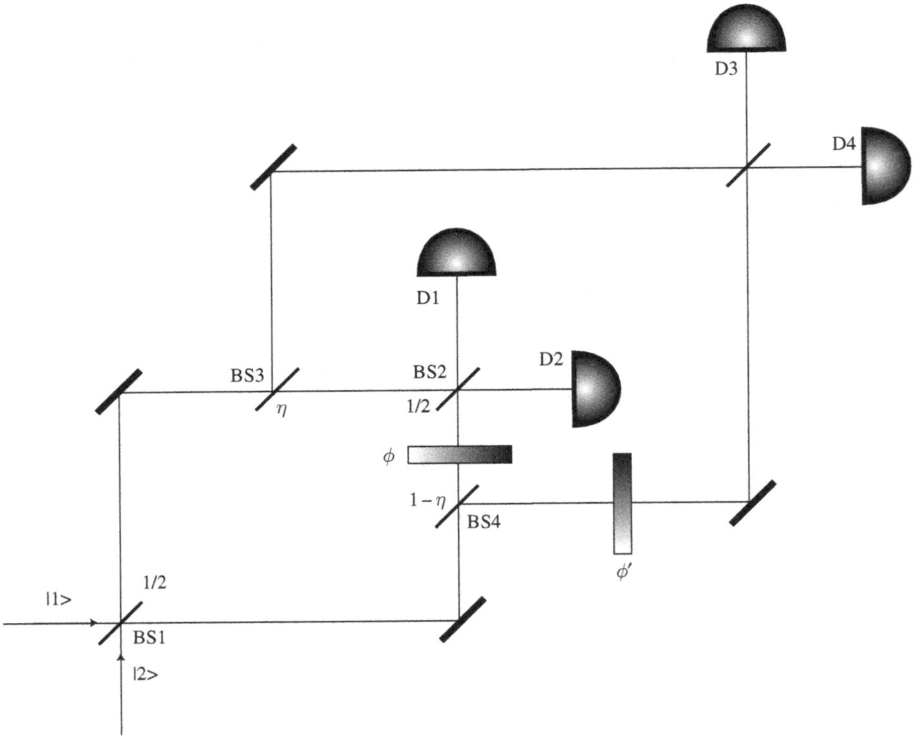

Figure 9.22 Beam splitters BS1 and BS2 are symmetric, while BS3 and BS4 have transmission parameters η and $1-\eta$, respectively. The two phase shifters produce phase shifts ϕ and ϕ'.

9.17 Calculate the probabilities (9.136), that is, when $\eta = 0$, by making use of the mean value of the projectors occurring therein, and show that they are identical to the results obtained by solving Prob. 9.15.

9.18 Find the explicit expressions for the effects (9.140). Then, compute probabilities (9.139) by making use of the mean value of these effects, and show that they are identical to the results obtained by solving Prob. 9.15.

9.19 Verify that the effects defined by Eqs. (9.140) fulfill the completeness requirement but do not commute with one another.

9.20 Describe the POVM performed by the arrangement shown in Fig. 9.22.

9.21 Solve the minimization problem of Subsec. 9.12.1: find the N functions $\wp_{\mathrm{H}_j}(\mathbf{D})$ that render the lhs of Eq. (9.168) a minimum.

9.22 Rewrite the classical decision theory of Subsec. 9.12.1 in terms of discrete data $\mathbf{D}$ as random variables.

(*Hint*: In the discrete case the probability density functions $\wp_k(\mathbf{D})$ have to be replaced by $\wp(\mathbf{D}|\mathrm{H}_k)$, i.e. by the probabilities of getting the data in the N hypotheses. Correspondingly, integrations over the continuous space must be replaced by summation.)

9.23 Rewrite the optimization equation (9.178a) in the case of binary decisions, i.e. when there are only two hypothesis, H_0 and H_1.

9.24 Show that the integral of Eq. (9.208) is equal to the identity operator.

9.25 Prove that the operator $\hat{\mathcal{R}}_j(\mathbf{c}) - \hat{L}$ in the lhs of Eq. (9.211) is non-negative definite.

Further reading

Measurement theory

Braginsky, V. B. and Khalili, F. Y., *Quantum Measurement*, Cambridge: Cambridge University Press, 1992.

Wheeler, J. A. and Zurek, W. (eds.), *Quantum Theory and Measurement*, Princeton: Princeton University Press, 1983.

Decoherence

Giulini, D., Joos, E., Kiefer, C., Kupsch, J., Stamatescu, I.-O., and Zeh, H. D. (eds.), *Decoherence and the Appearance of a Classical World in Quantum Theory*, Berlin: Springer, 1996.

Zurek, Wojciech H., Pointer basis of quantum apparatus: into what mixture does the wave packet collapse?" *Physical Review*, **D24** (1981), 1516–25.

Zurek, Wojciech H., Environment-induced superselection rules. *Physical Review*, **D26** (1982), 1862–80.

Zeno and anti-zeno effects

Kofman, A. G., Kurizki, G., and Opatrny, T., Zeno and anti-Zeno effects for photon polarization dephasing. *Physical Review*, **A63**, 042108.

Conditional measurements

Sherman, B. and Kurizki, G., Preparation and detection of macroscopic quantum superpositions by two-photon field–atom interactions. *Physical Review*, **A45** (1992), R7674–77.

POVM

Davies, E. B., *Quantum Theory of Open Systems*, London: Academic Press, 1976.

Holevo, A. S., *Probabilistic and Statistical Aspects of Quantum Theory* (engl. tr.), Amsterdam: North Holland, 1982.

Kraus, Karl, *States, Effects and Operations*, Berlin: Springer, 1983.

de Muynck, W. M., Stoffels, W. W., Martens, H., Joint measurement of interference and path observables in optics and neutron interferometry. *Physica*, **B175** (1991), 127–32.

QND measurement

Braginsky, V. B. and Khalili, F. Y., *Quantum Measurement*, Cambridge: Cambridge University Press, 1992.

Kimble, H. Jeffrey, in *Fundamental Systems in Quantum Optics*, J. Dalibard, J. M. Raymond, J. Zinn-Justin (eds.), Les Houches School Section LIII 1990, Amsterdam: Elsevier.

Detection and estimation theory

Helstrom, Carl W., *Quantum Detection and Estimation Theory*, New York: Academic, 1976.

PART III

MATTER AND LIGHT

10 Perturbations and approximation methods

In this chapter, we present some fundamental issues about approximation methods that are often used when a quantum-mechanical system is perturbed and about the relationship between classical and quantum mechanics. In Sec. 10.1 we introduce the stationary perturbation theory, while Sec. 10.2 is devoted to time-dependent perturbations. In Sec. 10.3 we briefly examine the adiabatic theorem. In Sec. 10.4 we introduce the variation method, an approximation method that is not based on perturbation theory. In Sec. 10.5 we discuss the classical limit of the quantum-mechanical equations, whereas in Sec. 10.6 we deal with the semiclassical approximation, in particular the WKB method. On the basis of the previous approximation methods in Sec. 10.7 we present scattering theory. Finally, in Sec. 10.8 we treat a method that has a wide range of applications: the path-integral method.

10.1 Stationary perturbation theory

Perturbation theory is a rather general approximation method that may be applied when a small additional force (the perturbation) acts on a system (the unperturbed system), whose quantum dynamics is fully known. If the disturbance is small, it modifies both the energy levels and the stationary states. This allows us to make an expansion in power series of a perturbation parameter, which is assumed to be small. Perturbation theory may be applied both to the case where the additional force is time-independent (in which case a stationary treatment suffices – the subject of the present section) as well as to the case where it explicitly depends on time.[1] In the former case we consider the perturbation as causing a modification of the states of the motion of the unperturbed system. In the latter, the perturbed system makes transitions from one state to another under the influence of the perturbation. In Sec. 10.2 we treat the dynamics of perturbed systems. Though this can be applied to both time-dependent and time-independent perturbations, the former case is obviously far more interesting. Given the fact that the number of quantum systems that are explicitly solvable is relatively small – free particle (Secs. 3.4 and 4.1), harmonic oscillator (Sec. 4.4), linear potentials (Subsec. 4.5.1) – we emphasize that quantum perturbation theory has a wide range of applications.

[1] See [*Dirac* 1930, 167–75].

10.1.1 Non-degenerate case

Stationary perturbation theory aims to find the changes in the discrete energy levels and eigenstates of a system when a small disturbance is applied. Our procedure is to split the total Hamiltonian into unperturbed and perturbated parts, according to the Dirac picture (see Subsec. 3.6.2), so that we have

$$\hat{H} = \hat{H}_0 + \zeta \hat{H}', \quad \hat{H}_0 \mid \psi_n\rangle = E_n^{(0)} \mid \psi_n\rangle, \quad \hat{H} \mid \varphi_n\rangle = E_n \mid \varphi_n\rangle, \tag{10.1}$$

where $\hat{H}_0$ is the Hamiltonian of the unperturbed system, $\mid \psi_n\rangle$ its eigenstates [see Subsec. 3.1.3], $E_n^{(0)}$ their corresponding energy levels, and ζ a real and positive parameter such that $\hat{H}$ is a well-defined Hamiltonian (see Subsec. 10.5.1). We assume that we are able to solve the unperturbed stationary Schrödinger equation and find $\{\mid \psi_n\rangle\}$ and $\{E_n^{(0)}\}$, while our goal is to find the perturbed states $\{\mid \varphi_n\rangle\}$ and their corresponding energy eigenvalues $\{E_n\}$.

For the time being, we also assume that all eigenvalues are different (non-degenerate case). In the next subsection we shall consider the degenerate case. In many situations, we can expand the perturbed eigenvalues as a Taylor power series in ζ (around $\zeta = 0$), such that the zero, first, etc., powers of ζ correspond to the zero, first, etc., orders of the perturbation. If ζ is a small parameter and the energy distances between the discrete unperturbed states are large compared to the energy-level modifications induced by the perturbation, we can rely on the fact that higher orders (i.e. terms with higher powers of ζ) of the perturbation expansion will be negligible. The existence of all the involved derivatives at $\zeta = 0$ does not imply that the function $E_n(\zeta)$ is analytic. If this is not the case, the Taylor series is not convergent but is asymptotic, i.e.

$$\sum_{k=0}^{N} \zeta^k E_n^{(k)} - E_n(\zeta) = O\left(\zeta^{N+1}\right). \tag{10.2}$$

Since the unperturbed eigenstates form an orthonormal basis on the system's Hilbert space, we may expand the eigenstates of the total Hamiltonian in terms of the eigenstates $\left| \psi_j \right\rangle$ of $\hat{H}_0$:

$$\mid \varphi_n\rangle = \sum_j c_{n,j} \left| \psi_j \right\rangle. \tag{10.3}$$

Now, we may explicitly write the Taylor-series expansion of the coefficients $c_{n,j}$ and of the eigenvalues E_n as functions of ζ around $\zeta = 0$, i.e.

$$c_{n,j} = c_{n,j}^{(0)} + \zeta c_{n,j}^{(1)} + \zeta^2 c_{n,j}^{(2)} + \cdots, \tag{10.4a}$$

$$E_n = E_n^{(0)} + \zeta E_n^{(1)} + \zeta^2 E_n^{(2)} + \cdots. \tag{10.4b}$$

Back-substituting the expansion (10.4a) into Eq. (10.3), we obtain

$$\mid \varphi_n\rangle = \sum_j c_{n,j}^{(0)} \left| \psi_j \right\rangle + \zeta \sum_j c_{n,j}^{(1)} \left| \psi_j \right\rangle + \zeta^2 \sum_j c_{n,j}^{(2)} \left| \psi_j \right\rangle + \cdots. \tag{10.5}$$

In the limit $\zeta \to 0$ the lhs of Eq. (10.5) becomes $|\psi_n\rangle$, which in turn implies that $c^{(0)}_{n,j} = \delta_{n,j}$ in the first term of the rhs of the same equation. We can then write

$$|\varphi_n\rangle = \left|\varphi_n^{(0)}\right\rangle + \zeta\left|\varphi_n^{(1)}\right\rangle + \zeta^2\left|\varphi_n^{(2)}\right\rangle + \cdots, \tag{10.6}$$

where

$$\left|\varphi_n^{(0)}\right\rangle = |\psi_n\rangle, \tag{10.7a}$$

$$\left|\varphi_n^{(k)}\right\rangle = \sum_j c^{(k)}_{n,j}\left|\varphi_j^{(0)}\right\rangle. \tag{10.7b}$$

Equation (10.6) is the Taylor expansion of the perturbed eigenkets in a power series of ζ. Using the expansion (10.6) in the perturbed stationary Schrödinger equation, we obtain

$$\left(\hat{H}_0 + \zeta\hat{H}'\right)\left(\left|\varphi_n^{(0)}\right\rangle + \zeta\left|\varphi_n^{(1)}\right\rangle + \zeta^2\left|\varphi_n^{(2)}\right\rangle + \cdots\right)$$
$$= \left(E_n^{(0)} + \zeta E_n^{(1)} + \zeta^2 E_n^{(2)}\cdots\right)\left(\left|\varphi_n^{(0)}\right\rangle + \zeta\left|\varphi_n^{(1)}\right\rangle + \zeta^2\left|\varphi_n^{(2)}\right\rangle + \cdots\right). \tag{10.8}$$

Since Eq. (10.8) must be valid for any (small) value of ζ, we can equate the coefficients of equal power of ζ on both sides so as to derive the infinite system of equations

$$\left(E_n^{(0)} - \hat{H}_0\right)\left|\varphi_n^{(0)}\right\rangle = 0, \tag{10.9a}$$

$$\left(E_n^{(0)} - \hat{H}_0\right)\left|\varphi_n^{(1)}\right\rangle + E_n^{(1)}\left|\varphi_n^{(0)}\right\rangle = \hat{H}'\left|\varphi_n^{(0)}\right\rangle, \tag{10.9b}$$

$$\left(E_n^{(0)} - \hat{H}_0\right)\left|\varphi_n^{(2)}\right\rangle + E_n^{(1)}\left|\varphi_n^{(1)}\right\rangle + E_n^{(2)}\left|\varphi_n^{(0)}\right\rangle = \hat{H}'\left|\varphi_n^{(1)}\right\rangle, \tag{10.9c}$$

$$\cdots = \cdots,$$

$$\left(E_n^{(0)} - \hat{H}_0\right)\left|\varphi_n^{(m)}\right\rangle + E_n^{(1)}\left|\varphi_n^{(m-1)}\right\rangle + \cdots + E_n^{(m)}\left|\varphi_n^{(0)}\right\rangle = \hat{H}'\left|\varphi_n^{(m-1)}\right\rangle. \tag{10.9d}$$

As expected, Eq. (10.9a) simply tells us that $\left|\varphi_n^{(0)}\right\rangle$ is an eigenket of $\hat{H}_0$ with eigenvalue $E_n^{(0)}$. Moreover, if we add a component $\eta\left|\varphi_n^{(0)}\right\rangle$, η being an arbitrary parameter, to any of the $\left|\varphi_n^{(j)}\right\rangle$ in each of the first terms of Eqs. (10.9), these terms – taking into account Eq. (10.9a) – are left unchanged. In order to simplify the calculation we can choose the arbitrary term in such a way that higher-order corrections are orthogonal to the unperturbed eigenket, i.e.

$$\left\langle\varphi_n^{(0)} \mid \varphi_n^{(j)}\right\rangle = 0, \quad \forall j > 0. \tag{10.10}$$

In other words, this implies that the sum in the rhs of Eq. (10.7b) has to be performed on $\forall j \neq n$, Using this property and the fact that the eigenvalue $E_n^{(0)}$ of the unperturbed Hamiltonian is non-degenerate, we can multiply each term of Eq. (10.9d) by $\left\langle\varphi_n^{(0)}\right|$ from the left so as to obtain the general expression

$$E_n^{(m)} = \left\langle\varphi_n^{(0)}\left|\hat{H}'\right|\varphi_n^{(m-1)}\right\rangle, \tag{10.11}$$

for any $m > 0$, which shows that the calculation of any E_n to a given order in ζ requires the knowledge of the eigenkets of $\hat{H}'$ to the next lower order.

Therefore, from Eq. (10.11) we immediately obtain that the first-order correction to the energy eigenvalues ($m = 1$) is given by

$$E_n^{(1)} = \left\langle \varphi_n^{(0)} \left| \hat{H}' \right| \varphi_n^{(0)} \right\rangle = \left\langle \psi_n \left| \hat{H}' \right| \psi_n \right\rangle = H'_{nn}, \tag{10.12}$$

which is the expectation value of the perturbing energy $\hat{H}'$ on the corresponding unperturbed state $|\psi_n\rangle$. In other words, the energy change in the first order is equal to the diagonal element of the perturbation Hamiltonian on the unperturbed eigenstate.

We may now proceed to calculate the first-order correction $\left|\varphi_n^{(1)}\right\rangle$ of the eigenket of the perturbed Hamiltonian. From Eq. (10.7b), we have

$$\left|\varphi_n^{(1)}\right\rangle = \sum_{j \neq n} c_{n,j}^{(1)} \, |\psi_j\rangle, \tag{10.13}$$

and reduce the problem to that of computing the coefficients $c_{n,j}^{(1)}$ for $j \neq n$. Substituting Eq. (10.13) into Eq. (10.9b) and rearranging the terms, we obtain

$$\sum_{j \neq n} c_{n,j}^{(1)} \left(E_n^{(0)} - E_j^{(0)}\right) |\psi_j\rangle = \left(\hat{H}' - E_n^{(1)}\right) |\psi_n\rangle. \tag{10.14}$$

Multiplying both sides of Eq. (10.14) from the left by $\langle\psi_k|$ ($k \neq n$), we obtain

$$c_{n,k}^{(1)} = \frac{\left\langle \psi_k \left| \hat{H}' \right| \psi_n \right\rangle}{E_n^{(0)} - E_k^{(0)}}, \tag{10.15}$$

which gives immediately the eigenket of the perturbed Hamiltonian to the first order. Moreover, Eq. (10.15) also determines the range of validity of the perturbation expansion. In fact, in order for the $\zeta c_{n,k}^{(1)}$ to be small, we must have

$$\zeta \left\langle \psi_k \left| \hat{H}' \right| \psi_n \right\rangle \ll E_n^{(0)} - E_k^{(0)}, \tag{10.16}$$

which is the formal expression of the condition we anticipated at the beginning of this subsection: the energy differences between the discrete unperturbed states have to be large relative to the energy changes induced by the perturbation.

The previous result allows us to calculate the second-order correction ($m = 2$) to the energy eigenvalues. Inserting Eq. (10.15) and (10.13) into Eq. (10.11), we obtain

$$E_n^{(2)} = \sum_{j \neq n} c_{n,j}^{(1)} \left\langle \psi_n \left| \hat{H}' \right| \psi_j \right\rangle = \sum_{j \neq n} \frac{\left\langle \psi_n \left| \hat{H}' \right| \psi_j \right\rangle \left\langle \psi_j \left| \hat{H}' \right| \psi_n \right\rangle}{E_n^{(0)} - E_j^{(0)}}, \tag{10.17}$$

from which we obtain

$$E_n^{(2)} = \sum_{j \neq n} \frac{\left|\left\langle \psi_n \left| \hat{H}' \right| \psi_j \right\rangle\right|^2}{E_n^{(0)} - E_j^{(0)}}. \tag{10.18}$$

In order to calculate the eigenkets $\left|\varphi_n^{(2)}\right\rangle$ of the perturbed Hamiltonian, let us again expand $\left|\varphi_n^{(2)}\right\rangle$ in terms of $|\psi_j\rangle$ (see Eq. (10.7b)):

$$\left|\varphi_n^{(2)}\right\rangle = \sum_{j \neq n} c_{n,j}^{(2)} \,|\psi_j\rangle. \tag{10.19}$$

Once again, we substitute Eqs. (10.19), (10.13), and (10.7a) into Eq. (10.9c), to obtain

$$\sum_{j \neq n} c_{n,j}^{(2)} \left(E_n^{(0)} - \hat{H}_0\right) |\psi_j\rangle + \sum_{j \neq n} c_{n,j}^{(1)} E_n^{(1)} \,|\psi_j\rangle + E_n^{(2)} \,|\psi_n\rangle = \sum_{j \neq n} c_{n,j}^{(1)} \hat{H}' \,|\psi_j\rangle. \tag{10.20}$$

Proceeding in the same way, starting from Eq. (10.14), we have ($k \neq n$)

$$c_{n,k}^{(2)} = \sum_{j \neq n} \frac{\left\langle \psi_k \left| \hat{H}' \right| \psi_j \right\rangle \left\langle \psi_j \left| \hat{H}' \right| \psi_n \right\rangle}{\left(E_n^{(0)} - E_k^{(0)}\right)\left(E_n^{(0)} - E_j^{(0)}\right)} - \frac{\left\langle \psi_k \left| \hat{H}' \right| \psi_n \right\rangle \left\langle \psi_n \left| \hat{H}' \right| \psi_n \right\rangle}{\left(E_n^{(0)} - E_k^{(0)}\right)^2}, \tag{10.21}$$

which immediately provides the explicit expression for the correction $\left|\varphi_n^{(2)}\right\rangle$ of the eigenket of the perturbed Hamiltonian to the second order.

Summarizing these results, we may write the energy eigenvalues and eigenstates of the perturbed Hamiltonian to the second-order in ζ as

$$E_n = E_n^{(0)} + \zeta H'_{nn} + \zeta^2 \sum_{j \neq n} \frac{\left|H'_{nj}\right|^2}{E_n^{(0)} - E_j^{(0)}} + O\left(\zeta^3\right), \tag{10.22a}$$

$$\begin{aligned} |\varphi_n\rangle = |\psi_n\rangle + \zeta \sum_{j \neq n} \Bigg[& \frac{H'_{jn}}{E_n^{(0)} - E_j^{(0)}} \left(1 - \zeta \frac{H'_{nn}}{E_n^{(0)} - E_j^{(0)}}\right) \\ & + \zeta \sum_{k \neq n} \frac{H'_{jk} H'_{kn}}{\left(E_n^{(0)} - E_k^{(0)}\right)\left(E_n^{(0)} - E_j^{(0)}\right)} \Bigg] |\psi_j\rangle. \end{aligned} \tag{10.22b}$$

We recall that $|\varphi_n\rangle$ is not normalized, so an extra computation must be performed in order to obtain the normalization factor (see Subsec. 2.2.2).

10.1.2 Degenerate case

So far, we have assumed that the eigenvalues corresponding to the unperturbed eigenstates $\left|\varphi_n^{(0)}\right\rangle = |\,\psi_n\rangle$ are non-degenerate. Suppose now that there are two states $|\,\psi_l\rangle$ and $\left|\,\psi_q\right\rangle$ (with $l \neq q$) that have the same unperturbed energy eigenvalue, i.e. $E_l^{(0)} = E_q^{(0)} = E^{(0)}$. As we know (see Subsec. 3.1.4), in this case $\{|\,\psi_l\rangle, \left|\,\psi_q\right\rangle\}$ is just a pair among the infinite possible pairs that span the two-dimensional subspace pertaining to the doubly degenerate energy eigenvalue $E^{(0)}$. We limit ourselves here to the case of doubly degenerate levels, but the generalization to higher-order degeneracies is straightforward.

The main difficulty arises in Eq. (10.15), where the denominator vanishes. Unless we simultaneously have[2] $\left\langle \psi_l \left| \hat{H}' \right| \psi_q \right\rangle = 0$, the coefficient $c_{q,l}^{(1)}$ is not defined, and the procedure described in the previous subsection cannot be valid. Moreover if we choose the basis of the unperturbed Hamiltonian in such a way that the perturbed Hamiltonian is diagonal in the space of degenerate eigenvalues, still no problems arise and the previous equations can be used without modifications.

Among all possible linear combinations of $|\,\psi_l\rangle$ and $\left|\,\psi_q\right\rangle$, let us introduce the particular superposition

$$\left|\varphi^{(0)}\right\rangle = d_l \,|\,\psi_l\rangle + d_q \left|\,\psi_q\right\rangle, \tag{10.23}$$

which removes the degeneracy in the first order. In order to reach this goal, we rewrite Eq. (10.9b) as

$$\left(E^{(0)} - \hat{H}_0\right)\left|\varphi^{(1)}\right\rangle = \left(\hat{H}' - E^{(1)}\right)\left(d_l \,|\,\psi_l\rangle + d_q \left|\,\psi_q\right\rangle\right), \tag{10.24}$$

where we have dropped the index n because we are implicitly referring to the subspace spanned by $\{|\,\psi_l\rangle, \left|\,\psi_q\right\rangle\}$. Then, we take the inner product of this equation with $\langle\psi_l\,|$ and $\left\langle\psi_q\,\right|$. We then find the following linear and homogeneous system of equations in d_l and d_q:

$$\left(\left\langle \psi_l \left| \hat{H}' \right| \psi_l \right\rangle - E^{(1)}\right) d_l + \left\langle \psi_l \left| \hat{H}' \right| \psi_q \right\rangle d_q = 0, \tag{10.25a}$$

$$\left\langle \psi_q \left| \hat{H}' \right| \psi_l \right\rangle d_l + \left(\left\langle \psi_q \left| \hat{H}' \right| \psi_q \right\rangle - E^{(1)}\right) d_q = 0. \tag{10.25b}$$

This system of equations in d_l and d_q can be solved if and only if the determinant of the coefficients vanishes, i.e.

$$\begin{vmatrix} H'_{ll} - E^{(1)} & H'_{lq} \\ H'_{ql} & H'_{qq} - E^{(1)} \end{vmatrix} = 0. \tag{10.26}$$

This a quadratic equation for $E^{(1)}$, whose solutions are given by

$$E^{(1)} = \frac{1}{2}\left[H'_{ll} + H'_{qq} \pm \sqrt{\left(H'_{ll} - H'_{qq}\right)^2 + 4|H'_{ql}|^2}\right]. \tag{10.27}$$

[2] If $\left\langle \psi_l \left| \hat{H}' \right| \psi_q \right\rangle = 0$, i.e. the perturbing Hamiltonian is zero in the subspace of the degenerate eigenvalues, there are no problems, because the numerator cancels with the zero in the denominator of Eq. (10.15), and we can just neglect these terms. If the Hamiltonian restricted to this subspace is proportional to the identity, we can just shift the Hamiltonian by adding an appropriate constant.

Since the diagonal matrix elements of $\hat{H}'$ are real, both values of $E^{(1)}$ are also real. They are equal if and only if

$$H'_{ll} = H'_{qq} \quad \text{and} \quad H'_{ql} = 0. \tag{10.28}$$

In this case we say that the degeneracy has not been removed in the first order (see Prob. 10.1). However, if at least one of the requirements (10.28) is not satisfied, then the values of $E_{(1)}$ are distinct and each can be used for computing d_l and d_q from Eqs. (10.25) and obtaining the desired pair of linear combinations of $|\psi_l\rangle$ and $|\psi_q\rangle$.

In order to find the first-order correction $|\varphi^{(1)}\rangle$ to $|\varphi^{(0)}\rangle$, we make use of the expansion

$$\left|\varphi^{(1)}\right\rangle = \sum_j c_j^{(1)} \left|\varphi_j^{(0)}\right\rangle, \tag{10.29}$$

and multiply from the left Eq. (10.24) by $\langle\psi_k|$, where $k \neq l, q$, so as to obtain

$$c_k^{(1)} = \frac{H'_{kl}d_l + H'_{kq}d_q}{\left(E^{(0)} - E_k^{(0)}\right)}, \tag{10.30}$$

which gives the desired $c_k^{(1)}$ for all $k \neq l, q$. We may impose $c_l^{(1)} = c_q^{(1)} = 0$ so as to satisfy the requirement (10.10), i.e. $\langle\varphi^{(0)} \mid \varphi^{(1)}\rangle = 0$. This computation may be then carried on to higher order by following the methodology of the non-degenerate case.

10.1.3 Perturbation of an oscillator

We may consider the example of the Hamiltonian

$$\hat{H} = \frac{1}{2}\left(\frac{\hat{p}_x^2}{m} + m\omega^2\hat{x}^2\right) + \zeta\hat{x}^4, \tag{10.31}$$

where

$$H_0 = \frac{1}{2}\left(\frac{\hat{p}_x^2}{m} + m\omega^2\hat{x}^2\right) \tag{10.32}$$

is the simple harmonic-oscillator Hamiltonian (see Sec. 4.4), and $\zeta H' = \zeta\hat{x}^4$ is a small anharmonic perturbation. A simple computation (see Eq. (4.72) and Prob. 10.2) shows that

$$E_n^{(0)} = \hbar\omega\left(n + \frac{1}{2}\right), \tag{10.33a}$$

$$\left\langle n \left| \hat{x}^4 \right| n \right\rangle = \left(\frac{\hbar}{2m\omega}\right)^2 (6n^2 + 6n + 3). \tag{10.33b}$$

This result implies that (see Eq. (10.12))

$$E_n = \hbar\omega\left(n + \frac{1}{2}\right) + \zeta\left(\frac{\hbar}{2m\omega}\right)^2\left(6n^2 + 6n + 3\right) + O(\zeta^2). \tag{10.34}$$

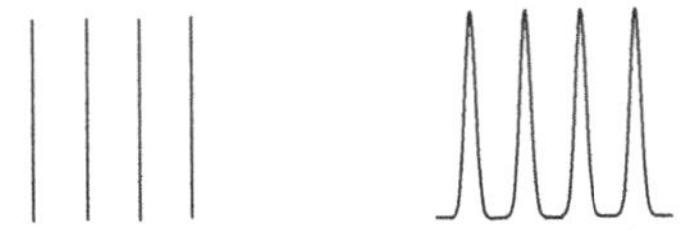

Figure 10.1 Stark effect. (a) The energy spectrum in an atom in absence of an external field is constituted by a series of delta functions centered at the energy eigenvalues of unperturbed Hamiltonian. (b) The perturbation induced by a small external electric field causes the delta function to become Breit–Wigner distributions.

The only matrix elements we need for the calculation of the first-order perturbed eigenstates of the anharmonic oscillator are (see Eq. (10.15) and again Prob. 10.2)

$$\left\langle n \left| \hat{x}^4 \right| n+2 \right\rangle = \left(\frac{\hbar}{2m\omega} \right)^2 (4n+6)\sqrt{(n+1)(n+2)}, \tag{10.35a}$$

$$\left\langle n \left| \hat{x}^4 \right| n+4 \right\rangle = \left(\frac{\hbar}{2m\omega} \right)^2 \sqrt{(n+1)(n+2)(n+3)(n+4)}. \tag{10.35b}$$

In conclusion, we may now compute

$$c_{n,m}^{(1)} = \frac{\langle n \left| \hat{x}^4 \right| m \rangle}{\hbar\omega(m-n)}, \tag{10.36}$$

whose only surviving terms are those for which we have

$$m = \{n-4, n-2, n+2, n+4\}. \tag{10.37}$$

The higher-order terms of the perturbation expansion can be computed in a similar way as explained in detail in Subsec. 10.1.1.

10.1.4 Stark effect for a rigid rotator

In this subsection we shall deal with a type of problem that occurs in the study of the polarization of diatomic molecules in an electric field.[3] We can consider two limiting cases: when there is no field and the electronic energy levels are represented by delta functions and when the external field is so strong that no energy level survives. The Stark effect is the situation in between these two extreme cases, when the energy levels, perturbed by the external field, begin to broaden (see Fig. 10.1). The density of the states in this case is given by a Breit–Wigner distribution,[4] which may be written as

$$\rho_{bw}(E) = \sum_j \frac{1}{2\pi} \frac{\Gamma_j}{(E-E_j)^2 + \left(\Gamma_j/2\right)^2}, \tag{10.38}$$

[3] This effect is strictly connected with the Paschen–Bach and Zeeman effects (see Sec. 11.3) that arise when an atom is subject to a static magnetic field.

[4] See [Breit/Wigner 1936].

where Γ_j is the width of the j-th energy level. In the limit $\Gamma_j \rightarrow 0$ we obviously obtain a series of δ-functions . If the electric field is not very large, the Γ_j are negligible and the only effect of the electric field is a shift in the energy levels. Here, we limit ourselves to computing only the effect for a small electric field.

Let us suppose, for the sake of simplicity, that the system may be treated as a rigid rotator (see Subsec. 6.2.1). The rigid rotator represents, e.g., the motion of the nuclei in a diatomic molecule,[5] and its only degrees of freedom are the angular variables (θ, ϕ). If $\hat{\mathbf{L}}$ is the rotator's angular momentum and $\mathrm{I} = mr_0^2$ its moment of inertia (see Eq. (6.78)), where m is the reduced mass and r_0 the mutual distance of the nuclei, the Hamiltonian, in the absence of electric field, is given by (see Eq. (6.80))

$$\hat{H}_0 = \frac{\hat{\mathbf{L}}^2}{2\mathrm{I}}. \tag{10.39}$$

Its eigenfunctions are represented by the spherical harmonics $Y_{lm}(\phi, \theta)$ and the corresponding eigenvectors are, as usually, $|\,l, m\rangle$ (see Subsec. 6.1.4). The corresponding energy eigenvalues depend only on l, so that we may write

$$\hat{H}_0\,|\,l, m\rangle = E_l^{(0)}\,|\,l, m\rangle\,, \quad E_l^{(0)} = \frac{\hbar^2}{2\mathrm{I}}l\,(l+1). \tag{10.40}$$

In the presence of an uniform electric field $\mathbf{E}$, directed along the z-axis, the Hamiltonian contains the additional term

$$\hat{H}^{'} = -E\,d\cos\theta, \tag{10.41}$$

where d is the electric dipole moment of the rotator. In the $\{|\,l, m\rangle\}$ representation nearly all of the matrix elements of $\hat{H}^{'}$ vanish. In order to have

$$\left\langle l_1, m_1 \left| \hat{H}^{'} \right| l_2, m_2 \right\rangle \neq 0, \tag{10.42}$$

we must have

$$m_1 = m_2, \quad l_1 = l_2 \pm 1. \tag{10.43}$$

When these conditions are fulfilled, we can deduce the matrix elements from the following formula:

$$\langle l, m\,|\cos\theta|\,l-1, m\rangle = \langle l-1, m\,|\cos\theta|\,l, m\rangle = \left(\frac{l^2 - m^2}{4l^2 - 1}\right)^{\frac{1}{2}}. \tag{10.44}$$

Except for the level $l = 0$, all of the unperturbed levels are degenerate. However, both $\hat{H}_0$ and $\hat{H} = \hat{H}_0 + \hat{H}^{'}$ commute with $\hat{L}_z$ and therefore one can separately solve the eigenvalue problem for $\hat{H}$ in each of the subspaces $\mathcal{H}_m$ of a given eigenvalue m of $\hat{L}_z$. In each of these subspaces, the spectrum of $\hat{H}_0$ is non-degenerate, and

$$\left\langle l, m \left| \hat{H}^{'} \right| l, m \right\rangle = 0. \tag{10.45}$$

[5] See Ch. 12 for a treatment of diatomic molecules.

Thus, the first-order correction to the energy levels vanishes. The second-order correction (see Eq. (10.18)) reduces to the two terms corresponding to $l \pm 1$, i.e.

$$E_{lm}^{(2)} = \frac{2\mathrm{I}(E\,d)^2}{\hbar^2} \sum_{l' \neq l} \frac{|\langle l, m\,|\cos\theta|\,l', m\rangle|^2}{l(l+1) - l'(l'+1)} \tag{10.46}$$

$$= \frac{2\mathrm{I}(E\,d)^2}{\hbar^2} \left[\frac{|\langle l, m\,|\cos\theta|\,l+1, m\rangle|^2}{l(l+1) - (l+1)(l+2)} + \frac{|\langle l, m\,|\cos\theta|\,l-1, m\rangle|^2}{l(l+1) - l(l-1)} \right].$$

Calculation of the rhs by means of Eq. (10.44) gives the final result:

$$E_{lm}^{(2)} = \frac{(E\,d)^2}{E_l^{(0)}} \frac{l(l+1) - 3m^2}{2(2l-1)(2l+3)}, \tag{10.47}$$

which also yields the right condition of applicability of a perturbation theory, namely

$$E\,d \ll E_l^{(0)}. \tag{10.48}$$

10.2 Time-dependent perturbation theory

When we wish to treat the dynamics of a perturbed system, we must make use of the time-dependent Schrödinger equation (3.8). As we have said, this assumes particular relevance when the perturbation is time-dependent. In other words, we assume the perturbation Hamiltonian $\hat{H}'$ depends on time. Our aim is to approximate the perturbed state vectors computing the "stationary" state vectors as linear superpositions of the eigenstates of the unperturbed system. As we have said we work with

$$\imath\hbar \frac{\partial}{\partial t} |\varphi\rangle = \hat{H} |\varphi\rangle, \tag{10.49}$$

with the same notation as in Eq. (10.1). As we know, the time evolution of the unperturbed eigenstates (see Eqs. (3.18) and (3.21)) is given by

$$|\psi_n(t)\rangle = e^{-\imath t \frac{E_n^{(0)}}{\hbar}} |\psi_n\rangle. \tag{10.50}$$

We now expand $|\varphi\rangle$ in terms of the $|\psi_n(t)\rangle$'s, with time-dependent expansion coefficients

$$|\varphi(t)\rangle = \sum_j c_j(t)\,|\psi_j\rangle\, e^{-\frac{\imath}{\hbar}E_j^{(0)}t}, \quad \sum_j |c_j(t)|^2 = 1. \tag{10.51}$$

Substitution of Eq. (10.51) in Eq. (10.49) gives

$$\sum_j \imath\hbar\dot{c}_j\,|\psi_j\rangle\, e^{-\frac{\imath}{\hbar}E_j^{(0)}t} + \sum_j c_j E_j^{(0)}\,|\psi_j\rangle\, e^{-\frac{\imath}{\hbar}E_j^{(0)}t} = \sum_j c_j\left[\hat{H}_0 + \zeta\hat{H}'(t)\right]|\psi_j\rangle\, e^{-\frac{\imath}{\hbar}E_j^{(0)}t}, \tag{10.52}$$

where we have assumed that $\hat{H}'(t)$ and $c_j(t)$ commute: this is correct under the hypothesis that $\hat{H}'(t)$ does not contain time derivatives. We replace $\hat{H}_0\,|\psi_j\rangle$ by $E_j^{(0)}\,|\psi_j\rangle$ on the rhs and multiply on the left by $\langle\psi_k\,|$, so as to obtain

$$\imath\hbar\dot{c}_k e^{-\frac{\imath}{\hbar}E_k^{(0)}t} = \sum_j c_j e^{-\frac{\imath}{\hbar}E_j^{(0)}t}\left\langle\psi_k\left|\zeta\hat{H}'(t)\right|\psi_j\right\rangle. \tag{10.53}$$

Making use of the Bohr frequency $\omega_{kj} = (E_k^{(0)} - E_j^{(0)})/\hbar$ (see Eq. (1.75)), we obtain

$$\dot{c}_k = \frac{1}{\imath\hbar}\sum_j\left\langle\psi_k\left|\zeta\hat{H}'(t)\right|\psi_j\right\rangle c_j e^{\imath\omega_{kj}t}. \tag{10.54}$$

The quantities

$$H'_{kj}(t) = \left\langle\psi_k\left|\zeta\hat{H}'(t)\right|\psi_j\right\rangle e^{\imath\omega_{kj}t} \tag{10.55}$$

are the matrix elements of the perturbation, which explicitly depend on time. The group of Eqs. (10.54) for all k is equivalent to the Schrödinger equation (10.49). Therefore, solving Eq. (10.54) – i.e. finding the time-dependent coefficients $c_j(t)$ – is equivalent to solving the original time-dependent Schrödinger equation. As in the previous section, we express the coefficient $c_k(t)$ as a formal Taylor expansion around $\zeta = 0$

$$c_k = c_k^{(0)} + \zeta c_k^{(1)} + \zeta^2 c_k^{(2)} + \cdots. \tag{10.56}$$

As usual, we have assumed that all the quantities have finite derivatives at $\zeta = 0$. Then, we substitute the expansion (10.56) in Eq. (10.54), we equate the coefficients of equal power of ζ, and obtain the set of equations

$$\dot{c}_k^{(s+1)} = \frac{1}{\imath\hbar}\sum_j \hat{H}'_{kj}(t)c_j^{(s)}, \tag{10.57}$$

where $\dot{c}_k^{(0)} = 0$. We shall assume that all but one $c_k^{(0)}$ are zero, so that the system is initially in a definite unperturbed energy eigenstate, let us say it is l. Integration of the first-order equation gives

$$c_k^{(1)}(t) = \frac{1}{\imath\hbar}\int_{-\infty}^{t} dt' H'_{kl}(t'). \tag{10.58}$$

Equation (10.58) takes a particularly simple form if the perturbation $\hat{H}'(t)$ depends harmonically on time except for being turned on at a certain time $t_0 = 0$ and off at a later time t_∞. Integrating Eq. (10.58) by parts, we obtain

$$c_k^{(1)}(t) = -\frac{1}{\hbar\omega_{kl}}\left\langle\psi_k\left|\hat{H}'(t')\right|\psi_l\right\rangle e^{\imath\omega_{kl}t'}\Big|_{-\infty}^{t} + \int_{-\infty}^{t} dt'\frac{e^{\imath\omega_{kl}t'}}{\hbar\omega_{kl}}\frac{\partial}{\partial t'}\left\langle\psi_k\left|\hat{H}'(t')\right|\psi_l\right\rangle. \tag{10.59}$$

The first term disappears in the lower limit whereas in the upper limit it coincides with the coefficients

$$\frac{H'_{kl}(t)}{E_l^{(0)} - E_k^{(0)}}. \tag{10.60}$$

If the perturbation acts for a finite range of time, the probability of transition from the state l to a state k after the perturbation has been switched off ($t > t_\infty$) is given by the square modulus of the second term:

$$\wp(l \to k) = \frac{1}{\hbar^2 \omega_{kl}^2} \left| \int_{-\infty}^{+\infty} dt' e^{\imath \omega_{kl} t'} \frac{\partial}{\partial t'} \left\langle \psi_k \left| \hat{H}'(t') \right| \psi_l \right\rangle \right|^2 . \tag{10.61}$$

Let us now assume that the perturbation has a harmonic time dependence, i.e.

$$H'_{kl}(t) = 2 \left\langle \psi_k \left| \hat{H}'_i \right| \psi_l \right\rangle e^{\imath \omega_{kl} t} \sin \omega t, \tag{10.62}$$

where $\left\langle \psi_k \left| \hat{H}'_i \right| \psi_l \right\rangle$ is independent of time and ω is positive. Substitution of Eq. (10.62) into Eq. (10.58) gives for the first-order amplitude at any time $t \geq t_\infty$ (see Prob. 10.3)

$$c_k^{(1)}(t \geq t_\infty) = -\frac{\left\langle \psi_k \left| \hat{H}'_i \right| \psi_l \right\rangle}{\imath \hbar} \left[\frac{e^{\imath(\omega_{kl}+\omega)t_\infty} - 1}{\omega_{kl} + \omega} - \frac{e^{\imath(\omega_{km}-\omega)t_\infty} - 1}{\omega_{kl} - \omega} \right]. \tag{10.63}$$

Equation (10.63) shows that the amplitude is bigger when the denominator of one or the other of the two terms is near zero.

Let us consider the situation in which the initial state $|\psi_l\rangle$ is a discrete bound state and the final state $|\psi_k\rangle$ is one of a continuous set of dissociated states. Then, $E_k > E_l$ and only the second term of Eq. (10.63) need to be considered. The first-order probability of finding the system in the state k after the perturbation is removed is (see Prob. 10.4)

$$\wp(l \to k) = |c_k^{(1)}(t \geq t_\infty)|^2 = \frac{4 \left| \left\langle \psi_k \left| \hat{H}'_i \right| \psi_l \right\rangle \right|^2 \sin^2 \frac{1}{2} (\omega_{kl} - \omega) t_\infty}{\hbar^2 (\omega_{kl} - \omega)^2}. \tag{10.64}$$

The transition probability per unit time is given by integrating Eq. (10.64) over all energy levels k of the continuous set and dividing by t_∞, i.e.

$$\wp_{\text{trans}} = \frac{1}{t_\infty} \int dE_k |c_k^{(1)}(t \geq t_\infty)|^2 \rho(k), \tag{10.65}$$

where $\rho(k) dE_k$ is the number of final states with energies between E_k and $E_k + dE_k$ and $\rho(k)$ is the energy density of the final states. Substitution of Eq. (10.64) into Eq. (10.65) yields

$$\wp_{\text{trans}} = \frac{2\pi}{\hbar} \rho(k) \left| \left\langle \psi_k \left| \hat{H}'_i \right| \psi_l \right\rangle \right|^2 , \tag{10.66}$$

where we have made use of the equality

$$\int_{-\infty}^{+\infty} dx \frac{\sin^2 x}{x^2} = \pi . \tag{10.67}$$

Equation (10.66) is often called the *Fermi golden rule* and tells us that in the presence of a weak resonant perturbation the population of the k-th state grows linearly with time.

In conclusion, let us consider Eq. (3.21) again. We see that, when the Hamiltonian is time-independent, the initial state evolves with time in such a way that the probabilities of the different energy eigenvalues are conserved, since the initial coefficients of the eigenstates of the Hamiltonian in the expansion are only multiplied by phase factors, and therefore only the relative phase is modified. This remains obviously true also in the case considered in the previous section. Instead, when the perturbation depends on time, the eigenstates of the perturbed Hamiltonian themselves are time-dependent and therefore the above probabilities are not conserved. This in turn means that, by the effect of the perturbation, the systems undergoes a progressive shift in the distribution of its components in the initial superposition.

10.3 Adiabatic theorem

As we have seen, it is very difficult in general to solve the Schrödinger equation with an explicitly time-dependent potential. In Sec. 3.1.3 we have solved the Schrödinger equation in the time-independent case. In the previous section we have presented an approximation method that allows us to deal with weak time-dependent perturbation. In this section we discuss the problem of the time evolution of a system with a Hamiltonian that explicitly depends on time but changes slowly – i.e. *adiabatically*.

Let us consider a Hamiltonian that is dependent on a parameter $\boldsymbol{\zeta}(t)$ that may be seen as a vector and that describes a closed loop in the parameter space.[6] It is then clear that, due to the time dependence of the parameter $\boldsymbol{\zeta}$, the Hamiltonian itself is time-dependent. The Schrödinger equation of the system may be written as

$$\imath\hbar\frac{d}{dt}\,|\,\psi(t)\rangle = \hat{H}\left[\boldsymbol{\zeta}(t)\right]|\,\psi(t)\rangle, \tag{10.68}$$

and the eigenvalue equation of the Hamiltonian is

$$\hat{H}\left[\boldsymbol{\zeta}(t)\right]\left|n\left[\boldsymbol{\zeta}(t)\right]\right\rangle = E_n\left[\boldsymbol{\zeta}(t)\right]\left|n\left[\boldsymbol{\zeta}(t)\right]\right\rangle, \tag{10.69}$$

where $\left|n\left[\boldsymbol{\zeta}(t)\right]\right\rangle$ is an instantaneous eigenstate of the Hamiltonian. Consider now an arbitrary initial state $|\,\psi(t_0 = 0)\rangle$ and expand it as

$$|\,\psi(t_0)\rangle = \sum_n \psi_n(t_0)\left|n\left[\boldsymbol{\zeta}(t_0)\right]\right\rangle, \tag{10.70}$$

where $\psi_n = \langle n\mid\psi\rangle$. We shall now try to describe the time evolution of this state. Suppose that we may write

$$|\,\psi(t)\rangle = \sum_n \psi_n(t)e^{-\imath\phi_n^{(d)}(t)}\left|n\left[\boldsymbol{\zeta}(t)\right]\right\rangle, \tag{10.71}$$

where

$$\phi_n^{(d)}(t) = \frac{1}{\hbar}\int_0^t dt' E_n\left[\boldsymbol{\zeta}(t')\right] \tag{10.72}$$

[6] See [*Schleich* 2001, 172–74].

is the dynamical phase for a time-dependent system.[7] Equation (10.72) may be also written as

$$\frac{d}{dt}\phi_n^{(d)}(t) = \frac{1}{\hbar} E_n\left[\boldsymbol{\zeta}(t)\right]. \tag{10.73}$$

By performing the time derivative of Eq. (10.71),

$$\imath\hbar\frac{d}{dt}\mid\psi(t)\rangle = \imath\hbar\sum_n e^{-\imath\phi_n^{(d)}}\left\{\left[\left(\frac{d}{dt}\psi_n\right) + \psi_n\left(-\imath\frac{d}{dt}\phi_n^{(d)}\right)\right]\mid n\rangle + \psi_n\left(\frac{d}{dt}\mid n\rangle\right)\right\}, \tag{10.74}$$

where, for the sake of notation, the argument $\boldsymbol{\zeta}(t)$ has been suppressed, and, by taking into account Eq. (10.73), we obtain

$$\imath\hbar\frac{d}{dt}\mid\psi(t)\rangle = \imath\hbar\sum_n e^{-\imath\phi_n^{(d)}}\left[\left(\frac{d}{dt}\psi_n\right)\mid n\rangle + \psi_n\left(\frac{d}{dt}\mid n\rangle\right)\right] + \sum_n E_n\psi_n e^{-\imath\phi_n^{(d)}}\mid n\rangle. \tag{10.75}$$

Considering the rhs of Eq. (10.68) and, by taking into account Eqs. (10.69) and (10.71), yields

$$\hat{H}\left[\boldsymbol{\zeta}(t)\right]\mid\psi(t)\rangle = \sum_n E_n\psi_n e^{-\imath\phi_n^{(d)}}\mid n\rangle. \tag{10.76}$$

Equating the rhs of Eqs. (10.75) and (10.76), we have

$$\sum_n e^{-\imath\phi_n^{(d)}}\left[\left(\frac{d}{dt}\psi_n\right)\mid n\rangle + \psi_n\left(\frac{d}{dt}\mid n\rangle\right)\right] = 0. \tag{10.77}$$

Projecting onto the m-th energy eigenstate and assuming the orthonormality condition of the eigenstates of the Hamiltonian, we obtain

$$\frac{d}{dt}\psi_m = -\left\langle m\left|\frac{d}{dt}\right|m\right\rangle\psi_m - \sum_{n\neq m} e^{-\imath(\phi_n^{(d)}-\phi_m^{(d)})}\left\langle m\left|\frac{d}{dt}\right|n\right\rangle\psi_n, \tag{10.78}$$

which obviously shows that different energy eigenstates are coupled with each other: even if the system begins in an initial state given by a single energy eigenstate, all the other energy eigenstates come into play due to the natural time-dependent Schrödinger evolution. Three factors determine the transition probability from the n-th to the m-th level:

- the matrix element $M_{m,n}(t) = \langle m\mid\frac{d}{dt}\mid n\rangle$;
- the phase difference $\phi_n^{(d)} - \phi_m^{(d)}$;
- the initial probability amplitude ψ_n of the n-th level.

Let us rewrite Eq. (10.78), indicating explicit time dependence and expanding the differentiation with respect to time as a scalar product between the gradient with respect to the vector parameter $\boldsymbol{\zeta}$ and the time derivative of the parameter. Then, we obtain

$$M_{m,n} = \left\langle m\left[\boldsymbol{\zeta}(t)\right]\left|\frac{d}{dt}\right|n\left[\boldsymbol{\zeta}(t)\right]\right\rangle = \langle m\left[\boldsymbol{\zeta}(t)\right]\mid\nabla_{\boldsymbol{\zeta}}\mid n\left[\boldsymbol{\zeta}(t)\right]\rangle\frac{d}{dt}\boldsymbol{\zeta}(t), \tag{10.79}$$

[7] This is just the obvious generalization of the phase factor $E_n t/\hbar$, which we have in the case of the time-independent Hamiltonian (see Eq. (3.21)).

which shows that the size of $M_{m,n}$ is proportional to the rate of change of ζ. Provided that this rate is small compared to the time scales involved, the coupling due to $M_{m,n}$ is also small. Moreover, each of the terms in the sum in the rhs of Eq. (10.78) has a phase factor which is given by the time-dependent differences $\phi_n^{(d)} - \phi_m^{(d)}$, whose oscillations contribute to make the coupling small. The *adiabatic theorem* then states that a system initially in the energy eigenstate $|j[\zeta(t_0)]\rangle$ remains in the same instantaneous (and slowly evolving in time) energy eigenstate, provided that the change in $\zeta(t)$ is adiabatic – however, as we shall see in Sec. 13.8, it also acquires a *geometric phase* factor.

10.4 The variational method

Generally speaking, the variational method, also called the method of Ritz, can be used for the approximate determination of the ground-state of the energy of a system when the perturbation method is inapplicable. If an arbitrary ket $|\varphi\rangle$ is expanded as

$$|\varphi\rangle = \sum_n c_n |\psi_n\rangle, \tag{10.80}$$

in the energy eigenkets $|\psi_n\rangle$ that form a complete orthonormal set, the expectation value of the Hamiltonian $\hat{H}$ (where $\hat{H}|\psi_n\rangle = E|\psi_n\rangle$) on the state described by the wave function $\varphi(\mathbf{r}) = \langle \mathbf{r} | \varphi\rangle$ is given by

$$\left\langle \hat{H} \right\rangle_\varphi = \int d\mathbf{r}\varphi^*(\mathbf{r})\hat{H}\varphi(\mathbf{r}) = \sum_n E|c_n|^2, \tag{10.81}$$

where the integration is extended over the entire range of all the coordinates of the system. For convenience, it is assumed that the energy eigenvalues are discrete in Eqs. (10.80) and (10.81). As we know, this can be accomplished by enclosing the system in a box (see Sec. 3.4).

A useful inequality may be derived from Eq. (10.81) by replacing each eigenvalue E in the summation on the rhs by the lowest energy eigenvalue E_0, i.e. the energy of the ground state, so that

$$\left\langle \hat{H} \right\rangle_\varphi \geq E_0 \sum_n |c_n|^2. \tag{10.82}$$

Since $\sum_n |c_n|^2 = 1$ for a normalized ket, then we also have

$$E_0 \leq \int d\mathbf{r}\varphi^*(\mathbf{r})\hat{H}\varphi(\mathbf{r}). \tag{10.83}$$

The variational method consists in evaluating the integral on the rhs of Eq. (10.83) with a trial ket $|\varphi\rangle$ that depends on a "small" number of parameters, and by varying these parameters until the expectation value of the energy reaches its minimum. The result is an upper limit for the ground-state energy of the system. Notice that, if $\varphi(\mathbf{r}) = \psi_0(\mathbf{r}) + \epsilon\delta\psi(\mathbf{r})$, with small ϵ, then

$$\int d\mathbf{r}\varphi^*(\mathbf{r})\hat{H}\varphi(\mathbf{r}) = E_0 + O(\epsilon^2). \tag{10.84}$$

This means that, with the variational method, the estimate of the wave function is inaccurate at order ϵ while that of the energy is inaccurate at the order ϵ^2.

A similar procedure may be applied to the estimate of the other eigenvalues E_n, with $E_{n-1} < E_n < E_{n+1}$. In this case, besides the extremum condition on the integral

$$\int d\mathbf{r}\psi_n^*(\mathbf{r})\hat{H}\psi_n(\mathbf{r}), \tag{10.85}$$

and the normalization condition

$$\int d\mathbf{r}\psi_n^*(\mathbf{r})\psi_n(\mathbf{r}) = 1, \tag{10.86}$$

we also have the orthogonality condition

$$\int d\mathbf{r}\psi_n^*(\mathbf{r})\psi_m(\mathbf{r}) = 0, \tag{10.87}$$

where $m = 0, 1, \ldots, n-1$.

10.5 Classical limit

As we have seen (Pr. 2.3: p. 72), the classical world may be understood in some sense as a limit of the quantum world when the physical systems become larger and larger, i.e. , formally, when $\hbar \to 0$. This is equivalent to taking the high-energy limit. In this section we shall see some consequences of this limit and try to give a physical and formal meaning to the transition from the quantum to the classical domain.

10.5.1 Lieb's theorem

Let us limit the following examination to the one-dimensional classical case, being the trivial generalization. Classically, in absence of dissipation, the time derivative of the position is given by $\dot{x} = p_x/m$, while the time derivative of the momentum can be written as a function of x, i.e. $\dot{p}_x = f(x)$ (see Sec. 1.1, and in particular Eqs. (1.7)). Since position and momentum variables classically commute, the second and third time derivatives of the position variable may be written as (see Prob. 10.5)

$$\ddot{x} = \frac{f(x)}{m} \quad \text{and} \quad \dddot{x} = \frac{\dot{f}(x)}{m} = \frac{p_x}{m^2} \cdot f'(x), \tag{10.88}$$

respectively, where $f'(x) = df(x)/dx$. In the quantum case, however, the third time derivative is given by (see Prob. 10.6)

$$\hat{\dddot{x}} = \frac{1}{2m^2}\left[\hat{p}_x f'(\hat{x}) + f'(\hat{x})\hat{p}_x\right]. \tag{10.89}$$

This result coincides with the classical one in the limit $\hbar \to 0$, because in this limit the observables commute. Let us come back on the problem of the relationship between classical Poisson brackets and quantum commutators. As we know (see Eq. (3.109) and remarks to Eq. (3.124)), in the classical limit – apart from a constant factor – commutators have to be replaced by the corresponding Poisson brackets. In general, there are strong similarities between the classical Poisson brackets and the quantum commutators provided that the Heisenberg commutation relations are valid. For instance,

$$\{x, p_x\} = 1, \quad \left[\hat{x}, \hat{p}_x\right] = \imath\hbar. \tag{10.90}$$

The previous equations can be generalized to (see Eq. (1.9) and Probs. 2.26–2.27)

$$\{f(x), p_x\} = f'(x), \quad \left[f(\hat{x}), \hat{p}_x\right] = \imath\hbar f'(\hat{x}) \tag{10.91a}$$

$$\{x, g(p_x)\} = g'(p_x), \quad \left[\hat{x}, g(\hat{p}_x)\right] = \imath\hbar g'(\hat{p}_x), \tag{10.91b}$$

where f and g are some functions. We recall here that the similarity between Poisson brackets and commutators also covers more formal aspects: for instance, both obey the Jacobi identity (see Eq. (2.96)). Notwithstanding, the correspondence between the classical and quantum formalism is not always so specular. For example, we have

$$\{f(x), p_x^2\} = 2p_x f'(x), \quad \left[f(\hat{x}), \hat{p}_x^2\right] = \imath\hbar\left[\hat{p}_x f'(\hat{x}) + f'(\hat{x})\hat{p}_x\right]. \tag{10.92}$$

In general, for two observables $\hat{\xi}$ and $\hat{\eta}$, we may write

$$\left[\hat{\xi}, \hat{\eta}\right] = \imath\hbar\{\xi, \eta\} + O(\hbar^2). \tag{10.93}$$

Therefore, we recover the classical case only if we neglect terms of order higher than $\hbar$.

Let us suppose that at time t_0 the quantum system is represented by a wave packet (see Box 2.6) – in the state $\psi(x)$ – centered around a given value ($\langle\hat{x}\rangle = x(t_0)$) of the position and of the momentum ($\langle\hat{p}_x\rangle = p(t_0)$) such that

$$\Delta_\psi x = O(\hbar^{1/2}), \quad \Delta_\psi p_x = O(\hbar^{1/2}). \tag{10.94}$$

This wave packet will evolve in such way that, neglecting all the terms of order higher than $\hbar$, $\langle\hat{x}\rangle_t = x(t)$ and $\langle\hat{p}_x\rangle_t = p(t)$ satisfy the classical equations of motion (see final discussion in Sec. 3.7), and the variances of the position and momentum always go to zero when $\hbar$ goes to zero. This conclusion appears to be natural because, formally, if we neglect the higher-order terms in the commutators, the operators do satisfy the classical equations of motion.

The previous discussion might suggest that it is always possible to perform the classical limit. As a matter of fact, this statement needs some specification, first for the different nature of the objects involved in classical and in quantum mechanics, second because, in the quantum case, if the Hamiltonian is well-defined, the quantum dynamics is also well-defined at all times, while a classically determined dynamics does not always exist: the passage to the classical limit is in general only possible if the classical counterpart is well defined. This is the content of Lieb's theorem,[8] which states that this passage is possible

[8] See [Lieb 1973].

only with the exception of "pathological cases." For instance, Lieb's theorem tells us that the previous formal development is actually correct.

On the other hand, as an example of "pathological system," let us consider the Hamiltonian

$$H = ap_x^2 + bx^4, \tag{10.95}$$

with $a > 0, b < 0$. When the energy of the particle is zero, we have $p_x^2 = -(b/a)x^4$. In this case, we also have

$$m\frac{dx}{dt} = \pm\sqrt{-\frac{b}{a}}, \tag{10.96}$$

from which it follows that

$$\frac{dx}{x^2} = \pm\frac{1}{m}\sqrt{-\frac{b}{a}}dt. \tag{10.97}$$

Integrating the previous equation from t_0 to t, we obtain

$$\frac{1}{x(t_0)} - \frac{1}{x(t)} = \pm\frac{1}{m}\sqrt{-\frac{b}{a}}\,(t - t_0). \tag{10.98}$$

As a consequence, the system goes to infinity in a finite time: if t_∞ is the time for which $x(t_\infty) = \infty$,

$$t_\infty = t_0 + m\frac{\sqrt{-\frac{a}{b}}}{x(t_0)}. \tag{10.99}$$

Classically, three possibilities arise: (1) the system is lost at infinity; (2) it bounces back; (3) it disappears at $+\infty$ and comes out from $-\infty$. Since the equations of motion do not allow us to decide which of these different possibilities must occur, the dynamics of the system, when passing the critical point t_∞ (which depends on the initial position), is classically not wholly determined. Therefore, the statement "the quantum evolution goes to the classical one" does not make sense as far as the classical evolution is not uniquely defined.

It is interesting to note that, for the example chosen, the Hamiltonian is also not unambiguously defined quantum-mechanically. In fact, it is not bounded from below (see Box 2.1) and it is not well-defined as a Hermitian operator, i.e. its definition as a Hermitian operator is not unique. Therefore, in this particular case, we could properly say that only when $t < t_\infty$ can the quantum system tend to its classical counterpart in the limit $\hbar \to 0$. However, in order to define the time evolution at any time, which is also much smaller than t_∞, in the quantum case we must make some choices on the behavior of the wave function at $x = \pm\infty$. In the quantum case, the resolution of the ambiguity is a must at any time because the probability of arriving at infinity is always different from zero, although very small for short times.

10.5.2 Amplitudes and action

Given the above limitation, we may now discuss the case when the passage at the limit $\hbar \to 0$ is possible. Here, we can safely assume that the classical dynamics is well-defined.

The probability for a quantum system $\mathcal{S}$ that is in an initial state $x(t_0)$ at time t_0 to be found in a position $x(t)$ at time t is $|\langle x(t) \mid x(t_0)\rangle|^2$ and its relative amplitude is given by the Green's function[9] (see Subsec. 3.5.5)

$$\mathrm{G}(x(t_0), x(t); t_0, t) = \langle x(t) \mid x(t_0)\rangle = e^{\frac{\iota}{\hbar}\mathrm{S}(x(t_0),x(t);t_0,t)}, \tag{10.100}$$

where the function $\mathrm{S}(x(t_0), x(t); t_0, t)$ will be defined below. Equation (10.100) is also a solution of the Schrödinger equation with respect to both variables x, t, which in the one-dimensional case is

$$-\iota\hbar\frac{d}{dt_0}\langle x(t) \mid x(t_0)\rangle = \int dx'(t_0)\left\langle x(t) \mid x'(t_0)\right\rangle\left\langle x'(t_0)\left|\hat{H}_{t_0}\right| x(t_0)\right\rangle. \tag{10.101}$$

One can prove that in the limit in which $\hbar \to 0$ (see again Pr. 2.3: p. 72) the function $\mathrm{S}(x(t_0), x(t); t_0, t)$ will coincide with the classical action evaluated along the classical trajectory going from $x(t_0)$ to $x(t)$ (see Eq. (1.14)). Then we have that

$$-\frac{\partial \mathrm{S}}{\partial t_0} = H^{\mathrm{c}}\left(x(t_0), p_x(t_0)\right), \tag{10.102}$$

where

$$p_x(t_0) = -\frac{\partial \mathrm{S}}{\partial x(t_0)}. \tag{10.103}$$

and H^{c} is the Hamiltonian of the corresponding classical system, which is formally identical to the above quantum Hamiltonian. To obtain the quantum analogue of the classical Lagrangian, we must consider an infinitesimal time interval $t = t_0 + \delta t$. We obtain then $\langle x(t_0 + \delta t) \mid x(t_0)\rangle$ as the analogue of $e^{\frac{\iota}{\hbar}L(t_0)\delta t}$. In this case, one should consider $L(t_0)$ as a function of the coordinate x at time $t_0 + \delta t$ and of the coordinate $x(t_0)$ at time t_0 rather than as a function of position and momentum, as is usually assumed. Finally, the quantum amplitude $\langle x(t) \mid x(t_0)\rangle$ may be written in the limit of a small $\hbar$ as

$$\langle x(t) \mid x(t_0)\rangle = e^{\frac{\iota}{\hbar}\int_{t_0}^{t} dt' L(t')} = e^{\frac{\iota}{\hbar}\mathrm{S}(x(t_0),x(t);t_0,t)}. \tag{10.104}$$

10.5.3 Hamilton–Jacobi equation

We consider now the case of a three-dimensional particle subject to a potential energy $V(\mathbf{r})$. Let us write a wave function describing the state of this particle as

$$\psi(\mathbf{r}) = \vartheta(\mathbf{r})e^{\frac{\iota}{\hbar}\phi(\mathbf{r})}, \tag{10.105}$$

where ϑ and ϕ are the amplitude and the phase of $\psi(\mathbf{r})$, respectively. Back-substituting Eq. (10.105) into the time-dependent Schrödinger equation (3.14) and separating the real and the imaginary parts, we obtain the equations (see Prob. 10.7)

$$\frac{\partial\phi}{\partial t} + \frac{(\nabla\phi)^2}{2m} + V = \frac{\hbar^2}{2m}\frac{\Delta\vartheta}{\vartheta}, \tag{10.106a}$$

$$m\frac{\partial\vartheta}{\partial t} + (\nabla\vartheta\cdot\nabla\phi) + \frac{1}{2}\vartheta\Delta\phi = 0, \tag{10.106b}$$

[9] See [*Dirac* 1930, 125–30].

where $\boldsymbol{\nabla}$ is the nabla operator and Δ the Laplacian. As we have already emphasized, the classical limit consists in taking the limit $\hbar \to 0$. Equation (10.106a) then becomes

$$\frac{\partial \phi}{\partial t} + \frac{(\boldsymbol{\nabla}\phi)^2}{2m} + V = 0, \tag{10.107}$$

which is the *Hamilton–Jacobi equation*, where the phase ϕ has to be interpreted as the classical action and therefore coincides with the function S introduced in the previous subsection.

In the classical limit, ψ describes a flux of non-interacting classical particles of mass m (i.e. a statistical mixture), which are subject to the potential energy $V(\mathbf{r})$. The density and the current density of this fluid at each point of space are at all times equal to the probability density ρ and the probability current density $\mathbf{J}$ of the quantum particle at that point (see Sec. 4.2). This can be shown as follows: multiplying both sides of Eq. (10.106b) by $2\vartheta/m$, we obtain

$$\frac{\partial}{\partial t}\vartheta^2 + \boldsymbol{\nabla}\left(\vartheta^2 \frac{\boldsymbol{\nabla}\phi}{m}\right) = 0, \tag{10.108}$$

which is the continuity equation (4.21) for the probability density

$$\rho(\mathbf{r}) = \vartheta^2(\mathbf{r}), \tag{10.109a}$$

and the current density

$$\mathbf{J}(\mathbf{r}) = \vartheta^2(\mathbf{r})\frac{\boldsymbol{\nabla}\phi(\mathbf{r})}{m}. \tag{10.109b}$$

Since the continuity equation of this fluid is satisfied, to make the analogy complete it is sufficient to show that the velocity field

$$\mathbf{v} = \frac{\mathbf{J}}{\rho} = \frac{\boldsymbol{\nabla}\phi}{\mathbf{m}} \tag{10.110}$$

of this fluid actually follows the law of motion of the classical fluid. Taking into account Eq. (10.110), Eq. (10.107) may be rewritten as

$$\frac{\partial \phi}{\partial t} + \frac{mv^2}{2} + V = 0. \tag{10.111}$$

Taking the gradient of both sides of Eq. (10.111), we obtain

$$\left[\frac{\partial}{\partial t} + (\mathbf{v}\cdot\boldsymbol{\nabla})\right] m\mathbf{v} + \boldsymbol{\nabla}V = 0, \tag{10.112}$$

from which we conclude that the particles of the fluid obey the classical equation of motion.

10.5.4 Spreading of the wave packets

A wave packet (see Box 2.6) may be considered the analogue of a classical particle if its "position" and "momentum" follow the laws of classical mechanics and if the dimension of the packet is sufficiently small at all times (see also Subsec. 10.5.1). The two requirements are clearly connected, as the Ehrenfest theorem shows (see Sec. 3.7). The importance of the second requirement is emphasized by the argument that follows. Let

$$\hat{H} = \frac{\hat{p}_x^2}{2m} + V(\hat{x}) \tag{10.113}$$

be the Hamiltonian of this one-dimensional packet. In the classical limit, the packet represents a particle whose position and momentum are

$$x^{\mathrm{c}} = \langle \hat{x} \rangle, \quad p_x^{\mathrm{c}} = \langle \hat{p}_x \rangle, \tag{10.114}$$

respectively. If the classical approximation is justified, the energy in the classical limit

$$E_{\mathrm{c}} = \frac{\langle \hat{p}_x \rangle^2}{2m} + V(\langle \hat{x} \rangle), \tag{10.115}$$

which does not necessarily coincide with $\langle \hat{H} \rangle$, is constant in time and so is also the difference $\langle \hat{H} \rangle - E_{\mathrm{c}}$. Since the extension of the wave packet should remain confined to a small interval, we can replace the operator $V(\hat{x})$ and its first derivative $V'(\hat{x})$ by their Taylor expansion about $\langle \hat{x} \rangle$,

$$V(\hat{x}) = V_{\mathrm{c}} + (\hat{x} - \langle \hat{x} \rangle)\, V'_{\mathrm{c}} + \frac{1}{2}(\hat{x} - \langle \hat{x} \rangle)^2\, V''_{\mathrm{c}} + \cdots, \tag{10.116a}$$

$$V'(\hat{x}) = V'_{\mathrm{c}} + (\hat{x} - \langle \hat{x} \rangle)\, V''_{\mathrm{c}} + \frac{1}{2}(\hat{x} - \langle \hat{x} \rangle)^2\, V'''_{\mathrm{c}} + \cdots, \tag{10.116b}$$

where the expressions $V_{\mathrm{c}}, V'_{\mathrm{c}}, \ldots$ indicate the values assumed by the potential–energy function V and its derivatives $V', V'', \ldots$ at the point $\hat{x} = \langle \hat{x} \rangle$. Taking the average values of Eqs. (10.116), we obtain

$$\langle V \rangle = V_{\mathrm{c}} + \frac{1}{2}\sigma_x^2 V''_{\mathrm{c}} + \cdots, \tag{10.117a}$$

$$\langle V' \rangle = V'_{\mathrm{c}} + \frac{1}{2}\sigma_x^2 V'''_{\mathrm{c}} + \cdots. \tag{10.117b}$$

where $\sigma_x^2 = \langle \hat{x}^2 \rangle - \langle \hat{x} \rangle^2$ is the square deviation of $\hat{x}$ from its mean (see Subsec. 2.3.1). Ehrenfest theorem provides us with the equations[10]

$$\frac{d}{dt}\langle \hat{x} \rangle = \frac{\langle \hat{p}_x \rangle}{m}, \tag{10.118a}$$

$$\frac{d}{dt}\langle \hat{p}_x \rangle = -\langle \hat{V}' \rangle, \tag{10.118b}$$

which coincide with the classical equations of motion if the rhs of Eq. (10.118b) is replaced by the first term in the rhs of Eq. (10.117b). In order for this approximation to be exact, the third and higher derivatives of $V(\hat{x})$ should be zero. This is the case when the potential energy is a polynomial of at most second degree in $\hat{x}$, as it is the case for the harmonic oscillator and the free particle (see Secs. 3.4 and 4.4). Alternatively, the approximation

[10] These equations are easily derived by simply taking the expectation values of both sides of Eqs. (3.126) and (3.128).

may still be reasonably valid when $V(\hat{x})$ varies slowly enough over the extension of the wave packet, so that $V'''(\hat{x})$ is small.

It can be shown (see Prob. 10.8) that a free-particle wave packet undergoes uniform rectilinear motion in which $\langle \hat{x} \rangle$ moves with velocity $\langle \hat{p}_x \rangle / m$. Moreover, the momentum square deviation $\sigma_p^2 = \langle \hat{p}_x^2 \rangle - \langle \hat{p}_x \rangle^2$ remains constant, while

$$\sigma_x^2(t) = \sigma_x^2(t_0) + \dot{\sigma}_x^2(t_0)(t - t_0) + \frac{\sigma_p^2(t_0)}{m^2}(t - t_0)^2. \tag{10.119}$$

This means that σ_x^2 grows indefinitely with time, causing the so-called spreading of the wave packet. As a consequence, a free quantum wave packet can be considered the analogue of a classical particle only for small times, such that the second and the third terms in Eq. (10.119) remain negligibly small.

As a matter of fact, the wave packet undergoes such a spreading in almost all the cases. A special exception is represented by the harmonic oscillator potential for which the wave packet does not spread at all (see Subsecs. 4.4.2 and 13.4.2, as well as Probs. 10.9 and 13.23).

10.6 Semiclassical limit and WKB approximation

10.6.1 Semiclassical limit

Here we shall introduce some general considerations about the classical limit before dealing with the WKB approximation. Let us consider the trace of the Green's function G (see again Subsec. 3.5.5) times some observable $\hat{O}$

$$\mathrm{Tr}\left(\hat{O}\mathrm{G}\right) = g_O(E) = \lim_{\epsilon \to 0} \sum_j \frac{\left\langle j \left| \hat{O} \right| j \right\rangle}{E - E_j + \imath\epsilon}, \tag{10.120}$$

where $\{| j \rangle\}$ is an eigenbasis of the Hamiltonian. Taking the imaginary part of expression (10.120) we obtain

$$\begin{aligned} \rho_O(E) &= -\frac{1}{\pi}\mathrm{Im}\left(g_O(E)\right) \\ &= \sum_j \left\langle j \left| \hat{O} \right| j \right\rangle \delta\left(E - E_j\right). \end{aligned} \tag{10.121}$$

The function $\rho_O(E)$ has poles at the quantum eigenvalues with residues given by the matrix elements of $\hat{O}$. When the observable $\hat{O}$ is the identity, the expression above reduces to

$$\rho(E) = \sum_j \delta(E - E_j), \tag{10.122}$$

and it expresses the density of energy levels of the system, i.e. the number of energy eigenvectors per unit interval of energy.

The number of energy levels with energy smaller than a certain value E is then given by[11]

$$N(E) = \int_{-\infty}^{E} dE' \rho(E'). \tag{10.123}$$

The semiclassical limit is given when $E \to \infty$. Then, in this case we have

$$N(E) \simeq N^{\mathrm{sc}}(E) = \frac{1}{h^3} \int d\mathbf{p} d\mathbf{r} \delta \left(E - H^{\mathrm{c}}(\mathbf{r}, \mathbf{p})\right), \tag{10.124}$$

where $H^{\mathrm{c}}(\mathbf{r}, \mathbf{p})$ is the classical Hamiltonian, and the integral is done in the region where the integrand is positive. The corrections of this limit are given by the formula

$$N(E) = N^{\mathrm{sc}}(E) + O\left(\hbar^2 V''\right) + \cdots, \tag{10.125}$$

where the dots represent oscillating terms vanishing at high energy. Therefore, in the mean we have $N(E) - N^{\mathrm{sc}}(E) \simeq 0$. In other words, while in the quantum case we have a step-wise profile of the energy, in the semiclassical limit we have a continuous function that goes up and down the classical curve of energy and therefore presents itself as a continuous oscillation around the classical curve.

10.6.2 WKB approximation

The method that constitutes the subject of this and the following subsections is known as *Wentzel–Kramer–Brillouin* or, in short, the *WKB approximation*.[12] It is very useful for dealing, for example, with a fundamental problem of the motion in one-dimensional case and provides a derivation of the Bohr–Sommerfeld quantization rules: in quantum mechanics, as we have said (in Subsec. 1.5.4), the Bohr–Sommerfeld quantization rule occupies an intermediate status between classical mechanics and quantum mechanics. The WKB method is an approximate treatment of the Schrödinger equation that can be used to derive the quantization rule.

As it is well known, the classical orbit of a one-dimensional harmonic oscillator is a circle in the phase space. When we perform the Fourier transform and the inverse Fourier transform from the position to the momentum representation and *vice versa* a problem arises: the points where the circle intersects the x-axis are turning points where the oscillator inverts its motion in the position representation, and the same occurs for the points where the circle intersects the p_x-axis in the momentum representation. From the modern point of view, the WKB method consists of making use of the usual Schrödinger equation in the x representation of the turning points and by performing explicit computations at the turning points only.

The method, in its simplest form, consists of introducing an expansion of the phase and amplitude of the wave function in powers of $\hbar$ and neglecting terms of order higher than $\hbar$. In this way, one can replace the Schrödinger equation in some regions of the space by

[11] See [Balian/Bloch 1971].

[12] See [Wentzel 1926] [Kramers 1926] [Brillouin 1926]. For a complete but very technical treatment see also [*Maslow* 1994].

its classical limit. However, this approximation may also be applied in regions where the classical limit is meaningless.

For the sake of simplicity, we shall consider the one-dimensional case. This is sufficient for deriving the Bohr–Sommerfeld quantization rule for the radial part of the wave functions. Let $\psi(x)$ be the wave function satisfying the stationary Schrödinger equation

$$\psi''(x) + \frac{2m}{\hbar^2}\left[E - V(x)\right]\psi(x) = 0. \tag{10.126}$$

Writing

$$\psi = e^{\frac{\imath\eta}{\hbar}}, \quad \eta = \phi + \frac{\hbar}{\imath}\ln\vartheta, \tag{10.127}$$

where ϑ and ϕ are the phase and amplitude of the wave function, respectively (see Eq. (10.105)), one obtains the system of equations (see Eqs. (10.106))

$$\phi'^2 - 2m\,(E - V) = \hbar^2\frac{\vartheta''}{\vartheta}, \tag{10.128a}$$

$$2\vartheta'\phi' + \vartheta\phi'' = 0. \tag{10.128b}$$

By integrating the continuity equation (10.128b), one obtains

$$\vartheta = C\left(\phi'\right)^{-\frac{1}{2}}, \tag{10.129}$$

where C is a constant. Substituting this expression into Eq. (10.128a), one has

$$\phi'^2 = 2m\,(E - V) + \hbar^2\left[\frac{3}{4}\left(\frac{\phi''}{\phi'}\right)^2 - \frac{1}{2}\frac{\phi'''}{\phi'}\right]. \tag{10.130}$$

This third-order differential equation is equivalent to the initial Schrödinger equation. By imposing that ϕ (as well as ϑ) be a even function of $\hbar$, we may expand ϕ in a power series of $\hbar^2$

$$\phi = \phi_0 + \hbar^2\phi_1 + \cdots, \tag{10.131}$$

substitute this expansion into Eq. (10.130), and keep only the zero-order terms

$$\phi'^2 \simeq \phi_0'^2 = 2m\left[E - V(x)\right]. \tag{10.132}$$

The integration of Eq. (10.132) is straightforward. One may distinguish between two cases: when $E > V(x)$ and when $E < V(x)$.

- In the first case, we define the wavelength (see Eq. (1.78))

$$\lambda(x) = \frac{\hbar}{\sqrt{2m\left[E - V(x)\right]}}, \tag{10.133}$$

which is inversely proportional to the classical momentum as a function of the position. Equation (10.132) is satisfied if $\phi' \simeq \pm\hbar/\lambda$. The WKB solution is a linear combination of oscillating functions

$$\psi(x) = \alpha\sqrt{\lambda}\cos\left(\int dx\frac{1}{\lambda} + \beta\right), \tag{10.134}$$

where, for the time being, α and β are arbitrary constants.

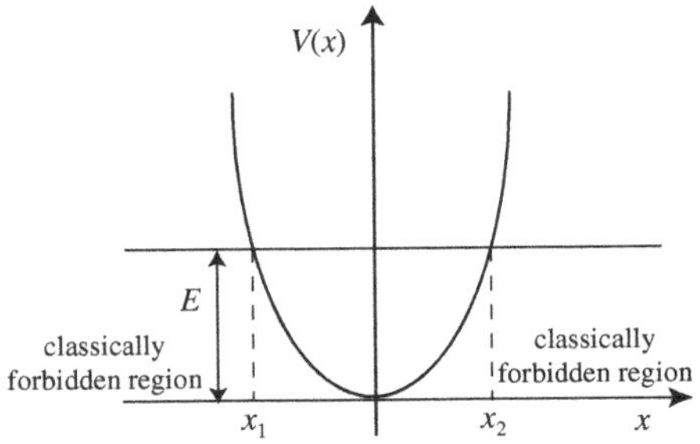

Figure 10.2 **WKB approximation: forbidden regions outside a potential well.**

- In the second case, i.e. when $E < V(x)$ (the region forbidden to classical particles), we may write

$$\lambda_q(x) = \frac{\hbar}{\sqrt{2m\,[V(x) - E]}}. \tag{10.135}$$

Equation (10.132) is satisfied if $\phi' \simeq \pm\imath\hbar/\lambda_q$. The WKB solution is a linear combination of real exponentials

$$\psi = (\lambda)^{\frac{1}{2}} \left[\gamma \exp\left(+ \int dx \frac{1}{\lambda} \right) + \delta \exp\left(- \int dx \frac{1}{\lambda} \right) \right], \tag{10.136}$$

where γ and δ are arbitrary constants. In the case of a parabolic potential well there are only two forbidden regions (the non-forbidden region is connected), and only one of the two coefficients of the exponential is different from zero, following the situation in which the system is in the classical forbidden region ($x > x_2$ or $x < x_1$) that lies either to the right or left of the allowed region (see Fig. 10.2).

As one would expect, for non-zero $\hbar$ the function λ is regular. It is also for non-zero $\hbar$ that this approximation breaks down near the *turning points*, i.e. those points for which $E = V(x)$.

10.6.3 Turning points

We have seen that the WKB approximation works everywhere but in the vicinity of the points for which $E = V(x)$. These are the turning points of the classical motion, i.e. points where the velocity of the particle vanishes and changes sign (see Sec. 4.3). Mathematically, the WKB approximation consists of replacing the Schrödinger equation

$$\psi'' + \frac{\psi}{\lambda^2} = 0, \tag{10.137}$$

by

$$\psi'' + \left(\frac{1}{\lambda^2} - \frac{1}{\sqrt{\lambda}} \left(\sqrt{\lambda} \right)'' \right) \psi = 0 \tag{10.138}$$

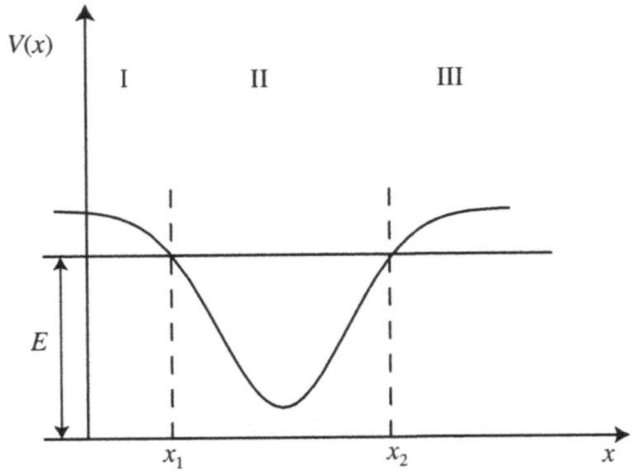

Figure 10.3 Schematic drawing of a one-dimensional potential well. Given the energy E of the particle, the turning points x_1 and x_2 divide the x-line into three regions: regions I and III are the classically forbidden ones.

both in the regions where $E > V(x)$ and where $E < V(x)$. Equation (10.138) has a second-order singularity of the type $(x-a)^{-2}$ at each point $x = a$, where $E = V(x)$.

Let us suppose that we have $E > V(x)$ or $E < V(x)$ according to whether $x > a$ or $x < a$. Making a careful analysis of the Schrödinger equation near the turning points (usually one uses an explicit solution that can be written in terms of the Airy function: indeed near the turning points in the generic case the potential can be approximated by a linear one), one finds that the general solution will be a linear combination of two solutions ψ_1 and ψ_2, whose asymptotic forms are

$$\text{for}\quad x \ll a: \qquad\qquad \text{for}\quad x \gg a:$$

$$\psi_1 \simeq \lambda_q^{\frac{1}{2}} \exp\left(+\int_x^a dx \frac{1}{\lambda_q}\right), \qquad \psi_1 \simeq -\lambda^{\frac{1}{2}} \sin\left(\int_a^x dx \frac{1}{\lambda} - \frac{\pi}{4}\right), \tag{10.139a}$$

$$\psi_2 \simeq \frac{1}{2}\lambda_q^{\frac{1}{2}} \exp\left(-\int_x^a dx \frac{1}{\lambda_q}\right), \qquad \psi_2 \simeq \lambda^{\frac{1}{2}} \cos\left(\int_a^x dx \frac{1}{\lambda} - \frac{\pi}{4}\right). \tag{10.139b}$$

This result may be derived by assuming that the potential is linear near the turning points. The same result could be obtained through less explicit computation by studying the solution of the Schrödinger equation in the complex plane.

10.6.4 Energy levels of a potential well

As an application we consider the potential well of Fig. 10.3 and calculate the energy levels of the discrete spectrum. For a given energy E there are two turning points x_1 and x_2 of the classical motion, which divide the x-axis into three regions. We look for the WKB solution that decreases exponentially in regions I and III, namely

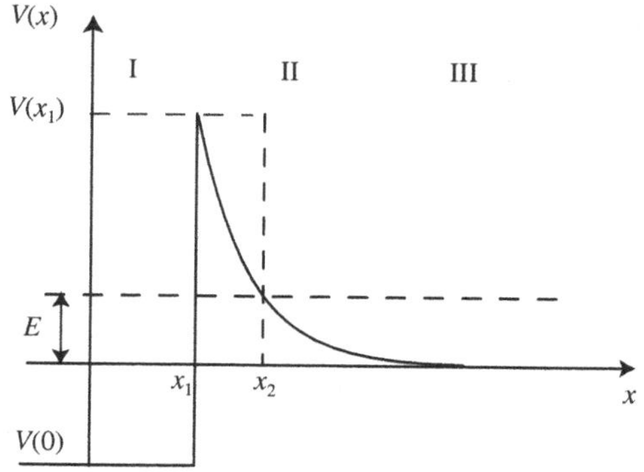

Figure 10.4 Schematic drawing of a one-dimensional potential barrier. Here, the classical forbidden region is represented by region II.

$$\psi_{\mathrm{I}} = \frac{1}{2} C \lambda_q^{\frac{1}{2}} \exp\left(- \int_x^{x_1} dx \frac{1}{\lambda_q} \right) \quad \text{for } x \ll x_1, \tag{10.140a}$$

$$\psi_{\mathrm{III}} = \frac{1}{2} C' \lambda_q^{\frac{1}{2}} \exp\left(- \int_{x_2}^{x} dx \frac{1}{\lambda_q} \right) \quad \text{for } x \gg x_2, \tag{10.140b}$$

where C and C' are adjustable constants. In accordance with Eqs. (10.139b), these functions are continued into region II (where $x_1 \ll x \ll x_2$) by the functions

$$\psi_{\mathrm{II}}^{(x_1)}(x) = C \lambda^{\frac{1}{2}} \cos\left(- \int_{x_1}^{x} dx \frac{1}{\lambda} - \frac{\pi}{4} \right), \tag{10.141a}$$

$$\psi_{\mathrm{II}}^{(x_2)}(x) = C' \lambda^{\frac{1}{2}} \cos\left(- \int_{x}^{x_2} dx \frac{1}{\lambda} - \frac{\pi}{4} \right), \tag{10.141b}$$

respectively. We have $\psi_{\mathrm{II}}^{(x_1)}(x) = \psi_{\mathrm{II}}^{(x_2)}(x)$ if

$$\int_{x_1}^{x_2} dx \frac{1}{\lambda} = \left(n + \frac{1}{2} \right) \pi, \tag{10.142}$$

where we have taken into account the expression (10.133).

The previous is valid in the one-dimensional case. In fact, the WKB method is suitable also for all completely integrable systems in higher dimensions. In this case, one is able to derive the Bohr–Sommerfeld quantization equation in the form

$$\oint dq_j dp_j = 2\pi\hbar(n_j + \delta_j), \tag{10.143}$$

where j goes from 1 to the number of dimensions of the system and q, p are appropriate generalized canonical position and momentum variables. Now, as the case may be, δ_j may assume several values, i.e. $0, 1/2, 1/4, 3/4$. The value of δ_j depends on the topology of phase space. For example, for the angular case $\delta_j = 0$. As a further example of application of the WKB method, one may consider (see Prob. 10.10) the calculation of the transmission probability through the potential barrier depicted in Fig. 10.4.

10.7 Scattering theory

In scattering theory we are interested in understanding what happens when a particle collides with other particles. Here, we shall discuss the simplest possible case: the scattering of a particle on a fixed potential, which for simplicity we suppose to have spherical symmetry.

In classical mechanics, we can compute the scattering angle as a function of the impact parameter b, i.e. the minimum distance that the unperturbed trajectory would have from the origin (the scattering center). In the quantum case things are more difficult as there is no unique trajectory.

We could take two different approaches:

- We can study the evolution of a wave packet – localized both in the position and momentum space – that is approaching the origin. We can compute the asymptotic form of the wave function at large times after scattering. It is clear that in this case we have to consider a time-dependent phenomenon.
- We can study what happens in a stationary regime. We have a constant flux of incoming particles localized in momentum space, but delocalized in position space (a plane wave): some of them are scattered by the potential and form a constant flux of outgoing particles. In this case, we need to study the solution of the time-independent Schrödinger equation.

The second approach has the advantage of being more directly related to a typical experimental situation: experiments are usually done using highly collimated (in momentum space) beams of particles and the resolution in position space is quite poor on the scale of the potential (the two resolutions usually differ by many orders of magnitude).

10.7.1 Plane-wave approach

Let us assume that the incoming beam moves with momentum $\mathbf{k}$ in the z direction (in units where $\hbar = 1$) and that the scattering potential has radial symmetry. The stationary wave function corresponding to this situation in absence of the potential would be given by (see Eq. (2.141))

$$\varphi_k(x) = e^{ik_z z}. \tag{10.144}$$

We impose that, in presence of the potential, the solution of the time-independent Schrödinger equation at distance far away from the origin is of the form

$$\psi(x) = e^{ik_z z} + \frac{e^{ik_r r}\vartheta(\theta)}{r} + O(r^{-3}), \tag{10.145}$$

where r is the distance from the origin and θ is the usual polar angle. The second (and new) term in the previous equation corresponds to the scattered wave. It is easy to check that, at large distances r, the previous form of the wave function, with an appropriate choice of the terms $O(r^{-3})$, is a solution of the free Schrödinger equation of a particle of mass m corresponding to the energy $k^2/(2m)$.

It is interesting to compute the flux $(\Phi(\theta)d\Omega)$ of particles outgoing from a sphere of a large radius R in an infinitesimal angular region $d\Omega$. This is given by the flux of the current density J defined in Subsec. 4.2.1. Generally speaking we find that

$$\Phi(\theta) = \Phi_{in}(\theta) + \Phi_{out}(\theta) + \Phi_I(\theta). \tag{10.146}$$

The first term corresponds to the unperturbed incoming particles and is given by

$$\Phi_{in}(\theta) = v R^2 \cos\theta, \tag{10.147}$$

where the velocity $\mathbf{v}$ is given by $\mathbf{k}/m$. It corresponds to a constant flux in the z direction: it is proportional to R^2 and changes sign in the same way as $\cos\theta$. Indeed, particles enter from below. The second term corresponds to the outgoing wave and is given by

$$\Phi_{out}(\theta) = v|\vartheta(\theta)|^2. \tag{10.148}$$

There is no R^2 term because the term proportional to the surface is compensated by the term r^{-1} in Eq. (10.145).

The ratio between the scattered flux (as function of θ) and the incoming flux per unit area is called the differential scattering cross section and it has (in the same way as the total cross section defined below) the dimension of a squared length. The differential scattering cross section is then given by

$$\sigma(\theta) = |\vartheta(\theta)|^2. \tag{10.149}$$

The total cross section is given by

$$\sigma = 2\pi \int d\theta \sin(\theta)\sigma(\theta) = 2\pi \int d\theta \sin(\theta)|\vartheta(\theta)|^2. \tag{10.150}$$

Up to now we have neglected the interference term $\Phi_I(\theta)$. When $\theta \neq 0$, one finds that the interference between the two terms in Eq. (10.145) oscillates with R and disappears if we average over R, thus not contributing to $\Phi_I(\theta)$. On the other hand, at $\theta = 0$ ($r = z$), the two terms have the same phase, so that the interference term becomes relevant. Therefore, the interference term must be of the form

$$\Phi_I(\theta)d\Omega = B\delta(\theta)d\theta. \tag{10.151}$$

As far as the total flux of particles entering the sphere is equal to zero, we must have that the interference term is proportional to the number of particles missing in the forward direction, because they have been scattered. We thus find that the conservation of the current implies that

$$B = -v\sigma. \tag{10.152}$$

A detailed computation[13] gives

$$mB = -4\pi\, \Im(\vartheta(0)), \tag{10.153}$$

which implies the so-called *optical theorem*

$$\Im(\vartheta(0)) = \frac{k}{4\pi}\sigma. \tag{10.154}$$

It is interesting to introduce the partial-wave scattering amplitude defined by

$$\vartheta(\theta) = \frac{1}{2ik} \sum_{l=0,\infty} (2l+1)\vartheta_l \Theta_{ll}(\cos\theta), \tag{10.155}$$

where $\Theta_{ll}(\cos\theta)$ are Legendre polynomials of order l (see Eq. (6.64)). The previous relation can be inverted; in this case, one gets

$$\vartheta_l = 2ik \int d\theta \, \sin(\theta)\Theta_{ll}(\cos\theta)\vartheta(\theta). \tag{10.156}$$

By imposing the conservation of the current in a more complex situation – when the incoming waves are a combination of plane waves – one arrives at a more detailed prediction for the partial-wave scattering amplitude. Indeed, one finds[14] that the partial wave amplitude ϑ_l must be of the form

$$\vartheta_l = \frac{e^{2\imath\delta_l} - 1}{2ik}, \tag{10.157}$$

where δ_l is called the *phase shift* (the factor 2 in the exponent in the definition of the phase shift is a convention). This equation implies the relation

$$k|\vartheta_l|^2 = \frac{1 - \cos(2\imath\delta_l)}{2ik} = \Im(\vartheta_l). \tag{10.158}$$

The previous relations are also called *unitary* relations, because conservation of the current is related to the conservation of probability and, therefore, to the unitarity of the time-evolution operator.

10.7.2 Perturbation theory

It is interesting to compute the scattering amplitude in the context of perturbation theory (Sec. 10.1). To this end we consider a scattering potential energy equal to $\zeta\, V(\mathbf{r})$ and write

$$\psi(\mathbf{r}) = \psi_0(\mathbf{r}) + \zeta\psi_1(\mathbf{r}) + \zeta^2\psi_2(\mathbf{r}) + \cdots, \tag{10.159}$$

where

$$\psi_0(\mathbf{r}) = e^{ik_z z}. \tag{10.160}$$

[13] The interested reader can find the explicit calculation in [*Landau/Lifshitz* 1976b, Ch. 17].
[14] See again [*Landau/Lifshitz* 1976b, Ch. 17] for details.

In order to compute the coefficients of the perturbation expansion, it may be convenient to write the Schrödinger equation in momentum space

$$\frac{\mathbf{p}^2}{2m}\tilde{\psi}(\mathbf{p}) + \zeta \int d\mathbf{p}' \tilde{V}(\mathbf{p} - \mathbf{p}')\tilde{\psi}(\mathbf{p}') = E\tilde{\psi}(\mathbf{p}), \tag{10.161}$$

where

$$\tilde{V}(\mathbf{p}) = (2\pi)^{-\frac{3}{2}} \int d\mathbf{r} e^{-\imath \mathbf{p}\cdot\mathbf{r}} V(\mathbf{r}), \tag{10.162}$$

or

$$\tilde{G}(\mathbf{p})^{-1}\tilde{\psi}(\mathbf{p}) + \zeta \int d\mathbf{p}' \tilde{V}(\mathbf{p} - \mathbf{p}')\tilde{\psi}(\mathbf{p}') = 0, \tag{10.163}$$

where

$$\tilde{G}(\mathbf{p}) = \left(\frac{\mathbf{p}^2}{2m} - E\right)^{-1}, \tag{10.164}$$

and $E = k^2/(2m)$. By taking advantage of the expansion (10.159) in momentum space

$$\tilde{\psi}(\mathbf{p}) = \tilde{\psi}_0(\mathbf{p}) + \zeta\tilde{\psi}_1(\mathbf{p}) + \zeta^2\tilde{\psi}_2(\mathbf{p}) + \cdots, \tag{10.165}$$

with

$$\tilde{\psi}_0(\mathbf{p}) = \delta(\mathbf{p} - \mathbf{k}), \tag{10.166}$$

where $\mathbf{k}$ is the vector with components $(0, 0, k)$, we may insert this expansion to the first order into Eq. (10.163) so as to obtain

$$\tilde{\psi}_1(\mathbf{p}) = -\tilde{G}(\mathbf{p}) \int d\mathbf{p}' \tilde{V}(\mathbf{p} - \mathbf{p}')\delta(\mathbf{p}' - \mathbf{k}) = -\tilde{G}(\mathbf{p})\tilde{V}(\mathbf{p} - \mathbf{k}). \tag{10.167}$$

A similar expression is obtained for $\tilde{\psi}_2(\mathbf{p})$ (with one more integration).

However the previous approach presents two drawbacks:

- The quantity $\tilde{G}(\mathbf{p})$ has a singularity in momentum space, and consequently its Fourier transform is ambiguous. In a nutshell, $\tilde{G}(\mathbf{p})$ is not well-defined.
- We would like $\tilde{\psi}_1(\mathbf{p})$ to contain only outgoing waves and no incoming waves.

Fortunately, we find ourselves in the happy situation where the two problems can be solved together. We redefine $\tilde{G}(\mathbf{p})$ as

$$\tilde{G}(\mathbf{p}) = \lim_{\epsilon\to 0^+} \left(\frac{\mathbf{p}^2}{2m} - E - i\epsilon\right)^{-1}. \tag{10.168}$$

Now, for $\epsilon \neq 0$ everything is well-defined. The limit $\epsilon \to 0^+$ is not divergent because the singularity is integrable but it differs from the limit $\epsilon \to 0^-$.

A simple computation shows that in position space this function is given by

$$G(\mathbf{r}) = \frac{C}{r}e^{ikr}, \tag{10.169}$$

where $C = m/(2\pi)$. Indeed, the inverse Fourier transform of $1/(p^2 + \mu^2)$ is given by

$$\frac{e^{-\mu r}}{4\pi r}, \tag{10.170}$$

Box 10.1 **Remarks on the Fourier transform**

We start by computing the Fourier transform $\tilde{f}(\mathbf{p})$ of $f(\mathbf{r}) = e^{-\mu r}/r$. Because of rotational invariance we have that $\tilde{f}(\mathbf{p})$ is a function of $|\mathbf{p}|$ only. We can thus compute it when $\mathbf{p}$ points in the z direction without loss of generality. We have that

$$\begin{aligned}
\tilde{f}(\mathbf{p}) &= (2\pi)^{-\frac{3}{2}} \int d\mathbf{r} e^{-\imath p z} \frac{e^{-\mu r}}{r} \\
&= (2\pi)^{-\frac{1}{2}} \int_{-1}^{1} d\cos\theta \int_{0}^{\infty} dr r^2 e^{-\imath p r \cos\theta} \frac{e^{-\mu r}}{r} \\
&= (2\pi)^{-\frac{1}{2}} (-\imath p)^{-1} \int_{0}^{\infty} dr (e^{-\imath p r} - e^{\imath p r}) e^{-\mu r} \\
&= (2\pi)^{-\frac{1}{2}} (-\imath p)^{-1} \left(\frac{1}{\imath p + \mu} + \frac{1}{\imath p - \mu} \right) = \sqrt{\frac{2}{\pi}} \frac{1}{p^2 + \mu^2}.
\end{aligned} \tag{10.171}$$

Conversely, we can look at the inverse Fourier transform:

$$\begin{aligned}
(2\pi)^{-\frac{3}{2}} \int d\mathbf{p} e^{\imath p_z z} \sqrt{\frac{2}{\pi}} \frac{1}{p_x^2 + p_y^2 + p_z^2 + \mu^2} &= \frac{1}{4\pi} \int dp_x dp_y \frac{e^{-|z|\sqrt{p_x^2+p_y^2+\mu^2}}}{\sqrt{p_x^2 + p_y^2 + \mu^2}} \\
&= \frac{1}{2} \int_0^{+\infty} dp^2 \frac{e^{-|z|\sqrt{p^2+\mu^2}}}{\sqrt{p^2 + \mu^2}},
\end{aligned} \tag{10.172}$$

where the first integral has been done by deforming the integration path in the complex p_z variable and picking the contribution of one of the two poles at $\pm\imath\sqrt{p_x^2 + p_y^2 + \mu^2}$ depending on the sign of z. The last integral can be done exactly. However, it is more instructive to consider the asymptotic limit for large $|z|$. In this case, the integral is dominated by the maximum at $p^2 = 0$ and can be written as

$$\frac{1}{2\mu} \int_0^{+\infty} dp^2 e^{-|z|\left(\mu + \frac{p^2}{2\mu}\right)} = \frac{e^{-\mu|z|}}{|z|}. \tag{10.173}$$

as can be seen by doing the integrals using polar coordinates. Consequently

$$\psi_1(\mathbf{r}) = -\frac{m}{2\pi} \int d\mathbf{r}' e^{\imath \mathbf{k}\cdot\mathbf{r}'} V(\mathbf{r}') G(|\mathbf{r} - \mathbf{r}'|). \tag{10.174}$$

We thus find that $\psi_1(\mathbf{r})$ is a superposition of outgoing waves with an amplitude proportional to $V(\mathbf{r})$. As far as the potential goes to zero at infinity, at large distances all these waves can be considered as outgoing from the origin.

The computation of the scattering amplitude $\vartheta(\theta)$ could also be done directly by writing Eq. (10.145) in momentum space. It is clear that the behavior at infinity of the wave function is connected to the singularity of its Fourier transform in momentum space. It can be shown that the equivalent of Eq. (10.145) in momentum space is

$$\tilde{\psi}(\mathbf{p}) = \delta(\mathbf{p} - \mathbf{k}) + \frac{\pi \vartheta(\theta(\mathbf{p}))}{E(\mathbf{p})}, \tag{10.175}$$

where

$$k^2 \cos(\theta(\mathbf{p})) = \mathbf{k} \cdot \mathbf{p}. \tag{10.176}$$

Indeed, the proof of Eq. (10.175) can be done by applying a procedure based on the guidelines shown in Box 10.1. One computes the inverse Fourier transform of Eq. (10.175) in two steps:

1. Compute the integral over the momentum parallel to $\mathbf{r}$ by picking the residuum of the pole, which is dominant at large distances r.
2. Integrate over the transverse momenta using the method of the point of maximum: one finds that the maximum is located at zero transverse momenta as in the previous case.

The function $\theta(\mathbf{p})$ matters only near the singularity where $p^2 = k^2$; different definitions would give identical results if they coincide for $p^2 = k^2$. By comparing the previous formulae we find that

$$\vartheta(\theta) = -\frac{m}{2\pi}\tilde{V}(\mathbf{p}(\theta)), \tag{10.177}$$

where $\mathbf{p}(\theta)$ is a vector whose squared length is given by

$$p(\theta)^2 = k^2\left((1-\cos\theta)^2 + \sin(\theta)^2\right) = (2k\sin(\theta/2))^2. \tag{10.178}$$

We finally find for the differential cross section that

$$\sigma(\theta) = \frac{m^2}{4\pi^2\hbar^4}|\tilde{V}(\mathbf{p}((\theta))|^2, \tag{10.179}$$

where we have written the result in the usual physical units. This formula is called the Born approximation, because it was derived by Born in 1926.[15]

We notice that, if the Fourier transform of $V(\mathbf{r})$ is finite, the differential cross section is also non-singular at $\theta = 0$ and the total cross section is finite. This is in contrast with the classical case, where, if the potential has non-compact support, no matter how fast it goes to zero at infinity, both the forward differential cross section $\sigma(0)$ and the total cross section σ are infinite. In quantum mechanics, the total cross section is convergent for all potentials going to zero as a power of $1/r$, when the power is greater than one. In the case of the Coulomb potential, for which the power is exactly equal to one, the total cross section is only logarithmically divergent.

10.8 Path integrals

In classical mechanics, a system follows a trajectory in phase space that is determined by imposing the least-action condition (see Eq. (1.14)). In other words, among all possible trajectories that connect the initial state to the final state, a classical system *chooses* precisely the one that minimizes the action S. The situation is pretty much different in quantum mechanics, where a principle of least action does not exist. The path-integral

[15] See [Born 1926].

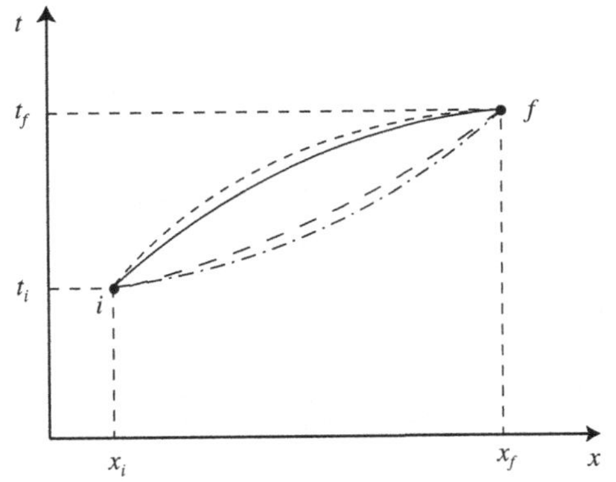

Figure 10.5 A pictorial representation of a few possible paths in one-dimensional configuration space. All the (infinite) possible paths connecting the initial position i to the final f contribute to the probability amplitude of moving from i to f.

method provides an interesting view-point that allows us to interpret quantum probabilities as a certain sum or integral over all possible paths of the system (see Subsec. 10.8.1). The importance of this method has been emphasized by a wide range of applications, particularly those involving perturbation theory (see Subsec. 10.8.2).

10.8.1 General features

Let us consider the transition of a quantum system from a certain initial point i of the configuration space to a certain final point f. As we know, each possible trajectory contributes with a different phase to the total probability amplitude of the transition $i \rightarrow f$. More precisely, the probability $\wp(f, i)$ is the absolute square of the Green's function $\mathrm{G}(f, i)$ (see Subsec. 3.5.5).[16] The total probability amplitude may then be interpreted as the sum of the contributions $\vartheta[x(t)]$ of each possible path connecting i and f in the configuration space[17] (see Fig. 10.5). In the one-dimensional case, the Green's function may be written as

$$\mathrm{G}(f, i) = \sum \vartheta\,[x(t)], \tag{10.180}$$

where $i = (x_i, t_i)$ and $f = (x_f, t_f)$, and the sum is taken of all possible paths. On the most general grounds, each path from i to f has equal probability, and therefore contributes by an equal amount to the total probability amplitude. However, the contribution $\vartheta[x(t)]$ of each path has a phase proportional to the action S (see Subsec.10.5.2)

$$\vartheta[x(t)] = C e^{\frac{i}{\hbar}\mathrm{S}[x(t)]}, \tag{10.181}$$

16 Note that the definition of the Green's functions introduced here is slightly different from that of Subsec. 3.5.5. Here, we have omitted the imaginary unity factor. This has obviously no consequence on the the absolute square of the Green's functions, even though it yields a slightly different form of Eq. (3.92).

17 [Feynman 1948]. Exposition in [Feynman/Hibbs 1965, 28–37].

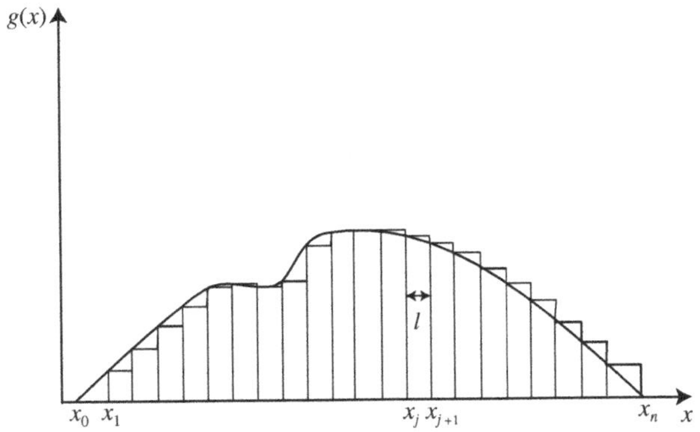

Figure 10.6 The integral of a continuous function, i.e. the area under the curve may be approximated by the product l times the sum of the ordinates. This approximation approaches the correct value as $l \to 0$.

where C is a constant. The number of paths is certainly infinite, with a high order of infinity. Therefore, it is natural to ask how to assign the correct measure to the space of these paths. To answer this question, consider the properties of the Riemann integral. The area A under a curve $g(x_j)$ is proportional to the sum of all its ordinates. Let us take a subset of these ordinates, e.g. those spaced at equal intervals (see Fig. 10.6),

$$l = \frac{x_n - x_0}{n}. \tag{10.182}$$

Then, through summation over the finite set of points x_j, we obtain

$$A \propto \sum_{j=0}^{n-1} g(x_j). \tag{10.183}$$

We may now define A as the limit of this sum. It is possible to pass to the limit in a smooth way by taking continuously smaller and smaller values of l (i.e. by increasing n). However, in this way every the sum will depend on n and, in order to obtain the limit, we must specify some normalization factor which should depend on l. For the Riemann integral this normalization factor is l itself, since each of the rectangles in Fig. 10.6 has area $lg(x_j)$. Therefore, the limit exists and we may write[18]

$$A = \lim_{l \to 0} \left[l \sum_{j=0}^{n-1} g(x_j) \right] = \int_{x_0}^{x_n} dx g(x). \tag{10.184}$$

We can follow an analogous procedure in defining the sum over all paths. Again, we choose a subset of the paths (see Fig. 10.7). For this purpose, we divide the independent time variable into steps of width ϵ. At each time t_j we select some point x_j (from the starting point x_i to arrival point x_f). We then build a path by connecting all positions that have been

[18] Here the limit $l \to 0$ is equivalent to the limit $n \to \infty$.

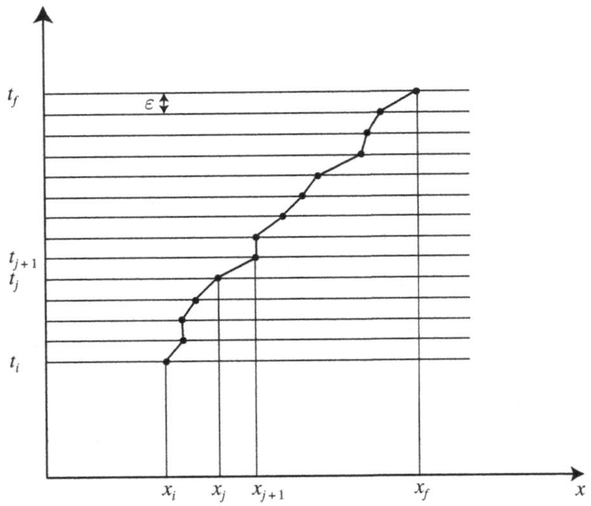

Figure 10.7 The sum over paths is defined as a limit, in which at first the path is specified by the coordinates at a large number of times separated by very small intervals ϵ. The sum over the paths is then an integral over all possible values of these intermediate coordinates. Then, in order to achieve the correct measure, the limit is taken for $\epsilon \to 0$.

found, that is, by connecting all the points (x_j, t_j) by straight lines. Given that we have n time intervals of length ϵ, i.e. $n\epsilon = t_f - t_i$, we may define a sum over all paths by taking a multiple integral over all possible values of x_j at each node:

$$\mathrm{G}(x_n, t_n; x_0, t_0) \simeq \int\int \cdots \int dx_1 dx_2 \cdots dx_{n-1} \vartheta[x(t)], \tag{10.185}$$

where $t_0 = t_i, t_n = t_f$ and $x_0 = x_i, x_n = x_f$. Of course, there is no need to integrate over x_0 and x_n, because these points are known and fixed. In the general case it is very difficult to compute the correct normalization factor. However, in all cases where the action is derived by integrating the Lagrangian

$$\mathrm{L}(\dot{x}, x, t) = \frac{m}{2}\dot{x}^2 - V(x, t), \tag{10.186}$$

it is possible to prove (see, e.g., Prob. 10.11) that the normalization factor is given by $\mathcal{N}^{-n}$, where

$$\mathcal{N} = \left(\frac{2\pi\imath\hbar\epsilon}{m}\right)^{\frac{1}{2}}. \tag{10.187}$$

With this factor the limit exists and it produces the correct value of the Green's function $\mathrm{G}(f, i)$. Hence, we can write

$$\mathrm{G}(x_n, t_n; x_0, t_0) = \lim_{\epsilon \longrightarrow 0} \frac{1}{\mathcal{N}} \int\int \cdots \int \frac{dx_1}{\mathcal{N}} \frac{dx_2}{\mathcal{N}} \cdots \frac{dx_{n-1}}{\mathcal{N}} e^{\frac{\imath}{\hbar}\mathrm{S}(f,i)}, \tag{10.188}$$

where

$$S(f,i) = \int_{t_0}^{t_n} dt L(\dot{x}, x, t), \tag{10.189}$$

and the integral is done using a trajectory $x(t)$ that is piecewise linear and passes through the points (x_j, t_j).

The action S for each trajectory is given by

$$S = \epsilon \sum_j \left[\frac{(x_j - x_{j-1})^2}{2\epsilon^2} m \right] - \int_0^1 du V\left[(1-u)x_j + u x_{j-1}\right], \tag{10.190}$$

where u parametrizes the trajectory between x_{j-1} and x_j. Assuming the trajectory to be continuous, when ϵ goes to zero the difference $x_j - x_{j-1}$ also does, and the action remains finite. In this case, neglecting terms that vanish when ϵ goes to 0, the integral in the previous formula may be approximated by $V(x_j)$ or, equivalently, by $(1/2)\left(V(x_j) + V(x_{j-1})\right)$.

This is not the only way to define a subset of all paths between i and f. For practical purposes, we might possibly need other formulations. Nevertheless, the concept of the sum (or integration) over all paths is rather general. Independently from the method we use for defining the integral, it is usually written in the following notation:

$$G(f,i) = \int_i^f d[x(t)] e^{\frac{i}{\hbar} S(f,i)}, \tag{10.191}$$

and is called the *path integral*, with the meaning given to it by expression (10.188).

Suppose now that we have two events in succession (say, a particle moving first from x_i at time t_i to x_c at time t_c, and then to x_f at time t_f), such that $S(f,i) = S(f,c) + S(c,i)$, where $c = (x_c, t_c)$. The action is an integral in time and the Lagrangian does not depend on derivatives higher than the velocity. Making use of Eq. (10.191) we can write

$$G(x_n, t_n; x_0, t_0) = \int d[x(t)] e^{\frac{i}{\hbar} S(f,c) + \frac{i}{\hbar} S(c,i)}. \tag{10.192}$$

It is possible to split any path into two parts (see Fig. 10.8), so that we can write

$$G(x_n, t_n; x_0, t_0) = \int dx_c \int_c^f d[x(t)] e^{\frac{i}{\hbar} S(f,c)} G(x_c, t_c; x_0, t_0), \tag{10.193}$$

where the integration is now performed not only over all possible paths from c to f, but also over the variable central point x_c. Then, we carry out the integration over paths between x_0 and an arbitrary x_c and between x_c and x_n, i.e.

$$G(x_n, t_n; x_0, t_0) = \int dx_c G(x_n, t_n; x_c, t_c) G(x_c, t_c; x_0, t_0). \tag{10.194}$$

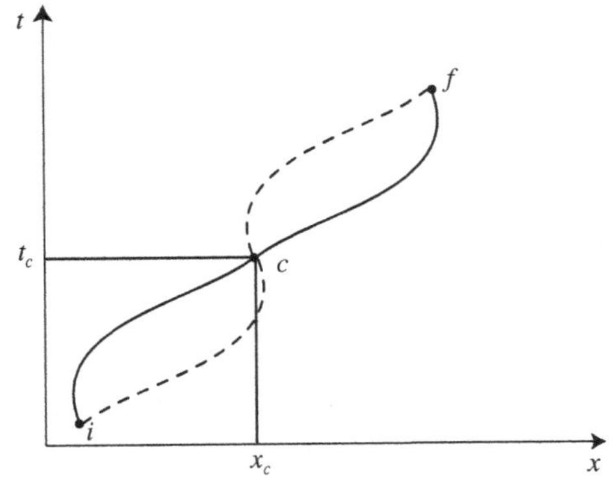

Figure 10.8 Two possible paths from *i* to *f* both passing through the same central point *c*.

This result can be summarized as follows: all alternative paths going from i to f can be labelled by specifying the position x_c through which they pass at time t_c. Then, the integral over the paths, or the *kernel* $\mathrm{G}(x_n, t_n; x_0, t_0)$ for a particle going from the point i to the point f can be computed according to the following rules:

- The kernel from i to f is the sum over all possible values of c of the amplitudes for the particle to go from i to f passing through c.
- The probability amplitude for a particle to go from i to c and then to f is given by the product of the kernel from i to c times the kernel from c to f.

Equation (10.194) can be proved independently from the path integral formalism and it is a direct consequence of the superposition principle. In fact, let us write an arbitrary eigenket of the position operator $\hat{x}$ at time t as $|x,t\rangle$ and an arbitrary eigenket of the position operator at time t' as $|x',t'\rangle$. In this case, the Green's function is simply given by the scalar product $\mathrm{G}(x,t;x',t') = \langle x',t' \mid x,t\rangle$. Now, for any time t_c such that $t < t_c < t'$ we also have

$$\begin{aligned}\int dx_c \mathrm{G}(x,t;x_c,t_c)\mathrm{G}(x_c,t_c;x',t') &= \int dx_c \langle x',t' \mid x_c,t_c\rangle \langle x_c,t_c \mid x,t\rangle \\ &= \mathrm{G}(x,t;x',t'). \end{aligned} \tag{10.195}$$

where we have made use of the identity

$$\hat{I} = \int dx_c \, |x_c,t_c\rangle \langle x_c,t_c|. \tag{10.196}$$

It is natural to ask, at this stage, whether the path-integral formulation of quantum mechanics is equivalent to the more traditional formalism, i.e. the Schrödinger picture (see Sec. 3.1). For this purpose, we must relate a path integral at one time to its value an infinitesimal time later, obtaining a differential equation for the path integral, and show that it is identical to the Schrödinger equation (see Prob. 10.11).

It is interesting to study the classical limit of the path integral approach, represented by the limit for the action being much larger than h. As we know, in classical mechanics only one trajectory exists, precisely the one that minimizes the action. In the classical limit, the phase $S/\hbar$ is very large. Therefore, small changes on the classical scale in the path will produce large variations in the phase contributions, making it a rapidly oscillating function. As a consequence, the total contribution resulting from the paths that are far from the classical path will add to zero. Instead, in the vicinity of the classical path – where the action is at a minimum – small variations in the path itself give rise to no change in the action in the first order, and the paths in that region will be in phase and give rise to a non-zero net contribution. In conclusion, in the classical limit the only path that needs to be considered is precisely the classical one.

To summarize, the path-integral method is a very powerful instrument for calculating probability amplitudes for evolutions between different experimental results – the events "i" and "f" in Eqs. (10.192)–(10.194) – because in between there are no "paths" in the classical sense.

10.8.2 The perturbation expansion revisited

Let us consider a one-dimensional particle moving in a potential $V(x,t)$.[19] The kernel for the motion between two points i and f is (see Eq. (10.191))

$$\mathrm{G}_V(f,i) = \int_i^f d[x(t)] e^{\frac{\imath}{\hbar}\int_{t_i}^{t_f} dt\left[\frac{m}{2}\dot{x}^2 - V(x,t)\right]}, \tag{10.197}$$

where the subscript V refers to the potential $V(x,t)$, whereas we shall use G_0 to indicate the kernel for the motion of a free particle (see Prob. 10.12). If the potential is at most quadratic in x (as it happens, e.g., for the harmonic oscillator), the kernel can be determined exactly, since the resulting calculations only involve Gaussian integrals. In the general case, if the potential varies sufficiently slowly, one may make use of the semiclassical approximation. The following method may be used when the potential is small and therefore can be treated as a small perturbation contribution to the free-particle evolution.

If the time integral of the potential along a path is small relative to $\hbar$, then the part of the exponential of Eq. (10.197) which depends upon the potential may be expanded as follows:

$$e^{-\frac{\imath}{\hbar}\int_{t_i}^{t_f} dt V(x,t)} = 1 - \frac{\imath}{\hbar}\int_{t_i}^{t_f} dt V(x,t) + \frac{1}{2!}\left(\frac{\imath}{\hbar}\right)^2 \left[\int_{t_i}^{t_f} dt V(x,t)\right]^2 + \cdots, \tag{10.198}$$

[19] See [Feynman/Hibbs 1965, 120–29].

which is defined along any path $x(t)$. Substituting the expression (10.198) into Eq. (10.197) yields

$$\mathrm{G}_V(f,i) = \mathrm{G}_0(f,i) + \mathrm{G}_1(f,i) + \mathrm{G}_2(f,i) + \cdots, \tag{10.199}$$

where

$$\mathrm{G}_0(f,i) = \int_i^f d[x(t)] e^{\frac{\imath}{\hbar}\int_{t_i}^{t_f} dt \frac{m}{2}\dot{x}^2}, \tag{10.200a}$$

$$\mathrm{G}_1(f,i) = -\frac{\imath}{\hbar}\int_i^f d[x(t)] e^{\frac{\imath}{\hbar}\int_{t_i}^{t_f} dt \frac{m}{2}\dot{x}^2} \int_{t_i}^{t_f} dt' V(x(t'),t') \tag{10.200b}$$

$$\mathrm{G}_2(f,i) = -\frac{1}{2\hbar^2}\int_i^f d[x(t)] e^{\frac{\imath}{\hbar}\int_{t_i}^{t_f} dt \frac{m}{2}\dot{x}^2} \int_{t_i}^{t_f} dt' V[x(t'),t'] \times \int_{t_i}^{t_f} dt'' V[x(st''),t''], \tag{10.200c}$$

$$\cdots\;\cdots\;\cdots\;\cdots$$

We wish now to evaluate the term $\mathrm{G}_1(f,i)$. For this purpose, we exchange the order of integration over the time variable t' and the path $x(t)$, writing

$$\mathrm{G}_1(f,i) = -\frac{\imath}{\hbar}\int_i^f dt' g(t'), \tag{10.201}$$

where

$$g(t') = \int_i^f d[x(t)] e^{\frac{\imath}{\hbar}\int_{t_i}^{t_f} dt \frac{m}{2}\dot{x}^2} V[x(t'),t']. \tag{10.202}$$

The path integral $g(t')$ is the sum over all paths of the amplitude of the free particle, each path being weighted by the potential $V[x(t'),t']$ evaluated at time t'. Before and after time t' the paths involved in $g(t')$ are the paths of an ordinary free particle. Using the same procedure that has led to Eq. (10.194), we cut each path in a part before and a part after $t = t'$. Therefore, the sum over all paths may be written as $\mathrm{G}_0(f,x')\mathrm{G}_0(x',i)$, where $x' = x(t')$. Then, we have

$$g(t') = \int_{-\infty}^{+\infty} dx' \mathrm{G}_0(f,x') V\left(x',t'\right) \mathrm{G}_0(x',i), \tag{10.203}$$

and, by substituting this expression into Eq. (10.201), we obtain

$$\mathrm{G}_1(f,i) = -\frac{\imath}{\hbar}\int_i^f dt' \int_{-\infty}^{+\infty} dx' \mathrm{G}_0(f,x') V\left(x',t'\right) \mathrm{G}_0(x',i). \tag{10.204}$$

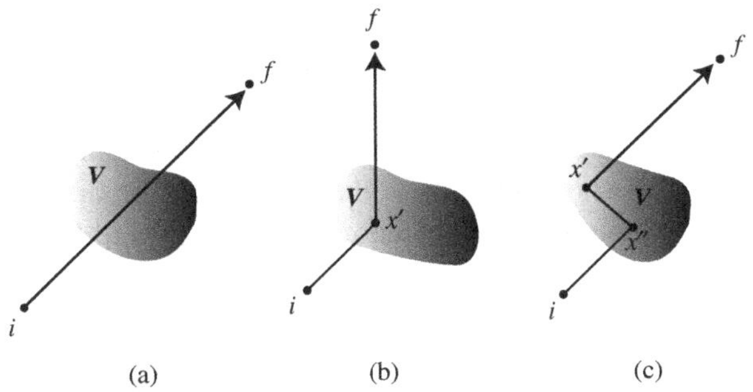

Figure 10.9 Pictorial representation of the motion of a particle from *i* to *f* through a region under the action of a potential $V(x, t)$. (a) A free particle is not scattered at all and is described by the kernel $G_0(f, i)$. (b) A single scattering event occurs in the shaded region, and the kernel is here $G_1(f, i)$. (c) Two scattering events occur, and the kernel is $G_2(f, i)$. The total amplitude for the motion is given by Eq. (10.199).

Equations (10.199) and (10.204) suggest an interesting interpretation of the process that describes the transition form i to f. We may think at the effect of the potential on the particle as a scattering process. Accordingly, we may interpret the kernels $G_j(f, i)$ ($j = 0, 1, 2, \ldots$) in the following terms:

- The particle may not be scattered at all, as described by $G_0(f, i)$.
- The particle may be scattered once, as described by $G_1(f, i)$.
- The particle may be scattered twice, and this is described by the kernel $G_2(f, i)$, and so on (see Fig. 10.9).

The final amplitude $G(f, i)$ is then a sum of all the kernels describing all these possibilities. For example, in the case of $G_1(f, i)$, for each possible scattering point x', the amplitude is given by

$$G_0(f, x')\left[-\frac{\imath}{\hbar}dt'dx'V(x', t')\right]G_0(x', i). \tag{10.205}$$

The kernel $G_1(f, i)$ is then obtained by summing up all these alternatives, i.e. by integrating over x' and t' (see Eq. (10.204)).

The kernel for a two-scattering process may be easily written as

$$\begin{aligned} G_2(f, i) = \left(-\frac{\imath}{\hbar}\right)^2 \int\int dx'dt' \int\int dx''dt'' G_0(f, x')V(x', t')G_0(x', x'')V(x'', t'') \\ \times G_0(x'', i). \end{aligned} \tag{10.206}$$

Summing up, the total amplitude for the motion from i to f is obtained as

$$\begin{aligned} G_V(f, i) = {} & G_0(f, i) - \frac{\imath}{\hbar}\int\int dx'dt' G_0(f, x')V(x', t')G_0(x', i) \\ & - \frac{1}{\hbar^2}\int\int dx'dt' \int\int dx''dt'' G_0(f, x')V(x', t')G_0(x', x'')V(x'', t'')G_0(x'', i) \\ & + \cdots . \end{aligned} \tag{10.207}$$

This equation is called the *Born expansion* for the amplitude. It involves a series of terms: at the n-th order, we have n scattering points. The motion before the first scattering point, after the last scattering point, and between two consecutive scattering points is free. It can be shown that Eq. (10.207) is equivalent to the perturbation expansion (10.6). We show that this is the case up to the first-order approximation. To this purpose, we note that the evolution described by Eq. (10.207) is unitary, so that it may be written

$$\left\langle f \left| e^{\imath E_n t} \right| i \right\rangle. \tag{10.208}$$

Summary

In this chapter we have dealt with some methods and problems concerning approximations and perturbations. In particular:

- We have introduced the *stationary perturbation theory*, in which a small time-independent perturbation is applied to a system, whose unperturbed dynamics is known. In this case, by making use of a Taylor-series expansion in power series of the small perturbation parameter we can find the perturbed energy eigenvalues and the coefficients of the perturbed eigenstates in terms of the unperturbed ones. Both the non-degenerate and the degenerate cases have been discussed. Two examples of application have also been presented: the perturbation of an oscillator and the Stark (electric) effect for a rigid rotator.
- We have also presented the *time-dependent perturbation theory*, where the perturbation explicitly depends on time. In this case, we must take into account the time-dependent Schrödinger equation. Here, the perturbed system undergoes a true modification of the probabilities of the perturbed energy eigenvalues.
- The *adiabatic theorem* deals with the problem of the time evolution of a system with a Hamiltonian that explicitly depends on time but changes slowly.
- *Scattering theory* has been summarized and the important concept of cross section has been introduced.
- We have also discussed the problem of the *classical limit*. In most cases is possible to interpret the classical world as an emergence from quantum physics when the Planck constant becomes very small relative to the system's action. However, it is not true that, whenever a quantum dynamics is well described, a classical dynamics also exists (Lieb's theorem). Several related problems have been presented, namely the concept of classical action, the Hamilton–Jacobi equation, and the spreading of wave packets. Moreover, the *semiclassical limit* and in particular the WKB approximation have been presented.
- An important and widely used technique for dealing with a vast class of problems is known as the *path integral* method. We have also shown that it is possible to deal with perturbation theory by making use of this method (*Born expansion*).

Problems

10.1 Develop the second-order approximation, when the degeneracy is not removed in the first order.

10.2 Compute the mean value $\langle n | \hat{x}^4 | n \rangle$ and use it to generalize the result to $\langle n | \hat{x}^4 | m \rangle$.

10.3 Derive Eq. (10.63).

10.4 Derive Eq. (10.64).

10.5 Show that for a one-dimensional classical system we have $\ddot{x} = f(x)/m$ and $\dddot{x} = p_x/m^2 \cdot f'(x)$.

10.6 Derive Eq. (10.89).

10.7 Follow the procedure described in Subsec. 10.5.3 to derive Eqs. (10.106).

10.8 Show that for a free particle the wave packet spreads according to the law (10.119).

10.9 Find the time evolution of the mean values $\langle \hat{x} \rangle$ and $\langle \hat{p}_x \rangle$ and of the square deviations σ_x^2 and σ_p^2 in the case of a one-dimensional harmonic oscillator. What is the condition for σ_x^2 and σ_p^2 to remain constant?

10.10 Using the WKB approximation, compute the transition probabilities for the wave packet of energy E impinging from the left on the potential barrier represented in Fig. 10.4.

10.11 Show that, for a one-dimensional particle, it is possible to derive the Schrödinger equation from the path-integral formalism.

10.12 Prove that the kernel for a one-dimensional free particle is given by

$$\mathrm{G}_0(f, i) = \left[\frac{2\pi \imath \hbar \left(t_f - t_i\right)}{m}\right]^{-\frac{1}{2}} e^{\frac{\imath m (x_f - x_i)^2}{2\hbar (t_f - t_i)}}. \tag{10.209}$$

(*Hint*: Start from Eq. (10.200a) and divide the time interval $t_f - t_i$ into n small intervals of duration ϵ. Then, the kernel is represented by a set of Gaussian integrals. Since the integral of a Gaussian is again a Gaussian, it is possible to carry out the integrations one after the other and, finally, take the limit $\epsilon \to 0$.)

Further reading

Scattering

Bjorken, J. D. and Drell, S. D., *Relativistic Quantum Mechanics*, New York: McGraw-Hill, 1964.

Schweber, Silvan S., *An Introduction to Relativistic Quantum Field Theory*, New York: Harper and Row, 1961.

Semiclassical methods

Maslow, V. P., *The Complex WKB Method for Nonlinear Equations 1: Linear Theory*, trans. from the Russian, Basel: Birkhäuser-Verlag, 1994.

Maslow, V. P. and Fedoriuk, M. V., *Semi-Classical Approximation in Quantum Mechanics*, trans. from the Russian, Dordrecht: Reidel, 1981.

Path integrals

Feynman, R. P. and Hibbs, A. R., *Quantum Mechanics and Path Integrals*, New York: McGraw-Hill, 1965.

11 Hydrogen and helium atoms

11.1 Introduction

As is well known, the concept of the atom appears for the first time as a hypothesis about the structure of matter by Democritus (*c*. 460 BC–*c*. 370 BC). Democritus also proposed a rudimentary mechanics: Atoms moved in straight lines and through their collisions bodies were formed and destroyed. It is also interesting to recall that later Epicurus (341 BC–270 BC) and Lucretius (*c*. 94 BC–*c*. 49 BC) introduced a random, reasonless, deviation (*clinamen*) from the straight line in order to account for the contingency of our world. Though atomism remained a speculative theory for more than 2000 years, in modern ages it was still alive among many scientists and philosophers (Newton, Locke, Spinoza, etc.).

At the beginning of the nineteenth century John Dalton (1766–1844) introduced the concept of the atom in a scientific framework in order to explain some chemical phenomena. He was one of the earliest scientists to work on the structure of matter. He supposed that atoms were the smallest units of an element that enter into chemical combinations, and that a chemical element was composed entirely of one type of atom. Then, compounds contain atoms of two or more different elements, where the relative number of atoms of each element in a particular compound is always the same. Experimentally, he observed that, for each gram of hydrogen found in water there were always 8 grams of oxygen. Thus, he concluded that, in every pure substance, the same elements are in the same proportions. This observation is known as the law of constant composition. Moreover, it was possible to assign a mass to each given element in such a way that the ratio of different atoms in an element was a rational number (e.g. in water for each atom of oxygen there were two atoms of hydrogen). However, the most important of Dalton's assumptions was that atoms do not change their identities (neither the number nor the type of atoms change) in chemical reactions, which simply change the way the atoms are joined together. As a consequence, he also showed that there was no change in mass during a chemical reaction. This is called the law of conservation of mass.

A further step in transforming the atomic theory into a scientific hypothesis was the introduction in 1869 by D. I. Mendeleev (1834–1907) of the table of the elements that was later completed by himself. The table showed considerable predictive power in anticipating the discovery of some chemicals. At first, the table was taken as a numerical extravagance. The discovery of new elements that filled the gaps in the table forced scientists to take more seriously the hypothesis of atoms as elementary particles, whose combination allowed for the existence of the different elements in a specific way.

However, atoms only began to be considered to be real after the discovery of Brownian motion, the disordered motion of small particles suspended in a fluid or gaseous solution.[1] It is well known that Albert Einstein (1905) succeeded in formalizing the problem. Once the experimental measurements of Jean Perrin (1908) confirmed the theory (his results also led to a better estimation of Avogadro's number) and Robert Millikan (1910) succeeded in measuring the elementary electric charge, as well as the first trajectories of elementary particles were observed in the Wilson chamber (1912), atomic theory began to be an integral part of modern physics.

The first hypothesis about the structure of the atom was formulated in 1902 by J. J. Thompson, who suggested that the neutral atom was formed of a positively charged homogeneous sphere, within which the electrons were suspended in such a number and such a way as to give rise to a zero total charge. In 1911, Geiger and Marsden performed an experiment involving the scattering of alpha particles (see the end of Sec. 4.3) by a thin metal film that falsified Thompson's model by showing that most of the alpha particles passed through the film undeflected, whereas the few that were scattered underwent large deflection. If the positive charge (and the mass) of the atom were distributed uniformly within its volume, we would observe a large number of deflections at small angles. Therefore, Ernest Rutherford inferred that the positive charge of the atom and nearly all of its mass was in a tiny region, the nucleus, whose radius we now know to be of the order of 10^{-15} m, while the radius of the entire atom was known to be the order of 10^{-10} m. It was the strong electrostatic repulsion between the nucleus and the alpha particles that caused the large deflections. According to Rutherford's model the number of electrons was equal to the atomic number Z: they orbited circularly around the nucleus. This model presented two main problems: first, since an accelerating electric charge, according to Larmor's law,[2] emits electromagnetic radiation, the electrons should progressively lose energy and fall, with a spiral trajectory, into the nucleus. The second problem was related to its inability to explain the existence of line spectra (see Subsec. 1.5.4).

A further step toward the quantum-mechanical model of the atom was made in 1913 when Bohr introduced the hypothesis of the quantization of atomic levels, extending and generalizing Planck's quantization rule to the hydrogen atom (see also Subsec. 1.5.1). We know now that Bohr's result only represents a partial solution of the problem,[3] whose completion can be obtained from the quantum-mechanical approach to the dynamics of the electrons. We shall see in this chapter the use of the Schrödinger equation in modern atomic theory.

After having introduced the general problem of the motion of quantum systems in the presence of a central potential, we shall introduce the theory of the hydrogen atom (in Sec. 11.2). In Sec. 11.3 we shall examine the problem of the atom in a magnetic field, while in Sec. 11.4 we shall introduce some relativistic corrections. However, in the present book we do not consider full relativistic effects as such and limit our exposition to approximation

[1] See [Parisi 2005a].

[2] See [*Jackson* 1962].

[3] The reason why Bohr's model has to be considered an *ad hoc* hypothesis lies in the fact that there is no explanation concerning the stability of the discrete stationary electronic levels.

methods. In Sec. 11.5 we shall make use of different approximation methods for treating the theory of the helium atom. Finally, in Sec. 11.6 we shall consider the Thomas–Fermi and Hartree–Fock methods for dealing with the multi-electron problem.

11.2 Quantum theory of the hydrogen atom

Generally speaking, an atomic system is represented by a heavy central nucleus made by Z protons and a certain number of neutrons (see Box 7.1). Z is a positive integer number and is called the *atomic number*. Indicating with N the number of neutrons, the sum

$$A = Z + N \tag{11.1}$$

is called *mass number*, because, as we shall see, the mass of the neutron is approximately equal to the mass of the proton, which in turn is almost 2000 times the mass of the electron (see also Subsec. 6.3.1). On the other hand, the charge of the proton is, apart from the sign, approximately equal to that of the electron. Combining these facts, we may conclude that the atom is an analogue of the solar system with the nucleus replacing the Sun and the planets represented by the electrons. There is, however, a certain number of differences that need to be emphasized:

- The atom is a microscopic system, and therefore quantum effects not only cannot be neglected but even determine the nature of the system. As a consequence, electrons do not follow classical trajectories.
- The force that binds the solar system is the gravitational interaction. In an atomic system gravitation is negligible, and the relevant interaction is represented by the Coulomb force.
- While in the solar system, different planets may have considerable differences of mass, the electrons' mass is the same.

In order for the atom to be electrically neutral, it must have the same number of electrons and neutrons. Otherwise, we have positive or negative ions.

In what follows we shall be mainly concerned with the hydrogen atom, since it undoubtedly the simplest atomic system, in that its nucleus is made of a proton (without neutrons) and there is only one electron. For this reason, it can be considered a prototype of the two–body Kepler problem. The helium atom is the next simplest problem, since it involves two electrons.

11.2.1 The quantum Kepler problem

The classical Kepler problem deals with the gravitational force among celestial bodies in the solar system.[4] As is well known from classical mechanics, when the number of

[4] See [*Goldstein* 1950, Ch. 3].

interacting bodies is equal to or larger than three, there is no close-form solution of the set of differential equations describing this physical situation. In order to solve such a problem, Kepler treated the bodies as point masses and reduced the problem to a set of equations describing one-to-one interactions. An equivalent treatment is also useful for dealing with the description of the hydrogen atom. On an abstract plane, let us first consider the case of two particles (say a and b) that interact with a conservative force, i.e. whose Hamiltonian is given by

$$\hat{H} = \frac{(\hat{\mathbf{p}}^a)^2}{2m_a} + \frac{(\hat{\mathbf{p}}^b)^2}{2m_b} + V(\hat{r}), \tag{11.2}$$

where $\hat{\mathbf{r}} = \hat{\mathbf{r}}^a - \hat{\mathbf{r}}^b$. Note that in this case V does not depend on the angular coordinates θ and ϕ. We may therefore apply the same considerations we have made in Subsec. 6.2.2 about central potentials. As in the classical case, we move to the center-of-mass reference frame and write

$$\hat{\boldsymbol{R}} = \frac{m_a\hat{\mathbf{r}}^a + m_b\hat{\mathbf{r}}^b}{M}, \tag{11.3}$$

where $M = m_a + m_b$. We now show that, given the expressions

$$\hat{\boldsymbol{P}} = M\dot{\hat{\boldsymbol{R}}} \quad \text{and} \quad \hat{\mathbf{p}} = m\dot{\hat{\mathbf{r}}}, \tag{11.4}$$

where

$$m = \frac{m_a m_b}{M} \tag{11.5}$$

is the system reduced mass, the Hamiltonian of the system can be written as

$$\hat{H} = \frac{\hat{\boldsymbol{P}}^2}{2M} + \frac{\hat{\mathbf{p}}^2}{2m} + V(\hat{r}) = \hat{H}_0 + \hat{H}_{\mathrm{I}}, \tag{11.6}$$

where $\hat{H}_0 = \hat{\boldsymbol{P}}^2/2M$. Since

$$\hat{\boldsymbol{P}} = m_a\dot{\hat{\mathbf{r}}}^a + m_b\dot{\hat{\mathbf{r}}}^b = \hat{\mathbf{p}}^a + \hat{\mathbf{p}}^b, \tag{11.7}$$

we have

$$\frac{\hat{\boldsymbol{P}}^2}{2M} + \frac{\hat{\mathbf{p}}^2}{2m} = \frac{1}{2M}\left[(\hat{\mathbf{p}}^a)^2 + (\hat{\mathbf{p}}^b)^2 + 2\hat{\mathbf{p}}^a\hat{\mathbf{p}}^b\right] + \frac{m}{2}\left[(\dot{\hat{\mathbf{r}}}^a)^2 + (\dot{\hat{\mathbf{r}}}^b)^2 - 2\dot{\hat{\mathbf{r}}}^a\dot{\hat{\mathbf{r}}}^b\right], \tag{11.8}$$

from which the result (11.6) follows (see Prob. 11.1).

It is possible to prove that $\hat{\boldsymbol{P}}, \hat{\boldsymbol{R}}$ and $\hat{\mathbf{p}}, \hat{\mathbf{r}}$ are two pairs of conjugate variables classically (see Prob. 11.2) and quantum-mechanically, and the variables of the first pair commute with those of the second pair. For the quantum case, it is clear that $\left[\hat{r}_j^a, \hat{p}_k^a\right] = \imath\hbar\delta_{jk}$, where $\hat{r}_j^a$ and $\hat{p}_k^a$ are the j-th component of position and the k-th component of momentum of the first particle, respectively, and the same relations hold for the second particle. Moreover,

$$\begin{aligned}\left[\hat{R}_j, \hat{P}_k\right] &= \left[\frac{m_a}{M}\hat{r}_j^a + \frac{m_b}{M}\hat{r}_j^b, \hat{p}_k^a + \hat{p}_k^b\right] \\ &= \left(\frac{m_a}{M} + \frac{m_b}{M}\right)\imath\hbar\delta_{jk} = \imath\hbar\delta_{jk},\end{aligned} \tag{11.9}$$

which is the desired result for the first pair of observables. For the second pair, we have

$$\left[\hat{r}_j, \hat{p}_k\right] = \left[\hat{r}^a_j - \hat{r}^b_j, \frac{m}{m_a}\hat{p}^a_k - \frac{m}{m_b}\hat{p}^b_k\right]$$
$$= \left(\frac{m}{m_a} + \frac{m}{m_b}\right)\imath\hbar\delta_{jk} = \imath\hbar\delta_{jk}, \tag{11.10}$$

which is again the result we wished to have. In a similar way, one can show that the observables of the first pair commute with the observables of the second pair, i.e. (see Prob. 11.3)

$$\left[\hat{R}_j, \hat{p}_k\right] = \left[\hat{r}_j, \hat{P}_k\right] = \left[\hat{r}_j, \hat{R}_k\right] = \left[\hat{p}_j, \hat{P}_k\right] = 0. \tag{11.11}$$

We have divided the Hamiltonian into two parts (Eq. (11.6)), so that, if the total wave function is of the form

$$\psi(\mathbf{r}^a, \mathbf{r}^b) = \Phi(\boldsymbol{R}) \cdot \varphi(\mathbf{r}), \tag{11.12}$$

and is an eigenfunction of the total Hamiltonian, we have

$$\hat{H}\psi(\mathbf{r}^a, \mathbf{r}^b) = \left(\hat{H}_0 + \hat{H}_{\mathrm{I}}\right)\Phi(\boldsymbol{R})\varphi(\mathbf{r})$$
$$= (E_R + E_r)\Phi(\boldsymbol{R})\varphi(\mathbf{r}) = E\psi(\mathbf{r}^a, \mathbf{r}^b), \tag{11.13a}$$

with

$$\hat{H}_0\Phi(\boldsymbol{R}) = E_R\Phi(\boldsymbol{R}), \tag{11.13b}$$
$$\hat{H}_{\mathrm{I}}\varphi(\mathbf{r}) = E_r\varphi(\mathbf{r}). \tag{11.13c}$$

Equation (11.13b) represents the trivial eigenvalue equation for a free particle, and we will focus our attention on Eq. (11.13c), which, with the substitution made in Eq. (6.85), i.e.

$$\varphi(\mathbf{r}) = \mathrm{R}(r)Y_{lm}(\phi, \theta), \tag{11.14}$$

can also be written as (see Eq. (6.87) and Prob. 6.12)

$$\left[-\frac{\hbar^2}{2mr^2}\frac{\partial}{\partial r}\left(r^2\frac{\partial}{\partial r}\right) + \frac{\hbar^2\hat{\mathbf{l}}^2}{2mr^2} + V(r) - E_r\right]\mathrm{R}(r) = 0. \tag{11.15}$$

Equation (11.14) is justified by the spherical symmetry of the problem, and the $Y_{lm}(\phi, \theta)$ are the spherical harmonics (6.65) (see Fig. 11.1).

11.2.2 The radial Schrödinger equation

When we deal with hydrogenoid atoms,[5] the potential energy has the form

$$V(r) = -\frac{Ze^2}{r}, \tag{11.16}$$

[5] Examples of hydrogenoid atoms are He^+, Li^{++}, Be^{+++}, ...

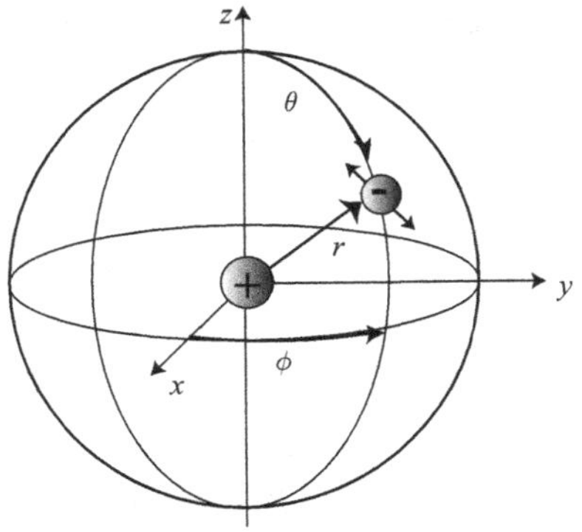

Figure 11.1 Electron coordinates in the atomic system. A negative electron is in motion around a positively charged nucleus. Even though it is not possible to define a quantum-mechanical trajectory of the electron, its position is individuated by the three spherical coordinates r, ϕ, θ: r represents the distance from the nucleus (see Fig. 6.3), and, via the energy eigenvalue, is associated with the *principal quantum number n* (see Fig. 6.9), ϕ is the azimuthal angle, and θ the polar angle (see Subsec. 6.1.3). Each (spin-1/2) electron may have a projection of the spin along the z-axis equal to $+\hbar/2$ or $-\hbar/2$ (see Sec. 6.3).

which represents the electrostatic interaction between a nucleus with charge Ze and only one electron with charge $-\mathrm{e}$, where

$$\mathrm{e} = 1.602\,176\,53 \times 10^{-19}\mathrm{C}, \tag{11.17}$$

C being the Coulomb charge unit. Then, Eq. (11.15) (see Eq. (6.92)) can be rewritten as

$$\xi'' + \frac{2m}{\hbar^2}\left[E_r - \frac{\hbar^2}{2m}\frac{l(l+1)}{r^2} + \frac{Ze^2}{r}\right]\xi = 0, \tag{11.18}$$

where

$$\mathrm{R}(r) = \frac{\xi(r)}{r}. \tag{11.19}$$

Figure 11.2 shows that the total potential resulting from the sum of the attractive Coulomb potential and the repulsive centrifugal term has a well shape (see Prob. 11.4).

Before looking for the solution of Eq. (11.18), we wish to introduce a general remark. We have here three natural constants, $\hbar$, e, and m, where

$$\frac{1}{m} = \frac{1}{m_e} + \frac{1}{m_n}, \tag{11.20}$$

and m_e is the electron mass; m_n is the nucleus mass, which is approximately equal to $Z\,m_p$; and m_p is the mass of the proton, which is about $1840\,m_e$.[6] Therefore, to an excellent

[6] The mass of a "free" proton and a "free" neutron is $1.672\,621\,58 \times 10^{-27}$ kg and $1.674\,927\,16 \times 10^{-27}$ kg, respectively. The sum amounts to $3.347\,548\,74 \times 10^{-27}$ kg. However, the mass of a "free" deuteron is only $3.343\,583\,09 \times 10^{-27}$ kg. The difference is $-0.003\,965\,65 \times 10^{-27}$ kg. This problem is called "mass defect." The reason is that the binding of the proton and neutron to form the deuteron releases a certain amount of energy which is equivalent to the mass defect.

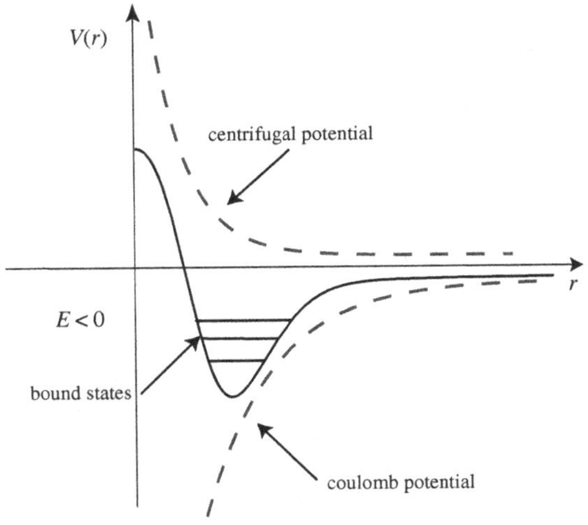

Figure 11.2 Resulting potential in the hydrogen atom (see also Fig. 4.9). The sum of the centrifugal potential $\hbar^2 \frac{l(l+1)}{2mr^2}$ and of the Coulomb potential $-Ze^2/r$ gives rise to the final well-shaped resulting potential. The negative-energy bound states are also schematically shown.

approximation (see Prob. 11.5), the reduced mass m is equal to the electron mass. With these constants we can build the quantities

$$r_0 = \frac{\hbar^2}{me^2} = 0.529\,177\,208\,59(36) \times 10^{-10}\,\mathrm{m}, \quad E_0 = \frac{me^4}{\hbar^2} = 27.2\,\mathrm{eV}, \tag{11.21}$$

where r_0 is called *Bohr's radius*, and has an uncertainty of $0.000\,000\,000\,36 \times 10^{-10}$ m. It is interesting to note that with the same constants it is also possible to define Bohr's magneton or the characteristic unity of magnetic momenta (see also Eq. (6.167)), i.e.

$$\mu_B = \frac{e\hbar}{2m}. \tag{11.22}$$

Let us now come back to the Schrödinger equation (11.18) in the case where $Z = 1$ (hydrogen atom). It is possible to write (see also Prob. 11.6)

$$r_0^2 \frac{d^2}{dr^2}\xi = \frac{d^2}{d\tilde{r}^2}\xi, \tag{11.23a}$$

$$\left[\frac{d^2}{d\tilde{r}^2} + \frac{2}{\tilde{r}} - \frac{l(l+1)}{\tilde{r}^2} + 2\tilde{E}\right]\xi(\tilde{r}) = 0, \tag{11.23b}$$

where

$$\tilde{r} = \frac{r}{r_0} \quad \text{and} \quad \tilde{E} = \frac{E}{E_0}. \tag{11.24}$$

Consider the change of variable

$$n^2 = -\frac{1}{2\tilde{E}}. \tag{11.25}$$

Box 11.1 Confluent hypergeometric functions

A confluent hypergeometric function has the general form [*Gradstein/Ryshik* 1981, 9.210, 9.216]

$$F(\alpha;\gamma;z) = 1 + \frac{\alpha}{\gamma}\frac{z}{1!} + \frac{\alpha(\alpha+1)}{\gamma(\gamma+1)}\frac{z^2}{2!} + \frac{\alpha(\alpha+1)(\alpha+2)}{\gamma(\gamma+1)(\gamma+2)}\frac{z^3}{3!} + \cdots, \tag{11.26}$$

where z is the (complex) variable and α and γ are two complex parameters. $F(\alpha;\gamma;z)$ is the solution of the differential equation

$$z\frac{d^2}{dz^2}f(z) + (\gamma - z)\frac{d}{dz}f(z) - \alpha f(z) = 0, \tag{11.27}$$

and its asymptotic behavior for $|z| \to \infty$ is given by

$$F(\alpha;\gamma;z) \to e^{-\imath\pi\alpha}\frac{\Gamma(\gamma)}{\Gamma(\gamma-\alpha)}z^{-\alpha} + \frac{\Gamma(\gamma)}{\Gamma(\alpha)}e^z z^{\alpha-\gamma}, \tag{11.28}$$

where Γ is the Euler gamma function [*Gradstein/Ryshik* 1981, 8.3]. Equation (11.28) is not valid for $\alpha = -n$, with $n = 0, 1, 2, \ldots$, in which case F becomes a polynomial of degree n in z. An important property of F is represented by *Kummer's transformation* [*Gradstein/Ryshik* 1981, 9.212]

$$F(\alpha;\gamma;z) = e^z F(\gamma - \alpha;\gamma;-z). \tag{11.29}$$

The asymptotic behavior of the solution of Eq. (11.23b) for $\tilde{r} \to \infty$ and for $\tilde{r} \to 0$ suggests the ansatz[7]

$$\xi(\tilde{r}) = \tilde{r}^{l+1}e^{-\frac{\tilde{r}}{n}}W(\tilde{r}). \tag{11.30}$$

Let us use the relations

$$\frac{\partial}{\partial\tilde{r}}\xi(\tilde{r}) = \left(\frac{l+1}{\tilde{r}} - \frac{1}{n} + \frac{W'}{W}\right)\tilde{r}^{l+1}e^{-\frac{\tilde{r}}{n}}W, \tag{11.31a}$$

$$\frac{\partial^2}{\partial\tilde{r}^2}\xi(\tilde{r}) = \left[-\frac{l+1}{\tilde{r}^2} + \frac{W''}{W} - \frac{(W')^2}{W^2} + \left(\frac{l+1}{\tilde{r}} - \frac{1}{n} + \frac{W'}{W}\right)^2\right]\tilde{r}^{l+1}e^{-\frac{\tilde{r}}{n}}W, \tag{11.31b}$$

in order to derive (see Prob. 11.7)

$$\left[\tilde{r}W'' + 2\left(1 + l - \frac{\tilde{r}}{n}\right)W' + 2\frac{n-l-1}{n}W\right]\xi(\tilde{r}) = 0, \tag{11.32}$$

where the derivatives of $W(\tilde{r})$ are taken with respect to $\tilde{r}$. With the change of variable

$$\eta = \frac{2\tilde{r}}{|n|}, \tag{11.33}$$

[7] Some textbooks incorrectly infer the form of the solution from the asymptotic behavior of Eq. (11.23b). We choose here to follow a more general approach (see [Loinger 2003]).

we may distinguish two cases:

- $n > 0$, i.e. $n = |n|$;
- $n < 0$, i.e. $n = -|n|$.

We shall show in the following that the two cases lead to the same result.

In the first case ($n = |n|$), we obtain

$$\eta W''(\eta) + [2(l+1) - \eta]\, W'(\eta) + (|n| - l - 1)\, W(\eta) = 0, \tag{11.34}$$

where the derivatives of $W(\eta)$ are now taken with respect to η. The solution of Eq. (11.34) is

$$W(\eta) = F\,(-|n| + l + 1; 2l + 2; \eta)\,, \tag{11.35}$$

where $F\,(\alpha; \gamma; z)$ is the confluent hypergeometric function (see Box 11.1), that is, a function of the form

$$\begin{aligned} F\,(-|n| + l + 1; 2l + 2; \eta) = 1 &+ \frac{(-n + l + 1)}{(2l + 2)} \frac{\eta}{1!} \\ &+ \frac{(-n + l + 1)[(-n + l + 1) + 1]}{(2l + 2)[(2l + 2) + 1]} \frac{\eta^2}{2!} \\ &+ \frac{(-n + l + 1)[(-n + l + 1) + 1][(-n + l + 1) + 2]}{(2l + 2)[(2l + 2) + 1][(2l + 2) + 2]} \frac{\eta^3}{3!} \\ &+ \cdots . \end{aligned} \tag{11.36}$$

This in turn justifies our ansatz. Then, in the limit $\tilde{r} \to \infty$ ($\eta \to \infty$), taking into account the expression (11.28), from Eq. (11.30) we have

$$\xi(\tilde{r}) = \tilde{r}^{l+1} e^{-\frac{\tilde{r}}{|n|}} F\,(-|n| + l + 1; 2l + 2; \eta) \to g(\tilde{r}) e^{\frac{\tilde{r}}{|n|}}, \tag{11.37}$$

where we recall substitution (11.33), and $g(\tilde{r}) \simeq \tilde{r}$.

In the second case ($n = -|n|$), we obtain the differential equation

$$\eta W'' + [2(l+1) + \eta]\, W' + (|n| + l + 1)\, W = 0, \tag{11.38}$$

whose solution is (see Prob. 11.8)

$$W(\eta) = F\,(|n| + l + 1; 2l + 2; -\eta)\,, \tag{11.39}$$

where F is again the confluent hypergeometric function with the form

$$\begin{aligned} F\,(|n| + l + 1; 2l + 2; -\eta) = 1 &+ \frac{(n + l + 1)}{(2l + 2)} \frac{-\eta}{1!} \\ &+ \frac{(n + l + 1)[(n + l + 1) + 1]}{(2l + 2)[(2l + 2) + 1]} \frac{(-\eta)^2}{2!} \\ &+ \frac{(n + l + 1)[(n + l + 1) + 1][(n + l + 1) + 2]}{(2l + 2)[(2l + 2) + 1][(2l + 2) + 2]} \frac{(-\eta)^3}{3!} \\ &+ \cdots . \end{aligned} \tag{11.40}$$

We make use now of Kummer's transformation

$$F\,(|n| + l + 1; 2l + 2; -\eta) = e^{-\eta} F\,(-|n| + l + 1; 2l + 2; \eta)\,, \tag{11.41}$$

and, for $\tilde{r} \to \infty$ ($\eta \to \infty$), we obtain

$$\begin{aligned}\xi(\tilde{r}) &= \tilde{r}^{l+1} e^{\frac{\tilde{r}}{|n|}} F\left(|n| + l + 1; 2l + 2; -\eta\right) \\ &= \tilde{r}^{l+1} e^{\frac{\tilde{r}}{|n|}} e^{-\frac{2\tilde{r}}{|n|}} F\left(-|n| + l + 1; 2l + 2; \eta\right) \to g(\tilde{r}) e^{\frac{\tilde{r}}{|n|}}. \end{aligned} \quad (11.42)$$

Equation (11.42), as expected, proves that the two cases ($n < 0$ and $n > 0$) lead to the same result. Since there is dependence on the sign of E (negative for bound states), but not on the sign of n, from now on we may take $n > 0$. From the expansion (11.36) we learn that $W(\eta)$ bears the general form

$$W(\eta) = \sum_{j=0}^{\infty} c_j \eta^j, \quad (11.43)$$

where

$$c_{j+1} = \frac{j + l + 1 - n}{(j+1)\left[j + 2(l+1)\right]} c_j. \quad (11.44)$$

The first three terms are given by

$$c_0 = 1, \quad c_1 = \frac{(l + 1 - n)}{2(l+1)}, \quad c_2 = \frac{(l+2) - n}{2(2l+3)} c_1. \quad (11.45)$$

11.2.3 Eigenvalues and eigenfunctions

Let us have a closer look at Eqs. (11.43) and (11.44). We have here two possibilities:

- One of the c_j, say c_k, is zero and consequently all other c_j with $j > k$ are also zero, and $W(\eta)$ is a polynomial.
- Otherwise, using the fact that

$$\frac{c_{j+1}}{c_j} \to \frac{1}{j}, \quad (11.46)$$

for $j \to \infty$, we would have $c_j \simeq 1/(j-1)!$, and consequently also

$$W(\eta) = \sum_{j=0}^{\infty} c_j \eta^j \propto \eta e^{\eta}. \quad (11.47)$$

This second possibility implies that $W(\eta)$ is not a polynomial and grows as e^{η} for $\eta \to \infty$. This alternative must be discarded, and we must therefore require that the series (11.43) or (11.36) be truncated at a certain value $j = k$. In order for this to happen, it is necessary that

$$n = l + 1 + k, \quad (11.48)$$

where k must be a non-negative integer. In other words, the radial wave function contains only a finite number of terms.Since l is an integer number, also n must be an integer num-

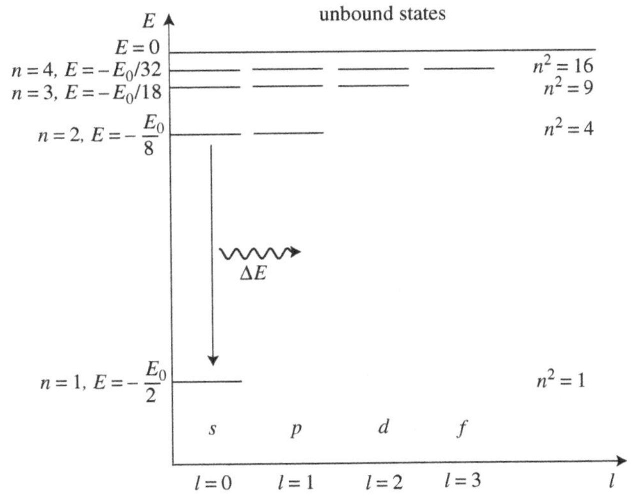

Figure 11.3 The Grotrian scheme: a schematic representation of the populated energy levels as a function of the two quantum numbers n and l. The arrow shows the transition from the $n = 2, l = 0$ state to the $n = 1, l = 0$ state, with emission of a photon of energy $\Delta E = 3E_0/8$.

ber. For each integer value of n, the corresponding value of energy, given the definitions (11.24), is then given by

$$E_n = E_0 \tilde{E} = -\frac{1}{2n^2} E_0, \tag{11.49}$$

where $n = 1, 2, \ldots$. It is then clear that the energy levels of the bound states become more and more dense as n goes to infinity, i.e. as the energy approaches the limit $E = 0$ (see Fig. 11.3). The number n is also called the *principal quantum number*. The energy levels with the same value of l may be grouped together. In this way one may build the Lymann ($l = 0$), Balmer ($l = 1$), Paschen ($l = 2$), Blackett ($l = 3$), and Pfund ($l = 4$) series (see also the end of Subsec. 6.1.4). The number of eigenstates for each given value of n is equal to n^2 (see Prob. 11.9). As we have seen, the energy, for a given n, does not depend on l. This may look bizarre, as l appears in the Hamiltonian. Such an l-degeneracy is only accidental.[8] As we shall see in the following, relativistic corrections may remove such a degeneracy.

The ground state $n = 1$ ($l = 0$) is obviously stable. However, transitions from upper levels to lower ones involve the emission of photons with energy equal to the difference between the energies of the these levels. For example, the transition from the $n = 2, l = 0$ state to the $n = 1, l = 0$ state involves the emission of a photon of energy:

$$\Delta E = -\frac{E_0}{8} - \left(-\frac{E_0}{2}\right) = \frac{3}{8} E_0. \tag{11.50}$$

[8] It occurs only for the Coulomb potential proportional to $1/r$. For instance, a different degeneracy is present in the case of the three-dimensional harmonic oscillator (see Subsec. 6.2.4).

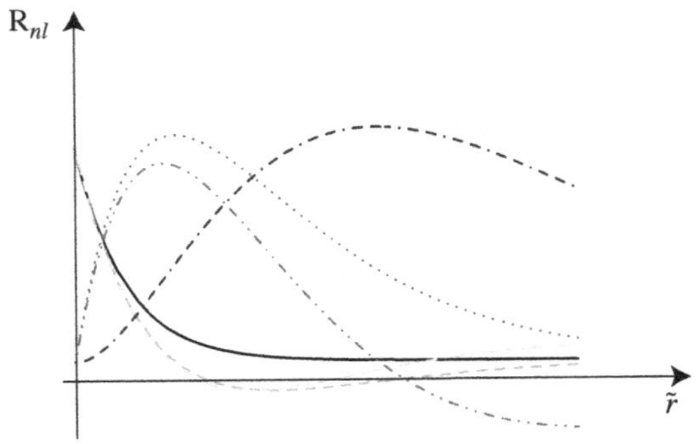

Figure 11.4 Schematic plot of the first few radial eigenfunctions of the hydrogenoid atom: R_{10} (solid line), R_{20} (dashed line), R_{21} (dotted line), R_{30} (dot-dashed line), R_{31} (dot-dot-dashed line), and R_{32} (dot-dash-dashed line).

For completeness in the following we present explicit expressions for the first few functions R_{nl} (see Fig. 11.4 and Prob. 11.10):

$$R_{10}(\tilde{r}) \propto e^{-\tilde{r}}, \tag{11.51a}$$

$$R_{20}(\tilde{r}) \propto e^{-\frac{\tilde{r}}{2}}\left(1-\frac{\tilde{r}}{2}\right), \tag{11.51b}$$

$$R_{21}(\tilde{r}) \propto \tilde{r}e^{-\frac{\tilde{r}}{2}}, \tag{11.51c}$$

$$R_{30}(\tilde{r}) \propto e^{-\frac{\tilde{r}}{3}}\left(1-\eta+\frac{1}{6}\eta^2\right) = e^{-\frac{\tilde{r}}{3}}\left(1-\frac{2}{3}\tilde{r}+\frac{2}{27}\tilde{r}^2\right), \tag{11.51d}$$

$$R_{31}(\tilde{r}) \propto \eta\left(1-\frac{1}{4}\eta\right)e^{-\frac{\eta}{2}} = \frac{2}{3}\tilde{r}\left(1-\frac{1}{6}\tilde{r}\right)e^{-\frac{\tilde{r}}{3}}, \tag{11.51e}$$

$$R_{32}(\tilde{r}) \propto \eta^2 e^{-\frac{\eta}{2}} = \frac{4}{9}\tilde{r}^2 e^{-\frac{\tilde{r}}{3}}. \tag{11.51f}$$

As expected (see property (ii) of Subsec. 3.2.3), R_{10} has no zeros, R_{20} has one zero (for $r = 2r_0$), R_{30} has two zeros, and so on. In order to find the correct normalization factors for the radial wave functions $R_{nl}(r)$, one has to impose the condition (see Prob. 11.11)

$$\int_0^{+\infty} dr\, |R_{nl}(r)|^2\, r^2 = 1. \tag{11.52}$$

It is particularly interesting to study the radial probability density for the electron, since it may give an idea on at what distance from the nucleus the probability of finding the electron is maximal (see Fig. 11.5). Such a probability density (see Prob. 11.12) is given by

$$\wp_{nl}(\tilde{r}) = |R_{nl}(r)|^2\, r^2. \tag{11.53}$$

11.2.4 Total eigenfunctions and hydrogenoid atoms

Recalling the definition of the total eigenfunction $\varphi(\mathbf{r})$ (Eq. (11.14)), we may now use the result for the radial wave function, which is given by Eqs. (11.19), (11.30), (11.43), and

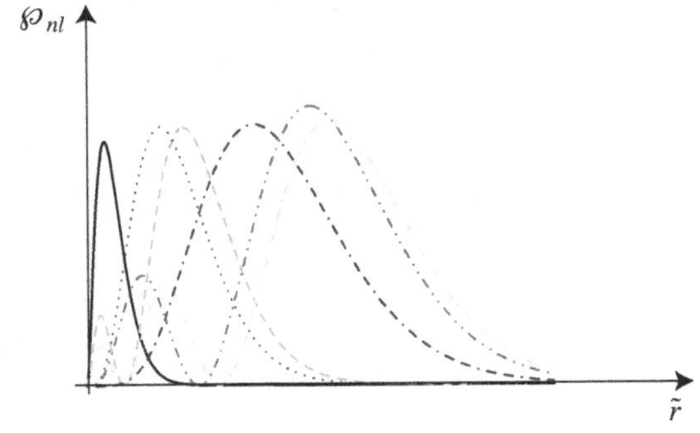

Figure 11.5 Schematic plot of the radial probability densities $\wp_{nl}(\tilde{r}) = |R_{nl}(\tilde{r})|^2 \tilde{r}^2$ for the first few radial eigenfunctions of the hydrogenoid atom (same convention as in Fig. 11.4).

(11.44), and Eq. (6.65) for the spherical harmonics in order to build the explicit form of

$$\varphi_{nlm}(r,\theta,\phi) = \mathrm{R}_{nl}(r)Y_{lm}(\theta,\phi). \tag{11.54}$$

In the case of hydrogenoid atoms, the zero-angular momentum total eigenfunctions $\varphi_{nlm}(r,\theta,\phi)$ for $n = 1, 2, 3$, are (see Probs. 11.14 and 11.15)

$$\varphi_{100}(\mathbf{r}) = \frac{1}{\sqrt{\pi}}\left(\frac{Z}{r_0}\right)^{\frac{3}{2}} e^{-\frac{Zr}{r_0}}, \tag{11.55a}$$

$$\varphi_{200}(\mathbf{r}) = \frac{1}{4\sqrt{2\pi}}\left(\frac{Z}{r_0}\right)^{\frac{3}{2}}\left(2 - \frac{Zr}{r_0}\right) e^{-\frac{Zr}{2r_0}}, \tag{11.55b}$$

$$\varphi_{300}(\mathbf{r}) = \frac{1}{81\sqrt{3\pi}}\left(\frac{Z}{r_0}\right)^{\frac{3}{2}}\left(27 - 18\frac{Zr}{r_0} + \frac{2Z^2r^2}{r_0^2}\right) e^{-\frac{Zr}{3r_0}}. \tag{11.55c}$$

11.3 Atom and magnetic field

As is natural to expect, the presence of a constant and homogenous magnetic field modifies the energy levels of an electron in an atomic system and may be considered as a perturbation. This effect is basically due the magnetic interaction between the external magnetic field and the intrinsic magnetic momentum of the electron. As we shall see, depending on the strength of the magnetic field, it is possible to distinguish between two regimes that correspond to two different effects: the Paschen–Bach effect (Subsec. 11.3.2) and the Zeeman effect (Subsec. 11.3.3).

Before entering into the detailed investigation of the effects produced by a magnetic field, it is necessary to consider another type of interaction that is always present in an atom (when $l \neq 0$) and goes under the name of the spin–orbit interaction. As we shall see in the next section, this interaction has a relativistic nature.

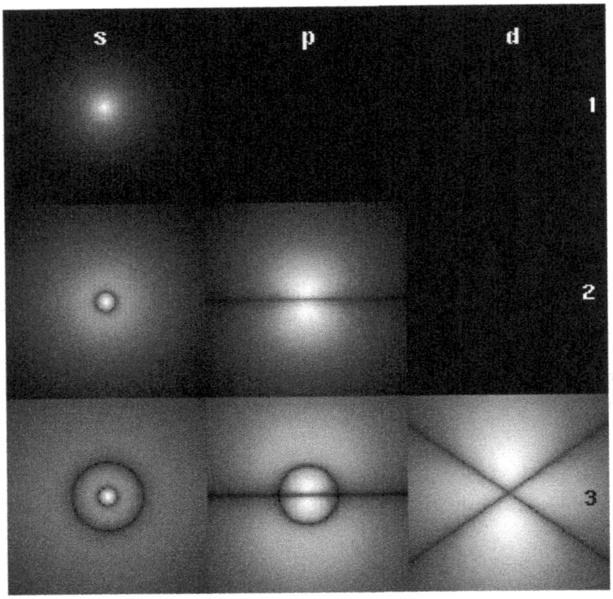

Figure 11.6 Pictorial representation (density plot) of the spatial probability density corresponding to the wave functions of an electron in a hydrogen atom possessing definite values of energy (increasing downward: $n = 1, 2, 3, \ldots$) and of angular momentum (increasing across: $s, p, d, \ldots$). Brighter areas correspond to higher probability density for a position measurement. Adapted from the web page www.physicsdaily.com/physics/Quantum_mechanics.

11.3.1 Spin–orbit interaction

We recall that Uhlenbeck and Goudsmit pointed out that certain spectroscopical lines, if observed with high precision (with a relative accuracy of the order of 10^{-5}), turn out to be doublets instead of single lines (see Fig. 6.9).[9] They made the hypothesis that this was due to the presence of a magnetic momentum intrinsic to the electron, i.e. the spin (see Sec. 6.3).

Let us consider the reference frame where the electron is at rest. Spin orbits effect, as all magnetic effects, is a relativistic effect. A careful study would involve relativistic considerations that are outside the aim of this book. Let use first consider a simplified approach. The proton's motion induces a current that generates a magnetic field which interacts with the electron. The electrostatic interaction between electron and proton is represented by the potential defined in Eq. (11.16), which corresponds to the electric field[10]

$$\mathbf{E} = \frac{Ze}{r^3}\mathbf{r}. \tag{11.56}$$

[9] See [Uhlenbeck/Goudsmit 1925, Uhlenbeck/Goudsmit 1926].

[10] We consider here electric and magnetic fields in unquantized form (and for this reason we have omitted the operatorial expression in the first few equations). The quantization of electromagnetic fields will be the subject of Ch. 13.

On the other hand, the "proton current" generates a magnetic field determined by (see Eq. (6.168))

$$\mathbf{B}_p = \mathbf{E} \times \mathbf{v} = \frac{Ze}{r^3}\mathbf{r} \times \mathbf{v} = \frac{Ze}{m_e}\frac{1}{r^3}\mathbf{L} = B_p \boldsymbol{k}, \tag{11.57}$$

where m_e is the electron mass, $\mathbf{L}$ its angular momentum, and $\boldsymbol{k}$ is the z-axis versor. Since the electron has an intrinsic magnetic momentum

$$\boldsymbol{\mu}_s = \frac{g}{2}\frac{e}{m_e}\mathbf{S}, \tag{11.58}$$

there will be an interaction potential energy (see Subsec. 6.3.3) that in the nucleus reference frame[11] has the form

$$V = -\boldsymbol{\mu}_s \cdot \mathbf{B}_p = \frac{Ze^2}{2m_e^2 r^3}\mathbf{L} \cdot \mathbf{S} = \frac{g}{2}\frac{Zc\alpha\hbar}{m_e^2}\frac{\mathbf{L} \cdot \mathbf{S}}{r^3}, \tag{11.59}$$

where

$$\alpha = \frac{e^2}{\hbar c} \simeq 137^{-1} \tag{11.60}$$

is the fine–structure constant and g is the gyromagnetic ratio of the electron that with high precision is equal to 2 (see Subsec. 6.3.3). However there is also an extra term of relativistic nature (called Thomas spin precession) so that the final result is ultimately the previous one but putting $(g - 1)$ at the place of g. Due to the smallness of the fine-structure constant, such a spin–orbit interaction may be treated as a weak perturbation. However, this perturbation is sufficient to render l and m_l as "bad" quantum numbers. The new good quantum numbers are then j and m_j (see Sec. 6.4). In fact, in the quantized framework, from Eq. (6.188) we obtain

$$\hat{J}^2 = \hat{L}^2 + \hat{S}^2 + 2\hat{\mathbf{L}} \cdot \hat{\mathbf{S}}, \tag{11.61}$$

and

$$\hat{\mathbf{L}} \cdot \hat{\mathbf{S}} = \frac{1}{2}\left(\hat{J}^2 - \hat{L}^2 - \hat{S}^2\right). \tag{11.62}$$

In order to find the first-order perturbational correction to the energy level (see Eq. (10.12)), we have to compute the mean value

$$\left\langle \varphi_{nlm}^{(0)} \left| V \right| \varphi_{nlm}^{(0)} \right\rangle = \frac{Ze^2}{2m_e^2}\left\langle \hat{\mathbf{L}} \cdot \hat{\mathbf{S}} \right\rangle_{\text{ang,spin}} \left\langle \frac{1}{r^3} \right\rangle_{\text{rad}}, \tag{11.63}$$

where $\langle \cdot \rangle_{\text{ang,spin}}$ denotes the mean value on the angular and spin variables, and

$$\left\langle \frac{1}{r^3} \right\rangle_{\text{rad}} = \frac{Z^3}{r_0^3 n^3 l \left(l + \frac{1}{2}\right)(l+1)}. \tag{11.64}$$

In conclusion, the first-order correction to the energy level is given by

[11] See also [*Jackson* 1962, 546].

$$E_{nlm}^{(1)} = \frac{Z^4 c\alpha e^6}{2\hbar^5 n^3 l\left(l+\frac{1}{2}\right)(l+1)} \left\langle \hat{\mathbf{L}} \cdot \hat{\mathbf{S}} \right\rangle_{\text{ang,spin}} = \kappa \left\langle \hat{\mathbf{l}} \cdot \hat{\mathbf{s}} \right\rangle_{\text{ang,spin}}, \tag{11.65}$$

where

$$\kappa = \frac{Z^4 c\alpha e^6 m_e}{2\hbar^3 n^3 l\left(l+\frac{1}{2}\right)(l+1)} = \frac{Z^4 \alpha^4 m_e c^4}{2n^3 l\left(l+\frac{1}{2}\right)(l+1)}, \tag{11.66}$$

and

$$\left\langle \hat{\mathbf{l}} \cdot \hat{\mathbf{s}} \right\rangle_{\text{ang,spin}} = \frac{1}{2}\left[j(j+1) - l(l+1) - s(s+1)\right]. \tag{11.67}$$

Since $\hat{\mathbf{s}} = 1/2$, we shall necessarily have $j = l \pm 1/2$ in Eq. (11.67). From Eq. (11.66) we expect the spin–orbit interaction to decrease as $1/n^3$ and $1/l^3$, so that the doublets reduce their separations as n and l increase (see Fig. 6.9 and Prob. 11.16).

Let us now go back to the problem of the behavior of an atom in an external homogeneous magnetic field $\mathbf{B}_{\text{ext}}$. Taking into account both the magnetic momenta induced by the orbital angular momentum and spin (see Eqs. (6.166) and (6.169)), and the spin–orbit interaction (Eq. (11.59)), the total Hamiltonian will be given by

$$\hat{H} = \hat{H}_A - \frac{e\hbar}{2m_e}\left(\hat{\mathbf{l}} + 2\hat{\mathbf{s}}\right) \cdot \mathbf{B}_{\text{ext}} - f(r)\hat{\mathbf{l}} \cdot \hat{\mathbf{s}}, \tag{11.68}$$

where $\hat{H}_A$ is the ordinary atomic Hamiltonian (11.2), and

$$f(r) = \frac{Ze^2\hbar^2}{2m_e^2 r^3}. \tag{11.69}$$

It should be noted that

$$\kappa = \langle f(r) \rangle_{\text{rad}}. \tag{11.70}$$

To find the solution of the general problem with Hamiltonian (11.68) is impossible. However, a perturbational approach (see Sec. 10.1) is viable in specific situations. In particular, Eq. (11.68) shows that we are in the presence of two perturbational terms in the Hamiltonian. We have necessarily to ask which of the two is the more relevant. If the magnetic term due to the external field is much larger than the spin–orbit interaction, the coupling between the orbital magnetic momentum and the spin magnetic momentum is negligible. In this case, $\hat{\mathbf{l}}$ and $\hat{\mathbf{s}}$ turn out to be decoupled, and m_l and m_s are again "good" quantum numbers. We denote this situation as the *Paschen–Bach effect*. In the opposite case, $\hat{\mathbf{l}}$ and $\hat{\mathbf{s}}$ are coupled, and the good quantum numbers are represented by j and m_j, giving rise to the *Zeeman effect*.

It should be noted that this distinction, given the strength B_{ext} of the magnetic field, depends on the excitation level of the atom: increasing n, the spin–orbit interaction decreases, and the highly excited levels (large n) feel the Paschen–Bach effect, whereas levels with small n feel the Zeeman effect.

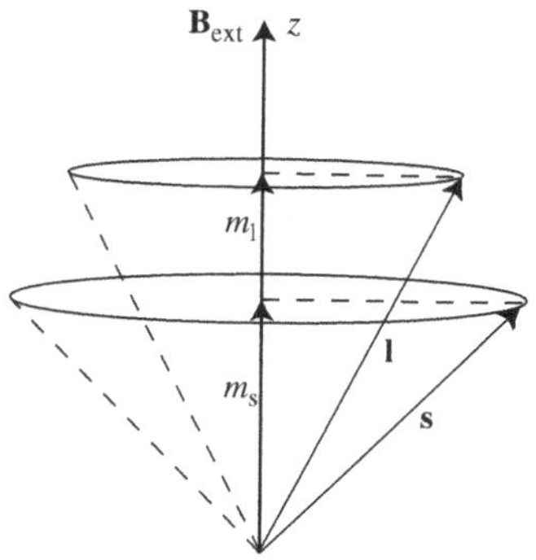

Figure 11.7 Landé semiclassical vectorial model for the Paschen–Bach effect. In order to compute the contribution of the spin–orbit correction, the spin and orbital angular momentum vectors can be thought of as precessing about the external magnetic field directed along the z-axis. As a consequence, only the time averaged components of $\hat{\mathbf{l}}$ and $\hat{\mathbf{s}}$ (directed along the z-axis) contribute to the scalar product in the third term of Eq. (11.71).

11.3.2 Paschen–Bach effect

Since the third term in Eq. (11.68) is negligible with respect to the second, we are allowed to separate the Hamiltonian according to the Dirac picture

$$\hat{H} = \hat{H}_A - \frac{e\hbar}{2m_e} B_{\text{ext}} \left(\hat{l}_z + 2\hat{s}_z\right) - f(r)\hat{\mathbf{l}} \cdot \hat{\mathbf{s}} = \hat{H}_0 + \hat{H}_I, \tag{11.71}$$

where

$$\hat{H}_0 = \hat{H}_A - \frac{e\hbar}{2m_e} B_{\text{ext}} \left(\hat{l}_z + 2\hat{s}_z\right), \tag{11.72}$$

and

$$\hat{H}_I = -f(r)\hat{\mathbf{l}} \cdot \hat{\mathbf{s}}. \tag{11.73}$$

We have assumed that the external magnetic field is along the z direction, i.e.

$$\mathbf{B}_{\text{ext}} = B_{\text{ext}}\boldsymbol{k}. \tag{11.74}$$

The eigenfunctions of $\hat{H}_A$ are also eigenfunctions of $\hat{l}_z$ and s_z. As a consequence, the solution of the eigenvalue equation for $\hat{H}_0$ is straightforward, i.e.

$$\begin{aligned} \hat{H}_0 \varphi_{nlm_l m_s} &= \left[\hat{H}_A - \frac{e\hbar}{2m_e} B_{\text{ext}} \left(\hat{l}_z + 2\hat{s}_z\right)\right] \varphi_{nlm_l m_s} \\ &= \left[E_{nl}^{(0)} - \mu_B B_{\text{ext}} \left(m_l + 2m_s\right)\right] \varphi_{nlm_l m_s}, \end{aligned} \tag{11.75}$$

where $E_{nl}^{(0)}$ is the unperturbed atomic energy level. We have now to add the spin–orbit perturbation. To this aim, we compute the first-order perturbational correction – given by $\hat{H}_I$ – to the energy level, i.e. we have to calculate the mean value of $-f(r)\hat{\mathbf{l}} \cdot \hat{\mathbf{s}}$ on the unperturbed states of the atom. In order to perform this calculation, we take advantage of the so-called (semiclassical) Landé's *vectorial model* (see Figs. 6.12 and 11.7): the spin

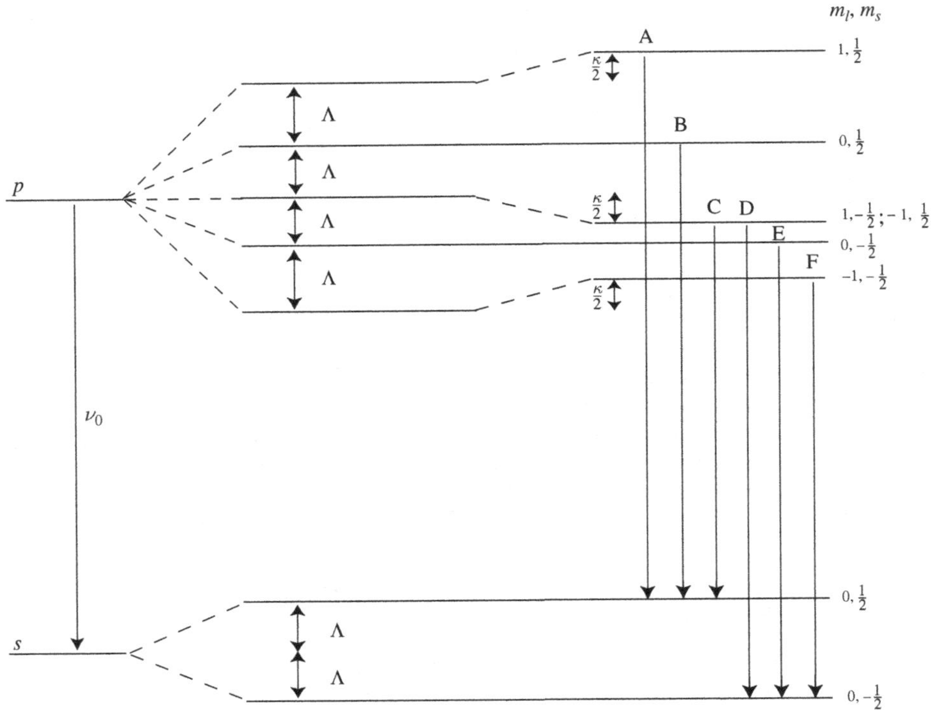

Figure 11.8 Diagram of the energy corrections to the s and p levels due to the Paschen–Bach effect. In the case of the s level, $l = 0$ and therefore there is no spin–orbit correction. However, the correction due to the external magnetic field gives rise to the doublet $(0, 1/2)$, $(0, -1/2)$, where the first number is m_l and the second one is m_s. The two components of the doublet are here separated by 2Λ. As for the p level ($m_l = -1, 0, 1$, $m_s = -1/2, 1/2$), there are five possibilities given by the five possible results of $m_l + 2m_s$. Apart from the $(0, 1/2)$ and $(0, -1/2)$ sublevels, for which the spin–orbit correction $\kappa\hat{\boldsymbol{l}} \cdot \hat{\boldsymbol{s}}$ is obviously zero, the sublevels are further corrected by the spin–orbit perturbation. The levels $(1, -1/2)$ and $(-1, 1/2)$ are degenerate, since both the magnetic and the spin–orbit corrections contribute exactly to the same extent.

and orbital angular momentum vectors can be thought of as precessing about the external magnetic field vector. Then, only $\hat{l}_z$ and $\hat{s}_z$ contribute to the scalar product, so that the final result for the correction is simply given by $\kappa m_l m_s$. To summarize, to first-order in the perturbation expansion, the energy eigenvalues in the case of the Paschen–Bach effect are given by

$$E_{nl}^{(0)} + E_{nlm_lm_s}^{(1)} = E_{nl}^{(0)} + \left(m_l + 2m_s\right)\Lambda + \kappa m_l m_s, \tag{11.76}$$

where $\Lambda = \mu_B B_{\text{ext}}$ is sometimes called the Lorentz parameter. In Fig. 11.8 we sketch a diagram of the s and p energy levels of a hydrogen atom with emphasis on the magnetic and spin–orbit corrections to the unperturbed energies (see Prob. 11.17).

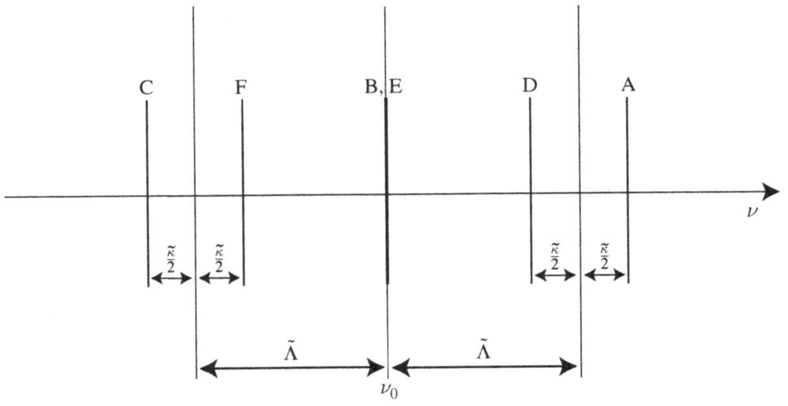

Figure 11.9 **Schematic diagram of the spectroscopical lines resulting from the allowed transitions A, B, C, D, E, F of Fig. 11.8. Here ν_0 is the central frequency, i.e. the frequency of the unperturbed $s \leftrightarrow p$ transition, $\tilde{\Lambda} = \Lambda/h$, and $\tilde{\kappa} = \kappa/h$.**

It is also evident from Fig. 11.8 that the $s \leftrightarrow p$ transition is profoundly affected by the magnetic and spin–orbit perturbations, even though not all the – virtually – possible transitions are actually allowed. In the electric-dipole approximation,[12] one may establish some rules that limit the permitted transitions. These are called *selection rules*, and in the present case are

$$\Delta l = \pm 1, \quad \Delta m_l = 0, \pm 1, \quad \Delta m_s = 0. \tag{11.77}$$

The first selection rule implies that transitions within the levels s and p are not allowed, whereas the second states that a photon emission or absorption may change the magnetic quantum number by at most 1. The third rule is justified by the fact that a photon exchange cannot invert the spin. In Fig. 11.9 we show the spectroscopical lines that stem from the allowed transitions depicted by vertical arrows in Fig. 11.8 (see Prob. 11.18).

11.3.3 Zeeman effect

Let us now go back to Eq. (11.68) and consider the case in which the magnetic field is weak, i.e. the second (magnetic field) term is much smaller than the third (spin–orbit) one.[13] In this case, the "good" quantum numbers are represented by $\{n, l, j, m_j\}$, and the total Hamiltonian, according to the Dirac Picture, may be separated into

$$\hat{H} = \hat{H}_0 + \hat{H}_I, \tag{11.78}$$

[12] For the classical treatment see [*Jackson* 1962, Chs. 4–5]. See also Sec. 13.7 for the general quantum treatment of the interaction between atom and electromagnetic field.

[13] For the history of the anomalous Zeeman effect see [*Mehra/Rechenberg* 1982–2001, I, 445–85].

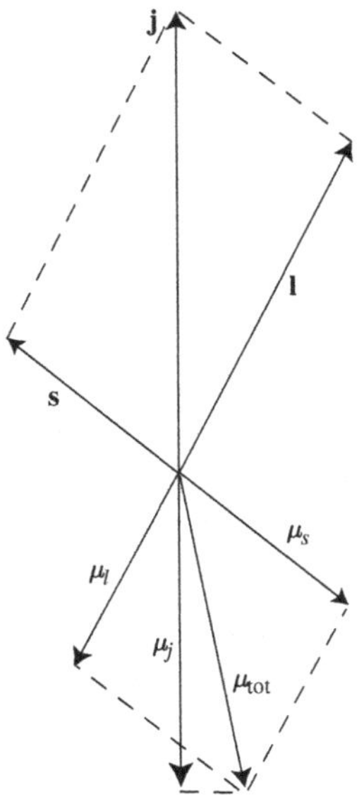

Figure 11.10 Semiclassical Landé vectorial model for the Zeeman effect. The orbital magnetic momentum μ_l (see Eq. (6.166)) is opposite to **l**, whereas the spin magnetic momentum μ_s (see Eq. (6.169)) is opposite to **s**. The total magnetic momentum $\mu_{tot} = \mu_l + \mu_s$, due to the different gyromagnetic ratio (see p. 224) in the two cases, is not parallel to **j**. The relevant quantity for our purposes is then represented by μ_j, i.e. the projection of μ_{tot} along the direction of **j**.

where

$$\hat{H}_0 = \hat{H}_A - f(r)\hat{\mathbf{l}} \cdot \hat{\mathbf{s}}, \tag{11.79}$$

and

$$\hat{H}_I = -\frac{e\hbar}{2m_e} B_{ext} \left(\hat{l}_z + 2\hat{s}_z\right). \tag{11.80}$$

In order to find the energy shift due to the magnetic perturbation, we have to evaluate the mean value

$$E^{(1)}_{nljm_j} = \left\langle \varphi_{nljm_j} \middle| \hat{H}_I \middle| \varphi_{nljm_j} \right\rangle. \tag{11.81}$$

To accomplish this task, we again resort to the Landé vectorial model (see Fig. 11.10). It turns out that the total magnetic momentum $\boldsymbol{\mu}_{tot}$, due to the different gyromagnetic ratio (see p. 224) of $\boldsymbol{\mu}_l$ and $\boldsymbol{\mu}_s$, is not parallel to $\mathbf{j}$.

Since the spin–orbit interaction is much larger than the magnetic one, the time scale of the "magnetic" dynamics is much larger than the corresponding spin–orbit interaction time scale. As a consequence, due to time averaging, the contribution to Eq. (11.81) of the component of $\boldsymbol{\mu}_{tot}$ orthogonal to $\mathbf{j}$ vanishes. The relevant quantity for our purposes is then represented by $\boldsymbol{\mu}_j$, i.e. the projection of $\boldsymbol{\mu}_{tot}$ along the direction of $\mathbf{j}$. In order to calculate $\boldsymbol{\mu}_j$, we start from

$$\mu_j = \mu_l \cos\theta + \mu_s \cos\phi = \frac{e\hbar}{2m_e}\left(l\cos\theta + 2s\cos\phi\right), \tag{11.82}$$

where θ is the angle between **l** and **j**, and ϕ is the angle between **s** and **j**. Making use of the Carnot theorem in the form

$$s^2 = l^2 + j^2 - 2lj\cos\theta, \tag{11.83}$$

and

$$l^2 = s^2 + j^2 - 2sj\cos\phi, \tag{11.84}$$

we obtain

$$l\cos\theta = \frac{j^2 + l^2 - s^2}{2j} \quad \text{and} \quad s\cos\phi = \frac{j^2 + s^2 - l^2}{2j}. \tag{11.85}$$

Substituting Eqs. (11.85) into Eq. (11.82), we get

$$\mu_j = \frac{e\hbar}{2m_e}\left(\frac{j^2 + l^2 - s^2}{2j} + 2\frac{j^2 + s^2 - l^2}{2j}\right) = \frac{e\hbar}{2m_e} g_L j, \tag{11.86}$$

where

$$g_L = 1 + \frac{j^2 + s^2 - l^2}{2j^2} = 1 + \frac{j(j+1) + s(s+1) - l(l+1)}{2j(j+1)} \tag{11.87}$$

is the *Landé g factor*. Such a factor should be equal to 2 from a relativistic viewpoint (and equal to 1 from a classical viewpoint). As a matter of fact, it turns out to be different from 2 (as well as from 1). This is the reason why this effect is sometimes called *anomalous* Zeeman effect.

We are now in the position to evaluate the energy shift (11.81), and write

$$E^{(1)}_{nljm_j} = \mu_B B_{\text{ext}} g_L m_j = \Lambda g_L m_j, \tag{11.88}$$

where $\Lambda = \mu_B B_{\text{ext}}$ is the already introduced Lorentz parameter. In Fig. 11.11 we draw the modifications due to the spin–orbit interaction and the Zeeman effect to the s and p levels (see Prob. 11.19), whereas in Fig. 11.12 we schematically depict the spectroscopical lines resulting from the transitions illustrated in the previous figure (see Prob. 11.20). In the Zeeman effect we have further selection rules some:

- If the spectroscopical light is observed orthogonal to $\mathbf{B}_{\text{ext}}$:

$$\Delta m_j = \pm 1 \quad \text{(for light polarized orthogonally to } \mathbf{B}_{\text{ext}}\text{)}, \tag{11.89a}$$
$$\Delta m_j = 0 \quad \text{(for light polarized parallel to } \mathbf{B}_{\text{ext}}\text{)}. \tag{11.89b}$$

- If the spectroscopical light is observed parallel to $\mathbf{B}_{\text{ext}}$:

$$\Delta m_j = \pm 1 \quad \text{(for light circularly polarized)}, \tag{11.90a}$$
$$\Delta m_j = 0 \quad \text{(is not allowed)}. \tag{11.90b}$$

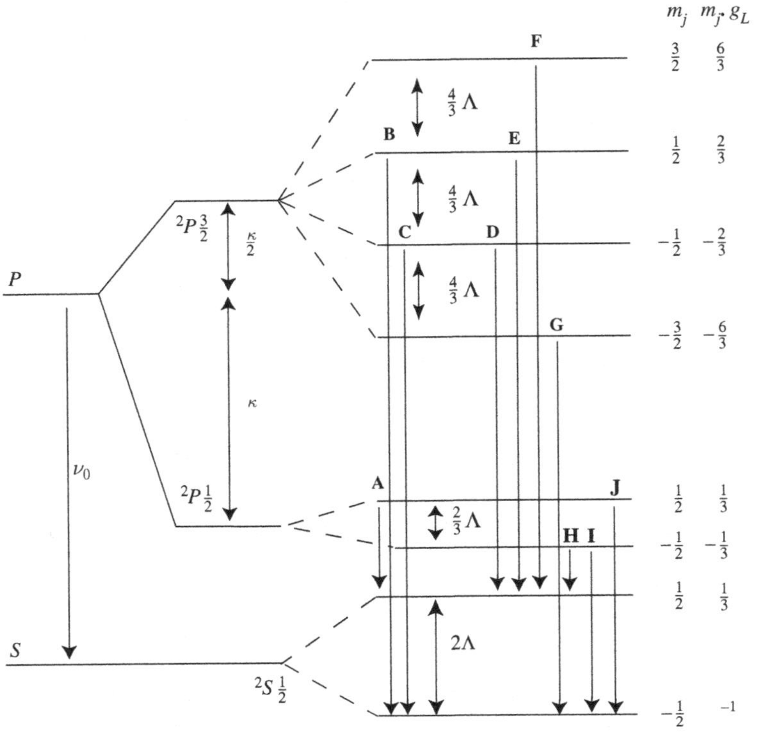

Figure 11.11 Modifications of the energy levels s and p due to the Zeeman effect. Proceeding from the left, the unperturbed levels are first affected by the spin–orbit correction, which leaves the s level unchanged. Then, each of the three resulting levels are further split due to the magnetic interaction with the external field B_{ext}. This splitting is ruled by the strength of the magnetic field (represented by Λ), the Landé g factor (which depends on l, s, j), and the eigenvalue m_j of $\hat{j}_z$. The ten possible transitions among the levels are denoted by the capital letters A, B, C, . . ., J.

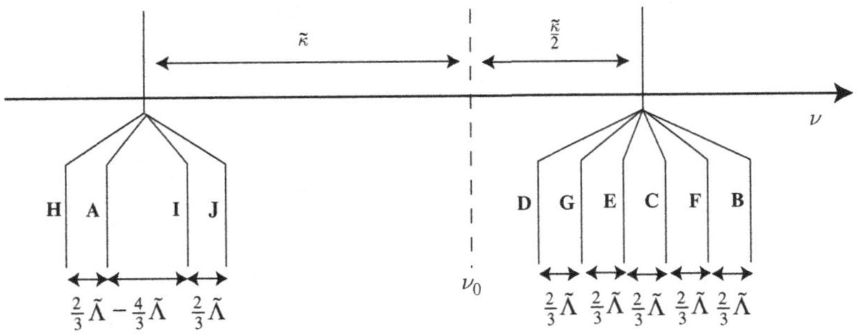

Figure 11.12 Schematic diagram of the spectroscopical lines resulting from the allowed transitions A, B, C, . . ., J of Fig. 11.11. Here ν_0 is the frequency of the unperturbed $s \leftrightarrow p$ transition, $\tilde{\Lambda} = \Lambda/h$, and $\tilde{\kappa} = \kappa/h$.

It should be noted that, as the strength of the external magnetic field increases, the Lorentz parameter Λ grows linearly up to the point where two levels (the levels ${}^2P_{3/2,-3/2}$ and ${}^2P_{1/2,1/2}$ in Fig. 11.11) overlap: this intermediate situation cannot be dealt with in perturbation theory. However, for magnetic field strengths much larger than this critical value, we fall back into the framework determined by the Paschen–Bach effect.

Finally, it is interesting to emphasize that the magnetic field completely removes the degeneracy with respect to the angular momentum direction, which is opposite to what happens in the electric case – the Stark effect (see Subsec. 10.1.4).[14]

11.4 Relativistic corrections

In Sec. 11.2 we have neglected relativistic effects. For a full treatment of relativistic effects one should use the relativistic extension of the Schrödinger equation, i.e. the Dirac equation. However, this goes beyond the aim of this book. First-approximation relativistic corrections may be obtained by using simple heuristic arguments as shown in the present section.

Beside the spin–orbit interaction, which should also be considered to be of relativistic nature, relativistic corrections to the hydrogen atom derive from two further effects:

- Relativistically, the kinetic energy presents the form

$$\hat{T} \simeq m_e c^2 \sqrt{1 + \frac{\hat{\mathbf{p}}^2}{m_e^2 c^2}} - m_e c^2. \tag{11.91}$$

  Since

$$\sqrt{1+x} \simeq 1 + \frac{1}{2}x - \frac{1}{8}x^2 + O(x^3), \tag{11.92}$$

  then, for small momenta, we obtain the expansion

$$\hat{T} = \frac{\hat{\mathbf{p}}^2}{2m_e} - \frac{1}{8}\frac{\hat{\mathbf{p}}^4}{m_e^3 c^2} = \hat{T}_{nr} + \Delta\hat{T}, \tag{11.93}$$

  where $\hat{T}_{nr}$ is the non-relativistic expression for the kinetic energy.
- Suppose that we wish to measure the electron position with great precision. Then, we must make use of high energy and this produces electron–positron couples. For position uncertainties,

$$\Delta x \simeq \frac{\hbar}{m_e c}, \tag{11.94}$$

[14] This is due to the fact that the magnetic field is an axial vector, i.e. it changes sign upon reflection on a plane that contains its direction, while the electric field is a polar vector (see Subsec. 6.1.1).

we must take into account the creation of virtual positron–electron pairs, since such an uncertainty is of the order of the Compton wavelength (1.68), where relativistic corrections become large. A detailed computation is needed to take care of this effect. However, in the non-relativistic limit we can treat this effect as a perturbation and consider, instead of the potential energy $V(\mathbf{r})$, a perturbed potential energy $V(\mathbf{r}+\delta\mathbf{r})$, where $\delta\mathbf{r}$ is the shift due to the uncertainty (11.94). The uncertainty in the electron's position is called *Zitterbewegung*, whereby the electron does not move smoothly but instead undergoes extremely rapid small-scale fluctuations, causing the electron to see a smeared-out Coulomb potential of the nucleus. The correction will then be given by

$$\begin{aligned} V(\mathbf{r}+\delta\mathbf{r}) - V(\mathbf{r}) = & \sum_{j=\{x,y,z\}} \partial_j V(\mathbf{r})\delta r_j + \frac{1}{2} \sum_{j,k=\{x,y,z\}} \partial_j \partial_k V(\mathbf{r})\delta r_j \delta r_k \\ & + O(\delta r^3) \\ \simeq & \, C \Delta V \delta r^2, \end{aligned} \tag{11.95}$$

where C is a numeric constant and the first-order terms as well as the second-order cross terms vanish in the averaging over $\delta\mathbf{r}$. From this we derive

$$\hat{H}_{\text{Darwin}} \simeq \frac{\hbar^2}{8m_e^2c^2}\Delta V = \frac{e\hbar^2}{8m_e^2c^2}\nabla\cdot\mathbf{E} = \frac{\pi\hbar^3\alpha Z}{2m_e^2c}\delta^3(\mathbf{r}), \tag{11.96}$$

where we have made use of the Gauss law

$$\nabla\cdot\mathbf{E} = 4\pi Q(\mathbf{r}), \tag{11.97}$$

where $Q(\mathbf{r})$ is the cubic charge density. For a point-like nucleus of charge Ze, we have

$$Q(\mathbf{r}) = Ze\delta^3(\mathbf{r}). \tag{11.98}$$

The Hamiltonian (11.96) is called the *Darwin term*, from the English physicist Charles Galton Darwin (1887–1962).

Since these corrections to the kinetic and potential energies are small, we can use first-order perturbation theory. In what follows we consider the joint action of both corrections, and try to estimate their relevance. First, we wish to calculate the mean value of $\hat{\mathbf{p}}^4/8m^3c^2$. Then, we have that

$$\hat{\mathbf{p}}^2 \left|\varphi_0\right\rangle = 2m_e\left(E_n^{(0)} + \frac{Ze^2}{r}\right)\left|\varphi_0\right\rangle, \tag{11.99}$$

where $\left|\varphi_0\right\rangle$ is the unperturbed eigenket of the Hamiltonian, so that

$$\begin{aligned} \left\langle\hat{\mathbf{p}}^4\right\rangle_{\varphi_0} &= \left\langle\varphi_0\left|\hat{\mathbf{p}}^2\hat{\mathbf{p}}^2\right|\varphi_0\right\rangle = 4m_e^2\left\langle\left(E_n^{(0)} + \frac{Ze^2}{r}\right)^2\right\rangle_{\varphi_0} \\ &= 4m_e^2\left[(E_n^{(0)})^2 + 2E_n^{(0)}\left\langle V\right\rangle_{\varphi_0} + \left\langle V^2\right\rangle_{\varphi_0}\right] \end{aligned}$$

$$= 4m_e^2\left[(E_n^{(0)})^2 + \frac{2E_n^{(0)}Ze^2}{r_0 n^2} + \frac{Z^2e^4}{r_0^2 n^3\left(l+\frac{1}{2}\right)}\right]. \tag{11.100}$$

From this calculation one obtains

$$-\frac{1}{8}\frac{\langle\hat{\mathbf{p}}^4\rangle}{m_e^3c^2} = -\frac{(E_n^{(0)})^2}{2m_ec^2}\left(\frac{4n}{l+\frac{1}{2}} - 3\right). \tag{11.101}$$

As for the second correction, i.e. the Darwin term (11.96), for hydrogenoid atoms we have

$$\left\langle \hat{H}_{\text{Darwin}}\right\rangle = \frac{\pi Z\alpha\hbar^3}{2m_e^2c}|\varphi_{nlm}(0)|^2. \tag{11.102}$$

The condition $\varphi_{nlm}(0) \neq 0$ is equivalent to $l = 0$ (see Eqs. (11.55a) and (11.51)) (and therefore also $m = 0$). As a consequence, the Darwin correction will never be present together with the spin–orbit contribution. Since

$$|\varphi_{n00}(0)|^2 = \frac{1}{\pi}\left(\frac{Ze^2m_e}{\hbar n}\right)^3, \tag{11.103}$$

then we have

$$\left\langle \hat{H}_{\text{Darwin}}\right\rangle = \frac{1}{2}m_ec^2\left(\frac{Z\alpha}{n}\right)^4 n\delta_{l0}. \tag{11.104}$$

Now we must include the spin–orbit contribution (see Eqs. (11.65) and (11.66)) and combine it with the other two relativistic correction terms. To this purpose we must distinguish three cases:

- When $l = 0$, we only have the kinetic and Darwin corrections, and, making use of the explicit expression (11.49) for the energy eigenvalues, which for hydrogenoid atoms are

$$E_n = -\frac{Z^2e^4m_e}{2n^2\hbar^2}, \tag{11.105}$$

  we have

$$m_ec^2\left(\frac{Z\alpha}{n}\right)^4\left(\frac{3}{8} - n + \frac{n}{2}\right) = -m_ec^2\left(\frac{Z\alpha}{n}\right)^4\frac{n}{2}\left(\frac{1}{j+\frac{1}{2}} - \frac{3}{4n}\right), \tag{11.106}$$

  because for $l = 0$, $j = 1/2$.
- When we consider the case in which $l = j - 1/2$, i.e. when the Darwin term is zero and the spin–orbit correction contributes, we obtain again

$$-m_ec^2\left(\frac{Z\alpha}{n}\right)^4\frac{n}{2}\left(\frac{1}{j+\frac{1}{2}} - \frac{3}{4n}\right) = -7,2\times 10^4\text{eV}\frac{Z^4}{n^3}\left(\frac{1}{j+\frac{1}{2}} - \frac{3}{4n}\right), \tag{11.107}$$

  where we have also give the explicit expression of the sum of the contributes.
- The same result holds also for $l = j + 1/2$.

This means that we always obtain the same correction either by adding the Darwin and the kinetic corrections or by summing the spin–orbit (in both cases $l = j \pm 1/2$) and the kinetic terms. For the ground state of the hydrogen ($Z = 1, n = 1$), the correction is $\simeq 10^{-3}$ eV and though very small it is easily detectable with spectroscopy.

There are also other relativistic aspects in the hydrogen atom that – as we have said – are beyond the scope of this book; we recall that some of them are due to quantum effects such as vacuum fluctuations: the most famous being the Lamb shift.[15]

11.5 Helium atom

The atomic Schrödinger equation can be solved only in its simplest case: the hydrogenoid atoms. In all other cases, one has to resort to approximation methods. In this section we consider the helium atom, with two electrons and $Z = 2$. For larger Z we have ionized atoms.[16] The Hamiltonian of the helium atom may be written as

$$\hat{H} = \frac{\hat{\mathbf{p}}_1^2}{2m_e} - \frac{Ze^2}{r_1} + \frac{\hat{\mathbf{p}}_2^2}{2m_e} - \frac{Ze^2}{r_2} + \frac{e^2}{r_{12}}, \tag{11.108}$$

where r_1, r_2 are the distances from the nucleus of the two electrons, and r_{12} is the distance between the electrons. We wish to study the first levels and in particular the ground state. In order to calculate the energy of the ground state, we have two methods: a perturbational approach and a variational procedure, which will be the subject of the next two subsections.

11.5.1 Perturbation method

Let us first rewrite Eq. (11.108) as

$$\hat{H} = \hat{H}_0 + \hat{H}_I, \tag{11.109}$$

where $\hat{H}_0 = \hat{H}_1 + \hat{H}_2$, and

$$\hat{H}_1 = \frac{\hat{\mathbf{p}}_1^2}{2m_e} - \frac{Ze^2}{r_1} \quad \text{and} \quad \hat{H}_2 = \frac{\hat{\mathbf{p}}_2^2}{2m_e} - \frac{Ze^2}{r_2}, \tag{11.110}$$

while $\hat{H}_I = e^2/r_{12}$. The first method consists considering the term $\hat{H}_I$ as a perturbation. Clearly, this is not a good approximation if Z is small, since, in this case, $\hat{H}_I$ is not much smaller than the two other Coulombian terms.

We consider first the case where $\hat{H}_I$ is neglected (zeroth order). In this case, the Hamiltonian is fully separable and the ground state is described by $\varphi_{100}(\mathbf{r}_1)\varphi_{100}(\mathbf{r}_2)$.

[15] For a relativistic treatment of these problems see [*Bjorken/Drell* 1964, 52–60].

[16] Examples of helioid atoms are $Li^+, Be^{++}, \ldots$.

As a consequence, the ground state energy will be the sum of the eigenvalues: $E_1^{(0)} = E_{100}^1 + E_{100}^2$, with (see Eq. (11.105) for $n = 1$)

$$E_{100}^1 = E_{100}^2 = -\frac{m_e c^2}{2}(Z\alpha)^2 . \tag{11.111}$$

For the sake of simplicity, we move to *atomic units*,[17] i.e. (see also Eq. (11.60))

$$r_0 = \frac{\hbar^2}{m_e e^2} = 1, E_0 = m_e c^2 \alpha^2 = 1, m_e = 1. \tag{11.112}$$

Therefore,

$$E_1^{(0)} = -\left(\frac{Z^2}{2} + \frac{Z^2}{2}\right) = -Z^2, \tag{11.113}$$

which for the helium atom is -4. Then, in this approximation, the binding energy in the ground state of the helium atom is eight times the corresponding one in the hydrogen atom, i.e.

$$E_1^{(0)} = 8 \cdot 13.6\,\text{eV} \simeq -109\,\text{eV}. \tag{11.114}$$

The first excited level is degenerate because we can get it by combining $\varphi_{100}\varphi_{200}$ and $\varphi_{100}\varphi_{21m}$. As a consequence, its energy in the zeroth approximation is given by

$$E_2^{(0)} = E_{100}^1 + E_{21m}^2 = -\left(\frac{Z^2}{2} + \frac{Z^2}{8}\right) = -\frac{5}{8}Z^2 = -5 \cdot 13.6 \simeq -68\,\text{eV}. \tag{11.115}$$

Within the same approximation it is also possible to compute the limiting energy E_l, above which the spectrum becomes a mix of a continuous and a discrete part. To this aim, it is sufficient to take the limit $n \to \infty$ for one of the two electrons (say, electron 2), obtaining

$$E_l^{(0)} = E_1^1 + E_\infty^2 = -\frac{Z^2}{2} = -2 \simeq -54.3\,\text{eV}. \tag{11.116}$$

In order to get a more precise estimation of the ground-state energy, we move to the first-order perturbation analysis, and insert the repulsive Coulombian correction, i.e.

$$\begin{aligned} E_1^{(1)} &= {}_1\langle\varphi_{100}|\,{}_2\langle\varphi_{100}|\frac{1}{r_{12}}|\varphi_{100}\rangle_1\,|\varphi_{100}\rangle_2 \\ &= \int d\mathbf{r}_1 d\mathbf{r}_2\, |\varphi_{100}(\mathbf{r}_1)|^2 \frac{1}{r_{12}} |\varphi_{100}(\mathbf{r}_2)|^2 , \end{aligned} \tag{11.117}$$

with (see Eq. (11.55a))

$$\varphi_{100}(\mathbf{r}) = \frac{1}{\sqrt{\pi}} Z^{\frac{3}{2}} e^{-Zr}. \tag{11.118}$$

After a direct computation (see Prob. 11.22), we finally obtain

$$E_1^{(1)} = \frac{5}{8}Z. \tag{11.119}$$

[17] This corresponds to the change of variables (11.21). On a practical side, the passage to atomic units may be formally effected by choosing $e = 1, m_e = 1, \hbar = 1$. Replacing e^2 by Ze^2, the atomic units become the so-called Coulombian units (see Prob. 11.21).

We find that, for $Z = 2$, $E_1^{(1)} \simeq -34\,\text{eV}$; this first-order perturbational result translates into a variation of the ground state from level $-109\,\text{eV}$ to level $-75\,\text{eV}$.

11.5.2 Variational method

A second approach involves the variational method (see Sec. 10.4). In this context, one must compute the expected value of the Hamiltonian on a "smart" wave function with a special choice of parameters. At the zeroth order we have

$$\varphi_{\text{He}}(\mathbf{r}_1, \mathbf{r}_2) = \varphi_{100}(\mathbf{r}_1)\varphi_{100}(\mathbf{r}_2) = \frac{Z^3}{\pi} e^{-Z(\mathbf{r}_1+\mathbf{r}_2)}. \tag{11.120}$$

Since one of the electrons is shielded from the proton by the other electron, we substitute Z with $\tilde{Z}$ and seek to find the $\tilde{Z}$ which minimizes $\langle \hat{H} \rangle$ (see Eq. (11.108)). Given that, for the ground state,

$$\varphi_{\text{He}}(\mathbf{r}_1, \mathbf{r}_2) = \frac{\tilde{Z}^3}{\pi} e^{-\tilde{Z}(\mathbf{r}_1+\mathbf{r}_2)}, \tag{11.121}$$

we have (see Eqs. (11.110) and (11.116))

$$\hat{H}_1(\tilde{Z}) \left| \varphi_{\text{He}} \right\rangle = \left(\frac{\hat{\mathbf{p}}_1^2}{2} - \frac{\tilde{Z}}{r_1} \right) \left| \varphi_{\text{He}} \right\rangle = \frac{\tilde{Z}^2}{2} \left| \varphi_{\text{He}} \right\rangle, \tag{11.122}$$

from which we obtain

$$\begin{aligned} \left\langle \hat{H}_1(\tilde{Z}) \right\rangle_{\varphi_{\text{He}}} &= \int d\mathbf{r}_1 \int d\mathbf{r}_2 \varphi_{\text{He}}(\mathbf{r}_1, \mathbf{r}_2) \left(\frac{\hat{\mathbf{p}}_1^2}{2} - \frac{\tilde{Z}}{r_1} \right) \varphi_{\text{He}}(\mathbf{r}_1, \mathbf{r}_2) \\ &= - \int d\mathbf{r}_1 \int d\mathbf{r}_2 \varphi_{\text{He}}(\mathbf{r}_1, \mathbf{r}_2) \frac{\tilde{Z}^2}{2} \varphi_{\text{He}}(\mathbf{r}_1, \mathbf{r}_2). \end{aligned} \tag{11.123}$$

It follows that

$$\left\langle \frac{\hat{\mathbf{p}}_1^2}{2} \right\rangle_{\varphi_{\text{He}}} = -\frac{\tilde{Z}^2}{2} + \tilde{Z} \left\langle \frac{1}{r_1} \right\rangle. \tag{11.124}$$

For the hydrogen atom we have found (see Prob. 11.13)

$$\left\langle \frac{1}{r} \right\rangle = \frac{1}{r_0} \frac{1}{n^2}, \tag{11.125}$$

with $r_0 = \hbar^2 / m_e Z e^2$. In atomic units we derive

$$\left\langle \frac{1}{r} \right\rangle = \frac{Z}{n^2}, \tag{11.126}$$

and, coming back to the helium atom, for $n = 1$, we have

$$\left\langle \frac{\hat{\mathbf{p}}_1^2}{2} \right\rangle_{\varphi_{\text{He}}} = \frac{\tilde{Z}^2}{2}. \tag{11.127}$$

Table 11.1 The table of the ground-state energy of the first few helioid atoms calculated according to the perturbation and variational methods and compared with their experimental determination

Element	Z	$(E_1^{(0)})^{\text{pert}}$	$(E_1^{(0)} + E_1^{(1)})^{\text{pert}}$	E_1^{var}	E_1^{exp}
He	2	−109	−75	−77.7	−78.6
Li^+	3	−245	−194	−196.7	−197.1
Be^{++}	4	−435	−367	−369.7	−370

Therefore,

$$\left\langle \frac{\hat{\mathbf{p}}_1^2}{2} + \frac{\hat{\mathbf{p}}_2^2}{2} \right\rangle_{\varphi_{\text{He}}} = \tilde{Z}^2. \tag{11.128}$$

On the other hand, for the potential energy of the interaction-free Hamiltonian $\hat{H}_0$ (see Eqs. (11.110)), we obtain

$$\left\langle -\frac{Z}{r_1} - \frac{Z}{r_2} \right\rangle_{\varphi_{\text{He}}} = -2Z\tilde{Z}, \tag{11.129}$$

where we have again made use of Eq. (11.126) with $\tilde{Z}$ in place of Z. Finally, since (see Eq. (11.117))

$$\left\langle -\frac{1}{r_{12}} \right\rangle_{\varphi_{\text{He}}} = \frac{5}{8}\tilde{Z}, \tag{11.130}$$

we also have

$$\left\langle \hat{H} \right\rangle_{\varphi_{\text{He}}} = \tilde{Z}^2 - 2Z\tilde{Z} + \frac{5}{8}\tilde{Z}. \tag{11.131}$$

Minimizing relatively to $\tilde{Z}$, i.e. imposing

$$\frac{\partial \left\langle \hat{H} \right\rangle_{\varphi_{\text{He}}}}{\partial \tilde{Z}} = 2\tilde{Z} - 2\left(Z - \frac{5}{16}\right) = 0, \tag{11.132}$$

we finally obtain

$$\tilde{Z} = Z - \frac{5}{16}, \tag{11.133}$$

and therefore

$$\left\langle \hat{H} \right\rangle_{\varphi_{\text{He}}} = -\left(Z - \frac{5}{16}\right)^2, \tag{11.134}$$

which for the ground-state energy yields −77.5 eV. We can summarize the results of this section in Table 11.1, which shows how the accuracy of the estimation of the ground-state energy of the first few helioid atoms increases as we move from zeroth-order perturbation theory to first-order perturbation theory to end up with the variational method.

11.5.3 The spin component

It should be noted that the Hamiltonian (11.108) does not depend on the spin. However, the two electrons are identical particles (fermions) and the exchange interaction (see Sec. 7.4) comes into play. In other words, one has to multiply all the wave functions considered in this section by the spin component. For the ground state of the helium atom we shall have

$$\Psi_g(\mathbf{r}_1, \mathbf{r}_2; s) = \frac{\tilde{Z}^3}{\pi} e^{-\tilde{Z}(r_1+r_2)} \varsigma(s), \tag{11.135}$$

and, since the spatial wave function is symmetric under the exchange of the two electrons, the spin wave function $\varsigma(s)$ must be antisymmetric, in order for the total wave function Ψ_g of the two fermions to be antisymmetric. As a consequence, $\varsigma(s)$ must be the wave function of a singlet state (see Subsec. 6.4.3) and the ground state of helium has always spin zero.

As for the wave function of the first excited level, in place of the simple product

$$\varphi_{100}(\mathbf{r}_1)\, \varphi_{200}(\mathbf{r}_2), \tag{11.136}$$

we shall have

$$\frac{1}{\sqrt{2}} \left[\varphi_{100}(\mathbf{r}_1)\, \varphi_{200}(\mathbf{r}_2) \pm \varphi_{100}(\mathbf{r}_2)\, \varphi_{200}(\mathbf{r}_1)\right] \varsigma(s), \tag{11.137}$$

and, instead of the product

$$\varphi_{100}(\mathbf{r}_1)\, \varphi_{21m}(\mathbf{r}_2), \tag{11.138}$$

we must write

$$\frac{1}{2\sqrt{2}} \left[\varphi_{100}(\mathbf{r}_1)\, \varphi_{21m}(\mathbf{r}_2) \pm \varphi_{100}(\mathbf{r}_2)\, \varphi_{21m}(\mathbf{r}_1)\right] \varsigma(s), \tag{11.139}$$

where this time $\varsigma(s)$ can be either symmetric (triplet state together with the minus sign in the spatial part of the wave function) or antisymmetric (singlet state with the plus sign for the spatial part of the wave function). The spin-zero helium atom is called *para*-helium, whereas the spin-one is called *ortho*-helium.[18]

Going to the estimation of the energy of the levels, one has to evaluate expectation values of the type

$$\left\langle \Psi \left| \frac{1}{r_{12}} \right| \Psi \right\rangle, \tag{11.140}$$

which, due to the forms (11.137) and (11.139), give rise to several spin-dependent terms: This is the effect of the exchange interaction that makes the energy eigenvalues depend on the spin even though the original Hamiltonian did not.

[18] This nomenclature is also used for all the two-fermion systems, such as, e.g., the electron-positron pair, called *positronium*.

11.6 Many-electron effects

The analytic solution of the Schrödinger equation is only possible in rather fortunate cases. In the one-dimensional case numerical techniques can be used with very high precision and it is relative easy to obtain results with 10 significant decimal digits. However, this is not so simple when we increase the number of dimensions. If we introduce a lattice of step a in a box of size l in D dimensions, the number $\mathcal{N}$ of points of the lattice is given by

$$\mathcal{N} = \left(\frac{1}{a}\right)^{D}. \tag{11.141}$$

This could still be managed in three dimensions also with non-spherically symmetric (but not too rough) potentials: 10^8 points can be used on a PC. Using our analytic command of elliptic differential equations, many efficient techniques may be found to extrapolate to the limit where $a \to 0$ and $\mathrm{l} \to \infty$. Unfortunately, for a system with N particles we have

$$\mathcal{N} = \left(\frac{1}{a}\right)^{ND}, \tag{11.142}$$

and, as soon as $N > 2$, the number of points becomes overwhelmingly too large.

A brute force discretization method therefore fails in studying the properties of many-particle states; people in the past have developed many smart tools in order to overcome these difficulties. While in the $N \to \infty$ limit, in certain cases, thermodynamic arguments may be used to simplify the computations, a very interesting and difficult problem is provided by atoms and moleculae, where N may be rather large (but also not too large).

On the top of this problem we have to deal with the statistics. Boson statistics is usually benign, but Fermi statistics (which is the relevant one for atoms) makes everything more difficult. There are many negative effects of Fermi statistics:

- The wave function may be not positive; integrals of non-positive functions are notoriously difficult to be computed numerically.
- For the same token, the path-integral representation of the Green functions (or of the partition function) of many fermions is not positive definite and this adds extra difficulties to the already difficult method of path integrals.
- The Pauli exclusion principle implies that, roughly speaking, different fermions occupy different regions of phase space: Therefore, when the number of fermions is high, we have to explore simultaneously a quite large region of phase space, e.g. both the region of large and small momenta.

11.6.1 Variational methods

A natural approach is provided by the variational method, which, as we have seen, gives very good results for the helium atom. In the most powerful version of the method one can use the following theorem: the ground state of a certain Hamiltionian has an energy eigenvalue that is greater than or equal to the eigenvalue of the ground state of the same Hamiltonian restricted to a subspace of the original space.

Therefore, we can consider a set of n vectors $|\psi_k\rangle$. If we are able to compute the following quantities

$$\vartheta_{j,k} = \langle\psi_j|\psi_k\rangle\,, \quad H_{j,k} = \langle\psi_j|\hat{H}|\psi_k\rangle\,, \tag{11.143}$$

the evaluation of the ground state of the Hamiltonian, restricted to the space spanned by the n vectors $|\psi_k\rangle$, involves only n-dimensional algebra and is relatively easy.

In this way, we can use as variational parameters the form of all the n vectors $|\psi_k\rangle$ and we have a quite large variational space. In some cases the quantities in Eq. (11.143) can be evaluated numerically and very good precision results are obtained.

However, the variational method needs a good starting point, and this stresses the importance of approximation methods. This problem will be studied in the rest of this sction for the fermionic case, neglecting for simplicity the complications due to spin.

11.6.2 The Hartree–Fock equation

The simplest way to write a wave function of N fermions is to use the Slater determinant (see Eq. (7.20)) in terms of N orthogonal wave functions. The Hartree–Fock method consists in combining the variational approach of the previous section with the Slater determinant. We shall see how this construction may be applied to the case of Z electrons in an atom of charge Z, neglecting the spins (or equivalently assuming that they have all the same spin).

The Hamiltonian of our problem is given by

$$\hat{H} = \sum_{k=1}^{Z}\left(\frac{\hat{\mathbf{p}}_k^2}{2m} - \frac{Ze^2}{r_k} + \frac{1}{2}\sum_{j=1}^{Z}{}' \frac{e^2}{r_{j,k}}\right), \tag{11.144}$$

where the primed sum indicates that the sum has to be taken for $i \neq k$, and Z is both the charge of the nucleus and the number of electrons: we are indeed considering neutral atoms.

Our aim is to compute the expectation of $\hat{H}$ in the state $|\Psi\rangle$ formed by the Slater determinant of Z vectors $|\psi_k\rangle$. We preliminary note that if $f(x)$ is a function of one variable we have that

$$\langle\Psi|f(x_1)|\Psi\rangle = \frac{1}{Z}\sum_{j=1}^{Z}\int dx\,\psi_j^*(x)\, f(x)\psi_j(x)\,,$$
$$\langle\Psi|\sum_{j=1}^{Z} f(x_j)|\Psi\rangle = \sum_{j=1}^{Z}\int dx\,\psi_j^*(x)\, f(x)\psi_j(x)\,. \tag{11.145}$$

In the case of a function of two variables $f(x, y)$ we have that

$$\langle\Psi| \sum_{j,k=1}^{Z}{}' f(x_j, x_k)|\Psi\rangle = \sum_{j,k=1}^{Z}{}' \int dx \int dy f(x, y) \Big[\psi_j^*(x)\psi_j(x)\psi_k^*(x)\psi_k(x) - \psi_k^*(x)\psi_j(x)\psi_j^*(x)\psi_k(x)\Big]. \tag{11.146}$$

We notice that in the case of non-interacting fermions, where the total Hamiltonian is the sum of single fermion Hamiltonians, if the $|\psi_k\rangle$ is an eigenvector of the single fermion Hamiltonian with eigenvalue E_k, then the state $|\Psi\rangle$ is an eigenvector of the total Hamiltonian $\hat{H}$ with energy

$$E = \sum_{k=1}^{Z} E_k . \tag{11.147}$$

If we apply the previous formulae to our Hamiltonian we get

$$\left\langle\Psi \left|\hat{H}\right| \Psi\right\rangle = \sum_{j=1}^{Z} \int dx \psi_j^*(x) \left(\frac{p^2}{2m} - \frac{Ze^2}{r}\right) \psi_j(x) + \frac{e^2}{2} \sum_{j,k=1}^{Z}{}' \int dx \int dy \frac{\psi_j^*(x)\, \psi_j(x)\, \psi_k^*(x)\, \psi_k(x) - \psi_k^*(x)\, \psi_j(x)\, \psi_j^*(x)\, \psi_k(x)}{|x-y|}. \tag{11.148}$$

The task now is to minimize the previous expression with respect to the number Z of the functions that enter into the definition of $|\Psi\rangle$.

This is a rather complex task. Explicit equations can be written, but finding the solution of these equations numerically leads to elaborate numerical computations. A sequence of approximations can be made in order to bring the problem to a more manageable formulation and also to a physical insight of the computation.

The first approximation that, as we shall see below, is physically motivated, consists introducing an effective potential energy $V(x)$ and assuming that all the wave functions $\psi_k(x)$ satisfy the equation

$$\left(-\frac{\hbar^2}{2m}\Delta + V(x)\right) \psi_k(x) = E_k\psi_k(x), \tag{11.149}$$

with the same potential energy $V(x)$. In this way we restrict the set of variational functions, and the ground state energy that we find may be larger than the one obtained without this restriction. The advantage is that we have only one function $V(x)$ that has to be computed by minimization (instead of Z functions, as in the original case); this leads to a strong simpification. Moreover, if we consider only spherically symmetric effective potentials, the complexity of the computation significantly decreases, because, for central potential, the numerical computation of the eigenfunctions can be performed in a very effective way. In other words, the mathematical problem is reduced to that of a functional minimization with only a one-dimensional function, namely, the effective radial potential energy $V(r)$.

If we neglect the exchange term, i.e. the last term in the double integral in Eq. (11.148) (namely the one proportional to $\psi_k^*(x)\psi_j(x)\psi_j^*(x)\psi_k(x)$), we find that the potential energy $V(x)$ that minimizes the energy must satisfy the self-consistent equation

$$V(x) = -\frac{Ze^2}{r} + e^2 \int dy \frac{\rho(y)}{|x-y|}, \tag{11.150}$$

where

$$\rho(x) = \sum_{k=1}^{Z} |\psi_k(x)|^2 . \tag{11.151}$$

Under this form the numerical computations are quite easy and can be done with high accuracy.

Similar computations can be done for molecules. However, radial symmetry cannot be used any longer. Sometimes, especially if high accuracy is not needed and qualitative results are sufficient, the computation can be done by considering the atoms as separate entities and treating the interaction among the atoms as a perturbation with the exception of the most external orbitals.

11.6.3 The Thomas–Fermi approximation

The first and simplest approximation that gives many of the characteristics of the atoms, and especially the overall dependence on Z, neglecting the oscillations of the periodic table, is the Thomas–Fermi approximation.[19] It is based on the semiclassical approximation (10.124).

Let us first study what happens to Z non-interacting fermions subject to the Hamiltonian

$$\hat{H}_{tot} = \sum_{k=1}^{Z} \left(\frac{\hat{p}_k^2}{2m} - V(x_k) \right) = \sum_{k=1}^{Z} \hat{H}(p_k, x_k) . \tag{11.152}$$

In order to simplify the formulae and to stress the basic principles let us present the computations in units where $h = 1$, $2m = 1$ and $e = 1$. In the semiclassical approximation, we find that the energy of the ground state is given by

$$E = \int_{H<E_F} dpdx\, H(p,x), \tag{11.153}$$

where E_F is known as the Fermi level of energy. The density of fermions in the point x, i.e. $\rho(x)$, is given, in the same approximation, by

$$\rho(x) = \int_{H<E_F} dp. \tag{11.154}$$

The total number of fermions is Z, and this implies that

$$\int dx \rho(x) = Z . \tag{11.155}$$

[19] See [Thomas 1927] [Fermi 1928].

In this case we find

$$\rho(x) = \int_{p^2 < E_F - V(x)} dp = \mathrm{V}\Delta(x)^{3/2} \tag{11.156}$$

where $\mathrm{V} = 4/3\pi$ (the volume of the unit sphere in three dimensions) and

$$\Delta(x) = V(x) - E_F \,. \tag{11.157}$$

The total energy can be written as

$$\begin{aligned} E &= \mathrm{V} \int dx \left(\frac{3}{5} \Delta(x)^{5/2} + V(x)\Delta(x)^{3/2} \right) \\ &= \int dx \left(\frac{3}{5} \mathrm{V}^{-2/3} \rho(x)^{5/3} + V(x)\rho(x) \right) . \end{aligned} \tag{11.158}$$

The first term is the kinetic energy and the second is the potential term.

We would like to stress the following point. If we consider the energy as a functional of ρ and we look for its stationary point, with the constraint given by Eq. (11.155), we find the equation

$$\frac{\delta E[\rho]}{\delta \rho(x)} \equiv \mathrm{V}^{-2/3} \rho(x)^{2/3} = \lambda \,, \tag{11.159}$$

where λ is a Lagrange multiplier that is needed to implement the constraint given by Eq. (11.155). With the identification of E_F with λ we recover Eq. (11.156).

Therefore we can say that the energy of Z fermions can be found using the variational principle where we have to minimize the expression (11.158).

The Thomas–Fermi equation can be derived in a similar way. We obtain for the free energy functional

$$E_{TF}[\rho] = \int dx \left(\frac{3}{5} \mathrm{V}^{-2/3} \rho(x)^{5/3} - \frac{Z\rho(x)}{r} \right) + \frac{1}{2} \int \int dx dy \frac{\rho(x)\rho(y)}{|x-y|} . \tag{11.160}$$

The variation of the previous functional leads to the equation

$$\mathrm{V}^{-2/3} \rho(x)^{2/3} = -V(x) \equiv \frac{Z\rho(x)}{r} + \int dy \frac{\rho(y)}{|x-y|} \,, \tag{11.161}$$

where we have set $E_F = 0$ in such a way that the charge density $\rho(x)$ extends up to infinity. By applying the Laplacian to the previous equation, we get

$$\mathrm{V}^{-2/3} \Delta \rho(x)^{2/3} = 4\pi \left(-Z\delta(x) + \rho(x) \right) , \tag{11.162}$$

which is the Thomas–Fermi equation.[20]

By introducing the function

$$\omega(x) = \rho(Z^{1/3} x) \,, \tag{11.163}$$

the previous equation becomes independent from Z and we get

$$\mathrm{V}^{-2/3} \Delta \omega(x)^{2/3} = -4\pi \left(\delta(x) + \omega(x) \right) . \tag{11.164}$$

[20] The Thomas–Fermi Equation is sometimes written for the function $\phi(x) \equiv \mathrm{V}^{-2/3} \rho(x)^{2/3}$.

A detailed study of the previous equation shows that there is a unique solution that goes to zero at infinity as x^{-3} and satisfies the normalization condition

$$\int dx\omega(x) = 1, \tag{11.165}$$

which is equivalent to $\int dx\rho(x) = Z$.

The possibility of arriving to a Z-independent equation implies the scaling of many quantities as function of Z. In particular, the radius (R) scales as $Z^{-1/3}$ and the total binding energy (E) as $Z^{7/3}$. If we use the numerical solution of the Thomas–Fermi equation and go back to the standard units,[21] we get

$$R = \frac{\hbar^2}{me^2 Z^{1/3}} = Z^{-1/3}\, 0.53\, \times\, 10^{-11}\,\mathrm{m}, \qquad E = Z^{7/3}\, 20.8\,\mathrm{eV}\,. \tag{11.166}$$

This scaling is correct, however the numerical prefactors are typically wrong by about 30%. Indeed, the repulsive correlation among the electrons (i.e. the exchange interactions) – which is neglected here – makes the atom larger and less bound.

This is the first example of the local density functional approach where one writes more precise and accurate forms for the functional $E[\rho]$. In this way one arrives to more accurate versions of the equations (11.149), where the exchange energy is taken into account with high precision. There are many books[22] and reviews on the local density functional approach (and also public-domain computer codes): the whole subject is quite active.

Summary

- First, we have shown how to apply to the hydrogen atom *Kepler's solution* for the problem of many interacting bodies, i.e. by only considering one-to-one interactions.
- The previous treatment has provided us with a *radial Schrödinger equation* that, with suitable variable changes, can be solved leading us to find the eigenvalues $E_n = -(1/n^2)E_0$ and the first total eigenfunctions.
- We have then studied the behavior of the atom in a magnetic field, by distinguishing the case in which the field is strong and the spin–orbit interaction can be treated as a small perturbation (*Paschen–Bach effect*), and the case when the external magnetic field is weak and can be treated as a perturbation (*Zeeman effect*).
- Apart form the spin–orbit interaction, two further *relativistic corrections* have been considered: the kinetic-energy term and the Darwin term. When the spin–orbit term is absent, the Darwin term is present and vice versa. However, we have shown that summation of one of the two with the kinetic-energy term gives the same result.

[21] See [*Landau/Lifshitz* 1976b].
[22] See [*Fiolhais et al.* 2003].

- We have then treated the *helium atom*, whose Schrödinger equation can be solved only by making use of approximation methods. We have in particular shown that perturbation and variational methods lead to similar results.
- Finally, we have considered *multielectron effects*, by making use of the Hartree–Fock method and of the Thomas–Fermi approximation.

Problems

11.1 Derive Eq. (11.6) from Eq. (11.2) and Eq. (11.8).

11.2 For the example of Subsec. 11.2.1, show that $\boldsymbol{P}$, $\boldsymbol{R}$ and $\mathbf{p}$, $\mathbf{r}$ are two pairs of classically conjugate variables.

11.3 Prove the commutators (11.11).

11.4 Study the behavior of the function

$$y(x) = \frac{a}{x^2} - \frac{b}{x},$$

with $a, b > 0$ and for $x > 0$.

11.5 Compute the relative deviation of the reduced mass from the electron mass in the case of the hydrogen atom.

11.6 Derive Eq. (11.23b).

11.7 Derive Eq. (11.32).

11.8 Prove the result (11.39).

11.9 Prove that the degree of degeneracy of the n-th energy eigenvalue of the hydrogen atom is equal to n^2.

11.10 Using Eqs. (11.43), (11.44), and (11.48), find the first radial wave functions of the hydrogen atom for $n = 1, n = 2, n = 3$.

11.11 Find the correct normalization factors for the radial wave functions listed in Eqs. (11.51).

11.12 Find the radial probability densities corresponding to the radial wave functions listed in Eqs. (11.51).

11.13 Compute the mean value of $(r/r_0)^j$ for $j = 2, 1, -1, -2, -3$ for the first eigenfunctions of an electron in a hydrogen atom.

11.14 Check the correctness of the normalization of Eqs. (11.55a).

11.15 Compute the eigenfunctions φ_{210}, $\varphi_{21,\pm1}$, φ_{310}, $\varphi_{31,\pm1}$, φ_{320}, $\varphi_{32,\pm1}$, and $\varphi_{32,\pm2}$ for a hydrogenoid atom.

11.16 Compute the energy-level corrections due to the spin–orbit interaction for the levels s, p, d, and f in the case of the hydrogen atom. Use the results of this calculation to draw a more refined version of Fig. 6.9.

11.17 Compute the energy-level corrections due to the Paschen–Bach effect (included in the spin–orbit term) for the levels s and p, which are necessary to draw Fig. 11.8. (*Hint*: First determine the magnetic correction as a function of m_l and m_s. Then, find the further spin–orbit modification, again as a function of m_l and m_s.)

11.18 Use the selection rules (11.77) to find the six allowed transitions among the levels depicted in Fig. 11.8. Compute the frequencies of the resulting transitions depicted in Fig. 11.9.

11.19 Compute the energy-level corrections due to the Zeeman effect (included the spin–orbit term) for the levels s and p, which are necessary to draw Fig. 11.11.
(*Hint*: First determine the spin–orbit correction (see Subsec. 11.3.1). Then, find the further magnetic modification due to the external field as a function of m_j and g_L.)

11.20 Use the selection rules (11.89a) to find the ten allowed transitions among the levels depicted in Fig. 11.12. Compute the frequencies of the resulting transitions depicted in Fig. 11.11.

11.21 From the definition of atomic units (11.112) deduce the corresponding units for the fundamental physical quantities of length, mass, time, and electronic charge.

11.22 Derive the result (11.119).

Further reading

Atkins, P. and Friedman, R., *Molecular Quantum Mechanics*, Oxford: Oxford University Press, 4th edn., 2005.

Brandsen, B. C. J., *Physics of Atoms and Molecules*, London: Longman, 1983.

Fiolhais, C. Nogueira, F., and Marques, M. (eds.), *A Primer in Density Functional Theory*, Berlin: Springer, 2003.

Slater, John C., *Quantum Theory of Matter*, New York: McGraw-Hill, 1951.

White, Harvey E., *Introduction to Atomic Spectra*, New York: McGraw-Hill, 1934.

12 Hydrogen molecular ion

In this chapter we move from the atomic to the molecular problem, i.e. we try to understand how quantum mechanics, besides explaining the structure and stability of a single atom, is able to account for the existence and the dynamics of molecules, the smallest building blocks of matter that determine and maintain the chemical properties of macroscopic substances. Molecular quantum physics is the field that deals with these types of problems and constitutes one of the most beautiful and successful applications of the basic principles of quantum mechanics. It is also of enormous relevance, because it lays the foundations of a consistent theory of condensed matter.

Our aim here is only to present the basic framework of molecular physics and to give a flavor of its power. A systematic treatment of the subject would go far beyond the scope of this book. In order to reach this goal, we take as our reference molecule the H_2^+ hydrogen molecular ion, which has the advantage of being simple enough to allow exact calculations. While it embodies all the main ingredients of any molecule, it avoids the difficulties due to multielectronic effects (see Sec. 11.6).

In Sec. 12.1 we present the H_2^+ molecule. In Sec. 12.2 we discuss the powerful Born–Oppenheimer approximation that allows us to deal with molecules with an arbitrary number of nuclei and electrons, a problem that would be otherwise unsolvable. In Sec. 12.3 we come back to the diatomic molecule for examining the three main aspects of the physical problem: the translational, the rotational, and the vibrational degrees of freedom. In Sec. 12.4 we present the Morse potential, which allows us to go beyond the harmonic approximation when considering the internuclear potential of diatomic molecules. Finally, in Sec. 12.5, we present two approximation methods useful for dealing with the molecular states, the linear combination of atomic orbitals (LCAO) and the valence bond method (VBM).

12.1 The molecular problem

The H_2^+ molecule consists of a single electron that moves within the field generated by two protons situated at relative fixed positions. As we shall see, the solution to this problem may be obtained thanks to a separation of variables. This separation may be effected within the framework of the so-called *spheroidal* (or elliptical) coordinates (see Fig. 12.1), where

$$\lambda = \frac{r_a + r_b}{r_{ab}} \quad \text{and} \quad \mu = \frac{r_a - r_b}{r_{ab}}, \tag{12.1}$$

with $1 \leq \lambda \leq +\infty$ and $-1 \leq \mu \leq +1$.

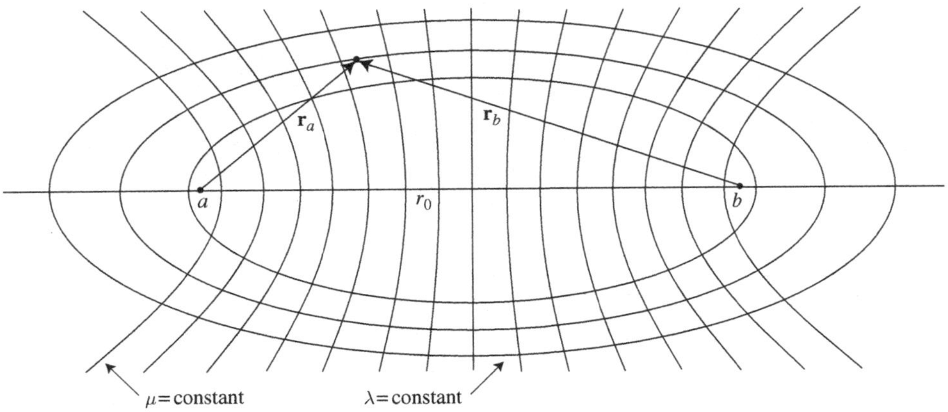

Figure 12.1 Spheroidal coordinates for the H_2^+ ion. The two hydrogen nuclei (i.e. the two protons) a, b are located at distance r_{ab} at the foci of the ellipses (λ = constant) and hyperbolas (μ = constant). The electron is located at distances r_a and r_b from the nuclei a and b, respectively. Besides λ and μ, the third coordinate is given by the rotation angle ϕ about the nuclear axis ab.

The Schrödinger equation for an electron in the field of two fixed protons in atomic units is given by

$$-\left(\frac{\nabla^2}{2} + \frac{1}{r_a} + \frac{1}{r_b}\right)\psi = E\psi, \tag{12.2}$$

where the Laplacian operator in spheroidal coordinates may be written as

$$\Delta = \frac{4}{r_{ab}^2(\lambda^2 - \mu^2)}\left[\frac{\partial}{\partial\lambda}\left(\lambda^2 - 1\right)\frac{\partial}{\partial\lambda} + \frac{\partial}{\partial\mu}\left(1 - \mu^2\right)\frac{\partial}{\partial\mu}\right] + \frac{1}{r_{ab}^2(\lambda^2 - 1)(1 - \mu^2)}\frac{\partial^2}{\partial\phi^2}. \tag{12.3}$$

Making the ansatz

$$\psi(\lambda, \mu, \phi) = \Lambda(\lambda)M(\mu)e^{\imath U\phi}, \tag{12.4}$$

Equation (12.2) can be changed into the set of two equations (see Prob. 12.1)

$$\frac{d}{d\lambda}\left[\left(\lambda^2 - 1\right)\frac{d\Lambda}{d\lambda}\right] + \left(\frac{Er_{ab}^2}{2}\lambda^2 + 2r_{ab}\lambda - \frac{U^2}{\lambda^2 - 1}\right)\Lambda = 0, \tag{12.5a}$$

$$\frac{d}{d\mu}\left[\left(1 - \mu^2\right)\frac{dM}{d\mu}\right] + \left(\frac{Er_{ab}^2}{2}\mu^2 - \frac{U^2}{1 - \mu^2}\right)M = 0. \tag{12.5b}$$

12.2 Born–Oppenheimer approximation

Let us consider for the time being the molecular problem in the most general terms, i.e. N_e electrons moving in the field generated by N_n nuclei. We denote by $\mathbf{r}_k^n$ and $\mathbf{p}_k^n$, respectively, the position and momentum of the k-th nucleus with charge $Z_k e$ and mass m_k

($k = 1, \ldots, N_n$), relative to the origin of the frame of reference. Similarly, the position and momentum of the j-th electron are denoted by $\mathbf{r}_j^e$ and $\mathbf{p}_j^e$, respectively ($j = 1, \ldots N_e$), again relative to the origin of the frame of reference. The total energy of the system will be given by

$$\hat{H} = \hat{T}_n + \hat{T}_e + \hat{V}_n + \hat{V}_e + \hat{V}_{ne}, \tag{12.6}$$

where the first two terms are the kinetic energies of the nuclei and electrons, respectively, i.e.

$$\hat{T}_n = \sum_{k=1}^{N_n} \frac{(\hat{\mathbf{p}}_k^n)^2}{2m_k}, \tag{12.7a}$$

$$\hat{T}_e = \frac{1}{2m_e} \sum_{j=1}^{N_e} (\hat{\mathbf{p}}_j^e)^2, \tag{12.7b}$$

and the three potential-energy terms are given by

$$\hat{V}_n = \sum_{k<l=1}^{N_n} \frac{(Z_k \mathrm{e})\,(Z_l \mathrm{e})}{\left|\mathbf{r}_k^n - \mathbf{r}_l^n\right|}, \tag{12.8a}$$

$$\hat{V}_e = \sum_{i<j=1}^{N_e} \frac{\mathrm{e}^2}{\left|\mathbf{r}_i^e - \mathbf{r}_j^e\right|}, \tag{12.8b}$$

$$\hat{V}_{ne} = \sum_{j=1}^{N_e} \sum_{k=1}^{N_n} \frac{Z_k\,(\mathrm{e})\,(-\mathrm{e})}{\left|\mathbf{r}_j^e - \mathbf{r}_k^n\right|}. \tag{12.8c}$$

The Born–Oppenheimer approximation separates the nuclear motion from the electronic one. In fact, we expect the electron dynamics to be much faster than the nuclear motion. As a consequence, we try to solve the molecular problem by first considering a "frozen" configuration of the nuclei, which induces a certain potential on each electron. Changing the nuclear configuration, we assume that the electrons will rearrange themselves almost instantaneously. Then, we shall look for solutions of the Schrödinger equation

$$\hat{H}\Psi(\mathbf{r}_a^e, \ldots, \mathbf{r}_j^e; \mathbf{r}_a^n, \ldots, \mathbf{r}_k^n) = E\Psi(\mathbf{r}_a^e, \ldots, \mathbf{r}_j^e; \mathbf{r}_a^n, \ldots, \mathbf{r}_k^n) \tag{12.9}$$

where the electronic states are defined in a parametric way relative to a given distribution of the nuclei, as we shall see in a moment. For example, in the case of a diatomic molecule, the parameter chosen is the distance between the nuclei. The solution of the general problem is given by the product

$$\Psi(\{\mathbf{r}_j^e\}; \{\mathbf{r}_k^n\}) = \psi(\{\mathbf{r}_k^n\})\varphi_{\mathbf{r}_k^n}(\{\mathbf{r}_j^e\}), \tag{12.10}$$

where $\psi(\{\mathbf{r}_k^n\})$ is the nuclear wave function. The electronic wave function $\varphi_{\mathbf{r}_k^n}(\{\mathbf{r}_j^e\})$ depends parametrically on the parameter set $\{\mathbf{r}_k^n\}$, and may be taken as a solution of the Schrödinger equation

$$\hat{H}_e \varphi_{\mathbf{r}_k^n}(\{\mathbf{r}_j^e\}) = E_e \varphi_{\mathbf{r}_k^n}(\{\mathbf{r}_j^e\}), \tag{12.11}$$

for a given nuclear configuration, where

$$\hat{H}_e = \hat{T}_e + \hat{V}_e + \hat{V}_{ne}. \tag{12.12}$$

Taking into account the above considerations, we may rewrite the lhs of the Schrödinger equation (12.9) as (see Prob. 12.2)

$$\begin{aligned}\hat{H}\Psi(\{\mathbf{r}_k^n\};\{\mathbf{r}_j^e\}) &= \left[-\sum_{k=1}^{N_n}\frac{\hbar^2}{2m_k}\nabla_k^2 + \hat{V}_n(\{\mathbf{r}_k^n\}) + \hat{H}_e\right]\psi(\{\mathbf{r}_k^n\})\varphi_{\mathbf{r}_k^n}(\{\mathbf{r}_j^e\}) \\ &= \varphi_{\mathbf{r}_k^n}(\{\mathbf{r}_j^e\})\left[-\sum_{k=1}^{N_n}\frac{\hbar^2}{2m_k}\nabla_k^2 + \hat{V}_n(\{\mathbf{r}_k^n\}) + E_e\right]\psi(\{\mathbf{r}_k^n\}) \\ &\quad -\sum_{k=1}^{N_n}\frac{\hbar^2}{2m_k}\left[\psi(\{\mathbf{r}_k^n\})\nabla_k^2\varphi_{\mathbf{r}_k^n}(\{\mathbf{r}_j^e\}) + 2\boldsymbol{\nabla}_k\psi(\{\mathbf{r}_k^n\})\boldsymbol{\nabla}_k\varphi_{\mathbf{r}_k^n}(\{\mathbf{r}_j^e\})\right],\end{aligned} \tag{12.13}$$

where

$$\nabla_k^2 = \frac{\partial^2}{\partial(x_k^n)^2} + \frac{\partial^2}{\partial(y_k^n)^2} + \frac{\partial^2}{\partial(z_k^n)^2}. \tag{12.14}$$

In Eq. (12.13) the two terms in the last square brackets may be neglected for reasons that are summarized in the following. In fact, the electronic wave function $\varphi_{\mathbf{r}_k^n}(\{\mathbf{r}_j^e\})$ may be thought of as a function of the difference $\mathbf{r}_j^e - \mathbf{r}_k^n$. As a consequence, denoting by

$$\left(\nabla_j^{'}\right)^2 = \frac{\partial^2}{\partial(x_j^e)^2} + \frac{\partial^2}{\partial(y_j^e)^2} + \frac{\partial^2}{\partial(z_j^e)^2} \tag{12.15}$$

the Laplacian relative to the electronic coordinates, we have

$$\nabla_k^2\varphi_{\mathbf{r}_k^n}(\{\mathbf{r}_j^e\}) \simeq \left(\nabla_j^{'}\right)^2\varphi_{\mathbf{r}_k^n}(\{\mathbf{r}_j^e\}), \tag{12.16}$$

from which it follows that the term

$$\frac{\hbar^2}{2m_k}\nabla_k^2\varphi_{\mathbf{r}_k^n}(\{\mathbf{r}_j^e\}) \tag{12.17}$$

in Eq. (12.13) is of the order of m_e/m_k times the kinetic energy of one electron. Moreover (see comments to Eq. (11.20)),

$$\frac{m_e}{m_k} \leq \frac{1}{1840}. \tag{12.18}$$

For the same reasons,

$$\boldsymbol{\nabla}_k\varphi_{\mathbf{r}_k^n}(\{\mathbf{r}_j^e\}) = -\boldsymbol{\nabla}^{'}_j\varphi_{\mathbf{r}_k^n}(\{\mathbf{r}_j^e\}), \tag{12.19}$$

and we also have

$$\hbar\boldsymbol{\nabla}_k\psi \simeq \hat{\mathbf{p}}_k^n \quad \text{and} \quad \hbar\boldsymbol{\nabla}_k\varphi_{\mathbf{r}_k^n} \simeq \hat{\mathbf{p}}_j^e, \tag{12.20}$$

where, for the sake of notation, we have dropped dependencies, from which it follows that

$$\frac{\hbar^2}{m_k}\boldsymbol{\nabla}_k\psi\boldsymbol{\nabla}_k\varphi_{\mathbf{r}_k^n} \simeq \frac{1}{m_k}\hat{\mathbf{p}}_k^n\hat{\mathbf{p}}_j^e \ll \frac{(\hat{\mathbf{p}}_j^e)^2}{2m_e}, \tag{12.21}$$

because, apart from (12.18), as we have said the dynamics of the electrons is much faster than that of the nuclei and therefore $p_k^n \ll p_j^e$. The last term in the rhs of Eq. (12.21) is of the order of the kinetic energy of one electron, and this finally justifies neglecting those two terms.

With these approximations, we may rewrite Eq. (12.9) as a Schrödinger equation for the nuclear wave function, i.e. (see Prob. 12.3)

$$\left[-\sum_{k=1}^{N_n}\frac{\hbar^2}{2m_k}\nabla_k^2+\hat{V}_n(\{\mathbf{r}_k^n\})+E_e(\{\mathbf{r}_k^n\})\right]\psi(\{\mathbf{r}_k^n\})=E\psi(\{\mathbf{r}_k^n\}), \tag{12.22}$$

where E is the total energy, and the electron energy eigenvalue (parametrically dependent on the position of the nuclei) $E_e(\{\mathbf{r}_k^n\})$, which appears in Eq. (12.11), may be considered as a correction to the nuclear potential energy $\hat{V}_n(\{\mathbf{r}_k^n\})$, giving rise to the total potential energy

$$\hat{V}(\{\mathbf{r}_k^n\})=\hat{V}_n(\{\mathbf{r}_k^n\})+E_e(\{\mathbf{r}_k^n\}). \tag{12.23}$$

In other words, the electrons with their motion generate a diffuse charge distribution that alters the nuclear Coulomb repulsion term $\hat{V}_n$. As a matter of fact, in the absence of electrons, the positively charged nuclei would repel each other, and no molecule could be built: the shield generated by the electron cloud binds the otherwise repelling nuclei – the essence of the chemical bond.

This framework that separates the nuclear motion from the electronic dynamics is precisely the essence of the Born–Oppenheimer approximation.[1] Such a method requires that one first solves the electronic Schrödinger equation (12.11) for any given possible configuration of the nuclei, finding the eigenvalues E_e as a function of the nuclear positions $\{\mathbf{r}_k^n\}$. Then, one uses these results in order to solve the Schrödinger equation (12.22) for the nuclei, with the potential energy (12.23).

12.3 Vibrational and rotational degrees of freedom

In the previous section we have laid the foundations of the quantum-mechanical treatment of the molecular problem in its general form. We now go back to the more specific case of a diatomic molecule, with particular reference to the H_2^+ molecular ion. As a guiding model for this analysis we take the rigid rotator model (see Subsec. 6.2.1). A diatomic molecule has three types of collective motion:

1. *Translation* of the center of mass of the molecule, which practically coincides with the center of mass of the nuclei. This motion gives rise to a continuous spectrum characteristic of the motion of a free particle. This motion can be neglected if we work in a reference system in which the center of mass is at rest.

[1] This approximation is sometimes called adiabatic approximation (see also Sec. 10.3), because it separates the fast electronic motion from the slow nuclear dynamics.

2. *Rotation* of the molecule about an axis passing through the center of mass. The pure rotational motion is obtained if the two nuclei stay at a fixed distance r_{ab} (see Figs. 6.5 and 12.1). The rotational spectrum of the molecule is given by Eqs. (6.81)–(6.82) (see Fig. 6.6).
3. *Vibration* along the axis connecting the two nuclei. If the two nuclei must stay at a fixed distance, the potential energy must have a minimum at $r = r_{ab}$. Under the harmonic approximation, the internuclear distance may oscillate around the equilibrium point r_{ab}, giving rise to vibrational energy levels and spectra.

Let us rewrite Eq. (12.22) for the case of a diatomic molecule, made up of atoms a and b,

$$\left[-\frac{\hbar^2}{2m_a}\nabla_a^2 - \frac{\hbar^2}{2m_b}\nabla_b^2 + \hat{V}_n(\mathbf{r}_a^n, \mathbf{r}_b^n) + E_e(\mathbf{r}_a^n, \mathbf{r}_b^n)\right]\psi(\mathbf{r}_a^n, \mathbf{r}_b^n) = E\psi(\mathbf{r}_a^n, \mathbf{r}_b^n). \tag{12.24}$$

Assuming that the nuclei are almost at rest relative to each other, and taking into account Eq. (12.23), the total potential energy will have the form

$$\hat{V}(\mathbf{r}_a^n, \mathbf{r}_b^n) = \hat{V}(\mathbf{r}_b^n - \mathbf{r}_a^n) = \hat{V}(\mathbf{r}_{ab}). \tag{12.25}$$

Defining the position of the center of mass as

$$\mathbf{r}_c = \frac{m_a\mathbf{r}_a^n + m_b\mathbf{r}_b^n}{m_a + m_b}, \tag{12.26}$$

and taking into account Eqs. (12.24)–(12.25), we apply the change of variables

$$\{\mathbf{r}_a^n, \mathbf{r}_b^n\} \to \{\mathbf{r}_c, \mathbf{r}_{ab}\} \tag{12.27}$$

in order to derive the Hamiltonian (see Prob. 12.4)

$$\hat{H} = -\frac{\hbar^2}{2(m_a + m_b)}\nabla_c^2 - \frac{\hbar^2}{2m}\nabla_0^2 + \hat{V}(\mathbf{r}_{ab}), \tag{12.28}$$

where m is the reduced mass and

$$\nabla_c^2 = \frac{\partial^2}{\partial x_c^2} + \frac{\partial^2}{\partial y_c^2} + \frac{\partial^2}{\partial z_c^2}, \tag{12.29a}$$

$$\nabla_0^2 = \frac{\partial^2}{\partial x_0^2} + \frac{\partial^2}{\partial y_0^2} + \frac{\partial^2}{\partial z_0^2}, \tag{12.29b}$$

being $\mathbf{r}_c = (x_c, y_c, z_c)$ and $\mathbf{r}_{ab} = (x_0, y_0, z_0)$.

In the Schrödinger equation (12.24), with the transformed Hamiltonian (12.28), we have separated the center-of-mass motion from the relative motion of the nuclei. This suggests the ansatz

$$\psi(\mathbf{r}_a^n, \mathbf{r}_b^n) = \zeta(\mathbf{r}_c)\eta(\mathbf{r}_{ab}). \tag{12.30}$$

Equation (12.28) will then give rise to the two Schrödinger equations (see Prob. 12.6)

$$-\frac{\hbar^2}{2(m_a + m_b)}\nabla_c^2\zeta(\mathbf{r}_c) = E_c\zeta(\mathbf{r}_c), \tag{12.31a}$$

$$\left[-\frac{\hbar^2}{2m}\nabla_0^2 + \hat{V}(\mathbf{r}_{ab})\right]\eta(\mathbf{r}_{ab}) = E_0\eta(\mathbf{r}_{ab}), \tag{12.31b}$$

where E_c is the energy eigenvalue of the center-of-mass degree of freedom and E_0 means in this context the energy eigenvalue pertaining to the relative motion. Equation (12.31a) represents, as we have said, a free-particle Schrödinger equation. Its solution will be a spherical wave of the form (see Eq. (2.147))

$$\zeta(\mathbf{r}_c) \propto e^{\imath \mathbf{k}\cdot\mathbf{r}_c}, \tag{12.32}$$

describing the translational motion. The corresponding eigenvalues will be given by

$$E_c = \frac{\hbar^2 \mathbf{k}^2}{2(m_a + m_b)}. \tag{12.33}$$

The solution of the second equation gives the rotational part of the motion. Since the potential $\hat{V}(\mathbf{r}_{ab})$ only depends on the distance between the two nuclei, we are in presence of a central potential (see Subsec. 6.2.2), i.e.

$$\hat{V}(\mathbf{r}_{ab}) = \hat{V}(r_{ab}), \tag{12.34}$$

and it is then convenient to translate this equation into spherical coordinates, where the angular part of the wave function $\eta(\mathbf{r}_{ab})$ will be given by spherical harmonics. The solution of the relative-motion equation therefore has the form (see Eq. (6.57))

$$\eta(\mathbf{r}_{ab}) = \mathrm{R}(r_{ab})Y(\phi, \theta), \tag{12.35}$$

where we recall that $\mathrm{R}(r_{ab})$ is the radial wave function. The spherical harmonics $Y_j(\phi, \theta)$ have eigenvalues (see Eq. (6.81))

$$E_j = \frac{\hbar^2}{2\mathrm{I}} j(j+1), \tag{12.36}$$

where $\mathrm{I} = m\bar{r}_{ab}^2$ is the moment of inertia (see also Eq. (6.78)) of the two nuclei, $\bar{r}_{ab}$ being the equilibrium distance between the two nuclei. Obviously, the eigenvalues of $\hat{j}^2$ and $\hat{j}_z$ are given by $j(j+1)$ and m_j (see Eqs. (6.58)), respectively.

For the solution of the radial wave function $\mathrm{R}(r_{ab})$, we first recall that

$$|r_{ab} - \bar{r}_{ab}| \ll \bar{r}_{ab}, \tag{12.37}$$

that is, the two nuclei will oscillate around the equilibrium position. This allows a Taylor series expansion of the total potential energy $\hat{V}(r_{ab})$ about $\bar{r}_{ab}$, i.e.

$$\hat{V}(r_{ab}) \simeq \hat{V}(\bar{r}_{ab}) + \frac{a}{2} \frac{\partial^2 \hat{V}(r_{ab})}{\partial r_{ab}^2}\bigg|_{r_{ab}=\bar{r}_{ab}} \cdot (r_{ab} - \bar{r}_{ab})^2 + \cdots, \tag{12.38}$$

where the first-order term is missing because $\bar{r}_{ab}$ is an equilibrium point,

$$\frac{\partial \hat{V}(r_{ab})}{\partial r_{ab}}\bigg|_{r_{ab}=\bar{r}_{ab}} = 0. \tag{12.39}$$

Up to the second order (see Fig. 12.2), the solutions will be of the harmonic-oscillator type, that is,

$$\hat{V}(r_{ab}) \simeq -D + \frac{\kappa}{2}(r_{ab} - \bar{r}_{ab})^2, \tag{12.40}$$

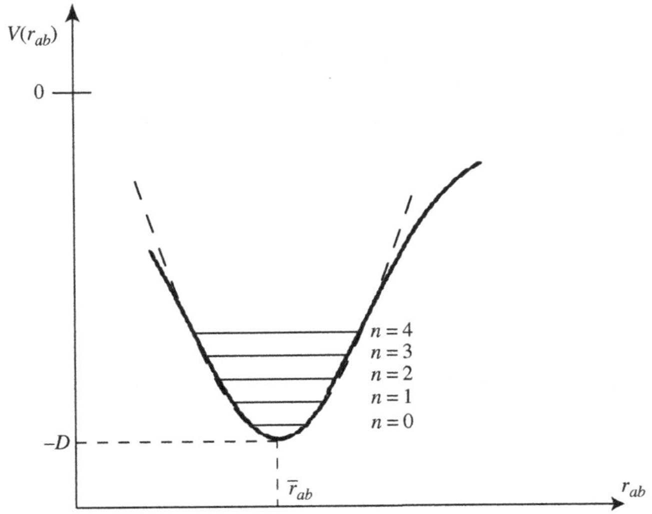

Figure 12.2 Diagram of the potential energy $\hat{V}(r_{ab})$ (solid line) of the nuclear motion of a diatomic molecule (see Eqs. (12.25), (12.38), and (12.40)). For $r_{ab} \ll \bar{r}_{ab}$, the Coulomb repulsion prevents the nuclei from getting close to each other, whereas for $r_{ab} \to \infty$ the molecule dissociates, D being the dissociation energy. The dashed line represents the harmonic approximation $\kappa(r_{ab} - \bar{r}_{ab})^2/2$ that is valid in the vicinity of the classical equilibrium point $\bar{r}_{ab}$ (harmonic region). In this region the vibrational energy levels ($n = 0, n = 1, n = 2, \ldots$) are harmonic and therefore equidistant.

where $D = -\hat{V}(\bar{r}_{ab})$ is the dissociation energy, or the energy needed to move the two nuclei to an infinite distance from each other. If we neglect the centrifugal term (6.93), making use of a rigid-rotator approximation, then we can write the radial Schrödinger equation as

$$\left[-\frac{\hbar^2}{2m} \frac{\partial^2}{\partial r_{ab}^2} + \frac{1}{2}\kappa \left(r_{ab} - \bar{r}_{ab} \right)^2 - D \right] \Phi(r_{ab}) = E^{\text{vib}} \Phi(r_{ab}), \tag{12.41}$$

which – apart from the constant term D – is the harmonic-oscillator equation (4.49), whose solutions will be given by the Hermite polynomials (4.98) with energy levels

$$E_n^{\text{vib}} = -D + \left(n + \frac{1}{2} \right) \hbar\omega_0, \tag{12.42}$$

where

$$\omega_0 = \sqrt{\frac{\kappa}{m}}. \tag{12.43}$$

Therefore, the total energy, relative to the nuclear part, is given by

$$\begin{aligned} E &= E_c + E_0 = E_c + E_j + E_n^{\text{vib}} \\ &= \frac{\hbar^2 \mathbf{k}^2}{2(m_a + m_b)} + \frac{\hbar^2}{2\mathrm{I}} j(j+1) - D + \left(n + \frac{1}{2} \right) \hbar\omega_0. \end{aligned} \tag{12.44}$$

It is useful to go back to the meaning of the Born–Oppenheimer approximation. Given a certain electronic state, the form of $E_e(\mathbf{r}_a^n, \mathbf{r}_b^n)$ in Eq. (12.24) is fixed, and Eq. (12.44)

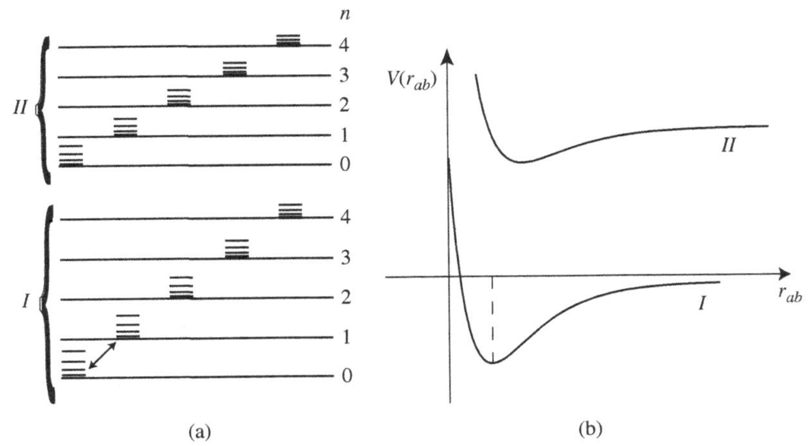

Figure 12.3 (a) First vibrational and rotational levels of two electronic states *I* and *II* (whose shapes are represented in (b)) in a diatomic molecule. The long horizontal lines depict the vibrational levels ($n = 0, n = 1, n = 2, \ldots$). To each vibrational level is associated a rotational band (short horizontal lines), whose eigenvalues are given by Eq. (12.36). The transitions between sublevels are regulated by the selection rules: $\Delta n = \pm 1$, $\Delta j = \pm 1$. For example, the transition from the level $n + 1, j + 1$ to the level n, j (represented by the double-headed arrow) will involve the emission of a photon of frequency (1.75), which, by making use of Eq. (12.44), gives $\nu = \omega_0/2\pi + 2B(j + 1)$, where B is the rotational constant defined by Eq. (6.83) (see Prob. 12.7). It should be noted that, due to the different curvature of the energy profiles around the minima, different electronic states have different vibrational frequencies. Moreover, due to the different values of $\bar{r}_{ab}$ and thus of the moment of inertia, also the separations between the rotational levels may be significantly different.

follows. Changing the electronic state will change the form of the potential $\hat{V}(r_{ab})$ given by Eq. (12.25) and thus the harmonic constant κ. As a consequence, there will be several electronic states, the lowest of which is the ground state. Each electronic state will possess many vibrational levels. In turn, every vibrational level will display rotational bands. This situation is schematically depicted in Fig. 12.3.

However, Eq. (12.44) is not the end of the story. As a matter of fact we have neglected two small effects: the anharmonicity of the potential $\hat{V}(r_{ab})$ and the centrifugal-barrier term. Adding these two effects, however, will only introduce two small additional perturbations in Eq. (12.44) and the physical picture will not be substantially modified.

12.4 The Morse potential

P. M. Morse[2] introduced a very interesting analytic form of the potential energy $\hat{V}(r_{ab})$ that represents a good approximation for the actual energy obtained from the stable molecular states of diatomic molecules. A further advantage of the Morse potential is that it leads to

[2] [Morse 1929].

a vibrational Schrödinger equation that is susceptible of an exact solution. The form of the Morse potential is (see Prob. 12.8)

$$\hat{V}(r_{ab}) = D\left[e^{-2a(r_{ab}-\bar{r}_{ab})} - 2e^{-a(r_{ab}-\bar{r}_{ab})}\right], \tag{12.45}$$

where D is again the dissociation energy, $\bar{r}_{ab}$ is the equilibrium internuclear distance, and a is a constant that, as we shall see, is related to the vibrational frequency.

If, for the time being, we ignore the rotation degree of freedom and concentrate upon the vibrational Schrödinger equation, under the hypothesis (12.45), we may first observe that the problem is equivalent to that of a central field with $l = 0$. Taking into account Eqs. (12.31b), (12.35), and (6.92), the vibrational Schrödinger equation for the product

$$\xi(r_{ab}) = \mathrm{R}(r_{ab})r_{ab} \tag{12.46}$$

may be written as

$$-\frac{d^2}{dr_{ab}^2}\xi + \frac{2m}{\hbar^2}\left(E^{\mathrm{vib}} - De^{-2ar_{ab}} + 2De^{-ar_{ab}}\right)\xi = 0, \tag{12.47}$$

where we have also performed a suitable (ideal) translation so as to put $\bar{r}_{ab} = r_{ab}$.

With a change of variable

$$\tilde{r} = \frac{2\sqrt{2mD}}{a\hbar}e^{-ar_{ab}}, \tag{12.48}$$

we obtain the Schrödinger equation (see Prob. 12.9)

$$a^2\tilde{r}^2\frac{d^2}{d\tilde{r}^2}\xi + a^2\tilde{r}\frac{d}{d\tilde{r}}\xi + \left(\frac{2m}{\hbar^2}E^{\mathrm{vib}} - \frac{a^2}{4}\tilde{r}^2 + \sqrt{2mD}\frac{a}{\hbar}\tilde{r}\right)\xi = 0. \tag{12.49}$$

We now substitute the notation[3]

$$\tilde{k} = \frac{\sqrt{-2mE^{\mathrm{vib}}}}{a\hbar}, \quad n = \frac{\sqrt{2mD}}{a\hbar} - \left(\tilde{k} + \frac{1}{2}\right), \tag{12.50}$$

so as to obtain (see Prob. 12.10)

$$\xi'' + \frac{1}{\tilde{r}}\xi' + \left(-\frac{\tilde{k}^2}{\tilde{r}^2} - \frac{1}{4} + \frac{n + \tilde{k} + \frac{1}{2}}{\tilde{r}}\right)\xi = 0. \tag{12.51}$$

The asymptotic behavior of the solution of Eq. (12.51) for $\tilde{r} \to \infty$ and for $\tilde{r} \to 0$ suggests the ansatz (see also Eq. (11.30))

$$\xi(\tilde{r}) = \tilde{r}^{\tilde{k}}e^{-\frac{\tilde{r}}{2}}W(\tilde{r}), \tag{12.52}$$

which in turn leads to the following equation for $W(\tilde{r})$:

$$\tilde{r}W'' + \left(2\tilde{k} + 1 - \tilde{r}\right)W' + nW = 0. \tag{12.53}$$

We may now proceed in a similar way as we have done in Sec. 11.2, finding the hypergeometric function

$$W(\tilde{r}) = F(-n, 2\tilde{k} + 1, \tilde{r}) \tag{12.54}$$

[3] Since we are looking for the discrete spectrum, we may assume $E^{\mathrm{vib}} < 0$.

as the solution to our problem. This solution is obtained for non-negative integers n, and the energy levels are given by (see Prob. 12.11)

$$E_n^{\mathrm{vib}} = -D\left[1 - \left(n + \frac{1}{2}\right)\frac{a\hbar}{\sqrt{2mD}}\right]^2, \tag{12.55}$$

up to the maximum value of n for which

$$n < \frac{\sqrt{2mD}}{a\hbar} - \frac{1}{2}. \tag{12.56}$$

On the other hand, when

$$2\sqrt{2mD} < a\hbar, \tag{12.57}$$

the discrete spectrum does not exist. The energy levels (12.55) should be compared with the eigenvalues (12.42) obtained in the harmonic approximation (see Prob. 12.12).

12.5 Chemical bonds and further approximations

As we have seen, in general it is not possible to find the exact solution (i.e. the exact energy eigenvalues and eigenfunctions) for the Schrödinger equation of a molecule. In this section, we shall describe the most popular methods used for the estimate of the electronic states in the diatomic molecules: the *linear combination of atomic orbitals* (LCAO) and the *valence bond method* (VBM). Although their starting points are slightly different, both these methods are approximate and their goal is to find the best estimate of the exact solution.

12.5.1 LCAO

The LCAO method assumes that the wave function of a set of atoms in a molecule can be appropriately dealt with as a linear combination of the original atomic wave functions with certain weights that have to be determined. In other words, according to this method the original wave functions are only slightly modified due to the formation of the molecule. In order to find the best estimate, one has then to apply a variational method (see Sec. 10.4 and Subsec. 11.5.2) and obtain the value of the parameters (the weights of the superposition) that minimize the energy.

Let us consider two atoms, say a and b, whose electronic states are described by the wave functions ψ_a and ψ_b, respectively. We then make the ansatz that the electronic state of the a–b molecule may be expressed as the superposition

$$\Psi = c_a\psi_a + c_b\psi_b, \tag{12.58}$$

or

$$\Psi = \psi_a + b\psi_b, \tag{12.59}$$

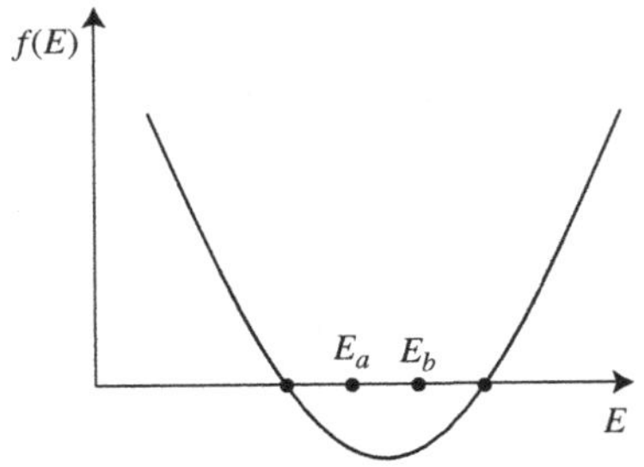

Figure 12.4 Schematic diagram of the function $f(E)$ (see Eq. (12.65)). The zeros of $f(E)$ represent the LCAO solutions for the energy of the molecule $a - b$.

where $b = c_b/c_a$ (with $c_a \neq 0$). In the latter case, if ψ_a and ψ_b are normalized, the resulting wave function Ψ is not normalized. Denoting by $\hat{H}$ the total Hamiltonian of the molecule, we have the Schrödinger equation

$$\hat{H}\Psi = E\Psi, \tag{12.60}$$

or

$$\hat{H}\psi_a + b\hat{H}\psi_b = E\psi_a + bE\psi_b. \tag{12.61}$$

Multiplying both sides first by ψ_a^* and then by ψ_b^* and integrating, we obtain the set of two equations

$$H_{aa} - E + b(H_{ab} - ES_{ab}) = 0, \tag{12.62a}$$

$$H_{ab} - ES_{ab} + b(H_{bb} - E) = 0, \tag{12.62b}$$

where

$$H_{ab} = \int d\mathrm{V}\psi_a^* \hat{H}\psi_b = \beta \tag{12.63}$$

is the so-called *resonance integral*, and

$$S_{ab} = \int d\mathrm{V}\psi_a^* \psi_b = S, \tag{12.64}$$

with $0 \leq S \leq 1$, is the so-called *overlapping integral*. By eliminating b, we obtain

$$(E_a - E)(E_b - E) - (\beta - ES)^2 = f(E) = 0, \tag{12.65}$$

where

$$E_a = H_{aa} = \int d\mathrm{V}\psi_a^* \hat{H}\psi_a, \tag{12.66a}$$

$$E_b = H_{bb} = \int d\mathrm{V}\psi_b^* \hat{H}\psi_b \tag{12.66b}$$

should not be confused with the energy eigenvalues of the atoms a and b, respectively. Equation (12.65) is an implicit quadratic equation for E, given E_a, E_b, β, and S. The function $f(E)$ is schematically depicted in Fig. 12.4, and its zeros represent the LCAO approximate solutions to the molecular energy.

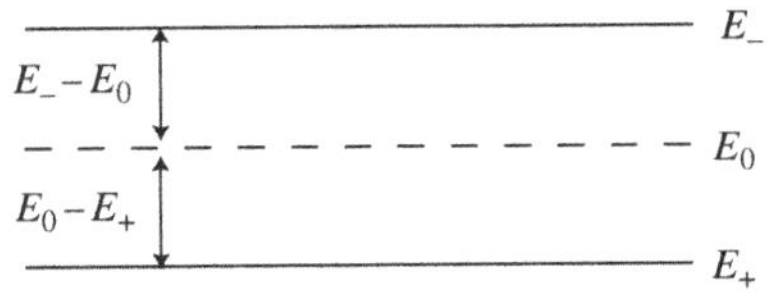

Figure 12.5 LCAO energy solutions for the H_2^+ molecular ion. Only the symmetric Ψ_+ state gives rise to a stable molecule with energy eigenvalue E_+. The antisymmetric state Ψ_-, on the other hand, with energy eigenvalue E_- is unstable.

In the case of the H_2^+ molecule, we have (see Fig. 12.1, and Eqs. (12.8a), (12.22)–(12.23))

$$\hat{H} = -\frac{\hbar^2}{2m}\nabla^2 - \frac{e^2}{\mathbf{r}_a} - \frac{e^2}{\mathbf{r}_b} + \frac{e^2}{\mathbf{r}_{ab}}. \quad (12.67)$$

Since the molecule is homonuclear, we have $|b|^2 = 1$, or $b = \pm 1$, which implies a symmetric and an antisymmetric wave function

$$\Psi_+ = \psi_a + \psi_b, \quad (12.68a)$$

$$\Psi_- = \psi_a - \psi_b, \quad (12.68b)$$

with

$$E_a = E_b = E_o, \quad (12.69)$$

from which it follows that, in the case of Ψ_+, Eqs. (12.62) become

$$(E_0 - E_+) + \beta - E_+ S = 0, \quad (12.70a)$$

$$\beta - E_+ S + (E_0 - E_+) = 0, \quad (12.70b)$$

which lead to

$$E_+ = \frac{E_0 + \beta}{1 + S}. \quad (12.71a)$$

Analogously, we have

$$E_- = \frac{E_0 - \beta}{1 - S}. \quad (12.71b)$$

The corresponding wave functions are given by

$$\psi_a(r) = C_a e^{-\frac{r}{r_0}} \quad (12.72a)$$

$$\psi_b(r) = C_b e^{-\frac{(r - r_{ab})}{r_0}}, \quad (12.72b)$$

where C_a and C_b are constants and r_0 is the Bohr's radius of the hydrogen atom (11.21). In the LCAO approximation, therefore, two energy levels for the molecule are possible, and are separated by an energy gap $E_- - E_+$ (see Fig. 12.5).

In Fig. 12.6 we show the graph of Ψ_+ and Ψ_- as a function of r_{ab}. As expected, only Ψ_+ has a minimum for $r_{ab} \simeq 2r_0$, with $E \simeq -0.56\mathrm{e}^2/r_0$. The excitation from the state Ψ_+ to the state Ψ_- is already sufficient to break the molecule.

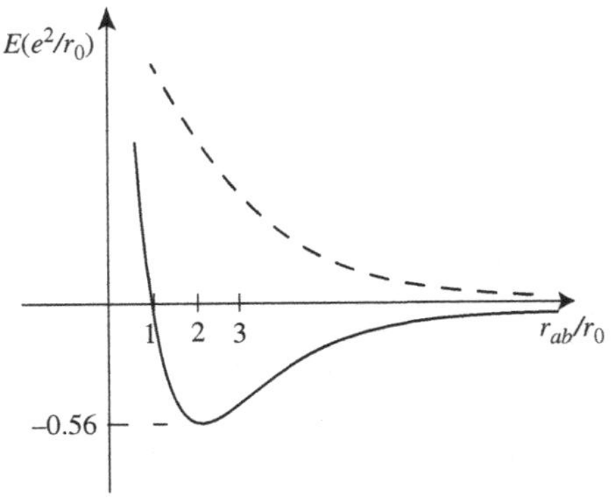

Figure 12.6 Graphic representation of the symmetric (solid line) and antisymmetric (dashed line) states of the ground level of the H_2^+ molecule.

Similar considerations can be developed for the case of other homonuclear diatomic molecules, in particular for the hydrogen molecule H_2. In the general case of heteronuclear diatomic molecules, it will be $b \neq \pm 1$, but, apart from this, similar calculations can be carried out as in Eqs. (12.58)–(12.71b), with $E_a \neq E_b$.

12.5.2 VBM

The valence bond method is under certain respects similar to the LCAO approximation. Nevertheless, it attempts to obtain the molecular wave function as a product – rather than a linear combination – of the original atomic wave function. While in the LCAO method we first build the molecular states that are successively filled with the existing electrons, the VBM makes use of the bonds between decoupled electrons, ignoring the internal electronic levels (complete shells).

Differently from other sections of the present chapter, we need here to consider the case of the hydrogen molecule H_2. The reason is that the VBM works with bonds, i.e. with molecules presenting at least two electrons. We start with the product wave function

$$\Psi(1,2) = \psi_a(1)\psi_b(2), \tag{12.73}$$

where 1 and 2 denote the two electrons. By applying the usual optimization procedure (see Subsec. 12.5.1), we would arrive at a dissociation energy of about 0.25 eV, that should be compared with the experimental value of about 4.8 eV. This approximation is therefore very bad. However, one should not be surprised of such a discrepancy, because the wave function (12.73) is not symmetrized. A more clever ansatz is given by

$$\Psi^{\pm}_{\rm cov}(1,2) = \psi_a(1)\psi_b(2) \pm \psi_a(2)\psi_b(1), \tag{12.74}$$

which is sometimes called the *covalent wave function*: $\Psi^{+}_{\rm cov}$ yields a dissociation energy of 3.15 eV. The approximation can be further improved by accepting a small probability of an *ionic* wave function

$$\Psi^{\pm}_{\rm ion}(1,2) = \psi_a(1)\psi_a(2) \pm \psi_b(1)\psi_b(2), \tag{12.75}$$

i.e. by writing the total wave function as

$$\Psi(1,2) = \Psi^{\pm}_{\rm cov}(1,2) + \alpha\Psi^{\pm}_{\rm ion}(1,2), \qquad (12.76)$$

where α should be calculated using the variational method. In the case of the hydrogen molecule, it turns out that $\alpha = 0.25$, i.e. the weight of the ionic bond is about 25%. With this improvement, we arrive at a value for the dissociation energy of about 4.1 eV. Other small improvements may be obtained if one accepts a certain polarization of the atomic orbitals due to the electrons. Moreover, with these simple procedures it is also possible to foresee the existence (or the non-existence) of some diatomic molecules, such as He_2 or Li_2.

As with the LCAO method, in the case of heteronuclear molecules, besides what we have seen above, one should also consider having a certain weight in front of the product wave function terms. An interesting example is given by the fluoride hydrogen HF, where $\alpha = 0.5$, due to the fact that fluorine is very electronegative.

Summary

- In this chapter we have first introduced a formalism for the simplest molecule – namely the molecular ion H_2^+ – making use of *spheroidal coordinates*.
- Next, we introduced the *Born–Oppenheimer approximation* in order to solve the molecular problem in the most general terms. Here, we made use of the fact that the electronic dynamics is much faster than the nuclear one, so that the nuclei may be treated as essentially static.
- We also presented the analysis of diatomic molecules by separating the *vibrational*, *rotational*, and *translational degrees of freedom*.
- We have also introduced the *Morse potential* approximation, which turns out to be a very good description for the vibrational motion of diatomic molecules.
- Finally, we have presented two further approximations, the *linear combination of atomic orbitals* (LCAO) and the *valence bond method* (VBM). These tools are very useful when dealing with more complex molecules.

Problems

12.1 Make use of the ansatz (12.4) and of the explicit expression (12.3) for the Laplacian in order to derive Eqs. (12.5) from the Schrödinger equation (12.2).

12.2 Justify Eq. (12.13).

12.3 Derive Eq. (12.22).

12.4 Starting from Eq. (12.24) and applying the change of variables (12.27), derive Eq. (12.28).

12.5 Consider the Schrödinger equation $\hat{H}\psi(\mathbf{r}_a, \mathbf{r}_b) = E\psi(\mathbf{r}_a, \mathbf{r}_b)$, with $\hat{H} = \hat{H}_A + \hat{H}_B$, and the ansatz $\psi(\mathbf{r}_a, \mathbf{r}_b) = \varphi_a(\mathbf{r}_a)\varphi_b(\mathbf{r}_b)$. Show that such Schrödinger equation is separable into two Schrödinger equations for $\varphi_a(\mathbf{r}_a)$ and $\varphi_b(\mathbf{r}_b)$.

12.6 Make use of the ansatz (12.30) into the Schrödinger equation (12.24), with the transformed Hamiltonian (12.28), in order to separate the center-of-mass motion from the relative motion of the nuclei (Eqs. (12.31a)).

12.7 Build the complete roto-vibrational spectrum of transitions between rotational sublevels belonging to adjacent vibrational excitations.

12.8 Study the function (12.45) representing the Morse potential. Find its extrema and draw its graph.

12.9 Perform the change of variable (12.48) and derive Eq. (12.49).

12.10 Derive Eq. (12.51).

12.11 Derive the eigenvalues (12.55).

12.12 What is the main difference between the eigenvalues (12.55) and (12.42)? Compare the two expressions and make a link between the vibrational frequency ω_0 and the constant a.

Further reading

Atkins, P. and Friedman, R., *Molecular Quantum Mechanics*, Oxford: Oxford University Press, 4th edn., 2005.

Brandsen, B. C. J., *Physics of Atoms and Molecules*, London: Longman, 1983.

Bohm, Arno, *Quantum Mechanics: Foundations and Applications*, New York: Springer, 1st edn., 1979; 2nd edn., 198; 3rd edn., 1993.

Harrison, Walter A., *Applied Quantum Mechanics*, Singapore: World Scientific, 2000.

Herzberg, Gerhard, *Molecular Spectra and Molecular Structure. I. Spectra of Diatomic Molecules*, New York: Van Nostrand Reinhold, 1950.

McWeeny, Roy (Ed.), *Coulson's Valence*, Oxford: Oxford University Press, 3rd edn., 1979.

Slater, John C., *Quantum Theory of Matter*, New York: McGraw-Hill, 1951.

White, Harvey E., *Introduction to Atomic Spectra*, New York: McGraw-Hill, 1934.

13 Quantum optics

The behavior of light and its interaction with matter is certainly one of the most interesting phenomena of physics. It has puzzled thinkers and scientists in ancient Greece (Archimedes) and through the Middle Ages, at least since the times of Robert Grosseteste (1168–1235), right up to the theories of Galileo Galilei and Isaac Newton.[1] The classical electromagnetic theory of light (see Tab. 13.1) was beautifully established in the second half of the nineteenth century with the Maxwell equations.[2] Quantum optics is nothing other than the quantum theory of light. It finds its roots in Planck's discovery (1900) that the assumption of energy quantization for a harmonic oscillator allowed us to solve the black-body problem (see Subsec. 1.5.1). The successive early steps that led to the quantization of the electromagnetic field include Einstein's explanation of the photoelectric effect with the introduction of the quantum of light (see Subsec. 1.2.1) and the interpretation of atomic spectra (see Subsec. 1.5.4).

Quantum optics is usually not covered by conventional textbooks on quantum mechanics. We include its discussion here for two main reasons: first, because it is a fundamental and advanced application of the basic framework of quantum mechanics, and, second, because in the last 30 years quantum optics has played (and is still playing) a major role in the advancement and understanding of quantum mechanics.

The main aim of this chapter is then to present the quantum properties of light and of its interaction with matter. We shall mainly focus on linear quantum optics and also include a short account of non-linear phenomena (such as parametric down-conversion). We start in Sec. 13.1 by illustrating how the classical equations of the electromagnetic field can be quantized, leading to the concept of the photon and to its genuinely quantum properties. In Sec. 13.2 we study the thermodynamic equilibrium of the radiation field. In Sec. 13.3 we derive the phase–number uncertainty relation. In Sec. 13.4 we consider three fundamentally quantum states of light: fock, coherent, and squeezed states. In Sec. 13.5 we present the main phase-space quasi-probability distributions: the Q-function, the characteristic function, the P-function, and the Wigner function, which enable us to illustrate the formalism of phase-space quantum optics. In Sec. 13.6 we discuss both first- and second-order coherence and an interesting optical detection technique, namely homodyne detection. In Sec. 13.7 we introduce the formalism of atom–cavity interaction, while, in Sec. 13.8, we finally discuss the geometric phase and the Aharonov–Bohm effect. Finally, in Sec. 13.9 we analyze the Casimir effect.

1 [*Newton* 1704].

2 [*Maxwell* 1873].

Table 13.1 Main components of the electromagnetic spectrum. For each component the typical transitions giving rise to radiation belonging to that part of the spectrum are indicated. The wave number is given by $\tilde{\nu} = E/hc = \nu/c = \lambda^{-1}$

Energy (ev)	Frequency (s^{-1})	Transitions	Radiation	Wave number (cm^{-1})	Wavelength (cm)
$E < 5 \times 10^{-6}$	$\nu < 1.2 \times 10^{9}$	Nuclear magnetic resonance	Radio waves	$\tilde{\nu} < 4 \times 10^{-2}$	$\lambda < 25$
5×10^{-6}	1.2×10^{9}	Spin orientation in magnetic field		4×10^{-2}	25
5×10^{-6}	1.2×10^{9}	Electron spin resonance	Microwaves (radar)	4×10^{-2}	25
$< E <$	$< \nu <$			$< \tilde{\nu} <$	$> \lambda >$
3.1×10^{-3}	7.5×10^{11}			25	4×10^{-2}
3.1×10^{-3}	7.5×10^{11}	Molecular rotations		25	4×10^{-2}
5×10^{-2}	1.2×10^{13}	Molecular vibrations	Infrared region	4×10^{2}	2.5×10^{-3}
$< E <$	$< \nu <$			$< \tilde{\nu} <$	$> \lambda >$
0.5	1.2×10^{14}			4×10^{3}	2.5×10^{-4}
1.55	3.8×10^{14}		Visible spectrum	12.5×10^{3}	8×10^{-5}
$< E <$	$< \nu <$			$< \tilde{\nu} <$	$> \lambda >$
3.1	7.5×10^{14}			25×10^{3}	4×10^{-5}
3.1	7.5×10^{14}	Valence electronic transitions		25×10^{3}	4×10^{-5}
$< E <$	$< \nu <$			$< \tilde{\nu} <$	$> \lambda >$
6.2	1.5×10^{15}			50×10^{3}	2×10^{-5}
6.2	1.5×10^{15}		Ultraviolet	50×10^{3}	2×10^{-5}
1,240	3×10^{17}	Inner shell electronic transitions		10^{7}	10^{-7}
$< E <$	$< \nu <$		X-rays	$< \tilde{\nu} <$	$> \lambda >$
1.24×10^{4}	3×10^{18}			10^{8}	10^{-8}
$1.24 \times 10^{4} < E$	$3 \times 10^{18} < \nu$	Nuclear transitions	Gamma rays	$10^{8} < \tilde{\nu}$	$10^{-8} > \lambda$

13.1 Quantization of the electromagnetic field

13.1.1 Classical equations of the electromagnetic field

The most natural and convenient way to describe how the electromagnetic field can be quantized is to start from the classical Maxwell's equations, which in free space and in absence of sources read as

$$\nabla \cdot \mathbf{B} = 0, \tag{13.1a}$$

$$\nabla \times \mathbf{E} = -\frac{\partial}{\partial t}\mathbf{B}, \tag{13.1b}$$

$$\nabla \cdot \mathbf{E} = 0, \tag{13.1c}$$

$$\nabla \times \mathbf{B} = \frac{1}{c^2}\frac{\partial}{\partial t}\mathbf{E}, \tag{13.1d}$$

where $\mathbf{B}$ is the magnetic field (see footnote 17, p. 224), $\mathbf{E}$ is the electric field,

$$c = \frac{1}{\sqrt{\epsilon_0 \mu_0}} \tag{13.2}$$

is the vacuum speed of light, and ϵ_0 and μ_0 are the vacuum electric permittivity and magnetic permeability, respectively. These equations could be written in a shorter and more compact way using the relativistic formalism of quadrivectors, but we omit these considerations in this textbook. As it is well known from classical electrodynamics, the properties of the electromagnetic field can be extracted from the vector potential $\mathbf{A}$ and the scalar potential U. In fact, $\mathbf{E}$ and $\mathbf{B}$ can be calculated from $\mathbf{A}$ and U thanks to the relations (see Eqs. (4.128))

$$\mathbf{E} = -\nabla \mathrm{U} - \frac{\partial}{\partial t}\mathbf{A}, \tag{13.3a}$$

$$\mathbf{B} = \nabla \times \mathbf{A}. \tag{13.3b}$$

Making use of the $\mathbf{A}$–U representation, Eqs. (13.1a) and (13.1b) are automatically satisfied, whereas Eqs. (13.1c) and (13.1d) become

$$\nabla \cdot \left(-\nabla \mathrm{U} - \frac{\partial}{\partial t}\mathbf{A}\right) = 0, \tag{13.4a}$$

$$\nabla \times (\nabla \times \mathbf{A}) = \frac{1}{c^2}\frac{\partial}{\partial t}\left(-\nabla \mathrm{U} - \frac{\partial}{\partial t}\mathbf{A}\right). \tag{13.4b}$$

This change of representation from the fields to the potentials presents a difficulty, since different scalar and vector potentials may lead to the same fields. In particular, since for any scalar function $f(\mathbf{r}, t)$,

$$\nabla \times (\nabla f) = 0, \tag{13.5}$$

the combined *gauge* transformations

$$\mathbf{A} \rightarrow \mathbf{A} + \nabla f, \tag{13.6a}$$

$$\mathrm{U} \rightarrow \mathrm{U} - \frac{\partial}{\partial t} f \tag{13.6b}$$

do not alter the values of **E** and **B**. Since the latter represent the measurable properties of the electromagnetic field, different potentials that lead to the same fields would describe the same physical situation. Then, if wish to translate Eqs. (13.1c) and (13.1d) in a univocal way, we have to add an extra constraint to Eqs. (13.3). For our purposes, it is convenient to adopt the so-called Coulomb or radiation gauge, for which

$$\nabla \cdot \mathbf{A} = 0. \tag{13.7}$$

Moreover, the scalar potential is a function of the spatial charge distribution and, in the case where there are no sources, can be eliminated from the problem with the choice

$$\mathrm{U} = 0. \tag{13.8}$$

Then, Eq. (13.3a) becomes

$$\mathbf{E} = -\frac{\partial}{\partial t}\mathbf{A}, \tag{13.9}$$

and Eq. (13.1c) is automatically satisfied. On the other hand, Eq. (13.4b) finally reads (see Prob. 13.2)

$$\nabla^2 \mathbf{A}(\mathbf{r}, t) = \frac{1}{c^2}\frac{\partial^2}{\partial t^2}\mathbf{A}(\mathbf{r}, t). \tag{13.10}$$

Equation (13.10) shows the important result that the vector potential satisfies the classical wave equation. It is known that the classical solution of Eq. (13.10) is given by the expansion

$$\mathbf{A} = \frac{1}{2\pi}\int d\mathbf{k}d\omega e^{\imath\omega t}e^{\imath\mathbf{k}\cdot\mathbf{r}}\delta(k^2 - c^2\omega^2), \tag{13.11}$$

where, as usual, **k** is the wave propagation vector and ω is the angular frequency. Note that we have expressed the vector potential as an integral ranging on two components that show a spatial and a temporal dependency, respectively.

13.1.2 Quantum modes of the field

In order to accomplish the quantization of the electromagnetic field, it is now required that we replace the classical vector potential **A** by a quantum-mechanical operator $\hat{\mathbf{A}}$.[3] In analogy with the classical case, we then expand the vector potential operator in a Fourier series as

$$\hat{\mathbf{A}}(\mathbf{r}, t) = \sum_{\mathbf{k}} c_{\mathbf{k}}\left[\hat{a}_{\mathbf{k}}\mathbf{u}_{\mathbf{k}}(\mathbf{r})e^{-\imath\omega_{\mathbf{k}}t} + \hat{a}^{\dagger}_{\mathbf{k}}\mathbf{u}^{*}_{\mathbf{k}}(\mathbf{r})e^{\imath\omega_{\mathbf{k}}t}\right], \tag{13.12}$$

[3] It is clear that we cannot make use here of all the tools needed to perform a proper quantization of the electromagnetic field. Indeed, this should be done in a relativistic setting, writing the Maxwell equation in Hamiltonain form, contolling that the procedure does not depend on the gauge, and, if we work in the Coulomb gauge, quantizing a system with these constraints. Here, we follow the heuristic shortcut of assuming that, in radiation gauge, the quantum electromagnetic field satisfies the same equations of the classical field.

where the $c_{\mathbf{k}}$'s are constants to be determined, the dimensionless amplitudes $\hat{a}_{\mathbf{k}}, \hat{a}_{\mathbf{k}}^{\dagger}$ are now quantum-mechanical operators (to be interpreted below), and $\mathbf{u}_{\mathbf{k}}(\mathbf{r})$ is a discrete set of orthogonal mode functions that, due to Eqs. (13.10) and (13.7), have to satisfy

$$\left(\nabla^2 + \frac{\omega_{\mathbf{k}}^2}{c^2}\right) \mathbf{u}_{\mathbf{k}}(\mathbf{r}) = \left(\nabla^2 + \mathbf{k}^2\right) \mathbf{u}_{\mathbf{k}}(\mathbf{r}) = 0, \tag{13.13a}$$

$$\nabla \cdot \mathbf{u}_{\mathbf{k}}(\mathbf{r}) = 0, \tag{13.13b}$$

where we have made use of the fact that $\omega_{\mathbf{k}} = c|\mathbf{k}|$. If, for simplicity, we choose to quantize the free field inside a cube of side l (the so-called *cavity* or electromagnetic resonator) with periodic boundary conditions, the wave (propagation) vector $\mathbf{k}$ has components

$$k_x = \frac{2\pi n_x}{\mathrm{l}}, \quad k_y = \frac{2\pi n_y}{\mathrm{l}}, \quad k_z = \frac{2\pi n_z}{\mathrm{l}}, \tag{13.14}$$

where n_x, n_y, and n_z are integers. Notice that in many applications, l will tend to infinity. Moreover, the mode functions may be expressed as plane waves, i.e.

$$\mathbf{u}_{\mathbf{k}}(\mathbf{r}) = \frac{\mathbf{e}}{\mathrm{l}^{\frac{3}{2}}} e^{\imath \mathbf{k} \cdot \mathbf{r}}, \tag{13.15}$$

and the polarization vector $\mathbf{e}$ must satisfy the condition $\mathbf{e} \cdot \mathbf{k} = 0$ as a consequence of the gauge condition (13.13b). This implies, as expected, that $\mathbf{e}$ is always orthogonal to the propagation direction, and therefore the polarization vector has two independent directions. In other words, this confirms the transverse nature of the electromagnetic waves: the space where $\mathbf{e} \cdot \mathbf{k} = 0$ is two-dimensional, so we have the unit polarization vectors $\mathbf{e}_\lambda$ ($\lambda = 1, 2$), for which

$$\mathbf{k} \cdot \mathbf{e}_\lambda = 0, \tag{13.16a}$$

$$\mathbf{e}_\lambda \cdot \mathbf{e}_{\lambda'} = \delta_{\lambda\lambda'}. \tag{13.16b}$$

Then, taking into account Eqs. (13.16), we rewrite the mode functions as

$$\mathbf{u}_{\mathbf{k},\lambda}(\mathbf{r}) = \frac{\mathbf{e}_\lambda}{\mathrm{l}^{\frac{3}{2}}} e^{\imath \mathbf{k} \cdot \mathbf{r}}, \tag{13.17}$$

as well as the operator $\hat{\mathbf{A}}$

$$\hat{\mathbf{A}}(\mathbf{r}, t) = \sum_{\mathbf{k},\lambda} c_{\mathbf{k}} \left[\hat{a}_{\mathbf{k},\lambda} \mathbf{u}_{\mathbf{k},\lambda}(\mathbf{r}) e^{-\imath \omega_{\mathbf{k}} t} + \hat{a}_{\mathbf{k},\lambda}^{\dagger} \mathbf{u}_{\mathbf{k},\lambda}^{*}(\mathbf{r}) e^{\imath \omega_{\mathbf{k}} t}\right]. \tag{13.18}$$

A complete calculation shows that the normalization factors have to be written as (see Prob. 13.3)

$$c_{\mathbf{k}} = \sqrt{\frac{\hbar}{2\omega_{\mathbf{k}} \epsilon_0}}. \tag{13.19}$$

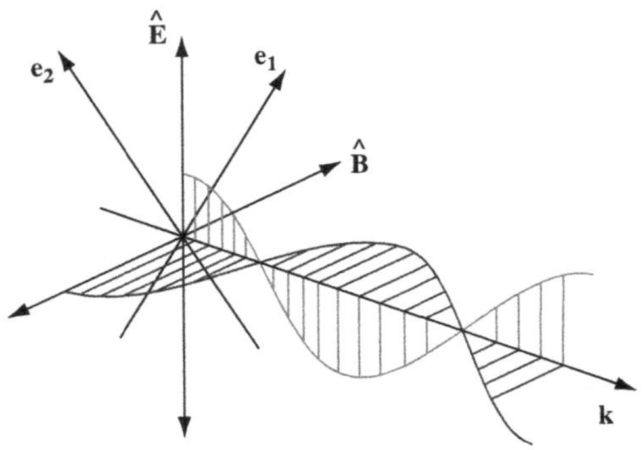

Figure 13.1 The three principal directions of the electromagnetic field. We assume that e_1 is parallel to the x direction, e_2 is parallel to the y direction, and k is in the z direction. The electric and magnetic fields oscillate along orthogonal directions.

Equation (13.18) together with Eqs. (13.9) and (13.3b) allows us to express the electric and magnetic fields as (see Prob. 13.4)

$$\hat{\mathbf{E}}(\mathbf{r},t) = \imath \sum_{\mathbf{k},\lambda} \left(\frac{\hbar\omega_{\mathbf{k}}}{2\epsilon_0}\right)^{\frac{1}{2}} \left[\hat{a}_{\mathbf{k},\lambda}\mathbf{u}_{\mathbf{k},\lambda}(\mathbf{r})e^{-\imath\omega_{\mathbf{k}}t} - \hat{a}^{\dagger}_{\mathbf{k},\lambda}\mathbf{u}^{*}_{\mathbf{k},\lambda}(\mathbf{r})e^{\imath\omega_{\mathbf{k}}t}\right], \tag{13.20a}$$

$$\hat{\mathbf{B}}(\mathbf{r},t) = \imath \sum_{\mathbf{k},\lambda} \left(\frac{\hbar k}{2c\mathrm{l}^3\epsilon_0}\right)^{\frac{1}{2}} \left[\hat{a}_{\mathbf{k},\lambda}e^{\imath(\mathbf{k}\cdot\mathbf{r}-\omega_{\mathbf{k}}t)} - \hat{a}^{\dagger}_{\mathbf{k},\lambda}e^{-\imath(\mathbf{k}\cdot\mathbf{r}-\omega_{\mathbf{k}}t)}\right]\mathbf{b}_\lambda, \tag{13.20b}$$

where

$$\mathbf{b}_\lambda = \mathbf{k} \times \mathbf{e}_\lambda \tag{13.21}$$

is a unit vector whose direction is orthogonal both to $\mathbf{k}$ and to $\mathbf{e}_\lambda$, in agreement with the fact that electric and magnetic fields oscillate along orthogonal directions, both remaining orthogonal to the wave vector (see Figs. 1.8 and 13.1).

13.1.3 Quantization of the field

In classical electrodynamics, $\mathbf{E}(\mathbf{r},t)$ and $\mathbf{B}(\mathbf{r},t)$ are real vectors and, as we have said, they represent the measurable properties of the electromagnetic field. Their quantum counterparts, as expected, become Hermitian operators. As we have said, quantization requires that $\hat{a}_{\mathbf{k}}$ and $\hat{a}^{\dagger}_{\mathbf{k}}$ be interpreted as quantum-mechanical operators. In order to correctly identify them, we resort to the classical expression of the energy for the electromagnetic field in a resonator, that is,[4]

$$H = \frac{1}{2}\int_{\mathrm{l}^3} d\mathbf{r}\left(\epsilon_0\mathbf{E}^2 + \frac{\mathbf{B}^2}{\mu_0}\right). \tag{13.22}$$

[4] [*Jackson* 1962, 237].

Substituting the expressions (13.20), we finally obtain (see Prob. 13.5)

$$\hat{H} = \sum_{\mathbf{k},\lambda} \hbar\omega_{\mathbf{k}} \left(\hat{a}^{\dagger}_{\mathbf{k},\lambda} \hat{a}_{\mathbf{k},\lambda} + \frac{1}{2} \right), \tag{13.23}$$

which shows (see Eq. (4.79)) that, if we interpret $\hat{a}_{\mathbf{k},\lambda}$ and $\hat{a}^{\dagger}_{\mathbf{k},\lambda}$ as the annihilation and creation operators, the Hamiltonian of the electromagnetic field describes a system of independent harmonic oscillators (see Sec. 4.4). Quantizing the electromagnetic field then amounts to quantizing each of the harmonic oscillators, one for each mode of the field and polarization direction. As a consequence, field quantization is accomplished by interpreting $\hat{a}_{\mathbf{k},\lambda}$ and $\hat{a}^{\dagger}_{\mathbf{k},\lambda}$ as the annihilation and creation operators of the **k**-th field mode with polarization direction λ, respectively. In particular, they obey the commutation relations

$$\left[\hat{a}_{\mathbf{k},\lambda}, \hat{a}_{\mathbf{j},\lambda'}\right] = \left[\hat{a}^{\dagger}_{\mathbf{k},\lambda}, \hat{a}^{\dagger}_{\mathbf{j},\lambda'}\right] = 0, \quad \left[\hat{a}_{\mathbf{k},\lambda}, \hat{a}^{\dagger}_{\mathbf{j},\lambda'}\right] = \delta_{\mathbf{kj}}\delta_{\lambda\lambda'}. \tag{13.24}$$

As we know from Subsec. 4.4.2, the creation (annihilation) operator $\hat{a}^{\dagger}_{\mathbf{k},\lambda}$ ($\hat{a}_{\mathbf{k},\lambda}$) applied to a state of definite energy "adds" ("removes") a quantum of energy $\hbar\omega_{\mathbf{k}}$ to (from) the mode $\mathbf{k}$, λ of the cavity field. These energy quanta need to be interpreted as light quanta or photons (see Subsecs. 1.2.1 and 1.5.2), whose wave vector is precisely $\mathbf{k}$. The number of photons in each mode of the cavity is given by the eigenvalue $n_{\mathbf{k},\lambda}$ of the corresponding number operator

$$\hat{N}_{\mathbf{k},\lambda} = \hat{a}^{\dagger}_{\mathbf{k},\lambda} \hat{a}_{\mathbf{k},\lambda}. \tag{13.25}$$

In order to specify the total field in the cavity, it is then necessary to indicate the number of photons (or *occupation number*) for each mode. A generic state of the total field can be written as

$$\left| n_{\mathbf{k}_1}, n_{\mathbf{k}_2}, n_{\mathbf{k}_3}, \ldots \right\rangle = \left| n_{\mathbf{k}_1} \right\rangle \otimes \left| n_{\mathbf{k}_2} \right\rangle \otimes \left| n_{\mathbf{k}_3} \right\rangle \otimes \cdots, \tag{13.26}$$

where, for the sake of simplicity, we have omitted the λ dependence, and the equality sign is a consequence of the fact that different cavity modes are independent. This also means that, for instance,

$$\hat{a}^{\dagger}_{\mathbf{k}_j} \left| n_{\mathbf{k}_1}, n_{\mathbf{k}_2}, n_{\mathbf{k}_3}, \ldots, n_{\mathbf{k}_j}, \ldots \right\rangle = \sqrt{n_{\mathbf{k}_j} + 1} \left| n_{\mathbf{k}_1}, n_{\mathbf{k}_2}, n_{\mathbf{k}_3}, \ldots, n_{\mathbf{k}_j} + 1, \ldots \right\rangle, \tag{13.27}$$

i.e. an operator pertaining to a certain mode $\mathbf{k}$ affects only the photons in that particular mode. Finally, we note that the total energy of a cavity field in the state (13.26), that is,

$$\left| \{n_{\mathbf{k}}\} \right\rangle = \left| n_{\mathbf{k}_1} \right\rangle \otimes \left| n_{\mathbf{k}_2} \right\rangle \otimes \left| n_{\mathbf{k}_3} \right\rangle \otimes \cdots, \tag{13.28}$$

is given by

$$E = \sum_{\mathbf{k}} \hbar\omega_{\mathbf{k}} \left(n_{\mathbf{k}} + \frac{1}{2} \right). \tag{13.29}$$

It should be noted from Eq. (13.29) that, even in the case when no excitations are present (i.e. when no photon is present in each of the modes **k**), the total energy does not vanish. In fact, the energy of this state, called the *vacuum state*, is

$$E_0 = \frac{\hbar}{2} \sum_{\mathbf{k}} \omega_{\mathbf{k}}, \tag{13.30}$$

and, since the sum is infinite and there is no upper bound for the allowed frequencies $\omega_{\mathbf{k}}$, it is also infinite. This energy is called the zero-point energy and constitutes a puzzling feature of quantum electrodynamics (see Sec. 13.9). However, if we neglect gravity, this feature does not create particular difficulties for calculations aimed at comparing theoretical results with experiments: practical experiments are only sensitive to *changes* in the total energy of the electromagnetic field, for which, of course, the zero-point energy cancels out. As a matter of fact, from a practical perspective, the zero-point energy represents a conceptual difficulty in quantum field theory but will not affect the results of the present chapter.[5]

13.2 Thermodynamic equilibrium of the radiation field

In the previous section, we have identified the quantized electromagnetic field with an ensemble of harmonic oscillators. In order to study the properties of the field in thermodynamic equilibrium, we then need to investigate the equilibrium properties of a single quantized harmonic oscillator, using the results of Sec. 5.4. In particular, Eq. (5.34) may be easily evaluated for a harmonic oscillator in the basis where $\hat{H}$ is diagonal. The elements of the density operator in this case are simply given by

$$\rho_{nm} = \langle n | \hat{\rho} | m \rangle = \frac{e^{-\beta\hbar\omega(n+1/2)}}{Z(\beta)} \delta_{nm}, \tag{13.31}$$

where $Z(\beta)$ is the canonical partition function and $\beta = (k_B T)^{-1}$. Equation (13.31) tells us that the equilibrium density operator $\hat{\rho}$ is diagonal in the energy (and number) representation. As expected, the thermodynamical equilibrium makes the off-diagonal elements in the energy representation vanish (see Sec. 9.4). In other words, the state at the equilibrium is a statistical mixture. In order to investigate its properties, let us first compute the partition function $Z(\beta)$ (see Eq. (5.31) and Prob. 13.6)

$$Z(\beta) = \mathrm{Tr}\left(e^{-\beta\hat{H}}\right) = \frac{e^{-\frac{1}{2}\beta\hbar\omega}}{1 - e^{-\beta\hbar\omega}}. \tag{13.32}$$

Using Eq. (5.32a), we may derive the mean energy (see Prob. 13.7)

$$\langle E \rangle = \frac{1}{2}\hbar\omega + \frac{\hbar\omega}{e^{\beta\hbar\omega} - 1}, \tag{13.33a}$$

[5] See [*Mandl/Shaw* 1984].

which, apart from the additive constant $(1/2)\hbar\omega$ is precisely Planck's formula (1.66) for the average energy of a quantized harmonic oscillator, a result which confirms and strengthens our correspondence, established in the previous section, between the modes of the electromagnetic fields and harmonic oscillators. In a similar way (see part (b) of Prob. 13.7), we may calculate the mean occupation number as

$$\langle \hat{N} \rangle = \sum_{n=0}^{\infty} n\rho_{nn} = \left(e^{\beta\hbar\omega} - 1\right)^{-1}. \tag{13.33b}$$

The last two results allow us to study the behavior of the oscillator in the low- and high-temperature limits. In particular, for $T \to 0$ we easily find

$$\langle E \rangle \to \frac{1}{2}\hbar\omega, \quad \langle \hat{N} \rangle \to 0, \tag{13.34}$$

which may be interpreted by saying that in the low-temperature limit all the oscillators will be in the ground state with a probability close to 1.

On the other hand, for $T \to \infty$, we have

$$\langle E \rangle \approx k_{\rm B}T, \quad \langle \hat{N} \rangle \approx \frac{k_{\rm B}T}{\hbar\omega}, \tag{13.35}$$

which show that the mean energy approaches the value predicted by the classical Maxwell–Boltzmann distribution. As a consequence, the mean occupation is the average energy available divided by the energy of a quantum.

The above limits are a clear signature of the Bose–Einstein distribution (see Subsec. 7.3.3). If we have many oscillators (as it is the case for the quantized radiation field), the total distribution may be written by multiplying many terms of the type (13.32), one for each mode of the field. This result comes about straightforwardly when substituting Eq. (13.23) into Eq. (13.32).

13.3 Phase-number uncertainty relation

In classical electrodynamics, it is usual to separate the complex fields into a product of a real amplitude and a phase factor in Fourier space (see Eq. (10.105)). If we want to proceed in a similar way for the quantized field, we need to write the operators $\hat{a}$ and $\hat{a}^\dagger$ in Eqs. (13.18) and (13.20) as a product of amplitude and phase operators. As we shall see, however, the introduction of the concepts of phase and of phase operator into the quantum-mechanical description of the field is not free from obstacles and difficulties. First of all, such a decomposition is not unique, which leaves a certain freedom in the definition of a phase operator. Nevertheless, any good candidate as phase operator should obviously be an Hermitian operator and give the correct phase properties of the classical fields in the classical limit.

As an attempt, following the suggestion of Eqs. (4.85), let us consider the phase operator $\widehat{e^{\imath\phi}}$ defined by the relation[6]

$$\hat{a} = \left(\hat{N}+1\right)^{\frac{1}{2}} \widehat{e^{\imath\phi}}, \tag{13.36a}$$

whose Hermitian conjugate reads as

$$\hat{a}^{\dagger} = \widehat{e^{-\imath\phi}} \left(\hat{N}+1\right)^{\frac{1}{2}}. \tag{13.36b}$$

From Eqs. (13.36) we may derive

$$\widehat{e^{\imath\phi}} = \left(\hat{N}+1\right)^{-\frac{1}{2}} \hat{a}, \tag{13.37a}$$

$$\widehat{e^{-\imath\phi}} = \hat{a}^{\dagger} \left(\hat{N}+1\right)^{-\frac{1}{2}}, \tag{13.37b}$$

which yield

$$\widehat{e^{\imath\phi}}\widehat{e^{-\imath\phi}} = \hat{I}. \tag{13.38}$$

It should be noted, however, that the product $\widehat{e^{-\imath\phi}}\widehat{e^{\imath\phi}}$ is not equal to the unity operator (see Prob. 13.8). Therefore, the previous expressions should not be considered as exponentials of a phase operator (see Th. 3.1: p. 123). This is the reason why the operator symbol refers to the entire expression rather than to ϕ. The action of the exponential phase operator on the basis states $|n\rangle$ can easily be determined using Eqs. (4.85) and (13.37) in order to obtain (see Prob. 13.9)

$$\widehat{e^{\imath\phi}}\,|n\rangle = \begin{cases} |n-1\rangle & \text{if } n \neq 0 \\ 0 & \text{if } n = 0 \end{cases}, \tag{13.39a}$$

$$\widehat{e^{-\imath\phi}}\,|n\rangle = |n+1\rangle\,. \tag{13.39b}$$

It can be shown that the two exponential phase operators are not Hermitian (see Prob. 13.11), and therefore cannot describe observable properties of the electromagnetic field. Nevertheless, they can be used to define another pair of operators (see Prob. 13.12)

$$\widehat{\cos\phi} = \frac{1}{2}\left(\widehat{e^{\imath\phi}} + \widehat{e^{-\imath\phi}}\right), \tag{13.40a}$$

$$\widehat{\sin\phi} = \frac{1}{2\imath}\left(\widehat{e^{\imath\phi}} - \widehat{e^{-\imath\phi}}\right), \tag{13.40b}$$

whose non-vanishing matrix elements are (see again Prob. 13.11)

$$\langle n-1\,|\widehat{\cos\phi}|\,n\rangle = \langle n\,|\widehat{\cos\phi}|\,n-1\rangle = \frac{1}{2}, \tag{13.41a}$$

$$\left\langle n-1\left|\widehat{\sin\phi}\right|n\right\rangle = \frac{1}{2\imath}, \quad \left\langle n\left|\widehat{\sin\phi}\right|n-1\right\rangle = -\frac{1}{2\imath}, \tag{13.41b}$$

which show that the operators $\widehat{\cos\phi}$ and $\widehat{\sin\phi}$ are Hermitian. Therefore, it is suitable to adopt them as representing the observable phase properties of the electromagnetic field.

6 See [Susskind/Glogower 1964].

It can also be shown (see Prob. 13.13) that the operators $\widehat{\cos\phi}$ and $\widehat{\sin\phi}$ do not commute with the number operator, i.e.

$$\left[\hat{N}, \widehat{\cos\phi}\right] = \imath \widehat{\sin\phi}, \quad \left[\hat{N}, \widehat{\sin\phi}\right] = -\imath \widehat{\cos\phi}. \tag{13.42}$$

This fact is a peculiar feature of the quantized electromagnetic field, and suggests that number and phase – similarly to position and momentum for a quantum particle – are two "incompatible" properties of the electromagnetic field: they cannot be simultaneously measured with arbitrary precision (see Subsec. 2.2.7). Rigorously speaking, it is possible to derive a set of number–phase uncertainty relations similar to the uncertainty relation between position and momentum. In fact, applying Eq. (2.200) to the above considered operators and making use of Eqs. (13.42) yields[7]

$$\Delta\hat{N}\Delta\widehat{\cos\phi} \geq \frac{1}{2}\left|\langle\widehat{\sin\phi}\rangle\right|, \quad \Delta\hat{N}\Delta\widehat{\sin\phi} \geq \frac{1}{2}\left|\langle\widehat{\cos\phi}\rangle\right|. \tag{13.43}$$

It should be noted that the latter two uncertainty relations do not suffer the problem pointed out (in Subsec. 6.5) in the context of angular momentum. In fact, when $\Delta\hat{N} = 0$, the rhs of Eqs. (13.43) also vanishes (see Prob. 13.14).

13.4 Special states of the electromagnetic field

As we have already pointed out (in Subsec. 13.1.3), each mode of the electromagnetic field may be characterized by the occupation number. In general, however, the quantum state of a field mode can be represented as an expansion in terms of occupation numbers, i.e.

$$|\Psi_F\rangle = \sum_n c_n |n\rangle, \tag{13.44}$$

c_n being complex numbers satisfying $\sum_n |c_n|^2 = 1$. Among all possible states, in the following we discuss three special classes:

- *Fock (number) states.*[8] These are precisely the states we referred to in Subsec. 13.1.3, i.e. eigenstates of the energy and of the number operator. For these states the number of photons is perfectly determined and the phase is completely unknown according to the uncertainty relations (13.43).
- *Coherent states*, which are eigenstates of the annihilation operator $\hat{a}$ and are minimum-uncertainty states.[9] As we shall see, these are the states most similar to classical states, i.e. to points in phase space. The origin of their name comes from the fact that an initially

[7] See [Louisell 1963].
[8] See [Fock 1932].
[9] See [Glauber 1963a, Glauber 1963b, Glauber 1963c].

coherent state of an oscillator remains coherent at all later times and retains its minimum-uncertainty character at all times, with its center moving in phase space along trajectories predicted by classical mechanics.[10]

- *Squeezed states*. This is the most general class of minimum-uncertainty states. In this respect, they represent a generalization of the coherent state to the case where the uncertainty product is not symmetric. Since they have no classical analogue, they constitute a genuine quantum feature of the radiation field and are very useful, for instance, for quantum noise reduction in communication.

13.4.1 Fock states

As we have seen, Fock states are states with definite energy of the radiation field (see also Subsec. 13.1.3). We have already mentioned the phase–number uncertainty properties (see Prob. 13.14). From a less formal and more physical viewpoint, an important property of a Fock state concerns the electric field. In fact, for a single-mode field, we have (see Eq. (13.20a))

$$\left\langle n \left| \hat{\mathbf{E}} \right| n \right\rangle = \imath \left(\frac{\hbar\omega}{2\epsilon_0} \right)^{\frac{1}{2}} \left\langle n \left| \hat{a}\mathbf{u}(\mathbf{r})e^{-\imath\omega t} - \hat{a}^\dagger \mathbf{u}^*(\mathbf{r})e^{\imath\omega t} \right| n \right\rangle = 0, \tag{13.45}$$

that is, the expectation value of the electric field vanishes. On the other hand,

$$\left\langle n \left| \hat{\mathbf{E}}^2 \right| n \right\rangle = \frac{\hbar\omega}{2\epsilon_0} \left\langle n \left| \hat{a}\hat{a}^\dagger + \hat{a}^\dagger\hat{a} \right| n \right\rangle |\mathbf{u}(\mathbf{r})|^2 = \frac{\hbar\omega}{\epsilon_0 l^3} \left(n + \frac{1}{2} \right), \tag{13.46}$$

where we have made use of the fact that $\langle n \left| \hat{a}^2 \right| n \rangle = \left\langle n \left| (\hat{a}^\dagger)^2 \right| n \right\rangle = 0$ and of Eq. (13.15). Equation (13.46) shows that the fluctuations in the electric field are given by

$$\Delta_n \hat{\mathbf{E}} = \sqrt{\left\langle \hat{\mathbf{E}}^2 \right\rangle_n - \left\langle \hat{\mathbf{E}} \right\rangle_n^2} = \left[\frac{\hbar\omega}{\epsilon_0 l^3} \left(n + \frac{1}{2} \right) \right]^{\frac{1}{2}}, \tag{13.47}$$

even though $\left\langle \hat{\mathbf{E}} \right\rangle_n = 0$. It is especially interesting to note that such fluctuations are present even in the case $n = 0$, i.e. in the vacuum state. Such *vacuum fluctuations* have many interesting consequences and provide explanations for several physical phenomena, among which spontaneous emission (see Subsec. 13.7.5) and the Lamb shift.

13.4.2 Coherent states

In Subsec. 4.4.2 we have introduced the concept of coherent state of the harmonic oscillator, defined as a state which minimizes the uncertainty product $\Delta\hat{x}\Delta\hat{p}_x$ and which does not spread under the harmonic oscillator potential. Thus, the first example of a coherent

[10] See [Glauber 1966].

state is the ground state of the harmonic oscillator, which in the language of quantum electrodynamics is the vacuum state. The vacuum is an eigenstate of the annihilation operator with eigenvalue 0,

$$\hat{a}\,|0\rangle = 0\,|0\rangle = 0. \tag{13.48}$$

Equation (13.48) suggests the existence of a class of states $|\alpha\rangle$ that are eigenstates of the annihilation operator, i.e. are defined by the relation

$$\hat{a}\,|\alpha\rangle = \alpha\,|\alpha\rangle, \tag{13.49a}$$

where $\alpha = \langle\hat{a}\rangle_\alpha$. It is also evident that we must have

$$\langle\alpha\,|\,\alpha^* = \langle\alpha\,|\,\hat{a}^\dagger. \tag{13.49b}$$

In order to obtain an explicit expression of a coherent state we write it as an expansion in terms of Fock states, i.e.

$$|\alpha\rangle = \sum_n |n\rangle\,\langle n\,|\,\alpha\rangle, \tag{13.50}$$

where the amplitudes $\langle n\,|\,\alpha\rangle$ are the coefficients of this expansion. Multiplying Eq. (13.49a) by $\langle n|$ from the left, we obtain

$$\langle n\,|\,\alpha\rangle = \frac{\langle n\,|\hat{a}|\,\alpha\rangle}{\alpha} = \frac{\sqrt{n+1}}{\alpha}\,\langle n+1\,|\,\alpha\rangle. \tag{13.51}$$

Using the adjoint of (see Prob. 4.13)

$$|n\rangle = \frac{\left(\hat{a}^\dagger\right)^n}{\sqrt{n!}}\,|0\rangle, \tag{13.52}$$

i.e.

$$\langle n+1| = \langle 0|\,\frac{\hat{a}^{n+1}}{\sqrt{(n+1)!}}, \tag{13.53}$$

yields

$$\langle n\,|\,\alpha\rangle = \frac{\langle 0\,|\hat{a}^{n+1}|\,\alpha\rangle}{\alpha\sqrt{n!}} = \frac{\alpha^n}{\sqrt{n!}}\,\langle 0\,|\,\alpha\rangle. \tag{13.54}$$

Substituting Eq. (13.54) into Eq. (13.50), we obtain

$$|\alpha\rangle = \sum_n \langle 0\,|\,\alpha\rangle\,\frac{\alpha^n}{\sqrt{n!}}\,|n\rangle. \tag{13.55}$$

By imposing the normalization condition for coherent states,

$$\begin{aligned} 1 = \langle\alpha\,|\,\alpha\rangle &= \sum_{n,m} |\,\langle 0\,|\,\alpha\rangle\,|^2 \frac{\alpha^n(\alpha^*)^m}{\sqrt{n!\,m!}}\,\langle m\,|\,n\rangle \\ &= |\,\langle 0\,|\,\alpha\rangle\,|^2 \sum_n \frac{|\alpha|^{2n}}{n!} \\ &= |\,\langle 0\,|\,\alpha\rangle\,|^2 e^{|\alpha|^2}, \end{aligned} \tag{13.56}$$

and choosing the arbitrary phase of $\langle 0 \mid \alpha\rangle$ equal to zero, we obtain

$$\langle 0 \mid \alpha\rangle = e^{-\frac{1}{2}|\alpha|^2}, \tag{13.57}$$

which, substituted into Eq. (13.55), yields

$$|\alpha\rangle = e^{-\frac{|\alpha|^2}{2}} \sum_{n=0}^{\infty} \frac{\alpha^n}{\sqrt{n!}} |n\rangle. \tag{13.58}$$

Equation (13.58) shows that coherent states display a Poisson-like distribution of photons. In fact, the probability of finding n photons is given by

$$\wp(n) = |\langle n \mid \alpha\rangle|^2 = e^{-|\alpha|^2} \frac{|\alpha|^{2n}}{n!}, \tag{13.59}$$

which is a Poisson distribution of the photon number with (see Prob. 13.16)

$$\langle n\rangle = \left\langle \hat{N} \right\rangle_\alpha = |\alpha|^2, \tag{13.60a}$$

$$\Delta_\alpha n = \sqrt{\left\langle \hat{N}^2 \right\rangle_\alpha - \left\langle \hat{N} \right\rangle_\alpha^2} = |\alpha|. \tag{13.60b}$$

In order to investigate the properties of coherent states it is convenient to introduce the so-called quadrature operators,[11] defined by

$$\hat{X}_1 = \frac{1}{\sqrt{2}} \left(\hat{a}^\dagger + \hat{a}\right), \quad \hat{X}_2 = \frac{\imath}{\sqrt{2}} \left(\hat{a}^\dagger - \hat{a}\right). \tag{13.61}$$

These operators may be viewed as the dimensionless electric and magnetic fields of a single mode (see Eqs. (13.20)). In the context of a one-dimensional particle subjected to a harmonic-oscillator potential, $\hat{X}_1$ and $\hat{X}_2$ represent the dimensionless position and momentum operators, respectively (see Eqs. (4.73) and Prob. 13.17). It trivially follows that the quadratures obey the commutation relation

$$\left[\hat{X}_1, \hat{X}_2\right] = \imath \hat{I}. \tag{13.62}$$

From Eqs. (13.62) and (2.200), we obtain the quadrature uncertainty relation

$$\Delta \hat{X}_1 \Delta \hat{X}_2 \geq \frac{1}{2}. \tag{13.63}$$

It is possible to show that the necessary (see Prob. 13.18) and sufficient condition for a state $|\beta\rangle$ to be coherent is

$$\Delta_\beta \hat{X}_1 = \Delta_\beta \hat{X}_2 = \frac{1}{\sqrt{2}}, \tag{13.64}$$

[11] See [Yuen 1976].

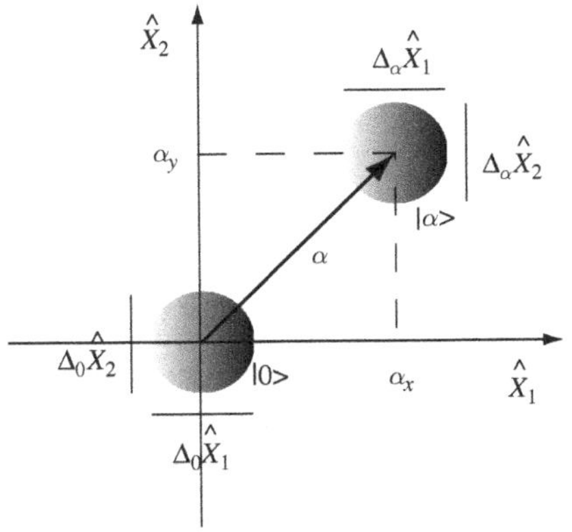

Figure 13.2 Pictorial representation of the coherent states and of the action of the displacement operator. Here the phase space is represented by the quadrature operators. Since $\langle \alpha | \hat{a} | \alpha \rangle = \alpha$ and $\langle \alpha | \hat{a}^\dagger | \alpha \rangle = \alpha^*$, it follows that for the vacuum state $\langle \hat{X}_1 \rangle_0 = \langle \hat{X}_2 \rangle_0 = 0$. Moreover, $\Delta_0 \hat{X}_1 = \Delta_0 \hat{X}_2 = 1/\sqrt{2}$, and so the vacuum state may be schematically depicted as a circle of radius $1/2\sqrt{2}$ centered around the origin. Similarly, any coherent state may be represented by a circle of radius $1/2\sqrt{2}$ centered around the point $\alpha = \alpha_x + \imath\alpha_y$ in the complex plane $(\hat{X}_1, \hat{X}_2) \equiv (\Re(\alpha), \Im(\alpha))$. Therefore, the action of the displacement operator $\hat{D}(\alpha)$ may be interpreted as that of displacing the vacuum state from the origin to the point $\alpha = (\alpha_x, \alpha_y)$ in the complex plane, as shown by the arrow.

i.e. coherent states are minimum-uncertainty states for which the quadratures are equally uncertain. Equation (13.64) may be seen as a third definition of coherent state.

Coherent states may also be viewed as the result of the action of a suitable operator on the vacuum state, which is the zero-amplitude coherent state. Making use again of the result of Eq. (13.52) in Eq. (13.58), we obtain

$$|\alpha\rangle = e^{-\frac{|\alpha|^2}{2}} \sum \frac{\alpha^n \left(\hat{a}^\dagger\right)^n}{n!} |0\rangle. \tag{13.65}$$

Since $\hat{a}\,|0\rangle = 0$, Eq. (13.65) may be rewritten as

$$|\alpha\rangle = e^{-\frac{1}{2}|\alpha|^2} e^{\alpha\hat{a}^\dagger} e^{-\alpha^*\hat{a}}\,|0\rangle. \tag{13.66}$$

The Baker–Hausdorff theorem (see Prob. 13.20) allows us to write the previous equation as

$$|\alpha\rangle = e^{\alpha\hat{a}^\dagger - \alpha^*\hat{a}}|0\rangle = \hat{D}(\alpha)\,|0\rangle, \tag{13.67}$$

where

$$\hat{D}(\alpha) = e^{\alpha\hat{a}^\dagger - \alpha^*\hat{a}} \tag{13.68}$$

is known as the *displacement operator*. Equation (13.67) shows that an arbitrary coherent state can be "generated" by displacing the vacuum state (see Fig. 13.2). It is easy to prove that this operator is unitary (see Prob. 13.21).

It is easy to show that the scalar product of two coherent states $|\alpha\rangle$ and $|\beta\rangle$ is given by (see Prob. 13.22)

$$\langle\alpha|\beta\rangle = e^{\alpha^*\beta-\frac{1}{2}|\alpha|^2-\frac{1}{2}|\beta|^2}, \tag{13.69}$$

which shows that two coherent states are never actually orthogonal to each other. However, since

$$|\langle\alpha \mid \beta\rangle|^2 = e^{-|\alpha-\beta|^2}, \tag{13.70}$$

if α and β are significantly different from each other,[12] i.e. $|\alpha-\beta| \gg 1$, then the two states are almost orthogonal. The last result is an indication of the fact that coherent states are *overcomplete*, i.e. they can be used as a basis for the whole Hilbert space and are also normalized, but notwithstanding they are not an orthogonal basis. In other words, any state may be expanded in terms of coherent states but this decomposition is not unique. The completeness relation for coherent states is given by

$$\frac{1}{\pi}\int d^2\alpha \, |\alpha\rangle \langle\alpha| = \hat{I}, \tag{13.71}$$

where $d^2\alpha = d\alpha_x d\alpha_y$ and the integral is over the whole complex plane. Equation (13.71) allows us to expand an arbitrary state in term of coherent states (see Prob. 13.24).

Concerning the phase properties, it is possible to show[13] that for large mean numbers of photons ($|\alpha|^2 \gg 1$) the phase uncertainty of a coherent state $|\alpha\rangle$ is given by

$$\Delta\widehat{\cos}\phi = \frac{\sin\theta}{2|\alpha|}, \tag{13.72}$$

where θ is the phase of the coherent excitation, i.e. $\theta = \arg\alpha$.[14] Making use of Eq. (13.60b) we arrive at the phase–number uncertainty relation for large-amplitude coherent states:

$$\Delta\hat{N}\,\Delta\widehat{\cos}\phi = \frac{1}{2}\sin\theta. \tag{13.73}$$

Equations (13.60) and (13.72) show that, for large mean photon number, the coherent state turns out to be better defined for what concerns both amplitude and phase. As a matter of fact, both the photon number fractional uncertainty ($\Delta n/ < n >$) and the phase uncertainty behave as $|\alpha|^{-1}$. This is one of the reasons why large-amplitude coherent states are considered to be a good approximation of classical states (see Fig. 13.3).

13.4.3 Squeezed states

Similarly to coherent states, squeezed states are also minimum uncertainty states for the quadratures $\hat{X}_1$ and $\hat{X}_2$. They are more general, however, since their minimum uncertainty condition

[12] In the representation of Fig. 13.2 this would correspond to the case where the distance between the centers of the two coherent states $|\alpha\rangle$ and $|\beta\rangle$ is much larger than their diameter.

[13] For a proof of the following result see [*Loudon* 1973, 148–50].

[14] Sometimes it is convenient to write $\alpha = |\alpha|e^{\iota\theta}$.

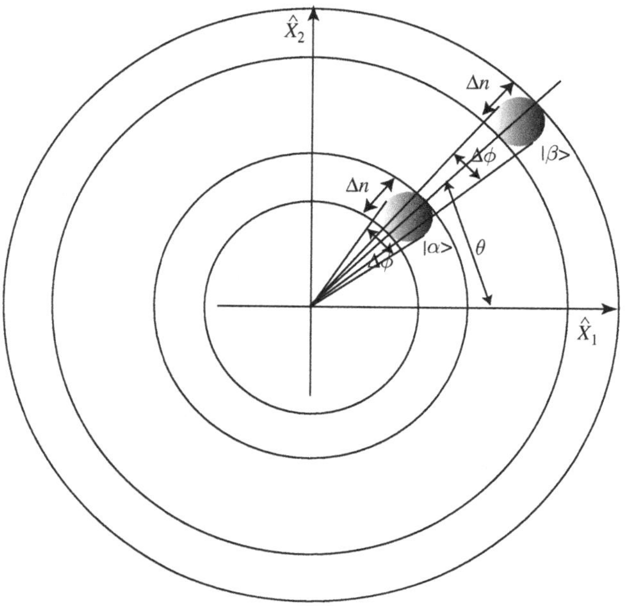

Figure 13.3 Phase-number uncertainty properties of coherent states. As the amplitude of the coherent state increases ($|\beta| > |\alpha|$ in the figure), it becomes better defined, both concerning the phase ($\Delta\phi$) and the number of photons ($\Delta n / < n >$).

$$\Delta\hat{X}_1\Delta\hat{X}_2 = \frac{1}{2} \tag{13.74}$$

does not require that $\Delta\hat{X}_1$ and $\Delta\hat{X}_2$ be equal: squeezed states are those states for which Eq. (13.74) is satisfied in spite of the fact that Eq. (13.64) is not satisfied. For this reason, we can conceive the coherent state as a limiting case of the squeezed states when $\Delta\hat{X}_1 = \Delta\hat{X}_2$. As a consequence, the phase space uncertainties of a squeezed state will be represented by an ellipse instead of a circle as in Fig. 13.4.

A squeezed state $|\alpha, \xi\rangle$ may be generated from the vacuum state as

$$|\alpha, \xi\rangle = \hat{D}(\alpha)\hat{S}(\xi)\,|0\rangle\,, \tag{13.75}$$

where $\hat{D}(\alpha)$ is the displacement operator (13.68), $\hat{S}(\xi)$ is the squeeze operator, defined as

$$\hat{S}(\xi) = e^{\frac{1}{2}(\xi^*\hat{a}^2 - \xi(\hat{a}^\dagger)^2)}, \tag{13.76}$$

and the complex number

$$\xi = re^{\imath\chi} \tag{13.77}$$

is the *squeezing parameter*. In other words, a squeezed state may be generated by first "squeezing" the vacuum along the direction at the angle $\chi/2$ (or $\chi/2 + \pi/2$) and then

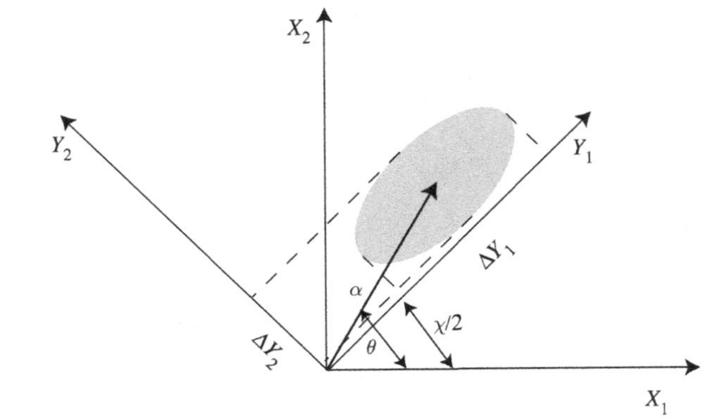

Figure 13.4 Phase convention for squeezed states. The phase of the coherent excitation is given by θ while the direction of squeezing is set by the parameter $\chi/2$. In the rotated frame $(\hat{Y}_1, \hat{Y}_2)$ the uncertainty $\Delta\hat{Y}_1$ and $\Delta\hat{Y}_2$ assume the simple form (13.81).

displacing the resulting squeezed vacuum state by α along the direction at the angle θ (see Fig. 13.5).[15] It can be immediately verified that

$$\hat{S}(-\xi) = \hat{S}^\dagger(\xi) = \hat{S}^{-1}(\xi). \tag{13.78}$$

As for a coherent state, the mean values of the annihilation and creation operators for a squeezed state are given by

$$\langle\hat{a}\rangle_{\alpha,\xi} = \langle\alpha,\xi\,|\hat{a}|\,\alpha,\xi\rangle = \alpha, \quad \left\langle\hat{a}^\dagger\right\rangle_{\alpha,\xi} = \left\langle\alpha,\xi\,\left|\hat{a}^\dagger\right|\,\alpha,\xi\right\rangle = \alpha^*. \tag{13.79}$$

In order to compute the quadrature uncertainties it is convenient to work in the rotated frame $(\hat{Y}_1, \hat{Y}_2)$ (see Fig. 13.4). Let us define the rotated quadratures $\hat{Y}_1$ and $\hat{Y}_2$ as

$$\left(\hat{Y}_1 + \imath\hat{Y}_2\right)e^{\imath\frac{\theta}{2}} = \hat{X}_1 + \imath\hat{X}_2 = \sqrt{2}\hat{a}. \tag{13.80}$$

In terms of these operators, we have (see Prob. 13.26)

$$\Delta_{\alpha,\xi}\hat{Y}_1 = \frac{1}{\sqrt{2}}e^{-r}, \tag{13.81a}$$

$$\Delta_{\alpha,\xi}\hat{Y}_2 = \frac{1}{\sqrt{2}}e^{r}, \tag{13.81b}$$

$$\Delta_{\alpha,\xi}\hat{Y}_1\Delta_{\alpha,\xi}\hat{Y}_2 = \frac{1}{2}. \tag{13.81c}$$

These equations show that r determines the amount of squeezing (and therefore is often called the *squeezing factor*), while χ determines the direction of squeezing. In certain contexts, one may talk about amplitude- and phase-squeezed states, depending on the quadrature that is squeezed (see Fig. 13.6).

[15] It should be noted that the operators $\hat{D}(\alpha)$ and $\hat{S}(\xi)$ do not commute with each other. Therefore, the state $|\,\tilde{\alpha},\xi\rangle = \hat{S}(\xi)\hat{D}(\alpha)\,|\,0\rangle$ is different from the state of Eq. (13.75) but is still a squeezed state.

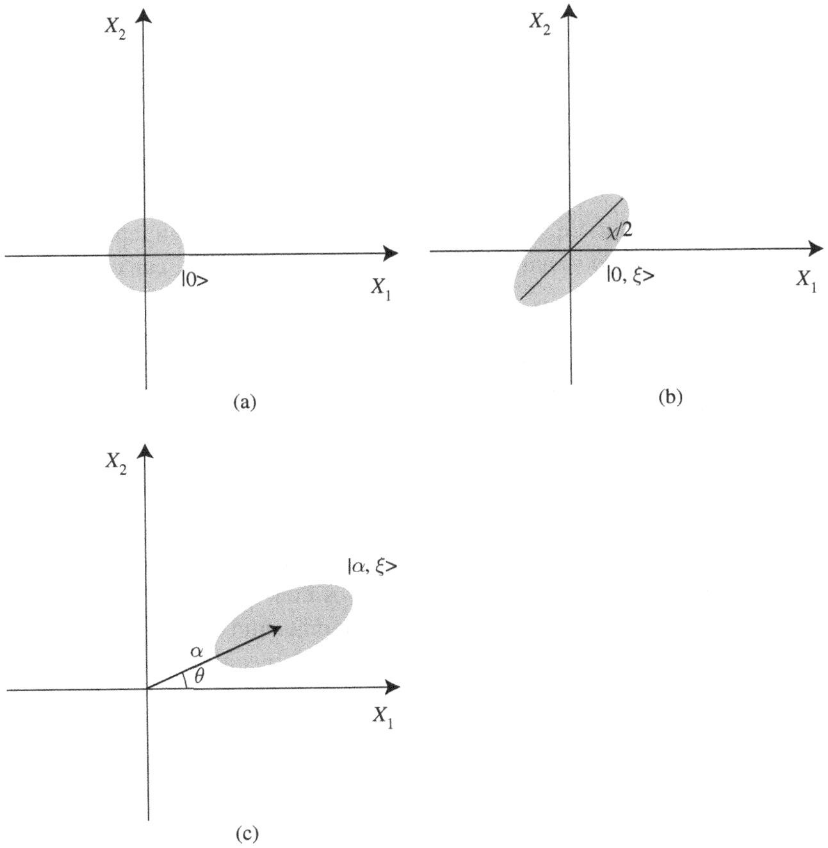

Figure 13.5 Generation of a squeezed state from the vacuum state $|0\rangle$. Starting from a vacuum state (a) we may first "squeeze" it through the squeeze operator in the phase space along the direction at the angle $\chi/2$, obtaining the squeezed state $|0, \xi\rangle$ (b). Second, we displace it by α through the displacement operator along the direction at the angle θ, obtaining the state $|\alpha, \xi\rangle$ (c).

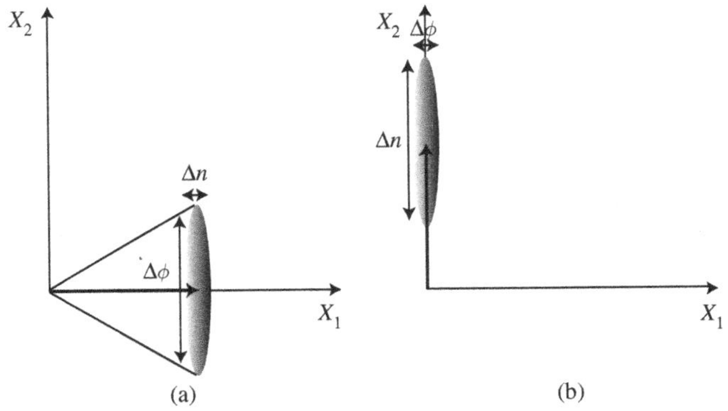

Figure 13.6 Phase-space of amplitude- and phase-squeezed states. (a) The quadrature carrying the coherent excitation is squeezed. This case corresponds to $\Im(\alpha) = 0$ and $\chi = \pi$. (b) The quadrature out of phase with the coherent excitation is squeezed. This case corresponds to $\Re(\alpha) = 0$ and $\chi = \pi$.

Finally, as regards the mean and variances of the number operator for a squeezed state, we have (see again Prob. 13.26)

$$\left\langle \hat{N} \right\rangle_{\alpha,\xi} = |\alpha|^2 + \sinh^2 r, \tag{13.82a}$$

$$\left(\Delta \hat{N}\right)^2_{\alpha,\xi} = \left|\alpha \cosh r - \alpha^* e^{\imath\chi} \sinh r\right|^2 + 2\cosh^2 r \sinh^2 r. \tag{13.82b}$$

In particular, Eq. (13.82a) shows the remarkable result that the mean number of photons in the squeezed vacuum state is larger (and in principle can even be much larger) than zero.

13.5 Quasi-probability distributions

Here we consider some phase-space methods which have widespread application in quantum optics. As we have seen (in Subsec. 2.3.3), quantum mechanics cannot be represented in a straightforward way in phase space as classical mechanics does. The greatest problem stems from the commutation relations, which seem to forbid a simultaneous perfect representation of momentum and position. In particular, it is not possible to define the joint probability that position and momentum (or, in the quantum optics language, the two quadrature operators) take on certain specified values. However, as we shall see, it is possible to build some very useful representations of quantum mechanical states in phase space.[16] The price that we must pay if we wish to avoid the use of negative probabilities is some "smoothing" of the phase-space distributions. In the following we introduce several quasi-probability distributions: the Q-function, the P-function, the Wigner function, and the characteristic function.[17] Even though, as we shall see, these have different properties, they may be seen as equivalent representations of the quantum state of a system. In other words, there is a one-to-one correspondence between each of these quasi-probability distributions and the density operator of the system.

Historically, the Wigner function was the first to be formulated. However, for the sake of simplicity we shall start with the Q-function.

13.5.1 Q-function

The Q-function[18] is essentially a coherent-state representation of a given state described by the density operator $\hat{\rho}$. It is defined as

$$\mathrm{Q}(\alpha, \alpha^*) = \frac{1}{\pi}\langle \alpha | \hat{\rho} | \alpha \rangle, \tag{13.83}$$

[16] In the following we shall specifically consider states of the quantized radiation field, but the same formalism can be translated to a generic quantum state.

[17] See [Cahill/Glauber 1969] [Hillery *et al.* 1984] for a review.

[18] It was introduced by Husimi [Husimi 1937].

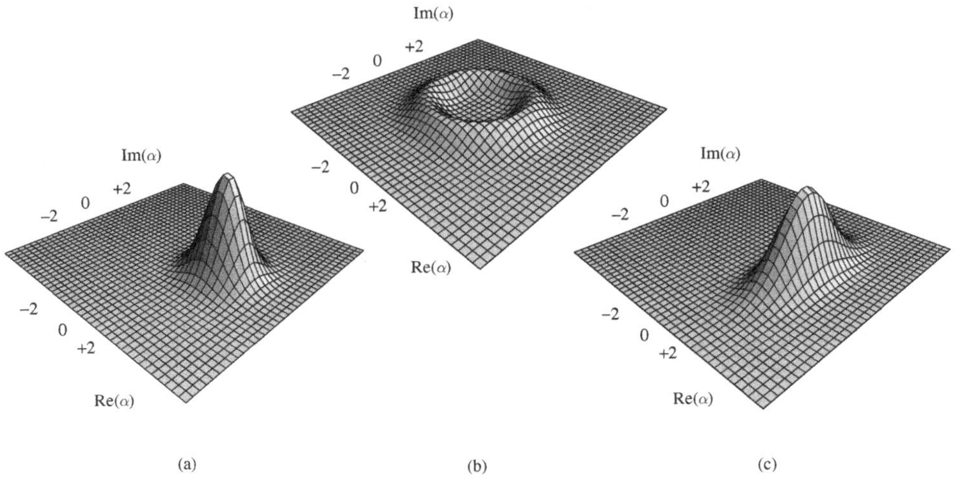

Figure 13.7 (a) Representation of the Q-function of a coherent state $|\alpha\rangle = |2\rangle$. As it may be seen from Eq. (13.84), it is a bidimensional Gaussian centered at the point $\alpha_0 = (2, 0)$ in the complex plane $(\Re(\alpha), \Im(\alpha))$. (b) Representation of the Q-function of a number state $|n\rangle$ with $n = 4$, i.e. with the number of photons equal to the mean number of photons in the coherent state (a). Its annular shape shows the phase-invariance of the number state. (c) Q-function representation of a squeezed state $|\alpha, \xi\rangle$ with $\alpha = 2$ and $\xi = 0.8$. For the analytic treatment of the Q-representation of a squeezed state see [*Scully/Zubairy* 1997] [*Walls/Milburn* 1994].

which represents the expectation value of the density matrix on a coherent state $|\alpha\rangle$, and is always real and positive. The simplest example of a Q-function is when the density matrix $\hat{\rho} = |\alpha_0\rangle\langle\alpha_0|$ describes a coherent state. In fact, Eq. (13.83) for a coherent state $|\alpha_0\rangle$ gives (see Eq. (13.70))

$$Q(\alpha, \alpha^*) = \frac{1}{\pi} e^{-|\alpha-\alpha_0|^2}, \tag{13.84}$$

which is represented in Fig. 13.7(a). The normalization of the density operator ensures the normalization of the Q-function. In fact,

$$1 = \mathrm{Tr}\left(\hat{\rho}\right) = \mathrm{Tr}\left(\frac{1}{\pi}\int d^2\alpha\,|\alpha\rangle\langle\alpha|\,\hat{\rho}\right) = \frac{1}{\pi}\int d^2\alpha\left\langle\alpha\left|\hat{\rho}\right|\alpha\right\rangle, \tag{13.85}$$

from which it follows that

$$\int d^2\alpha Q(\alpha, \alpha^*) = 1. \tag{13.86}$$

A similar procedure shows that

$$\begin{aligned}\left\langle\hat{a}^r\left(\hat{a}^\dagger\right)^s\right\rangle &\equiv \mathrm{Tr}\left[\hat{a}^r\left(\hat{a}^\dagger\right)^s\hat{\rho}\right] = \mathrm{Tr}\left[\frac{1}{\pi}\int d^2\alpha\hat{a}^r\,|\alpha\rangle\langle\alpha|\,(\hat{a}^\dagger)^s\hat{\rho}\right] \\ &= \frac{1}{\pi}\int d^2\alpha\alpha^r\alpha^{*s}\left\langle\alpha\left|\hat{\rho}\right|\alpha\right\rangle \\ &= \int d^2\alpha\alpha^r\left(\alpha^*\right)^s Q(\alpha, \alpha^*),\end{aligned} \tag{13.87}$$

i.e. the antinormally ordered quantum moments are given by simple moments of the Q-function.[19]

Though the Q-function behaves as a probability function (i.e. it is always positive and its integral is equal to one), not all positive normalizable Q-functions correspond to positive definite normalizable density operators, that is, it is a quasi-probability. The concept of *-probability* takes its roots in the fact that all the phase-space functions introduced in the present section may be seen as quantum counterparts to classical distribution functions. However, the quantum phase-space functions do not always express "bona fide" probabilities and in certain cases (as happens for the Wigner function) they may also have negative values, and this explains the term *quasi-*.

We note that the Q-function is also a bounded function (see Prob. 13.27), i.e.

$$Q(\alpha, \alpha^*) \leq \frac{1}{\pi}. \tag{13.88}$$

Making use of Eq. (13.58) we may write the Q-function in terms of an absolutely convergent power series in α and α^* as follows:

$$Q(\alpha, \alpha^*) = e^{-\alpha\alpha^*} \sum_{n,m} \frac{\langle n | \hat{\rho} | m \rangle}{\pi \sqrt{n!\, m!}} \alpha^m (\alpha^*)^n. \tag{13.89}$$

The double power series is absolutely convergent[20] since $e^{-\alpha\alpha^*}$ has an absolutely convergent power series and because (see Prob. 13.28)

$$| \langle n | \hat{\rho} | m \rangle | \leq 1. \tag{13.90}$$

In fact, the product of two absolutely convergent series is also absolutely convergent. Let us write the power series (13.89) in the form

$$Q(\alpha, \alpha^*) = \sum_{n,m} Q_{n,m} \alpha^m (\alpha^*)^n, \tag{13.91}$$

then an alternative expression for the density operator $\hat{\rho}$ in terms of the Q-function is represented by

$$\hat{\rho} = \pi \sum_{n,m} Q_{n,m} \left(\hat{a}^\dagger\right)^n \hat{a}^m. \tag{13.92}$$

In fact, Eqs. (13.49) show that $\langle \alpha | \hat{\rho} | \alpha \rangle / \pi$, with $\hat{\rho}$ given by the previous expression, is identical to the power series (13.91). As a second example, it is immediate to write the Q-function of a number state as (see Fig. 13.7(b))

19 A product of the creation and annihilation operators is said to be *normally* ordered when all the annihilation operators appear to the right of the creation operators. On the other hand, it is said to be *antinormally* ordered when all the annihilation operators appear to the left of the creation operators.

20 For the definition of absolutely convergent series see [*Apostol* 1969, I, ch. 10].

$$Q(\alpha, \alpha^*) = \frac{1}{\pi} | \langle \alpha \mid n \rangle |^2 = \frac{1}{\pi} e^{-|\alpha|^2} \frac{|\alpha|^{2n}}{n!}. \tag{13.93}$$

Finally, the Q-function of a squeezed state is presented for comparison in Fig. 13.7(c).

13.5.2 Characteristic function

We can write the quantum characteristic function for the electromagnetic field (see Eq. (2.75)) as

$$\chi(\eta, \eta^*) = \mathrm{Tr}\left[\hat{\rho} e^{\eta \hat{a}^\dagger} e^{-\eta^* \hat{a}}\right]. \tag{13.94}$$

This function is particularly useful because it helps to establish a relationship between different quantum-optical phase-space representations. As in classical mechanics (see Eq. (2.77)), the normally ordered moments are given by the derivatives of the characteristic function at $\eta = \eta^* = 0$, that is,

$$\left\langle \left(\hat{a}^\dagger\right)^r \hat{a}^s \right\rangle = (-)^s \frac{\partial^{r+s}}{\partial^r \eta \partial^s \eta^*} \chi(\eta, \eta^*) \bigg|_{\eta=\eta^*=0}, \tag{13.95}$$

where

$$\frac{\partial}{\partial \eta} \eta^* = \frac{\partial}{\partial \eta^*} \eta = 0, \tag{13.96}$$

which shows that η and η^* must be considered as two independent variables. Making use of the completeness condition (13.71) of the coherent states and of the Baker–Hausdorff theorem (see Prob. 13.20), we find

$$\chi(\eta, \eta^*) = \mathrm{Tr}\left[\frac{1}{\pi} \int d^2\alpha \hat{\rho} e^{|\eta|^2} e^{-\eta^* \hat{a}} |\alpha\rangle \langle \alpha | e^{\eta \hat{a}^\dagger}\right], \tag{13.97}$$

from which it follows that the characteristic function is related to the Q-function by the relationship

$$\chi(\eta, \eta^*) = e^{|\eta|^2} \int d^2\alpha e^{\eta\alpha^* - \alpha\eta^*} Q(\alpha, \alpha^*), \tag{13.98}$$

i.e. the characteristic function is a two-variable Fourier transform of the Q-function. One can also define an antinormally ordered characteristic function, that is,

$$\chi_A(\eta, \eta^*) = \mathrm{Tr}[\hat{\rho} e^{-\eta^* \hat{a}} e^{\eta \hat{a}^\dagger}]. \tag{13.99}$$

The characteristic function for the coherent state $\hat{\rho} = |\beta\rangle\langle\beta|$ is given by (see Prob. 13.29)

$$\chi(\eta, \eta^*) = e^{\eta\beta^* - \beta\eta^*}, \tag{13.100}$$

while the characteristic function for the number state $\hat{\rho} = |n\rangle\langle n|$ is given by

$$\chi(\eta, \eta^*) = \sum_{m=0}^{n} \frac{n!}{(m!)^2(n-m)!}(-|\eta|^2)^m. \qquad (13.101)$$

13.5.3 P-function

The P-function[21] is a kind of coherent-state expansion of the density operator $\hat{\rho}$. In other words, let us assume that the density operator may be written in the form

$$\hat{\rho} = \int d^2\alpha \mathrm{P}(\alpha, \alpha^*)|\alpha\rangle\langle\alpha|, \qquad (13.102)$$

then $\mathrm{P}(\alpha, \alpha^*)$ is the so-called Glauber–Sudarshan representation of $\hat{\rho}$. Since we have

$$1 = \mathrm{Tr}\left(\hat{\rho}\right) = \int d^2\alpha \mathrm{P}(\alpha, \alpha^*) \sum_n \langle n \mid \alpha\rangle \langle\alpha \mid n\rangle = \int d^2\alpha \mathrm{P}(\alpha, \alpha^*) \langle\alpha \mid \alpha\rangle, \qquad (13.103)$$

it follows that

$$\int d^2\alpha \mathrm{P}(\alpha, \alpha^*) = 1, \qquad (13.104)$$

which is the normalization condition for the P-function. Using this expression and making use of the cyclic property of the trace, it is easy to calculate that

$$\begin{aligned}\left\langle\left(\hat{a}^\dagger\right)^r \hat{a}^s\right\rangle &= \mathrm{Tr}\left[\left(\hat{a}^\dagger\right)^r \hat{a}^s \hat{\rho}\right] = \mathrm{Tr}\left[\hat{a}^s \hat{\rho}\left(\hat{a}^\dagger\right)^r\right] \\ &= \int d^2\alpha \mathrm{P}(\alpha, \alpha^*) \sum_n \langle n\left|\hat{a}^s\right|\alpha\rangle \langle\alpha \mid \left(\hat{a}^\dagger\right)^r \mid n\rangle, \end{aligned} \qquad (13.105)$$

from which, using Eqs. (13.49), we deduce

$$\left\langle\left(\hat{a}^\dagger\right)^r \hat{a}^s\right\rangle = \int d^2\alpha \left(\alpha^*\right)^r \alpha^s \mathrm{P}(\alpha, \alpha^*), \qquad (13.106)$$

i.e. the P-function straightforwardly gives the normally ordered quantum moments. Using Eqs. (13.94) and (13.102), we may immediately write the relationship between the P-function and the characteristic function as

$$\chi(\eta, \eta^*) = \int d^2\alpha e^{\eta\alpha^* - \alpha\eta^*} \mathrm{P}(\alpha, \alpha^*), \qquad (13.107)$$

21 This representation was independently introduced by Glauber [Glauber 1963b] and Sudarshan [Sudarshan 1963].

which shows that the characteristic function is the Fourier transform of the P-function. By Fourier transforming expressions (13.107) and (13.98) we obtain the direct relationship between the P- and Q-functions:

$$Q(\alpha,\alpha^*) = \frac{1}{\pi}\int d^2\beta e^{-|\alpha-\beta|^2} P(\beta,\beta^*). \tag{13.108}$$

The P-representation may present in practice some difficulties. In fact, the P-function may correspond to a characteristic function which does not have an ordinary Fourier transform. For example, for a coherent state $\hat{\rho} = |\beta\rangle\langle\beta|$ we have

$$P(\alpha,\alpha^*) = \delta^{(2)}(\alpha-\beta). \tag{13.109}$$

It may even happen that the P-function is given by derivatives of δ functions. For a Fock state $\hat{\rho} = |n\rangle\langle n|$, for instance, we have

$$P(\alpha,\alpha^*) = \frac{1}{n!}e^{|\alpha|^2}\left(\frac{\partial^2}{\partial\alpha\partial\alpha^*}\right)^n \delta^{(2)}(\alpha). \tag{13.110}$$

13.5.4 Wigner function

The Wigner distribution function[22] may be written as the Fourier transform of the Wigner characteristic function $\chi_W(\eta,\eta^*)$

$$W(\alpha,\alpha^*) = \frac{1}{\pi^2}\int d^2\alpha e^{-\eta\alpha^*+\eta^*\alpha}\chi_W(\eta,\eta^*), \tag{13.111}$$

where (see also Eq. (13.99))

$$\chi_W(\eta,\eta^*) = \mathrm{Tr}\left[\hat{\rho}e^{\eta\hat{a}^\dagger-\eta^*\hat{a}}\right] = e^{\frac{1}{2}|\eta|^2}\chi_A(\eta,\eta^*) = e^{-\frac{1}{2}|\eta|^2}\chi(\eta,\eta^*). \tag{13.112}$$

It is also interesting to express the Wigner function in terms of position and momentum,[23] which is given by

$$W(x,p_x) = \frac{1}{\pi h}\int_{\Re} dx'\langle x+x'|\hat{\rho}|x-x'\rangle e^{2\imath\frac{p_x x'}{h}}. \tag{13.113}$$

If the W-function is integrated with respect to $\hat{p}_x$, it gives the correct probability distribution (marginal distribution) of $\hat{x}$ and vice versa (see Prob. 13.30). However, the W-function can also assume negative values, which is a signature of the fact that it is not a true probability distribution.

The relationship between W-function and P-function is given by

$$W(\alpha) = \frac{2}{\pi}\int d^2\beta e^{-2|\beta-\alpha|^2} P(\beta,\beta^*). \tag{13.114}$$

[22] It is actually, from a historical view point, the first quasi-probability and was introduced in [Wigner 1932]. For an extensive exposition see [Dragoman 1998].

[23] This was the original Wigner's formulation.

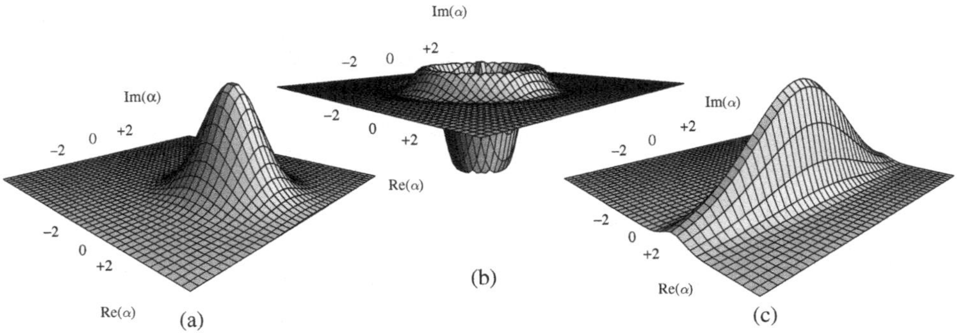

Figure 13.8 (a) Representation of the W-function of a coherent state $|\alpha\rangle = |2\rangle$. As it may be seen from Eq. (13.115), it is a bidimensional Gaussian centered at the point $\alpha_0 = (2, 0)$ in the complex plane $(\Re(\alpha), \Im(\alpha))$. (b) Representation of the W-function of a number state $|n\rangle$ with $n = 4$, i.e. with the number of photons equal to the mean number of photons in the coherent state (a). Its annular shape shows the phase-invariance of the number state. Note that there are regions where the function becomes negative. (c) W-function representation of a squeezed state $|\alpha, \xi\rangle$ with $\alpha = 2$ and $\xi = 0.8$. Note also that the three figures do not have the same z-scale. It is interesting to observe the main differences between the Q-function and the W-function (see Fig. 13.7): the latter, in the representation of the number state, shows negative values and oscillations that are absent in the former.

That is, the W-function is a Gaussian convolution of the P-function, just like the Q-function – but with a different Gaussian weight (see Eq. (13.108)). In other terms, the Q-function is a less detailed average (it is smoother) than the W-function (because of the factor 2 in the exponential), and hence, unlike the W-function, it is never negative. For this reason, the W-function is a better candidate than the Q-function for distinguishing the true quantum features of a certain state, without presenting the difficulties of definition typical of the P-function.

Let us now consider two examples (see Fig. 13.8). First, the Wigner function for a coherent state $|\alpha_0\rangle$. Using Eqs. (13.100), (13.112), and (13.111), we have

$$W(\alpha, \alpha^*) = \frac{2}{\pi} e^{-2|\alpha - \alpha_0|^2}. \tag{13.115}$$

The second example is the W-function for a number state $|n\rangle$, which may be written as (see Prob. 13.31)

$$W(\alpha, \alpha^*) = (-1)^n \frac{1}{2\pi} e^{-2|\alpha|^2} L_n\left(4|\alpha|^2\right), \tag{13.116}$$

where $L_n(x)$ is the Laguerre polynomial of degree n.

In conclusion, the different phase-space representations have advantages and limitations. Since we have studied the tranformations from a representation to the other, their use can be chosen depending on practical reasons determined by the specific necessity of the problem under consideration.

13.6 Quantum-optical coherence

13.6.1 First- and second-order coherence

When we have a single mode of the field, Eq. (13.20a) may be written as

$$\hat{\mathbf{E}}(\mathbf{r},t) = \hat{\mathbf{E}}^{(+)}(\mathbf{r},t) + \hat{\mathbf{E}}^{(-)}(\mathbf{r},t), \tag{13.117}$$

where

$$\hat{\mathbf{E}}^{(+)}(\mathbf{r},t) = \imath \left(\frac{\hbar\omega}{2\epsilon_0}\right)^{\frac{1}{2}} e^{\imath(\mathbf{k}\cdot\mathbf{r}-\omega t)}\hat{a}, \tag{13.118a}$$

$$\hat{\mathbf{E}}^{(-)}(\mathbf{r},t) = -\imath \left(\frac{\hbar\omega}{2\epsilon_0}\right)^{\frac{1}{2}} e^{-\imath(\mathbf{k}\cdot\mathbf{r}-\omega t)}\hat{a}^\dagger, \tag{13.118b}$$

where, for the sake of simplicity, we have taken $l = 1$. These expressions show that a coherent state is also an eigenstate of the positive component of the field (13.118a), that is (see Eq. (13.49a)),

$$\hat{\mathbf{E}}^{(+)} \left|\alpha\right\rangle = E^{(+)} \left|\alpha\right\rangle, \tag{13.119}$$

where

$$E^{(+)} = \imath \left(\frac{\hbar\omega}{2\epsilon_0}\right)^{\frac{1}{2}} e^{\imath(\mathbf{k}\cdot\mathbf{r}-\omega t)}\alpha \tag{13.120}$$

is the eigenvalue of the field operator $\hat{\mathbf{E}}^{(+)}$, as well as the bra $\langle\alpha\,|$ is the eigenstate of the negative component (see Eq. (13.49b)), that is,

$$\langle\alpha\,|\,\hat{\mathbf{E}}^{(-)} = \langle\alpha\,|\,E^{(-)}, \tag{13.121}$$

where similarly

$$E^{(-)} = -\imath \left(\frac{\hbar\omega}{2\epsilon_0}\right)^{\frac{1}{2}} e^{-\imath(\mathbf{k}\cdot\mathbf{r}-\omega t)}\alpha^*. \tag{13.122}$$

Now, let us calculate the mean value of the product of the field operators (13.118) on the coherent state $\left|\alpha\right\rangle$, i.e.

$$\left\langle\alpha\left|\hat{\mathbf{E}}^{(-)}\hat{\mathbf{E}}^{(+)}\right|\alpha\right\rangle = \frac{\hbar\omega}{2\epsilon_0}\left\langle\alpha\left|\hat{a}^\dagger\hat{a}\right|\alpha\right\rangle = \frac{\hbar\omega}{2\epsilon_0}\left\langle\hat{N}\right\rangle_\alpha = \frac{\hbar\omega}{2\epsilon_0}\left|\alpha\right|^2, \tag{13.123}$$

which shows that, for a coherent state, such a mean value is proportional to the intensity of light. By performing a similar calculation with field operators for two different space–time points

$$\hat{\mathbf{E}}^{(-)}(\mathbf{r}_1,t_1) = \hat{\mathbf{E}}_1^{(-)} = -\imath \left(\frac{\hbar\omega}{2\epsilon_0}\right)^{\frac{1}{2}} e^{-\imath(\mathbf{k}\cdot\mathbf{r}_1-\omega t_1)}\hat{a}, \tag{13.124a}$$

$$\hat{\mathbf{E}}^{(+)}(\mathbf{r}_2,t_2) = \hat{\mathbf{E}}_2^{(+)} = \imath \left(\frac{\hbar\omega}{2\epsilon_0}\right)^{\frac{1}{2}} e^{\imath(\mathbf{k}\cdot\mathbf{r}_2-\omega t_2)}\hat{a}^\dagger, \tag{13.124b}$$

we obtain

$$G_{12}^{(1)} = \left\langle \alpha \left| \hat{\mathbf{E}}_1^{(-)} \hat{\mathbf{E}}_2^{(+)} \right| \alpha \right\rangle = \frac{\hbar\omega}{2\epsilon_0} \left\langle \hat{N} \right\rangle_\alpha e^{\imath \mathbf{k}\cdot(\mathbf{r}_2 - \mathbf{r}_1)} e^{\imath\omega(t_1 - t_2)}. \tag{13.125}$$

This is the typical interference that is produced in basic interferometry experiments, where differences in paths and even time delays are produced. The function $G^{(1)}$ is called *first-order coherence*, as in classical wave interferometry. In a similar way we may define the second-order interference function as the mean value of the product of the two intensities taken at the two space–time points,[24] i.e.

$$G_{12}^{(2)} = \left\langle \alpha \left| \hat{\mathbf{E}}_1^{(-)} \hat{\mathbf{E}}_1^{(+)} \hat{\mathbf{E}}_2^{(-)} \hat{\mathbf{E}}_2^{(+)} \right| \alpha \right\rangle = \left\langle \alpha \left| \hat{\mathbf{E}}_1^{(-)} \hat{\mathbf{E}}_2^{(-)} \hat{\mathbf{E}}_2^{(+)} \hat{\mathbf{E}}_1^{(+)} \right| \alpha \right\rangle, \tag{13.126}$$

where we have made use of the fact that operators taken in two different space–time points commute. If we move from a single-mode field to the general case of a multimode field (see Eq. (13.20a)), we must rewrite Eq. (13.118a) as

$$\hat{\mathbf{E}}^{(+)}(\mathbf{r}, t) = \imath \sum_{\mathbf{k},\lambda} \left(\frac{\hbar\omega_{\mathbf{k}}}{2\epsilon_0} \right)^{\frac{1}{2}} e^{\imath(\mathbf{k}\cdot\mathbf{r} - \omega_{\mathbf{k}} t)} \hat{a}_{\mathbf{k},\lambda}(t), \tag{13.127}$$

and similarly for the negative field operator. Since the photons of each mode are noninteracting, the state for the multimode field can be written as the product of states for each mode (see Eqs. (13.26)–(13.28)), i.e.,

$$|\alpha\rangle = \left|\alpha_{\mathbf{k}_1,\lambda}\right\rangle \left|\alpha_{\mathbf{k}_2,\lambda}\right\rangle \ldots = \left|\{\alpha_{\mathbf{k}_j,\lambda}\}\right\rangle. \tag{13.128}$$

Using this formalism, we may write the first-order correlation function for the multimode field as

$$G^{(1)} = \left\langle \{\alpha_{\mathbf{k}_j,\lambda}\} \left| \hat{\mathbf{E}}_1^{(-)} \hat{\mathbf{E}}_2^{(+)} \right| \{\alpha_{\mathbf{k}_j,\lambda}\} \right\rangle, \tag{13.129}$$

and similarly for $G^{(2)}$. Detailed calculations[25] show that thermal light from a single source may exhibit first-order coherence – this is the common interference effect of a Young's double slit experiment. However, it can never be second-order coherent. Thermal light is random, Gaussian (i.e. chaotic) light coming from an incoherent source (for instance, a light bulb). It may be described as a mixture of coherent states $|\alpha\rangle$, where the probability of finding a particular state $|\alpha\rangle$ is given by the Gaussian distribution

$$\wp(\alpha) = \frac{1}{\pi\bar{n}} e^{-\frac{|\alpha|^2}{\bar{n}}}. \tag{13.130}$$

Perfect laser (i.e. coherent-state) light, on the other hand, exhibits both first-order and second-order coherence.

[24] Second-order interference plays a crucial role, for instance, in the Hanbury Brown and Twiss experiment, where the intensity of a plane wave is measured by photodetectors at two different space–time points. One of the two outputs is delayed by a small time, and the two intensities are multiplied together and averaged. See [Hanbury Brown/Twiss 1956, Hanbury Brown/Twiss 1958] [Lewenstein 2007].

[25] See [*Loudon* 1973] [*Goldin* 1982].

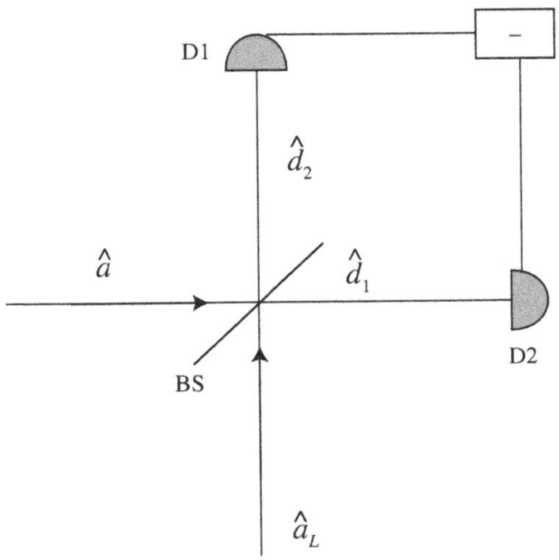

Figure 13.9 Scheme of homodyne detection in the balanced configuration. The input mode $\hat{a}$ is mixed to a strong local oscillator $\hat{a}_L$ thanks to a beam splitter, whose output ports are directly detected by the photodetectors D1 and D2. The resulting photocurrents are then subtracted each other, so that the final signal is proportional to the difference between the number of photons in mode $\hat{d}_1$ and mode $\hat{d}_2$.

13.6.2 Homodyne detection

In simple one-dimensional systems, one often faces the problem of measuring the position or the momentum of a particle (see Sec. 2.3). As we know, in the case of the harmonic oscillator (see Sec. 4.4), these operators are linear combinations of the creation and annihilation operators, and correspond to the quadrature operators (see Eq. (13.61)). It is natural to ask how one can measure these quadrature operators in a quantum-optical system. The answer to this question comes from the so-called homodyne detection. As we shall see, this technique plays a crucial role in quantum-state measurement (see Ch. 15).

In a typical quantum-optical homodyne detection scheme, the input light signal (a single mode of the electromagnetic field) is "superposed" to a strong local oscillator field thanks to a lossless beam splitter, whose transmission and reflection coefficients are given by T and R, respectively. The light emerging from the two output modes of the beam splitter is then measured by two distinct photodetectors D1 and D2 (see Fig. 13.9). If we denote by $\hat{a}$ and $\hat{a}_L$ the annihilation operators of the input mode and of the local oscillator mode, respectively, the annihilation operators of the output modes ($\hat{d}_1$ and $\hat{d}_2$) are then given by the transformation

$$\begin{pmatrix} \hat{d}_1 \\ \hat{d}_2 \end{pmatrix} = \begin{bmatrix} \mathrm{R} & \mathrm{T} \\ \mathrm{T} & \mathrm{R} \end{bmatrix} \begin{pmatrix} \hat{a}_L \\ \hat{a} \end{pmatrix}, \tag{13.131}$$

where the unitarity of the coupling matrix is ensured by the relations

$$|\mathrm{R}|^2 + |\mathrm{T}|^2 = 1 \quad \text{and} \quad \mathrm{T}^*\mathrm{R} + \mathrm{R}^*\mathrm{T} = 0. \tag{13.132}$$

The second condition is satisfied if

$$\arg \mathrm{R} - \arg \mathrm{T} = \frac{\pi}{2}. \tag{13.133}$$

In the *balanced* homodyne detection, the reflection and transmission coefficients satisfy $|\mathrm{R}| = |\mathrm{T}| = 2^{-\frac{1}{2}}$, and the output signal is given by the difference of the photocurrents of the two detectors. In this configuration (see again Fig. 13.9), the measured signal is represented by the operator

$$\hat{d}_1^\dagger \hat{d}_1 - \hat{d}_2^\dagger \hat{d}_2 = \imath \left(\hat{a}^\dagger \hat{a}_L - \hat{a}_L^\dagger \hat{a} \right), \tag{13.134}$$

i.e. the difference between the measured number of photons in mode $\hat{d}_1$ and mode $\hat{d}_2$. Its expectation value is given by (see Prob. 13.32)

$$\left\langle \hat{N}_{12} \right\rangle = \left\langle \hat{d}_1^\dagger \hat{d}_1 - \hat{d}_2^\dagger \hat{d}_2 \right\rangle = 2|\alpha_L| \left\langle \hat{X}_\theta \right\rangle, \tag{13.135}$$

where α_L is the coherent amplitude of the local oscillator (assumed to be in a coherent state), and

$$\hat{X}_\theta = \hat{X}_\theta^\dagger = \frac{1}{\sqrt{2}} \left(\hat{a}^\dagger e^{\imath\theta} + \hat{a} e^{-\imath\theta} \right) \tag{13.136}$$

is the quadrature operator at angle (see Eqs. (13.61) and (13.80))

$$\theta = \arg \mathrm{R} - \arg \mathrm{T} + \phi_L = \frac{\pi}{2} + \phi_L, \tag{13.137}$$

ϕ_L being the phase of α_L. It results from Eq. (13.135) that the balanced configuration has allowed us to remove the contributions due to the local oscillator and to the signal only. If we assume that the intensity of the local oscillator is much stronger than that of the input signal, and we take into account the quantum efficiency η of the photodetectors, the final expression for the expectation value of the photocounts in the balanced homodyne detection is given by

$$\left\langle \hat{N}_{12} \right\rangle = 2\eta|\alpha_L| \left\langle \hat{X}_\theta \right\rangle, \tag{13.138}$$

which shows that the output signal is proportional to the mean value of the operator $\hat{X}_\theta$, i.e. that homodyne detection is indeed an effective way to measure a quadrature operator. Moreover, the balanced configuration has a great advantage because it completely removes the noise contribution of the local oscillator.

13.7 Atom–field interaction

The interaction of an atom with an electromagnetic field is one of the richest and most important phenomena in optical physics. Such interaction bears a lot of interesting physical consequences, some of which will be discussed in the following sections and chapters.

In the present section, we start from the general quantum theory of atom–field interaction, but our aim and focus is the treatment of the simplest case: the interaction between a single mode of the quantized electromagnetic field in a cavity and a single two-level atom, which is described by the so-called *Jaynes–Cummings model*.[26]

13.7.1 The interaction between the electromagnetic field and an atom

Let us first begin with some general considerations. In order to give the quantized treatment of the interaction, it would be necessary to quantize the electron field of the atom. This, however, goes far beyond the scope of this book and we shall limit ourselves to a non-relativistic treatment. The Hamiltonian which describes the interaction of a radiation field and an atom is given by

$$\hat{H} = \frac{1}{2m}\left(\hat{\mathbf{p}} - \frac{\mathrm{e}}{c}\hat{\mathbf{A}}(\mathbf{r})\right)^2 + \mathrm{eU}(\mathbf{r}) + \hat{H}_F, \tag{13.139}$$

where $\hat{\mathbf{p}}$ is the electron's momentum, $\mathbf{r}$ its position, $\hat{\mathbf{A}}$ is the vector potential of the external field, U is the Coloumb interaction term, and $\hat{H}_F$ is the Hamiltonian of a free field, which has the form (13.22) or (13.23). The justification of the first term of the Hamiltonian (13.139) is given by the fact that it provides the correct equations of motion (see Eq. (4.133)).[27]

The Hamiltonian (13.139) can be simplified with the help of the so-called *electric-dipole approximation*.[28] This approximation consists of the assumption that the spatial behavior of the mode function of the vector field varies more slowly than the electronic wave function. In the optical region (see Tab. 13.1), for instance, this is justified by the fact that the wavelength of the photon (of the order of 10^{-7} m) is much larger than the dimension of the atom (of the order of 10^{-10} m). As a consequence, the vector potential (and also the electric field) does not change appreciably within the size of the atom. Then, $\hat{\mathbf{A}}(\mathbf{r})$ in the previous equation may be replaced by $\hat{\mathbf{A}}(\mathbf{R})$, where $\mathbf{R}$ is the center-of-mass position. A detailed analysis[29] shows that Eq. (13.139) may be written as

$$\hat{H} = \hat{H}_F + \hat{H}_A + \hat{H}_{\mathrm{I}}, \tag{13.140}$$

where

$$\hat{H}_A = \frac{\hat{\mathbf{p}}^2}{2m} + \mathrm{eU}(\mathbf{r}) \tag{13.141}$$

is the Hamiltonian of the free atom,

$$\hat{H}_{\mathrm{I}} = -\mathrm{e}\hat{\mathbf{r}} \cdot \hat{\mathbf{E}} \tag{13.142}$$

is the interaction Hamiltonian, and $\hat{\mathbf{E}}$ is the electric radiation field. Equation (13.140) can be justified by the following argument. The first term in Eq. (13.139) gives rise to three different components: the term proportional to $\hat{\mathbf{p}}^2$, which, together with the Coloumb term,

26 See [Cummings 1965] [Jaynes/Cummings 1963].
27 See, e.g. [*Mandl/Shaw* 1984, 17].
28 For the classical treatment of the multipole-field expansion see [*Jackson* 1962, Ch. 16].
29 See [*Schleich* 2001, pp. 382–402].

constitutes the Hamiltonian $\hat{H}_A$ of the matter (of the free atom); the term proportional to $\hat{\mathbf{A}}^2$, which is usually small and can be neglected; and finally the term proportional to $\hat{\mathbf{p}} \cdot \hat{\mathbf{A}}$.[30] This last term is particularly important when one considers transition amplitudes between different states of the atom and the field which involve the emission or absorbtion of one photon. In this case, one has to expand the vector potential according to Eqs. (13.18) and (13.15) and, following the electric-dipole approximation, replace the exponential functions $e^{\imath \mathbf{k} \cdot \mathbf{r}}$ by 1. As we have said, the $\hat{\mathbf{p}} \cdot \hat{\mathbf{A}}$ term is equivalent to $\hat{H}_{\rm I}$, defined in Eq. (13.142).

In the following, we explicitly limit ourselves to a two-level atom,[31] but the results can be generalized to the multilevel case, and adopt the center-of-mass reference frame.[32] In order to make explicit $\hat{H}_A$ and the interaction term $\hat{H}_{\rm I}$ in the full Hamiltonian (13.140), we have to introduce quantized atomic levels. If $|g\rangle$ and $|e\rangle$ denote the ground and excited levels of the atom, respectively, the operators

$$\hat{\sigma}_+ = |e\rangle\langle g|, \quad \hat{\sigma}_- = |g\rangle\langle e| \tag{13.143}$$

may be considered as atomic raising and lowering operators.[33] In other words, they satisfy the relations (see Prob. 13.33)

$$\hat{\sigma}_+|e\rangle = 0, \quad \hat{\sigma}_+|g\rangle = |e\rangle, \tag{13.144a}$$

$$\hat{\sigma}_-|e\rangle = |g\rangle, \quad \hat{\sigma}_-|g\rangle = 0, \tag{13.144b}$$

$$\hat{\sigma}_+ = \frac{1}{2}\left(\hat{\sigma}_x + \imath\hat{\sigma}_y\right), \quad \hat{\sigma}_- = \frac{1}{2}\left(\hat{\sigma}_x - \imath\hat{\sigma}_y\right), \tag{13.144c}$$

and the following commutation relations:

$$\left[\hat{\sigma}_+, \hat{\sigma}_-\right] = \hat{\sigma}_z, \quad \left[\hat{\sigma}_\pm, \hat{\sigma}_z\right] = \mp 2\hat{\sigma}_\pm, \quad \left[\hat{\sigma}_+, \hat{\sigma}_-\right]_+ = \hat{I}, \tag{13.145}$$

where $\hat{\sigma}_x$, $\hat{\sigma}_y$, and $\hat{\sigma}_z$ are the Pauli spin matrices defined by Eqs. (6.154).

In terms of these pseudo-spin atomic operators, the free Hamiltonian of the atom can be derived from the eigenvalue equations

$$\hat{H}_A|g\rangle = E_g|g\rangle, \quad \hat{H}_A|e\rangle = E_e|e\rangle, \tag{13.146}$$

where E_g and E_e are the energies of the ground and excited levels, respectively. Then, we have

$$\hat{H}_A = E_g|g\rangle\langle g| + E_e|e\rangle\langle e|, \tag{13.147}$$

which may be rewritten as

$$\hat{H}_A = \frac{1}{2}\left(E_g + E_e\right) + \frac{1}{2}\hbar\omega_A\hat{\sigma}_z, \tag{13.148}$$

where

$$\omega_A = \frac{E_e - E_g}{\hbar} \tag{13.149}$$

30 Note that in the radiation gauge, where $\nabla \cdot \mathbf{A} = 0$ (see Eq. (13.7)), the term $\hat{\mathbf{A}} \cdot \hat{\mathbf{p}}$ may be replaced by $\hat{\mathbf{p}} \cdot \hat{\mathbf{A}}$.

31 Two-level atoms clearly do not exist in nature, but physical situations can be created by optical pumping such that only two levels are effectively involved.

32 See, e.g., [*Scully/Zubairy* 1997].

33 These operators coincide with the spin-1/2 operators introduced in Eq. (6.151). This coincidence displays the analogy between a two-level atom and a spin-1/2 particle in a magnetic field.

is the "frequency" of the atomic transition (see Eq. (1.75)) and we have made use of the identities

$$|e\rangle\langle e|+|g\rangle\langle g|=\hat{I} \quad \text{and} \quad |e\rangle\langle e|-|g\rangle\langle g|=\hat{\sigma}_z. \tag{13.150}$$

Finally, we have now to evaluate the interaction term $\hat{H}_{\rm I}$ in the Hamiltonian (13.140). Let us consider a single mode of the cavity with frequency ω. In the electric-dipole approximation, the electric field may be evaluated at the origin (i.e. at the position of the nucleus or of the center of mass) and, taking into account the expansion (13.20a), for a single mode in the Schrödinger picture (where the observables do not evolve and therefore the exponentials $e^{\pm\imath\omega t}$ are not present) it may be written as

$$\hat{\mathbf{E}}=\imath\left(\frac{\hbar\omega}{2\epsilon_o \mathrm{l}^3}\right)^{\frac{1}{2}}\mathbf{e}\left(\hat{a}-\hat{a}^\dagger\right), \tag{13.151}$$

where, for the sake of simplicity, $\mathbf{e}$ denotes the linear polarization vector, and the exponentials $e^{\pm\imath\mathbf{k}\cdot\mathbf{r}}$ vanish at the origin. Let us concentrate our attention on the atom's position operator $\hat{\mathbf{r}}$, according to the expression for the interaction Hamiltonian (13.142). Since the atomic potential energy – the Coulomb term in Eq. (13.141) – is even (with respect to x, y, and z), the wave functions of the energy eigenstates are either even or odd (see property (iv) in Subsec. 3.2.3). As a consequence, the diagonal elements of the position operator vanish. On the other hand, the off-diagonal elements of the dipole operator

$$\hat{\mathbf{d}}=\mathrm{e}\hat{\mathbf{r}} \tag{13.152}$$

may be written as

$$\mathbf{d}=\mathrm{e}\langle e|\hat{\mathbf{r}}|g\rangle, \tag{13.153a}$$

$$\mathbf{d}^*=\mathrm{e}\langle g|\hat{\mathbf{r}}|e\rangle. \tag{13.153b}$$

Therefore, the dipole operator takes the form

$$\mathrm{e}\hat{\mathbf{r}}=\mathbf{d}|e\rangle\langle g|+\mathbf{d}^*|g\rangle\langle e|=\mathbf{d}\hat{\sigma}_+ +\mathbf{d}^*\hat{\sigma}_-, \tag{13.154}$$

and describes transitions from the ground state $|g\rangle$ to the excited state $|e\rangle$ and vice versa. Back-substituting this result into Eq. (13.142), we are left with the complex scalar products

$$\mathbf{d}\cdot\mathbf{e}=|\mathbf{d}\cdot\mathbf{e}|e^{\imath\phi}, \tag{13.155a}$$

$$\mathbf{d}^*\cdot\mathbf{e}=|\mathbf{d}\cdot\mathbf{e}|e^{-\imath\phi}. \tag{13.155b}$$

Taking $\phi=\pi/2$, the interaction term (13.142) can be written as

$$\hat{H}_{\rm I}=\hbar\varepsilon_0\left(\hat{\sigma}_+-\hat{\sigma}_-\right)\left(\hat{a}-\hat{a}^\dagger\right), \tag{13.156}$$

where the coupling constant

$$\varepsilon_0=\left(\frac{\omega}{2\epsilon_0\hbar \mathrm{l}^3}\right)^{\frac{1}{2}}|\mathbf{d}\cdot\mathbf{e}|, \tag{13.157}$$

is the so-called *vacuum Rabi frequency*. Equation (13.156) gives rise to four terms. The term $\hat{\sigma}_+\hat{a}$ ($\hat{\sigma}_-\hat{a}^\dagger$) describes the physical process by which the atom is (de-)excited and a photon is absorbed (emitted). The other two terms, i.e. $\hat{\sigma}_+\hat{a}^\dagger$ and $\hat{\sigma}_-\hat{a}$, describe processes by which either the atom is excited and a photon is emitted or the atom is de-excited and a photon is absorbed. As a consequence, these are non-energy conserving terms and, in this context, have no physical meaning. In the so-called *rotating-wave approximation* these terms can therefore be neglected. This heuristic argument may be made rigorous by moving into the interaction picture. When doing so,[34] one arrives at an interaction Hamiltonian

$$\hat{H}_{\mathrm{I}}^{\mathrm{I}} = \hbar\varepsilon_0 \left(\hat{\sigma}_-\hat{a}^\dagger e^{\imath(\omega-\omega_A)t} + \hat{\sigma}_+\hat{a}e^{-\imath(\omega-\omega_A)t} - \hat{\sigma}_-\hat{a}e^{-\imath(\omega+\omega_A)t} - \hat{\sigma}_+\hat{a}^\dagger e^{\imath(\omega+\omega_A)t}\right), \tag{13.158}$$

where the non-energy conserving terms are multiplied by oscillatory terms that involve the sum of the frequencies of the cavity and the atomic transitions and therefore, on averaging, vanish, yielding

$$\hat{H}_{\mathrm{I}}^{\mathrm{I}} \simeq \hbar\varepsilon_0 \left(\hat{\sigma}_-\hat{a}^\dagger e^{\imath\Delta t} + \hat{\sigma}_+\hat{a}e^{-\imath\Delta t}\right), \tag{13.159}$$

where $\Delta = \omega - \omega_A$ is the so-called *detuning*.

Finally, dropping the zero-point energy in $\hat{H}_F$ and the constant energy term in $\hat{H}_A$, we arrive, at the total Hamiltonian in the Schrödinger picture,

$$\hat{H}_{JC} = \hbar\omega\hat{a}^\dagger\hat{a} + \frac{1}{2}\hbar\omega_A\hat{\sigma}_z + \hbar\varepsilon_0\left(\hat{\sigma}_+\hat{a} + \hat{\sigma}_-\hat{a}^\dagger\right), \tag{13.160}$$

which describes the interaction of a single mode of the electromagnetic field with a single two-level atom in the rotating-wave and dipole approximations. This is also called the Jaynes–Cummings model and is particularly relevant in quantum optics, because it is an exactly solvable model of the matter–field interaction and displays some very interesting and genuine quantum features.

13.7.2 Jaynes-Cummings model

In order to solve the Jaynes–Cummings model in the simplest case where the detuning is zero, i.e. when ω coincides with the atomic frequency ω_A (resonant case), we first note that the free eigenstates undergo an obvious degeneracy: A situation in which n photons are present in the cavity and the atom is excited is energetically equivalent to the situation in which we have $n+1$ photons in the cavity and the atom is in the ground state.[35] In other words, the combined field–atom eigenkets

$$|n,e\rangle = |n\rangle_F \otimes |e\rangle_A \quad \text{and} \quad |n+1,g\rangle = |n+1\rangle_F \otimes |g\rangle_A \tag{13.161}$$

of the free Hamiltonian $\hat{H}_0 = \hat{H}_F + \hat{H}_A$ share the same energy $\hbar\omega(n+1/2)$. Let us work in the interaction picture (see Subsec. 3.6.2) and recast the Jaynes–Cummings Hamiltonian (13.160) as

[34] See [*Schleich* 2001, pp. 407–409].

[35] The only exception is the combined ground state $|0,g\rangle$, where there is no photon and the atom is in the ground state, and for this reason there can be no further evolution.

$$\hat{H}_{JC} = \hat{H}_0 + \hat{H}_{\rm I}, \tag{13.162a}$$

$$\hat{H}_0 = \hbar\omega\left(\hat{a}^\dagger\hat{a} + \frac{1}{2}\hat{\sigma}_z\right), \tag{13.162b}$$

$$\hat{H}_{\rm I} = \hbar\varepsilon_0\left(\hat{a}\hat{\sigma}_+ + \hat{a}^\dagger\hat{\sigma}_-\right). \tag{13.162c}$$

Now, it is possible to show that $\hat{H}_0$ and $\hat{H}_{\rm I}$ commute. Indeed, we write

$$\left[\hat{H}_0, \hat{H}_{\rm I}\right] = \hbar^2\omega\varepsilon_0\left\{\hat{\sigma}_+\left[\hat{a}^\dagger\hat{a}, \hat{a}\right] + \hat{\sigma}_-\left[\hat{a}^\dagger\hat{a}, \hat{a}^\dagger\right] + \frac{1}{2}\left[\hat{\sigma}_z, \hat{\sigma}_+\right]\hat{a} + \frac{1}{2}\left[\hat{\sigma}_z, \hat{\sigma}_-\right]\hat{a}^\dagger\right\}, \tag{13.163}$$

and, since we also have that (see Prob. 4.11)

$$\left[\hat{a}^\dagger\hat{a}, \hat{a}\right] = -\hat{a} \quad \text{and} \quad \left[\hat{a}^\dagger\hat{a}, \hat{a}^\dagger\right] = \hat{a}^\dagger, \tag{13.164}$$

taking into account the second of Eqs. (13.145), we immediately obtain the desired result, i.e.

$$\left[\hat{H}_0, \hat{H}_{\rm I}\right] = 0. \tag{13.165}$$

From this fact two consequences follow:

(i) $\hat{H}_{\rm I} = \hat{H}_{\rm I}^{\rm I}$, i.e. the interaction Hamiltonian in the interaction picture coincides with $\hat{H}_{\rm I}$.
(ii) The eigenstates of $\hat{H}_{\rm I}$ are linear combinations of eigenstates of $\hat{H}_0$ (see Th. 2.4: p. 66).

The first consequence is due to the fact that we have (see Eq. (3.118))

$$\hat{H}_{\rm I}^{\rm I} = e^{\frac{\iota}{\hbar}\hat{H}_0 t}\hat{H}_{\rm I}e^{-\frac{\iota}{\hbar}\hat{H}_0 t}, \tag{13.166}$$

from which it also follows that

$$|\psi(t)\rangle_{\rm I} = e^{-\frac{\iota}{\hbar}\hat{H}_{\rm I}t}|\psi(0)\rangle\,. \tag{13.167}$$

The second consequence enables us to calculate the eigenkets of $\hat{H}_{\rm I}$. Considering the interaction Hamiltonian $\hat{H}_{\rm I}$ as a perturbation relative to the free Hamiltonian $\hat{H}_0$, its eigenvalue equation may be written as

$$\hat{H}_{\rm I}|\psi_n\rangle_{\rm I} = E_n^{\rm I}|\psi_n\rangle_{\rm I}, \tag{13.168}$$

where $|\psi_n\rangle_{\rm I}$ are the perturbed eigenkets, while $E_n^{\rm I}$ are the eigenvalues of the perturbed Hamiltonian. As we have said, any eigenket of $\hat{H}_{\rm I}$ may be expanded in terms of the degenerate eigenstates of the free Hamiltonian, i.e.

$$|\psi_n\rangle_{\rm I} = c_n|n, e\rangle + c_n^{'}|n+1, g\rangle\,. \tag{13.169}$$

Now, it is easy to see that we have

$$\hat{H}_{\rm I}|\psi_n\rangle_{\rm I} = \hbar\varepsilon_n\left(c_n|n+1, g\rangle + c_n^{'}|n, e\rangle\right), \tag{13.170}$$

where $\varepsilon_n = \varepsilon_0\sqrt{n+1}$. Since $|\psi_n\rangle_{\rm I}$ has to be an eigenstate of $\hat{H}_{\rm I}$, we must have that $c_n = c_n^{'} = 1/\sqrt{2}$, in order for the state to be correctly normalized. Therefore,

$$|\psi_n^+\rangle_{\rm I} = \frac{1}{\sqrt{2}}\left(|n, e\rangle + |n+1, g\rangle\right) \tag{13.171a}$$

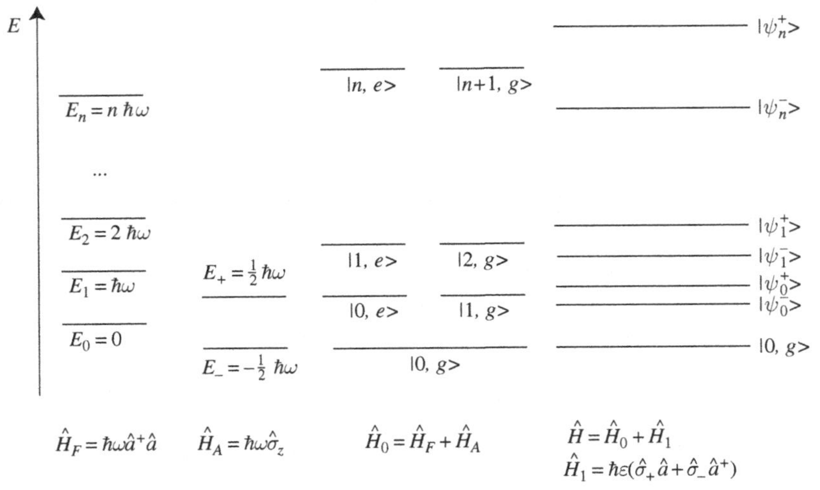

Figure 13.10 **Resonant Jaynes–Cummings energy levels. In the first two columns, the separated energy levels for the free field and for the free atom are shown. In the third column, the total free energy is represented: exactly on resonance, apart from the ground state, all levels are doubly degenerate. In the fourth column, the total energy levels are represented, in which the degeneracy is removed.**

is one of the two eigenstates of $\hat{H}_\mathrm{I}$ for a given n, with energy $E^\mathrm{I}_{n,+} = \hbar\varepsilon_n$. The other one can be written as

$$|\,\psi_n^-\rangle_\mathrm{I} = \frac{1}{\sqrt{2}}\left(|\,n,e\rangle - |\,n+1,g\rangle\right), \tag{13.171b}$$

which corresponds to the energy eigenvalue $E^\mathrm{I}_{n,-} = -\hbar\varepsilon_n$. In other words, the perturbation has removed the energy degeneracy of the unperturbed Hamiltonian (see Fig. 13.10).

13.7.3 Rabi oscillations

In order to understand the dynamics of a two-level atom resonantly interacting with a single mode of an electromagnetic field, it is instructive to study the time evolution of an initially excited atom when n photons are present in the field. In this case, the initial state of the system is $|\,\psi(0)\rangle = |\,n,e\rangle$ and, according to Eq. (13.167), the evolved state at time t in the interaction picture can be expressed as

$$|\,\psi(t)\rangle = \frac{1}{\sqrt{2}}\left[e^{-\imath\varepsilon_n t}\,|\,\psi_n^+\rangle + e^{\imath\varepsilon_n t}\,|\,\psi_n^-\rangle\right], \tag{13.172}$$

where for the sake of simplicity we have dropped the subscript I to the states in the interaction picture. In obtaining Eq. (13.172) we have made use of the identity

$$|\,n,e\rangle = \frac{1}{\sqrt{2}}\left(|\,\psi_n^+\rangle + |\,\psi_n^-\rangle\right). \tag{13.173}$$

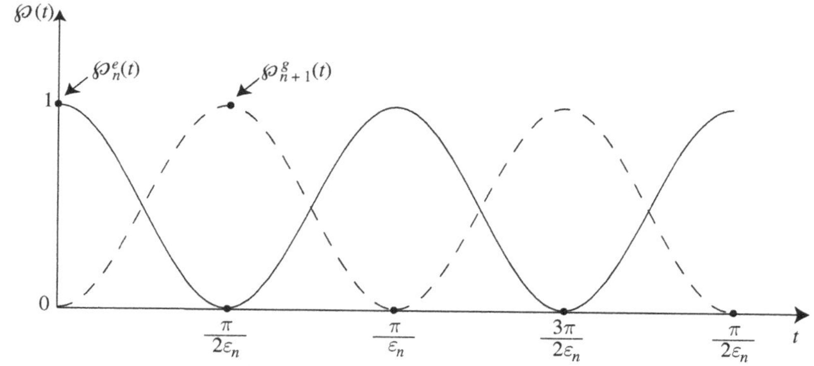

Figure 13.11 Rabi oscillations. The probabilities $\wp_n^e(t)$ (solid line) and $\wp_{n+1}^g(t)$ (dashed line) oscillate sinusoidally in time in the resonant Jaynes–Cummings model when the initial state at $t = 0$ is $|n, e\rangle$. The sum of the two probabilities is obviously equal to 1 at any time.

After some simple calculations, and transforming back the perturbed eigenstates according to Eqs. (13.171), we finally arrive at

$$|\psi(t)\rangle = \cos(\varepsilon_n t)|n, e\rangle - \imath \sin(\varepsilon_n t)|n+1, g\rangle. \tag{13.174}$$

Equation (13.174) shows an interesting quantum phenomenon: the evolved state *oscillates* between the degenerate states $|n, e\rangle$ (initial state) and $|n+1, g\rangle$. In particular, at times $t = \pi/2\varepsilon_n$ the state has entirely evolved into $|n+1, g\rangle$, while at time $t = \pi/\varepsilon_n$ it has come back to the initial state (Fig. 13.11).[36] It is possible to calculate the probability of finding simultaneously exactly n photons inside the cavity and the atom in the excited state at a generic time t, that is,

$$\wp_n^e(t) = |\langle n, e \mid \psi(t)\rangle|^2 = \cos^2(\varepsilon_n t). \tag{13.175}$$

This is the so-called *Rabi oscillation* and ε_n is called the Rabi frequency. When $n = 0$, we have the *vacuum* Rabi frequency ε_0.

13.7.4 Collapses and revivals

It is particularly interesting to investigate the dynamics of our system (atom plus field) when initially the atom is excited and the field is in a coherent state, i.e.

$$|\psi(0)\rangle = |\alpha, e\rangle = e^{-\frac{|\alpha|^2}{2}} \sum_{n=0}^{\infty} \frac{\alpha^n}{\sqrt{n!}} |n, e\rangle. \tag{13.176}$$

[36] At first glance, this result may appear in contradiction with the statement given in Prob. 3.12. However, a more careful analysis shows that it is instead a beautiful confirmation of that statement (see also Subsec. 3.5.3).

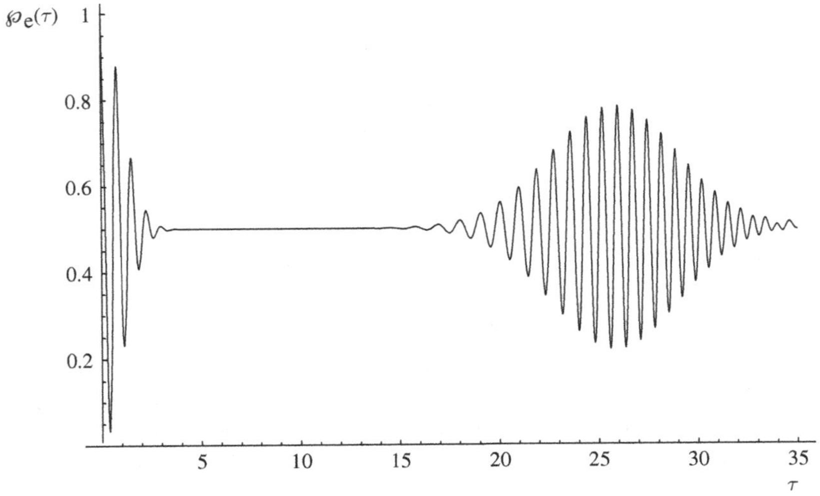

Figure 13.12 **The phenomenon of collapses and revivals. The probability $\wp_e(\tau)$ from Eq. (13.181) is plotted against the dimensionless time τ, for $|\alpha| = 4$. After a time of the order of $\tau = 1$, the envelope of the oscillations collapses, but the oscillations revive after $\tau \simeq 25$.**

The dynamics in this case can be easily derived from Eq. (13.174) and yields

$$| \psi(t)\rangle = e^{-\frac{|\alpha|^2}{2}} \sum_{n=0}^{\infty} \frac{\alpha^n}{\sqrt{n!}} \left[\cos(\varepsilon_n t) \,|n, e\rangle - \imath \sin(\varepsilon_n t) \,|n+1, g\rangle\right]. \tag{13.177}$$

Again, it is interesting to calculate the probability of finding at time t the atom in the excited state, but regardless of the state of the field. To this end we have first to obtain the atom–field density matrix

$$\hat{\rho}_{AF}(t) = |\psi(t)\rangle \langle \psi(t)| . \tag{13.178}$$

Then, we have to perform the partial trace (see Subsec. 5.5.2) over the field degrees of freedom

$$\hat{\tilde{\rho}}_A(t) = \mathrm{Tr}_F \left[\hat{\rho}_{AF}(t)\right], \tag{13.179}$$

so as to obtain the reduced density matrix of the atom alone, regardless of the state of the field. Finally, the desired probability is found as (see Eq. (9.100))

$$\wp_e(t) = \left\langle e \left| \hat{\tilde{\rho}}_A(t) \right| e \right\rangle . \tag{13.180}$$

The final result is (see Prob. 13.34)

$$\wp_e(t) = \frac{1}{2}\left[1 + \sum_{n=0}^{\infty} \frac{e^{-|\alpha|^2} |\alpha|^{2n}}{n!} \cos(2\varepsilon_n t)\right]. \tag{13.181}$$

This probability is plotted in Fig. 13.12 as a function of the dimensionless time $\tau = \varepsilon_0 t$ and represents a nice illustration of the experimentally observed phenomenon known as collapses and revivals: at the beginning, $\wp_e$ oscillates rapidly, but such oscillations are damped out after some time (called the collapse time, t_c, of the order of ε_0^{-1}). After another characteristic time (the revival time, t_r, of the order of $2\pi |\alpha|/\varepsilon_0$), the oscillations in the probability revive.

This striking phenomenon is deeply connected with the discrete nature of the quantized electromagnetic field – which reflects into the sum in Eq. (13.181) – and to the factor $\sqrt{n+1}$: after t_c, the contributions from the sum are out of phase with respect to each other. After t_r, however, the contributions come back in phase, giving rise to a revival of the oscillations.

13.7.5 Spontaneous emission

If we investigate the dynamics of our system (atom plus field) in the case where the atom is initially excited but there are no photons in the field, we immediately realize that Rabi oscillations occur even when the field is in the vacuum state. In fact, in this case the initial state is

$$| \psi(0)\rangle = | 0, e\rangle , \tag{13.182}$$

and the calculation performed in Subsec. 13.7.3 yields (see Eq. (13.174))

$$| \psi(t)\rangle = \cos(\varepsilon_0 t) \, | 0, e\rangle - \imath \sin(\varepsilon_0 t) \, | 1, g\rangle . \tag{13.183}$$

In other words, the initially excited atom may emit a photon (and make a transition to the ground state) even when no photons are present in the field. This is the simplest case of a phenomenon known as *spontaneous emission*, in contrast to the so-called *stimulated emission*, which in turn occurs in presence of photons (see Box 13.1). Spontaneous emission may also be regarded as a result of the perturbation of the vacuum fluctuations of the field onto the excited atom. Moreover, spontaneous emission is an entirely random process which occurs isotropically, whereas stimulated emission has a preferred direction, which is precisely that of the stimulating photon.

Let us very briefly summarize the history of the laser (see Box 13.1). In 1954 it was first shown in [Gordon *et al.* 1954, Gordon *et al.* 1955] that coherent radiation can be generated in the radio-frequency domain. Their experiments, performed in ammonia, gave rise to the first MASER (Microwave Amplification by Stimulated Emission of Radiation). This phenomenon was later extended in [Schawlow/Townes 1958, Prokhorov 1958] to optical radiation (LASER). Today it is possible to experimentally produce single-atom laser beams, which show a more orderly photon stream than even the quietest "ordinary" laser.[37] By making use of other techniques, in particular of a specific application of the Airy function, it is also possible to produce beams that do not show diffraction effects but preserve the intensity profile (are "propagation invariant").[38] Indeed, a laser cannot preserve such a coherence over long distances.

[37] See [McKeever *et al.* 2003] [Carmichael/Orozco 2003].
[38] See [Siviloglou *et al.* 2007] [Dholakia 2008].

Box 13.1 LASER

In Sec. 13.7 we have investigated the dynamics of the interplay between matter and light. Ordinary light is dominated by spontaneous emission, with photons emitted randomly in all directions, and represents just *optical noise*. In contrast, a LASER (Light Amplification by Stimulated Emission of Radiation) is a device that amplifies the process of stimulated emission so as to obtain coherent and highly directional light beams. In order to understand how a laser works, we have first to study the thermodynamic equilibrium of a two-level atom in the presence of radiation (see Fig. 13.13). Let us consider an atomic medium, N_g and N_e being the number of ground-state and excited atoms, respectively. The rate of the spontaneous emission from $|e\rangle$ to $|g\rangle$ is given by A and is independent from the photon number. However, atoms can also make a transition from $|e\rangle$ to $|g\rangle$ through stimulated emission, and in this case the corresponding rate will be given by $B\rho$, where ρ is the electromagnetic energy density in the medium. Finally, atoms may be excited from $|g\rangle$ to $|e\rangle$ through (obviously, stimulated) absorption of photons, again with a rate $B\rho$. A and B are the so-called *Einstein coefficients*.[39] At the thermodynamic equilibrium at temperature T, the number of excited atoms will be much smaller than the number of the atoms in the ground state, as their ratio is given by the *Boltzmann factor*, i.e.

$$\frac{N_e}{N_g} = e^{-\frac{\hbar\omega}{k_B T}}, \tag{13.184}$$

where $\hbar\omega$ is the energy separations of the two levels, ω is the photon frequency (i.e. the energy separation of the two levels $|e\rangle$ and $|g\rangle$ divided by $\hbar$), and k_B the Boltzmann constant. Therefore, spontaneous emission dominates the relaxation process and photons are emitted incoherently in all possible directions. Laser operation, on the other hand, may occur only when the system is far from equilibrium, i.e. when the number of excited atoms is much larger than the number of atoms in ground state. This *population inversion* cannot be reached just by increasing the temperature due to condition (13.184). A possible way to bypass this problem is provided by the scheme shown in Fig. 13.14, where the medium is made of three-level atoms: A strong *pump* (which, for example, may be of optical nature) excites the atoms from level $|0\rangle$ to $|e\rangle$, while Einstein's A and B coefficients rule the transitions between $|e\rangle$ and $|g\rangle$. Atoms may also make a transition from $|g\rangle$ to $|0\rangle$ through a relaxation process. In this case, the condition (13.184) forces N_e to be much smaller than N_0. However, N_e is much larger than N_g and population inversion is achieved. We may then imagine a laser as schematically made of a three-level active medium contained in a long thin rod with mirrors at the two ends (see Fig. 13.15).

13.7.6 Parametric down-conversion

Spontaneous parametric down-conversion (SPDC) is a very important process in quantum optics that has relevant fields of application. A non-linear crystal splits incoming

[39] See [Einstein 1917].

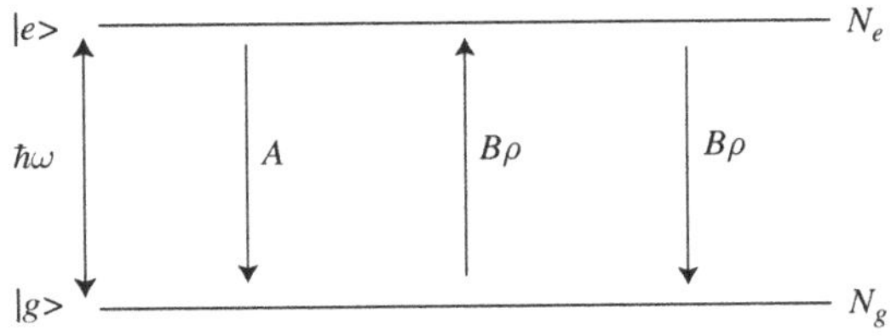

Figure 13.13 Schematic diagram representing stimulated and spontaneous transitions between atomic levels. The rate A of the spontaneous process is independent of the field energy density ρ, whereas the rate for the stimulated processes is given by $B\rho$.

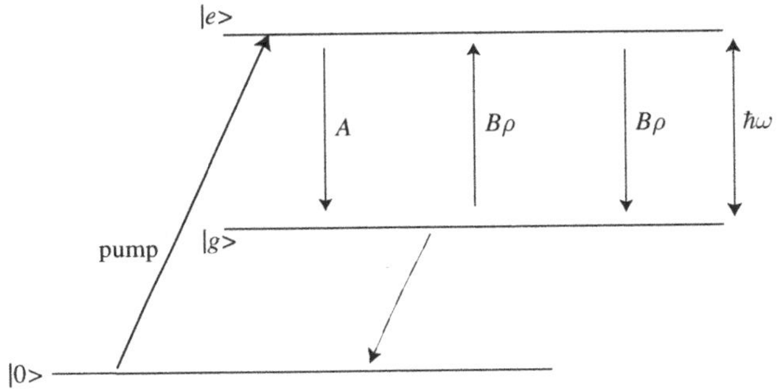

Figure 13.14 Three-level atomic configuration able to achieve population inversion. Initially, all atoms are in the level $|0\rangle$ and are pumped to the excited level $|e\rangle$. Spontaneous and stimulated transitions between levels $|e\rangle$ and $|g\rangle$ occur as in Fig. 13.13, while relaxation process brings the atom back from $|g\rangle$ to the $|0\rangle$ level.

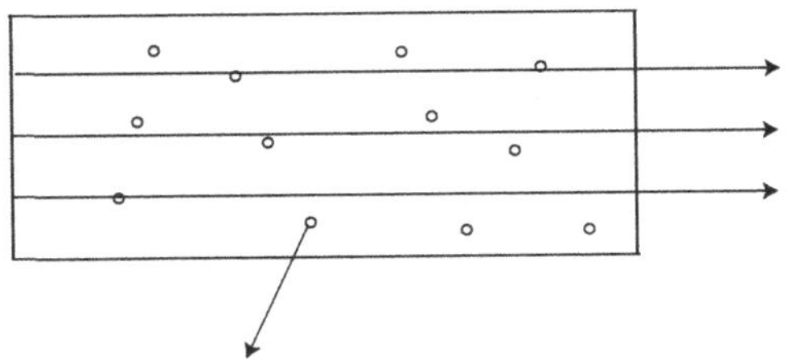

Figure 13.15 A very schematic diagram of a laser. A long thin rod with two end-mirrors contains the active medium shown in Fig. 13.14, which is brought to population inversion by a strong external pump. Coherent photons emitted in the same direction (the axis of the rod) by stimulated emission are further amplified through repeated passage of the photons through the medium, while spontaneous photons emitted in different directions are simply lost. The process is initiated by a spontaneously emitted photon which will in turn encounter other excited atoms and force them to emit in the same direction. The phase of a stimulated photon is equal to the phase of the stimulating one. As a result, laser light is simultaneously coherent, higly directional, and has a very narrow distribution of frequencies. Among the processes which perturb a perfectly coherent, unidirectional, and monochromatic laser beam, we may include spontaneous emission, imperfections or inhomogeneities of the medium, and Doppler effect.

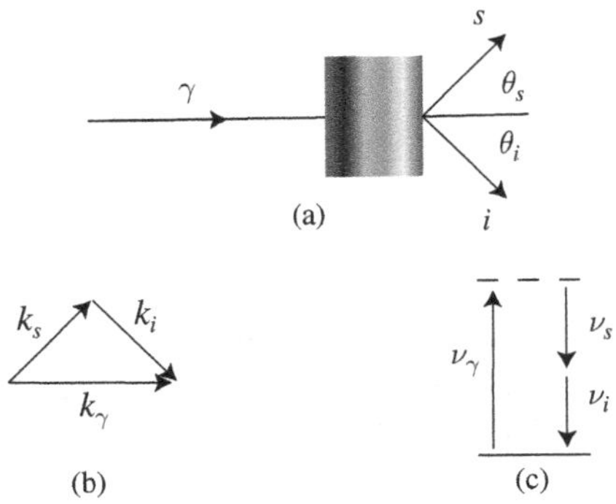

Figure 13.16 Parametric down-conversion (PDC). (a) Generation of the idler and signal photons. (b) Momentum conservation. (c) Energy conservation.

photons (a coherent laser beam) into pairs of photons of lower energy whose combined energy and momentum is equal to the energy and momentum of the original photon (see Fig. 13.16). If the frequency of the laser is in the near ultraviolet (wavelength about 300 nanometers), then the light that emerges takes the form of a conical rainbow of visible light (wavelengths in the range 400–800 nm) (see also Tab. 13.1). "Parametric" refers to the fact that the process does not change the state of the crystal, which is the reason why energy and momentum are conserved.[40] We obtain a more correct view of this process, once we realize that the non-linear crystal brings about an interaction between the laser field and the zero-point field, and that, as a consequence of this interaction, there is a secondary field emitted by the crystal. In fact, the process is "spontaneous" in the same sense as spontaneous emission – it is stimulated by random vacuum fluctuations. Consequently, the photon pairs are created at random times. However, if one photon of the pair (the "signal") is detected at any time, then one knows with certainty that its partner (the "idler") is present. This allows for the creation of optical fields containing (to a good approximation) a single photon. As a matter of fact, this is the predominant mechanism for experimentalists to create single photons (Fock states of the electromagnetic field). The single photons as well as the photon pairs are often used in quantum information experiments and applications such as quantum cryptography and the Bell test experiments (see Chs. 16–17).

SPDC is a special case of parametric amplification. The main difference lies in the fact that in the latter a laser field drives the crystal to squeeze an input state (see Subsec. 13.4.3), whereas in the former the input is simply given by the vacuum state. As a consequence, the output of SPDC is a squeezed vacuum, with an even number of photons. Apart from the zero-photon term (the vacuum itself), the leading-order contribution comes exactly from the two-photon process.

[40] This is related to the so-called *phase-matching condition* of nonlinear optics: in type-I (or degenerate) phase matching the two photons bear the same polarization, whereas in type-II they have orthogonal polarizations.

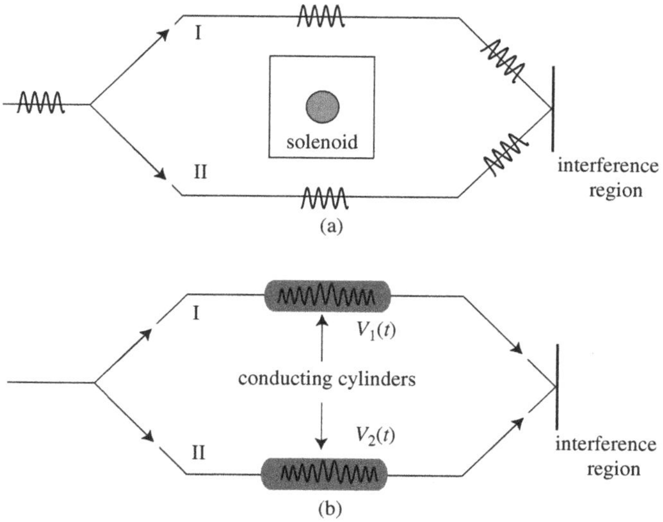

Figure 13.17 (a) Magnetic AB effect. The axis of the solenoid is perpendicular to the page. The wave function is a split plane wave. (b) Electric AB effect. $V_1 = V_2 = 0$ except when the wave packet is shielded from the electric field.

13.8 Geometric phase

13.8.1 Aharonov–Bohm effect

The Aharonov–Bohm (AB) effect[41] consists of the discovery that the electromagnetic potentials, and not only the fields, may have physical effects. The easiest way to detect this striking phenomenon is to let two beams of particles pass to the right and left of an isolated field. As we shall see, they will present a relative phase shift (see Fig. 13.17).

In classical electrodynamics the situation is quite different, since scalar and vector potentials are useful mathematical tools that enable us to calculate the fields (see Subsec. 13.1.1), but are deprived of direct physical significance. The AB effect manifests itself in quantum mechanics through the electromagnetic scalar and vector potentials U and $\mathbf{A}$. As a consequence, the AB effect has two forms, magnetic and electric.

Let us consider the schematic setup depicted in Fig. 13.17(a), where the magnetic field $\mathbf{B}$ is essentialy confined within the solenoid. On the other hand, the vector potential cannot be zero everywhere outside the solenoid, because the total magnetic flux Φ_M through every circuit containing the solenoid is given by

$$\Phi_M = \int \mathbf{B} \cdot d\Sigma = \oint \mathbf{A}(\mathbf{r}) \cdot d\mathbf{r} \neq 0, \qquad (13.185)$$

[41] Introduced in quantum mechanics in [Aharonov/Bohm 1959]. See also [*Peshkin/Tonomura* 1989].

where $d\Sigma$ is the surface element of the area delimited by the circuit and $d\mathbf{r}$ is the linear element of the circuit.

Now let

$$\psi(x,t) = \psi_1^0(x,t) + \psi_2^0(x,t) \tag{13.186}$$

be the wave function of the electron beams when the magnetic field is switched off ($\psi_1^0(x,t)$ and $\psi_2^0(x,t)$ represent the components that pass through the upper and lower paths of the apparatus, respectively). When the magnetic field is present, the Hamiltonian is given by (see Eq. (4.133))

$$\hat{H} = \frac{1}{2m}\left(\hat{\mathbf{p}} - \frac{\mathrm{e}}{c}\hat{\mathbf{A}}(\mathbf{r})\right)^2, \tag{13.187}$$

where e is the electron charge, and the total wave function modifies into

$$\psi(x,t) = \psi_1(x,t) + \psi_2(x,t), \tag{13.188}$$

where

$$\psi_1(x,t) = \psi_1^0(x,t)e^{\frac{\imath}{\hbar}\mathrm{S}_{M,1}}, \tag{13.189a}$$

$$\psi_2(x,t) = \psi_2^0(x,t)e^{\frac{\imath}{\hbar}\mathrm{S}_{M,2}}, \tag{13.189b}$$

and

$$\mathrm{S}_{M,1} = \frac{\mathrm{e}}{c}\int_I \mathbf{A}(\mathbf{r})\cdot d\mathbf{r}, \tag{13.190a}$$

$$\mathrm{S}_{M,2} = \frac{\mathrm{e}}{c}\int_{II} \mathbf{A}(\mathbf{r})\cdot d\mathbf{r} \tag{13.190b}$$

are the actions involved, where $d\mathbf{r}$ is the linear element of the circuit. Equations (13.190) may be justified with the following argument: the square in Eq. (13.187) gives rise to three terms. As we have already seen (in Subsec. 13.7.1), the first term is the usual kinetic-energy term (common to the two paths), the second is proportional to $\hat{\mathbf{A}}^2$ and can be neglected, while the third term is the only important term that makes a difference between the two paths. Its contribution to the unitary time evolution operator may be written as

$$\exp\left(\frac{\imath\mathrm{e}}{\hbar c}\int_0^t dt\frac{\hat{\mathbf{p}}}{m}\cdot\hat{\mathbf{A}}(\mathbf{r})\right) = \exp\left(\frac{\imath\mathrm{e}}{\hbar c}\int_0^t dt\hat{\mathbf{A}}(\mathbf{r})\cdot\mathbf{v}\right) = \exp\left(\frac{\imath\mathrm{e}}{\hbar c}\int_{I/II}\hat{\mathbf{A}}(\mathbf{r})\cdot d\mathbf{r}\right), \tag{13.191}$$

It is then clear that the interference pattern generated in the interference region will depend on the phase difference

$$\frac{\mathrm{S}_{M,1} - \mathrm{S}_{M,2}}{\hbar} = \frac{\mathrm{e}}{\hbar c}\oint_{I+II}\mathbf{A}(\mathbf{r})\cdot d\mathbf{r} = \frac{\mathrm{e}}{\hbar c}\Phi_M, \tag{13.192}$$

where the integral is intended along the circuit given by the paths I and II (in a clockwise direction). The most striking feature of this effect is that its existence does not depend on the magnetic forces acting on the electrons. Indeed, the above result is also valid if we surround the solenoid by a potential barrier that reflects the electrons perfectly.

The electric version of the AB effect is similar to the one we have just described (see Fig. 13.17(b)). Again, a single coherent electron beam is split into two parts (I and II) and each part is then allowed to enter along a cylindrical metal pipe that shields the electrons from any electric field. Each of the pipes is connected to an external generator, which makes the potential on the tubes to alternate in time, but only when the electron wave packet is deep inside the cylinder. This ensures that the electron beam does not experience any local field. The potential will add to the Hamiltonian of the particle a term $\mathrm{U}(x,t)$, which is, for the region inside the pipe, a function of time only, so that the global Hamiltonian may be written as

$$\hat{H} = \hat{H}_0 - \mathrm{eU}(x,t), \tag{13.193}$$

where $\hat{H}_0$ is the Hamiltionian of the free particle. The wave functions for the electrons in the two beams are then given by

$$\psi_1(x,t) = \psi_1^0(x,t)e^{-\frac{i}{\hbar}\mathrm{S}_{E,1}(x,t)}, \tag{13.194a}$$
$$\psi_2(x,t) = \psi_2^0(x,t)e^{-\frac{i}{\hbar}\mathrm{S}_{E,2}(x,t)}, \tag{13.194b}$$

where the action $\mathrm{S}_{E,j}$ $(j = 1,2)$ is given by

$$\mathrm{S}_{E,1}(x,t) = -\mathrm{e}\int_0^t dt'\mathrm{U}^1(x,t'), \tag{13.195a}$$
$$\mathrm{S}_{E,2}(x,t) = -\mathrm{e}\int_0^t dt'\mathrm{U}^2(x,t'). \tag{13.195b}$$

When the two beams meet again at the interference region, their relative phase is then shifted by an amount

$$\delta\phi = \frac{1}{\hbar}\left[\mathrm{S}_{E,2}(x,t) - \mathrm{S}_{E,1}(x,t)\right]. \tag{13.196}$$

How can we interpret the AB effect? Aharonov and Bohm supposed the physical reality of the vector and scalar potentials and this reality was understood as more fundamental then that of the field strength. In other words, Bohm saw in the AB effect the physical manifestation of a potential which cannot be accounted for in classical terms.[42] However, we can also explain the AB effect through the geometric phase formalism, as we shall see below.

13.8.2 Geometric phase

In Sec. 10.3 we have discussed the adiabatic theorem, which applies to systems subject to a Hamiltonian which is a function of a time-dependent parameter whose rate of change is small. In that context, we have concluded that, under the adiabatic condition, an initial eigenstate of a time-dependent Hamiltonian remains an instantaneous eigenstate. In

[42] Bohm was developing in those years a hidden-variable variant of quantum mechanics in which the vector potential played an important role (see Sec 16.3).

fact, the adiabaticity condition allows us to neglect the coupling to other energy eigenstates $|n[\mathbf{r}(t)]\rangle$ in Eq. (10.78), so that the latter equation reduces to (see also Eq. (10.79))

$$\frac{d}{dt}\psi_m(t) \simeq -\langle m[\mathbf{r}(t)]|\nabla_{\mathbf{r}}|m[\mathbf{r}(t)]\rangle \cdot \frac{d\mathbf{r}(t)}{dt}\psi_m(t), \tag{13.197}$$

where, for practical purposes of this section, we have substituted the parameter $\boldsymbol{\zeta}$ of Sec. 10.3 with $\mathbf{r}$. Now, we wish to show that, after a cycle in the parameter space, the system acquires a phase factor that adds up to its dynamical phase.[43] This is called the *geometric phase*. In fact, making use of the orthonormality condition $\langle m(\mathbf{r}) \mid n(\mathbf{r})\rangle = \delta_{mn}$, we have

$$\begin{aligned} 0 = \nabla_{\mathbf{r}}\langle m(\mathbf{r}) \mid m(\mathbf{r})\rangle &= \langle \nabla_{\mathbf{r}} m(\mathbf{r}) \mid m(\mathbf{r})\rangle + \langle m[\mathbf{r}]|\nabla_{\mathbf{r}}|m[\mathbf{r}]\rangle \\ &= \langle m[\mathbf{r}]|\nabla_{\mathbf{r}}|m[\mathbf{r}]\rangle^* + \langle m[\mathbf{r}]|\nabla_{\mathbf{r}}|m[\mathbf{r}]\rangle \\ &= 2\Re\{\langle m[\mathbf{r}]|\nabla_{\mathbf{r}}|m[\mathbf{r}]\rangle\}, \end{aligned} \tag{13.198}$$

which shows that $\langle m[\mathbf{r}]|\nabla_{\mathbf{r}}|m[\mathbf{r}]\rangle$ is purely imaginary. Therefore, we may integrate Eq. (13.197) to find

$$\psi_m(t) = e^{-\imath\phi_m^{(g)}(t)}\psi_m(0), \tag{13.199}$$

where

$$\phi_m^{(g)}(t) = \int_{\mathbf{r}(0)}^{\mathbf{r}(t)} d\mathbf{r}'\Im\{\langle m[\mathbf{r}']|\nabla_{\mathbf{r}'}|m[\mathbf{r}']\rangle\} \tag{13.200}$$

is just the geometric phase.[44] Notice that the presence of the factor $d\mathbf{r}(t)/dt$ in Eq. (13.197) changes the integration in dt into an integration in $d\mathbf{r}'$. If the parameter $\mathbf{r}(t)$ describes a close loop C in the parameter space, Eq. (13.200) may be rewritten as

$$\phi_m^{(g)} = \oint_C d\mathbf{r}\Im\{\langle m[\mathbf{r}]|\nabla_{\mathbf{r}}|m[\mathbf{r}]\rangle\}. \tag{13.201}$$

In Eq. (13.201) the vector $\nabla_{\mathbf{r}} m[\mathbf{r}]$ is tangent to the vector $|m[\mathbf{r}]\rangle$. Let us take as an example a state vector on the Poincaré sphere (see Subsec. 1.3.3). Then, the vector $\nabla_{\mathbf{r}} m[\mathbf{r}]$ is tangent to the surface and $\mathbf{r}$ may be taken as the vector pointing to the surface, whose motion causes the state $|m[\mathbf{r}]\rangle$ to cycle on the surface of the sphere (the so-called parallel transport) (see Fig. 13.18).

13.8.3 The AB effect as a geometric phase

As we have seen in the previous subsection, the geometric phase can be understood as a measure of the curvature of the parameter space. In fact, by applying Stokes theorem to transform the line integral in Eq. (13.201) into a surface integral, we have

43 See also [*Schleich* 2001, 174–76].

44 The introduction and the development of this concept is especially due to Sir Michael Berry [Berry 1984, Berry 1987, Berry 1989], but there were already some contributions in [Pancharatnam 1956]. See also [Aharonov/Anandan 1987].

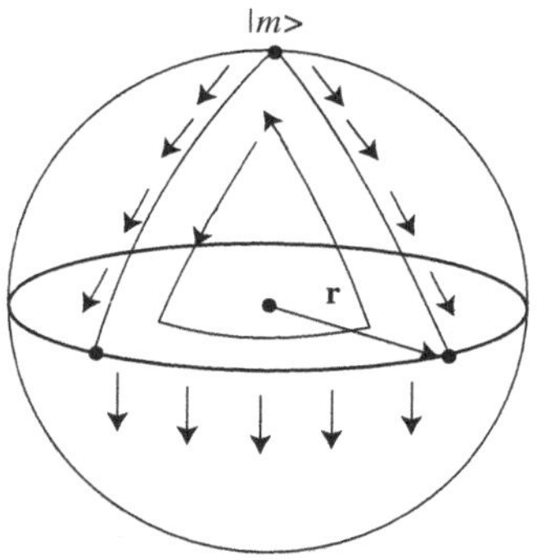

Figure 13.18 Parallel transport. The geometric phase takes its name from the fact that its expression (13.200) bears elements of differential geometry, in particular of the concept of measuring the curvature of a surface by parallel transport of a tangent vector. Here, we transport a vector (represented by the small arrows) that is tangent to a curved surface along a path on this surface. After a cycle on the closed path, the vector is not identical to the initial one: the two vectors have a non-vanishing angle between them, which is a measure for the curvature of the surface.

$$\phi_m^{(g)} = \int_{\mathbf{S}} d\Sigma \cdot \mathcal{B}_m(\mathbf{r}), \tag{13.202}$$

where $\mathbf{S}$ is the area enclosed by the path $\mathcal{C}$, $d\Sigma$ is a surface element and $\mathcal{B}_m(\mathbf{r})$ is a "field," given by

$$\mathcal{B}_m(\mathbf{r}) = \nabla_{\mathbf{r}} \times \Im\{\langle m\,[\mathbf{r}]\,|\nabla_{\mathbf{r}}|\,m\,[\mathbf{r}]\rangle\}, \tag{13.203}$$

whose "flux" outogoing through the surface is measured by $\phi_m^{(g)}$. Then, the formalism of the geometric phase is particularly useful when applied to the interpretation of the AB effect. In the case of the magnetic AB effect (see Eq. (13.192)), for example, we may write

$$\int_{\mathcal{C}} d\mathbf{r} \cdot \mathbf{A}(\mathbf{r}) = \int_{\mathbf{S}} d\Sigma \cdot \mathbf{B}(\mathbf{r}) = \Phi_M = \frac{\hbar c}{\mathrm{e}} \phi^{(g)}, \tag{13.204}$$

where this time the flux is that of the magnetic field. This confirms that the AB effect is only dependent on abstract geometric properties that have nothing to do with the particular nature of forces and fields involved.

13.9 The Casimir effect

The Casimir effect deals with the variation of the vacuum energy of the electromagnetic field inside a cavity limited by a perfectly conducting surface. In the case of two parallel plates the vacuum energy increases by increasing the distance and this leads to an attractive force between the plates; however, in some other geometries the force may be repulsive. There are folkloric statements that the Casimir force plays a crucial role in emulsions like mayonnaise; however, it has a definitive impact on nanotechnological problems.

In the case of the electromagnetic field in a region with given boundary conditions the vacuum energy is given by (see also Eq. (13.30))

$$E_R = \sum_{n=0}^{\infty} \hbar c E_n^{1/2}, \tag{13.205}$$

where E_n are the eigenvalues of the Laplacian inside the region: the factor 1/2 of the zero-point energy compensates with the factor 2 arising from the number of the polarization states producing a net result equal to 1. If the boundaries are conducting, zero boundary conditions should be used for the Laplacian.

In the case of a region of size $L \times M \times M$ the eigenvalues of the Laplacian are

$$E(n_x, n_y, n_z) = \frac{\pi^2 n_x^2}{L^2} + \frac{\pi^2 n_y^2}{M^2} + \frac{\pi^2 n_z^2}{M^2}, \tag{13.206}$$

where n_x, n_y, and n_z run over all positive integer numbers.

It is quite evident that the vacuum energy as defined in Eq. (13.205) is infinite: The sum over the variables n is not convergent. However, in the physical world, we have vacuum also outside the cavity and this also contributes to the total energy. Moreover, another infinite term is produced by creating a perfect conducting surface in the space. Clearly, we are not interested in computing the total energy of vacuum fluctuations that cannot be measured directly anyway. On the contrary, we would rather like to control the amount of energy that comes from objects moving in space and that contributes to deforming the cavity, because these energy variations produce observable forces.

If, for the time being, we restrict ourselves to a not too rigorous level and do not pay too much attention to problems possibly arising from the interchange of summation and integration limits, we could first compute the limit $M \to \infty$ in order to obtain the energy per unit area. If we proceed in this way, we find

$$E_L \equiv \lim_{M\to\infty} \frac{E_{L\times M\times M}}{M^2} = \frac{\hbar c}{(2\pi)^2} \sum_{n=1}^{\infty} \int dk_y dk_z \sqrt{\frac{\pi^2 n^2}{L^2} + k_y^2 + k_z^2}. \tag{13.207}$$

At the front of the previous equation we face the serious problem that the integral is infinite: we have two ways around this in order to obtain the final result, applying some wizardry or carefully extracting the divergent terms.

13.9.1 A useful wizardry: analytic continuation

Let us first discuss the wizardry. We start by contemplating the following equation:

$$\sum_{k=0}^{\infty} 2^k = \left[\sum_{k=0}^{\infty} s^k \right]_{s=2} = \left. \frac{1}{1-s} \right|_{s=2} = -1. \tag{13.208}$$

It is clear that the divergent sum of the positive numbers on the rhs cannot be equal to –1: we have used for $s = 2$ a formula that is valid only for $|s| < 1$.

In the same way we could write

$$\int_0^\infty dx\, 2^x = \left[\int_0^\infty dx\, s^x\right]_{s=2} = -\left.\frac{1}{\ln(s)}\right|_{s=2} = -\,[\ln(2)]^{-1}, \tag{13.209}$$

which does not make sense for the same reasons.

However, there may be contexts where it could make sense to write

$$\sum_{k=0}^\infty 2^k - \int_0^\infty dx\, 2^x = F(s)|_{s=2} = -1 + [\ln(2)]^{-1}, \tag{13.210}$$

where

$$F(s) = \frac{1}{1-s} + \frac{1}{\ln(s)}. \tag{13.211}$$

In the general case, also Eq. (13.210) is meaningless. However, it would be correct if we could prove the following:

- The problem under study depends on a parameter s and for $s < 1$ the solution is given by the lhs of Eq. (13.210), which in this region is correctly given by the function $F(s)$.
- The solution of the problem is an analytic function of s, for positive s. Notice that the function $F(s)$ is an analytic function of s at $s = 1$, because the poles at $s = 1$ in each of the two components cancel exactly.

Now, what could be the equivalent of the parameter s in the case of the Casimir effect? A natural candidate might be the dimension of the space.[45]

If we redo our computation in a space of dimension D, we find that the Casimir energy (neglecting trivial proportionality factors) is given by

$$C_L(D) = \sum_{n=1}^\infty \left[\prod_{\nu=1}^{D-1}\left(\int dk_\nu\right)\sqrt{\frac{\pi^2 n^2}{L^2} + \mathbf{k}^2}\,\right]. \tag{13.212}$$

We can now go into polar coordinates. If we denote by $S(D)$ the area of surface of the unit sphere in D dimensions,[46] after integration over the angular variables we obtain

$$C_L(D) = \sum_{n=1}^\infty S(D-1)\int_0^\infty dk\, k^{D-2}\sqrt{\frac{\pi^2 n^2}{L^2} + k^2} \tag{13.213}$$

$$= \frac{1}{2}\sum_{n=1}^\infty S(D-1)\int_0^\infty \frac{dz}{z} z^{\frac{D-1}{2}}\sqrt{\frac{\pi^2 n^2}{L^2} + z}. \tag{13.214}$$

The integral over z in the previous equation is not convergent. In fact, it is divergent near 0, for $D \le 1$. However, these divergences may be eliminated by integration by parts,

45 Of course, space dimensions are always integer, but with a little of imagination we could consider non-integer dimensions. See [Parisi 2003].

46 We do not need its analytic expression, which we only write for completeness, i.e. $2\pi^{D/2}/\Gamma(D/2)$. We only need that $S(D)$ is an entire function.

exposing the divergences in an explicit algebraic form that is compensated by the zeros of $S(D-2)$.[47] We are not going to discuss this point in greater details. At any rate, the dangerous divergence at infinity disappears for $D < 0$.

We shall make use of the formula[48]

$$\int_0^\infty dz z^{\mu-1}(a+z)^{-\nu} = a^{\mu-\nu}\frac{\Gamma(\nu)}{\Gamma(\mu)\Gamma(\nu-\mu)}. \tag{13.215}$$

If we do not pay too much attention to the conditions for convergence, we obtain

$$C_L(D) = \frac{S(D-1)\,\Gamma\left(\frac{D-2}{2}\right)}{2\Gamma\left(-\frac{1}{2}\right)\Gamma\left(-\frac{1}{2}-\frac{D-1}{2}\right)}\sum_{n=1}^{\infty} A(n)^{\frac{D-1}{2}+\frac{1}{2}}, \tag{13.216}$$

where

$$A(n) \equiv \frac{\pi^2 n^2}{L^2}. \tag{13.217}$$

The previous formula can also be written for $L = 1$ (neglecting π factors) as

$$\frac{1}{\Gamma\left(-\frac{1}{2}\right)\Gamma\left(-\frac{D}{2}\right)}\sum_{n=1}^{\infty} n^D = \frac{1}{\Gamma\left(-\frac{1}{2}\right)\Gamma\left(-\frac{D}{2}\right)}\zeta\left(-D\right), \tag{13.218}$$

where $\zeta\,(s)$ is the Riemann Zeta function. If we use the previous equations for $D = 3$ we get

$$C_L(3) = \frac{1}{2}\left(\frac{\pi}{L}\right)^3 S(2)\frac{\Gamma\left(\frac{1}{2}\right)}{\Gamma\left(-\frac{1}{2}\right)\Gamma\left(-\frac{3}{2}\right)}\zeta\,(-3). \tag{13.219}$$

Putting everything together, and using the fact that $\zeta(-3) = -1/120$, we obtain

$$E_L = -\frac{\hbar c \pi^2}{720\,L^3}, \tag{13.220}$$

which leads to an attractive force (per unit area) equal to

$$\frac{\hbar c \pi^2}{240\,L^4}. \tag{13.221}$$

The computation has a chance to be correct as far as the divergent terms – coming from the modes outside the cavity – may only give a contribution to the energy of the form $a + bL$. These terms do not contribute to the final form of the Casimir force – they could only add a constant term. However, this force should go to zero when L goes to infinity, as happens in our result, which is in fact correct.

[47] See [*Marinari/Parisi* 2008].
[48] See [*Gradstein/Ryshik* 1981, 3.1.94.4].

13.9.2 A sketch of the physical approach

In order to perform a proper computation we should first introduce a cutoff at large momenta and consider the following function

$$E_L^\Lambda = \frac{\hbar c}{(2\pi)^2}\sum_{n=1}^{\infty}\int dk_y\, dk_z\, F^\Lambda\left(\frac{\pi^2 n^2}{L^2} + k_y^2 + k_z^2\right), \tag{13.222}$$

where the function $F^\Lambda(k^2)$ is equal to k for $k \ll \Lambda$ and goes to zero very fast for $k \gg \Lambda$. The introduction of a momentum cutoff is natural from a physical point of view as far as all cavities become transparent at high energy.

Manipulating the previous equation and after some complex computations, described in [*Parisi* 1988], one obtains for large Λ the result

$$E_L^\Lambda = -\frac{\hbar c \pi^2}{720\, L^3} + a(\Lambda) + b(\Lambda)L, \tag{13.223}$$

where $a(\Lambda)$ and $b(\Lambda)$ are divergent functions when the volume goes to infinity, and whose detailed form depends on F^Λ.

The computation for a generic cavity can be done in the following way.[49] One writes

$$E_R = \hbar c \int d\lambda \left(\rho(\lambda) - s_V V \lambda^{\frac{3}{2}} - s_S A \lambda^{\frac{1}{2}}\right)\lambda^{\frac{1}{2}}, \tag{13.224}$$

where $\rho(\lambda)$ is the spectral function of the Laplacian with zero boundary condition, which for large values of λ behaves as

$$\rho(\lambda) \simeq s_V V \lambda^{\frac{3}{2}} - s_S A \lambda^{\frac{1}{2}}, \tag{13.225}$$

as can be seen from a semiclassical analysis[50] (we have neglected terms oscillating around zero).

It turns out that if we add to the naive form

$$E_R = \hbar c \int d\lambda \rho(\lambda)\lambda^{\frac{1}{2}}, \tag{13.226}$$

only terms proportional to the volume and to the surface of the cavity, the result is finite.[51] The term proportional to the volume should be compensated from the mode outside the cavity and is therefore irrelevant. Here, we consider only deformations of the cavity that keep the area of the cavity constant. There is also a physical term proportional to the area, which depends on the conduction characteristic of the surface, but this point will not be investigated further in this context.

[49] See [Balian/Duplantier 1977, Balian/Duplantier 1978].

[50] See [Balian/Bloch 1970, Balian/Bloch 1972, Balian/Bloch 1974]. In these papers, the quantities V and A, which are respectively the volume and the area of the cavity, as well as s_V and s_S are computed.

[51] See [Balian/Duplantier 1977, Balian/Duplantier 1978].

The integral in Eq. (13.224) is convergent and thus we have found an explicit form for the Casimir energy of a generic cavity. Of course, we have to compute the spectrum of the Laplacian inside the cavity, but only the lowest eigenvalues will be relevant, whereas for high λ the integral is convergent and the spectral density can be estimated by the subleading terms in a semiclassical expansion. Clearly, this method gives the same results as those of the previous subsection, as can be checked with some pain by a diligent reader.

Approximated methods based on a multipole expansion can also be found in the literature[52] for computing the force among two objects of arbitrary shape.

Summary

- Starting from the Maxwell equations, we have shown how to quantize the electromagnetic field (the so-called *second quantization*). This amounts to quantizing an infinite set of harmonic oscillators, one for each mode of the field. Quantization is then accomplished by interpreting the amplitude operators $\hat{a}_{\mathbf{k}}$ and $\hat{a}^{\dagger}_{\mathbf{k}}$ as the annihilation and creation operators of the $\mathbf{k}$-th field mode, respectively.
- Using the classical methods of statistical mechanics, we have studied the *thermodynamic equilibrium* of the radiation field.
- The *phase-number uncertainty relation* has been derived.
- We have successively investigated three special states of the electromagnetic field: (1) the *Fock (number) states*: these are eigenstates of both the energy and of the number operators. For these states, the number of photons is perfectly determined and the phase is completely unknown according to the phase–number uncertainty relation. (2) *Coherent states*, which are eigenstates of the annihilation operator $\hat{a}$. These are most similar to the classical states, i.e. to points in phase space. (3) *Squeezed states*. These are the most general class of minimum uncertainty states. In this respect, they represent a generalization of the coherent state in the case where the uncertainty product is not symmetric. They constitute a genuine quantum feature of the radiation field.
- Then, some *quasi-probability distributions* have been presented, as functions in the phase space. In particular, we have presented the Q-function, the characteristic function, the P-function, and the Wigner function.
- Then, we have studied the atom–field interaction, with particular attention to the *Jaynes–Cummings model* for two-level atoms and to *Rabi oscillations*. In this context, we have shown that quantum systems may present a collapse-and-revival behavior.
- We have discussed the *geometric phase* and interpreted the *Aharonov–Bohm effect* as a useful application.
- Finally, we have analyzed the *Casimir effect*.

[52] See [Emig *et al.* 2007].

Problems

13.1 Prove that

$$\nabla \cdot (\nabla \times \mathbf{V}) = 0,$$
$$\nabla \times (\nabla f) = 0,$$
$$\nabla \cdot (\nabla f) = \nabla^2 f = \Delta f,$$

where f is an arbitrary scalar function and $\mathbf{V}$ an arbitrary vector.

13.2 Using Maxwell's equations, the definition of the vector potential, and the condition for the Coulomb gauge, show that $\mathbf{A}$ satisfies the wave equation (13.10).

13.3 Derive the coefficient (13.19).

13.4 Derive the explicit expressions (13.20) for the electric and magnetic fields.

13.5 Using Eqs. (13.20) and (13.22) derive the Hamiltonian (13.23) for the quantized electromagnetic field.

13.6 Compute the partition function for a quantized harmonic oscillator so as to obtain the result (13.32).

13.7 (a) Applying Eq. (5.32a), derive the mean energy (13.33a) for a harmonic oscillator at the equilibrium.

(b) Obtain the same result by using the number distribution (13.31) of the state.

13.8 Show that the operator $\widehat{e^{\imath\phi}}$ admits the number state expansion

$$\widehat{e^{\imath\phi}} = \sum_{n=0}^{\infty} |n\rangle\langle n+1|. \tag{13.227}$$

Use this result to prove the commutation relation

$$\left[\widehat{e^{\imath\phi}}, \widehat{e^{-\imath\phi}}\right] = |0\rangle\langle 0|, \tag{13.228}$$

which shows that $\widehat{e^{\imath\phi}}$ is not unitary.

(*Hint*: Take advantage of Eq. (13.39a) and of the resolution of identity for number states $\sum_{n=0}^{\infty} |n\rangle\langle n| = \hat{I}$.)

13.9 Prove Eqs. (13.39).

13.10 Verify that the states

$$\left|e^{\imath\phi}\right\rangle = \sum_{n=0}^{\infty} e^{\imath n\phi}\,|n\rangle \quad (-\pi < \phi \le \pi) \tag{13.229}$$

are eigenstates of the exponential phase operator $\widehat{e^{\imath\phi}}$, and compute the corresponding eigenvalues. Are they orthogonal? Why? Write down their resolution of identity.

13.11 Show that the only non-vanishing matrix elements of the exponential phase operators are given by

$$\left\langle n-1\left|\widehat{e^{\imath\phi}}\right|n\right\rangle = 1, \tag{13.230a}$$
$$\left\langle n+1\left|\widehat{e^{-\imath\phi}}\right|n\right\rangle = 1. \tag{13.230b}$$

Use these results to prove that $\widehat{e^{\imath\phi}}$ and $\widehat{e^{-\imath\phi}}$ are not Hermitian operators.

13.12 Derive the uncertainty relation between $\widehat{\cos\phi}$ and $\widehat{\sin\phi}$.

13.13 Use the results of Prob. 4.11 to show that

$$\left[\hat{N}, \widehat{e^{\imath\phi}}\right] = -\widehat{e^{\imath\phi}} \tag{13.231a}$$

$$\left[\hat{N}, \widehat{e^{-\imath\phi}}\right] = \widehat{e^{-\imath\phi}}. \tag{13.231b}$$

Take advantage of these latter results to prove the commutation relations (13.42).

13.14 Evaluate the uncertainty relation (13.43) for the eigenstates of the number operator. Use this result to obtain the phase uncertainties $\Delta\widehat{\cos\phi}$ and $\Delta\widehat{\sin\phi}$.

13.15 Write the commutation relation betweeen magnetic and electric field $[\hat{\mathbf{E}}, \hat{\mathbf{B}}]$ of a single mode and use it to derive the corresponding uncertainty relation. Evaluate the uncertainty product for the vacuum state.

13.16 Prove Eqs. (13.60) by using the properties of the number operator and of the coherent states. Obtain the same results exploiting directly the Poisson distribution (13.59) for the photon number.

13.17 By exploiting the analogy between a one-dimensional particle subjected to a harmonic oscillator potential and a single mode of the electromagnetic field, identify the dimensionless position and momentum operators corresponding to the quadratures.

13.18 Prove Eq. (13.64).
(*Hint*: Make use of Eqs. (13.49), (13.61), and (13.24).)

13.19 Prove that, if $\hat{\xi}$ and $\hat{\pi}$ are two non-commuting operators, then

$$\begin{aligned} e^{-\eta\hat{\xi}}\hat{\pi}e^{\eta\hat{\xi}} &= \hat{\pi} - \eta\left[\hat{\xi}, \hat{\pi}\right] + \frac{\eta^2}{2!}\left[\hat{\xi}, \left[\hat{\xi}, \hat{\pi}\right]\right] + \cdots \\ &= \sum_n \frac{(-\eta)^n}{n!}\left[\hat{\xi}, \hat{\pi}\right]_n, \end{aligned} \tag{13.232}$$

where η is a complex parameter,

$$\left[\hat{\xi}, \hat{\pi}\right]_n = \left[\hat{\xi}, \left[\hat{\xi}, \hat{\pi}\right]_{n-1}\right],$$

and

$$\left[\hat{\xi}, \hat{\pi}\right]_1 = \left[\hat{\xi}, \hat{\pi}\right].$$

First, give the proof up to the second order in η; then, prove the general case.
(*Hint*: For the general derivation, make use of the derivatives of Eq. (13.232) with respect to η.)

13.20 Prove the Baker–Hausdorff(–Campbell) theorem, which states that, given any two operators $\hat{O}$ and $\hat{O}'$ for which

$$\left[\hat{O}, \left[\hat{O}, \hat{O}'\right]\right] = \left[\hat{O}', \left[\hat{O}, \hat{O}'\right]\right] = 0,$$

when in particular $\left[\hat{O}, \hat{O}'\right]$ is a c-number, we have

$$e^{\hat{O}+\hat{O}'} = e^{\hat{O}}e^{\hat{O}'}e^{-\frac{1}{2}[\hat{O},\hat{O}']} = e^{\hat{O}'}e^{\hat{O}}e^{\frac{1}{2}[\hat{O},\hat{O}']}. \tag{13.233}$$

(*Hint*: Make use of the results of the previous problem.)

13.21 Prove that $\hat{D}(\alpha)\hat{D}^\dagger(\alpha) = \hat{I}$.

13.22 Prove Eq. (13.69).

13.23 Show that a free-field single-mode coherent state remains coherent under time evolution.

13.24 Coherent states are overcomplete. This means that any coherent state may be written as an expansion in terms of the other coherent states. Show that this is indeed the case.
(*Hint*: Use the completeness relation (13.71).)

13.25 Make use of

$$e^{-\hat{O}}\hat{O}'e^{\hat{O}} = \sum_n \frac{(-1)^n}{n!}\left[\hat{O},\hat{O}'\right]_n, \tag{13.234}$$

where, as in Prob. 13.19, $\left[\hat{O},\hat{O}'\right]_n = \left[\hat{O},\left[\hat{O},\hat{O}'\right]_{n-1}\right]$ is the iterated commutator, to prove the equalities

$$\hat{S}^\dagger(\xi)\hat{a}\hat{S}(\xi) = \hat{a}\cosh r - \hat{a}^\dagger e^{\imath\chi}\sinh r, \tag{13.235a}$$
$$\hat{S}^\dagger(\xi)\hat{a}^\dagger\hat{S}(\xi) = \hat{a}^\dagger\cosh r - \hat{a}e^{-\imath\chi}\sinh r. \tag{13.235b}$$

13.26 Use the results of problem 13.25 to prove Eqs. (13.81) and (13.82).

13.27 Prove that the Q-function is bounded by $1/\pi$.

13.28 Prove Eq. (13.90).
(*Hint*: Make use of the fact that for any x and y we have $2xy \le x^2 + y^2$.)

13.29 Derive the characteristic functions for a coherent state $\hat{\rho} = |\beta\rangle\langle\beta|$ (see Eq. (13.100)) and a number state $\hat{\rho} = |n\rangle\langle n|$ (see Eq. (13.101)).

13.30 Prove that integration of the W-function with respect to p_x gives the correct probability distribution of the position.

13.31 Derive Eq. (13.116).

13.32 Using the explicit expression (13.131), derive Eq. (13.135).
(*Hint*: Derive first Eq. (13.134) and then assume that the local oscillator is in a coherent state $|\alpha_L\rangle$.)

13.33 Prove the relations (13.144) and (13.145).

13.34 Derive the result (13.181).

Further reading

Quantization of the electromagnetic field

Loudon, Rodney, *The Quantum Theory of Light*, London: Clarendon, 1973, 2nd edn., 1983.

General quantum optics

Barnett, S. M. and Radmore, P. M., *Methods in Theoretical Quantum Optics*, Oxford: Oxford University Press, 2002.

Goldin, Edwin, *Waves and Photons: An Introduction to Quantum Optics*, New York: Wiley, 1982.

Loudon, R. and Knight, P. L., Squeezed light. *Journal of Modern Optics*, **34** (1987), 709–59.

Scully, M. O. and Zubairy, M. S., *Quantum Optics*, Cambridge: Cambridge University Press, 1997.

Walls, D. F. and Milburn, G. J., *Quantum Optics*, Berlin: Springer, 1994.

Special states of light

Ma'nko, V. I. and Dodonov, V. V., *Theory of Non-classical States of Light*, Oxford: CRC Press, 2003.

Quantum optics in phase space

Cahill, K. E. and Glauber, R. J., Density operators and quasiprobability distributions. *Physical Review*, **177** (1969), 1882–902.

Gardiner, Crispin W., *Quantum Noise*, Berlin: Springer, 1991.

Hillery, M. A., O'Connell, R. F., Scully, M. O., and Wigner, E., Distribution functions in physics: fundamentals. *Physics Reports*, **106** (1984), 121–67.

Schleich, Wolfgang P., *Quantum Optics in Phase Space*, Berlin: Wiley-VCH, 2001.

Optical coherence

Mandel, L. and Wolf, E., *Optical Coherence and Quantum Optics*, Cambridge: Cambridge University Press, 1995.

Geometric phase

Bohm, A., Mostafazadeh, A., Koizumi, H., Niu, Q., and Zwanziger, J., *The Geometric Phase in Quantum Systems*, Berlin: Springer, 2003.

Shapere, A. and Wilczek, F. (eds.), *Geometric Phases in Physics*, Singapore: World Scientific, 1989.

Aharonov–Bohm effect

Peshkin, M. and Tonomura, A., *The Aharonov–Bohm Effect*, Berlin: Springer, 1989.

PART IV

QUANTUM INFORMATION: STATE AND CORRELATIONS

14 Quantum theory of open systems

In this chapter we shall deal with the quantum dynamics of an *open* system. By open quantum system we mean here a system which interacts with an environment (see Sec. 9.4): since we are not interested in the dynamics of the environment, we shall have to describe the evolution of the system in some "effective" way. In particular, if we consider only the evolution of the system, it will be non-unitary, and this will represent the subject of this chapter. As we know, Hamiltonian quantum dynamics is unitary and changes pure states into pure states. On the other hand, non-unitary dynamics changes an initially pure state into a mixture, which must be described by a density matrix (see Ch. 5). In the case of macroscopic systems, the coupling with the environment may be arbitrarily reduced and therefore its influence can be made correspondingly small (see Sec. 1.1). Microscopic systems, however, always couple to the environment and this coupling *cannot* be considered negligible. This is the reason why the quantum theory of open systems is one of the most important and fundamental chapters of quantum mechanics that, though born in quantum optics, has many implications in almost all fields of physics. The present chapter can be seen as a further development of the measurement theory (see Ch. 9), as open systems manifest a decoherent dynamics (see in particular Sec. 9.4).

One of the simplest and most effective ways to describe the interaction of a system with an environment is the so-called *reservoir* or *heat bath* approach. In this approach, the microscopic quantum system is assumed to interact with a much larger system endowed with a large number of degrees of freedom. This concept has been mutuated from statistical mechanics: the reservoir may be the water tank at a certain temperature used in the Carnot engine, or a bath of harmonic oscillators representing the radiation field viewed as an external perturbation acting on the dynamics of an atom. In some cases, the separation into "system" and "bath" is not physically evident, but only convenient for the characteristics of the system's dynamics. In general, however, due to the large number of degrees of freedom of the bath, the system does not influence the reservoir, whereas the presence of the reservoir and its coupling to the system induces *dissipation* (i.e. energy loss) and *fluctuations* (i.e. random noise).

In Sec. 14.1 we shall discuss the general effect of a reservoir on a quantum system in terms of decoherence and dephasing and introduce the concept of superoperator. In Sec. 14.2 the concept of master equation is introduced. In this section we shall also consider a first example of superoperator, the Lindblad superoperator. Finally, methods for finding the solution of the master equation are presented. In Sec. 14.3 we shall provide a generalization of the master equation formalism. In Sec. 14.4 the second example of superoperator, the jump superoperator, is presented and the quantum Monte-Carlo method is briefly reviewed. Finally, in Sec. 14.5 the issue of the Schrödinger cats is again discussed by introducing the most recent methods developed in quantum optics.

14.1 General considerations

14.1.1 Decoherence and dephasing

Let us consider a quantum system $\mathcal{S}$. We shall study here the simplest case, i.e. a generic two-level system described by the ket

$$| \psi \rangle = c_0' \, | 0 \rangle + c_1' | 1 \rangle , \tag{14.1}$$

$| 0 \rangle$ and $| 1 \rangle$ representing an arbitrary basis. $\mathcal{S}$ interacts with a large reservoir $\mathcal{R}$, whose state can be expanded in the orthonormal basis $\{| r_n \rangle\}$. Initially, the state of the combined system $(\mathcal{S} + \mathcal{R})$ is simply given by the factorized state

$$| \Psi \rangle = | \psi \rangle \otimes | R \rangle , \tag{14.2}$$

where $| R \rangle = \sum_n a_n \, | r_n \rangle$ is some initial state of the bath. In other words, the initial state is separable, i.e. it may be represented as a direct product of the two initial state vectors. In general, after the interaction, the state of the combined system will no longer be factorized, i.e. we may write

$$\left| \Psi' \right\rangle = c_0 \, | 0 \rangle \, | R_0 \rangle + c_1 \, | 1 \rangle \, | R_1 \rangle , \tag{14.3}$$

where $| R_0 \rangle$ and $| R_1 \rangle$ are some other (normalized but not necessarily orthogonal, i.e. $\langle R_1 \mid R_0 \rangle \neq 0$) states of the reservoir. The density matrix corresponding to the state described in Eq. (14.3) is given by

$$\begin{aligned} \hat{\rho}_{\mathcal{SR}} = {} & |c_0|^2 \, | 0 \rangle \, \langle 0 \mid R_0 \rangle \, \langle R_0 | + |c_1|^2 \, | 1 \rangle \, \langle 1 \mid R_1 \rangle \, \langle R_1 | \\ & + c_0 c_1^* \, | 0 \rangle \, \langle 1 \mid R_0 \rangle \, \langle R_1 | + c_1 c_0^* \, | 1 \rangle \, \langle 0 \mid R_1 \rangle \, \langle R_0 | , \end{aligned} \tag{14.4}$$

which is a pure state. In order to obtain the reduced density matrix of the system alone, we have to trace the reservoir out, so that (see Prob. 14.1)

$$\hat{\hat{\rho}}_{\mathcal{S}} = \begin{bmatrix} |c_0|^2 & c_0 c_1^* \, \langle R_1 \mid R_0 \rangle \\ c_0^* c_1 \, \langle R_0 \mid R_1 \rangle & |c_1|^2 \end{bmatrix} . \tag{14.5}$$

We see from Eq. (14.5) that the diagonal elements are not affected by this transformation, whereas the off-diagonal elements are reduced by the factor $| \langle R_1 \mid R_0 \rangle | < 1$. The more the two states $| R_1 \rangle$ and $| R_0 \rangle$ become orthogonal, the smaller the off-diagonal elements will be. This is therefore a simple example of how decoherence may occur in an open quantum system.

Another "phenomenological" way to show the effects of the reservoir's random fluctuations is provided by the so-called *dephasing*. In this case we assume that the state of the system $\mathcal{S}$ after interaction with $\mathcal{R}$ may be written as

$$| \psi \rangle = c_0 \, | 0 \rangle + c_1 e^{\imath \phi} | 1 \rangle , \tag{14.6}$$

where ϕ represents a random phase. When the fluctuations are large enough, the off-diagonal terms again tend to vanish, since $\langle e^{\imath\phi}\rangle = 0$. Similar considerations may be developed in the case of an n-level system, with $n > 2$.

14.1.2 Operators and superoperators

We want now to develop the appropriate formalism to study the dynamics described in a qualitative way in the previous subsection. Generally speaking, any operator $\hat{O}$ may be represented in the following form (see Eq. (5.5)):

$$\hat{O} = \sum_{j,k} | j\rangle \langle j | \hat{O} | k\rangle \langle k |, \tag{14.7}$$

where $\{| n\rangle \}$ is an arbitrary basis on the underlying Hilbert space. In this way, operators may be considered as – not necessarily normalized – "vectors" in a super Hilbert space,[1] which is the direct product of the original Hilbert space $\mathcal{H}$ and its dual $\mathcal{H}^*$ (see also Subsec. 1.3.2). In fact, Eq. (14.7) may be rewritten as

$$\hat{O} = \sum_{j,k} O_{jk} | j\ \rangle \langle k |, \tag{14.8}$$

where

$$O_{jk} = \left\langle j \left| \hat{O} \right| k \right\rangle. \tag{14.9}$$

Therefore, we may associate to any operator $\hat{O}$ a S-ket $|\hat{O}\}$ and a S-bra $\{\hat{O}|$, defined by

$$\left| \hat{O} \right\} = \sum_{j,k} O_{jk} | j,k\}, \tag{14.10a}$$

$$\left\{ \hat{O} \right| = \sum_{j,k} O^*_{jk} \{ j,k |, \tag{14.10b}$$

where

$$| j,k\} = || j\ \rangle \langle k |\}, \tag{14.11a}$$

$$\{ j,k | = \{| j\ \rangle \langle k || \tag{14.11b}$$

represent the basis in which the S-ket (and the S-bra) is expanded. Their scalar product may be represented as

$$\{ l,m | j,k\} = \langle l \mid j\rangle \langle k \mid m\rangle = \delta_{l,j}\delta_{k,m}, \tag{14.12}$$

from which the generalized scalar product follows (see Prob. 14.2)

$$\left\{ \hat{O} \middle| \hat{O}' \right\} = \mathrm{Tr}\left(\hat{O}^\dagger \hat{O}' \right). \tag{14.13}$$

From Eq. (14.13) it follows that we may reformulate Eq. (14.9) as

$$\left\langle j \left| \hat{O} \right| k \right\rangle = \{ j,k | \hat{O}\}. \tag{14.14}$$

[1] See [Royer 1989].

As a consequence of the introduction of S-kets and S-bras, the operators acting on the S-kets and S-bras are called *superoperators*, which will be symbolized with a double hat in the following. We shall consider in this chapter two examples: the Lindblad superoperator (in Subsec. 14.2.2) and the jump superoperator (in Sec. 14.4). In the next subsections, when a superoperator $\hat{\hat{\mathcal{L}}}$ acts on an operator $\hat{O}$, if not explicitly expressed, it will be understood that the superoperator acts on the S-ket $\left|\hat{O}\right\}$.

14.2 The master equation

In the open-system context, the *master equation* is a first-order differential equation for the reduced density operator of the system. It incorporates all the "ingredients" of the system–reservoir dynamics, so that the knowledge of the initial density matrix and the ability to solve the master equation allow us to determine the density operator of the system at any time t.

14.2.1 General formalism of the master equation

In this subsection we shall derive a rather general form of the master equation. In order to reach this goal we shall use several assumptions and approximations, which will be emphasized during the derivation.

Let us consider the Hamiltonian of the system plus reservoir. This may be written as

$$\hat{H}_{\mathcal{SR}} = \hat{H}_{\mathcal{S}} + \hat{H}_{\mathcal{R}} + \hat{H}_{\mathrm{I}} = \hat{H}_0 + \hat{H}_{\mathrm{I}}, \tag{14.15}$$

where $\hat{H}_0 = \hat{H}_{\mathcal{S}} + \hat{H}_{\mathcal{R}}$ is the free part and $\hat{H}_I$ is the interaction part of the Hamiltonian. In the Schrödinger picture, the dynamics of the total system is ruled by the von Neumann equation (5.28), which, in this case, is

$$\frac{d}{dt}\hat{\rho}_{\mathrm{S}}^{\mathcal{SR}} = \frac{\imath}{\hbar}\left[\hat{\rho}_{\mathrm{S}}^{\mathcal{SR}}, \hat{H}_{\mathcal{SR}}\right], \tag{14.16}$$

where the subscript S refers to the Schrödinger picture. Our aim is to find an equation for the reduced density operator of the system $\mathcal{S}$

$$\hat{\rho}_{\mathrm{S}}^{\mathcal{S}}(t) = \mathrm{Tr}_{\mathcal{R}}\left[\hat{\rho}_{\mathrm{S}}^{\mathcal{SR}}(t)\right], \tag{14.17}$$

where the superscript $\mathcal{S}$ refers to the system. First, we move to the Dirac picture (see Subsec. 3.6.2 and Eq. (5.30)) and transform the density operator $\hat{\rho}_{\mathrm{S}}^{\mathcal{SR}}$ according to

$$\hat{\rho}_{\mathrm{I}}^{\mathcal{SR}}(t) = e^{\frac{\imath}{\hbar}\hat{H}_0 t}\hat{\rho}_{\mathrm{S}}^{\mathcal{SR}}(t)e^{-\frac{\imath}{\hbar}\hat{H}_0 t}. \tag{14.18}$$

In the interaction picture, the von Neumann equation for the total system reads

$$\frac{d}{dt}\hat{\rho}_{\mathrm{I}}^{\mathcal{SR}}(t) = \frac{\imath}{\hbar}\left[\hat{\rho}_{\mathrm{I}}^{\mathcal{SR}}(t), \hat{H}_{\mathrm{I}}^{\mathrm{I}}(t)\right], \tag{14.19}$$

where the interaction Hamiltonian in the interaction picture is given by

$$\hat{H}_I^I(t) = e^{\frac{i}{\hbar}\hat{H}_0 t}\hat{H}_I(t)e^{-\frac{i}{\hbar}\hat{H}_0 t}. \tag{14.20}$$

As a consequence, making use of Eqs. (14.17) and (14.18), the reduced density operator of the system $\mathcal{S}$ may be written as

$$\hat{\tilde{\rho}}_S^S(t) = \text{Tr}_{\mathcal{R}}\left[e^{-\frac{i}{\hbar}\hat{H}_0 t}\hat{\rho}_I^{S\mathcal{R}}(t)e^{\frac{i}{\hbar}\hat{H}_0 t}\right], \tag{14.21}$$

from which, taking into account that $\hat{H}_0 = \hat{H}_S + \hat{H}_{\mathcal{R}}$, that the trace is over the reservoir degrees of freedom, and the cyclic property of the trace (see Box 3.1), it follows that

$$\hat{\tilde{\rho}}_S^S(t) = e^{-\frac{i}{\hbar}\hat{H}_S t}\hat{\tilde{\rho}}_I^S(t)e^{\frac{i}{\hbar}\hat{H}_S t}, \tag{14.22}$$

where

$$\hat{\tilde{\rho}}_I^S(t) = \text{Tr}_{\mathcal{R}}\left[\hat{\rho}_I^{S\mathcal{R}}(t)\right]. \tag{14.23}$$

At this point we may integrate Eq. (14.19), so as to obtain

$$\hat{\rho}_I^{S\mathcal{R}}(t) = \hat{\rho}_I^{S\mathcal{R}}(0) - \frac{i}{\hbar}\int_0^t dt'\left[\hat{H}_I^I(t'), \hat{\rho}_I^{S\mathcal{R}}(t')\right]. \tag{14.24}$$

If we perform a second iteration by substituting back $\hat{\rho}_I^{S\mathcal{R}}(t')$ into Eq. (14.24), we obtain

$$\hat{\rho}_I^{S\mathcal{R}}(t) = \hat{\rho}_I^{S\mathcal{R}}(0) - \frac{i}{\hbar}\int_0^t dt'\left[\hat{H}_I^I(t'), \hat{\rho}_I^{S\mathcal{R}}(0)\right]$$
$$- \frac{1}{\hbar^2}\int_0^t dt'\int_0^{t'} dt''\left[\hat{H}_I^I(t'), \left[\hat{H}_I^I(t''), \hat{\rho}_I^{S\mathcal{R}}(t'')\right]\right]. \tag{14.25}$$

One could keep on iterating this way, deriving an infinite series. This process, however, would not be very helpful and we prefer to differentiate Eq. (14.25). This calculation gives

$$\frac{d}{dt}\hat{\rho}_I^{S\mathcal{R}}(t) = -\frac{i}{\hbar}\left[\hat{H}_I^I(t), \hat{\rho}_I^{S\mathcal{R}}(0)\right]$$
$$- \frac{1}{\hbar^2}\int_0^t dt'\left[\hat{H}_I^I(t), \left[\hat{H}_I^I(t'), \hat{\rho}_I^{S\mathcal{R}}(t')\right]\right]. \tag{14.26}$$

Now we perform the trace over the reservoir degrees of freedom. By taking into account Eq. (14.23), this yields

$$\frac{d}{dt}\hat{\tilde{\rho}}_I^S(t) = -\frac{1}{\hbar^2}\int_0^t dt'\text{Tr}_{\mathcal{R}}\left\{\left[\hat{H}_I^I(t), \left[\hat{H}_I^I(t'), \hat{\rho}_I^{S\mathcal{R}}(t')\right]\right]\right\}, \tag{14.27}$$

where, without restrictions, we have assumed

$$\text{Tr}_{\mathcal{R}}\left\{\left[\hat{H}_I^I(t), \hat{\rho}_I^{S\mathcal{R}}(0)\right]\right\} = 0. \tag{14.28}$$

Now, we impose that the reservoir and the system $\mathcal{S}$ are initially uncoupled, i.e.

$$\hat{\rho}_S^{S\mathcal{R}}(0) = \hat{\rho}_S^S(0) \otimes \hat{\rho}_S^{\mathcal{R}}, \tag{14.29}$$

and that the interaction energy is much smaller than the energy of the system and of the reservoir (*weak coupling assumption*), so that the reservoir density matrix is practically not affected by the interaction with $\mathcal{S}$, i.e.

$$\hat{\rho}_{\mathrm{I}}^{\mathcal{R}}(t) \simeq \hat{\rho}_{\mathrm{I}}^{\mathcal{R}}(0) = \hat{\rho}^{\mathcal{R}}, \tag{14.30}$$

from which it follows that

$$\hat{\rho}_{\mathrm{I}}^{\mathcal{SR}}(t) \simeq \hat{\rho}_{\mathrm{I}}^{\mathcal{S}}(t) \otimes \hat{\rho}^{\mathcal{R}}. \tag{14.31}$$

In the weak coupling approximation, Eq. (14.27) takes the form of an integro-differential equation for $\hat{\tilde{\rho}}_{\mathrm{I}}^{\mathcal{S}}(t)$

$$\frac{d}{dt}\hat{\tilde{\rho}}_{\mathrm{I}}^{\mathcal{S}}(t) = -\frac{1}{\hbar^2}\int_0^t dt' \mathrm{Tr}_{\mathcal{R}}\left\{\left[\hat{H}_{\mathrm{I}}^{\mathrm{I}}(t), \left[\hat{H}_{\mathrm{I}}^{\mathrm{I}}(t'), \hat{\tilde{\rho}}_{\mathrm{I}}^{\mathcal{S}}(t') \otimes \hat{\rho}^{\mathcal{R}}\right]\right]\right\}. \tag{14.32}$$

In order to reduce the previous equation into a true differential equation, we consider the case where the correlation functions of the reservoir vary at a time scale much shorter than the characteristic time of the dynamics of $\mathcal{S}$. In this case, we can make use of the so-called *Markov approximation*, and say that $\hat{\tilde{\rho}}_{\mathrm{I}}^{\mathcal{S}}(t)$ changes by a negligible amount over the time scale on which the reservoir correlation functions in Eq. (14.32) tend to vanish. With such an approximation, we may replace $\hat{\tilde{\rho}}_{\mathrm{I}}^{\mathcal{S}}(t')$ by $\hat{\tilde{\rho}}_{\mathrm{I}}^{\mathcal{S}}(t)$ and let the lower limit in the integral go to minus infinity in Eq. (14.32). After defining $\tau = t - t'$, we finally obtain

$$\frac{d}{dt}\hat{\tilde{\rho}}_{\mathrm{I}}^{\mathcal{S}}(t) = -\frac{1}{\hbar^2}\int_0^\infty d\tau \mathrm{Tr}_{\mathcal{R}}\left\{\left[\hat{H}_{\mathrm{I}}^{\mathrm{I}}(t), \left[\hat{H}_{\mathrm{I}}^{\mathrm{I}}(t-\tau), \hat{\tilde{\rho}}_{\mathrm{I}}^{\mathcal{S}}(t) \otimes \hat{\rho}^{\mathcal{R}}\right]\right]\right\}. \tag{14.33}$$

As in the theory of stochastic processes, we see that the Markov approximation has produced exactly the desired result: the knowledge of $\hat{\tilde{\rho}}_{\mathrm{I}}^{\mathcal{S}}(0)$ and of the interaction Hamiltonian (both of the state of the system and of the "force" acting on it) is sufficient to determine $\hat{\tilde{\rho}}_{\mathrm{I}}^{\mathcal{S}}(t)$ at all future times. In the following subsections we shall see how these general considerations apply to particular cases.

14.2.2 The Lindblad master equation

Equation (14.33) has a rather general form and makes no assumptions on the type of system–reservoir interaction. Let us now write the interaction Hamiltonian in the Schrödinger picture in the form

$$\hat{H}_{\mathrm{I}} = \hbar \sum_j \left(\hat{\Lambda}_j^\dagger \hat{\Gamma}_j + \hat{\Lambda}_j \hat{\Gamma}_j^\dagger\right), \tag{14.34}$$

where $\hat{\Gamma}_k$ are reservoir operators and the system operators $\hat{\Lambda}_k$ satisfy the quite general property[2]

$$\left[\hat{H}_{\mathcal{S}}, \hat{\Lambda}_j^\dagger\right] = \hbar\omega_j \hat{\Lambda}_j^\dagger. \tag{14.35}$$

Equation (14.34) is sufficiently general and, for example, includes the case when $\hat{\Gamma}_j$ are annihilation operators of the reservoir (supposed to be a bath of harmonic oscillators) and $\hat{\Lambda}_j$ are annihilation operators of the system. In such a case, the master equation will describe the dynamics of the damped harmonic oscillator, and the Hamiltonian (14.34) expresses the exchange of quanta between the system and the bath: it is precisely this exchange which represents the energy loss due to dissipation. Using Eqs. (14.34) and (14.20) in Eq. (14.33), we obtain $4^2 = 16$ terms: the four terms given by the double commutator, in each of which $\hat{H}_{\mathrm{I}}^{\mathrm{I}}(t)$ is expanded following Eq. (14.34). For example, one of the terms has the form

$$-\int_0^t dt' \sum_{j,l} e^{-\imath\omega_j t}\hat{\Lambda}_j^\dagger e^{\imath\omega_l t}\hat{\Lambda}_l\, \hat{\tilde{\rho}}_{\mathrm{I}}^{\mathcal{S}}(t')\mathrm{Tr}_{\mathcal{R}}\left[\hat{\Gamma}_j(t)\hat{\Gamma}_l^\dagger(t')\hat{\rho}^{\mathcal{R}}\right], \tag{14.36}$$

where

$$\hat{\Gamma}_k(t) = e^{\frac{\imath}{\hbar}\hat{H}_{\mathcal{R}}t}\hat{\Gamma}_k e^{-\frac{\imath}{\hbar}\hat{H}_{\mathcal{R}}t}, \tag{14.37}$$

and we have used the formula (see Prob. 13.19)

$$e^{\frac{\imath}{\hbar}\hat{H}_{\mathcal{S}}t}\hat{\Lambda}_j e^{-\frac{\imath}{\hbar}\hat{H}_{\mathcal{S}}t} = e^{\imath\omega_j t}\hat{\Lambda}_j. \tag{14.38}$$

The other three terms have a similar form. Now, we assume, as we have already seen in the previous subsection, that the state of the reservoir is stationary, i.e. that the last term in the trace in Eq. (14.36) only depends on the difference $t' - t$. As a consequence, the terms in the sum for which $\omega_j \neq \omega_l$ are quickly oscillating terms with respect to the dynamical time scale of $\hat{\tilde{\rho}}_{\mathrm{I}}^{\mathcal{S}}(t)$, and can be neglected. This is another manifestation of the so-called rotating wave approximation, first introduced in Subsec. 13.7.1.

Next, we use the Markov approximation of the previous subsection, and therefore we are allowed to replace $\hat{\tilde{\rho}}_{\mathrm{I}}^{\mathcal{S}}(t')$ by $\hat{\tilde{\rho}}_{\mathrm{I}}^{\mathcal{S}}(t)$. The term (14.36) can be written as

$$-\sum_j \hat{\Lambda}_j^\dagger\hat{\Lambda}_j \int_0^\infty d\tau e^{\imath\omega_j\tau}\hat{\tilde{\rho}}_{\mathrm{I}}^{\mathcal{S}}(t)\mathrm{Tr}_{\mathcal{R}}\left[\hat{\Gamma}_j(\tau)\hat{\Gamma}_j^\dagger(0)\hat{\rho}^{\mathcal{R}}\right], \tag{14.39}$$

where $\tau = t - t'$ and we let t go to infinity. In conclusion, since the terms involving the correlation functions of products as $\hat{\Gamma}(\tau)\hat{\Gamma}(0)$ or $\hat{\Gamma}^\dagger(\tau)\hat{\Gamma}^\dagger(0)$ are for simplicity assumed to be negligible if, for $\tau \neq 0$, the reservoir is in a thermal state,[3] all the remaining terms can be cast in one of the following forms

[2] When, for some operators $\hat{\Lambda}_j$ and $\hat{\Lambda}_j^\dagger$, Eq. (14.35) holds, we say that $\hat{\Lambda}_j$ and $\hat{\Lambda}_j^\dagger$ are *eigenoperators* of the system's Hamiltonian. This property is general because it can be proved that any system's operator can be expanded into eigenoperators of the system's Hamiltonian.

[3] These terms, however, are not necessarily zero if the reservoir is in a squeezed state (see Subsec. 13.4.3).

$$\int_0^\infty d\tau e^{\imath\omega_j\tau}\mathrm{Tr}_{\mathcal{R}}\left[\hat{\Gamma}_j(\tau)\hat{\Gamma}_j^\dagger(0)\hat{\rho}^{\mathcal{R}}\right]=A_j+\imath B_j, \tag{14.40a}$$

$$\int_0^\infty d\tau e^{-\imath\omega_j\tau}\mathrm{Tr}_{\mathcal{R}}\left[\hat{\Gamma}_j(0)\hat{\Gamma}_j(\tau)\hat{\rho}^{\mathcal{R}}\right]=A_j-\imath B_j, \tag{14.40b}$$

$$\int_0^\infty d\tau e^{\imath\omega_j\tau}\mathrm{Tr}_{\mathcal{R}}\left[\hat{\Gamma}_j^\dagger(\tau)\hat{\Gamma}_j(0)\hat{\rho}^{\mathcal{R}}\right]=C_j+\imath D_j, \tag{14.40c}$$

$$\int_0^\infty d\tau e^{-\imath\omega_j\tau}\mathrm{Tr}_{\mathcal{R}}\left[\hat{\Gamma}_j^\dagger(0)\hat{\Gamma}_j(\tau)\hat{\rho}^{\mathcal{R}}\right]=C_j-\imath D_j, \tag{14.40d}$$

where A_j, B_j, C_j, and D_j are real numbers. The master equation in the interaction picture then takes the final form

$$\begin{aligned}\frac{d}{dt}\hat{\rho}_{\mathrm{I}}^{\mathcal{S}}(t) = &-\imath\sum_j\left[B_j\hat{\Lambda}_j^\dagger\hat{\Lambda}_j + D_j\hat{\Lambda}_j\hat{\Lambda}_j^\dagger,\ \hat{\rho}_{\mathrm{I}}^{\mathcal{S}}\right]\\ &+\sum_j A_j\left(2\hat{\Lambda}_j^\dagger\hat{\rho}_{\mathrm{I}}^{\mathcal{S}}\hat{\Lambda}_j - \hat{\Lambda}_j^\dagger\hat{\Lambda}_j\hat{\rho}_{\mathrm{I}}^{\mathcal{S}} - \hat{\rho}_{\mathrm{I}}^{\mathcal{S}}\hat{\Lambda}_j^\dagger\hat{\Lambda}_j\right)\\ &+\sum_j C_j\left(2\hat{\Lambda}_j\hat{\rho}_{\mathrm{I}}^{\mathcal{S}}\hat{\Lambda}_j^\dagger - \hat{\Lambda}_j\hat{\Lambda}_j^\dagger\hat{\rho}_{\mathrm{I}}^{\mathcal{S}} - \hat{\rho}_{\mathrm{I}}^{\mathcal{S}}\hat{\Lambda}_j\hat{\Lambda}_j^\dagger\right),\end{aligned} \tag{14.41}$$

where the first commutator term just represents a small perturbing Hamiltonian term (e.g. the Lamb and Stark shift terms) and is usually neglected.

Equation (14.41) is an example of a general class of equations known as *Lindblad's master equation*,[4] which, in the Schrödinger picture, can be written in the form

$$\frac{d}{dt}\hat{\rho}(t) = \hat{\hat{\mathcal{L}}}\hat{\rho}(t), \tag{14.42}$$

where $\hat{\hat{\mathcal{L}}}$ is the *Lindblad superoperator* for the generalized Liouville transformation given by

$$\hat{\hat{\mathcal{L}}}\,|\hat{\rho}\} = \left(\hat{\hat{\mathcal{L}}}_{\mathrm{d}} + \hat{\hat{\mathcal{L}}}_{\mathrm{nd}}\right)|\hat{\rho}\}, \tag{14.43a}$$

where the term (for the sake of notation, we drop the S-ket formalism, as already announced)

$$\hat{\hat{\mathcal{L}}}_{\mathrm{nd}}\hat{\rho} = -\frac{\imath}{\hbar}\left[\hat{H},\hat{\rho}\right] \tag{14.43b}$$

represents the unitary evolution given by the quantum counterpart of the Liouville equation (5.28), while the term $\hat{\hat{\mathcal{L}}}_d$ represents the dissipative part given by

[4] See [*Lindblad* 1983].

$$\hat{\mathcal{L}}_{\rm d}\hat{\rho} = \frac{1}{2}\sum_j \left(\left[\hat{\Lambda}_j\hat{\rho}, \hat{\Lambda}_j^\dagger\right] + \left[\hat{\Lambda}_j, \hat{\rho}\hat{\Lambda}_j^\dagger\right]\right), \tag{14.43c}$$

where, in this context, the set of operators $\{\hat{\Lambda}_j\}$ is also known as *Lindblad operators*.

14.2.3 The damped harmonic oscillator

Let us consider the simplest case of a harmonic oscillator interacting with a bath of harmonic oscillators (the heat bath), which we have already mentioned in the previous subsection. This model may represent, for instance, a single mode of the electromagnetic field interacting with its environment. The total Hamiltonian in this case would be

$$\hat{H} = \hat{H}_{\mathcal{S}} + \hat{H}_{\mathcal{B}} + \hat{H}_{\rm I}, \tag{14.44}$$

where

$$\hat{H}_{\mathcal{S}} = \hbar\omega\hat{a}^\dagger\hat{a} \tag{14.45}$$

is the harmonic-oscillator type Hamiltonian of the system,

$$\hat{H}_{\mathcal{B}} = \hbar\sum_n \omega_n\hat{b}_n^\dagger\hat{b}_n \tag{14.46}$$

is the bath Hamiltonian, $\hat{b}_n$ being the annihilation operator of the n-th oscillator of the bath, and

$$\hat{H}_{\rm I} = \hat{a}\hat{\Gamma}^\dagger + \hat{a}^\dagger\hat{\Gamma} \tag{14.47}$$

is the interaction Hamiltonian, where $\hat{\Gamma}$ is a bath operator defined as

$$\hat{\Gamma} = \sum_n g_n\hat{b}_n, \tag{14.48}$$

and g_n are the system–bath coupling constants. As we have seen, in this approximation the bath is stationary. Its correlation functions (14.40) can be shown to be[5]

$$\left\langle\hat{\Gamma}^\dagger(t)\hat{\Gamma}(t')\right\rangle = 2\gamma N\delta(t-t'), \tag{14.49a}$$

$$\left\langle\hat{\Gamma}(t)\hat{\Gamma}^\dagger(t')\right\rangle = 2\gamma\,(N+1)\,\delta(t-t'), \tag{14.49b}$$

$$\left\langle\hat{\Gamma}(t)\hat{\Gamma}(t')\right\rangle = 0, \tag{14.49c}$$

$$\left\langle\hat{\Gamma}^\dagger(t)\hat{\Gamma}^\dagger(t')\right\rangle = 0, \tag{14.49d}$$

[5] See, e.g., [*Gardiner* 1991, 336].

where the δ functions reflect the fact that the dynamics of the system are very fast, γ is known as *damping constant* and is a function of the coupling constants, and

$$N = \frac{1}{\exp\left(\frac{\hbar\omega}{k_B T}\right) - 1}, \tag{14.50}$$

where T is the temperature of the bath that we assume to be constant (the bath is both stationary and in equilibrium). As we have already mentioned, such an interaction induces exchange of quanta between the system and the bath. Using the above results into Eq. (14.41), we may derive the master equation for the reduced density matrix of the system in the interaction picture. For a zero-temperature heat bath, we have $N = 0$ and the master equation can be written as (see Prob. 14.3)

$$\frac{d\hat{\rho}}{dt} = \gamma \left(2\hat{a}\hat{\rho}\hat{a}^\dagger - \hat{a}^\dagger\hat{a}\hat{\rho} - \hat{\rho}\hat{a}^\dagger\hat{a}\right). \tag{14.51}$$

14.2.4 Solution of the master equation

The solution of the master equation, for instance of Eq. (14.51), is often difficult. In some cases, it is even impossible to solve an operatorial master equation directly and find $\hat{\rho}(t)$ in operatorial form. However, alternative methods of solution are available. For example, one can derive equations of motion for certain relevant expectation values and solve these to obtain the desired operator averages. Alternatively, one may take matrix elements of the master equation in a given representation and obtain equations of motion for the matrix elements of $\hat{\rho}$ (see for instance Subsec. 14.3.2). Another possible way out is to adopt numerical techniques to integrate the master equation. Finally, it is also possible to translate the master equation into an equivalent c-number partial differential equation for the Q, W, or P functions (see Sec. 13.5). This yields a Fokker–Planck-type equation for the corresponding quasi-probability distribution.[6] It is not our intention here to give a complete treatment of the methods which allow the conversion of the master equation into a partial differential equation. We shall only sketch briefly the steps which are necessary to accomplish this task for the Q-function and apply the method to the damped harmonic oscillator case.

If we want to turn Eq. (14.51) into a Fokker–Planck equation for the Q-function, it is sufficient to compute the expectation value of both sides of the master equation onto a coherent state $|\alpha\rangle$. The lhs of Eq. (14.51) then becomes (see also Eq. (13.83))

$$\frac{1}{\pi}\left\langle \alpha \left| \frac{d\hat{\rho}}{dt} \right| \alpha \right\rangle = \frac{\partial}{\partial t} \mathrm{Q}(\alpha, \alpha^*; t). \tag{14.52}$$

[6] See [Fokker 1914] and [Planck 1917]. Originally, the Fokker–Planck equation described the time evolution of the probability density function of the position and velocity of a particle, but it can be generalized to any other observable, too, It applies to systems that can be described by a small number of "macrovariables," where other parameters vary so rapidly with time that they can be treated as noise. The first use of the Fokker–Planck equation was the statistical description of Brownian motion of a particle in a fluid.

On the rhs, it is first necessary to normally order (see footnote 19 of Ch. 13) all the terms, i.e. to move all the creation operators $\hat{a}^\dagger$ to the left of the annihilation operators $\hat{a}$, using the commutation relation $[\hat{a}, \hat{a}^\dagger] = \hat{I}$, and then compute the corresponding expectation value.

For instance, considering the last term in brackets on the rhs of Eq. (14.51), we have (see Prob. 14.5)

$$\left\langle \alpha \left| \hat{\rho}\hat{a}^\dagger \hat{a} \right| \alpha \right\rangle = \alpha \left\langle \alpha \left| \hat{a}^\dagger \hat{\rho} + \frac{d\hat{\rho}}{d\hat{a}} \right| \alpha \right\rangle = \pi \left(|\alpha|^2 + \alpha \frac{\partial}{\partial \alpha} \right) \mathrm{Q}(\alpha, \alpha^*; t), \tag{14.53}$$

where we have taken advantage of the fact that $|\alpha\rangle$ is an eigenstate of the destruction operator with eigenvalue α. In a similar way,

$$\left\langle \alpha \left| \hat{a}^\dagger \hat{\rho}\hat{a} \right| \alpha \right\rangle = |\alpha|^2 (\alpha, \alpha^*; t), \tag{14.54}$$

and, for the first term in brackets on the rhs of Eq. (14.51), we have (see Prob. 14.6)

$$\left\langle \alpha \left| \hat{a}\hat{\rho}\hat{a}^\dagger \right| \alpha \right\rangle = |\alpha|^2 \mathrm{Q} + \alpha^* \frac{\partial}{\partial \alpha^*} \mathrm{Q} + \frac{\partial}{\partial \alpha} (\alpha \mathrm{Q}) + \frac{\partial^2}{\partial \alpha \partial \alpha^*} \mathrm{Q}. \tag{14.55}$$

In conclusion, one can establish an operatorial correspondence between the operators $\hat{a}, \hat{a}^\dagger$ (which act on the density matrix) and the differential operators $\partial/\partial\alpha, \partial/\partial\alpha^*$ (which act on the corresponding Q-function):

$$\hat{a}\hat{\rho} \leftrightarrow \left(\alpha + \frac{\partial}{\partial \alpha^*} \right) \mathrm{Q}(\alpha, \alpha^*), \tag{14.56a}$$

$$\hat{a}^\dagger \hat{\rho} \leftrightarrow \alpha^* \mathrm{Q}(\alpha, \alpha^*), \tag{14.56b}$$

$$\hat{\rho}\hat{a} \leftrightarrow \alpha \mathrm{Q}(\alpha, \alpha^*), \tag{14.56c}$$

$$\hat{\rho}\hat{a}^\dagger \leftrightarrow \left(\alpha^* + \frac{\partial}{\partial \alpha} \right) \mathrm{Q}(\alpha, \alpha^*). \tag{14.56d}$$

Proceeding in this way for all the operator products in the rhs of the master equation, one obtains the corresponding partial differential equation for the Q-function. In particular, if one applies this procedure to the master equation (14.51), one obtains (see Prob. 14.7)

$$\frac{\partial}{\partial t} \mathrm{Q} = \gamma \left(2 \frac{\partial^2}{\partial \alpha \partial \alpha^*} + \frac{\partial}{\partial \alpha} \alpha + \frac{\partial}{\partial \alpha^*} \alpha^* \right) \mathrm{Q}. \tag{14.57}$$

14.3 A formal generalization

In the present section, we want to make a formal connection between the master equation formalism, developed in the preceding subsections, and the POVM formalism, described in Sec. 9.10.

14.3.1 The master equation as a non-unitary transformation

Any time that a system and its environment interact, we may write the transformation $\mathcal{T}$ (see Subsec. 9.10.1), which occurs on the system as a result of the interaction,[7] as

$$\mathcal{T}\left(\hat{\rho}^{\mathcal{S}}\right) = \mathrm{Tr}_{\mathcal{E}}\left[\hat{U}\left(\hat{\rho}^{\mathcal{S}} \otimes \hat{\rho}^{\mathcal{E}}\right)\hat{U}^{\dagger}\right], \tag{14.58}$$

where, as usual, $\hat{\rho}^{\mathcal{S}}$ and $\hat{\rho}^{\mathcal{E}}$ stand for the density matrix of the system and the environment, respectively, and $\hat{U}$ is the evolution operator for the combination system plus environment. Equation (14.58) may also be written in terms of the transformation superoperator $\hat{\hat{\mathcal{T}}}$ as (see also Eq. (9.116a))

$$\begin{aligned}\hat{\hat{\mathcal{T}}}\left|\hat{\rho}^{\mathcal{S}}\right\} &= \sum_j \left\langle e_j \left| \hat{U}\left(\hat{\rho}^{\mathcal{S}} \otimes | e_0 \rangle \langle e_0 |\right)\hat{U}^{\dagger}\right| e_j\right\rangle \\ &= \sum_j \hat{\vartheta}_j \hat{\rho}^{\mathcal{S}} \hat{\vartheta}_j^{\dagger},\end{aligned} \tag{14.59}$$

where $| e_0 \rangle$ is the initial state of the environment and $\{| e_j \rangle\}$ is an orthogonal basis for the environment, and where

$$\hat{\vartheta}_k = \left\langle e_k \left| \hat{U} \right| e_0 \right\rangle \tag{14.60}$$

are the amplitude operators (here for the environment) introduced in Eqs. (9.123) and (9.124) (see also Eq. (9.150)). Let us now consider the probability that the system is in the state $\hat{\rho}_k^{\mathcal{S}}$, written as

$$\begin{aligned}\hat{\rho}_k^{\mathcal{S}} &\propto \mathrm{Tr}_{\mathcal{E}}\left[\hat{P}_k \hat{U}\left(\hat{\rho}^{\mathcal{S}} \otimes | e_0 \rangle \langle e_0 |\right)\hat{U}^{\dagger}\hat{P}_k\right] \\ &= \hat{\vartheta}_k \hat{\rho}^{\mathcal{S}} \hat{\vartheta}_k^{\dagger},\end{aligned} \tag{14.61}$$

where $\hat{P}_k = | e_k \rangle \langle e_k |$. The state $\hat{\rho}_k^{\mathcal{S}}$ can be normalized as

$$\hat{\rho}_k^{\mathcal{S}} = \frac{\hat{\vartheta}_k \hat{\rho}^{\mathcal{S}} \hat{\vartheta}_k^{\dagger}}{\mathrm{Tr}_{\mathcal{S}}\left(\hat{\vartheta}_k \hat{\rho}^{\mathcal{S}} \hat{\vartheta}_k^{\dagger}\right)}, \tag{14.62}$$

so that we may finally write the probability that the system is in the state $\hat{\rho}_k^{\mathcal{S}}$ as

$$\wp_k = \mathrm{Tr}\left(\hat{\vartheta}_k \hat{\rho}^{\mathcal{S}} \hat{\vartheta}_k^{\dagger}\right) \tag{14.63}$$

and the transformation given by Eqs. (14.58) and (14.59) as

$$\mathcal{T}\left(\hat{\rho}^{\mathcal{S}}\right) = \sum_k \wp_k \hat{\rho}_k^{\mathcal{S}}. \tag{14.64}$$

Let us now consider the special case in which we perform a measurement on the environment and project it onto to the state $| e_m \rangle$. In this case, the final state of the compound system is given by (see Eq.(9.101))

[7] See [*Nielsen/Chuang* 2000, 356–73, 386–89].

$$\hat{\rho}_f^{\mathcal{S}+\mathcal{E}} = \frac{\hat{P}_m \hat{U}\left(\hat{\rho}^{\mathcal{S}} \otimes \hat{\rho}^{\mathcal{E}}\right) \hat{U}^\dagger \hat{P}_m}{\mathrm{Tr}\left[\hat{P}_m \hat{U}\left(\hat{\rho}^{\mathcal{S}} \otimes \hat{\rho}^{\mathcal{E}}\right) \hat{U}^\dagger\right]}, \tag{14.65}$$

where $\hat{P}_m = | e_m\rangle \langle e_m |$. The net transformation acting on the object system may then be written as

$$\begin{aligned} \hat{\mathcal{T}}_m \left|\hat{\rho}^{\mathcal{S}}\right\} &= \mathrm{Tr}_{\mathcal{E}}\left[\hat{P}_m \hat{U}\left(\hat{\rho}^{\mathcal{S}} \otimes \hat{\rho}^{\mathcal{E}}\right) \hat{U}^\dagger \hat{P}_m\right] \\ &= \sum_{jk} w_j \mathrm{Tr}_{\mathcal{E}}\left[\hat{P}_k \hat{P}_m \hat{U}\left(\hat{\rho}^{\mathcal{S}} \otimes | e_j\rangle\langle e_j |\right) \hat{U}^\dagger \hat{P}_m \hat{P}_k\right] \\ &= \sum_{jk} \hat{\vartheta}_{jk} \hat{\rho}^{\mathcal{S}} \hat{\vartheta}_{jk}^\dagger, \end{aligned} \tag{14.66}$$

where

$$\hat{\rho}^{\mathcal{E}} = \sum_j w_j \, | e_j\rangle\langle e_j |, \tag{14.67}$$

the w_j being weights, i.e. real and positive numbers, and

$$\hat{\vartheta}_{jk} = \sqrt{w_j}\left\langle e_k \left| \hat{P}_m \hat{U} \right| e_j\right\rangle. \tag{14.68}$$

Now, we can show the connection of this formalism with that of the master equation. Let us consider, for the sake of simplicity, the specific example of a two-level atom coupled to the vacuum radiation field and undergoing spontaneous emission. The coherent part of the atom's evolution is described by the Hamiltonian

$$\hat{H} = \frac{1}{2}\hbar\omega\hat{\sigma}_z, \tag{14.69}$$

where $\hbar\omega$ is the energy difference of the atomic levels (see Subsec. 13.7.1). The emission of a photon, causing the atom to make a transition from the excited level $| e\rangle$ to the ground state $| g\rangle$, is described by the operator

$$\hat{\Lambda} = \sqrt{2\gamma}\hat{\sigma}_-, \tag{14.70}$$

where $\hat{\sigma}_- = |g\rangle\langle e|$ and γ is the rate of spontaneous emission. The other possible process, namely the absorption of a photon (which would be described by the $\hat{\sigma}_+$ operator), is not allowed since initially no photons are present in the radiation field (and the atom is supposed to be in the excited state). From Eqs. (14.43) and making use of Eq. (14.70), we immediately obtain the master equation ruling this process as

$$\hat{\mathcal{L}}\left|\hat{\rho}\right\} = \frac{d\hat{\rho}}{dt} = -\frac{\imath}{\hbar}\left[\hat{H}, \hat{\rho}\right] + \gamma\left(2\hat{\sigma}_-\hat{\rho}\hat{\sigma}_+ - \hat{\sigma}_+\hat{\sigma}_-\hat{\rho} - \hat{\rho}\hat{\sigma}_+\hat{\sigma}_-\right). \tag{14.71}$$

Now, we show that this master equation may be reduced to Eq. (14.59) by performing the change of variables

$$\hat{\rho}'(t) = e^{\frac{\imath}{\hbar}\hat{H}t}\hat{\rho}(t)e^{-\frac{\imath}{\hbar}\hat{H}t}. \tag{14.72}$$

In fact, the master equation for $\hat{\rho}'(t)$ is

$$\hat{\mathcal{L}}\left|\hat{\rho}'\right\} = \gamma\left(2\hat{\sigma}'_-\hat{\rho}'\hat{\sigma}'_+ - \hat{\sigma}'_+\hat{\sigma}'_-\hat{\rho}' - \hat{\rho}'\hat{\sigma}'_+\hat{\sigma}'_-\right), \tag{14.73}$$

where (see Prob. 13.19)

$$\hat{\sigma}'_- = e^{\imath \hat{H} t} \hat{\sigma}_- e^{-\imath \hat{H} t} = e^{-\imath \omega t} \hat{\sigma}_-, \tag{14.74a}$$

$$\hat{\sigma}'_+ = e^{\imath \hat{H} t} \hat{\sigma}_+ e^{-\imath \hat{H} t} = e^{\imath \omega t} \hat{\sigma}_+. \tag{14.74b}$$

Thus, Eq. (14.73) may be rewritten as

$$\hat{\hat{\mathcal{L}}} \left| \hat{\rho}' \right\} = \gamma \left(2\hat{\sigma}_- \hat{\rho}' \hat{\sigma}_+ - \hat{\sigma}_+ \hat{\sigma}_- \hat{\rho}' - \hat{\rho}' \hat{\sigma}_+ \hat{\sigma}_- \right), \tag{14.75}$$

which is formally similar to Eq. (14.51) – even though there we are faced with a mode of the electromagnetic field and here with an atom. In order to solve Eq. (14.75) we need to use the Bloch-vector representation for $\hat{\rho}'$.

14.3.2 Bloch-sphere representation of a two-dimensional system

It is interesting to note that there is a formal similarity between a two–level atom interacting with the electromagnetic field and a spin-1/2 magnetic dipole precessing in a magnetic field. In order to bring out this analogy in the most clear form, let us write the density matrix of the atom in the form

$$\hat{\rho} = \begin{bmatrix} \rho_{ee} & \rho_{eg} \\ \rho_{ge} & \rho_{gg} \end{bmatrix}, \tag{14.76}$$

where $\rho_{jk} = \langle j | \hat{\rho} | k \rangle$, with $\{ | j \rangle, | k \rangle \} = \{ | e \rangle, | g \rangle \}$. We may then introduce the *Bloch vector*

$$\mathbf{s} \equiv \left(s_x, s_y, s_z \right), \tag{14.77}$$

defined by its (real) components

$$s_x = \rho_{eg} + \rho_{ge}, \tag{14.78a}$$

$$s_y = \imath \left(\rho_{eg} - \rho_{ge} \right), \tag{14.78b}$$

$$s_z = \rho_{ee} - \rho_{gg}. \tag{14.78c}$$

The first two components are then linked to the coherences (the off-diagonal terms) of the density matrix while the third component is the so-called population inversion of the atom (see Box 13.1). The norm of the vector **s** cannot be larger than 1 (see Prob. 14.8). The Bloch vector is therefore a three-dimensional vector contained within a sphere with radius 1. Moreover, any pure state can be described as a Bloch vector pointing to the surface of the sphere (see Fig. 14.1). The Bloch-sphere representation is then analogous to the Poincaré-sphere representation (see Subsec. 1.3.3 and Sec. 5.6).

In conclusion, any density matrix (pure or mixed) $\hat{\rho}$ of a two-dimensional system may be written as

$$\hat{\rho} = \frac{1}{2} \left(\hat{I} + \mathbf{s} \cdot \hat{\sigma} \right), \tag{14.79}$$

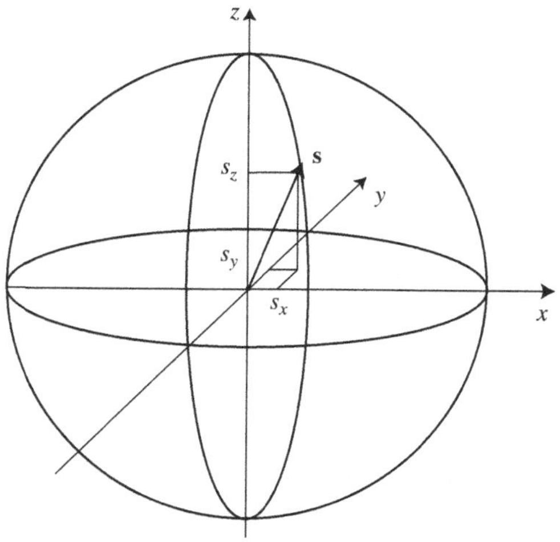

Figure 14.1 **Bloch-sphere representation of states of a two-level quantum system.**

where $\hat{\sigma}$ is the two-dimensional "spin" operator and

$$\mathbf{s} \cdot \hat{\sigma} = \sum_j s_j \hat{\sigma}_j, \tag{14.80}$$

where $j \in \{x, y, z\}$ (see Eqs. (6.154)). In explicit matricial form Eq. (14.79) may be written as

$$\hat{\rho} = \frac{1}{2} \begin{bmatrix} 1 + s_z & s_x - \imath s_y \\ s_x + \imath s_y & 1 - s_z \end{bmatrix}. \tag{14.81}$$

It is easy to verify that, by substituting expressions (14.78) into the matrix (14.81), we obtain the density matrix (14.76).

In order to illustrate the formalism of the Bloch vector, let us go back to the master equation (14.75) for a spontaneously emitting two-level atom. The equation of motion for the three components of the Bloch vector can then be written as (see Prob. 14.9)

$$\dot{s}_x = -\gamma s_x, \tag{14.82a}$$
$$\dot{s}_y = -\gamma s_y, \tag{14.82b}$$
$$\dot{s}_z = -2\gamma\,(1 + s_z)\,. \tag{14.82c}$$

These equations are formally similar to the Bloch equations for a magnetic dipole in a magnetic field and are known as the optical Bloch equations. Their solution (see again Prob. 14.9) is

$$s_x(t) = e^{-\gamma t} s_x(0), \tag{14.83a}$$
$$s_y(t) = e^{-\gamma t} s_y(0), \tag{14.83b}$$
$$s_z(t) = e^{-2\gamma t} s_z(0) + e^{-2\gamma t} - 1. \tag{14.83c}$$

Defining $\gamma' = 1 - e^{-2\gamma t}$, this evolution is equivalent to

$$\hat{\rho}'(t) = \hat{\hat{\mathcal{T}}} \left|\hat{\rho}'(0)\right\} = \hat{\vartheta}_0\hat{\rho}'(0)\hat{\vartheta}_0^\dagger + \hat{\vartheta}_1\hat{\rho}'(0)\hat{\vartheta}_1^\dagger, \tag{14.84}$$

where

$$\hat{\vartheta}_0 = \begin{bmatrix} 1 & 0 \\ 0 & \sqrt{1-\gamma'} \end{bmatrix}, \quad \hat{\vartheta}_1 = \begin{bmatrix} 0 & \sqrt{\gamma'} \\ 0 & 0 \end{bmatrix}, \tag{14.85}$$

where $\hat{\vartheta}_0$ and $\hat{\vartheta}_1$ are amplitude operators. In this specific case, the process (14.84) describes *amplitude damping* and γ' represents the probability of spontaneous emission, which tends to 1 as t goes to infinity.

It is interesting to note that the amplitude-operator formalism is more general than that of the master equation. In fact, the amplitude-operator formalism may describe processes that are not Markovian, i.e. describe state changes without the assumption of a continuous time evolution.

14.4 Quantum jumps and quantum trajectories

In this section we shall present an entirely different way of viewing the master equation. This approach is based on an analogy with classical statistical physics, where two descriptions of the dynamical evolution of a system are possible: first, one may describe the system by using a probability distribution; in this case, as we know, the evolution is generated by a Fokker–Planck-type equation. Alternatively, one may describe the system as an ensemble of stochastic trajectories, each of which is generated by a set of stochastic differential equations. In the quantum mechanics of open systems, quasi-probability distributions (see Sec. 13.4) may be used in the place of the classical probability distributions, whose evolution is given by the corresponding Fokker–Planck-type equation (see Subsec. 14.2.4). It is then natural to ask: can we envisage a description of a quantum system which is the analogue of the classical stochastic-trajectory method?

In order to answer this question, let us suppose for the time being that we are able to monitor in a perfect way our open quantum system. Even though we know that this is not physically possible (see Subsec. 2.3.3 and also Ch. 15), this procedure will help us in the derivation of the new formalism of *quantum trajectories*. If our monitoring is perfect, then we are able to detect any quanta lost by the system. For example, in the case of the electromagnetic field inside a cavity, we would be able to detect any single photon lost by the cavity. We can then record the times at which the quanta are released. Therefore, we may assume that between two successive detections the system evolves in a continuous way (without emission of quanta) (see also Sec. 9.8). In this hypothetical, perfectly monitored, quantum trajectory, the dynamics of the system would then consist of a succession of continuous evolutions and discrete emissions of quanta, called *quantum jumps*. Needless to say, these hypothetical trajectories have no physical meaning. In spite of this, the evolved density operator at a certain time can be determined as the weighted ensemble average of all possible quantum trajectories. This is the essence of the quantum-jump approach to the dynamics of open quantum systems.

Let us now briefly illustrate how this formalism works. Consider an open quantum system described by the reduced density matrix $\hat{\rho}$. Its dynamics is ruled by the master equation (14.42)

$$\frac{d}{dt}\hat{\rho}(t) = \hat{\hat{\mathcal{L}}}\hat{\rho}(t), \tag{14.86}$$

where (see also Eqs. (14.43))

$$\hat{\hat{\mathcal{L}}}\hat{\rho} = \left(\hat{\hat{\mathcal{L}}}_{\mathrm{d}} + \hat{\hat{\mathcal{L}}}_{\mathrm{nd}}\right)\hat{\rho} \tag{14.87}$$

is the superoperator pertaining to the system and its environment. Equation (14.86) can be formally solved as

$$\hat{\rho}(t) = e^{\hat{\hat{\mathcal{L}}}t}\hat{\rho}(0). \tag{14.88}$$

The jump superoperator, i.e. the superoperator describing the loss of a quantum by the system, can be defined as

$$\hat{\hat{\mathcal{J}}}\hat{\rho} = \hat{a}\hat{\rho}\hat{a}^{\dagger}, \tag{14.89}$$

since this term precisely accounts for the emission of a quantum.[8] We may then add and subtract the superoperator $\hat{\hat{\mathcal{J}}}$ to $\hat{\hat{\mathcal{L}}}$ in the exponent in the rhs of Eq. (14.88) to obtain

$$\hat{\rho}(t) = e^{[(\hat{\hat{\mathcal{L}}}-\hat{\hat{\mathcal{J}}})+\hat{\hat{\mathcal{J}}}]t}\hat{\rho}(0). \tag{14.90}$$

By making use of the identity (see Prob. 14.10)

$$e^{(\hat{O}+\eta\hat{O}')\xi} = \sum_{k=0}^{\infty}\eta^k\int_0^{\xi}d\xi_k\int_0^{\xi_k}d\xi_{k-1}\ldots\int_0^{\xi_2}d\xi_1 e^{\hat{O}(\xi-\xi_k)}\hat{O}'e^{\hat{O}(\xi_k-\xi_{k-1})}\hat{O}'\ldots\hat{O}'e^{\hat{O}'\xi_1}, \tag{14.91}$$

where the term for $k=0$ is defined as

$$e^{\hat{O}(\xi)}, \tag{14.92}$$

we arrive at

$$\begin{aligned}\hat{\rho}(t) &= e^{[(\hat{\hat{\mathcal{L}}}-\hat{\hat{\mathcal{J}}})+\hat{\hat{\mathcal{J}}}]t}\hat{\rho}(0)\\ &= \sum_{m=0}^{\infty}\int_0^t dt_m\int_0^{t_m}dt_{m-1}\cdots\int_0^{t_2}dt_1\\ &\quad\times e^{(\hat{\hat{\mathcal{L}}}-\hat{\hat{\mathcal{J}}})(t-t_m)}\hat{\hat{\mathcal{J}}}e^{(\hat{\hat{\mathcal{L}}}-\hat{\hat{\mathcal{J}}})(t_m-t_{m-1})}\hat{\hat{\mathcal{J}}}\ldots\hat{\hat{\mathcal{J}}}e^{(\hat{\hat{\mathcal{L}}}-\hat{\hat{\mathcal{J}}})t_1}\hat{\rho}(0).\end{aligned} \tag{14.93}$$

[8] Truly speaking, the choice of $\hat{\hat{\mathcal{J}}}$ is not unique and may well depend on how the system is thought to be monitored. This remark, however, does not alter the essence of the following argument.

The expression inside the integrals may be considered as the unnormalized conditioned density operator

$$\hat{\hat{\rho}}_c(t) = e^{(\hat{\hat{\mathcal{L}}}-\hat{\hat{\mathcal{J}}})(t-t_m)}\hat{\hat{\mathcal{J}}}e^{(\hat{\hat{\mathcal{L}}}-\hat{\hat{\mathcal{J}}})(t_m-t_{m-1})}\hat{\hat{\mathcal{J}}}\cdots\hat{\hat{\mathcal{J}}}\,e^{(\hat{\hat{\mathcal{L}}}-\hat{\hat{\mathcal{J}}})t_1}\hat{\rho}(0). \tag{14.94}$$

Moving from right to left, this term can be interpreted in the following manner: the initial density operator evolves in the time interval between $t = 0$ and $t = t_1$ (when there are no loss of quanta) under the propagator $e^{(\hat{\hat{\mathcal{L}}}-\hat{\hat{\mathcal{J}}})t_1}$, jumps under the action of $\hat{\hat{\mathcal{J}}}$ at the time of the first emission (t_1), evolves during the next interval ($t_2 - t_1$) under the propagator $e^{(\hat{\hat{\mathcal{L}}}-\hat{\hat{\mathcal{J}}})(t_2-t_1)}$, jumps again at t_2 under the action of $\hat{\hat{\mathcal{J}}}$, and so on. In this way, we are building a step-wise trajectory for the conditioned density operator. Two ingredients are necessary for this procedure: first, we need two types of evolution, one without jumps ruled by the superoperator $\hat{\hat{\mathcal{L}}} - \hat{\hat{\mathcal{J}}}$, and a jump ruled by the superoperator $\hat{\hat{\mathcal{J}}}$; second, we need the specific set of times for the jumps. Because neither $\hat{\hat{\mathcal{J}}}$ nor $e^{(\hat{\hat{\mathcal{L}}}-\hat{\hat{\mathcal{J}}})t}$ preserve the trace, we have to introduce a normalization by hand, and define the conditioned density operator as

$$\hat{\rho}_c(t) = \frac{\hat{\hat{\rho}}_c(t)}{\mathrm{Tr}\left[\hat{\hat{\rho}}_c(t)\right]}, \tag{14.95}$$

where, as defined above, $\hat{\hat{\rho}}_c(t)$ is the unnormalized density operator. This way of proceeding defines a decomposition of the quantum dynamics into an infinite number of quantum trajectories, which are generated by the times at which jumps happen, and between which jumps, although watched for, do not occur. Therefore, this method has a deep analogy with the path-integral method developed by Feynman (see Sec. 10.8). The main difference between the two methods lies in the fact that the former is based on the master equation (and therefore on an analogy to classical statistical mechanics), while the latter represents a decomposition of the Schrödinger equation (and therefore represents an analogy to deterministic classical mechanics).

The most powerful application of the quantum-jump approach is that it allows us to perform numerical (Monte Carlo) simulations, which often turn out to be more efficient than numerical integrations of the master equation or of the Fokker–Planck equation. In fact, this is the quantum analogue of the classical *Monte-Carlo method.* As it is well known, several numerical techniques are available in order to integrate a one-dimensional function between a and b (see Fig. 14.2). For instance, we may partition the interval $[a, b]$ into N subintervals and calculate the area of each rectangle separately and then sum them up. We can also generate N random points from a uniform distribution in the rectangle of the $x - y$ plane containing the area we wish to integrate, and count the number N_u of those falling below the curve. In the latter case, the quantity $(b - a)cN_u/N$, where c is the height of the rectangle, represents an estimate of the desired integral. For two-dimensional spaces the two methods are equivalent and for $N \longrightarrow \infty$ both converge with high accuracy to the integral of the function. However, for n-dimensional spaces ($n \geq 2$), in order to obtain the

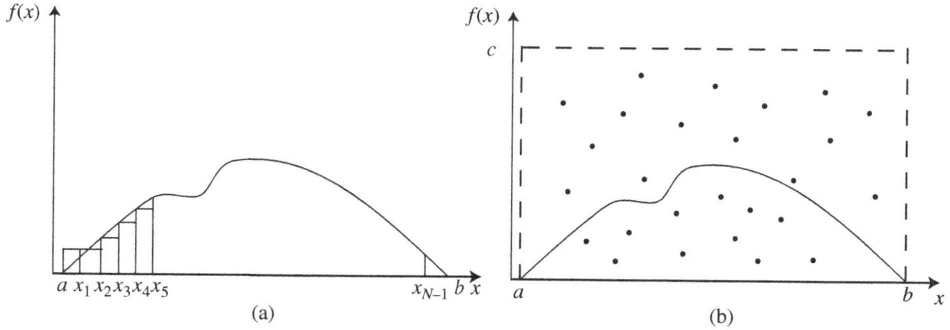

Figure 14.2 Two different numerical techniques for evaluating the integral $\int_a^b dx f(x)$. (a) We subdivide the interval $[a, b]$ in N subintervals, calculate the area of each rectangle, and finally sum them up. (b) We generate N uniform random points in the dashed rectangle and count those falling below the curve.

same degree of accuracy, while the first method requires N^n calculations of $f(x)$, the Monte-Carlo method only requires a number of calculations of $f(x)$ that is polynomial in N (see also Sec. 17.8).

In the following we wish to illustrate a practical method that is a direct consequence of what we have derived above and is known as *stochastic Monte-Carlo wave-functions* or simply *quantum trajectories*. We shall proceed now in an operational manner and make use of numerical simulations. Since the jumps occur at random times, we must build this randomness in the theory in a way that is statistically correct.[9] The probability that an emission occurs in the time interval $[t, t+\Delta t)$ for the conditioned density operator $\hat{\rho}_c(t)$ is given by

$$\wp_c(t) = \mathrm{Tr}\left[\hat{\tilde{\mathcal{J}}}\hat{\rho}_{\hat{c}}(t)\right]\Delta t, \tag{14.96}$$

i.e. simply the product of the conditioned mean flux of quanta times the time interval Δt. Suppose that $\hat{\rho}_c(t)$ may be written as the pure state

$$\hat{\rho}_c(t) = |\psi_c(t)\rangle\langle\psi_c(t)|, \tag{14.97}$$

and that, for the density operator (14.94), we also have

$$\hat{\bar{\rho}}_c(t) = \left|\bar{\psi}_c(t)\right\rangle\left\langle\bar{\psi}_c(t)\right|. \tag{14.98}$$

Then, we may write the propagation without emissions of quanta over a time Δt as

$$\left|\bar{\psi}_c(t+\Delta t)\right\rangle = e^{-\frac{i}{\hbar}\hat{\tilde{H}}\Delta t}\left|\bar{\psi}_c(t)\right\rangle, \tag{14.99}$$

where, for reasons that will be seen in short, $\hat{\tilde{H}}$ is not a (Hermitian) Hamiltonian. At the time of a jump, the unnormalized state undergoes the collapse

$$\left|\bar{\psi}_c(t)\right\rangle \leadsto \hat{C}\left|\bar{\psi}_c(t)\right\rangle, \tag{14.100}$$

[9] See [*Carmichael* 1993, 122–30].

where $\hat{C}$ is a collapse operator. For example, in the case $\hat{\hat{\mathcal{J}}}$ is defined as in Eq. (14.89), $\hat{C} = \hat{a}$. Now, given the ket $| \psi_c(t_n)\rangle$ (where $t_n = n\Delta t$), we may define a discrete-time operational procedure to calculate the ket $| \psi_c(t_{n+1})\rangle$. In fact, first we have to evaluate the probability

$$\wp_c(t_n) = \left\langle \psi_c(t_n) \left| \hat{C}^\dagger \hat{C} \right| \psi_c(t_n) \right\rangle \Delta t. \tag{14.101}$$

Then, we generate a random number r_n distributed uniformly on the interval [0, 1]. Finally, we compare $\wp_c(t_n)$ with r_n and calculate $| \psi_c(t_{n+1})\rangle$ according to

$$| \psi_c(t_{n+1})\rangle = \begin{cases} \dfrac{\hat{C}| \psi_c(t_n)\rangle}{\left(\left\langle \psi_c(t_n) \left| \hat{C}^\dagger \hat{C} \right| \psi_c(t_n)\right\rangle\right)^{\frac{1}{2}}} & \text{if } \wp_c(t_n) \geq r_n \\ \dfrac{e^{-\frac{\imath}{\hbar}\tilde{\hat{H}}\Delta t}| \psi_c(t_n)\rangle}{\left(\left\langle \psi_c(t_n) \left| e^{\frac{\imath}{\hbar}(\tilde{\hat{H}}^\dagger - \tilde{\hat{H}})\Delta t} \right| \psi_c(t_n)\right\rangle\right)^{\frac{1}{2}}} & \text{if } \wp_c(t_n) < r_n. \end{cases} \tag{14.102}$$

The result is a stochastic quantum mapping between the times t_k (possibly separated by many Δt) at which the collapses occur

$$| \psi_c(t_{k+1})\rangle = \frac{\hat{C}e^{-\frac{\imath}{\hbar}\tilde{\hat{H}}\tau_{k+1}}| \psi_c(t_k)\rangle}{\left(\left\langle \psi_c(t_k) \left| e^{\frac{\imath}{\hbar}\tilde{\hat{H}}^\dagger \tau_{k+1}} \hat{C}^\dagger \hat{C} e^{-\frac{\imath}{\hbar}\tilde{\hat{H}}\tau_{k+1}} \right| \psi_c(t_n)\right\rangle\right)^{\frac{1}{2}}}, \tag{14.103}$$

where $\tau_{k+1} = t_{k+1} - t_k$ is a random time, which depends on the jump statistics (see Eq. (14.101)). The central point (whose proof goes beyond our goal) is that this procedure leads to the same ensemble averages for observables as the master-equation approach.

One could be tempted to interpret this formalism as a description of real trajectories. However, as we have already said, since the single stochastic quantum trajectories do not bear any physical meaning, the quantum jumps formalism has been shown to be rather a pure numerical technique. This can also be seen if we apply this formalism to a simple case, the master equation for the damped harmonic oscillator (see Eqs. (14.51) and (14.89)). In this case, in the Schrödinger picture, we would have

$$\hat{\hat{\mathcal{J}}}\,\hat{\hat{\rho}}_c = 2\gamma \hat{a}\hat{\hat{\rho}}_c\hat{a}^\dagger, \tag{14.104a}$$

$$\left(\hat{\hat{\mathcal{L}}} - \hat{\hat{\mathcal{J}}}\right)\hat{\hat{\rho}}_c = -\imath\omega\left[\hat{a}^\dagger\hat{a}, \hat{\hat{\rho}}_c\right] - \gamma\left(\hat{a}^\dagger\hat{a}\hat{\hat{\rho}}_c + \hat{\hat{\rho}}_c\hat{a}^\dagger\hat{a}\right), \tag{14.104b}$$

where $\hat{\hat{\rho}}_c$ is unnormalized. It is easy to see that, in this case, the evolution operator for the relative ket $\left| \tilde{\psi}_c \right\rangle$ is $e^{-\frac{\imath}{\hbar}\tilde{\hat{H}}t}$, where (see Prob. 14.11)

$$\tilde{\hat{H}} = \hbar\omega\hat{a}^\dagger\hat{a} - \imath\hbar\gamma\hat{a}^\dagger\hat{a}. \tag{14.105}$$

As a consequence, also $\tilde{\hat{H}}$ is not Hermitian, which implies that the propagator $e^{-\frac{\imath}{\hbar}\tilde{\hat{H}}t}$ cannot be unitary (see Stone's theorem: p. 123).

14.5 Quantum optics and Schrödinger cats

In the present section we would like to show one of the many possible connections between quantum optics and fundamental quantum mechanics, via the theory of open systems. As we know (see Sec. 9.3), Schrödinger cats are at the same time paradoxical states and a fundamental key to a deep understanding of quantum mechanics. As we have already mentioned in the introduction of Ch. 13, quantum optics has been playing a major role in testing some fundamental (and sometimes puzzling) results of quantum theory. The Schrödinger cat business is precisely one of these examples.

14.5.1 The anharmonic oscillator model

Let us consider a very simple quantum-optical model which is also a fundamental example of quantum-mechanical dynamics, first pointed to in this context by Yurke and Stoler.[10] It is essentially an initial coherent state of the single-mode electromagnetic field evolving under the influence of an anharmonic-oscillator Hamiltonian (see Eq. (13.23)), given by

$$\hat{H} = \hbar\left(\omega\hat{N} + \omega_a\hat{N}^2\right), \tag{14.106}$$

where ω is the energy-level splitting of the unperturbed harmonic oscillator (see Sec. 4.4) and ω_a is the strength of the anharmonic perturbation.[11] Rigorously speaking, the anharmonic perturbation term in the Hamiltonian should be taken proportional to $\hat{x}^4$ (see Subsec. 10.1.3). However, it can be shown (see Probs. 10.2 and 14.12) that, except for a suitable frequency shift, the two formulations are equivalent. In the interaction picture, where the anharmonic term is considered as the interaction part of the Hamiltonian, an initial coherent state $|\,\alpha\rangle$ will evolve as

$$|\,\alpha, t\rangle = e^{-\imath\omega_a t\left(\hat{a}^\dagger\hat{a}\right)^2}\,|\,\alpha\rangle = e^{-\frac{|\alpha|^2}{2}}\sum_{j=0}^{\infty}\alpha^j\frac{e^{-\imath\omega_a t j^2}}{\sqrt{j!}}\,|\,j\rangle\,. \tag{14.107}$$

This equation immediately tells us that the state vector is periodic with period $2\pi/\omega_a$ and becomes particularly interesting for certain values of t: at time $t = \pi/\omega_a$, for example, $e^{-\imath\omega_a t j^2} = (-1)^j$ and the state $|\,\alpha\rangle$ has evolved to the state $|-\alpha\rangle$. Furthermore, when $t = \pi/2\omega_a$, we have (see Prob. 14.13)

$$\left|\,\alpha, \frac{\pi}{2\omega_a}\right\rangle = \frac{1}{\sqrt{2}}\left(e^{-\imath\frac{\pi}{4}}\,|\,\alpha\rangle + e^{\imath\frac{\pi}{4}}\,|-\alpha\rangle\right), \tag{14.108}$$

i.e. the initial coherent state has evolved towards a coherent superposition of the coherent states $|\,\alpha\rangle$ and $|-\alpha\rangle$, which are 180° out-of-phase with respect to each other and, as a consequence, are macroscopically distinguishable when $|\alpha|$ is large ($|\alpha| \gg 1$). The state

[10] See [Yurke/Stoler 1986].

[11] The anharmonic term may, in general, be written as $\hbar\omega_a\hat{N}^k$ with $k > 1$. Here we restrict our analysis to $k = 2$ even though many properties of the system are still present for any given $k \geq 2$.

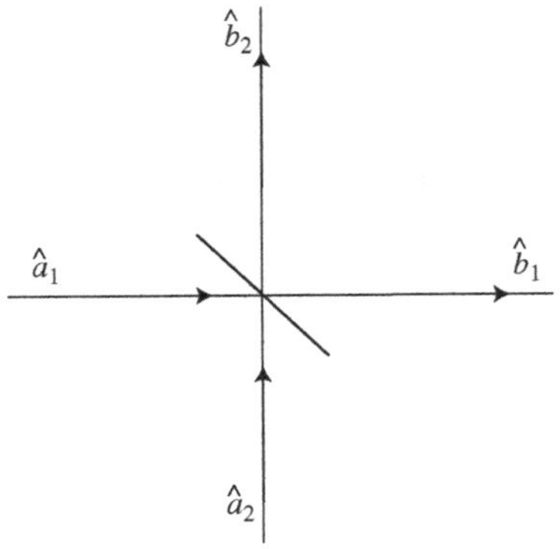

Figure 14.3 Beam-splitter model used to account for losses and dissipation in the anharmonic-oscillator model. The input state (14.110) is injected into the first port of the beam splitter (mode $\hat{a}_1$). At the output, one observes the light emerging from $\hat{b}_1$, *independently* of what emerges from $\hat{b}_2$.

(14.108) is a prototype of the quantum-optical Schrödinger cat state and, as shown by several authors,[12] is tremendously sensitive to decoherence and dissipation (which have not been considered up to now) and therefore very difficult to experimentally realize and detect.

In order to exploit such a sensitivity, it is very interesting to evaluate what happens when the state (14.108) is injected into a beam splitter that may be used to model losses due to the medium or the detector (see Fig. 14.3). If we denote by η the (real) transmission coefficient of the beam splitter (see Subsec. 13.6.2), then the loss will be proportional to $1 - \eta$. In this simple model, loss is represented by the fact that, at the output modes, one only observes light emerging from $\hat{b}_1$, independently from what emerges from $\hat{b}_2$. The beam-splitter transformation may be written as

$$\begin{pmatrix} \hat{b}_1 \\ \hat{b}_2 \end{pmatrix} = \begin{bmatrix} \sqrt{\eta} & \sqrt{1-\eta} \\ -\sqrt{1-\eta} & \sqrt{\eta} \end{bmatrix} \begin{pmatrix} \hat{a}_1 \\ \hat{a}_2 \end{pmatrix}, \tag{14.109}$$

while the input state is represented by

$$| \, \mathrm{in} \rangle = \left| \alpha, \frac{\pi}{2\omega_a} \right\rangle_1 | 0 \rangle_2 , \tag{14.110}$$

i.e. the input of the second port of the beam splitter is the vacuum. A direct calculation shows that the output state is given by (see Prob. 14.14)

$$| \, \mathrm{out} \rangle = \frac{1}{\sqrt{2}} \left[e^{-\frac{\iota\pi}{4}} \left| \sqrt{\eta}\alpha \right\rangle_1 \left| -\sqrt{1-\eta}\alpha \right\rangle_2 + e^{\frac{\iota\pi}{4}} \left| -\sqrt{\eta}\alpha \right\rangle_1 \left| \sqrt{1-\eta}\alpha \right\rangle_2 \right], \tag{14.111}$$

where the definition (13.58) of a coherent state has been used. Now, taking advantage of homodyne detection (see Subsec. 13.6.2), we may observe the light coming out from the output $\hat{b}_1$ of the beam splitter. In this way, we measure he operator (see also Eq. (13.136))

$$\hat{X}_\theta = \frac{1}{\sqrt{2}} \left(e^{\iota\theta} \hat{b}_1 + e^{-\iota\theta} \hat{b}_1^\dagger \right), \tag{14.112}$$

[12] Besides the article of Yurke and Stoler, see [Walls/Milburn 1985] [Milburn 1986].

where θ s the local-oscillator phase of the homodyne-detection scheme. Introducing the operator (see also Eq. (13.80))

$$\hat{Y}_\theta = \frac{1}{\sqrt{2}}\left(e^{\imath\theta}\hat{b}_2 + e^{-\imath\theta}\hat{b}_2^\dagger\right), \tag{14.113}$$

we may reconstruct the wave function $\psi_{\text{out}}(x, y) = \langle x, y \mid \text{out}\rangle$ corresponding to the state (14.111) in the representation x, y of the eigenvalues of $\hat{X}_\theta$ and $\hat{Y}_\theta$, respectively, that is,

$$\psi_{\text{out}}(x, y) = \frac{1}{\sqrt{2}}\left[e^{-\frac{\imath\pi}{4}}\psi_\gamma(x)\psi_{-\delta}(y) + e^{\frac{\imath\pi}{4}}\psi_{-\gamma}(x)\psi_\delta(y)\right], \tag{14.114}$$

where

$$\gamma = \alpha\sqrt{\eta} \quad \text{and} \quad \delta = \alpha\sqrt{1-\eta}, \tag{14.115}$$

while

$$\psi_\beta(x) = \frac{1}{\pi^{\frac{1}{4}}}\exp\left[-\frac{x^2}{2} + \frac{2x\beta e^{\imath\theta}}{\sqrt{2}} - \left(\frac{\beta e^{\imath\theta}}{\sqrt{2}}\right)^2 - \frac{|\beta|^2}{2}\right]. \tag{14.116}$$

The probability distribution for the outgoing photocurrent x, coming out from the homodyne detector, may be obtained by integrating $|\psi_{\text{out}}(x, y)|^2$ over all possible values of y, i.e.

$$\begin{aligned}\wp(x) &= \int_{-\infty}^{+\infty} dy\,\psi^*_{\text{out}}(x, y)\psi_{\text{out}}(x, y)\\ &= \frac{1}{2\sqrt{\pi}}\left\{e^{-[x-\sqrt{2\eta}|\alpha|\cos(\theta+\phi)]^2} + e^{-[x+\sqrt{2\eta}|\alpha|\cos(\theta+\phi)]^2}\right.\\ &\quad\left. + 2e^{-2(1-\eta)|\alpha|^2}e^{-x^2-2\eta|\alpha|^2\cos^2(\theta+\phi)}\sin\left[2\sqrt{2\eta}|\alpha|\sin(\theta+\phi)x\right]\right\},\end{aligned} \tag{14.117}$$

where $\alpha = |\alpha|e^{\imath\phi}$. This probability distribution may be immediately interpreted as an interference signal (see, e.g., Fig. 9.7): the first two terms represent two Gaussian bells centered, respectively, at

$$x = \pm\sqrt{2\eta}|\alpha|\cos(\theta + \phi), \tag{14.118}$$

whereas the third term represents the interference due to the coherence (or superposition) of the states $|\alpha\rangle$ and $|-\alpha\rangle$ in Eq. (14.108). One may then interpret the combination of the beam splitter plus the ideal detector as a model for a real, inefficient detector ($\eta < 1$), or even for lossy medium.

From a pedagogical point of view, the beauty of this model lies in the fact that, by varying the local-oscillator phase θ, it is first possible to verify that the state (14.108) has two macroscopically distinguishable components (see Fig. 14.4(a)), and, successively, that such a state is a coherent rather than a statical mixture. Moreover, the number of interference fringes is proportional to $|\alpha|$ (see Fig. 14.4 (b)), i.e. to the distance between the components (see Fig. 14.4.(c)). However, as soon as the detection efficiency becomes smaller than 1 or, equivalently, the loss enters into play (here represented by the the fraction $1 - \eta$ of the light coming out from the second port of the beam splitter), the interference fringes rapidly vanish. From an analytical point of view, this is made clear

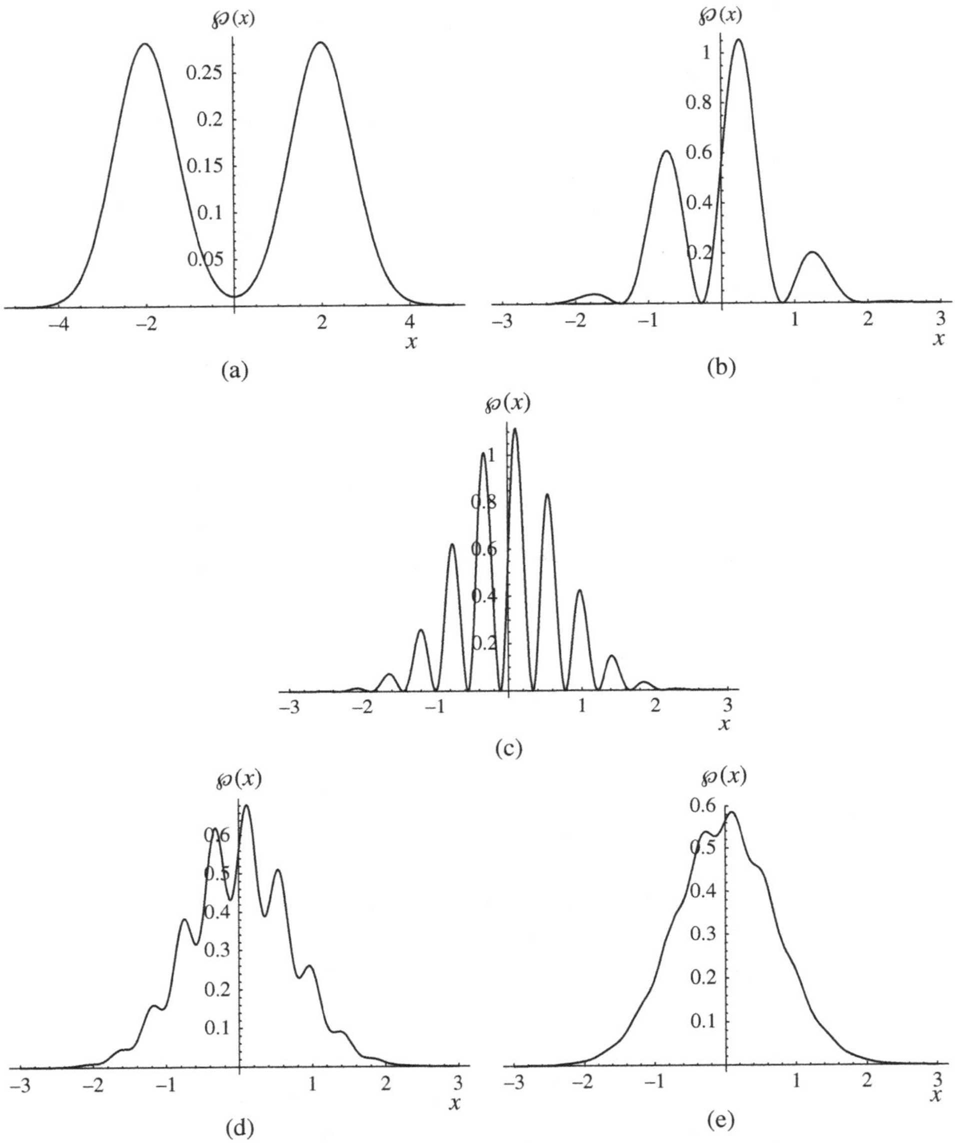

Figure 14.4 Interference fringes and their sensitivity to losses in the Yurke–Stoler model. (a) The probability distribution (14.117) is plotted for $|\alpha| = 2$ and $\eta = 1$. The local oscillator phase of the homodyne detector is set such that $\cos(\theta + \phi) = 1$. Here, the two distinguishable components are clearly visible. (b) As in (a) but with $\sin(\theta + \phi) = 1$: the interference fringes arise. (c) As in (b) but with $|\alpha| = 5$: the number of interference fringes increases. (d) As in (c) but with $\eta = 0.97$: even small losses partly destroy interference. (e) As in (c) but with $\eta = 0.94$: a slightly larger loss almost washes out interference fringes.

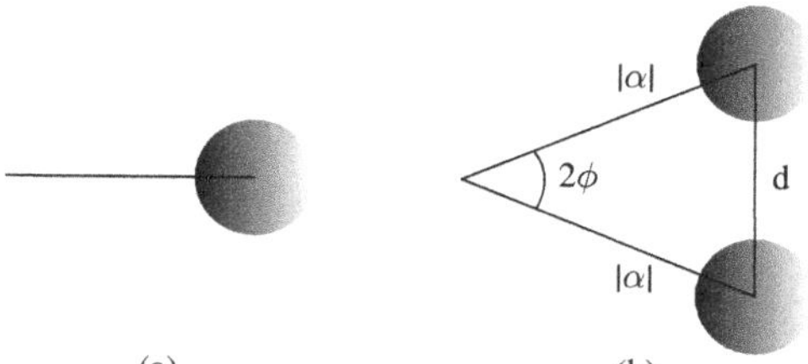

Figure 14.5 (a) Pictorial representation in phase space of a coherent state of a quantum oscillator. (b) The two coherent-state components separated by a distance d from a Schrödinger cat.

Box 14.1 **Coherent states and macroscopic distinguishability**

It is possible to conceive of a Schrödinger cat in a very intuitive way [Brune *et al.* 1996]. Consider a two-level atom (ground and excited state: $|g\rangle, |e\rangle$) coupled to an apparatus $\mathcal{A}$ represented by a quantum oscillator in a coherent state. The state vector $|\alpha\rangle$ defining it has a circular Gaussian distribution of radius unity due to quantum fluctuations which make the tip uncertain (see Figs. 13.2 and 14.5(a)). Consider an ideal measurement in which the atom–oscillator interaction entangles the phase of the oscillator $\pm\phi$ to the internal state of the atom leading to:

$$|\Psi\rangle = \frac{1}{\sqrt{2}}\left(|e, \alpha e^{\imath\phi}\rangle + |g, \alpha e^{-\imath\phi}\rangle\right). \tag{14.119}$$

When the distance[13] $d = 2\sqrt{N}\sin\phi$ (where $N = |\alpha|^2$ is the mean photon number) is larger than one, a Schrödinger cat is obtained (see Fig. 14.5(b)).

Practically, the biggest difficulty one encounters to generate and detect a Schrödinger cat is the preservation of the quantum coherence when enlarging sufficiently the "distance" between the component states in order to obtain a quantum behavior on, at least, a mesoscopic scale.

by the factor $e^{-2(1-\eta)|\alpha|^2}$ in front of the interference term of Eq. (14.117): when the loss $1-\eta$ become larger than $1/2\,|\alpha|^2$, such a factor becomes much smaller than 1 (see Figs. 14.4.(d)–14.4(e)). When α is large, then, a detection efficiency slightly smaller than 1 is sufficient to wash out interference fringes, making the probability distribution (14.117) indistinguishable from that corresponding to a statistical mixture of the states $|\alpha\rangle$ and $|-\alpha\rangle$.

[13] Here, we use the concept of distance between quantum states in an intuitive, heuristic way. More formal definitions of this concept may be introduced (Probs. 14.15–14.16). A complete discussion of this subject goes beyond the aim of this book. For more details see, e.g., [Knöll/Orlowski 1995] .

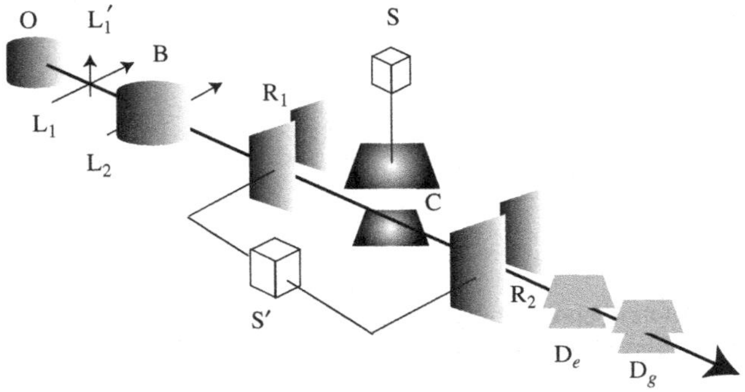

Figure 14.6 Experimental setup of Haroche's experiment. The cavity C is made by two superconducting niobium mirrors. The rubidium atoms effusing from the oven O are velocity selected by two laser beams L_1 and L_1' and are then excited into state $|e\rangle$ in box B. Each circular atom is prepared in a quantum superposition of $|e\rangle$ and $|g\rangle$ by a resonant microwave $\pi/2$ pulse in a low-Q- cavity R_1. It then crosses the high-Q cavity C in which a small coherent field with average photon number N varying from 0 to 10 is injected by a pulsed source S. The field, which evolves freely while each atom crosses C, relaxes to a vacuum before being regenerated for the next atom. The field is left coherent. After leaving C, each atom undergoes another $\pi/2$ pulse in a cavity R_2 identical to R_1. R_1 and R_2 are fed by a continuous-wave source S′. The atoms in states $|e\rangle$ ($|g\rangle$) are finally counted by detectors D_e (D_g).

14.5.2 A cavity QED model

A further example of Schrödinger cat realization in quantum optics is given by the combination of high-Q cavities[14] and circular Rydberg atoms.[15] In this kind of system, especially prepared two-level atoms are sent at a controlled speed through the cavity, where the electromagnetic field is usually prepared in an initially coherent state $|\alpha\rangle$, generating entangled states of the atom–field combined system.[16] Let us consider the apparatus shown in Fig. 14.6. It is schematically made of a high-Q cavity (C), two auxiliary low-Q cavities (R_1 and R_2), an atomic source (O, L_1 and L_1', and B), and atomic detectors (D_e and D_g). The Rydberg atoms may be in either of two possible states (ground, $|g\rangle$, or excited, $|e\rangle$).

The $|e\rangle \rightsquigarrow |g\rangle$ transition and the cavity frequency are slightly off resonance (detuning δ), so that the atom and field cannot exchange energy but only undergo $1/\delta$ dispersive frequency shifts. The atom–field coupling then produces an atomic-level dependent dephasing of the field and generates an entangled state.

[14] The quality factor of an electromagnetic field cavity (a resonator) is given by $Q = \omega/\gamma$, where ω is the resonance frequency of the cavity and γ its dissipation constant. Clearly, the higher the quality factor of the cavity, the smaller is the influence of dissipation on the dynamics of the electromagnetic field inside the cavity.

[15] A Rydberg atom is an excited atom with one or more electrons that have a very high quantum principal number. These atoms have a number of peculiar properties, including an exaggerated response to electric and magnetic fields, and long decay times. When also the orbital quantum number is high, their electron wave functions approximate classical (circular) orbits about the nucleus.

[16] See [Brune *et al.* 1996].

The initial state of the atom–cavity system is given by

$$|\Psi(0)\rangle = |e\rangle \otimes |\alpha\rangle = |e,\alpha\rangle . \tag{14.120}$$

When passing through R_1 the atom receives a $\pi/2$ pulse[17] and the state becomes

$$|\Psi(t_1)\rangle = \frac{1}{\sqrt{2}}\left(|g,\alpha\rangle + |e,\alpha\rangle\right). \tag{14.121}$$

As said above, when passing through the cavity C, the state of the field undergoes a phase shift, which is different for the two atomic states. As a consequence, the state of the system becomes

$$|\Psi(t_2)\rangle = \frac{1}{\sqrt{2}}\left(|g,\alpha e^{\imath\phi_g}\rangle + |e,\alpha e^{\imath\phi_e}\rangle\right). \tag{14.122}$$

Finally, when the atom crosses R_2, it receives a second $\pi/2$ pulse so that the combined state may be cast into the form

$$|\Psi(t_3)\rangle = \frac{1}{2}\left(|g,\alpha e^{\imath\phi_g}\rangle - |e,\alpha e^{\imath\phi_g}\rangle + |g,\alpha e^{\imath\phi_e}\rangle + |e,\alpha e^{\imath\phi_e}\rangle\right). \tag{14.123}$$

It is then clear that, detecting the atomic state in $|e\rangle$ or $|g\rangle$, respectively, leaves the cavity field into the state

$$|\psi_F\rangle = \frac{1}{\sqrt{2}}\left(|\alpha e^{\imath\phi_e}\rangle \mp |\alpha e^{\imath\phi_g}\rangle\right). \tag{14.124}$$

If $\phi_e \simeq -\phi_g \simeq \phi$, Eq. (14.124) may be rewritten as

$$|\psi_F\rangle = \frac{1}{\sqrt{2}}\left(|\alpha e^{\imath\phi}\rangle + |\alpha e^{-\imath\phi}\rangle\right), \tag{14.125}$$

which represents a coherent quantum superposition of two (distinct) states of the single-mode electromagnetic field inside the cavity. These two coherent components are (macroscopically) distinguishable if $|\alpha|$ is large and ϕ is of the order of $\pi/2$ (see Box 14.1).

The coherence between the two components of the state was revealed by a subsequent two-atom correlation experiment. While a first atom creates a superposition state involving the two field components, a second atom (the probe) crosses C with the same velocity after a short delay τ_2 and dephases the field again by an angle $\pm\phi$. The two field components then turn into three, with phases $+2\phi$, -2ϕ, and 0. The zero component may be obtained via two different paths, since the atoms may have crossed C either in the (e, g) configuration or in the (g, e) configuration (i.e. the second atom undoes the phase shift of the first one). Since the atomic states are mixed after C in R_2, the (e, g) and (g, e) "paths" are indistinguishable. As a consequence, there is an interference term in the joint probabilities $\wp^{(2)}_{ee}, \wp^{(2)}_{eg}, \wp^{(2)}_{ge}, \wp^{(2)}_{gg}$.

The experimental results confirm that mesoscopic Schrödinger cat states may be actually created in this system and that their subsequent decoherence may be observed and monitored: Fig. 14.7(a) shows the signal obtained when C is empty. The final probability distribution $\wp^{(1)}_g(\nu)$ (of finding an atom in $|g\rangle$ as a function of the continuous-wave

[17] This amounts to the transformation $|e\rangle \rightsquigarrow \frac{1}{\sqrt{2}}(|g\rangle + |e\rangle)$ and $|g\rangle \rightsquigarrow \frac{1}{\sqrt{2}}(|g\rangle - |e\rangle)$ (see Sec. 13.7 and Prob. 8.3).

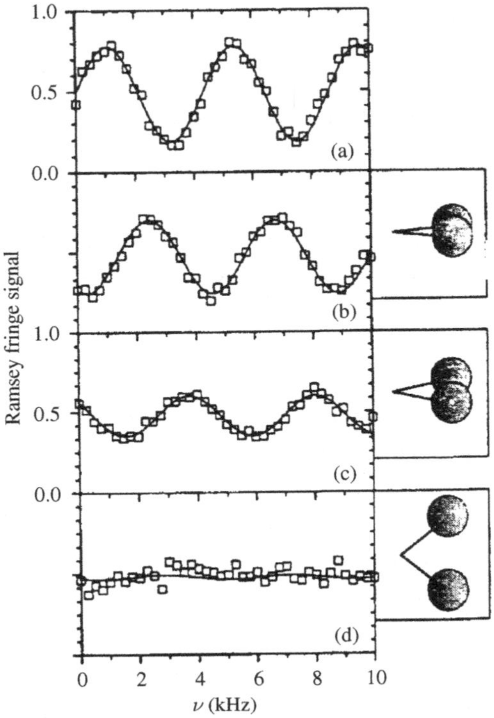

Figure 14.7 $\wp_g^{(1)}(\nu)$ signal exhibits Ramsey fringes.(a) C empty.(b)–(d) C stores a coherent field. Insets show the phase space representation of the field components left in C. Adapted from [Brune *et al.* 1996, 4888].

frequency ν) exhibits the so-called *Ramsey fringes* (interference) typical of atoms subjected to successive pulses. In fact, transitions $|e\rangle \rightsquigarrow |g\rangle$ occur either in R_1 or in R_2. Since the two "paths" are indistinguishable this leads to an interference term between the corresponding probability amplitudes.

In Fig. 14.7(b)–(d) we have represented the fringes when there is a coherent field in C (with an average number of photons $N = 9.5$ and the coherent amplitude $|\alpha| = 3.1$). When δ is reduced, the contrast of the fringes decreases (and the "paths" of the atoms become partially distinguishable) and their phase is shifted. The fringe contrast reduction demonstrates the separation of the field state into two components.

Summary

- In this chapter we have dealt with a very important problem, that of the analysis of open systems, mainly by using the methods developed in quantum optics. In particular, we have derived the general form of the so-called *master equation*.

- We have presented the most famous and general formulation, the *Lindblad master equation*. We have then applied this formalism to the case of the damped harmonic oscillator and have found a quantum analogue of the Fokker–Planck equation.
- After having shown the connection between master equation and POVM, we have presented the method of *quantum jumps*, an analogue of the classical use of stochastic trajectories for studying the evolution of a system.
- We have presented more details about some *Schrödinger cat experiments*: the anharmonic-oscillator model of a coherent single-mode electromagnetic field and a cavity-QED experiment.

Problems

14.1 Expand the two states $| R_0\rangle$ and $| R_1\rangle$ of the reservoir as $| R_0\rangle = \sum_j a_j^0 \, | r_j\rangle$ and $| R_1\rangle = \sum_j a_j^1 \, | r_j\rangle$, in order to derive the result (14.5).

14.2 Prove Eqs. (14.13) and (14.14).

14.3 Follow the procedure sketched in Subsec. 14.2.3 and explicitly derive the master equation for a damped harmonic oscillator, Eq. (14.51).

14.4 From Eq. (14.51) show that the diagonal matrix element $\wp(n) = \langle n \,|\hat{\rho}|\, n\rangle$, representing the probability of n quanta being in the system, obeys the equation

$$\frac{d\wp(n)}{dt} = 2\gamma \left[(n+1)\,\wp(n+1) - n\wp(n)\right]. \tag{14.126}$$

14.5 Derive Eq. (14.53), by first proving that

$$\hat{\rho}\hat{a}^\dagger = \hat{a}^\dagger \hat{\rho} + \frac{d\hat{\rho}}{d\hat{a}}. \tag{14.127}$$

14.6 Derive Eq. (14.55).

14.7 Taking into account the results of the previous two problems and starting from Eq. (14.51), prove Eq. (14.57).

14.8 Prove that in the case of a pure state the norm of the Bloch vector is equal to 1 while in the case of a mixed state it is strictly smaller then 1.

14.9 From the master equation (14.75) derive the equations of motion for ρ_{ee}, ρ_{gg}, and ρ_{eg}. After solving these equations, obtain and solve the corresponding Bloch equations for s_x, s_y, and s_z.

14.10 Prove Eq. (14.91).

14.11 Consider a non-unitary Schrödinger equation

$$\imath\hbar\frac{d}{dt}\,|\,\psi\rangle = \hat{\tilde{H}}\,|\,\psi\rangle\,, \tag{14.128}$$

where $\hat{\tilde{H}} = \hat{H}_0 + \hat{H}'$ is a non-Hermitian Hamiltonian, such that $\hat{H}_0 = \hat{H}_0^\dagger$ and $(\hat{H}')^\dagger = -\hat{H}'$. Show that the Louville equation in this case would be written as

$$\imath\hbar\frac{d}{dt}\hat{\rho} = \left[\hat{H}_0, \hat{\rho}\right] + \left[\hat{H}', \hat{\rho}\right]_+. \tag{14.129}$$

14.12 Show that the Hamiltonian (10.31) may be rewritten in the form of Eq. (14.106), with a suitable redefinition of the frequency.
(*Hint*: Take advantage of the results of Prob. 10.2 and of Eq. (4.78).)

14.13 Derive Eq. (14.108).

14.14 Derive Eq. (14.111).
(*Hint*: Rewrite the input state (14.110) so as to explicitly show the dependence on the creation operator $\hat{a}_1^\dagger$. Then, resolve Eq. (14.109) for $\hat{a}_1$ and substitute into Eq. (14.109).)

14.15 The simplest definition of distance between two pure quantum states $\mid \psi_1\rangle$ and $\mid \psi_2\rangle$ is the so-called Fubini–Study distance [Bargmann 1954], defined as

$$\mathrm{d}_{fs} = \sqrt{1 - |\langle \psi_1 \mid \psi_2 \rangle|^2}. \tag{14.130}$$

Verify that it satisfies the property of a distance and calculate the distance between the coherent states $\left|\alpha e^{-\imath\phi}\right\rangle$ and $\left|\alpha e^{\imath\phi}\right\rangle$ introduced in Box 14.1.

14.16 One of the most common ways to generalize the concept of distance to arbitrary (mixed) states $\hat{\rho}_1$ and $\hat{\rho}_2$ is the so-called Hilbert–Schmidt distance, defined as

$$\begin{aligned}\mathrm{d}_{hs}(\hat{\rho}_1, \hat{\rho}_2) &= \frac{1}{\sqrt{2}} \parallel \hat{\rho}_1 - \hat{\rho}_2 \parallel_2 = \frac{1}{\sqrt{2}} \left\{ \mathrm{Tr}\left[(\hat{\rho}_1 - \hat{\rho}_2)^2 \right] \right\}^{\frac{1}{2}} \\ &= \frac{1}{\sqrt{2}} \left[\mathrm{Tr}\left(\hat{\rho}_1^2\right) + \mathrm{Tr}\left(\hat{\rho}_2^2\right) - 2\mathrm{Tr}\left(\hat{\rho}_1\hat{\rho}_2\right) \right]^{\frac{1}{2}}. \end{aligned} \tag{14.131}$$

Verify that this distance reduces to the Fubini–Study distance in the case of pure states.

Further reading

General features

Davies, E. B., *Quantum Theory of Open Systems*, London: Academic Press, 1976.
Louisell, William H., *Quantum Statistical Theory of Radiation*, New York: Wiley, 1973.

Master equation

Gardiner, Crispin W., *Quantum Noise*, Berlin: Springer, 1991.
Lindblad, Göran, *Non-Equilibrium Entropy and Irreversibility*, Dordrecht: Reidel, 1983.

Fokker–Planck equation

Carmichael, Howard J., *Statistical Methods in Quantum Optics*, Berlin: Springer, 1999.
Risken, Hannes, *The Fokker–Planck Equation: Methods of Solution and Application*, Berlin: Springer, 1996.

Quantum jumps

Carmichael, Howard J., *An Open Systems Approach to Quantum Optics*, Heidelberg: Springer, 1993.

Distance between quantum states

Dodonov, V. V., Man'ko, O. V., Man'ko, V. I. and Wünsche, A., Energy-sensitive and "classical-like" distances between quantum states. *Physica Scripta*, **59** (1999), 81–89.

Knöll, L. and Orlowski, A., Distance between density operators: applications to the Jaynes–Cummings model. *Physical Review*, **A51** (1995), 1622–30.

15 State measurement in quantum mechanics

In classical mechanics, as we know, all observables commute with each other and therefore belong to the same "commuting set." This means that – in principle – we can measure all observables simultaneously. Then, the state manifests itself as a collection of properties (of values of observables) (see Sec. 1.1), and as such it is observable. On the other hand, quantum-mechanical observables may not commute (see Subsec. 2.1.5). As a consequence, the state is not just a collection of observable values (see also Sec. 2.3) so that we may ask ourselves whether it is observable or not, and therefore raise the following question: *What is the nature of the wave function?* Is it simply a mathematical tool, i.e. does it only represent the sum of all that we can know about a system? Or does it have an ontological status? Since the early days of quantum mechanics, this question has been very popular. As a first attempt, de Broglie tried to interpret the wave function as describing, in classical terms, a wave field that envelopes a classical particle (see also Subsec. 16.3.2). This theory should predict the existence of empty waves, but several experiments have shown negative results (see also the end of Sec. 9.6). More recently, many physicists have tried to propose some methods in order to measure the wave function. Here we shortly report on some of these proposals and discuss to what extent the quantum state can be measured.

In Sec. 15.1 we shall discuss Aharonov and co-workers' proposal to measure the wave function and show why this is ultimately a measurement of an observable of the system. In Sec. 15.2 we shall see an important consequence of the non-measurability of the state in quantum mechanics: the no-cloning theorem, i.e. the impossibility of cloning a quantum state. In Sec. 15.3 we shall return to an important problem (see also Sec. 9.5): the relationship between reversibility and irreversibility in quantum mechanics. In Sec. 15.4 we shall discuss the theory and the new techniques for measuring the Wigner function of a system. Finally, in Sec. 15.5, we shall develop some concluding remarks about how the quantum state should be understood.

15.1 Protective measurement of the state

One of the most interesting attempts at interpreting the wave function in ontological terms is due to Aharonov and coworkers.[1] The main idea is that of *protective measurement*, i.e. a

[1] See [Aharonov *et al.* 1993] [Aharonov/Vaidman 1993].

measurement of the wave function during which it is prevented from changing noticeably by means of another interaction which it undergoes at the same time.

Let us assume that we wish to measure an observable $\hat{O}$ on a system in the state $|\varsigma\rangle = \sum c_j |s_j\rangle$, where the states $|s_j\rangle$ are eigenkets of $\hat{O}$, and that the interaction between the apparatus $\mathcal{A}$ and the system $\mathcal{S}$ is described by the Hamiltonian

$$\hat{H}_{\mathcal{A}+\mathcal{S}} = \hat{H}_0 + \hat{H}_{\mathcal{AS}} + \hat{H}_{\mathcal{A}}, \tag{15.1}$$

where

$$\hat{H}_{\mathcal{AS}} = \hat{H} = \varepsilon(t)\hat{x}_{\mathcal{A}}\hat{O}, \tag{15.2}$$

is the interaction Hamiltonian, $\hat{H}_0$ is the free Hamiltonian of the system, $\hat{H}_{\mathcal{A}}$ is the Hamiltonian of the apparatus, $\hat{x}_{\mathcal{A}}$ is the one-dimensional pointer observable, and ε represents the coupling function (see Eq. (9.10)), i.e. $\varepsilon(t)$ is non-zero only in the interval $[0, \tau]$ (duration of the interaction). In general, such an interaction leads to an entangled state (see Sec. 9.1), which may be written as

$$|\Psi(\tau)\rangle = \sum_j c'_j |s_j\rangle |a_j\rangle, \tag{15.3}$$

where

$$|a_j\rangle = e^{-(\imath/\hbar)\varepsilon o_j \hat{x}_{\mathcal{A}}} |\mathcal{A}\rangle \tag{15.4}$$

are states of the apparatus $\mathcal{A}$ which, for sufficiently large ε are orthogonal for distinct eigenvalues a_j's of the pointer observable. The apparatus $\mathcal{A}$ is in the initial state $|\mathcal{A}\rangle$.

Let us now consider the case in which no entanglement takes place. Therefore, in place of Eq. (15.3), we write the factorized state

$$|\varsigma(0)\rangle|\mathcal{A}(0)\rangle \mapsto |\varsigma(t)\rangle|\mathcal{A}(t)\rangle, \; t > 0. \tag{15.5}$$

In this case, there is no reduction of the wave function. Instead, we would have the equation of motion

$$\frac{d}{dt}\langle\varsigma(t)\,|\langle\mathcal{A}(t)\,|\hat{p}_x^{\mathcal{A}}|\varsigma(t)\rangle|\mathcal{A}(t)\rangle = -\varepsilon(t)\langle\varsigma(t)|\hat{O}|\varsigma(t)\rangle, \tag{15.6}$$

where $\hat{p}_x^{\mathcal{A}}$ is the observable canonically conjugate to $\hat{x}_{\mathcal{A}}$ and, in the Heisenberg picture,

$$\frac{d}{dt}\hat{p}_x^{\mathcal{A}} = \frac{\imath}{\hbar}\left[\hat{H}, \hat{p}_x^{\mathcal{A}}\right] = -\varepsilon(t)\hat{O}. \tag{15.7}$$

Equation (15.7) shows that $\hat{p}_x^{\mathcal{A}}$ changes by different amounts for distinct eigenvalues o_j, and by Eq. (15.6) we can determine $\langle\varsigma(t)|\hat{O}|\varsigma(t)\rangle$ by the change in the apparatus' momentum.

A protective measurement can be made in two different ways. (i) If $|\varsigma(t)\rangle$ is a non-degenerate eigenstate of the Hamiltonian $\hat{H}$, then the interaction is assumed to be sufficiently weak and $\hat{H}$ changes slowly so that $|\varsigma(t)\rangle$ is nearly equal to $|\varsigma(0)\rangle$ up to a phase factor for $t \in [0, \tau]$. Then, following the adiabatic theorem (see Sec. 10.3), $|\varsigma(t)\rangle$ remains an eigenstate of the Hamiltonian and no entanglement takes place.

(ii) If we have an arbitrary evolution, so that $|\varsigma(t)\rangle$ is not necessarily an eigenstate of the Hamiltonian, we can operate in the following manner. If $|\varsigma_0(t)\rangle$ is the evolution of $|\varsigma\rangle$

determined by the unperturbed Hamiltonian $\hat{H}_0$ of the system $\mathcal{S}$, then one can measure an observable $\hat{O}'(t)$, for which $|\varsigma_0(t)\rangle$ is a non-degenerate eigenstate, a large number of times which are dense in the interval $[0,\tau]$ – say at times $t_n = (n/N)\tau, n = 1, 2, \ldots N$, where N is an arbitrarily large number. Then, $|\varsigma(t)\rangle$ does not noticeably depart from $|\varsigma_0(t)\rangle$ – it is a sort of quantum Zeno effect (see Sec. 9.8). Now, consider the branch of combined system evolution in which each measurement of $\hat{O}'(t_n)$ results in the state $|\varsigma_0(t_n)\rangle$ of $\mathcal{S}$

$$\begin{aligned}
|\Psi(\tau)\rangle_0 &= |\varsigma_0(t_N)\rangle\langle\varsigma_0(t_N)|e^{-\frac{\iota}{\hbar}\frac{\tau}{N}\hat{H}(t_N)}\cdots|\varsigma_0(t_2)\rangle\langle\varsigma_0(t_2)| \\
&\quad\times e^{-\frac{\iota}{\hbar}\frac{\tau}{N}\hat{H}(t_2)}|\varsigma_0(t_1)\rangle\langle\varsigma_0(t_1)|e^{-\frac{\iota}{\hbar}\frac{\tau}{2}\hat{H}(t_1)}|\varsigma(0)\rangle|\mathcal{A}(0)\rangle \\
&= |\varsigma_0(t_N)\rangle\langle\varsigma_0(t_N)|e^{-\frac{\iota}{\hbar}\frac{\tau}{N}\varepsilon(t_N)\hat{x}_{\mathcal{A}}\hat{O}}\cdots|\varsigma_0(t_3)\rangle\langle\varsigma_0(t_2)| \\
&\quad\times e^{-\frac{\iota}{\hbar}\frac{\tau}{N}\varepsilon(t_2)\hat{x}_{\mathcal{A}}\hat{O}}|\varsigma_0(t_2)\rangle\langle\varsigma_0(t_1)|e^{-\frac{\iota}{\hbar}\frac{\tau}{N}\varepsilon(t_1)\hat{x}_{\mathcal{A}}\hat{O}}|\varsigma_0(t_1)\rangle|\mathcal{A}_0(\tau)\rangle,
\end{aligned} \tag{15.8}$$

where $|\mathcal{A}_0(\tau)\rangle$ is the state of $\mathcal{A}$ when it evolves under the Hamiltonian $\hat{H}_{\mathcal{A}}$. We now calculate explicitly the last expectation value in Eq. (15.8) up to the second order in $1/N$ and find

$$\begin{aligned}
\langle\varsigma_0(t_1)|e^{-\frac{\iota}{\hbar}\frac{\tau}{N}\varepsilon(t_1)\hat{x}_{\mathcal{A}}\hat{O}}|\varsigma_0(t_1)\rangle &= 1 - \frac{\iota}{\hbar}\frac{\tau}{N}\varepsilon(t_1)\hat{x}_{\mathcal{A}}\left\langle\hat{O}\right\rangle - \frac{1}{2\hbar^2}\frac{\tau^2}{N^2}\varepsilon(t_1)^2\hat{x}_{\mathcal{A}}^2\left\langle\hat{O}^2\right\rangle \\
&= 1 - \frac{\iota}{\hbar}\frac{\tau}{N}\varepsilon(t_1)\hat{x}_{\mathcal{A}}\left\langle\hat{O}\right\rangle \\
&\quad - \frac{1}{2\hbar^2}\frac{\tau^2}{N^2}\varepsilon(t_1)^2\hat{x}_{\mathcal{A}}^2\left\langle\hat{O}\right\rangle^2 - \frac{1}{2\hbar^2}\frac{\tau^2}{N^2}\varepsilon(t_1)^2\hat{x}_{\mathcal{A}}^2\Delta\hat{O}^2 \\
&= e^{-\frac{\iota}{\hbar}\frac{\tau}{N}\varepsilon(t_1)\hat{x}_{\mathcal{A}}\left\langle\hat{O}\right\rangle}\left[1 - \frac{1}{2\hbar^2}\frac{\tau^2}{N^2}\varepsilon(t_1)^2\hat{x}_{\mathcal{A}}^2\Delta\hat{O}^2\right],
\end{aligned} \tag{15.9}$$

where we have made use of the fact that

$$\Delta\hat{O}^2 = \left\langle\hat{O}^2\right\rangle - \left\langle\hat{O}\right\rangle^2. \tag{15.10}$$

In the limit $N \to \infty$, where the product of the factors in the term containing $\Delta\hat{O}^2$ approaches 1, Eq. (15.8) reads

$$|\Psi(\tau)\rangle_0 = |\varsigma_0(\tau)\rangle \exp\left(-\frac{\iota}{\hbar}\int_0^\tau dt\varepsilon(t)\hat{x}_{\mathcal{A}}\left\langle\hat{O}\right\rangle\right)|\mathcal{A}_0(\tau)\rangle. \tag{15.11}$$

In this limit, the considered branch undergoes a unitary evolution and therefore the contribution from other branches – giving rise to states different from $|\varsigma_0(t)\rangle$ – vanishes. From the exponential operator in Eq. (15.11), the momentum of the apparatus is shifted by an amount (see also Eq. (15.7))

$$\Delta\hat{p}_x^{\mathcal{A}} = -\int_0^\tau dt\left\langle\hat{O}\right\rangle\varepsilon(t). \tag{15.12}$$

Therefore, by measuring $\hat{p}_x^{\mathcal{A}}$, $\langle \hat{O} \rangle$ can be determined. Then, according to the present approach, by repeating this experiment with different observables, the wave function of a single system may be determined up to an overall phase factor. Moreover, since a protective measurement as proposed by Aharonov and co-workers should not give rise to entanglement between the system and the apparatus and neither lead to a collapse, it could allow us in principle to distinguish between two non-orthogonal states, provided that both are protected.

In conclusion, this proposal aims at measuring the state of a quantum system and therefore to consider it as an observable. In other words, Aharonov and co-workers' try to consider the quantum state in classical terms.

Aharonov and co-workers' proposal has been criticized[2] (see Prob. 15.1) by pointing out that it only proved that, *if* the wave vector of a system $\mathcal{S}$ is known beforehand to be the eigenstate of the unknown Hamiltonian of $\mathcal{S}$, then it is possible to determine the properties of that eigenstate. In other words, one can determine some of the properties of an unknown Hamiltonian of $\mathcal{S}$, if one knows that $\mathcal{S}$ is in an eigenstate of that observable. In fact, the main condition of their model is a protective measurement, i.e. the system $\mathcal{S}$ interacts with an apparatus $\mathcal{A}$ or with the rest of the world in such a way that its wave function remains unchanged after the measurement but affects $\mathcal{A}$, so that a succession of measurements can completely determine it. This in turn means that, if $\mathcal{S}$ is in an energy eigenstate and if the interaction between $\mathcal{S}$ and the rest of the world is adiabatic, then the state vector after the measurement would still represent the same energy eigenstate. While the state vector is unchanged, the rest of the world has been changed in a manner dependent on the specific state of $\mathcal{S}$. However, if so, what we have obtained is only the measurement of an observable (the Hamiltonian) and not of the wave function as such. The problem is that we can force the wave function to be the eigenstate of an observable, but we cannot force the observable to have the unknown wave function as its eigenstate.

On the other hand, we know that the density matrix describing a pure state is a projector, i.e. it is an observable (see Subsec. 2.1.1 and Eq. (5.26)). So, why one can measure a projector but cannot obtain information about the state? The question is, what are the possible values that we would obtain by measuring a projector? Obviously, 0 or 1. If we obtain 0, we know that the system has not passed a certain test (say a vertical polarization filter), whereas if we obtain 1, we know that it has passed it (see also Subsec. 1.3.2). However, if the system before the test was in a superposition state of, say, vertical and horizontal polarization, we have a non-zero probability that it passes the test and a non-zero probability that it does not. Therefore, if we obtain a 0, we are not able to distinguish whether the system before the measurement was in a horizontal polarization state or in a superposition of vertical and horizontal polarization, and, similarly, if we obtain 1, we cannot distinguish between a previous vertical or superposed polarization state. In conclusion, the measurement of a projector (which, of course, is always possible) is not able to discriminate between non-orthogonal states. In other words, given an unknown state, we cannot decide *which* projector, if measured, would allow us to determine it.

[2] See [Unruh 1994].

15.2 Quantum cloning and unitarity violation

An important problem that is strictly connected with the question whether it is possible to measure the state of a single system is the possibility of *cloning* the state of a quantum system. By cloning we mean here the ability to make a perfect copy of a certain state without disturbing the original state. This issue is deeply connected with the problem of state measurement: as a matter of fact, if we had a *quantum copying machine*, then we would be able to make an arbitrarily large number of copies of the state to be measured, and then repeatedly measure different and incompatible observables on different copies, so as to extract enough information to determine the state of a single quantum system.

15.2.1 No-cloning theorem

The central question, which is deeply connected with the state measurement in quantum mechanics is the following: given a quantum system $\mathcal{S}$ in a certain (but in principle unknown) state $|\psi\rangle$, is it possible to make a perfect copy of it, i.e. is it possible to have a second system $\mathcal{S}'$ to be in the same state $|\psi\rangle$ while leaving $\mathcal{S}$ in the original state? We know that classically this is certainly possible. Ordinary photocopy machines do exactly the right job. Quantum-mechanically, the answer is more delicate. Yuen has shown that it is possible to duplicate a state known a priori to be any one of an orthogonal set of state vectors.[3] Is this also true for an arbitrary set of states (that is, not necessarily mutually orthogonal)? As a matter of fact, the following theorem may be proved:[4]

Theorem 15.1 (No-cloning) *In quantum mechanics no state can be cloned.*

Proof

Suppose that such a cloning is possible. For the sake of simplicity, let us restrict ourselves to a two-level system, e.g. the polarization state of a photon. Then, considering the apparatus being in an initial "ready" state $|\mathcal{A}_0\rangle$, an incoming photon with vertical polarization $|\updownarrow\rangle$, and a third system (the one onto which we want to make a copy of the photon's state) in a generic state $|0\rangle$, we would have the evolution

$$|\mathcal{A}_0\rangle|0\rangle|\updownarrow\rangle \leadsto |\mathcal{A}_\updownarrow\rangle|\updownarrow\rangle|\updownarrow\rangle . \tag{15.13a}$$

Similarly, for the same initial state of the apparatus and of the third system, but with an incoming photon in horizontal polarization, we would have

$$|\mathcal{A}_0\rangle|0\rangle|\leftrightarrow\rangle \leadsto |\mathcal{A}_\leftrightarrow\rangle|\leftrightarrow\rangle|\leftrightarrow\rangle . \tag{15.13b}$$

[3] See [Yuen 1986].
[4] See [Wootters/Zurek 1982].

Now, given the assumptions (15.13), and considering the case of an incoming photon in a linear superposition $c'|\updownarrow\rangle + c''|\leftrightarrow\rangle$, due to linearity, the result of the interaction will be

$$|\mathcal{A}_0\rangle|0\rangle\left(c'|\updownarrow\rangle + c''|\leftrightarrow\rangle\right) \rightsquigarrow c'|\mathcal{A}_\updownarrow\rangle|\updownarrow\rangle|\updownarrow\rangle + c''|\mathcal{A}_\leftrightarrow\rangle|\leftrightarrow\rangle|\leftrightarrow\rangle . \tag{15.14}$$

Now, it is easy to recognize that this result is in no way a clone of the superposition represented by the state of the incoming photon in the lhs of Eq. (15.14). In fact, such a cloning should, instead, be represented by

$$|\mathcal{A}_0\rangle|0\rangle\left(c'|\updownarrow\rangle + c''|\leftrightarrow\rangle\right) \rightsquigarrow |\mathcal{A}_?\rangle\left(c'|\updownarrow\rangle + c''|\leftrightarrow\rangle\right)\left(c'|\updownarrow\rangle + c''|\leftrightarrow\rangle\right). \tag{15.15}$$

Q.E.D

15.2.2 D'Ariano–Yuen theorem

The previous proof shows that cloning would represent a violation of the superposition principle (i.e. of the linearity of quantum mechanics). The proof has been generalized by Yuen and D'Ariano[5] by showing that cloning of two non-orthogonal states would represent a violation of the unitarity of the quantum-mechanical state evolution. Their main result may be summarized as:

Theorem 15.2 (D'Ariano–Yuen) *The possibility of discriminating between two non-orthogonal quantum-mechanical states contradicts the unitarity of quantum-mechanical transformations.*

Proof

A quantum-cloning machine that produces $n > 1$ copies of a generic state $|\psi\rangle$ from a given set of possible states, must effect a unitary evolution of the form

$$|\mathcal{A}\rangle \otimes |\psi\rangle \otimes |b_1\rangle \otimes \cdots \otimes |b_{n-1}\rangle \mapsto |\mathcal{A}'(\psi)\rangle \otimes |\psi\rangle \otimes \cdots \otimes |\psi\rangle, \tag{15.16}$$

where $|\mathcal{A}\rangle$ represents the state of the apparatus or the environment, the $|\psi\rangle$ states are present n times on the rhs, and $|b_1\rangle \otimes \cdots \otimes |b_{n-1}\rangle$ are the state preparation of the modes which support the clones. $|\mathcal{A}\rangle$ is the initial state of sufficiently enough other modes (environment and others), so that the transformation is unitary. Now, consider two non-orthogonal states $|\varphi\rangle, |\varsigma\rangle$ (with $0 < |\langle\varphi|\varsigma\rangle| < 1$), and suppose we know a priori that the system is in any one of them. We know that the transformation (15.16) must preserve the scalar product in order to be unitary (see Th. 8.1: p. 262). Let us write Eq. (15.16) for $|\varphi\rangle$,

$$|\mathcal{A}\rangle \otimes |\varphi\rangle \otimes |b_1\rangle \otimes \cdots \otimes |b_{n-1}\rangle \mapsto |\mathcal{A}'(\varphi)\rangle \otimes |\varphi\rangle \otimes \cdots \otimes |\varphi\rangle, \tag{15.17}$$

[5] See [D'Ariano/Yuen 1996] .

and for $|\varsigma\rangle$,

$$|\mathcal{A}\rangle \otimes |\varsigma\rangle \otimes |b_1\rangle \otimes \cdots \otimes |b_{n-1}\rangle \mapsto |\mathcal{A}'(\varsigma)\rangle \otimes |\varsigma\rangle \otimes \cdots \otimes |\varsigma\rangle, \tag{15.18}$$

and take the scalar product of the lhs and rhs of these two equations, so that

$$\langle\varphi|\varsigma\rangle = \langle\mathcal{A}'(\varphi)|\mathcal{A}'(\varsigma)\rangle(\langle\varphi|\varsigma\rangle)^n \, . \tag{15.19}$$

From this expression it immediately follows that

$$\langle\mathcal{A}'(\varphi)|\mathcal{A}'(\varsigma)\rangle(\langle\varphi|\varsigma\rangle)^{n-1} = 1, \tag{15.20}$$

which would in turn require that

$$|\langle\mathcal{A}'(\varphi)|\mathcal{A}'(\varsigma)\rangle| > 1 \tag{15.21}$$

for $n > 1$.

So far, we have proved that the cloning of two non-orthogonal states contradicts unitarity. Since the possibility to discriminate between two non-orthogonal states would in turn imply the possibility – through quantum-mechanical unitary transformations – of generating clones of the two original states, then also the discrimination of two non-orthogonal states would violate the unitarity of quantum-mechanical transformations.

Q.E.D

The above theorem proves the impossibility of measuring the state of a single system through a single measurement. Moreover, the same authors have shown that any succession of repeated measurements performed on a single system gives exactly the same probability distribution of an appropriately chosen single measurement with the output state independent of the input one – this is entirely different from what happens in classical mechanics, where successive measurements on the same system do increment the information about the system. This completes the picture and allows us to establish on the most general grounds the impossibility of measuring the wave function of a single system.[6]

15.3 Measurement and reversibility

An interesting question, which is to a certain extent complementary to the ones expressed at the beginning of this chapter and of the previous section, is whether a measurement can be made reversible (see also Sec. 9.5). Here we shall show that an ideal measurement is reversible if and only if no information about the initial state is obtained, and then, if so, it cannot be considered a true measurement (see also Subsec. 9.11.3).

[6] An interesting study of Hillery and Bužek [Bužek/Hillery 1996] shows to what extent one can copy a state imperfectly.

15.3.1 A "unitarily reversible" measurement

In a pioneering work, Ueda and Kitagawa[7] showed that a measurement can be made reversible if it is unsharp (see Sec. 9.10), and if it is sensitive to the vacuum field fluctuations. However, they did not consider a conserved quantity and, for this reason, the argument was not definitive.

A conserved quantity, the photon-number operator, was considered in a paper by Imamoḡlu,[8] who investigated the possibility of repetitive (logically reversible) measurements. He has demonstrated that, under certain conditions, the state of the field before and after the measurement is unchanged. However, those conditions contradict the results of the previous section.

More recently, Mabuchi and Zoller[9] have shown under which conditions reversibility is possible in the case of a system coupled to an environment. In short, they prove that the action of a jump superoperator (see Subsec. 14.4) cannot in general be inverted as such. Inversion is possible if one considers the system as pertaining to a subset of the original Hilbert space and the jump as unitary – in this case, however, no new information is obtained.

Let us discuss the problem on an abstract level (the authors also propose a concrete experiment). Let the evolution of some system – in the formalism of the Monte-Carlo wave-function approach – be subjected to the action of a collapse operator $\hat{C}_{j_r}$ (see again Subsec. 14.4)

$$|\psi_{\mathcal{S}}(t_r + dt)\rangle = \hat{C}_{j_r}|\psi_{\mathcal{S}}(t_r)\rangle, \tag{15.22}$$

where t_r is the observation time of the jump j_r (j denotes the j-th channel, for example the j-th harmonic oscillator of a bath reservoir). Between consecutive counts (jumps), the system state vector evolves according to

$$|\psi_{\mathcal{S}}(t)\rangle = e^{-\frac{\imath}{\hbar}\tilde{\hat{H}}(t-t_r)}|\psi_{\mathcal{S}}(t_r)\rangle, \tag{15.23}$$

where

$$\tilde{\hat{H}} = \hat{H}_{\mathcal{E}+\mathcal{S}} - \frac{\imath}{2}\sum_k \hat{C}_k^\dagger \hat{C}_k \tag{15.24}$$

is an effective non-Hermitian Hamiltonian. It is certainly true that the quantum collapse operator is not invertible on the entire Hilbert space of the system. In principle, it may be invertible on a restricted subspace of the same Hilbert space. If we consider the presence of a feedback mechanism, i.e. the action of the output of a system on its input, described by a unitary operator $\hat{U}_j$, such that

$$|\psi_{\mathcal{S}}(t_r + dt)\rangle = \hat{U}_j\hat{C}_{j_r}|\psi_{\mathcal{S}}(t_r)\rangle, \tag{15.25}$$

then the condition $\hat{C}_j = c_j\hat{U}_j^\dagger$, where c_j is a complex number, ensures that the final state is proportional to the initial state. If between two quantum jumps the unitary dynamics is not distorted by damping factors, i.e.

[7] See [Ueda/Kitagawa 1992].
[8] See [Imamoḡlu 1993].
[9] See [Mabuchi/Zoller 1996].

$$|\psi_{\mathcal{S}}(t)\rangle = e^{-\frac{i}{\hbar}\tilde{\hat{H}}t}|\psi_{\mathcal{S}}(0)\rangle = e^{-\frac{1}{2}\sum_j |c_j|^2 t} e^{-\frac{i}{\hbar}t\hat{H}_{\mathcal{E}+\mathcal{S}}}|\psi_{\mathcal{S}}(0)\rangle, \qquad (15.26)$$

where the state vectors pertain to the considered subspace and the Hamiltonian $\hat{H}_{\mathcal{E}+\mathcal{S}}$ is Hermitian, then in principle the feedback can undo the effect of the quantum collapse operator, as follows:

$$\begin{aligned} |\Psi_{\mathcal{E}+\mathcal{S}}(t)\rangle &= \frac{1}{\mathcal{N}} e^{-\frac{i}{\hbar}\tilde{\hat{H}}(t-t_n)} \hat{U}_{j_n}\hat{C}_{j_n} \cdots \hat{U}_{j_1}\hat{C}_{j_1} e^{-\frac{i}{\hbar}\tilde{\hat{H}}t_1}|\Psi_{\mathcal{E}+\mathcal{S}}\rangle \\ &= e^{-\frac{i}{\hbar}t\hat{H}_{\mathcal{E}+\mathcal{S}}}|\Psi_{\mathcal{E}+\mathcal{S}}\rangle, \end{aligned} \qquad (15.27)$$

where $\mathcal{N}$ is some normalization constant.

15.3.2 Is new information gained?

It can be shown that the scheme chosen by Mabuchi and Zoller is not a measurement in the most general sense[10] (see Subsec. 9.11.3). In fact they consider the action of an annihilation operator $\hat{a} = \hat{C}$ (which causes the jump by photon absorption, for example) and which acts as an unitary operator on a subspace $\mathcal{H}_1$ of the original Hilbert space, i.e. we have

$$\hat{a}|\psi\rangle = \hat{U}|\psi\rangle, \quad \langle\psi|\hat{a}^\dagger = \langle\psi|\hat{U}^\dagger. \qquad (15.28)$$

Due to the unitarity of $\hat{U}$, we also have

$$1 = \langle\psi|\hat{U}^\dagger\hat{U}|\psi\rangle = \langle\psi|\hat{a}^\dagger\hat{a}|\psi\rangle = \langle\psi|\hat{N}|\psi\rangle, \qquad (15.29)$$

where $\hat{N}$ is the number operator, which means that the expectation value of the photon number is equal to unity for an arbitrary state from the specified subset $|\psi\rangle \in \mathcal{H}_1$. If we expand $|\psi\rangle$ as follows (see Eq. (13.44))

$$|\psi\rangle = c_0|0\rangle + c_1|1\rangle + \cdots + c_n|n\rangle \cdots, \qquad (15.30)$$

where $|n\rangle$ is a (normalized) state with n photons, then Eq. (15.29) can be written as

$$\wp_1 + 2\wp_2 + 3\wp_3 + \cdots + n\wp_n + \cdots = 1, \qquad (15.31)$$

with non-negative numbers $\wp_n = |c_n|^2$. For this equation to be fulfilled, at least one of the numbers $\wp_1, \wp_2, \ldots$ must be non-zero. As a consequence, the state $|\psi\rangle \in \mathcal{H}_1$ cannot be the vacuum. Since a jump (annihilation operator) diminishes the number of photons by unity, the fact that a jump occurred gives the information that the initial state was not the vacuum – information that was already contained in the assumption that the initial state belongs to the subset $\mathcal{H}_1$. Since the fact that $|\psi\rangle \in \mathcal{H}_1$ depends on the preparation, no new information is gained by such a reversible quantum jump (see also Subsec. 3.5.3). The same argument can be extended to a double jump operator and in fact to any number of jumps (see Prob. 15.2).

[10] See [Mensky 1996].

Hence we may conclude the following:

- In general, the action of a jump operator (which may describe certain dissipative processes and certain kinds of measurement) is not unitary and cannot be inverted.
- If we restrict the possible states to a properly chosen subset of the original Hilbert space of the system, the action of a jump operator may become unitary and therefore, under those conditions, reversible.
- However, in all those cases in which the action of a jump operator is reversible, the measurement described by the jump operator itself gives no new information besides that already included in the preparation of the initial state.

In this context, a recent result due to Zurek is of particular relevance.[11] He has shown that, since a measurement consists in a transfer of information about a system to the apparatus, even imperfect copying essential in such situations restricts possible unperturbed outcomes to an orthogonal subset of all possible states of the system, thus breaking the unitary symmetry of its Hilbert space implied by the quantum superposition principle. Preferred outcome states emerge as a result. They provide therefore a framework for the so-called Òwave-packet collapse, Ó designating terminal points of quantum jumps and defining the measured observable by specifying its eigenstates.

15.3.3 A generalization on reversible measurements

Nielsen and Caves[12] provided a powerful formal generalization of the previous analysis. We recall here the formula (9.116a) for operations, i.e. (see also Sec. 14.3)

$$\mathcal{T}\hat{\rho} = \sum \hat{\vartheta}_k \hat{\rho} \hat{\vartheta}_k^\dagger. \tag{15.32}$$

If we have a single (not necessarily unitary) operator such that

$$\mathcal{T}\hat{\rho} = \hat{\vartheta}\hat{\rho}\hat{\vartheta}^\dagger \quad \text{and} \quad \hat{\vartheta}^\dagger\hat{\vartheta} = \hat{E}, \tag{15.33}$$

then we speak of an *ideal* measurement (perfect readout of the state of the apparatus). We then have a unitarily reversible measurement – on a subspace $\mathcal{H}_0$ of the state space $\mathcal{H}$ of the original problem – if there exists a unitary operator $\hat{U}$ acting on $\mathcal{H}_0$ such that (see Eq. (9.114))

$$\hat{\rho} = \hat{U} \frac{\mathcal{T}\hat{\rho}}{\mathrm{Tr}[\mathcal{T}\hat{\rho}]} \hat{U}^\dagger \tag{15.34}$$

for all $\hat{\rho}$ whose support lies in $\mathcal{H}_0$. Now, consider the following generalized form of measurement:

$$\mathcal{T}_j\hat{\rho} = \sum_k \hat{\vartheta}_{jk} \hat{\rho} \hat{\vartheta}_{jk}^\dagger, \tag{15.35}$$

[11] See [Zurek 2007].

[12] See [Nielsen/Caves 1997].

where j labels the outcome of the measurement and (see Eq. (9.115))

$$\sum_{jk} \hat{\vartheta}^{\dagger}_{jk} \hat{\vartheta}_{jk} = \hat{I}. \tag{15.36}$$

Supposing that j occurs, then the unnormalized state after the measurement is $\mathcal{T}_j\hat{\rho}$. The probability that the outcome j occurs is

$$\wp(j) = \mathrm{Tr}\left[\mathcal{T}_j\hat{\rho}\right] = \mathrm{Tr}\left[\hat{\rho}\sum_k \hat{\vartheta}^{\dagger}_{jk}\hat{\vartheta}_{jk}\right], \tag{15.37}$$

where use has been made of the cyclic property of the trace (see Prob. 5.4). It may be proved that the following statements are equivalent (see Prob. 15.3)

- The ideal quantum operation $\mathcal{T}_j\hat{\rho} = \hat{\vartheta}_j\hat{\rho}\hat{\vartheta}^{\dagger}_j$ is unitarily reversible on a subspace $\mathcal{H}_0$ of the total Hilbert space $\mathcal{H}$.
- The POVM $\hat{E} = \hat{\vartheta}^{\dagger}\hat{\vartheta}$, when restricted to $\mathcal{H}_0$, is a positive multiple of the identity operator on $\mathcal{H}_0$, i.e.

$$\hat{P}_{\mathcal{H}_0}\hat{E}\hat{P}_{\mathcal{H}_0} = \eta^2\hat{P}_{\mathcal{H}_0}, \tag{15.38}$$

  where η is a real constant satisfying $0 < \eta \leq 1$ and $\hat{P}_{\mathcal{H}_0}$ is the projector onto $\mathcal{H}_0$.
- We have

$$\mathrm{Tr}\left[\hat{\rho}\hat{\vartheta}^{\dagger}\hat{\vartheta}\right] = \mathrm{Tr}\left[\hat{\rho}\hat{E}\right] = \eta^2 \tag{15.39}$$

  for all density operators whose support lies in $\mathcal{H}_0$, and η takes the meaning of the probability of occurrence of the result represented by $\hat{\vartheta}$.
- The operator $\hat{\vartheta}$ can be written as

$$\hat{\vartheta} = \eta\hat{U}\hat{P}_{\mathcal{H}_0} + \hat{\vartheta}\hat{P}_{\mathcal{H}_0^{\perp}}, \tag{15.40}$$

  where $\hat{U}$ is some unitary operator acting on the whole $\mathcal{H}$ and $\hat{P}_{\mathcal{H}_0^{\perp}}$ projects onto the orthogonal complement of $\mathcal{H}_0$, that is $\hat{P}_{\mathcal{H}_0} + \hat{P}_{\mathcal{H}_0^{\perp}} = \hat{I}$. This condition makes formally clear why an ideal operation described by $\hat{\vartheta}$ can be unitarily reversed on $\mathcal{H}_0$: when restricted to $\mathcal{H}_0$, $\hat{\vartheta}$ acts as the unitary operator $\hat{U}$, except for rescaling by the real constant η, which accounts for the probability of obtaining the result corresponding to $\hat{\vartheta}$.

Summarizing, an ideal measurement is reversible if and only if no new information about the prior state is obtained from the measurement. Given the result as stated by Eq. (15.39), each state is equally likely. As a consequence, a reversible ideal measurement cannot be considered as a true measurement.

15.4 Quantum state reconstruction

In this section, we shall discuss a few methods which allow the reconstruction of the quantum state on a large set of identical systems. The following results are not in contrast with what we have stated in the previous sections (and actually confirm the conclusion of the previous examination), that is, the impossibility of measuring the state of a single system. In fact, the measurement of the Wigner function or any other of the methods discussed in the present section are only possible if one performs a large number of measurements, each one on a single element of a set of identically prepared systems.

15.4.1 Measurement of the Wigner function

We have seen that the Wigner function (see Subsec. 13.5.4) is, among others, a phase space representation of the density operator or, equivalently, a quasi-probability distribution for conjugate variables. It is then clear that measuring the Wigner function is completely equivalent to the measurement of the density operator. Indeed, there are circumstances where a direct measurement of the Wigner function is simply more convenient.

Royer[13] analyzed the problem in general terms by working out the premeasurement techniques developed by Lamb.[14] The problem can be cast as follows: given a well-defined preparation procedure and a certain number of identical systems, is it possible to determine experimentally (to measure) the state which such a procedure forces the systems to be in? Due to the one-to-one correspondence between the W-function and the density matrix of a system (see Subsec. 13.5.4), this is possible if one is able to determine the W-function.

Let us limit ourselves to a one-dimensional system whose phase state is represented by position $\hat{x}$ and momentum $\hat{p}_x$. Making use of the results of Subsec. 14.1.2, we introduce the S-vectors

$$|\hat{x}\,\hat{p}_x\} = \sqrt{\frac{2}{\pi\hbar}}\,|\hat{\Pi}_{xp}\}, \tag{15.41}$$

where

$$\begin{aligned}
\hat{\Pi}_{xp} &= \frac{\hbar}{2}\int_{-\infty}^{+\infty} dx' e^{\imath x' p_x}\left|x + \frac{1}{2}\hbar x'\right\rangle\left\langle x - \frac{1}{2}\hbar x'\right| \\
&= \frac{\hbar}{2}\int_{-\infty}^{+\infty} dp'_x e^{\imath x p'_x}\left|p_x + \frac{1}{2}\hbar p'_x\right\rangle\left\langle p_x - \frac{1}{2}\hbar p'_x\right| \\
&= \frac{\hbar}{4\pi}\int_{-\infty}^{+\infty} dp'_x \int_{-\infty}^{+\infty} dx' e^{\imath p'_x(\hat{x}-x)-\imath x'(\hat{p}_x-p_x)},
\end{aligned} \tag{15.42}$$

$\left|x + \frac{1}{2}\hbar x'\right\rangle$ and $\left|p_x + \frac{1}{2}\hbar p'_x\right\rangle$ being eigenkets of position and momentum, respectively, and $-\infty < x < +\infty, -\infty < p_x < +\infty$. We may expand the operator $\hat{\Pi}_{xp}$ about the phase-space point (x, p_x) by making use of the displacement operator (see Eq. (13.68) and Prob. 15.4)

$$\hat{D}_{xp} = e^{\frac{\imath}{\hbar}(p_x\hat{x}-x\hat{p}_x)}, \tag{15.43}$$

[13] See [Royer 1985, Royer 1989].

[14] See [Lamb 1969].

so that we have

$$\hat{\Pi}_{xp} = \hat{D}_{xp}\hat{\Pi}\hat{D}_{xp}^{-1}, \tag{15.44}$$

where the parity operator about the origin is given by

$$\hat{\Pi} = \int_{-\infty}^{+\infty} dx|-x\rangle\langle x| = \int_{-\infty}^{+\infty} dp_x|-p_x\rangle\langle p_x|. \tag{15.45}$$

It follows from Eqs. (15.44)–(15.45) that

$$\hat{\Pi}_{xp}(\hat{x}-x)\hat{\Pi}_{xp} = -(\hat{x}-x), \hat{\Pi}_{xp}(\hat{p}_x - p_x)\hat{\Pi}_{xp} = -(\hat{p}_x - p_x), \tag{15.46}$$

that is, $\hat{\Pi}_{xp}$ is the parity operator about the phase-space point (x, p_x). The key point of the following discussion is that the W-function is the expectation value of the parity operator $\hat{\Pi}_{xp}$ (see Eqs. (13.113) and (14.13))

$$W_{\hat{\rho}}(x, p_x, t) = \left\{\hat{\Pi}_{xp}|\hat{\rho}(t)\right\} = \frac{1}{\pi\hbar}\left\langle\hat{\Pi}_{xp}\right\rangle_{\hat{\rho}(t)}, \tag{15.47}$$

where $\hat{\Pi}_{xp}$ is Hermitian (see Prob. 15.5). Since obviously $\hat{\Pi}_{xp}^2 = \hat{I}$, $\hat{\Pi}_{xp}$ is an observable whose eigenvalues are ± 1 (see Prob. 15.6): a complete set of eigenstates $|\psi_{xp}^n\rangle, n = 1, 2, \ldots$, satisfying

$$\hat{\Pi}_{xp}|\psi_{xp}^n\rangle = (-1)^n|\psi_{xp}^n\rangle, \tag{15.48}$$

may be obtained by displacing in phase space any complete orthogonal set of kets $|\psi^n\rangle$ of definite parity about the origin. Thus (see also Probs. 2.8 and 15.7),

$$\psi^n(-x) = (-1)^n\psi^n(x), \tag{15.49a}$$

$$\hat{\Pi}_{xp} = \sum_n (-1)^n|\psi_{xp}^n\rangle\langle\psi_{xp}^n|, \tag{15.49b}$$

$$|\psi_{xp}^n\rangle = \hat{D}_{xp}|\psi^n\rangle, \tag{15.49c}$$

so that Eq. (15.47) can be rewritten as

$$W_{\hat{\rho}}(x, p_x, t) = \frac{1}{\pi\hbar}\sum_n (-1)^n\langle\psi_{xp}^n|\hat{\rho}(t)|\psi_{xp}^n\rangle. \tag{15.50}$$

We try now to measure $W_{\hat{\rho}}(x, p_x, t)$ at some definite time (e.g. $t = 0$). This can be done by measuring each transition probability $\langle\psi_{xp}^n|\hat{\rho}(0)|\psi_{xp}^n\rangle$ following the method introduced by Lamb: a simple approach is possible if we choose the $|\psi^n\rangle$s to be eigenstates of the Hamiltonian

$$\hat{H} = \frac{\hat{p}_x^2}{2m} + V(\hat{x}), \tag{15.51}$$

where $V(-x) = V(x)$ is a symmetric potential. Then, the $|\psi_{xp}^n\rangle$'s are eigenstates of the displaced Hamiltonian

$$\hat{H}_{xp} = \hat{D}_{xp}\hat{H}\hat{D}_{xp}^{-1} = \frac{(\hat{p}_x - p_x)^2}{2m} + V(\hat{x} - x), \tag{15.52}$$

so that measuring the set $\langle\psi^n_{xp}|\hat{\rho}(0)|\psi^n_{xp}\rangle$ (or $\hat{\Pi}_{xp}$) becomes equivalent to measuring the Hamiltonian $\hat{H}_{xp}$. A suitable method to measure $\hat{H}_{xp}$ almost in the strict sense is as follows: first, we place ourselves in a reference frame moving with uniform speed $v = p^f_x/m$ relative to the preparation apparatus $\mathcal{A}$. By virtue of the Galilei transformations, the observed density operator (for $t \leq 0$) is

$$\hat{\rho}^f(t) = \hat{D}^{-1}_{vt,p^f_x}\hat{\rho}(t)\hat{D}_{vt,p^f_x}. \tag{15.53}$$

At time $t = 0$ we turn on the potential $V(x - x^f)$ in the moving frame. The eigenstates of

$$\hat{H}_{x^f,0} = \frac{\left(\hat{p}^f_x\right)^2}{2m} + V(\hat{x}^f - x^f) \tag{15.54}$$

are

$$\hat{D}_{x^f,0}|\psi^n\rangle = |\psi^n_{x^f,0}\rangle, \tag{15.55}$$

with corresponding energies E_n. Then, at times $t \geq 0$ we obtain

$$\begin{aligned}
\hat{\rho}^f(t) &= e^{-\frac{\imath}{\hbar}t\hat{H}_{x^f,0}}\hat{\rho}^f(0)e^{\frac{\imath}{\hbar}t\hat{H}_{x^f,0}} \\
&= \sum_{m,n} e^{-\frac{\imath}{\hbar}(E_n - E_m)t}|\psi^n_{x^f,0}\rangle\langle\psi^m_{x^f,0}|\langle\psi^n_{x^f,0}|\hat{\rho}^f(0)|\psi^m_{x^f,0}\rangle \\
&= \sum_{m,n} e^{-\frac{\imath}{\hbar}(E_n - E_m)t}|\psi^n_{x^f,0}\rangle\langle\psi^m_{x^f,0}|\langle\psi^n_{x^f,p^f_x}|\hat{\rho}(0)|\psi^m_{x^f,p^f_x}\rangle,
\end{aligned} \tag{15.56}$$

where we have made use of the transformation (15.53) for $\hat{\rho}(0)$. Now, the transition probabilities (see Prob. 15.8)

$$\langle\psi^n_{x^f,0}|\hat{\rho}^f(0)|\psi^n_{x^f,0}\rangle = \langle\psi^n_{x^f p^f_x}|\hat{\rho}(0)|\psi^n_{x^f p^f_x}\rangle \tag{15.57}$$

are time-independent, so that we have a long time available to perform a measurement of $\hat{H}_{x^f,0}$ referring to the set $\{|\psi^n_{x^f,0}\rangle\}$ and "find" the particle in one of the states pertaining to this set. Repeating the measurement many times will allow to build the distribution (15.57), from which $W_{\hat{\rho}}(x^f, p^f_x, 0)$ can be deduced by means of Eq. (15.50). What has been done is a measurement of $W_{\hat{\rho}}(x^f, p^f_x, 0)$ by measuring

$$W_{\hat{\rho}^f}(x^f, 0, t) = W_{\hat{\rho}}(x^f + vt, p^f_x, t) \tag{15.58}$$

at $t = 0$ in the moving frame. In conclusion, applying the same procedure over and over again with different values x^f and p^f_x, it is in principle possible to reconstruct the Wigner function on any relevant region of the phase space.

15.4.2 Quantum tomography

Perhaps the most effective – and certainly the first – method that has been experimentally used for the reconstruction of the W-function is the so-called *quantum state tomography*. In this approach, one takes advantage of the inversion of a set of measured probability distributions of quadrature amplitudes (see Eq. (13.61)) in order to reconstruct the W-function.[15]

The starting key point is the pioneering contribution of Vogel and Risken,[16] who have shown that measuring appropriate probability distributions of certain quadratures, it is possible to obtain the W-function. In fact, we recall here the definition of the W-function (see Eq. (13.111))

$$W(\alpha,\alpha^*) = \frac{1}{\pi^2}\int d^2\xi e^{-\xi\alpha^*+\xi^*\alpha}\chi(\xi), \tag{15.59}$$

expressed as the Fourier transform of the characteristic function

$$\chi(\xi) = \mathrm{Tr}\left[\hat{\rho}e^{\xi\hat{a}^\dagger-\xi^*\hat{a}}\right]. \tag{15.60}$$

The homodyne detector (see Subsec. 13.6.2) measures the rotated quadrature operator $\hat{X}_\theta$. The complete information for calculating any single-time expectation value of $\hat{X}_\theta$ is given by the probability distribution $\wp(X,\theta)$, which is the Fourier transform

$$\wp(X,\theta) = \frac{1}{2\pi}\int d\zeta\, \chi_\wp(\zeta,\theta)e^{-\imath\zeta X} \tag{15.61}$$

of the characteristic function

$$\chi_\wp(\zeta,\theta) = \mathrm{Tr}\left[e^{\imath\zeta\hat{X}_\theta}\hat{\rho}\right]. \tag{15.62}$$

We try now to find a one-to-one correspondence between the W-function $W(\alpha,\alpha^*)$ and the probability distribution $\wp(X,\theta)$. Recalling the definition (13.136) of $\hat{X}_\theta$, we may rewrite

$$\chi_\wp(\zeta,\theta) = \chi\left(\imath\zeta\frac{e^{\imath\theta}}{\sqrt{2}}\right). \tag{15.63}$$

Making use of the notation $\xi = \xi_r + \imath\xi_i$ and $\chi(\xi) = \chi(\xi_r,\xi_i)$, $\chi_\wp(\zeta,\theta)$ may be written as

$$\chi_\wp(\zeta,\theta) = \chi\left(-\frac{1}{\sqrt{2}}\zeta\sin\theta, \frac{1}{\sqrt{2}}\zeta\cos\theta\right). \tag{15.64}$$

It is then clear that, if $\chi_\wp(\zeta,\theta)$ is known for all ζ values in the range $-\infty < \zeta < +\infty$ and for all θ values in the range $0 \le \theta < \pi$, then the characteristic function $\chi(\xi_r,\xi_i)$ is known in the whole complex plane ξ, and therefore we also obtain a one-to-one correspondence between the characteristic functions (15.60) and (15.62) as well as between the W-function (15.59) and the probability distribution (15.61).

In order to see this explicitly, let us make the inverse Fourier transform of Eq. (15.61), use Eq. (15.64), and insert the result into Eq. (15.59). After a proper change of variables, we obtain (see Prob. 15.9)

$$W(\alpha_r,\alpha_i) = \frac{1}{2\pi^2}\int_{-\infty}^{+\infty} dX \int_{-\infty}^{+\infty} d\zeta \int_0^\pi d\theta \wp(X,\theta)e^{\imath\zeta\left[X-\sqrt{2}(\alpha_r\cos\theta+\alpha_i\sin\theta)\right]}|\zeta|, \tag{15.65}$$

[15] See [Freyberger *et al.* 1997] for a beautiful review of the topic of quantum state measurement.

[16] See [Vogel/Risken 1989]. See also [Bertrand/Bertrand 1987].

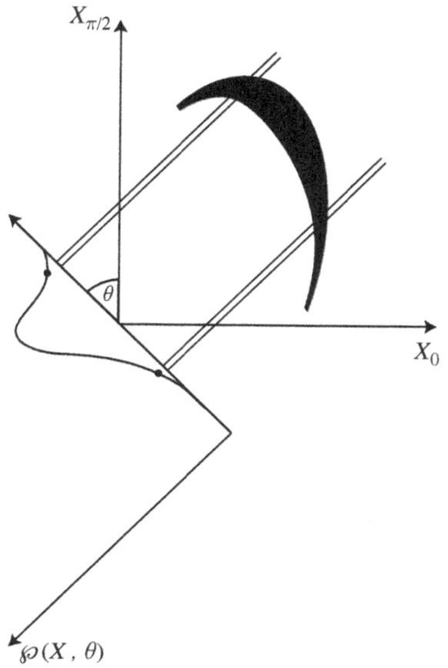

Figure 15.1 **Tomographic method for reconstructing the W-function. The probability distributions $\wp(X,\theta)$ of the observable $\hat{X}_\theta$ are measured. Pictorially, these correspond to overlaps between the W-function and infinitesimally thin stripes in phase space. A full reconstruction of the W-function requires $\wp(X,\theta)$ to be measured for all angles θ between 0 and π.**

where $\alpha = \alpha_r + \imath\alpha_i$. Equation (15.65) is the so-called inverse Radon transform[17] and reveals the essence of the tomographic method: one performs many measurements of entire "slices" through the Wigner function, which correspond to measurements of the observable $\hat{X}_\theta$, obtaining the marginal probability distribution $\wp(X,\theta)$ for any values of $0 \leq \theta \leq \pi$ (see Fig. 15.1). From these marginal distributions, the W–function can be reconstructed by means of the inverse Radon transform. This technique is similar to the method used in modern medicine to take the "picture" of humans organs – hence the name of *quantum tomography*.

Optical homodyne tomography of a single mode of the electromagnetic field has been experimentally realized[18] (see Fig. 15.2). An ensemble of repeated measurements of the quadrature for various phases relative to the local oscillator of the homodyne detector have been performed.[19]

There are of course other sampling methods, which use the same principle.[20] For example, in the *simultaneous method* the phase space is sampled with circular discs rather than cutting it into slices. In so doing one measures simultaneously two conjugate variables and, in accordance with the uncertainty relation, the values can be known with limited accuracy

[17] See [*Natterer* 1986].
[18] See [Smithey *et al.* 1993].
[19] See also [Freyberger/Herkommer 1994] [Wallentowitz/Vogel 1995].
[20] See [Freyberger *et al.* 1997].

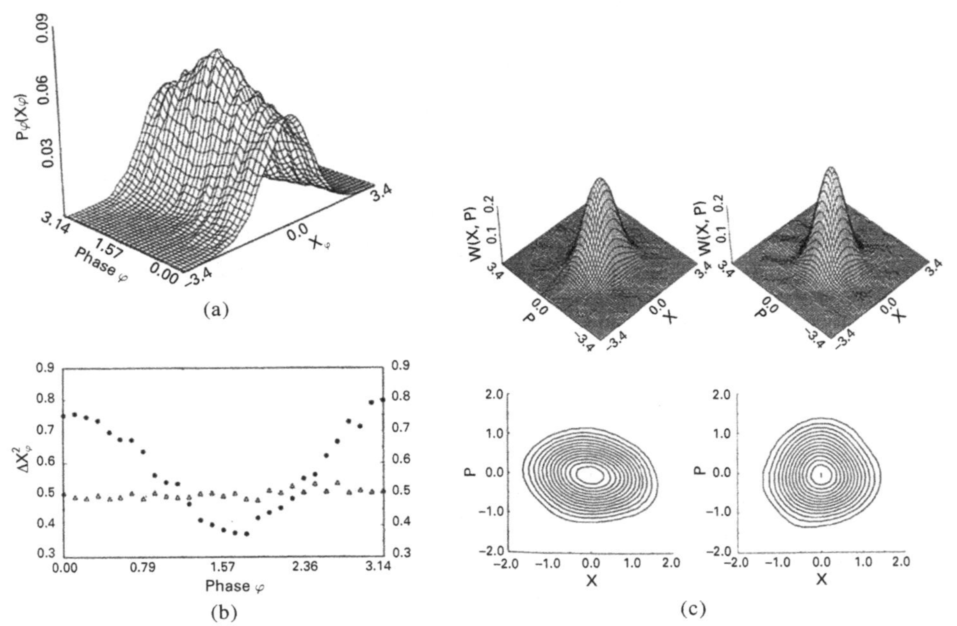

Figure 15.2 **Tomographic measurements of quantum states as experimentally performed by Raymer and co-workers. (a) Measured quadrature-amplitude distributions at various values of local-oscillator phase. Since these distributions are normalized, a decreasing width of a particular distribution is accompanied by an increase in its peak height. (b) Variances of quadrature amplitudes versus local oscillator phase. Circles denote squeezed state, triangles vacuum state. (c) Measured Wigner distributions for a squeezed state (left) and a vacuum state (right) viewed in 3D (top) and as contour graph (bottom). Adapted from [Smithey *et al.* 1993].**

(see also Fig. 2.8). In spite of the fuzzy character of such a measurement, it still fully characterizes a quantum state and directly yields the Q-function (see Subsec. 13.5.1).

A third sampling method is the so-called *ring method*, where, graphically, phase space is scanned with rings, each of which represents one energy eigenstate – Fock state (see Subsec. 13.4.1). Then, for each value of the energy level n, we may build the probability distributions $\wp_n(X_0, X_{\pi/2})$. The W-function can then be obtained by the simple sum

$$W(X_0, X_{\pi/2}) \propto \sum_{n=0}^{\infty} (-1)^n \wp_n(X_0, X_{\pi/2}). \tag{15.66}$$

The tomographic method has also been extended in invariant form, by relating the homodyne output distribution directly to the density matrix (without the intermediate step of reconstructing the W-function) in a quadrature-component basis[21] and in the Fock basis[22] In practice, the reconstruction of the W-function can be a very difficult task: the knowledge of the W-function is equivalent to the knowledge of all independent moments of the system operators (see Subsec. 2.1.3); in the case of the harmonic oscillator it amounts to the

[21] See [Kühn *et al.* 1994] [Zucchetti *et al.* 1996].

[22] See [D'Ariano *et al.* 1994, D'Ariano *et al.* 1995] [Schiller *et al.* 1996]. See also [Leonhardt 1995].

knowledge of all moments of the creation and annihilation operators. But the state under consideration is generally characterized by an infinite number of independent moments, i.e. we would need an infinite time in order to reconstruct the W-function. Hence, in the general case, we can only partially reconstruct the W-function.[23] Wallentowitz and Vogel[24] have proposed a simple alternative technique: a mixing of signal and coherent fields with controlled amplitude at a beam splitter may serve the reconstruction of the W-function and other distribution functions by photon counting alone.

One of the most amazing features of the tomographic method and its variants is that its application is not limited to the electromagnetic field. As a matter of fact, it has also been experimentally realized for obtaining the reconstruction of the quantum state of matter, in particular of atoms, molecules, ions in a Pauli trap, and even of an atomic beam.

15.4.3 Quantum state endoscopy of a cavity field

Schleich and co-workers[25] have proposed an alternative method to probe the quantum state of a single mode of electromagnetic field inside a cavity, called *quantum state endoscopy*. This technique has the advantage that it does not require us to take the field outside the cavity, as is the case for optical homodyne tomography.

The probe consists of a beam of two-level atoms that pass through a resonant quantized cavity field (see Sec. 13.7). The atoms enter the cavity after having been prepared in the coherent superposition

$$| \psi_A(0)\rangle = \frac{1}{\sqrt{2}} \left(| e\rangle + e^{\imath\phi} | g\rangle \right), \tag{15.67}$$

where $| g\rangle$ and $| e\rangle$ are the atomic ground and excited states, respectively. The method requires the measurement of the number of outgoing atoms that are found in the excited state as a function of the interaction time for four distinct internal phases $\phi = -\pi/2, 0, \pi/2, \pi$ of the initial state of the atoms. As we shall see, from these data it is possible to infer the initial field state

$$| \psi_F(0)\rangle = \sum_{n=0}^{\infty} c_n | n\rangle, \tag{15.68}$$

where $| n\rangle$ denotes a Fock state (see Subsec. 13.4.1) with n photons. In fact, we may calculate the probability $\wp_e(t, \phi)$ of finding the atom in the excited state after it has interacted for a time t with the cavity field. To this end, we make use of the Jaynes–Cummings atom–field interaction model (see Subsec. 13.7.2) and, in the rotating-wave approximation, we write the interaction Hamiltonian as (see Eq. (13.160))

$$\hat{H}_I = \hbar\varepsilon_0 \left(\hat{\sigma}_+ \hat{a} + \hat{\sigma}_- \hat{a}^\dagger \right), \tag{15.69}$$

[23] On this point see [Bužek *et al.* 1996], where quantifications of the precision in the reconstruction of the W-function and the minimal bound in order to discriminate between pure states and corresponding mixtures are also given.

[24] See [Wallentowitz/Vogel 1996].

[25] See [Bardroff *et al.* 1995] .

where $\hat{\sigma}_+$ and $\hat{\sigma}_-$ are atomic raising and lowering operators, respectively, and ε_0 is the vacuum Rabi frequency (13.157). The initial state of the compound system is then given by

$$|\Psi_{AF}(0)\rangle = \frac{1}{\sqrt{2}}\sum_{n=0}^{\infty} c_n \left(|e\rangle\,|n\rangle + e^{\imath\phi}\,|g\rangle\,|n\rangle\right). \tag{15.70}$$

After the interaction, the state of the combined system reads (see Prob. 15.10)

$$|\Psi_{AF}(t)\rangle = \sum_{n=0}^{\infty}\left(\psi_{e,n}(t)\,|e\rangle\,|n\rangle + \psi_{g,n+1}(t)\,|g\rangle\,|n+1\rangle\right) + \psi_{g,0}(t)\,|g\rangle\,|0\rangle, \tag{15.71}$$

where the probability amplitudes are

$$\psi_{e,n}(t) = \frac{1}{\sqrt{2}}\left[\cos(\varepsilon_n t)\,c_n - \imath\sin(\varepsilon_n t)\,e^{\imath\phi}c_{n+1}\right] \tag{15.72a}$$

and

$$\psi_{g,n+1}(t) = \frac{1}{\sqrt{2}}\left[-\sin(\varepsilon_n t)\,c_n + \cos(\varepsilon_n t)\,e^{\imath\phi}c_{n+1}\right], \tag{15.72b}$$

where $\varepsilon_n = \varepsilon_0\sqrt{n+1}$. From Eq. (15.71) we obtain the probability to find the atom, after the interaction time t, in the state $|e\rangle$, independent of the final field state, i.e. (see Prob. 15.11)

$$\wp_e(t,\phi) = \frac{1}{2}\left[1 - \frac{1}{2}|c_0|^2 + \frac{1}{2}\sum_{n=0}^{\infty}\cos(2\varepsilon_n t)\left(|c_n|^2 - |c_{n+1}|^2\right)\right.$$
$$\left. - \sum_{n=0}^{\infty}\sin(2\varepsilon_n t)\,\Im\left(c_n c_{n+1}^* e^{-\imath\phi}\right)\right]. \tag{15.73}$$

Our problem now is to infer the initial state of the field from the measured quantity $\wp_e(t,\phi)$. In order to do this, consider that the contributions to $\wp_e$ are (i) two time-independent terms, (ii) a contribution of cosines, and (iii) sines of the Rabi frequencies ε_n. The cosine contribution contains only the absolute values squares of the field probability amplitudes. Due to the initial superposition of the atom, the sine contribution brings in the relative phase between neighboring photon probability amplitudes. Therefore, in order to reconstruct the phases of c_n we have to obtain the value of the product $c_n c_{n+1}^*$ from $\wp_e(t,\phi)$. For this purpose, we use the above expression for different values of ϕ and define the complex function

$$f_c(t) = \wp_e\left(t,\frac{\pi}{2}\right) - \wp_e\left(t,-\frac{\pi}{2}\right) + \imath\left[\wp_e(t,\pi) - \wp_e(t,0)\right]$$
$$= \sum_{n=0}^{\infty}\sin(2\varepsilon_n t)\,c_n c_{n+1}^*, \tag{15.74a}$$

which only contains cross terms $c_n c_{n+1}^*$, and the real function

$$f_r(t) = \wp_e\left(t,-\frac{\pi}{2}\right) + \wp_e(t,0) + \wp_e\left(t,\frac{\pi}{2}\right) + \wp_e(t,\pi) - 2$$
$$= \sum_{n=0}^{\infty}\cos(2\varepsilon_n t)\left(|c_n|^2 - |c_{n+1}|^2\right) - |c_0|^2, \tag{15.74b}$$

which in turn only contains the photon number probabilities. Finally, the problem is reduced to a mathematical inversion of Eqs. (15.74) in order to obtain

$$e_n = c_n c_{n+1}^* \quad \text{and} \quad g_n = |c_n|^2 - |c_{n+1}|^2 \tag{15.75}$$

from the measured functions $f_c(t)$ and $f_r(t)$, and then to derive from Eqs. (15.75) the complex numbers c_n. This allows us to completely reconstruct the initial cavity-field state (15.68).

15.4.4 Informational completeness

The problem of quantum state reconstruction raises the question of the extent to which an observable's measurement informs us about the state of a given system. This, in turn, brings us back to the relationship between sharp and unsharp observables (see Sec. 9.10). Stated in other terms, we may ask ourselves whether the probability distributions of a certain set of observables are sufficient to determine the state of a quantum system, i.e. to discriminate between different states. Such a question leads naturally to the concept of *informational completeness*: a family of self-adjoint operators $\{\hat{O}_k\}$ is said to be informationally complete if

$$\text{Tr}\left[\hat{\rho}\hat{O}_k\right] = \text{Tr}\left[\hat{\rho}'\hat{O}_k\right], \tag{15.76}$$

$\forall k$, implies that $\hat{\rho} = \hat{\rho}'$ on a Hilbert space $\mathcal{H}$. It is possible to show that sharp observables are not informationally complete.

In fact, take as an example the relationship between the momentum $\hat{p}_x$ and the position $\hat{x}$ of a one-dimensional system. The famous *Pauli problem*[26] is summarized in the following question: do the position and momentum distributions determine uniquely the wave function? We can now reformulate it in the language of informational completeness of the canonically conjugate position and momentum observables: are sharp complementary observables informationally complete? As we shall prove now by making use of a counterexample, this is not the case.

Take[27] a normalized wave function $\psi(x) \in L^2(x)$, with

$$\psi(x) = |\psi(x)|e^{\imath\phi(x)}, \tag{15.77}$$

where

$$|\psi(x)| = |\psi(-x)|. \tag{15.78}$$

The function $\phi(x)$ satisfies

$$0 \le \phi(x) < 2\pi \quad \text{and} \quad \phi(x) + \phi(-x) \neq \text{constant} \, (\text{mod}\, 2\pi). \tag{15.79}$$

Then, the wave function

$$\psi'(x) = \psi^*(-x) = |\psi(x)|e^{-\imath\phi(-x)} \tag{15.80}$$

[26] See [*Pauli* 1980, 17]

[27] See [Prugovečki 1977].

represents a state different from $\psi(x)$ (i.e. $\hat{P}_{\psi'} \neq \hat{P}_{\psi}$) but with

$$|\psi'(x)|^2 = |\psi(x)|^2 \quad \text{and} \quad |\tilde{\psi}'(k_x)|^2 = |\tilde{\psi}(k_x)|^2, \tag{15.81}$$

where $\tilde{\psi}(k_x)$ is the Fourier transform of $\psi(x)$. Therefore, the pair $\hat{p}_x$, $\hat{x}$ is not able to distinguish between the states ψ and ψ'.

We recall, instead, that unsharp observables are the result of a smearing operation on sharp ones. Now, this smearing operation on sharp observables, say $\hat{\mathbf{p}}$ and $\hat{\mathbf{r}}$, can be understood as a coarse-graining operation on a set of sharp observable. This operation can have an informationally complete refinement as a result. This is nicely illustrated by a careful inspection of Eq. (15.65): the knowledge of $\wp(X,\theta)$ for only two (sharp) values of θ, namely $\theta = 0, \pi/2$, is not sufficient to reconstruct the W-function. In order to perform the inverse Radon transform, it is necessary to know $\wp(X,\theta)$ for a sufficiently large number of θ values.[28] In other words, the probability distributions of sharp momentum and position (or relative quadratures) do not cover the whole phase space, whereas they do if they are taken to be unsharp. Summarizing, the following theorem can be proved:

Theorem 15.3 (Informational completeness of unsharp observables) *A set of unsharp complementary observables can be informationally complete.*

15.5 The nature of quantum states

What we have learnt from this chapter is that we cannot have a direct evidence of, i.e. directly measure, a quantum state of a single system. Our experience is only connected with the experimental values of observables, and any time we measure an observable we can only have a partial experience of a system under a certain perspective but we can never have a complete experience that would be represented by an observation of the state vector, which is – in a quantum-mechanical sense (see Sec. 16.1) – a complete description of the system (see Subsec. 3.1.2). In other words, the quantum state is *not* an observable in the classical sense (see also Sec. 5.2). However, since this feature of the quantum state is not due to subjective ignorance but rather to an intrinsic characteristic of the microscopic world (see Sec. 1.4 and Ch. 16), there are no definitive reasons to deny the reality of a quantum state. We have indeed already seen (in Sec. 9.7) that the wave function can be considered as a form of reality that is not reducible to the events. We understand with *event*, in the simplest case, the click of a detector, that is, a local result of a physical interaction that can be directly experienced.

[28] The exact reconstruction may be achieved only when $\wp(X,\theta)$ is known for any θ, i.e. for an infinite number of values. However, finite-resolution quantum tomography may be achieved even when $\wp(X,\theta)$ is known for a finite number of θ values, if this number is sufficiently large.

We may therefore ask the following question: how is it possible to assign an ontological reality to something (the state or the wave function) which goes beyond the observables that we measure or beyond the events that actually happen?[29]

Indeed,[30] any quantum state can be conceived of as the result of a preparation. As a matter of fact, both classically and quantum-mechanically, it is an equivalence class of preparations, in the same way that any observable is an equivalence class of premeasurements (see Subsec. 9.10.1). Moreover, any property is an equivalence class of events. This is again true both classically and quantum-mechanically. In other words, properties are only assigned to a system conditionally upon the occurrence of certain events (the only reality that we directly experience). The substantial difference, is that classically we also assume that the state is a property, whereas quantum-mechanically we cannot. The reason is that classically the state is a collection of properties and can therefore be thought of as a property itself. We may again conclude that reality is an interplay between events and states, from which events, under certain environmental conditions and contexts (see Ch. 14), emerge.

Summary

In this chapter we have dealt with the problem of the measurement of the wave function or of the state of the system:

- The *protective measurement* of the wave function introduced by Aharonov and coworkers is actually the measurement of the Hamiltonian of the system.
- We have proved the *impossibility to clone* a quantum state and that any cloning or measurement of a state would represent a violation of the unitarity of quantum evolution.
- Moreover, we have shown that any measurement is an *irreversible process* and that a reversible process cannot extract information from a given system.
- We have then presented a general formalism for *measuring the Wigner function* and introduced some techniques (optical homodyne tomography and quantum state endoscopy) that enable us to obtain this result.
- Finally, we have proposed that, though a quantum state cannot be measured, it should be conceived as *a form of reality* even though not at the same ontological level as events.

Problems

15.1 Restate the criticism of the measurement of the wave function presented in Sec. 15.1 in the same mathematical form as that originally employed by Aharonov and coworkers.

[29] See [Auletta/Tarozzi 2004a, Auletta/Tarozzi 2004b].
[30] See [Auletta *et al.* 2008].

15.2 Why cannot the argument presented in Subsecs. 15.3.1 and 15.3.2 be developed by making use of a creation operator?

15.3 Prove that the statements presented in Subsec. 15.3.3 are equivalent.

15.4 Find the relation between the displacement operators defined in Eqs. (13.68) and (15.43).
(*Hint*: Make use of the explicit expressions (4.73) of the annihilation and creation operators.)

15.5 Prove that $\hat{\Pi}_{xp}$ is a Hermitian operator.

15.6 Prove that $\hat{\Pi}_{xp}^2 = \hat{I}$.

15.7 Prove Eq. (15.49c).

15.8 Prove Eq. (15.57).
(*Hint*: Make use of Eq. (15.53) written for $t = 0$.)

15.9 Derive Eq. (15.65).
(*Hint*: Make use of the change of variable $\xi_r = -\frac{1}{\sqrt{2}}\zeta \sin\theta$ and $\xi_i = \frac{1}{\sqrt{2}}\zeta \cos\theta$.)

15.10 Derive Eq. (15.71).
(*Hint*: Take advantage of Eq. (13.174) and generalize the procedure outlined in Subsec. 13.7.3.)

15.11 Derive Eq. (15.73).

15.12 Prove that $|\tilde{\psi}'(k_x)|^2 = |\tilde{\psi}(k_x)|^2$ in Eq. (15.81).

Further reading

Freyberger, M., Bardroff, P. J., Leichtle, C., Schrade, G., and Schleich, W. P., The art of measuring quantum states. *Physics World*, **10.11** (1997), 41–45.

Leonhardt, Ulf, *Measuring the Quantum State of Light*, Cambridge: Cambridge University Press, 1997.

Schleich, W. P. and Raymer, M. (eds.), *Quantum State Preparation and Measurement*, *Journal of Modern Optics*, **44** (1997), issues 11–12.

16 Entanglement: non-separability

In classical mechanics, two (or more) systems that do not interact – that do not exert any kind of *force* on each other – are completely separated (see Sec. 1.1), i.e. experiments performed locally on one of them cannot in any way influence experiments performed locally on the other. Quantum mechanics, on the other hand, admits that there can be a form of interdependence even in absence of physical interaction. This type of correlation goes under the name of *entanglement* (see Def. 5.1: p. 183). We already know that entanglement means non-factorizability of the state vector describing a system with at least two different degrees of freedom. In this chapter, we shall see that this may be the case also for systems that are very far away (space-like distant) from each other and we shall learn many interesting consequences of this state of affairs. Recent developments show that one may entangle even particles that have never interacted with each other.

The starting point of the following analysis is necessarily represented by the pioneering paper of Einstein, Podolsky, and Rosen,[1] who inaugurated an incredible series of theoretical and experimental investigations right up to the present day that have confirmed the predictions of quantum mechanics. The "irony" of this history is that Einstein, Podolsky, and Rosen, EPR for short, aimed to present a definitive proof of the inconsistency of quantum mechanics relative to classical physical theory. In fact, they initiated one of the most prolific fields of modern science.

After discussing the EPR argument (in Sec. 16.1) and the answers that have been given by Bohr and Schrödinger, in Sec. 16.2 we explain how to experimentally produce a relatively simple entangled state – Bohm's version of the EPR state. In Sec. 16.3 we introduce hidden-variable theories. In Sec. 16.4 we review Bell's important contributions about hidden-variable theories and the quantum non-separability problem (the so-called Bell inequalities). In Sec. 16.5 we describe the main experimental tests on the violations of Bell inequalities. In Sec. 16.6 we introduce an important result: the possibility of entangling particles coming from different sources, while in Sec. 16.7 we show how a conflict between quantum mechanics and the separability principle may arise without making use of Bell-like inequalities. In Sec. 16.8 we point out that, even though quantum mechanics violates the separability principle, it is not in contrast with relativistic locality. In Sec. 16.9 a few further developments about the exact determination of quantum violation of separability bounds are introduced, and, finally, in Sec. 16.10 some concluding remarks follow.

[1] See [Einstein *et al.* 1935].

16.1 EPR

In their ground-breaking paper, EPR formulated some general principles that any physical theory should satisfy. In particular, EPR distinguished between *objective reality* and *physical concepts*, the latter having to correspond to the former. When judging a physical theory that makes use of physical concepts, one should enquire about both the correctness and completeness of the theory.

16.1.1 The logical structure of the EPR argument

Following EPR, the *correctness* consists of the degree of agreement between the conclusions of the theory and human experience – the objective reality. However, EPR are mainly interested in the *completeness* of quantum mechanics, and hence formulated the following definition:

Definition 16.1 (Completeness) *A theory is complete if every element of objective reality has a counterpart in it.*

It is evident that correctness together with completeness (in analogy with the completeness theorem of mathematical logic) establishes an isomorphism between physical theory and objective reality.

The aim of the EPR article, however, is rather to show the *incompleteness* of quantum mechanics in the sense of its inability to give a satisfactory explanation of entities which are considered fundamental – in a word, it a "disproof" and not a positive proof.[2]

The core of the argument is the *separability principle*, which we can express as follows (see also Sec. 1.1):

Principle 16.1 (Separability) *Two dynamically independent systems possess their own separate state.*

The separability principle is the principle of individuation for physical systems.[3] Practically, it identifies the terms *insulated* and *dynamically independent*: in other words, it does not acknowledge a form of interdependence between systems other than the dynamical or

[2] In general, it is impossible to find a positive proof of an empirical theory (one can instead accumulate a certain amount of evidence in favor of it). In particular, a positive proof of the completeness of a certain physical theory would require a total knowledge of reality, and, as a consequence, an infinite time. On the other hand, theories can be disproved by experience and (even thought) experiments. This type of epistemology is the so-called *falsificationism*, and is due to the contributions of Peirce and Popper.

[3] See [Howard 1985].

the causal local form of interaction. Therefore, it is important to distinguish the problem of locality carefully – i.e. the existence of bounds in the transmission of signals and physical effects – from that of separability, which concerns only the impossibility of a correlation between separated systems *without dynamical and causal connections*. Part of the EPR argument is that, in the absence of physical interactions, the systems have no relation at all.

Furthermore, making use of the separability principle, EPR state a sufficient condition for the reality of observables, which can be formulated as follows:

Principle 16.2 (Criterion of physical reality) *If, without in any way disturbing a system, we can predict with probability equal to unity the value of a physical quantity, then, independently from our measurement procedure, there exists an element of the physical reality corresponding to this physical quantity.*

It is evident that, being a sufficient condition of physical reality, this criterion is strictly related to the definition of correctness, in the sense that only a correct theory can perform predictions of this type.

After having defined completeness and correctness, the aim of EPR is to show that, assuming separability and the sufficient condition of reality, quantum mechanics is not complete (see also Subsec. 9.2.2). In logical terms, according to the EPR argument, for quantum mechanics the following statement holds:

$$[(\text{Suff. Cond. Reality}) \wedge (\text{Separability})] \Longrightarrow \neg\text{Completeness}, \qquad (16.1)$$

where $\wedge$ is the logical AND, $\neg$ is the logical negation, and the arrow is the sign for logical implication.[4] Before entering into details, it is very important to understand the abstract logic form of the argument. According to EPR, the incompleteness of quantum mechanics would be a consequence of both separability and sufficient condition of reality. The argument, as said, has the logical structure of an implication. In order to invalidate it, it suffices to show that *at least one* of the two assumptions is false. In fact, if one of the two assumptions is false, also their joint assertion (Suff. Cond. Reality $\wedge$ Separability) is false, i.e. the antecedent of the implication is false. However, if the antecedent of an implication is false, its consequent (i.e. $\neg$Completeness) may be indifferently true or false, being, under this condition, the implication always true. It this case, the argument would prove neither the incompleteness, nor the completeness of quantum mechanics, and be finally inconclusive. As we shall see below, Schrödinger argued against the principle of separability, while Bohr tried to reject the sufficient condition of reality.

The argument of EPR is then structured as follows. From (i) Def. 16.1, (ii) Pr. 16.2, and (iii) the fact that, according to quantum mechanics, two non-commuting observables cannot simultaneously have definite values, it follows that:

[4] The implication $p \Longrightarrow q$, where p and q are arbitrary propositions, may be defined as: p is false OR q is true. In other words, the implication is false only in the case in which p is true AND q is false.

either (1) the quantum-mechanical description of reality given by the wave function is not complete, *or* (2) when the operators corresponding to two physical quantities do not commute, the two quantities cannot have simultaneous reality.[5]

At this stage, the separability concept is still hidden inside the words "[...] without in any way disturbing [...]," contained in Pr. 16.2 The importance of this concept will be made clear in what follows. As we shall see below, assuming that the wave function is a complete description of reality, EPR show that two non-commuting observables can have simultaneous reality and from this conclude that quantum mechanics cannot be complete.

16.1.2 A counterexample: the EPR state

For the structure of the argument, EPR must provide an example of wave function for which two non-commuting observables can have simultaneous reality, and it is here that separability comes into play. Let us, for this purpose, consider a one-dimensional system $\mathcal{S}$ made of two subsystems $\mathcal{S}_1$ and $\mathcal{S}_2$, say two particles which interact during the time interval between t_1 and t_2, after which they no longer interact. We take into account the momenta

$$\hat{p}_x^{(1)} = -\imath\hbar\frac{\partial}{\partial x_1} \quad \text{and} \quad \hat{p}_x^{(2)} = -\imath\hbar\frac{\partial}{\partial x_2} \tag{16.2}$$

on $\mathcal{S}_1$, with eigenfunctions $\varphi_p(x_1)$, where x_1 is a position variable used to describe system $\mathcal{S}_1$, and on $\mathcal{S}_2$, with eigenfunctions $\psi_p(x_2)$, x_2 being the position variable describing $\mathcal{S}_2$, respectively. Let us assume that the compound system is described by the wave function

$$\Psi(x_1, x_2) = \int_{-\infty}^{+\infty} dp\,\psi_p(x_2)\varphi_p(x_1). \tag{16.3}$$

We wish to stress that EPR are considering time-independent wave functions, that is, a system at a certain (not specified) instant of time. It is clear that $\mathcal{S}_1$ and $\mathcal{S}_2$ are entangled (see Subsec. 5.5.1), even if, as we shall see below, the concept of entanglement was formally introduced by Schrödinger after the publication of the EPR article. Indeed, the eigenfunctions of $\hat{p}_x^{(1)}$ with eigenvalue p in the position representation are (see Sec. 2.2)

$$\varphi_p(x_1) = \frac{1}{\sqrt{2\pi}} e^{\frac{\imath}{\hbar} p x_1}, \tag{16.4}$$

while for $\mathcal{S}_2$ we have

$$\psi_p(x_2) = \frac{1}{\sqrt{2\pi}} e^{-\frac{\imath}{\hbar}(x_2 - x_0)p}, \tag{16.5}$$

where x_0 is a constant and $\psi_p(x_2)$ is the eigenfunction in the position representation of the momentum $\hat{p}_x^{(2)}$, corresponding to the eigenvalue $-p$ of the second particle's momentum.

[5] This argument has the logical structure of an XOR: the propositions (1) and (2) can never be both true, as well as never both false.

Now we proceed as follows:

(a) We locally measure the momentum on $\mathcal{S}_1$ and suppose that we find the eigenvalue p'.
(b) Therefore, the state (16.3) reduces to

$$\psi_{p'}(x_2)\varphi_{p'}(x_1). \tag{16.6}$$

(c) Then, it is evident that $\mathcal{S}_2$ must be in state $\psi_{p'}$ and this result can be predicted with absolute certainty.
(d) However, we have obtained this result by not disturbing $\mathcal{S}_2$ (assumption of separability).
(e) Then, as a consequence of (c) and (d) and of the sufficient condition of reality, $\hat{p}_x^{(2)}$ is an element of reality.

Note that steps (a)–(c) are purely quantum-mechanical. Only steps (d)–(e) are connected to the EPR argument.

However, if we had chosen to consider another observable pertaining to $\mathcal{S}_1$, say $\hat{x}_1$, whose eigenfunctions are $\varphi_x(x_1)$, whereas $\psi_x(x_2)$ are the eigenfunctions of the observable $\hat{x}_2$ of $\mathcal{S}_2$, then we would have written Ψ as

$$\Psi(x_1, x_2) = \frac{1}{\sqrt{2\pi}} \int_{-\infty}^{+\infty} dx\, \psi_x(x_2)\varphi_x(x_1), \tag{16.7}$$

where the eigenfunction φ_x corresponding to the eigenvalue x of $\hat{x}_1$ is

$$\varphi_x(x_1) = \delta(x_1 - x), \tag{16.8}$$

and (see Prob. 16.1)

$$\psi_x(x_2) = \frac{1}{\sqrt{2\pi}} \int_{-\infty}^{+\infty} dp\, e^{-\frac{i}{\hbar}(x - x_2 + x_0)p} = h\delta(x - x_2 + x_0), \tag{16.9}$$

corresponding to the eigenvalue $x + x_0$ of $\hat{x}_2$. Let us now repeat the previous procedure for the position measurement:

(a′) We locally measure the position on $\mathcal{S}_1$ and find the eigenvalue x'.
(b′) Now it is clear that the state (16.7) reduces to

$$\psi_{x'}(x_2)\varphi_{x'}(x_1). \tag{16.10}$$

(c′) Then, it is evident that $\mathcal{S}_2$ must be in state $\psi_{x'}$ and this result can be predicted with absolute certainty.
(d′) However, we have obtained this result by not disturbing $\mathcal{S}_2$ (assumption of separability).
(e′) Then, as a consequence of (c′) and (d′) and of the sufficient condition of reality, $\hat{x}^{(2)}$ is an element of reality.

Conclusions (e) and (e′) look incompatible on the basis of the fact that position and momentum observables of particle 2 do not commute. However, going back to the alternatives (1) and (2) of the previous subsection, EPR have in this way shown that, assuming that (1) is false (the quantum-mechanical description of reality is not complete), (2) is proved to be false as well (both $\hat{p}_x^{(2)}$ and $\hat{x}^{(2)}$ therefore have simultaneous reality). Then, the previous assumption must be rejected, and (1) must be true. Therefore, according to the EPR argument, quantum mechanics cannot be complete, i.e. the wave functions (16.3) and (16.7) cannot be considered a complete description of the compound system.

16.1.3 Bohr's and Schrödinger's criticism of EPR

As we have said, due to the abstract logical structure (16.1) of the argument, it is evident that, in order to reject the conclusion that quantum mechanics is incomplete, one needs to show the inconsistency of separability with quantum mechanics or the failure of the sufficient condition of reality in a quantum framework. In fact, they are the only non quantum-mechanical assumptions in steps (a)–(e) and (a′)–(e′). EPR themselves anticipated a possible objection [Einstein *et al.* 1935]. In their words:

> One could object to this conclusion on the grounds that our criterion of reality is not sufficiently restrictive. Indeed, one would not arrive at our conclusion if one insisted that two or more physical quantities can be regarded as simultaneous element of reality *only when they can be simultaneously measured or predicted.* On this point of view, since either one or the other, but not both simultaneously, of the quantities $\hat{p}$ and $\hat{x}$ can be predicted, they are not simultaneously real. This makes the reality of $\hat{p}$ and $\hat{x}$ depend upon the process of measurement carried out on the first system, which does not disturb the second system in any way. No reasonable definition of reality could be expected to permit this.

In the same year (1935), Bohr rejected the sufficient condition of reality, while Schrödinger rejected the separability principle.

Bohr,[6] as partly anticipated by EPR themselves, criticized the EPR argument by pointing out that, even if the EPR thought-experiment excludes any direct physical interaction of the system with the measuring apparatus, the measurement process has an essential influence on the conditions on which the very definition of the physical observables in question rests. And these conditions must be considered as an inherent element of any phenomenon to which the term "physical reality" can be unambiguously applied. Bohr acknowledged that it is possible to determine experimental arrangements such that the measurement of the position or of the momentum of one particle automatically determines the position or the momentum of the other. However, such experimental arrangements for measuring momentum and position are incompatible with each other. As we see, the central point of Bohr's criticism is that it is not possible to assign a reality to observables of quantum systems *independently* of the experimental context in which we decide to interact with them. We wish to point out that the experimental procedures through which observables are determined – we have indeed defined observables as equivalence classes of premeasurements (see Subsec. 9.10.1 and Sec. 15.5) – are physical operations. Therefore, nothing prevents us from considering observables as elements of reality notwithstanding the fact that they may not commute. In this way, our interpretation assumes an important instance of EPR

[6] See [Bohr 1935a, Bohr 1935b] and also [*Jammer* 1974, 195–97].

and frames it in a wider ontological context, in which state, observables and properties show an increasing degree of determination. We are always allowed to assume the reality of the state of a quantum system, with all provisos about the word "reality" we have already spoken about and provided that, at least in principle, it is possible to prepare the system in that state. Moreover, it is also allowed to speak of observables as elements of reality, at least provided that we are able to show a (possible) context of measurement, that is, a premeasurement (a suitable coupling). Here, we are also somehow accepting Bohr's instance, even though, in our opinion, it is not sufficient to demolish the EPR argument. Indeed, the fact that a certain observable is an element of reality does not imply that also one of the properties (that is, one of its eigenvalues) is also a reality. In order to make this inference, we need an event (see again Sec. 15.5). This is the essence of the developments of this chapter, starting from Schrödinger's contribution, which we now briefly sketch.

In a series of articles,[7] Schrödinger answered EPR by introducing the concept of entanglement in quantum mechanics. As we shall see in this chapter, it is entanglement that (in the absence of events) prevents us in some situations attributing properties to a system or its subsystems. But, provided that there is an event, it is still entanglement that allows the attribution of properties in a way that is classically unknown.[8] However, we wish also to stress that Schrödinger also thought that entanglement was not a phenomenon with ontological pregnancy – in his opinion it was too far away from our common sense to be real. We do not wish now to enter into an analysis of Schrödinger's position. We only recall that his answer to EPR is developed in the same series of papers where he proposes for the first time the *Gedankenexperiment* of the Schrödinger cat as a possible bizarre consequence of entanglement. Since we have already discussed this particular aspect of quantum theory (see Secs. 9.3 and 14.5), in the following we wish to develop a deeper understanding of entanglement with particular regard to the separability problem.

16.2 Bohm's version of the EPR state

As a consequence of Bohr's and Schrödinger 's answers outlined above, we have two possible ways of solving what has been called the EPR paradox, i.e. we either accept the incompleteness of quantum mechanics (according to EPR) or we admit a violation of separability (following the pioneering work of Schrödinger). If these different solutions to this apparent paradox have a physical meaning, they must be somehow distinguishable and therefore lead to different predictions that can be experimentally tested. As we shall see, this is in fact the case. However, in order to arrive at a clear determination of the predictions to be tested, there was a long way to go: as a matter of fact, it took almost 40 years from the EPR paper to the first experimental tests of the above alternative. In this

[7] See [Schrödinger 1935].

[8] Incidentally, in finding entanglement as a solution of a possible conflict between a thought-experiment and quantum-mechanical laws, Schrödinger was performing an *abduction* in the meaning given to it by [*Peirce CP*, 2.96].

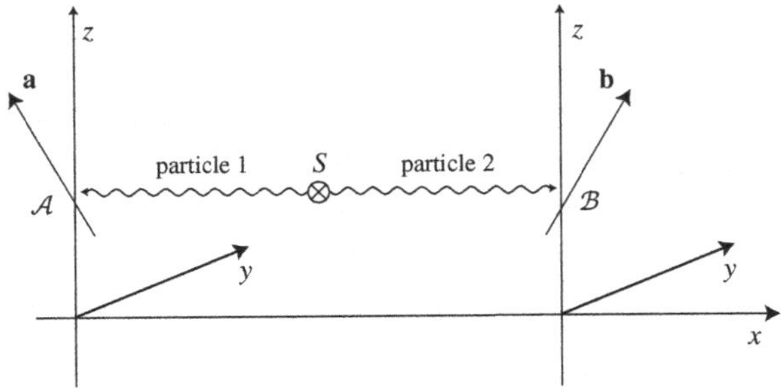

Figure 16.1 Schematic overview of the EPR – Bohm experiment. Two particles are generated (e.g. by decay of a spin-zero particle) from a common source $\mathcal{S}$, and their spin is measured with two apparata $\mathcal{A}$ and $\mathcal{B}$ at different angles with the z direction, **a** and **b**.

respect, David Bohm has given important contributions, leading both to a better definition of the problem and to an attempt to find a theory that could overcome the supposed incompleteness of quantum mechanics (a so-called hidden-variable theory). In this section, we shall consider the first contribution, while Sec. 16.3 will be devoted to hidden-variable theories.

16.2.1 Theory

The formulation of the original EPR thought experiment proposed by Bohm[9] deals with discrete observables, instead of continuous ones such as position and momentum. This step was originally understood as a further simplification of the EPR argument. It is an important point because it clearly shows that the non-local features of quantum theory are a consequence of entanglement, and because it is also experimentally realizable.

Consider two particles with spin-1/2 that are in a state in which the total spin is zero (singlet state). They can be produced by a single particle decay. After a time t_0 the two particles begin to separate and at time t_1 they no longer interact (see Fig. 16.1). On the hypothesis that they are not disturbed, the law of angular momentum conservation guarantees that they remain in a singlet state. Considering the projection of the spin along the z direction, the singlet state may be written in the form (see Eq. (6.194))

$$|\Psi_0\rangle = \frac{1}{\sqrt{2}}(|\uparrow\rangle_1 \otimes |\downarrow\rangle_2 - |\downarrow\rangle_1 \otimes |\uparrow\rangle_2), \tag{16.11}$$

where the subscripts 1 and 2 refer to the particles. Now, as we know, if a measurement of the spin component along the z direction of particle 1 leads to a result $+1/2$, that of particle 2 along the same direction must give the value $-1/2$, and vice versa. This means that $|\Psi_0\rangle$ is an eigenket of the operator $\hat{\sigma}_{1z}\hat{\sigma}_{2z}$ (see Prob. 16.2). This result remains true for the three

[9] See [*Bohm* 1951, 614–23].

possible directions of the spin. In other words, the state $|\Psi_0\rangle$ is rotationally invariant. In order to see this, let us write it in terms of the z-component eigenspinors as

$$|\Psi_0\rangle = \frac{1}{\sqrt{2}}\left[\begin{pmatrix}1\\0\end{pmatrix}_1 \otimes \begin{pmatrix}0\\1\end{pmatrix}_2 - \begin{pmatrix}0\\1\end{pmatrix}_1 \otimes \begin{pmatrix}1\\0\end{pmatrix}_2\right]. \tag{16.12}$$

Then, $|\Psi_0\rangle$ turns out to be also an eigenvector of $\hat{\sigma}_{1x}\hat{\sigma}_{2x}$ and $\hat{\sigma}_{1y}\hat{\sigma}_{2y}$. For example,

$$\begin{aligned}(\hat{\sigma}_{1y}\hat{\sigma}_{2y})\,|\Psi_0\rangle &= \frac{1}{\sqrt{2}}\begin{bmatrix}0 & -\imath\\ \imath & 0\end{bmatrix}_1\begin{bmatrix}0 & -\imath\\ \imath & 0\end{bmatrix}_2\\ &\times\left[\begin{pmatrix}1\\0\end{pmatrix}_1 \otimes \begin{pmatrix}0\\1\end{pmatrix}_2 - \begin{pmatrix}0\\1\end{pmatrix}_1 \otimes \begin{pmatrix}1\\0\end{pmatrix}_2\right]\\ &= \frac{1}{\sqrt{2}}\left[\begin{pmatrix}0\\ \imath\end{pmatrix}_1 \otimes \begin{pmatrix}-\imath\\0\end{pmatrix}_2 - \begin{pmatrix}-\imath\\0\end{pmatrix}_1 \otimes \begin{pmatrix}0\\ \imath\end{pmatrix}_2\right]\\ &= -|\Psi_0\rangle.\end{aligned} \tag{16.13}$$

Therefore, if the y component of the spin of particle 1 is measured, the y component of the spin of particle 2 must have the opposite value relative to the spin of the first particle. This state is perhaps the simplest discrete version of the EPR states (16.3) and (16.7), where the role of position and momentum is played by different components of the spin. Moreover, it is an ideal candidate for the possible realization of experimental tests.

16.2.2 Preparing a singlet state

An interesting question, which is preliminary to all further investigation, is: how can one prepare a singlet state? It is instructive to see how such a state can be prepared in a different context – two two-level atoms interacting with a single-mode of the electromagnetic field in a cavity.[10] One can prepare a singlet state by allowing the two atoms – 1 and 2 – initially in their excited ($|\uparrow\rangle_1$) and ground ($|\downarrow\rangle_2$) states, respectively, to interact with the resonant cavity mode in the vacuum state. After the preparation of the entangled atomic state, the cavity mode is left in its original state. The preparation takes place in two steps. First, we send atom 1, initially in the excited state $|\uparrow\rangle_1$ through the cavity in its vacuum state $|0\rangle_c$ in a direction $\mathbf{r}_1$. After 1 has left the cavity, atom 2, prepared in its ground state $|\downarrow\rangle_2$, is sent through the cavity in a different direction $\mathbf{r}_2$ (see Fig. 16.2). The interaction Hamiltonian in a rotating frame at the cavity mode frequency and in the rotating wave approximation is given by the Jaynes–Cummings model (see Eq. (13.162c))

$$\hat{H}_I = \hbar\varepsilon(\hat{\sigma}_+\hat{a} + \hat{a}^\dagger\hat{\sigma}_-), \tag{16.14}$$

where

$$\hat{\sigma}_+ = (\hat{\sigma}_-)^\dagger = |\uparrow\rangle\langle\downarrow| \tag{16.15}$$

[10] See [Cirac/Zoller 1994]. See also [Phoenix/Barnett 1993].

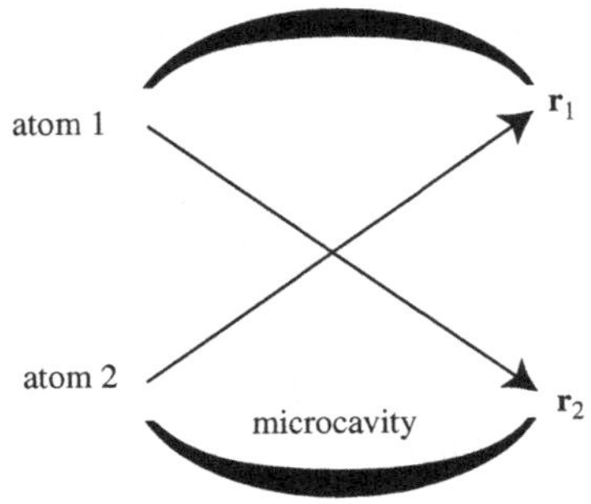

Figure 16.2 Preparation of a singlet state. Two atoms (1 and 2) are successively sent through an empty cavity in their initial excited and ground states, respectively. The velocity of the two atoms is selected in such a way that, after the interaction, the cavity is left in its vacuum state.

is the atomic raising operator, valid for each atom. The coupling parameter between the cavity and the atoms is ε. Here we take advantage of an important property of the resonant Jaynes–Cummings model, which is the conservation of the excitation number. In other words, the operator (see Prob. 16.3)

$$\hat{N}_e = \hat{a}^\dagger \hat{a} + |\uparrow\rangle\langle\uparrow| \tag{16.16}$$

is a conserved quantity (see Subsec. 8.1.2). In particular, the state $|\downarrow\rangle|0\rangle_c$ does not change during the interaction, and the states $|\uparrow\rangle|0\rangle_c$ and $|\downarrow\rangle|1\rangle_c$ (where $|1\rangle_c$ describes the presence of one photon in the cavity) will experience vacuum Rabi oscillations (see Subsec. 13.7.3), which will depend on the interaction time t_j of particle j ($j = 1, 2$) with the cavity field and therefore on the atomic velocity $v_j = L_c/t_j$—where L_c is the effective length of the cavity. We suppose that v_1 has been selected in such a way that atom 1 undergoes $1/4$ of a Rabi oscillation (a so-called $\pi/2$ pulse).

Therefore, the state of the combined system (atom 1 + atom 2 + cavity mode), after the first atom has crossed the cavity, will be given by

$$|\Psi\rangle = \frac{1}{\sqrt{2}}(|\uparrow\rangle_1|0\rangle_c - |\downarrow\rangle_1|1\rangle_c)|\downarrow\rangle_2. \tag{16.17}$$

The state of the system after atom 2 has crossed the cavity can be calculated if one takes into account that $|0\rangle_c|\downarrow\rangle_2$ remains unchanged during the interaction. By selecting v_2 in such a way that the state $|1\rangle_c|\downarrow\rangle_2$ performs half a Rabi cycle (a π pulse), the final state of the system may be written as

$$\begin{aligned} |\Psi\rangle_f &= \frac{1}{\sqrt{2}}(|\uparrow\rangle_1|\downarrow\rangle_2|0\rangle_c - |\downarrow\rangle_1|\uparrow\rangle_2|0\rangle_c) \\ &= |\Psi_0\rangle_{12}|0\rangle_c, \end{aligned} \tag{16.18}$$

and therefore the singlet state (16.11) for the two-atom system has been prepared.

16.3 HV theories

16.3.1 Preliminary definitions

As we have said, the aim of EPR was to show that quantum mechanics is incomplete and to urge for a new deterministic theory able to integrate the supposed statistical formulation of quantum theory. In other words, EPR implicitly believed that there exist some variables that are presently unknown to us and that are able to explain the (apparently random) phenomena described by quantum mechanics. These variables have been therefore called *hidden variables*.

First, we give a basic definition of hidden variables (HV):[11] if a given theory T contains a set of observables $\{\hat{O}\}$ which describe a physical system $\mathcal{S}$, and there are some variables $\{\lambda_{\rm HV}\}$ about $\mathcal{S}$ which are not experimentally accessible within the framework of T and the values of each $\hat{O}$ can be obtained by some averaging operation on the values of some $\lambda_{\rm HV}$, then the $\{\lambda_{\rm HV}\}$ are called HVs with respect to T.

Now we define an HV theory for quantum mechanics:[12] if for each quantum system there exists a measure space Ω together with a finite measure μ (normalized, so that $\mu(\Omega) = 1$) on Ω such that every state $\hat{\rho}$ of an arbitrary quantum system can be represented as a mixture

$$\hat{\rho} = \int_{\Omega} d\mu(\lambda_{\rm HV})\hat{\rho}_{\lambda_{\rm HV}}(\hat{P}) \tag{16.19}$$

of dispersion-free states $\hat{\rho}_{\lambda_{\rm HV}}(\hat{P})$ for all PVM $\hat{P}$, then we call the theory which admits such mixtures a quantum HV theory. A *dispersion-free state* may be defined as a state such that all relevant observables have zero variance (see Subsec. 2.3.1). Therefore, the expectation value of any observable coincides with one of its eigenvalues. If this were the case, all observables of a given system would be simultaneously measurable, and the quantum commutation relations and their consequence – the uncertainty principle – should be rejected.

In other words, an HV theory supposes that the state of a system and its observables are all – in a classical sense – perfectly determined, and hence also their dynamics. In this sense, it is a classically deterministic theory. An HV theory is also a complete theory in the sense of EPR (see Def. 16.1: p. 568): it would be characterized by the predictive exhaustivity of all possible measurement outcomes.

The development of a mature HV theory is mainly due to David Bohm. In so doing, Bohm also further developed some ideas that had been proposed for the first time by de Broglie (see also end of Sec. 9.6). In fact, de Broglie proposed a deterministic integration of quantum mechanics. Let us then briefly examine de Broglie's proposal.

[11] See [*Jammer* 1974, 256–57, 262].
[12] See [*Jauch* 1968, 116].

16.3.2 De Broglie's proposal

Two issues have to be spelled out here before proceeding further into our analysis:

- the pilot wave;
- the double solution.

The theory of the *pilot wave* is a proposal de Broglie put forward in order to understand the basic ontology of the microworld as composed of two different entities existing simultaneously: a wave and a particle. According to this theory, the wave ψ is a field which moves wave-like in the space and "pilots" a particle embedded in the field. The particle is sensible to all wave superpositions of the field. In the interferometer experiment example, the particle, although both paths are open, actually always goes through one of them, and the diffraction pattern is only due to the wave-like nature of the field (see again Sec. 9.6). From this perspective, there is no complementarity between wave and particle and no "uncertainty" at all.

The *double solution theory* is a mathematical aspect of the same idea: the correlation between particle and wave is a phase correlation, such that the particle is a singularity of the field, which differs from ψ only in amplitude, and which represents another, non-linear solution of the wave equation.

De Broglie published his results in a series of articles.[13] But, on the occasion of an exposition to a large scientific auditorium at the Fifth Physical Conference of the Solvay Institute in Brussels (October 1927), he presented only a simplified version, namely the pilot-wave theory.[14] The many important criticisms to his proposal pushed de Broglie to abandon the theory: in a public lecture at the university of Hamburg in early 1928 he embraced the complementarity principle.[15] Later (1955–56) he returned to his old proposal in a more systematic way, i.e. in the form of a double-solution theory.[16]

16.3.3 Bohm's developments

Bohm was the first proponent of a systematic HV theory for quantum mechanics. Bohm's proposal[17] originated from a sharp criticism – in the sense of EPR – of the Copenhagen interpretation. According to Bohm, the wave function cannot be the best description of microreality because it allows several forms of indeterminism.

Bohm's proposal is substantially similar to that of de Broglie about the simultaneous presence of waves and particles. However, in order to account for their reciprocal relationships and to face the measurement problems which quantum mechanics poses, Bohm

[13] See [de Broglie 1927a, de Broglie 1927b, de Broglie 1927c].
[14] See [de Broglie 1955].
[15] There is a historical reconstruction in [*Jammer* 1974, 110–14].
[16] See [*de Broglie* 1956]. It is worth mentioning that the double-solution theory (in its non-linear consequences) has also been experimentally rejected [*Auletta* 2000, Sec. 28.2].
[17] See [Bohm 1952, 371–75].

developed the concept of *quantum potential*. Bohm started with the usual wave definition (see Eq. (10.105))

$$\psi(\mathbf{r}) = \vartheta(\mathbf{r})e^{\frac{i}{\hbar}\phi(\mathbf{r})}, \tag{16.20}$$

and wrote the equations for the amplitude ϑ and the phase ϕ (both assumed to be real) in the following form (see Eqs. (10.106)):

$$\frac{\partial\phi}{\partial t} = -\left[\frac{(\nabla\phi)^2}{2m} + V(\mathbf{r}) - \frac{\hbar^2}{2m}\frac{\nabla^2\vartheta}{\vartheta}\right], \tag{16.21a}$$

$$\frac{\partial\vartheta}{\partial t} = -\frac{1}{2m}\left[\vartheta\nabla^2\phi + 2\nabla\vartheta\cdot\nabla\phi\right]. \tag{16.21b}$$

In the classical limit, the phase ϕ is a solution of the Hamilton–Jacobi equation (10.107). In this case, we can consider the probability density $\wp(\mathbf{r}) = \vartheta^2(\mathbf{r})$ as a classical (stochastic) probability density for ensembles of particles (see Eq. (10.109a)).

Instead of neglecting the third term of the rhs of Eq. (16.21a) – as in the case for the classical approximation – we derive a quantum equivalent of the Hamilton–Jacobi equation for an ensemble of particles, the so-called Bohm–Jacobi equation

$$\frac{\partial\phi}{\partial t} + \frac{(\nabla\phi)^2}{2m} + V(\mathbf{r}) - \frac{\hbar^2}{4m}\left[\frac{\nabla^2\wp}{\wp} - \frac{1}{2}\frac{(\nabla\wp)^2}{\wp^2}\right] = 0, \tag{16.22}$$

which implies that the particles are acted upon not only by the "classical" potential energy $V(\mathbf{r})$, but also by the quantum-potential energy

$$V_Q(\mathbf{r}) = \frac{-\hbar^2}{4m}\left[\frac{\nabla^2\wp}{\wp} - \frac{1}{2}\frac{(\nabla\wp)^2}{\wp^2}\right] = -\frac{\hbar^2}{2m}\frac{\nabla^2\vartheta}{\vartheta}. \tag{16.23}$$

A systematic analysis of Bohm's interpretation shows that its central features are:

- Each quantum particle has a well-defined, continuous, and causally determined position.
- The particle is always embedded in a new type of fundamental quantum field, the ψ-field, which satisfies the Schrödinger equation and represents a further development of de Broglie's position.
- The equation of motion of the particle is given by (see Eq. (1.17))

$$m\frac{d\mathbf{v}}{dt} = -\nabla V(\mathbf{r}) - \nabla V_Q(\mathbf{r}), \tag{16.24}$$

  which implies that, besides the classical forces $-\nabla V(\mathbf{r})$, the quantum force $-\nabla V_Q(\mathbf{r})$ also acts on the particle.
- The particle's momentum is equal to $\mathbf{p} = \nabla\phi$.

According to Bohm, the most important evidence for the existence of the quantum potential is given by the Aharonov–Bohm effect[18] (see Sec. 13.8 and also Figs. 16.3–16.4). Though the AB effect can be explained in pure quantum-mechanical terms, its discovery is one of the most important contributions of Bohm. It is another evidence for the circumstance that the struggle for interpreting a theory can give rise to very important results.

[18] See [*Bohm/Hiley* 1993, 51–52] .

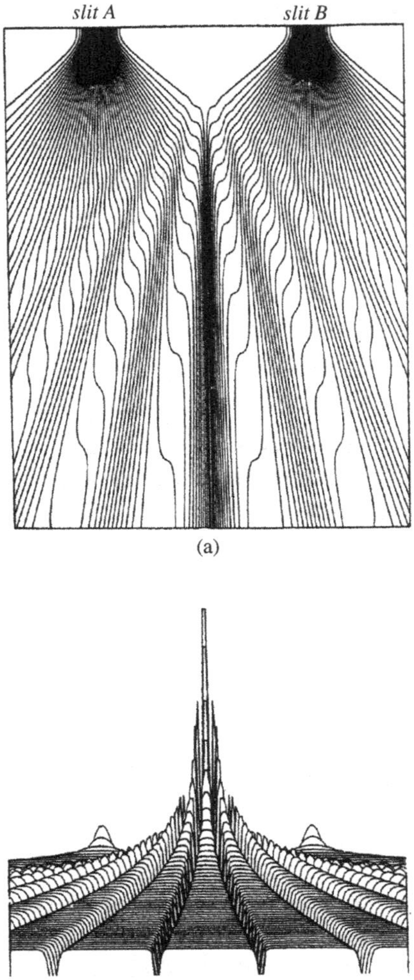

Figure 16.3 **(a) Particle trajectories for a two-Gaussian slit system after Bohm's model, and (b) the corresponding quantum potential. It is interesting to note that similar plots are used by Bohm and Hiley for the description of the AB effect. Adapted from [*Bohm/Hiley* 1993, 33–34].**

In the quantum-potential formula (16.23), the wave or field ψ appears both in the numerator and in the denominator, so that the quantum potential is not changed at all if ψ is multiplied by an arbitrary constant, i.e. the effect of the quantum potential is independent of the strength of the field.[19] If this is correct, then it is impossible to conceive the quantum potential as a form of physical energy. For this reason, the quantum potential was later interpreted by its proponents as a special type of information, an *active information* – so-called in order to distinguish it from the potential information represented by entropy (see Ch. 17). It is evident that this interpretation is suggested by de Broglie's concept of the pilot

[19] See [*Bohm/Hiley* 1993, 31–37].

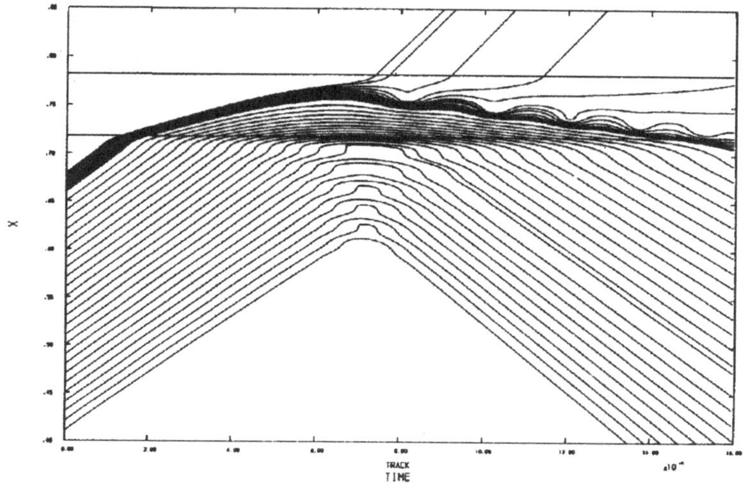

(a)

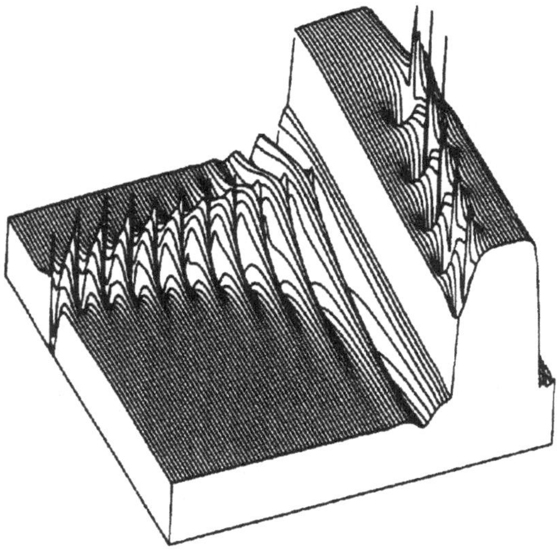

(b)

Figure 16.4 **(a) Trajectories for a potential barrier ($E = V/2$) after Bohm's model, and (b) the corresponding quantum potential. Adapted from [*Bohm/Hiley* 1993, 76–77].**

wave. The active information, according to Bohm, is a form of action where the strength of the signal is not important but only the form, or the structure, of the driving "message" is significant. Now, since the strength of the action does not decrease with distance, Bohm was forced to affirm a strong form of non-locality.[20]

[20] See [*Bohm/Hiley* 1993, 57, 62]. In the latter work, Bohm has proposed a theory based on the violation of the Lorentz invariance [*Bohm/Hiley* 1993, 282–85, 289–93, 350]. See also [Bell 1981, 133].

One should also mention that some more technical objections can be advanced. Englert, Scully, Süssman, and Walther[21] proved that Bohm's trajectories can differ from observed ones, and Dewdney, Hardy, and Squires[22] proved that in Bohm's model a particle can excite a detector on one path of an interferometer, even if it takes the other path as a consequence of the quantum potential.

16.4 Bell's contribution

After the EPR paper, developments were mainly in the discussion of HV theories, in particular due to the contribution of Bohm. However, until the mid 1960s the discussion about the issues of separability and the completeness of quantum mechanics remained on a rather abstract level. It was with the work of John S. Bell that this discussion entered a quantitative ground and could, in ten years, be translated into experimental tests.

Bell worked mainly in two areas: the problem whether quantum mechanics could be completed with a hidden variable theory and the question of separability and non-locality. Although the two problems have a common root and are deeply connected, they deserve a distinct discussion. In (Subsec. 16.4.1) we shall start by examing the first point and then present Bell's contribution to the original EPR problem (in Subsec. 16.4.2). In Subsec. 16.4.3 we shall finally introduce further developments of Bell's work.

16.4.1 Dispersion-free states

As we have seen (in Subsec. 16.3.1), dispersion-free states are states in which all observables are completely determined. In principle, one could interpret quantum-mechanical uncertainties as being valid only on a statistical level but admitting a deeper (hidden) level, where states are purely classical and are explained by adding hidden parameters to the quantum state vector. In 1932, von Neumann had given a proof of the contradiction between the hypothesis of dispersion-free states and quantum mechanics, with the evident aim of showing the completeness of the theory.[23] However, von Neumann's argument was not completely satisfactory because, at that time, it was not generally proved that any expectation value of a given observable could be expressed as the trace of the product between the operator representing the observable and the density matrix describing the system (see Eq. (5.18)).

In the 1950s, Gleason[24] was able to show that, for quantum-mechanical systems whose states are represented by a vector in a Hilbert space with a dimension of at least three,

[21] See [Englert *et al.* 1993].
[22] See [Dewdney *et al.* 1993].
[23] See [*von Neumann* 1932, 163–71] [*von Neumann* 1955, 324–25].
[24] See [Gleason 1957].

any expectation value may be expressed in terms of Eq. (5.18). Bell made use of Gleason theorem to prove the following corollary:[25]

Corollary 16.1 (Bell dispersion-free) *In a Hilbert space for a quantum system with at least three dimensions, the existence of dispersion-free states is incompatible with the additivity requirement for expectation values of commuting operators.*

We shall prove Cor. 16.1 by making use of another theorem, formulated by Bell himself. To this purpose, let $\hat{P}_\varsigma$ be the projector on the Hilbert space vector $|\varsigma\rangle$, i.e. acting on an arbitrary vector $|\psi\rangle$ in the following manner:

$$\hat{P}_\varsigma|\psi\rangle = \frac{|\varsigma\rangle\langle\varsigma|\psi\rangle}{\langle\varsigma|\varsigma\rangle}. \tag{16.25}$$

If $\{|\varsigma_j\rangle\}$ is a complete set of orthogonal vectors, we have (see Eq. (1.41a))

$$\sum_j \hat{P}_{\varsigma_j} = \hat{I}. \tag{16.26}$$

It is clear that, for any normalized state $|\varphi\rangle$, we also have the additivity requirement for expectation values of projectors pertaining to the same set, which are commuting operators due to the requirement (1.41b), i.e.

$$\sum_j \left\langle\varphi\left|\hat{P}_{\varsigma_j}\right|\varphi\right\rangle = \sum_j \left\langle\hat{P}_{\varsigma_j}\right\rangle_\varphi = 1. \tag{16.27}$$

We know that the expectation value of a projector is non-negative, since the eigenvalues may only assume values 0 or 1. Now, it is possible to prove the following lemma (see Prob. 16.4):

Lemma 16.1 (Bell I) *If some arbitrary vector $|\varsigma\rangle$ is such that $\left\langle\hat{P}_\varsigma\right\rangle_\varphi = 1$ for a given state $|\varphi\rangle$, then, for that state, $\left\langle\hat{P}_\psi\right\rangle_\varphi = 0$ for any $|\psi\rangle$ orthogonal to $|\varsigma\rangle$.*

If $|\psi_1\rangle$ and $|\psi_2\rangle$ represent another orthogonal basis for the subspace spanned by $|\varsigma_1\rangle$ and $|\varsigma_2\rangle$, then from Eq. (16.27) we have (see Prob. 16.5)

$$\left\langle\hat{P}_{\psi_1}\right\rangle_\varphi + \left\langle\hat{P}_{\psi_2}\right\rangle_\varphi = 1 - \sum_{j\neq 1,2}\left\langle\hat{P}_{\varsigma_j}\right\rangle_\varphi, \tag{16.28}$$

or

$$\left\langle\hat{P}_{\psi_1}\right\rangle_\varphi + \left\langle\hat{P}_{\psi_2}\right\rangle_\varphi = \left\langle\hat{P}_{\varsigma_1}\right\rangle_\varphi + \left\langle\hat{P}_{\varsigma_2}\right\rangle_\varphi. \tag{16.29}$$

[25] For a parallel and independent proof of the impossibility to have in quantum mechanics dispersion-free states see [Kochen/Specker 1967].

Since $|\psi_1\rangle$ and $|\psi_2\rangle$ can be any linear combination of $|\varsigma_1\rangle$ and $|\varsigma_2\rangle$, the following lemma can also be proved (see Prob. 16.6):

Lemma 16.2 (Bell II) *If, for a given state* $|\varphi\rangle$

$$\left\langle \hat{P}_{\varsigma_1} \right\rangle_\varphi = \left\langle \hat{P}_{\varsigma_2} \right\rangle_\varphi = 0 \tag{16.30}$$

for some pairs of orthogonal vectors $|\varsigma_1\rangle$ *and* $|\varsigma_2\rangle$*, then*

$$\left\langle \hat{P}_{c_1\varsigma_1 + c_2\varsigma_2} \right\rangle_\varphi = 0 \tag{16.31}$$

for all complex scalars c_1, c_2.

Next, we develop Bell's central idea, i.e. a theorem by which we shall prove Cor. 16.1:

Theorem 16.1 (Bell I) *Let* $|\varsigma\rangle$ *and* $|\psi\rangle$ *be some vectors in a Hilbert space with dimension* $d \geq 3$ *such that, for a given state* $|\varphi\rangle$*, we have*

$$\left\langle \hat{P}_\psi \right\rangle_\varphi = 1, \tag{16.32a}$$

$$\left\langle \hat{P}_\varsigma \right\rangle_\varphi = 0. \tag{16.32b}$$

Then, $|\varsigma\rangle$ *and* $|\psi\rangle$ *cannot be arbitrarily close. In fact,*

$$\| \, |\varsigma\rangle - |\psi\rangle \, \| > \frac{1}{2} \| \, |\psi\rangle \, \| . \tag{16.33}$$

Proof

In the case in which $|\varsigma\rangle$ and $|\psi\rangle$ are orthogonal, the theorem is obviously proved. If not, let us normalize $|\psi\rangle$ as

$$|\psi'\rangle = \frac{|\psi\rangle}{\| \, |\psi\rangle \, \|}, \tag{16.34}$$

and write

$$|\varsigma'\rangle = |\psi'\rangle + \epsilon |\psi_\perp\rangle, \tag{16.35}$$

where

$$|\varsigma'\rangle = \frac{|\varsigma\rangle}{\| \, |\psi\rangle \, \|}, \tag{16.36}$$

$|\psi_\perp\rangle$ is orthogonal to $|\psi\rangle$ and normalized, ϵ is a real and positive number (indeed any phase factor can be eventually reabsorbed in the definition of $|\psi_\perp\rangle$). Let $|\psi_{\perp\perp}\rangle$ now be a

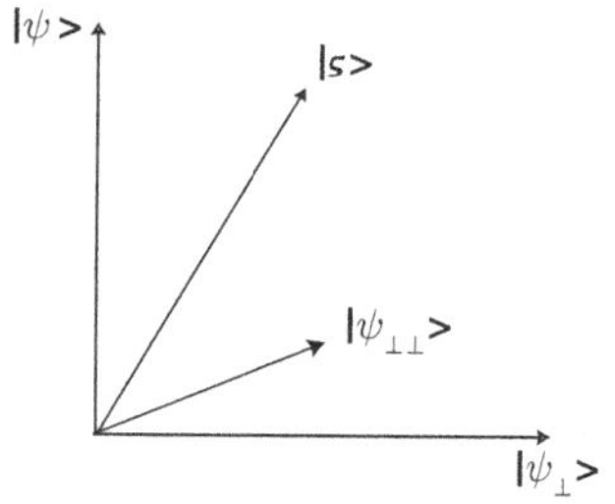

Figure 16.5 The three-dimensional Hilbert space proposed by Bell. The state vector $|\varsigma\rangle$ lies on the plane $|\psi\rangle$-$|\psi_\perp\rangle$, while $|\psi_{\perp\perp}\rangle$ is orthogonal to this plane.

normalized vector orthogonal to both $|\psi\rangle$ and $|\psi_\perp\rangle$ (see Fig. 16.5). By the orthogonality of $|\psi_\perp\rangle$ and $|\psi_{\perp\perp}\rangle$ to $|\psi\rangle$ and by Eq. (16.32a), we have (Lemma 16.1)

$$\left\langle \hat{P}_{\psi_\perp} \right\rangle_\varphi = 0 \quad \text{and} \quad \left\langle \hat{P}_{\psi_{\perp\perp}} \right\rangle_\varphi = 0. \tag{16.37}$$

However, from Eq. (16.32b) and Lemma 16.2 it follows

$$\left\langle \hat{P}_{\varsigma' + a^{-1}\epsilon \psi_{\perp\perp}} \right\rangle_\varphi = 0, \tag{16.38}$$

where a is a real number and we have taken $c_1 = 1$ and $c_2 = a^{-1}\epsilon$, $|\varsigma_1\rangle = |\varsigma'\rangle$, and $|\varsigma_2\rangle = |\psi_{\perp\perp}\rangle$. Again, by Lemma 16.2 and Eqs. (16.37) we also have

$$\left\langle \hat{P}_{-\epsilon\psi_\perp + a\epsilon\psi_{\perp\perp}} \right\rangle_\varphi = 0. \tag{16.39}$$

The vectors

$$|\varsigma'\rangle + a^{-1}\epsilon\, |\psi_{\perp\perp}\rangle \tag{16.40a}$$

and

$$-\epsilon\, |\psi_\perp\rangle + a\epsilon\, |\psi_{\perp\perp}\rangle \tag{16.40b}$$

are mutually orthogonal, so that, making use of Eq. (16.35) and again of Lemma 16.2, we obtain

$$\left\langle \hat{P}_{\psi' + \epsilon(a + a^{-1})\psi_{\perp\perp}} \right\rangle_\varphi = 0. \tag{16.41}$$

Now, if ϵ is smaller than $1/2$, there is a real number a such that (see Prob. 16.7)

$$\epsilon(a + a^{-1}) = \pm 1, \tag{16.42}$$

and therefore we also have

$$\left\langle \hat{P}_{\psi' + \psi_{\perp\perp}} \right\rangle_\varphi = \left\langle \hat{P}_{\psi' - \psi_{\perp\perp}} \right\rangle_\varphi = 0. \tag{16.43}$$

The vectors $|\psi'\rangle \pm |\psi_{\perp\perp}\rangle$ are orthogonal to each other: Adding them and again using Lemma 16.2, we have

$$\left\langle \hat{P}_{\psi'} \right\rangle_\varphi = 0, \tag{16.44}$$

which contradicts the assumption (16.32a). Therefore, we must take $\epsilon > 1/2$ and

$$\| \, | \varsigma' \rangle - | \psi' \rangle \, \| > \frac{1}{2} \, \| \, \psi_\perp \, \| \, . \tag{16.45}$$

By multiplying the lhs and the rhs by $\| \, | \, \psi \rangle \, \|$, since the norm of $| \, \psi_\perp \rangle$ is =1, we obtain Eq. (16.33).

Q.E.D

Now, using Th. 16.1 we can also prove Cor. 16.1.

Proof

In a dispersion-free state any projector has an expectation value which is given by either 0 or 1. Let us consider the case in which the projector $\hat{P}_\psi$ has expectation value 1 on a given state $| \, \varphi \rangle$. Then, there is at least one projector, corresponding to the state $| \, \varsigma \rangle$ that is arbitrarily close to $| \, \psi \rangle$, whose expectation value on $| \, \varphi \rangle$ is 0. However, this contradicts Th. 16.1. Therefore, there are no dispersion free states satisfying the additivity requirement (16.27).

Q.E.D

It should be noted that the previous result could be criticized in principle by rejecting Lemma 16.2 on the following ground.[26] $\hat{P}_{c_1 \varsigma_1 + c_2 \varsigma_2}$ commutes with $\hat{P}_{\varsigma_1}$ and $\hat{P}_{\varsigma_2}$ only if, respectively, $c_2 = 0$ or $c_1 = 0$. Thus, in general, the measurement of $\hat{P}_{c_1 \varsigma_1 + c_2 \varsigma_2}$ requires a distinct experimental arrangement. In other words, in Bell's proof it is implicitly assumed that the measurement of an observable must yield the same value independently of whichever other measurements may be made simultaneously, i.e. that measurement of an observable $\hat{O}$ yields the same value if it is measured as part of set $\hat{O}, \hat{O}'_1, \hat{O}''_1, \ldots$ of mutually commuting observables or of a second set $\hat{O}, \hat{O}'_2, \hat{O}''_2, \ldots$ of mutually commuting observables, though in general some observables of the second set fail to commute with some of the first set. This objection brings some analogy with Bohr's answer to EPR (see Subsec. 16.1.3).

However, any local HV theory should be non-contextual,[27] because it assumes that the value of any observable is already determined by the hidden parameters, independently of measurements on other observables. Then, if the non-contextuality of HV theories is accepted, there is no reason to deny the validity of the proof.

Obviously, the problem of whether HV contextual theories are possible as such remains open.[28] However, the second Bell theorem, as we shall see in the next subsections, forces these theories to assume a form of non-locality, so that a potential supporter of HV theories would be forced to accept a non-local HV contextual theory.

[26] See [Bell 1966, 8–9].

[27] Contextually physical theories assume that a result of any operation on a system may depend on any other operation that is simultaneously performed on the same system.

[28] For recent examination of this point see [Weihs 2007].

16.4.2 Non-locality

As shown in the previous subsection, the problem of quantum non-locality has for a long time been deeply connected with that of the existence of HVs. As we shall see now, Bell was able to prove the following striking result: no deterministic local HV theory can make predictions compatible with quantum mechanics. This achievement was of particular relevance because it moved the discussion from a qualitative level to a strict quantitative (and therefore experimentally testable) ground. The incompatibility between a local HV theory and quantum mechanics raises the difficult question of what type of locality is violated by quantum mechanics. This discussion must be postponed (see Sec. 16.8) and for the time being we will use the expression "quantum non-locality" as a generic term that could cover two very different possibilities, i.e. the violation of separability (see Pr. 16.1: p. 568) and or violation of Einstein's locality dictated by special relativity.

The *Gedankenexperiment* proposed by Bell[29] was a further refinement of the EPR–Bohm model (see Subsec. 16.2.1, in particular Fig. 16.1). He assumed the existence of a hidden parameter λ_{HV} such that, given λ_{HV}, the result $A_{\mathbf{a}}$ obtained by measuring the spin of the first particle along a chosen direction $\mathbf{a}$ (i.e. the observable $\hat{\sigma}_1 \cdot \mathbf{a}$), depends only on λ_{HV} and on $\mathbf{a}$. Similarly, the result $B_{\mathbf{b}}$ of measuring the spin of the second particle along a chosen direction $\mathbf{b}$ (i.e. $\hat{\sigma}_2 \cdot \mathbf{b}$), depends only on $\mathbf{b}$ and λ_{HV}. In fact, the separability principle denies that there can be a form of interdependence between two systems if they do not dynamically interact. This hypothesis can be formulated mathematically in a more rigorous way in terms of the factorization rule

$$(A_{\mathbf{a}} B_{\mathbf{b}})(\lambda_{HV}) = A_{\mathbf{a}}(\lambda_{HV}) B_{\mathbf{b}}(\lambda_{HV}), \tag{16.46}$$

where $A_{\mathbf{a}}$ and $B_{\mathbf{b}}$ are two deterministic functions of the hidden parameter. Equation (16.46) expresses the fact that the probability distributions for the two particles are mutually independent. In order to remain on a practical level, we assume that the result of each measurement can be either $+1$ (spin up) or -1 (spin down), that is

$$A_{\mathbf{a}}(\lambda_{HV}) = \pm 1, \quad B_{\mathbf{b}}(\lambda_{HV}) = \pm 1. \tag{16.47}$$

Following Eq. (16.46), if $\rho(\lambda_{HV})$ denotes the probability distribution of λ_{HV}, then the expectation value of the product of the two components $\hat{\sigma}_1 \cdot \mathbf{a}$ and $\hat{\sigma}_2 \cdot \mathbf{b}$ is

$$\left\langle (\hat{\sigma}_1 \cdot \mathbf{a})\,(\hat{\sigma}_2 \cdot \mathbf{b}) \right\rangle \equiv \langle \mathbf{a}, \mathbf{b} \rangle = \int_{\Lambda_{HV}} d\lambda_{HV}\, \rho(\lambda_{HV}) A_{\mathbf{a}}(\lambda_{HV}) B_{\mathbf{b}}(\lambda_{HV}), \tag{16.48}$$

where Λ_{HV} represents the set of all possible values of λ_{HV}. Since we do not know the values of the hidden parameters λ_{HV}, we must integrate over the possible values $\lambda_{HV} \in \Lambda_{HV}$.[30] Because $\rho(\lambda_{HV})$ is supposed to be a normalized probability distribution, we have

$$\int d\lambda_{HV}\, \rho(\lambda_{HV}) = 1, \tag{16.49}$$

29 See [Bell 1964, 15–19].

30 In the following we shall normally drop the explicit mention of integration over the entire space Λ_{HV}.

and

$$-1 \leq \langle \mathbf{a}, \mathbf{b} \rangle \leq +1. \tag{16.50}$$

Our aim is to compare the prediction of a deterministic HV theory as expressed by Eq. (16.48) with the quantum-mechanical expectation value, which for the singlet state $|\Psi_0\rangle$ (see Eq. (16.11)) is given by (see Prob. 16.8)

$$\langle \mathbf{a}, \mathbf{b} \rangle_{\Psi_0} = \langle \Psi_0 \left| (\hat{\sigma}_1 \cdot \mathbf{a}) \, (\hat{\sigma}_2 \cdot \mathbf{b}) \right| \Psi_0 \rangle = -\mathbf{a} \cdot \mathbf{b}. \tag{16.51}$$

When the two orientations **a** and **b** are parallel, quantum-mechanical calculations (see Eq. (16.13)) show that

$$\langle \mathbf{a}, \mathbf{a} \rangle_{\Psi_0} = -1, \tag{16.52}$$

as it should be since there is a perfect *anticorrelation* between the results of the two measurements.

Since the value given by Eq. (16.52) for perfect anticorrelation is an experimental fact, also a HV theory must satisfy this requirement. On the other hand, $\langle \mathbf{a}, \mathbf{a} \rangle = -1$ holds if and only if we also have

$$A_{\mathbf{a}}(\lambda_{\mathrm{HV}}) = -B_{\mathbf{a}}(\lambda_{\mathrm{HV}}), \tag{16.53}$$

for any direction **a**, and except for, at most, a set of points λ_{HV} of zero measure in Λ_{HV}. In this case, Eq. (16.48) reaches the minimum value (see also Eq. (16.50)). Under this assumption, we can rewrite Eq. (16.48) as

$$\langle \mathbf{a}, \mathbf{b} \rangle = -\int d\lambda_{\mathrm{HV}} \rho(\lambda_{\mathrm{HV}}) A_{\mathbf{a}}(\lambda_{\mathrm{HV}}) A_{\mathbf{b}}(\lambda_{\mathrm{HV}}). \tag{16.54}$$

Now we consider two possible orientations, say **b** and **c**, of the spin measurement of particle 2. Dropping the subscript HV for the sake of simplicity, we may write

$$\begin{aligned} \langle \mathbf{a}, \mathbf{b} \rangle - \langle \mathbf{a}, \mathbf{c} \rangle &= -\int d\lambda \rho(\lambda) [A_{\mathbf{a}}(\lambda) A_{\mathbf{b}}(\lambda) - A_{\mathbf{a}}(\lambda) A_{\mathbf{c}}(\lambda)] \\ &= \int d\lambda \rho(\lambda) A_{\mathbf{a}}(\lambda) A_{\mathbf{b}}(\lambda) [A_{\mathbf{b}}(\lambda) A_{\mathbf{c}}(\lambda) - 1], \end{aligned} \tag{16.55}$$

because, due to the property (16.47), for any orientation **a** we have $[A_{\mathbf{a}}(\lambda)]^2 = 1$. Then, from Eq. (16.55) we may prove (see Prob. 16.9) the inequality

$$|\langle \mathbf{a}, \mathbf{b} \rangle - \langle \mathbf{a}, \mathbf{c} \rangle| \leq \int d\lambda \rho(\lambda) [1 - A_{\mathbf{b}}(\lambda) A_{\mathbf{c}}(\lambda)], \tag{16.56}$$

so that we finally obtain

$$|\langle \mathbf{a}, \mathbf{b} \rangle - \langle \mathbf{a}, \mathbf{c} \rangle| \leq 1 + \langle \mathbf{b}, \mathbf{c} \rangle. \tag{16.57}$$

This formula is the first of a family of inequalities, collectively called *Bell inequalities*. Its importance lies in the fact that it sets precise quantitative bounds on the prediction of *any* local deterministic HV theory. It is now possible to formulate the second Bell theorem, also known simply as "Bell's theorem."

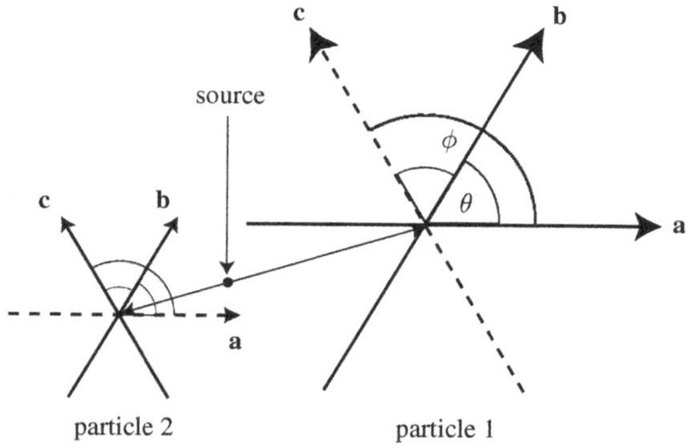

Figure 16.6 Scheme of the orientations used in the counterexample proposed for proving the second Bell theorem. Orientation **a** is chosen for particle 1 while **b** and **c** are the alternative orientations for particle 2. Dashed lines denote orientations used for the other particle.

Theorem 16.2 (Bell II) *A deterministic HV theory, which acknowledges the separability principle* (Pr. 16.1: p. 568) *must satisfy an inequality of type* (16.57). *The predictions of quantum mechanics, on the other hand, violate such an inequality.*

Proof

The first part of the theorem has been already proved. In order to prove the second part, it suffices to show a contradiction between Eq. (16.57) and Eq. (16.51) by means of a counterexample.[31] We take **a**, **b** and **c** to be coplanar, with **c** making an angle ϕ of $2\pi/3$ with **a**, and **b** making an angle θ of $\pi/3$ with both **a** and **c** (see Fig. 16.6).

Then, according to the elementary formula

$$cos\theta = \frac{\mathbf{a} \cdot \mathbf{b}}{\| \mathbf{a} \| \| \mathbf{b} \|}, \tag{16.58}$$

and by choosing unitary vectors, i.e. $\| \mathbf{a} \| = \| \mathbf{b} \| = \| \mathbf{c} \| = 1$, we have

$$\mathbf{a} \cdot \mathbf{b} = \mathbf{b} \cdot \mathbf{c} = \frac{1}{2}; \quad \mathbf{a} \cdot \mathbf{c} = -\frac{1}{2}, \tag{16.59}$$

from which, taking into account Eq. (16.51), it follows

$$\left| \langle \mathbf{a}, \mathbf{b} \rangle_{\Psi_0} - \langle \mathbf{a}, \mathbf{c} \rangle_{\Psi_0} \right| = 1, \tag{16.60}$$

while we have

$$1 + \langle \mathbf{b}, \mathbf{c} \rangle_{\Psi_0} = \frac{1}{2}. \tag{16.61}$$

These values do not satisfy inequality (16.57).

Q.E.D

[31] See [Clauser/Shimony 1978, 1888–90].

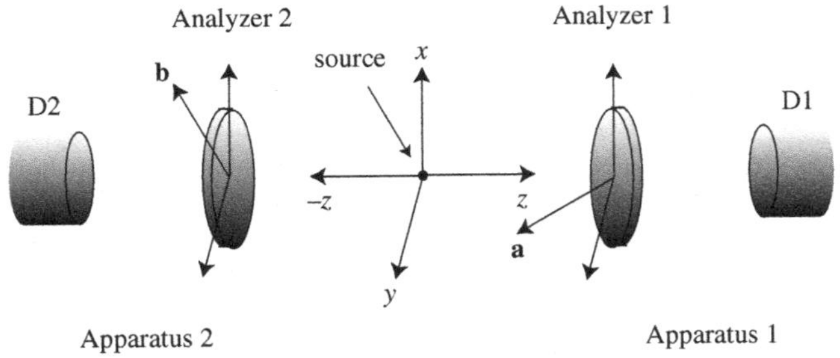

Figure 16.7 Ideal experimental configuration used in the proof by CHSH and by Clauser and Horne (CH). A source emitting particle pairs is analyzed by two apparata. Each apparatus consists of an analyzer and an associated detector. The two analyzers are characterized by the vector parameters **a** and **b**, respectively, which can be adjusted by the experimenter. In the above example **a** and **b** make some angles with a fixed reference axis on the *x*–*y* plane. Coincidence detections are counted.

16.4.3 Other inequalities

Many physicists have further worked on this subject brought to the general attention of the scientific community by the pioneering work of Bell, so that other similar inequalities have been successively derived. These further developments usually go either in the direction of generalizing Bell's result or in the direction of simplifying its scheme and proof. In a paper by Clauser, Horne, Shimony, and Holt (CHSH) Bell theorem was generalized in the following manner:[32]

- The assumption of determinism was left aside so that the authors were able to show an incompatibility between quantum mechanics and stochastic HV theories that acknowledge the separability principle.
- The authors conceived a more realistic *Gedankenexperiment* so that a considerable step toward factual experiments was made.

We suppose here correlated pairs of particles – that need neither be spin-1/2 particles nor perfectly anticorrelated – such that one particle enters apparatus $1_{\mathbf{a}}$ and the other one enters apparatus $2_{\mathbf{b}}$, where **a** and **b** are two generic vector parameters. This apparatus consists of two analyzers and a detector beyond each analyzer – it is an idealized coincidence count experiment (see Fig. 16.7).

Under the HV assumption, in each apparatus a particle must select one of two "channels," labelled $+1$ and -1, respectively. The results are represented by $A_{\mathbf{a}}$ and $B_{\mathbf{b}}$, each of which equals $+1$ or -1. We also assume the independence of $A_{\mathbf{a}}(\lambda)$ from **b** and of $B_{\mathbf{b}}(\lambda)$ from **a** (see Eq. (16.46)). As before, the normalized probability distribution $\rho(\lambda)$ is independent of both **a** and **b**. Let us write again the correlation function (16.54) for the results of the two measurements:

$$\langle \mathbf{a}, \mathbf{b} \rangle = \int_{\Lambda} d\lambda \rho(\lambda) A_{\mathbf{a}}(\lambda) B_{\mathbf{b}}(\lambda). \tag{16.62}$$

[32] See [Clauser *et al.* 1969].

We prefer, in this context, to remain on an abstract level, without specifying the nature of the involved particles and observables. As we shall see, in the next section we shall concentrate on possible realizations of this kind of experiments, where the particles are represented by photons and the observables by polarization orientations. Taking into account Eqs. (16.55) and (16.56), we also have

$$\begin{aligned} \mid \langle \mathbf{a}, \mathbf{b} \rangle - \langle \mathbf{a}, \mathbf{c} \rangle \mid &\leq \int_{\Lambda} d\lambda \rho(\lambda) |A_{\mathbf{a}}(\lambda) B_{\mathbf{b}}(\lambda) - A_{\mathbf{a}}(\lambda) B_{\mathbf{c}}(\lambda)| \\ &= \int_{\Lambda} d\lambda \rho(\lambda) |A_{\mathbf{a}}(\lambda) B_{\mathbf{b}}(\lambda)| [1 - B_{\mathbf{b}}(\lambda) B_{\mathbf{c}}(\lambda)] \\ &= 1 - \int_{\Lambda} d\lambda \rho(\lambda) B_{\mathbf{b}}(\lambda) B_{\mathbf{c}}(\lambda). \end{aligned} \tag{16.63}$$

Suppose now that for some pair of orientations $\mathbf{a}'$ and $\mathbf{b}$ (where $\mathbf{a}'$ is an alternative orientation for apparatus 1) we have

$$\left\langle \mathbf{a}', \mathbf{b} \right\rangle = 1 - \delta, \tag{16.64}$$

where $0 \leq \delta \leq 1$. In this way we overcome Bell's experimentally unrealistic condition that for some pair $\mathbf{b}$ and $\mathbf{a}'$ there is perfect (anti-)correlation. Partitioning the state space Λ into two regions Λ^+ and Λ^-, such that

$$\Lambda^{\pm} = \{\Lambda \mid A_{\mathbf{a}'}(\lambda) = \pm B_{\mathbf{b}}(\lambda)\}, \tag{16.65}$$

we have (see Prob. 16.10)

$$\int_{\Lambda^-} d\lambda \rho(\lambda) = \frac{1}{2}\delta, \tag{16.66}$$

and therefore (see Prob. 16.11)

$$\begin{aligned} \int_{\Lambda} d\lambda \rho(\lambda) B_{\mathbf{b}}(\lambda) B_{\mathbf{b}'}(\lambda) &\geq \int_{\Lambda} d\lambda \rho(\lambda) A_{\mathbf{a}'}(\lambda) B_{\mathbf{b}'}(\lambda) - 2 \int_{\Lambda^-} d\lambda \rho(\lambda) |A_{\mathbf{a}'}(\lambda) B_{\mathbf{b}'}(\lambda)| \\ &= \left\langle \mathbf{a}', \mathbf{b}' \right\rangle - \delta, \end{aligned} \tag{16.67}$$

where $\mathbf{b}'$ is an alternative orientation for apparatus 2 and corresponds to the $\mathbf{c}$ orientation in Eq. (16.63). In conclusion, making use of Eqs. (16.63), (16.64), and (16.67), we obtain

$$\mid \langle \mathbf{a}, \mathbf{b} \rangle - \left\langle \mathbf{a}, \mathbf{b}' \right\rangle \mid \leq 2 - \left\langle \mathbf{a}', \mathbf{b} \right\rangle - \left\langle \mathbf{a}', \mathbf{b}' \right\rangle, \tag{16.68}$$

which is the second inequality of the Bell's family and is called the *CHSH inequality*. As with Bell's inequality, Eq. (16.68) also conflicts with quantum mechanics, as we shall see below.

It is interesting to emphasize that inequality (16.68) implies inequality (16.57) as a special case, so that the CHSH inequality is a more general instance of Bell's results (see Prob. 16.12).

Stimulated by the progress of the research, Bell returned to the problem in 1971.[33] Differently from CHSH, Bell implicitly proposed a device with an auxiliary apparatus (*event–ready detectors*) to measure the number of pairs emitted by the source, and collocated it before the analyzers. Bell reformulated the CHSH inequality as (see Prob. 16.13)

[33] See [Bell 1971, 37].

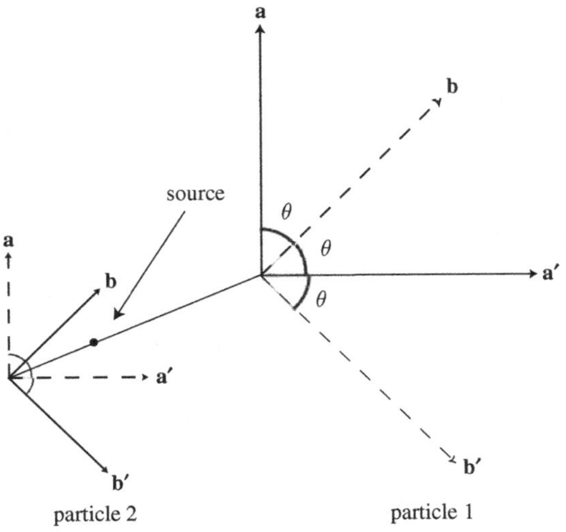

Figure 16.8 **Optimal orientations a, a′, b, and b′ for testing the CHSH inequality (16.69).**

$$|\,\langle \mathbf{a}, \mathbf{b}\rangle - \langle \mathbf{a}, \mathbf{b}'\rangle\,| + |\,\langle \mathbf{a}', \mathbf{b}'\rangle + \langle \mathbf{a}', \mathbf{b}\rangle\,| \leq 2. \tag{16.69}$$

It is easy to find a quantum counterexample to inequality (16.69).[34]

Proof

Taking into account the imperfections of instruments, the quantum-mechanical expectation values have the form

$$\langle \mathbf{a}, \mathbf{b}\rangle_{\mathrm{QM}} = -\eta\, \mathbf{a}\cdot\mathbf{b}, \tag{16.70}$$

where η (representing the efficiency of the detector) is some coefficient which is 1 only in the idealized case. If we take $\mathbf{a}$, $\mathbf{a}'$, $\mathbf{b}$, and $\mathbf{b}'$ to be coplanar and the angle between $\mathbf{a}$ and $\mathbf{b}$, between $\mathbf{b}$ and $\mathbf{a}'$, and between $\mathbf{a}'$ and $\mathbf{b}'$ to be $\theta = \pi/4$, we have (see Fig. 16.8)

$$\begin{aligned}\left[\left|\langle \mathbf{a}, \mathbf{b}\rangle - \langle \mathbf{a}, \mathbf{b}'\rangle\right| + \left|\langle \mathbf{a}', \mathbf{b}'\rangle + \langle \mathbf{a}', \mathbf{b}\rangle\right|\right]_{\mathrm{QM}} &= \left[\left|-\frac{\sqrt{2}}{2} + \left(-\frac{\sqrt{2}}{2}\right)\right| + \left|-\frac{\sqrt{2}}{2} - \frac{\sqrt{2}}{2}\right|\right]\eta \\ &= 2\sqrt{2}\eta,\end{aligned} \tag{16.71}$$

with $\| \mathbf{a} \| = \| \mathbf{b} \| = \| \mathbf{a}' \| = \| \mathbf{b}' \| = 1$ so that $\mathbf{a}\cdot\mathbf{b} = \cos\theta$. The necessary and sufficient condition to observe the quantum violation is that the efficiency $\eta > \sqrt{2}/2$.

Q.E.D

[34] See [Bell 1981, 152–53] [Clauser/Shimony 1978, 1893–94].

The previous proofs were generalized by Clauser and Horne[35] (CH) by eliminating the event-ready detectors of Bell's 1971 proof: another step was thus made towards the realization of real experiments.

The CH's apparatus is the same as that of CHSH (see Fig. 16.7). The source during a fixed time emits N pairs of particles. Let us denote with $N_1(\mathbf{a})$ and $N_2(\mathbf{b})$ the number of counts at detector 1 and 2, respectively, and with $N_{12}(\mathbf{a}, \mathbf{b})$ the number of simultaneous counts (coincidence counts). When N is sufficiently large, the frequencies

$$\pi_1(\mathbf{a}) = \frac{N_1(\mathbf{a})}{N}, \quad \pi_2(\mathbf{b}) = \frac{N_2(\mathbf{b})}{N}, \quad \pi_{12}(\mathbf{a}, \mathbf{b}) = \frac{N_{12}(\mathbf{a}, \mathbf{b})}{N}, \tag{16.72}$$

may be taken as the corresponding probabilities. Assuming a HV parameter λ, as before, the requirement of separability (see also Eq. (16.46)) is now

$$\pi_{12}(\lambda, \mathbf{a}, \mathbf{b}) = \pi_1(\lambda, \mathbf{a})\pi_2(\lambda, \mathbf{b}). \tag{16.73}$$

The ensemble average probabilities of Eqs. (16.72) are

$$\wp_k(\mathbf{j}) = \int d\rho \ \pi_k(\lambda, \mathbf{j}), \quad k = 1, 2, \quad \mathbf{j} = \mathbf{a}, \mathbf{b}, \tag{16.74a}$$

$$\wp_{12}(\mathbf{a}, \mathbf{b}) = \int d\rho \pi_1(\lambda, \mathbf{a})\pi_2(\lambda, \mathbf{b}). \tag{16.74b}$$

The probabilities for orientations $\mathbf{a}$ and $\mathbf{a}'$ of analyzer 1 and $\mathbf{b}$ and $\mathbf{b}'$ of analyzer 2, respectively, must obviously satisfy the inequalities

$$0 \leq \pi_k(\lambda, \mathbf{j}) \leq 1. \tag{16.75}$$

CH were finally able to derive (see Prob. 16.14)

$$\frac{\wp_{12}(\mathbf{a}, \mathbf{b}) - \wp_{12}(\mathbf{a}, \mathbf{b}') + \wp_{12}(\mathbf{a}', \mathbf{b}) + \wp_{12}(\mathbf{a}', \mathbf{b}')}{\wp_1(\mathbf{a}') + \wp_2(\mathbf{b})} \leq 1, \tag{16.76}$$

which is the *Clauser–Horne inequality*, the third of the Bell family, and involves only quantities which are independent of N. This is of relevance because – as we shall see in Subsec. 16.5.1 – the CH inequality can be expressed entirely in terms of a ratio of observable count rates.

Assuming now[36] $\mathbf{a}$ and $\mathbf{b}$ to be orientation angles relative to some reference axis in a fixed plane, cylindrical symmetry about a line normal to the fixed plane (see also Subsec. 16.2.1), and reflection symmetry with respect to the fixed plane, quantum mechanics allows the following predictions:

$$\left[\wp_k(\mathbf{j})\right]_{\text{QM}} = \wp_k, \quad k = 1, 2, \quad \mathbf{j} = \mathbf{a}, \mathbf{b}, \tag{16.77a}$$

$$\left[\wp_{12}(\mathbf{a}, \mathbf{b})\right]_{\text{QM}} = g(\widehat{\mathbf{ab}}), \tag{16.77b}$$

$$\langle \mathbf{a}, \mathbf{b} \rangle_{\text{QM}} = h(\widehat{\mathbf{ab}}), \tag{16.77c}$$

[35] See [Clauser/Horne 1974, 527] [Clauser/Shimony 1978, 1894–95].

[36] See [Clauser/Horne 1974, 528–29]. See also [Clauser/Shimony 1978, 1896–97, 1901–902].

where g and h are some functions and $\widehat{\mathbf{ab}}$ is the angle between the vectors $\mathbf{a}$ and $\mathbf{b}$. Stochastic HV theories that acknowledge the separability principle should satisfy similar symmetries, i.e.

$$\wp_k(\mathbf{j}) = \wp_k, \quad k = 1, 2, \quad \mathbf{j} = \mathbf{a}, \mathbf{b}, \tag{16.78a}$$

$$\wp_{12}(\mathbf{a}, \mathbf{b}) = \wp_{12}(\widehat{\mathbf{ab}}), \tag{16.78b}$$

$$\langle \mathbf{a}, \mathbf{b} \rangle = h'(\widehat{\mathbf{ab}}). \tag{16.78c}$$

Suppose now that we take $\mathbf{a}$, $\mathbf{a}'$, $\mathbf{b}$ and $\mathbf{b}'$ so that (see again Fig. 16.8)

$$\widehat{\mathbf{ab}} = \widehat{\mathbf{a'b}} = \widehat{\mathbf{a'b'}} = \frac{1}{3}\widehat{\mathbf{ab'}} = \theta. \tag{16.79}$$

We can reformulate Eq. (16.76) (see also the solution to Prob. 16.14) in the following manner:

$$-1 \leq 3\wp_{12}(\theta) - \wp_{12}(3\theta) - \wp_1 - \wp_2 \leq 0, \tag{16.80}$$

or

$$f(\theta) \leq 1, \tag{16.81}$$

where

$$f(\theta) = \frac{3\wp_{12}(\theta) - \wp_{12}(3\theta)}{\wp_1 + \wp_2}. \tag{16.82}$$

We now prove that quantum predictions can be in conflict with the CH inequality.

Proof

On general grounds, consider an experiment of the type shown in Fig. 16.7 with the following quantum predictions:

$$\left[\wp_{12}(\theta)\right]_{\mathrm{QM}} = \frac{1}{4}\eta_1\eta_2\wp_1\wp_{2/1}[\epsilon_+^1\epsilon_+^2 + \epsilon_-^1\epsilon_-^2 \mathrm{E}\cos n\theta], \tag{16.83a}$$

$$(\wp_1)_{\mathrm{QM}} = \frac{1}{2}\eta_1\wp_1\epsilon_+^1, \tag{16.83b}$$

$$(\wp_2)_{\mathrm{QM}} = \frac{1}{2}\eta_2\wp_2\epsilon_+^2, \tag{16.83c}$$

where η_j represents the effective quantum efficiency of detector j and $\epsilon_+^j = \epsilon_M^j + \epsilon_m^j$, $\epsilon_-^j = \epsilon_M^j - \epsilon_m^j$, ϵ_M^j and ϵ_m^j being the maximum and the minimum transmissions of the respective analyzers. Moreover, the functions $\wp_1$ and $\wp_2$ are the probabilities that an emission enters apparatus 1 or 2, respectively, while $\wp_{2/1}$ is the conditional probability that, if emission 1 enters apparatus 1 then emission 2 enters apparatus 2. E is a measure of the initial-state purity. The possible values of n are 1 or 2 depending upon whether the experiment is performed with fermions or bosons, respectively.

Now the quantum predictions (16.83) for the function $f(\theta)$ (Eq. (16.82)) yield

$$f_{\mathrm{QM}}(\theta) = \frac{1}{4}\eta\wp_{2/1}\left\{2\epsilon_+ + \mathrm{E}[3\cos n\theta - \cos 3n\theta]\left[\frac{(\epsilon_-)^2}{\epsilon_+}\right]\right\}, \tag{16.84}$$

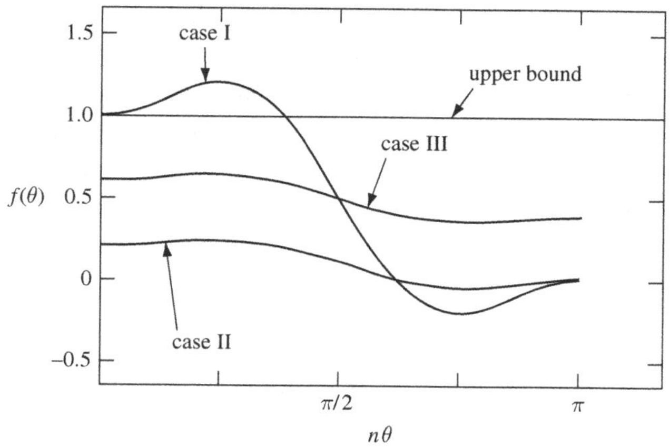

Figure 16.9 Typical dependence of $f(\theta)$ (Eq. (16.82)) upon $n\theta$ in three cases (I–III). The upper bound for $f(\theta)$ set by inequality (16.81) is $+1$. Case I (nearly ideal). We have $\eta \simeq \wp_{2/1} \simeq \mathrm{E} \simeq \epsilon_+ \simeq \epsilon_- \simeq 1$ (see Eqs. (16.83)). Case II. Experiments have nearly ideal parameters $\mathrm{E} \simeq \epsilon_+ \simeq \epsilon_- \simeq 1$, but have $\eta \ll 1$ and/or $\wp_{2/1} \ll 1$. Case III. Experiments have nearly ideal parameters $\eta \simeq \wp_{2/1} \simeq 1$ but have $\mathrm{E} \ll 1$ and/or $\epsilon_-/\epsilon_+ \ll 1$. Adapted from [Clauser/Shimony 1978, 1903].

where, for simplicity, $\eta \equiv \eta_1 = \eta_2$, $\wp_1 = \wp_2$, $\epsilon_+ \equiv \epsilon_+^1 = \epsilon_+^2$, $\epsilon_- \equiv \epsilon_-^1 = \epsilon_-^2$. Selecting the value $\theta = \pi/4n$, one finds that the condition for violation of inequality (16.81) is (see also Fig. 16.9)

$$\eta \wp_{2/1} \epsilon_+ \left[\sqrt{2} \left(\frac{\epsilon_-}{\epsilon_+} \right)^2 \mathrm{E} + 1 \right] > 2. \tag{16.85}$$

In particular, in the ideal case $\eta = \wp_{2/1} = \epsilon_+ = \epsilon_- = \mathrm{E} = 1$, $f_{\mathrm{QM}}(\theta)$ assumes the value $(1 + \sqrt{2})/2$, which is larger than 1.

Q.E.D

The lesson we learn from the above proof is that idealized conditions are not necessary for testing violations of Bell-like inequalities, even though, of course, if experimental conditions are too far from the ideal case, such violations may remain unrevealed.

16.5 Experimental tests

We now wish to discuss the transition from the theory in the form of *Gedankenexperiments* to performed experiments. In doing this, we necessarily have to leave aside a number of important and beautiful experiments, for many of which we refer to the literature.

From the nature of the Bell inequalities it follows that there are two different questions to test:

- Are Bell inequalities experimentally violated?
- Does the violation conform to the predictions of quantum mechanics?

Different tests of the Bell inequalities are possible. Some of these take advantage, e.g., of two-particle polarization correlation (spin entanglement), two-particle energy–time entanglement, time entanglement for fields, or photon–number correlation. These kinds of experiments can be performed either with massless particles (e.g. photon correlation) or with massive particles (e.g., proton–proton scattering and positron annihilation).

The experiment we choose to test the theoretical predictions of the previous section is of a photon-correlation type with linear polarizers on the line of CHSH or CH instead of massive spin-1/2 particles.[37]

16.5.1 From theory to experiment

Before entering into the details of the tests of Bell inequalities, we wish to discuss a few problems connected with this issue. CHSH assumed that, given a pair of photons emerging from the polarizers, the probability of their joint detection was independent of the polarizers' orientations **a** and **b**. This assumption is a bit problematic because there is no way to test it.[38] CH made another assumption which leads to the same result: for every pair emitted (i.e. for each value of λ), the probability of a count with a polarizer in place is less than or equal to the corresponding probability with the polarizer removed[39] – in fact very often only a small number of photons is actually detected. This assumption appears reasonable because the insertion of a polarization analyzer imposes an obstacle between the source of the emission and the detector, and it is natural to think that an obstacle cannot increase the detection probability.[40] However, it is very difficult to prove positively CH's second assumption because it requires that the probability be diminished upon the insertion of a polarizer *for all* λ. On the other hand, attempts at falsifying this assumption as a means of invalidating experimental tests of non-locality or non-separability[41] do not appear to be as reasonable as the assumption itself. We therefore assume CH's hypothesis and look for the general conditions for its test.[42]

We denote with ∞ an apparatus configuration in which the analyzer is absent. $\pi_k(\lambda, \infty)$ denotes the probability of a count from detector k when analyzer k is absent and the state of emission is λ. CH's assumption may then be restated as

$$0 \leq \pi_k(\lambda, \mathbf{j}) \leq \pi_k(\lambda, \infty) \leq 1, \quad k = 1, 2, \quad \mathbf{j} = \mathbf{a}, \mathbf{b}. \tag{16.86}$$

From Eq. (16.86) and following arguments similar to those used in the derivation of Eq. (16.76) (see Prob. 16.14), we have that

$$\begin{aligned} -\wp_{12}(\infty, \infty) \leq \wp_{12}(\mathbf{a}, \mathbf{b}) - \wp_{12}(\mathbf{a}, \mathbf{b}') + \wp_{12}(\mathbf{a}', \mathbf{b}) + \wp_{12}(\mathbf{a}', \mathbf{b}') \\ - \wp_{12}(\mathbf{a}', \infty) - \wp_{12}(\infty, \mathbf{b}) \leq 0. \end{aligned} \tag{16.87}$$

[37] For a review of first experimental attempts in quantum optics see [Reid/Walls 1986].

[38] See [Clauser/Shimony 1978, 1912–13].

[39] See [Clauser/Horne 1974, 530].

[40] Though this statement is not true in general in quantum mechanics (see Subsec. 2.1.5), it appears to be valid in this context.

[41] For example by Caser [Caser 1992, 24–25].

[42] See [Clauser/Shimony 1978, 1905–906] [Fry 1995, 234–35].

We again use the rotational invariance argument (Eqs. (16.78) and (16.79)), so that

$$\wp_{12}(\mathbf{j}, \infty) = \wp_{12}(\infty), \quad \mathbf{j} = \mathbf{a}, \mathbf{b}, \tag{16.88a}$$

$$\wp_{12}(\mathbf{a}, \mathbf{b}) = \wp_{12}(\theta), \quad \theta = \widehat{\mathbf{ab}}, \tag{16.88b}$$

so that we may rewrite inequality (16.87) in the following form (see inequality (16.80)):

$$-\wp_{12}(\infty, \infty) \leq 3\wp_{12}(\theta) - \wp_{12}(3\theta) - \wp_{12}(\mathbf{a}', \infty) - \wp_{12}(\infty, \mathbf{b}) \leq 0. \tag{16.89}$$

Since the emission rates in all experiments would be held constant, we can write the ratios of probabilities as ratios of the corresponding counting rates

$$\frac{\wp_{12}(\mathbf{a}', \infty)}{\wp_{12}(\infty, \infty)} = \frac{R_1}{R_0}, \quad \frac{\wp_{12}(\infty, \mathbf{b})}{\wp_{12}(\infty, \infty)} = \frac{R_2}{R_0}, \quad \frac{\wp_{12}(\theta)}{\wp_{12}(\infty, \infty)} = \frac{R(\theta)}{R_0}, \tag{16.90}$$

so that we can write Eq. (16.89) in the form

$$-R_0 \leq 3R(\theta) - R(3\theta) - R_1 - R_2 \leq 0. \tag{16.91}$$

If we take $\theta = \pi/8$ for the upper–limit violation, we have

$$-R_0 \leq 3R(\pi/8) - R(3\pi/8) - R_1 - R_2 \leq 0. \tag{16.92}$$

On the other hand, if we take $\theta = 3\pi/8$ for the lower-limit violation, using the fact that, in this context, $9\pi/8$ is the same angle as $\pi/8$, we have

$$-R_0 \leq 3R(3\pi/8) - R(\pi/8) - R_1 - R_2 \leq 0. \tag{16.93}$$

Dividing both inequalities by R_0 and subtracting (*cum grano salis*) the second inequality from the first, and then the first from the second, we obtain

$$\frac{|R(\pi/8) - R(3\pi/8)|}{R_0} \leq \frac{1}{4}. \tag{16.94}$$

If we take, in the ideal case,[43] pairs of photons propagating in opposite directions from the source along the z-axis with total angular momentum 0 and total parity $+1$, for the polarization part of the wave function we have

$$\begin{aligned} | \Psi_0 \rangle &= \frac{1}{\sqrt{2}} \left(| x \rangle_1 \otimes | x \rangle_2 + | y \rangle_1 \otimes | y \rangle_2 \right) \\ &= \frac{1}{\sqrt{2}} \left[\begin{pmatrix} 1 \\ 0 \\ 0 \end{pmatrix}_1 \otimes \begin{pmatrix} 1 \\ 0 \\ 0 \end{pmatrix}_2 + \begin{pmatrix} 0 \\ 1 \\ 0 \end{pmatrix}_1 \otimes \begin{pmatrix} 0 \\ 1 \\ 0 \end{pmatrix}_2 \right], \end{aligned} \tag{16.95}$$

where the kets represent polarization vectors along the x-axis and the other two along the y-axis.

The projection operator for linear polarization along an axis lying in the xy-plane and making angle ϕ with the x-axis is given by

$$\hat{P}(\phi) = | \phi \rangle \langle \phi | = \begin{bmatrix} \cos^2 \phi & \cos \phi \sin \phi & 0 \\ \cos \phi \sin \phi & \sin^2 \phi & 0 \\ 0 & 0 & 0 \end{bmatrix}. \tag{16.96}$$

[43] See [Shimony 1971, 82–85] [Clauser/Horne 1974, 530] [Clauser/Shimony 1978, 1906–907].

The vector

$$|\phi\rangle = \begin{pmatrix} \cos\phi \\ \sin\phi \\ 0 \end{pmatrix}, \qquad (16.97)$$

representing linear polarization in that chosen direction, is an eigenvector of the projector (16.96) with eigenvalue 1, while the vector

$$|\phi_\perp\rangle = \begin{pmatrix} -\sin\phi \\ \cos\phi \\ 0 \end{pmatrix}, \qquad (16.98)$$

representing linear polarization perpendicular to the chosen direction, is again an eigenvector of $\hat{P}(\phi)$ but with eigenvalue 0.

The quantum predictions (using again $\theta = \widehat{\mathbf{ab}}$) are (see Prob. 16.15)

$$\left[\frac{R(\theta)}{R_0}\right]_{\Psi_0} = \left\langle \Psi_0 | \hat{P}(\mathbf{a}) \otimes \hat{P}(\mathbf{b}) | \Psi_0 \right\rangle = \frac{1}{4}(1 + \cos 2\theta), \qquad (16.99)$$

from which we find

$$\left[\frac{R(\pi/8)}{R_0} - \frac{R(3\pi/8)}{R_0}\right]_{\Psi_0} = \frac{1}{4}\sqrt{2}, \qquad (16.100)$$

which obviously violates Eq. (16.94). As it has already been pointed out, we cast the following experimental tests into two classes:

- the test for stochastic HV predictions;
- the test for quantum predictions.

Freedman and Clauser[44] observed that the 5513 $\mathring{A}$ and 4227 $\mathring{A}$ pairs of photons produced by $4p^{2\,1}S_0 \rightsquigarrow 4s4p^1P_1 \rightsquigarrow 4s^{2\,1}S_0$ cascade in calcium (see Fig. 16.10). Calcium atoms in a beam from an oven were excited by resonance absorption to the $3d4p^1P_1$ level, from which a considerable fraction decayed to the $4p^{2\,1}S_0$ state at the top of the cascade (see Fig. 16.11).[45]

The average ratios for approximately 200 hours of running time were

$$\left\langle \frac{R(\pi/8)}{R_0} \right\rangle = 0.400 \pm 0.007, \qquad (16.101a)$$

$$\left\langle \frac{R(3\pi/8)}{R_0} \right\rangle = 0.100 \pm 0.003, \qquad (16.101b)$$

$$\left\langle \frac{R(\pi/8) - R(3\pi/8)}{R_0} \right\rangle = 0.300 \pm 0.008, \qquad (16.101c)$$

and clearly violate inequality (16.94) by about six standard deviations.

[44] See [Freedman/Clauser 1972].

[45] More technical details can be found in the original paper.

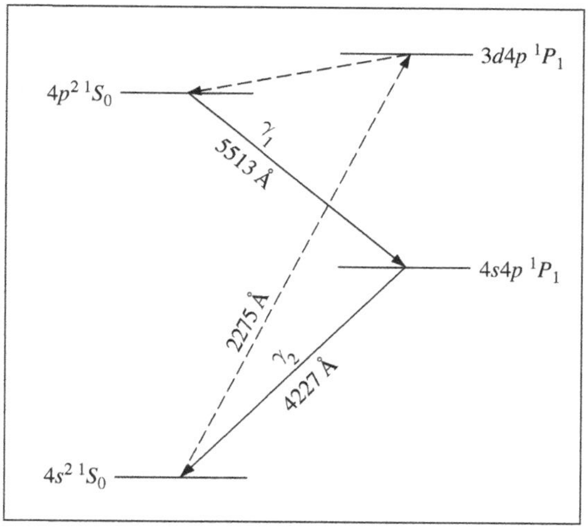

Figure 16.10 Partial Grotrian diagram of atomic calcium for Freedman and Clauser's experiment. Adapted from [Freedman *et al.* 1976, 53].

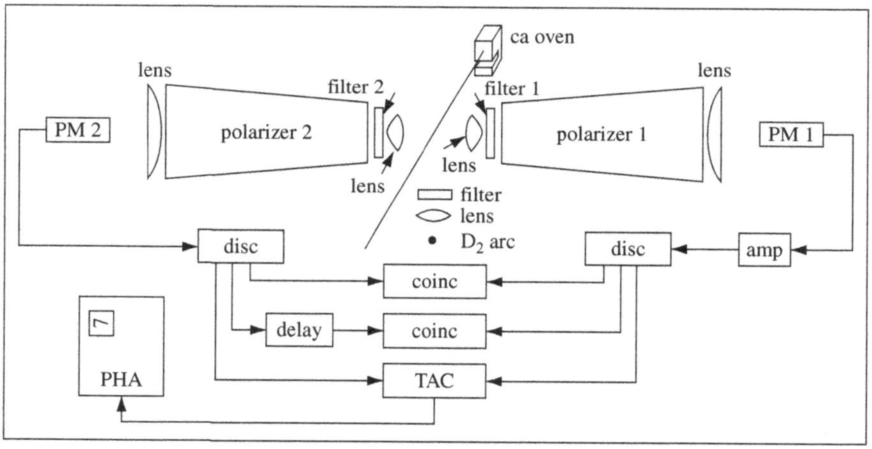

Figure 16.11 Schematic diagram of apparatus and associated electronics of the experiment by Freedman and Clauser. Scalers (not shown) monitored the outputs of the discriminators and coincidence circuits. Adapted from [Freedman/Clauser 1972].

We now test the *quantum predictions*. With all the necessary corrections from an ideal case, quantum prediction (16.100) has to be modified as

$$\frac{[R(\pi/8) - R(3\pi/8)]}{R_0} = (0.401 \pm 0.005) - (0.100 \pm 0.005) = 0.301 \pm 0.007, \quad (16.102)$$

which agrees exceptionally well with the experimental results (Eq. (16.101)). Freedman and Clauser's experiment is thus an important step toward the rejection of HV stochastic theories and a strong confirmation of quantum mechanics on one of its most counterintuitive playgrounds, namely the violation of the separability principle.

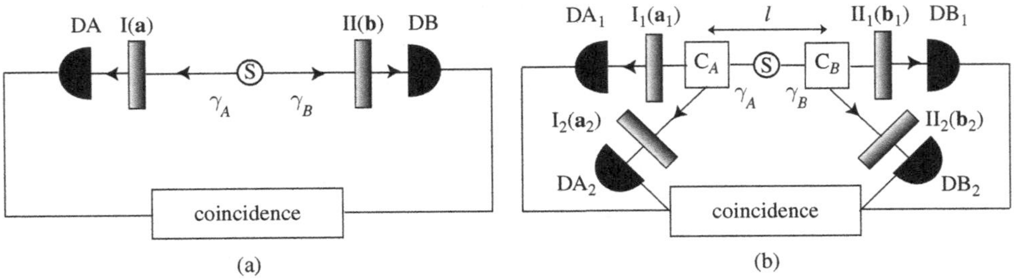

Figure 16.12 (a) Friedman–Clauser experiment: the correlated photons γ_A, γ_B coming from the source S impinge upon the linear polarizers I, II oriented in directions **a**, **b**, respectively. The rate of joint detection by the photomultipliers is monitored for various combinations of orientations. (b) Experiment proposed by Aspect: the optical commutator C_A directs the photon γ_A either towards polarizer I_1 with orientation $\mathbf{a}_1$ or to polarizer I_2 with orientation $\mathbf{a}_2$. Similarly for C_B. The two commutators work independently (the time intervals between two commutations are taken to be stochastic). The four joint detection rates are monitored and the orientations $\mathbf{a}_1, \mathbf{a}_2, \mathbf{b}_1, \mathbf{b}_2$ are not changed for the whole experiment. l is the separation between the switches.

16.5.2 Loopholes

The test we have discussed so far appears rather convincing and a clear confirmation of the validity of quantum mechanics against any HV theory. However, for the sake of completeness, one has to admit the possibility of some loopholes. A *loophole* may here be understood as an implicit assumption which undermines the full validity of an experimental test. This assumption is often connected to technical limits that have not been taken into account or to the introduction of undesired parameters. In the following we discuss some possible loopholes, which, at least in principle, may affect the type of test discussed in the previous subsection.

Locality loophole

The *first loophole* we consider is the locality loophole. In all experiments, one should consider the possibility that the result of a measurement obtained by using a certain polarizer direction depend on the orientation of the other polarizer. In other words, it could be the case that quantum mechanics violates not only separability but also locality, which we understand in this context as a superluminal connection between the two polarizers. We shall discuss the issue of locality in more detail later (see Subsec. 16.8.1). Here, we are not interested in excluding the non-locality as a possibility, but only in excluding its possible influence on the results of the experiments performed in order to test the separability as such.

In 1976 Aspect[46] proposed for the first time an experiment in which, instead of a fixed apparatus[47] as is shown in Fig. 16.12(a), an apparatus with optical commutators is used

[46] See [Aspect 1976].
[47] See also [Aspect *et al.* 1982a].

(see Fig. 16.12(b)). This configuration allows us to ensure the validity of Bell's locality condition (16.46). As a matter of fact, this condition, though reasonable, is not a consequence of any fundamental physical law.[48] Using time-variable analyzers – as in the experiment we are currently discussing – the locality condition becomes a consequence of Einstein's causality, which forbids superluminal influences. The experiment was later performed by Aspect, Dalibard, and Roger.[49] The switching between the two channels occurs about every 10 ns. Since this delay, as well as the lifetime of the intermediate level of the cascade (5 ns), is small compared with l/c (40 ns) (see Fig. 16.12(b)), a detection event and the corresponding change in the orientation of the polarizer on the other side of the apparatus are separated by space-like intervals.

Testing a Clauser–Horne type inequality (Eq. (16.76)) by choosing angles between the orientations $\mathbf{a}$ ($\mathbf{a}_1$) and $\mathbf{b}$ ($\mathbf{b}_1$), $\mathbf{b}$ and $\mathbf{a}'$ ($\mathbf{a}_2$), and $\mathbf{a}'$ and $\mathbf{b}'$ ($\mathbf{b}_2$) to be equal to 22.5°, and the angle between $\mathbf{a}$ and $\mathbf{b}'$ equal to 67.5°, the experimental result for the expression in Eq. (16.87) was 0.101 ± 0.020, clearly violating the corresponding inequality by five standard deviations.

Due to these results, we may conclude that the second Bell theorem excludes not only non-contextual HV theories, because they satisfy the requirement of separability, but also contextual ones (see the conclusions of Subsec. 16.4.1). In fact, if the two sides of the apparatus are space-like separated, a supporter of a HV theory is forced to accept a strong form of non-locality, i.e. an action-at-a-distance, in order to explain how it is possible to obtain Bell inequalities violations both with fixed polarizers and with random switches.[50]

Angular-correlation loophole

The problem of a possible correlation between polarizers is not the only difficulty in the experiments which have been performed to test Bell inequalities. Another difficulty (*second loophole*) concerns the angular correlation:[51] because of the cosine-squared angular correlation of the directions of the photons emitted in an atomic cascade, an inherent polarization decorrelation is present. Hence the very polarization correlation which could result in a violation of one of the Bell inequalities is reduced for non-collinear photons.

The problem can be overcome[52] by using SPDC sources (see Subsec. 13.7.6) instead of atomic cascade ones. Pairs of photons resulting from SPDC can have an angular correlation of better than 1 mrad, although in general they need not be collinear. Initially experiments involving SPDC were limited by low quantum efficiency, but more recently efficiencies larger than 90% have been reached.

We report here a SPDC experiment performed for the first time by Alley and Shih and successively improved by Ou and Mandel.[53] While Alley and Shih obtained a violation of

48 See [Bell 1964].
49 See [Aspect *et al.* 1982b].
50 See [Shimony 1984, 109–116].
51 See [Santos 1991] [Santos 1992].
52 See [Kwiat *et al.* 1994, 3210].
53 See [Shih/Alley 1988] [Ou/Mandel 1988].

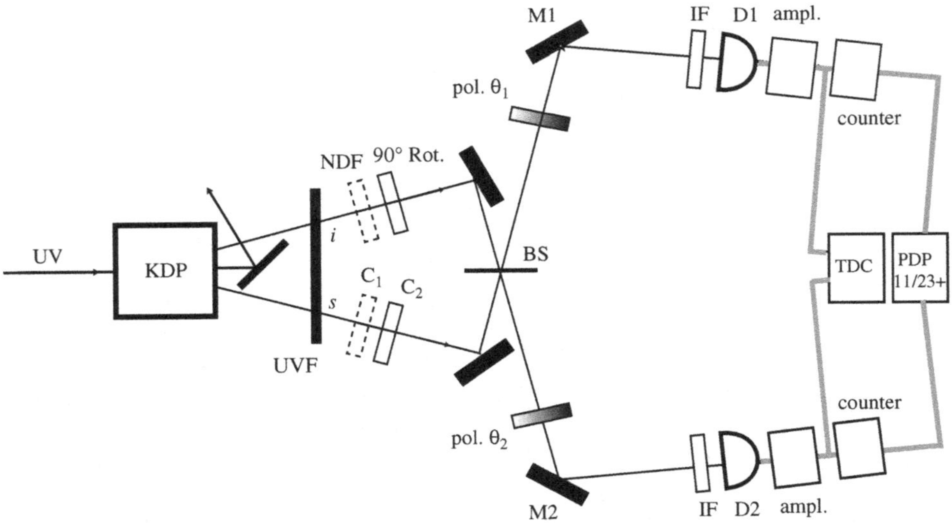

Figure 16.13 Outline of the Alley–Shih and Ou–Mandel's experiment. Light from the 351.1 nm line of an argon-ion laser falls on a non-linear crystal of potassium dihydrogen phosphate (KDP), where down-converted photons of wavelength of about 702 nm are produced. When the condition for degenerate phase matching is satisfied, down-converted, linearly polarized signal and idler photons emerge at angles of about ±2° relative to the ultraviolet (UV) pump beam with the electric vector in the plane of the diagram. The *idler* (i) photons pass through a 90° polarization rotator, while the *signal* (s) photons traverse a compensating glass plate C_1 producing an equal time delay. S-photons and i-photons are then directed from opposite sides towards a beam splitter (BS). The input to the BS consists of an x-polarized s-photon and of a rotated y-polarized i-photon. The light beams emerging from BS, consisting of a mixing of i-photons and s-photons, pass through linear polarizers set at adjustable angles θ_1 and θ_2, through similar interference filters (IF) and finally fall on two photodetectors D_1 and D_2. The photoelectric pulses from D_1 and D_2 are amplified and shaped and fed to the start and stop inputs of a time-to-digital converter (TDC) under computer control which functions as a coincidence counter.

Bell inequality by three standard deviations, the experiment performed by Ou and Mandel obtained violations as large as six standard deviations (see Fig. 16.13).

If we denote by $\wp(\theta_1, \theta_2)$ the joint probability of detecting two photons for a setting θ_1, θ_2 of the two linear polarizers, we may rewrite the Clauser–Horne inequality (16.87) as

$$\mathrm{C} = \wp(\theta_1, \theta_2) - \wp(\theta_1, \theta_2') + \wp(\theta_1', \theta_2') + \wp(\theta_1', \theta_2) - \wp(\theta_1', \infty) - \wp(\infty, \theta_2) \leq 0, \tag{16.103}$$

where again ∞ stands for the absence of analyzer.

Quantum mechanics describes the output state as the following linear superposition state of the two channels 1 and 2 (towards detectors D1 and D2):

$$\begin{aligned} |\Psi\rangle = {} & \mathrm{T}_x \mathrm{T}_y \,|x\rangle_1 \,|y\rangle_2 + \mathrm{R}_x \mathrm{R}_y \,|y\rangle_1 \,|x\rangle_2 \\ & - \imath \mathrm{R}_y \mathrm{T}_x \,|x, y\rangle_1 \,|0\rangle_2 + \imath \mathrm{R}_x \mathrm{T}_y \,|0\rangle_1 \,|x, y\rangle_2 , \end{aligned} \tag{16.104}$$

where $R_x^2, R_y^2, T_x^2, T_y^2$ are the polarization-dependent beam-splitter reflectivities and transmissivities with $R_x^2 + T_x^2 = 1$ and $R_y^2 + T_y^2 = 1$ (for the sake of simplicity we have assumed that all coefficients are real). Using polarized scalar fields at the two detectors, it is possible to calculate the probability

$$\wp(\theta_1, \theta_2) = \eta \left(T_x T_y \cos\theta_1 \sin\theta_2 + R_x R_y \sin\theta_1 \cos\theta_2 \right)^2, \qquad (16.105)$$

where η is characteristic of the detectors efficiency. Equation (16.105) reduces to

$$\wp(\theta_1, \theta_2) = \frac{1}{4}\eta \sin^2(\theta_1 + \theta_2) \qquad (16.106)$$

if $R_x^2 = T_x^2 = 1/2$ and $R_y^2 = T_y^2 = 1/2$. If the polarizer angles are chosen so that $\theta_1 = \pi/8, \theta_2 = \pi/4, \theta_1' = 3\pi/8, \theta_2' = 0$, one sees that the quantum correlation function (see Prob. 16.16)

$$C = \frac{1}{4}\eta(\sqrt{2} - 1) > 0 \qquad (16.107)$$

violates inequality (16.103) for any $\eta > 0$. If we express the function C in terms of coincidence rates instead of probabilities, Ou and Mandel have found the experimental result

$$\begin{aligned} C_{\text{exp}} &= R(22.5°, 45°) - R(22.5°, 0°) + R(67.5°, 45°) + R(67.5°, 0°) \\ &\quad - R(67.5°, \infty) - R(\infty, 45°) \\ &= (11.5 \pm 2.0)\,\text{min}^{-1}. \end{aligned} \qquad (16.108)$$

Therefore, C_{exp} is positive with an accuracy of about six standard deviations, in violation of inequality (16.103) (see Fig. 16.14).[54]

Detection loophole

Up to now we have considered the loopholes concerning possible non-local correlations and the angular (de-)correlation. A further issue (*third loophole*) is represented by the detection loophole.[55] In fact, we may raise the question of how high the detection efficiencies (the parameters η_j $(j = 1, 2)$ in Eqs. (16.83)) must be for the experimental confirmation of the quantum theoretical predictions. With experiments like that of Aspect and co-workers a sufficient value for the efficiency is $\eta = 83\%$.[56] But if the inequality is optimized changing the angle of the settings after the correction for $\eta < 1$, then a lower requirement may be sufficient, varying from 66.7% to 100% depending on the variation of the background level from 0.00% to 10.36%.[57]

[54] In successive experiments the number of standard deviations for the violation has been strongly improved. For instance, Kwiat and co-workers [Kwiat *et al.* 1999] have obtained a 242–σ violation of Bell's inequality.

[55] See [Ferrero *et al.* 1990, 686–87]. See also [Kwiat *et al.* 1994, 3209].

[56] See [Mermin 1986].

[57] See [Eberhard 1993].

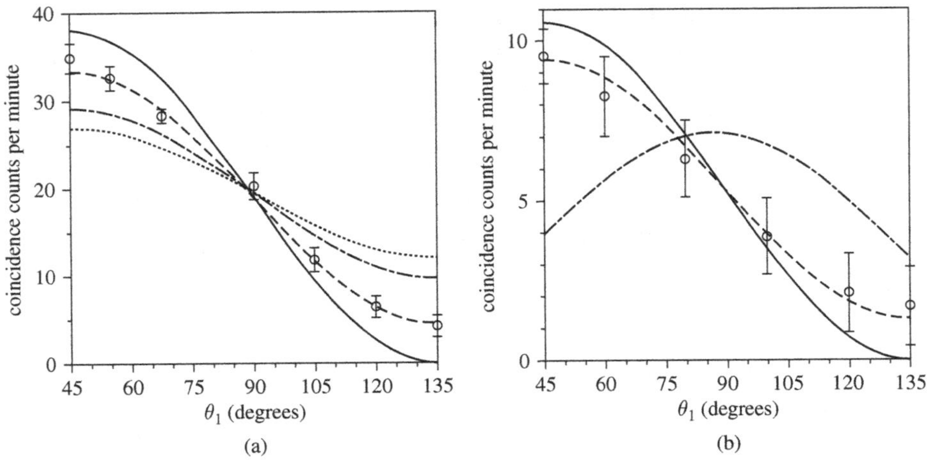

Figure 16.14 Confirmation of the quantum-mechanical predictions in the Ou–Mandel experiment. (a) Measured coincidence counting rate as a function of the polarizer angle θ_1, with θ_2 fixed at 45°. The solid line represents quantum prediction and the dash-dotted curve the classical prediction. The dashed and dotted lines are the preceding curves with some corrections added to reduce modulation caused by imperfect alignment. (b) Measured coincidence counting rate as a function of the polarizer angle θ_1 with θ_2 fixed at 45°, when a 8:1 attenuator is inserted into the idler beam. The curves are the same as in (a) except that the dotted curve is absent. Adapted from [Ou/Mandel 1988].

Instead, the problem with SPDC-type experiments is that, even with high detection efficiency, one must discard part of the counts, since, if we have an output state of the type (16.104), we are obliged to discard all events where both photons are in the same channel, represented by the third and fourth components of the rhs of Eq. (16.104), and one could pose the question whether this selection might represent a bias. Even though this is a remote possibility, in order to exclude any ambiguity a more refined solution is required. A possibility is to directly produce a pair of photons in singlet-type state, thus avoiding any post-selection. One of the first proposals for doing this is shown and summarized in Fig. 16.15. By means of this apparatus it is possible to produce output photons in the state

$$|\Psi\rangle \simeq |v\rangle_3 |h\rangle_4 + e^{\imath\phi}(|h\rangle_3 |v\rangle_4). \tag{16.109}$$

Finally, we briefly discuss a related problem connected to Clauser and Horne's assumption: with polarizers in place the probability of a count is not larger than without polarizers.[58] This assumption has been questioned on the basis of a possible form of *enhancement* in the detection process. It has been shown that, even in this hypothesis, there is a detectable difference between HV theories and quantum mechanics in the case of experiments with three polarizers.[59]

[58] See [Marshall *et al.* 1983].

[59] See [Garuccio/Selleri 1984]. For the sake of completeness, we cite a paper by Marshall and Santos [Marshall/Santos 1985] that does not strictly exclude a form of enhancement which would be able to invalidate

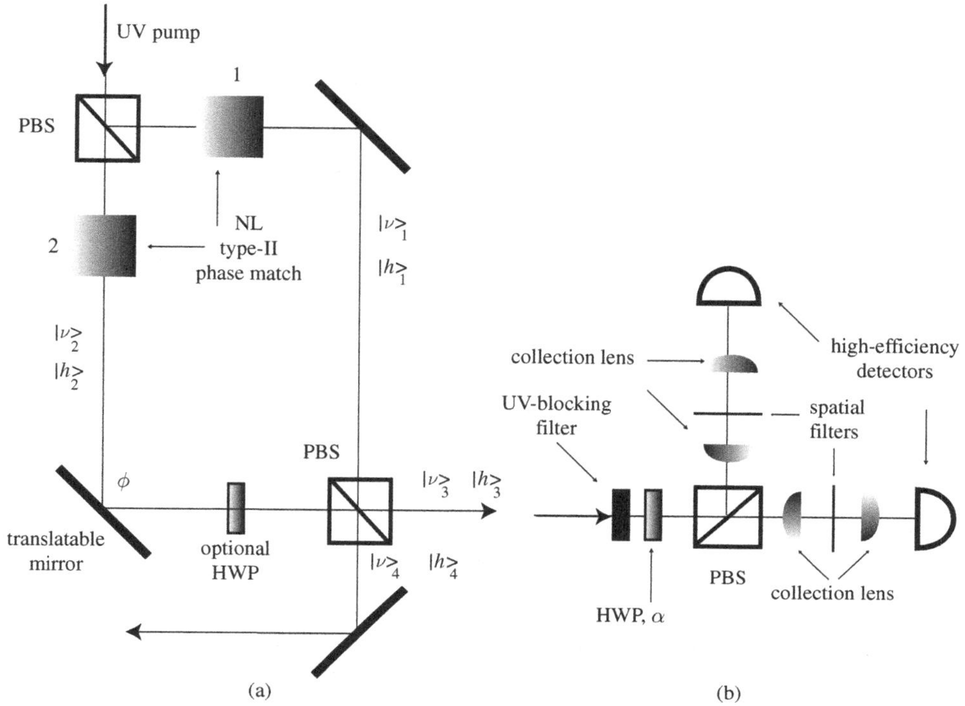

Figure 16.15 Proposed experiment for solving the detection loophole. (a) An ultraviolet pump photon may be spontaneously down-converted in either of two non-linear crystals, producing a pair of collinear orthogonally polarized photons at half the frequency (type-II phase matching). The outputs are directed toward a second PBS [see Subsec. 3.5.2]. When the outputs of both crystals are combined with an appropriately relative phase ϕ, a true singlet- or triplet-like state may be produced. By using a half-wave plate (HWP) to effectively exchange the polarizations of photons originating in crystal 2, one overcomes several problems arising from non-ideal phase matching. An additional mirror is used to direct the photons into opposite direction towards separated analyzers. (b) A typical analyzer, including an HWP to rotate by θ the polarization component selected by the analyzing BS, and precision spatial filters to select only conjugate pairs of photons. In an advanced version of the experiment, the HWP could be replaced by an ultrafast polarization rotator (such as Pockels or Kerr cells) to also close the locality loophole.

16.6 Bell inequalities with homodyne detection

We schematically present here a different way to test Bell inequalities, but we will not go into any details of the experimental testing. However, we include it here because we think that it may be very instructive.

at least some of the performed experiments – see also [Haji-Hassan *et al.* 1987] for more details. However, such a form of enhancement does not seem very plausible.

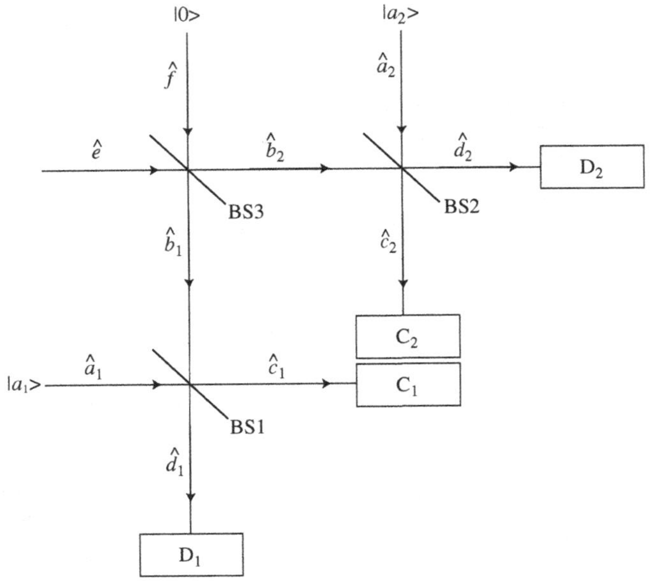

Figure 16.16 "Entanglement" with vacuum. Here, we consider a pair of homodyne detectors (see Subsec. 13.6.2), each of which consists of a symmetric beam splitter (BS1 or BS2), a coherent local oscillator with amplitude $a_k = \alpha e^{\imath \phi_k}$, and two photodetectors as output. One of the two inputs to these homodyne detectors is derived from a third symmetric beam splitter (BS3).

16.6.1 Photon-intensities correlation

An interesting variation of the self-interference experiment is represented by the "entanglement" of photons with a vacuum[60] (see Fig. 16.16). This can also be considered to be an interesting generalization of beam splitting. The transformations between the mode operators shown in Fig. 16.16 are (see Subsec. 3.5.2)

$$\begin{pmatrix} \hat{c}_k \\ \hat{d}_k \end{pmatrix} = \frac{1}{\sqrt{2}} \begin{bmatrix} 1 & \imath \\ \imath & 1 \end{bmatrix} \begin{pmatrix} \hat{a}_k \\ \hat{b}_k \end{pmatrix}; \tag{16.110a}$$

$$\begin{pmatrix} \hat{b}_1 \\ \hat{b}_2 \end{pmatrix} = \frac{1}{\sqrt{2}} \begin{bmatrix} 1 & \imath \\ \imath & 1 \end{bmatrix} \begin{pmatrix} \hat{f} \\ \hat{e} \end{pmatrix}. \tag{16.110b}$$

Summing up,

$$\begin{pmatrix} \hat{c}_1 \\ \hat{d}_1 \\ \hat{c}_2 \\ \hat{d}_2 \end{pmatrix} = \begin{bmatrix} \frac{1}{\sqrt{2}} & \frac{\imath}{2} & 0 & -\frac{1}{2} \\ \frac{\imath}{\sqrt{2}} & \frac{1}{2} & 0 & \frac{\imath}{2} \\ 0 & -\frac{1}{2} & \frac{1}{\sqrt{2}} & \frac{\imath}{2} \\ 0 & \frac{\imath}{2} & \frac{\imath}{\sqrt{2}} & \frac{1}{2} \end{bmatrix} \begin{pmatrix} \hat{a}_1 \\ \hat{f} \\ \hat{a}_2 \\ \hat{e} \end{pmatrix}. \tag{16.111}$$

60 See [Tan *et al.* 1991].

Now we calculate the corresponding coincidence probabilities. First, let us consider the case of vacuum inputs to the modes $\hat{f}, \hat{e}$. The local oscillators are assumed to be in coherent states: $|\alpha e^{\imath\phi_1}\rangle, |\alpha e^{\imath\phi_2}\rangle$. The intensities at all detectors are equal to

$$\langle I_{C_1}\rangle = \langle I_{C_2}\rangle = \langle I_{D_1}\rangle = \langle I_{D_2}\rangle = \frac{1}{2}\alpha^2, \tag{16.112a}$$

where the the detectors C_k, D_k correspond to mode operators $\hat{c}_k, \hat{d}_k$. The two-photon correlation functions are also equal, i.e.

$$\langle I_{C_1} I_{C_2}\rangle = \langle I_{D_1} I_{D_2}\rangle = \langle I_{C_1} I_{D_2}\rangle = \langle I_{D_1} I_{C_2}\rangle = \frac{1}{4}\alpha^4. \tag{16.112b}$$

Consider now the input of a single photon in mode $\hat{e}$ while the mode $\hat{f}$ is in the vacuum state. The total state after the first BS can be written as

$$|\Psi\rangle = \frac{1}{\sqrt{2}}\left(\imath|1\rangle_{b_1}|0\rangle_{b_2} + |0\rangle_{b_1}|1\rangle_{b_2}\right), \tag{16.113}$$

that looks as an entangled state of one-photon and the vacuum.[61] The photon intensities at each detector are given by

$$\langle I_{C_1}\rangle = \langle I_{C_2}\rangle = \langle I_{D_1}\rangle = \langle I_{D_2}\rangle = \frac{1}{2}\alpha^2 + \frac{1}{4}, \tag{16.114a}$$

and are increased by $1/4$ relatively to calculation (16.112a). Moreover,

$$\langle I_{C_1} I_{C_2}\rangle = \langle I_{D_1} I_{D_2}\rangle = \frac{1}{4}\left\{\alpha^4 + \alpha^2[1 + \sin(\phi_1 - \phi_2)]\right\}, \tag{16.114b}$$

$$\langle I_{C_1} I_{D_2}\rangle = \langle I_{D_1} I_{C_2}\rangle = \frac{1}{4}\left\{\alpha^4 + \alpha^2[1 - \sin(\phi_1 - \phi_2)]\right\}, \tag{16.114c}$$

where, if we set $\phi_1 - \phi_2 = -\pi/2$, we get the minimum of coincidence rate $\alpha^4/4$ for detector pairs (C_1, C_2) and (D_1, D_2), and the maximum value $\alpha^4/4 + \alpha^2/2$ for pairs (C_1, D_2) and (D_1, C_2). The classical wave description of light also shows a similar non-local behavior. However, it is possible to distinguish between the quantum and the classical predictions.

In order to show this, we calculate, according to quantum mechanics, an intensity correlation coefficient as follows:

$$C(\phi_1, \phi_2) = \frac{\langle (I_{D_1} - I_{C_1})(I_{D_2} - I_{C_2})\rangle}{\langle (I_{D_1} + I_{C_1})(I_{D_2} + I_{C_2})\rangle}. \tag{16.115}$$

By evaluating the previous expression in the case of a single-photon input, we obtain

$$C(\phi_1, \phi_2) = \left[\frac{1}{\alpha^2 + 1}\right]\sin(\phi_1 - \phi_2). \tag{16.116}$$

[61] Strictly speaking, the state (16.113) cannot be considered an entangled state, since it contains only one excitation. However, this form is useful for our purposes here.

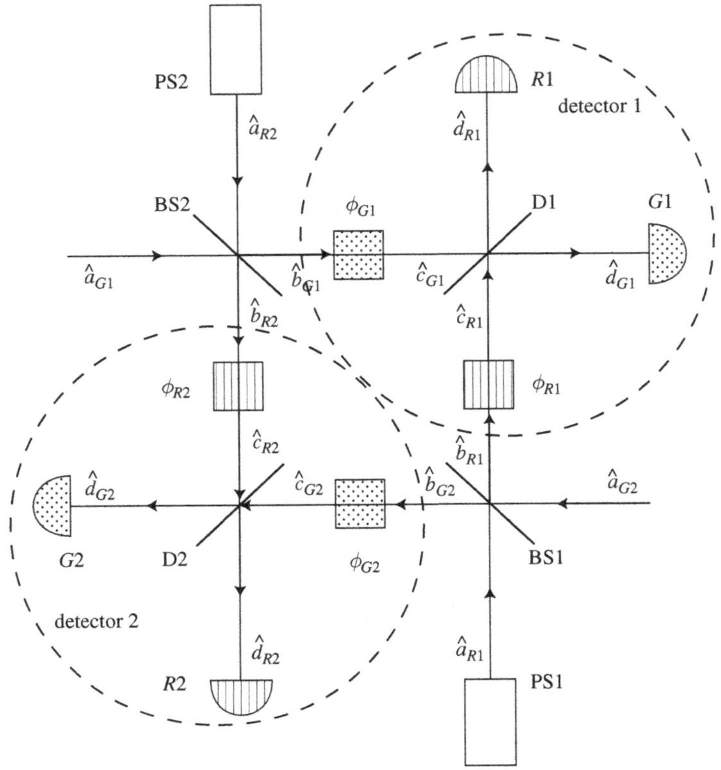

Figure 16.17 Schematics of Yurke and Stoler's thought-experiment. Particles enter by pumps PS1 and PS2 (lines), vacuum by the other input port (dots), before beam splitters BS1 and BS2. Each arm consists of a phase shifter ϕ_{Gj} (j = 1, 2) or ϕ_{Rj}(*j* = 1, 2), a beam merger (D1 or D2) and a particle counter *Gj* (*j* = 1, 2) or *Rj* (*j* = 1, 2). Detector *j* consists of phase shifters ϕ_{Gj} and ϕ_{Rj}, of beam splitter Dj, and of particle counters *Gj* and *Rj*. The original color version of this figure is on the Cambridge University Press website, at www.cambridge.org/9780521869638.

If the coefficient of $\sin(\phi_1 - \phi_2)$ is greater than $1/\sqrt{2}$, there is a non-local behavior that directly translates into a violation of a Bell-type inequality.

16.6.2 Particles from different sources

Until the end of the 1980s the general belief was that entanglement was a consequence of the fact that the particles involved originated from the same source. However, Yurke and Stoler, by means of a thought-experiment, showed[62] that entanglement can also originate with two or three photons deriving from different sources. The scheme of the experiment is depicted in Fig. 16.17, where *R* stands for "red" (represented by vertical lines) and *G* for "green" (represented by dots). The involved input–output transformations induced by the beam splitters are

62 See [Yurke/Stoler 1992a, Yurke/Stoler 1992b].

$$\begin{pmatrix} \hat{d}_{R1} \\ \hat{d}_{G1} \end{pmatrix} = \frac{1}{\sqrt{2}} \begin{bmatrix} 1 & \imath \\ \imath & 1 \end{bmatrix} \begin{pmatrix} \hat{c}_{R1} \\ \hat{c}_{G1} \end{pmatrix}, \tag{16.117a}$$

$$\begin{pmatrix} \hat{d}_{R2} \\ \hat{d}_{G2} \end{pmatrix} = \frac{1}{\sqrt{2}} \begin{bmatrix} 1 & \imath \\ \imath & 1 \end{bmatrix} \begin{pmatrix} \hat{c}_{R2} \\ \hat{c}_{G2} \end{pmatrix}, \tag{16.117b}$$

and

$$\begin{pmatrix} \hat{b}_{R1} \\ \hat{b}_{G2} \end{pmatrix} = \frac{1}{\sqrt{2}} \begin{bmatrix} 1 & \imath \\ \imath & 1 \end{bmatrix} \begin{pmatrix} \hat{a}_{R1} \\ \hat{a}_{G2} \end{pmatrix}, \tag{16.118a}$$

$$\begin{pmatrix} \hat{b}_{R2} \\ \hat{b}_{G1} \end{pmatrix} = \frac{1}{\sqrt{2}} \begin{bmatrix} 1 & \imath \\ \imath & 1 \end{bmatrix} \begin{pmatrix} \hat{a}_{R2} \\ \hat{a}_{G1} \end{pmatrix}. \tag{16.118b}$$

The possible events for each detector (1 or 2) are elements of the set $\{0, R, G, R^2, G^2, E\}$, where 0 represents the event in which nothing is detected, R (G) the event for the photon counter labelled R (G) firing once, R^2 (G^2) the event for the photon counter labelled R (G) firing twice, and E the event in which each photon counter of the detector counts a single photon. Let us now consider possible joint events of the two detectors. In particular, $A = \{RR, GG\}$ be the event in which both of the R photon counters or both of the G photon counters fire, $B = \{RG, GR\}$ the event where each detector counts a single photon and only one of the R photon counters fires, $C = \{0E, E0\}$ the event in which both counters of one detector fire, and finally $D = \{0R^2, 0G^2, R^20, G^20\}$ the event in which one counter of one detector fires twice. However, due to the geometry of the apparatus (in particular to the destructive interference), events C do not occur. Suppose also that the detector phase ϕ_1 of detector 1, given by $\phi_1 = \phi_{R1} - \phi_{G1}$, and the detector phase ϕ_2 of detector 2, given by $\phi_2 = -\phi_{R2} + \phi_{G2}$, can only take one of the three values: ϕ_a, ϕ_b, ϕ_c. For brevity, we shall refer to these settings as 1, 2, and 3, respectively. For the sake of simplicity, assume also that we always have $\phi_1 = \phi_2$, so that the detector phase settings are 11, 22, or 33. The probabilities for the possible events ξ_1 and ξ_2 for detectors 1 and 2 are given by

$$\frac{1}{4}\cos^2(\phi_1 - \phi_2) \quad \text{if} \quad \xi_1, \xi_2 \in A \tag{16.119a}$$

$$\frac{1}{4}\sin^2(\phi_1 - \phi_2) \quad \text{if} \quad \xi_1, \xi_2 \in B \tag{16.119b}$$

$$\frac{1}{8} \quad \text{if} \quad \xi_1, \xi_2 \in D. \tag{16.119c}$$

Since the detector phases can be changed randomly up to the instant the photon enters the detector, one concludes that, from a local realist point of view, there is a set of instructions of the form $\lambda_1, \lambda_2, \lambda_3; \lambda'_1, \lambda'_2, \lambda'_3$, such that the λ_j's (with $j \in \{1, 2, 3\}$) are elements of the set $\{R, G\}$, and λ_j is the instruction to the detector 1 telling it which counter has to fire when the detector switch for the phase is j. The λ'_j's play a similar role for detector 2. For instance, the set of instructions (RGR; RGR), means that, for detector 1:

- when the switch position is on 1 (phase ϕ_a), the particle counter R will fire;
- when the switch position is on 2 (phase ϕ_b), the particle counter G will fire;
- when the switch position is on 3 (phase ϕ_c), the particle counter R will fire.

The same instruction set is, in this case, sent to detector 2.

By the geometry of the apparatus, and limiting ourselves to the set B of events, we have

$$\wp(R, G, \phi_a, \phi_b) = \wp(RGR; RGR) + \wp(RGG; RGG), \quad (16.120a)$$

$$\wp(R, G, \phi_a, \phi_c) = \wp(RRG; RRG) + \wp(RGG; RGG), \quad (16.120b)$$

$$\wp(G, R, \phi_b, \phi_c) = \wp(RGR; RGR) + \wp(GGR; GGR), \quad (16.120c)$$

where $\wp(R, G, \phi_a, \phi_b)$ is the probability that detector 1 reports the event R and detector 2 the event G, given that the detector phase ϕ_1 of detector 1 is set to ϕ_a and the detector phase ϕ_2 of detector 2 is set to ϕ_b, and $\wp(RGR; RGR)$ is the probability that instruction set $(RGR; RGR)$ is sent. From Eq. (16.120b) and (16.120c) one obtains

$$\wp(R, G, \phi_a, \phi_c) \geq \wp(RGG; RGG), \quad (16.121a)$$

$$\wp(G, R, \phi_b, \phi_c) \geq \wp(RGR; RGR). \quad (16.121b)$$

From Eqs. (16.121) and (16.120a) one obtains the following Bell inequality (see Inequality (16.57))

$$\wp(R, G, \phi_a, \phi_b) \leq \wp(R, G, \phi_a, \phi_c) + \wp(R, G, \phi_b, \phi_c), \quad (16.122)$$

which can be rewritten as

$$\sin^2(\phi_a - \phi_b) \leq \sin^2(\phi_a - \phi_c) + \sin^2(\phi_b - \phi_c). \quad (16.123)$$

Taking $\theta = \phi_a - \phi_c = \phi_c - \phi_b$ we obtain

$$\sin^2(2\theta) \leq 2\sin^2\theta, \quad (16.124)$$

which is violated when $0 < |\theta| < \pi/4$.

16.6.3 Entanglement swapping

In the previous subsection we have seen that is possible to entangle particles coming from different sources. Here, we make another step towards the absolute generalization of the concept of entanglement: it is possible to entangle systems that have never directly interacted before. This was proposed by Zeilinger and co-workers[63] and is known as *entanglement swapping*. Consider two pairs of entangled photons emitted by two independent sources as shown in Fig. 16.18. The state of each photon pair will be given by

$$|\psi\rangle_I = \frac{1}{\sqrt{2}}(|h\rangle_1|v\rangle_2 - |v\rangle_1|h\rangle_2), \quad (16.125a)$$

$$|\psi\rangle_{II} = \frac{1}{\sqrt{2}}(|h\rangle_3|v\rangle_4 - |v\rangle_3|h\rangle_4), \quad (16.125b)$$

while the total four-particle state is factorized, i.e.

$$|\Psi\rangle = |\psi\rangle_I \otimes |\psi\rangle_{II}. \quad (16.126)$$

[63] See [Żukowski *et al.* 1993].

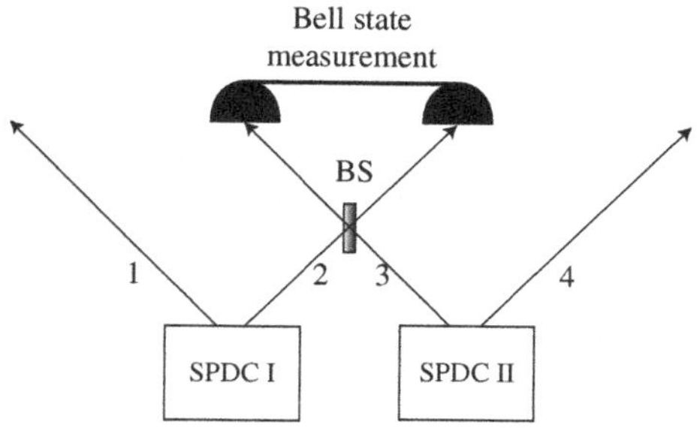

Figure 16.18 Scheme of entanglement swapping. Two cw pumped down-conversion sources (SPDC I and SPDC II) each emit a photon pair (1–2 and 3–4, respectively).

In other words, there is no entanglement of any of the photons 1 or 2 with any of the photons 3 or 4.

If we now perform a particular type of joint measurement on photons 2 and 3, we are able to project photons 1 and 4 onto a different entangled state, which depends on the result of the measurement of photons 2 and 3. To be specific, let us use the Bell states[64] (see also Sec. 17.5)

$$\left|\Psi^-\right\rangle_{23} = \frac{1}{\sqrt{2}}\left(|\,v\rangle_2\,|\,h\rangle_3 - |\,h\rangle_2\,|\,v\rangle_3\right), \tag{16.127a}$$

$$\left|\Psi^+\right\rangle_{23} = \frac{1}{\sqrt{2}}\left(|\,v\rangle_2\,|\,h\rangle_3 + |\,h\rangle_2\,|\,v\rangle_3\right), \tag{16.127b}$$

$$\left|\Phi^-\right\rangle_{23} = \frac{1}{\sqrt{2}}\left(|\,v\rangle_2\,|\,v\rangle_3 - |\,h\rangle_2\,|\,h\rangle_3\right), \tag{16.127c}$$

$$\left|\Phi^+\right\rangle_{23} = \frac{1}{\sqrt{2}}\left(|\,v\rangle_2\,|\,v\rangle_3 + |\,h\rangle_2\,|\,h\rangle_3\right), \tag{16.127d}$$

which represent a complete orthonormal basis of the Hilbert spaces of particles 2 and 3. In order to evaluate the effect of the joint measurement onto one of the states (16.127), we rewrite the state (16.126) in terms of the previous basis

$$\begin{aligned} |\,\Psi\rangle &= \frac{1}{2}\Big(\left|\Psi^+\right\rangle_{14}\left|\Psi^+\right\rangle_{23} + \left|\Psi^-\right\rangle_{14}\left|\Psi^-\right\rangle_{23} \\ &\quad + \left|\Phi^+\right\rangle_{14}\left|\Phi^+\right\rangle_{23} + \left|\Phi^-\right\rangle_{14}\left|\Phi^-\right\rangle_{23}\Big). \end{aligned} \tag{16.128}$$

A close inspection of Eq. (16.128) tells us that, in all cases, the projection onto the Bell basis of particles 2 and 3 also projects particles 1 and 4 on to an entangled state, and with precisely the same form.[65]

A slightly different experimental proposal, which allows for the realization of event-ready detectors, employs SPDC sources pumped by cw lasers.[66] One of the most stringent

[64] First introduced in [Braunstein *et al.* 1992a].
[65] For the first experimental realization of entanglement swapping, see [Pan *et al.* 1998].
[66] See [Żukowski *et al.* 1993].

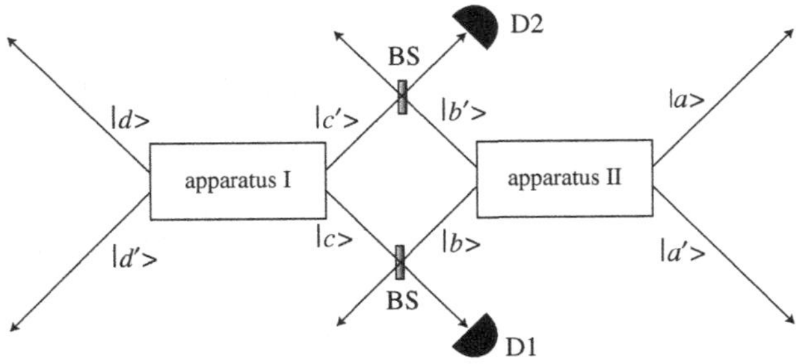

Figure 16.19 **Variation of entanglement swapping. Two cw pumped down-conversion sources each emit a photon pair (1–2 and 3–4, respectively), which after suitable beam splitters and mirrors results in the two states (16.129). The initially independent signal photons are entangled by an ultracoincident registration of the idlers (for sake of simplicity $\gamma_i = i$) at D1, D2.**

technical requirements for the realization of this proposal is the narrow filtering of i-photons and their detection in ultracoincidence (i.e. a coincidence window narrower than the filter bandwidth time) (see Fig. 16.19).[67]

Let the two pumped pairs of particles be described by the states

$$| \psi \rangle_{\mathrm{I}} = \frac{1}{\sqrt{2}}(|a\rangle_1 |b\rangle_2 + |a'\rangle_1 |b'\rangle_2), \tag{16.129a}$$

$$| \psi \rangle_{\mathrm{II}} = \frac{1}{\sqrt{2}}(|c\rangle_3 |d\rangle_4 + |c'\rangle_3 |d'\rangle_4), \tag{16.129b}$$

where both $| \psi \rangle_{\mathrm{I}}$ and $| \psi \rangle_{\mathrm{II}}$ represent, say, an entanglement of path degree of freedom of the two involved photons (1 is entangled with 2 while 3 is entangled with 4), so that for instance, photons 1 and 2 are emitted either in the joint state $|a\rangle_1 |b\rangle_2$ or in the joint state $|a'\rangle_1 |b'\rangle_2$. Then, the initial four-photon state can be written as

$$\begin{aligned} |\Psi\rangle &= | \psi \rangle_{\mathrm{I}} \otimes | \psi \rangle_{\mathrm{II}} \\ &= \frac{1}{2} \left(|a\rangle_1 |b\rangle_2 + |a'\rangle_1 |b'\rangle_2\right) \left(|c\rangle_3 |d\rangle_4 + |c'\rangle_3 |d'\rangle_4\right), \end{aligned} \tag{16.130}$$

where the photons 1, 4 can be called s-photons, and 2, 3 i-photons. In order to entangle uncorrelated s-photons (1 and 4) and to obtain

$$|\varphi\rangle_s = \frac{1}{\sqrt{2}} \left(|a\rangle_1 |d'\rangle_4 + |a'\rangle_1 |d\rangle_4\right), \tag{16.131}$$

we need to project the i-photons into an entangled state. This projection can be done after overlapping their modes at two BSs in such a way that any photon exiting a BS goes through the path leading to the detector (constructive interference) and by observing the i-photons at detectors D1, D2. If the two i-photons are indistinguishable, the joint detection projects the i-state into (see Prob. 16.17)

[67] In [Żukowski *et al.* 1995] these restrictions are abandoned.

$$|\varphi\rangle_i = \frac{1}{\sqrt{2}}\left(|b\rangle_2|c'\rangle_3 + |b'\rangle_2|c\rangle_3\right). \tag{16.132}$$

The consequence is that the registration of photons 2, 3 can operationally entangle the pair 1, 4. However, the joint detection of i-photons must happen in coincidence, and this poses the experimental requirements for this purpose.

Entanglement swapping of photons is not the only phenomenon of genuine quantum interference of systems from different sources. Recently, it has been shown that electrons also show a similar behavior.[68]

16.7 Bell theorem without inequalities

In the next two subsections we analyze two independent approaches that share with Bell's inequalities the same final goal: to prove that quantum mechanics as such is incompatible with at least separability. As we shall see, this statement can also be proved without making use of inequalities, but, instead, exploiting two-particle and three-particle entanglement.

16.7.1 Two-particle entanglement

We begin with the original formulation of the Stapp theorem:[69]

Theorem 16.3 (Stapp) *No theory can:*

- *give contingent predictions of the individual results of measurements;*
- *be compatible with the statistical predictions of quantum mechanics (to within a certain confidence level);*
- *satisfy* Pr. 16.1.

The *first requirement* may be understood in terms of the following *Gedankenexperiment*. Consider an experimental arrangement similar to that described in Figs. 16.6 and 16.8, where the input state is the usual EPRB state (see Eq. (16.11)), with two SGM apparata, whose axes can be rotated, allowing different alternative settings. The word "contingent" means that the theory gives predictions for various possible *alternative* settings. We denote

[68] See [Neder *et al.* 2007].
[69] See [Stapp 1971]. See also [Bell 1981, 132].

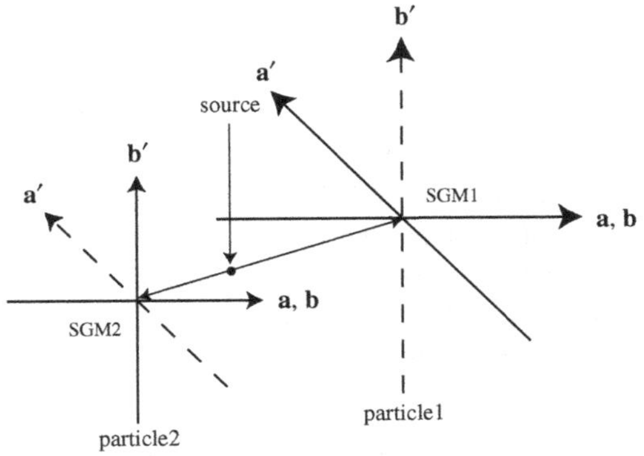

Figure 16.20 Schematic experimental arrangement and setting orientations used for the proof of Stapp's theorem.

the directions of the axes of SGM1 and SGM2 by α and β, respectively. They are both normal to the line of flight and $\theta(\alpha, \beta)$ is the angle between them. Two different settings $\alpha = \mathbf{a}, \mathbf{a}'$ and $\beta = \mathbf{b}, \mathbf{b}'$ of SGM1 and SGM2, respectively, are considered (see Fig. 16.20). Let i ($i = 1, 2$) label the SGMs and j label the individual runs of the experiments. Then, $n_{ij}(\alpha, \beta) = \pm 1$ according to whether the theory predicts that the particle from the j pair that passes through the SGMi is deflected up or down when the settings are α and β. Hence, the first condition of the theorem implies that, for each individual pair j, the numbers $n_{1j}(\alpha, \beta)$ and $n_{2j}(\alpha, \beta)$, are perfectly defined for all four combinations of arguments α and β.

According to quantum mechanics, the following relation holds with increasing accuracy as the number N of runs increases (see Eq. (16.51)):

$$\frac{1}{N}\sum_{j=1}^{N} n_{1j}(\alpha, \beta) n_{2j}(\alpha, \beta) = -\cos\theta(\alpha, \beta), \tag{16.133}$$

so that the second requirement of Stapp theorem amounts to say that Eq. (16.133) holds in the limit $N \to \infty$. Now let us choose the directions $\mathbf{a}$, $\mathbf{a}'$, $\mathbf{b}$, and $\mathbf{b}'$ so that

$$\cos\theta(\mathbf{a}, \mathbf{b}) = 1, \quad \cos\theta(\mathbf{a}, \mathbf{b}') = 0, \tag{16.134a}$$

$$\cos\theta(\mathbf{a}', \mathbf{b}) = -\frac{1}{\sqrt{2}}, \quad \cos\theta(\mathbf{a}', \mathbf{b}') = \frac{1}{\sqrt{2}}. \tag{16.134b}$$

The third condition of Stapp's theorem (separability) can be expressed as

$$n_{1j}(\mathbf{a}, \mathbf{b}) = n_{1j}(\mathbf{a}, \mathbf{b}') = n_{1j}(\mathbf{a}), \tag{16.135a}$$

$$n_{1j}(\mathbf{a}', \mathbf{b}) = n_{1j}(\mathbf{a}', \mathbf{b}') = n_{1j}(\mathbf{a}'), \tag{16.135b}$$

$$n_{2j}(\mathbf{a}, \mathbf{b}) = n_{2j}(\mathbf{a}', \mathbf{b}) = n_{2j}(\mathbf{b}), \tag{16.135c}$$

$$n_{2j}(\mathbf{a}, \mathbf{b}') = n_{2j}(\mathbf{a}', \mathbf{b}') = n_{2j}(\mathbf{b}'). \tag{16.135d}$$

Proof

Inserting Eqs. (16.134) and (16.135) into Eq. (16.133) we obtain

$$\frac{1}{N}\sum_{j=1}^{N} n_{1j}(\mathbf{a})n_{2j}(\mathbf{b}) = -1, \tag{16.136a}$$

$$\frac{1}{N}\sum_{j=1}^{N} n_{1j}(\mathbf{a})n_{2j}(\mathbf{b}') = 0, \tag{16.136b}$$

$$\frac{1}{N}\sum_{j=1}^{N} n_{1j}(\mathbf{a}')n_{2j}(\mathbf{b}) = \frac{1}{\sqrt{2}}, \tag{16.136c}$$

$$\frac{1}{N}\sum_{j=1}^{N} n_{1j}(\mathbf{a}')n_{2j}(\mathbf{b}') = -\frac{1}{\sqrt{2}}. \tag{16.136d}$$

From Eq. (16.136a) we have

$$n_{1j}(\mathbf{a}) = -n_{2j}(\mathbf{b}), \tag{16.137}$$

which, combined with Eq. (16.136b), gives

$$\frac{1}{N}\sum_{j=1}^{N} n_{2j}(\mathbf{b})n_{2j}(\mathbf{b}') = 0. \tag{16.138}$$

Subtraction of Eq. (16.136d) from Eq. (16.136c) yields

$$\frac{1}{N}\sum_{j=1}^{N} n_{1j}(\mathbf{a}')\left[n_{2j}(\mathbf{b}) - n_{2j}(\mathbf{b}')\right] = \sqrt{2}. \tag{16.139}$$

Using the fact that $n_{2j}(\mathbf{b}')n_{2j}(\mathbf{b}') = 1$ (because the allowed values are only ± 1), we obtain

$$\sqrt{2} = \frac{1}{N}\sum_{j=1}^{N} n_{1j}(\mathbf{a}')n_{2j}(\mathbf{b}')\left[n_{2j}(\mathbf{b})n_{2j}(\mathbf{b}') - 1\right] \tag{16.140a}$$

$$\leq \frac{1}{N}\sum_{j=1}^{N} \left|n_{2j}(\mathbf{b})n_{2j}(\mathbf{b}') - 1\right| \tag{16.140b}$$

$$= \frac{1}{N}\sum_{j=1}^{N} \left[1 - n_{2j}(\mathbf{b})n_{2j}(\mathbf{b}')\right] \tag{16.140c}$$

$$= 1 - \frac{1}{N}\sum_{j=1}^{N} n_{2j}(\mathbf{b})n_{2j}(\mathbf{b}'), \tag{16.140d}$$

$$= 1 \tag{16.140e}$$

which is clearly impossible, so that quantum predictions are not compatible with a theory that gives contingent predictions and acknowledges separability.

Q.E.D

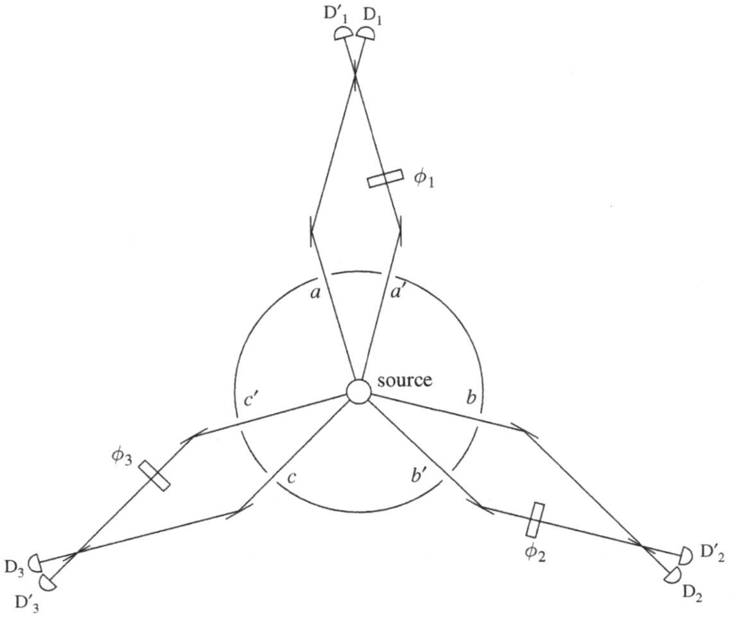

Figure 16.21 The GHSZ *Gedankenexperiment* with a three-particle interferometer. The source emits a triple of particles, 1, 2, and 3, in six beams, with the state given by Eq. (16.141). A phase shift ϕ_1 is imparted to beam a' of particle 1, and beams a, a' are brought together on a BS before illuminating detectors D_1, D'_1. Likewise for particles 2 with beams b, $b'$and for particle 3 with beams c, c'.

16.7.2 Three-particle entanglement

Greenberger, Horne, Shimony, and Zeilinger (GHSZ) have been able to prove that quantum mechanics violates separability in the spirit of Bell proofs but without using inequalities at all.[70]

Consider a particle with zero mean momentum that decays into three photons.[71] If all three daughter particles have the same energy, by momentum conservation they must be emitted 120° apart from each other. The central source is surrounded by an array of six apertures: a, b, and c at 120° separation, and a', b', and c' also at 120° separation, with respect to each other (see Fig. 16.21). Because of the placement of apertures, the three particles 1, 2, and 3 must emerge either through a, b, and c or through a', b', and c'. Thus, the state of the three particles beyond the apertures will be given by the superposition (GHSZ state)

$$|\varphi\rangle = \frac{1}{\sqrt{2}}\left(|a\rangle_1|b\rangle_2|c\rangle_3 + |a'\rangle_1|b'\rangle_2|c'\rangle_3\right), \tag{16.141}$$

70 In making use of a three-particle entanglement we follow here a later development by GHSZ of an idea proposed originally in [Greenberger *et al.* 1989], where the authors make use of a four-particle experiment, giving rise to a state known as the GHZ state.

71 See [Greenberger *et al.* 1990].

where $|a\rangle_1$ denotes the particle 1 in beam a, and so on. Beyond the apertures, beams $|a\rangle_1$ and $|a'\rangle_1$ are totally reflected so as to overlap at a 50/50 BS, and the two outgoing beams are monitored by detectors D_1 and D_1'. A similar arrangement is in place for the other beams. Suppose that $|a'\rangle_1$ passes through a phase plate which causes a phase shift ϕ_1. Consequently, we have the evolutions

$$|a\rangle_1 \mapsto \frac{1}{\sqrt{2}}(|D_1\rangle + \imath|D_1'\rangle), \tag{16.142a}$$

$$|a'\rangle_1 \mapsto \frac{1}{\sqrt{2}}e^{\imath\phi_1}(|D_1'\rangle + \imath|D_1\rangle), \tag{16.142b}$$

where $|D_1\rangle$ and $|D_1'\rangle$ denote the states of the particle emerging towards detectors D_1 and D_1', respectively. Particles 2 and 3 are subjected to similar treatment with detectors D_2 and D_2' for particle 2 and detectors D_3 and D_3' for particle 3.

The initial state $|\varphi\rangle$ of the three particles evolves then into

$$\begin{aligned}|\Psi\rangle = \frac{1}{4}\Big[&(1 - \imath e^{\imath(\phi_1+\phi_2+\phi_3)})|D_1\rangle|D_2\rangle|D_3\rangle + (\imath - e^{\imath(\phi_1+\phi_2+\phi_3)})|D_1\rangle|D_2\rangle|D_3'\rangle \\ &+ (\imath - e^{\imath(\phi_1+\phi_2+\phi_3)})|D_1\rangle|D_2'\rangle|D_3\rangle + (-1 + \imath e^{\imath(\phi_1+\phi_2+\phi_3)})|D_1\rangle|D_2'\rangle|D_3'\rangle \\ &+ (\imath - e^{\imath(\phi_1+\phi_2+\phi_3)})|D_1'\rangle|D_2\rangle|D_3\rangle + (-1 + \imath e^{\imath(\phi_1+\phi_2+\phi_3)})|D_1'\rangle|D_2\rangle|D_3'\rangle \\ &+(-1 + \imath e^{\imath(\phi_1+\phi_2+\phi_3)})|D_1'\rangle|D_2'\rangle|D_3\rangle + (-\imath + e^{\imath(\phi_1+\phi_2+\phi_3)})|D_1'\rangle|D_2'\rangle|D_3'\rangle\Big].\end{aligned} \tag{16.143}$$

The probability for detection of the three particles by the respective detectors D_1, D_2, and D_3 is

$$\wp^{\Psi}_{D_1D_2D_3}(\phi_1, \phi_2, \phi_3) = \frac{1}{16}\left|1 - \imath e^{\imath(\phi_1+\phi_2+\phi_3)}\right|^2 = \frac{1}{8}[1 - \sin(\phi_1 + \phi_2 + \phi_3)]. \tag{16.144a}$$

Likewise,

$$\wp^{\Psi}_{D_1'D_2D_3}(\phi_1, \phi_2, \phi_3) = \frac{1}{8}[1 + \sin(\phi_1 + \phi_2 + \phi_3)], \tag{16.144b}$$

and so on for the remaining six possible outcomes. The sum of the probabilities for all eight possible outcomes is of course 1.

Given a parameter λ that determines the state of the whole multiparticle system, we may define three functions $\alpha_\lambda(\phi_1)$, $\beta_\lambda(\phi_2)$, and $\gamma_\lambda(\phi_3)$ which represent the measurement result at the detector pairs 1, 2, and 3, respectively. For the sake of concreteness, we assign to each of these functions the value $+1$ when a particle enters an unprimed detector and -1 when it enters a primed one. There is an implicit assumption in the introduction of these three functions (see Eqs. (16.46) and (16.135)): $\alpha_\lambda(\phi_1)$ does not depend on ϕ_2 and ϕ_3, and so on. Now we calculate the expectation value on the state $|\Psi\rangle$ of the product of the three measurement outcomes at the three arms of the interferometer, given that the relative phase is tuned to ϕ_1, ϕ_2, and ϕ_3 (see Prob. 16.18),

$$
\begin{aligned}
C_\Psi(\phi_1,\phi_2,\phi_3) &= \wp^\Psi_{D_1D_2D_3}(\phi_1,\phi_2,\phi_3) - \wp^\Psi_{D'_1D'_2D'_3}(\phi_1,\phi_2,\phi_3) \\
&\quad + \wp^\Psi_{D_1D'_2D'_3}(\phi_1,\phi_2,\phi_3) + \wp^\Psi_{D'_1D_2D'_3}(\phi_1,\phi_2,\phi_3) + \wp^\Psi_{D'_1D'_2D_3}(\phi_1,\phi_2,\phi_3) \\
&\quad - \wp^\Psi_{D'_1D_2D_3}(\phi_1,\phi_2,\phi_3) - \wp^\Psi_{D_1D'_2D_3}(\phi_1,\phi_2,\phi_3) - \wp^\Psi_{D_1D_2D'_3}(\phi_1,\phi_2,\phi_3) \\
&= \sin(\phi_1+\phi_2+\phi_3).
\end{aligned}
\tag{16.145}
$$

Let us now show that a contradiction arises between Eq. (16.145) and the EPR-like requirement that $\alpha_\lambda(\phi_1)$, $\beta_\lambda(\phi_2)$, and $\gamma_\lambda(\phi_3)$ possess definite values once λ, ϕ_1, ϕ_2, and ϕ_3 are specified. For $\phi_1+\phi_2+\phi_3 = \pi/2$ we obtain $C_\Psi = +1$ and for $\phi_1+\phi_2+\phi_3 = 3\pi/2$ we obtain $C_\Psi = -1$. Stated in terms of the three functions $\alpha_\lambda(\phi_1)$, $\beta_\lambda(\phi_2)$, and $\gamma_\lambda(\phi_3)$, we have that

$$
\alpha_\lambda(\phi_1)\beta_\lambda(\phi_2)\gamma_\lambda(\phi_3) = \begin{cases} +1 & \text{if} \quad \phi_1+\phi_2+\phi_3 = \frac{1}{2}\pi \\ \\ -1 & \text{if} \quad \phi_1+\phi_2+\phi_3 = \frac{3}{2}\pi \end{cases} . \tag{16.146}
$$

Consider three different choices of the phase angles that satisfy the former assignment, i.e. $(\pi/2, 0, 0)$, $(0, \pi/2, 0)$, $(0, 0, \pi/2)$, and one choice that satisfies the latter, i.e. $(\pi/2, \pi/2, \pi/2)$. In the first three cases, we may write the product of the outcomes as

$$\alpha_\lambda(\pi/2)\beta_\lambda(0)\gamma_\lambda(0) = 1, \tag{16.147a}$$

$$\alpha_\lambda(0)\beta_\lambda(\pi/2)\gamma_\lambda(0) = 1, \tag{16.147b}$$

$$\alpha_\lambda(0)\beta_\lambda(0)\gamma_\lambda(\pi/2) = 1. \tag{16.147c}$$

Multiplying the three Eqs. (16.147) we have

$$\alpha_\lambda(\pi/2)\beta_\lambda(\pi/2)\gamma_\lambda(\pi/2) = 1, \tag{16.148}$$

because the other factors are equal to one, since $\alpha^2_\lambda(\phi) = \beta^2_\lambda(\phi) = \gamma^2_\lambda(\phi) = 1$ for any ϕ. This result clearly contradicts that, for $\phi_1+\phi_2+\phi_3 = 3\pi/2$, we have $C_\Psi = -1$.

Note that the observed count rates for coincidences (for instance, among detectors D_1, D_2, D_3), will depend on the phases. That is, if $\phi_1+\phi_2+\phi_3$ is varied linearly in time, then the three-particle coincidence rate (the three-particle interference) will vary sinusoidally (see Eq. (16.144a)). However, there will be no two-particle interference fringes (see Prob. 16.19).

The GHSZ thought experiment shows that there is an intrinsic contradiction between the assumption of perfect correlation and the other EPR assumptions, namely separability and reality, at the level of three – or more – particle systems. Moreover, the GHSZ thought experiment demonstrates that such contradiction arises even for an individual system rather than for the statistical properties of an ensemble of identically prepared systems.[72]

Finally, let us conclude this section by considering an interesting consequence of the GHSZ state: entanglement, in a given compound system, is strongly affected by measurements performed on one of its constituents.[73] For instance, consider the case of three spin-1/2 particles (a GHSZ-like state), defined by

[72] A three-particle GHZ-or GHSZ-like entangled state has also been experimentally realized [Bouwmeester *et al.* 1999].

[73] See [Krenn/Zeilinger 1996],

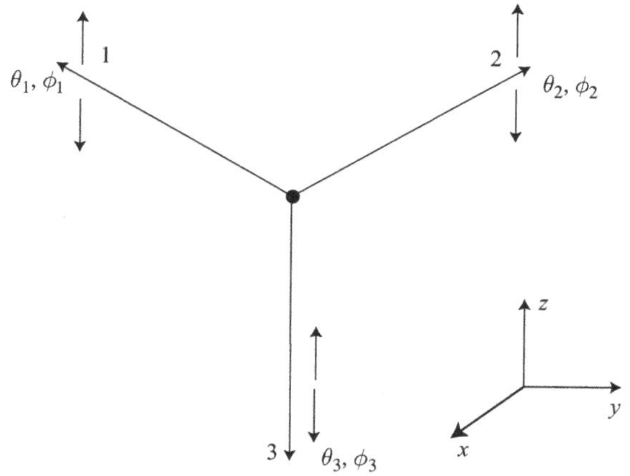

Figure 16.22 Conceptual scheme allowing investigation of the conditional entanglement arising between two out of three particles in a GHSZ state

$$|\Psi\rangle = \frac{1}{\sqrt{2}}(|\uparrow\rangle_1|\uparrow\rangle_2|\uparrow\rangle_3 + |\downarrow\rangle_1|\downarrow\rangle_2|\downarrow\rangle_3), \tag{16.149}$$

where $|\uparrow\rangle_j$ and $|\downarrow\rangle_j$ represent the particle j's up and down states of the spin component along the z-direction, respectively. Now, we look for two-particle entanglement between particles 1 and 2 when the spin of particle 3 is measured along the component defined by the spherical angles (θ_3, ϕ_3) (see Fig. 16.22). If one performs spin measurements within the xy-plane on particles 1 and 2 ($\theta_1 = \theta_2 = \pi/2$), two subensembles are generated depending on the result (up or down) of the measurement performed on particle 3. Then, the two-particle correlation functions for these subensembles are given by

$$\mathrm{C}^{\pm}_{12} = \pm\sin(\theta_3)\cos(\phi_1 + \phi_2 + \phi_3). \tag{16.150}$$

Equation (16.150) tells us that, after having measured the third particle, the other two still remain entangled, unless the third is measured along the z-direction ($\theta_3 = 0$), because in this case we obviously obtain $\mathrm{C}^{+}_{12} = \mathrm{C}^{-}_{12} = 0$.

16.8 What is quantum non-locality?

We have seen that quantum mechanics is not compatible with an HV theory that acknowledges the separability principle. However, in principle it could raise the problem of whether quantum mechanics violates relativistic locality by some form of superluminal communication between entangled entities. Here, by *locality* we mean the principle of absence of *action-at-a-distance* and the existence of bounds on the speed of transmission of signals or physical effects. By *relativistic locality* we mean that these bounds, for particles of real non-negative rest mass, are represented by the speed of light in vacuum. In this section, we shall see that quantum mechanics does not violate locality because it does not imply a superluminal transmission of signals. As a matter of fact, quantum dynamics is strictly local while quantum correlations exhibit non-local features.

16.8.1 Eberhard's theorem

Since 1978 Eberhard focused on the problem of separability and locality.[74] Let $\hat{O}_1$ and $\hat{O}_2$ be two observables on subsystems $\mathcal{S}_1$ and $\mathcal{S}_2$ of a system $\mathcal{S}$, respectively, and $\wp(o_a, \mathbf{a}; o_b, \mathbf{b})$ be the probability that the results of a measurement of $\hat{O}_1$ and $\hat{O}_2$ on $\mathcal{S}_1$ and $\mathcal{S}_2$ yield o_a and o_b when certain settings of the measurement apparata are **a** and **b**, respectively.

We now give the following definition of locality: the probability distribution of $\hat{O}_1$ ($\hat{O}_2$), independently of the measurement outcome on $\hat{O}_2$ ($\hat{O}_1$), obtained by integrating $\wp(o_a, \mathbf{a}; o_b, \mathbf{b})$ over o_b (o_a), is independent of the other setting **b** (**a**), that is

$$\sum_{o_b} \wp(o_a, \mathbf{a}; o_b, \mathbf{b}) = \wp(o_a, \mathbf{a}); \quad \sum_{o_a} \wp(o_a, \mathbf{a}; o_b, \mathbf{b}) = \wp(o_b, \mathbf{b}). \tag{16.151}$$

If locality were violated, we would have a causal interdependence between the two subsystems, because, by changing the setting **a** (**b**), we would be able to act on the result of the other measurement, and, hence, if we performed experiments on subsystems that are space-like separated, we would be able to transmit a message with superluminal or even infinite speed. We can now prove the following theorem:

Theorem 16.4 (Eberhard) *Quantum-mechanical correlations do not imply a violation of locality as expressed by* Eq. (16.151).

Proof

Let

$$\hat{P}_{o_a,\mathbf{a}} = |\, o_a, \mathbf{a}\rangle \langle o_a, \mathbf{a}\,| \quad \text{and} \quad \hat{P}_{o_b,\mathbf{b}} = |\, o_b, \mathbf{b}\rangle \langle o_b, \mathbf{b}\,| \tag{16.152}$$

be the projectors on the state $|\, o_a, \mathbf{a}\rangle$ of subsystem $\mathcal{S}_1$ when the setting is **a** and on the state $|\, o_b, \mathbf{b}\rangle$ of subsystem $\mathcal{S}_2$ when the setting is **b**, respectively, and $\hat{\rho}$ be a density matrix which represents the compound state of $\mathcal{S} = \mathcal{S}_1 + \mathcal{S}_2$. The probability $\wp(o_a, \mathbf{a})$ that, by measuring the observable $\hat{O}_1$ on $\mathcal{S}_1$, we obtain o_a, is

$$\wp(o_a, \mathbf{a}) = \mathrm{Tr}\left[\hat{P}_{o_a,\mathbf{a}}\hat{\rho}\right]. \tag{16.153}$$

After a measurement of $\hat{O}_1$ when the setting is **a** with result o_a we obtain the transformation (see Eq. (9.101))

$$\hat{\rho} \leadsto \hat{\rho}' = \frac{\hat{P}_{o_a,\mathbf{a}}\hat{\rho}\hat{P}_{o_a,\mathbf{a}}}{\wp(o_a, \mathbf{a})}. \tag{16.154}$$

74 See [Eberhard 1978]. A proof of a the following result can also be found in [Jordan 1983].

If we perform a second measurement on the second subsystem, and calculate the conditional probability of obtaining o_b by measuring $\hat{O}_2$ when the setting is $\mathbf{b}$, we have

$$\wp'(o_b, \mathbf{b}|o_a, \mathbf{a}) = \mathrm{Tr}\left[\frac{\hat{P}_{o_b,\mathbf{b}}\hat{P}_{o_a,\mathbf{a}}\hat{\rho}\hat{P}_{o_a,\mathbf{a}}}{\wp(o_a, \mathbf{a})}\right]. \tag{16.155}$$

We now compute the joint probability of obtaining the two results o_a and o_b given the settings $\mathbf{a}$ and $\mathbf{b}$, respectively, i.e.

$$\begin{aligned}\wp(o_a, \mathbf{a}; o_b, \mathbf{b}) &= \wp(o_a, \mathbf{a})\wp'(o_b, \mathbf{b}|o_a, \mathbf{a}) \\ &= \wp(o_a, \mathbf{a})\frac{\mathrm{Tr}\left[\hat{P}_{o_b,\mathbf{b}}\hat{P}_{o_a,\mathbf{a}}\hat{\rho}\hat{P}_{o_a,\mathbf{a}}\right]}{\wp(o_a, \mathbf{a})} \\ &= \mathrm{Tr}\left[\hat{P}_{o_b,\mathbf{b}}\hat{P}_{o_a,\mathbf{a}}\hat{\rho}\hat{P}_{o_a,\mathbf{a}}\right].\end{aligned} \tag{16.156}$$

Finally we have (see Eq. (16.151))

$$\sum_{o_a}\wp(o_a, \mathbf{a}; o_b, \mathbf{b}) = \mathrm{Tr}\sum_{o_a}\left(\hat{P}_{o_b,\mathbf{b}}\hat{P}_{o_a,\mathbf{a}}\hat{\rho}\hat{P}_{o_a,\mathbf{a}}\right) = \mathrm{Tr}\left[\hat{P}_{o_b,\mathbf{b}}\hat{\rho}\right] = \wp(o_b, \mathbf{b}), \tag{16.157}$$

where we have made use of the cyclic property of the trace (see Prob. 5.4), of the fact that $\hat{P}_{o_a,\mathbf{a}}$ and $\hat{P}_{o_b,\mathbf{b}}$ commute because they pertain to different subsystems, of the fact that $\hat{P}^2_{o_a,\mathbf{a}} = \hat{P}_{o_a,\mathbf{a}}$ (see Eq. (1.41b)), and of the property $\sum_{o_a}\hat{P}_{o_a,\mathbf{a}} = \hat{I}$ for any orthogonal set of projectors $\{\hat{P}_{o_a,\mathbf{a}}\}$ (see Eq. (1.41a)). We may proceed in a similar way starting from the conditional probability $\wp'(o_a, \mathbf{a}|o_b, \mathbf{b})$ in order to derive the first equality in Eq. (16.151).

Q.E.D

Eberhard's theorem is of particular relevance because it shows that, though the probability distributions of possible outcomes of measurement on two entangled systems are not independent, there is no way – by changing the setting for the measurement on one subsystem – to influence the probability distributions of the outcomes on the other subsystem, something which in turn would imply the possibility of exchanging superluminal signals. In the following we shall use the term *non-locality* as a short-hand way to refer to all separability-violating quantum correlations based on entanglement. However, we should never forget that in quantum mechanics there is no violation of locality *stricto sensu*.

16.8.2 A necessary condition for separability

As we have seen, certain entangled states violate Bell inequalities. However, it is not true that every violation of separability implies a violation of Bell inequalities. In other words, while a separable system always satisfies Bell inequalities, the converse is not necessarily true. We seek now a necessary condition for separability and show that it is more stringent than that implied by Bell inequalities.[75] Let us consider two separable subsystems $\mathcal{S}'$ and $\mathcal{S}''$ and write the density operator of the composite system as (see Eq. (5.39))

[75] See [Peres 1996].

$$\hat{\rho} = \sum_j w_j \hat{\rho}'_j \otimes \hat{\rho}''_j, \tag{16.158}$$

where the weights w_j are non-negative and satisfy $\sum_j w_j = 1$.

In order to derive the separability criterion, let us explicitly rewrite Eq. (16.158) in terms of the matrix elements, i.e.

$$\rho_{m\mu,n\nu} = \sum_j w_j (\rho'_j)_{mn} (\rho''_j)_{\mu\nu}, \tag{16.159}$$

where Latin indices refer to the first subsystem and Greek indices to the second one (in general the dimensions of the two subsystems can be different). Note that this equation can always be satisfied if we replace the density matrices by Liouville functions, which have to be non-negative. In the quantum-mechanical case, however, we require the non-negativity of the eigenvalues. Let us define a new density matrix

$$\hat{\varrho} = \sum_j w_j (\hat{\rho}'_j)^{\mathrm{T}} \hat{\rho}''_j, \tag{16.160}$$

where only the density matrix referring to the first subsystem has been transposed. From Eq. (16.160) it is clear that

$$\varrho_{m\mu,n\nu} = \rho_{n\mu,m\nu}. \tag{16.161}$$

The transformation $\hat{\rho} \mapsto \hat{\varrho}$ is non-unitary, but $\hat{\varrho}$ is still Hermitian. Since a transposed matrix of the form $(\hat{\rho}'_j)^{\mathrm{T}} = (\hat{\rho}'_j)^*$ is a non-negative matrix with unit trace, it follows that none of the eigenvalues of $\hat{\varrho}$ is negative: this is the necessary condition for separability.

For example, take a pair of spin-1/2 particles in a Werner state (impure singlet state), which, for the bidimensional case, has the form

$$\hat{\rho}_{\mathrm{W}} = w \hat{P}_{\Psi_0} + (1 - w)\hat{I}, \tag{16.162}$$

where $\hat{P}_{\Psi_0}$ is the projector on the singlet state (6.194). In terms of matrix elements we have

$$(\rho_{\mathrm{W}})_{m\mu,n\nu} = w\rho^0_{m\mu,n\nu} + (1 - w)\frac{\delta_{mn}\delta_{\mu\nu}}{4}, \tag{16.163}$$

where the density matrix elements for a pure singlet state are given by

$$\rho^0_{01,01} = \rho^0_{10,10} = -\rho^0_{01,10} = -\rho^0_{10,01} = \frac{1}{2}, \tag{16.164}$$

and all other components of $\hat{\rho}^0$ vanish. An explicit calculation (see Prob. 16.20) shows that $\hat{\varrho}$ (where the density matrix of the first subsystem has been transposed) has three eigenvalues equal to $(1 + w)/4$ and the fourth one equal to $(1 - 3w)/4$. The condition for the lowest one to be non-negative (a necessary requirement for separability) is $w \le 1/3$. It is worth noting that this criterion is a stronger test for separability than that represented by the Bell inequalities, which hold for $w \le 1/\sqrt{2}$ (see Fig. 16.23 and Prob. 16.21).

Finally, it should be noted that, in the particular case chosen above, this necessary separability criterion is also sufficient – as it is for composite systems having dimensions 2×2 and 2×3. However, for higher dimensions it does not represent a sufficient condition.[76]

[76] As proved in [Horodecki *et al.* 1996].

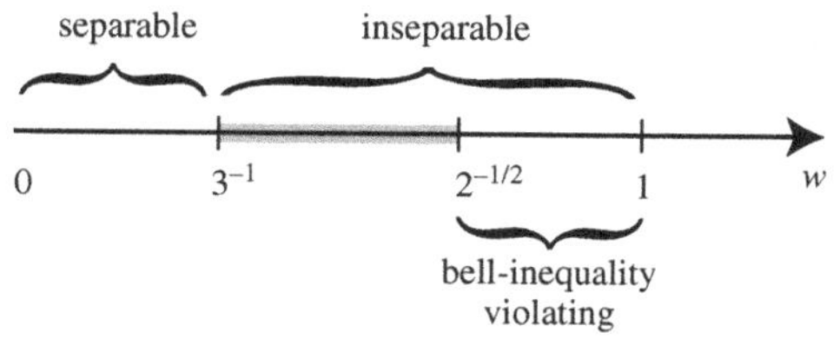

Figure 16.23 Simple representation of separability for a Werner state as a function of the weight w. It is interesting to note that the grey area represents all those Werner states which *do not* violate the Bell inequality but are still not separable.

16.9 Further developments about inequalities

The investigation of the non-local properties of quantum mechanics has been developed and generalized in many directions. In the present section we first wish to recall some developments, and then to stress in particular one of them, the Tsirelson theorem. The main further developements may be summarized as follows:

- Several generalized forms of inequalities, of which the original Bell inequalities represent a special case. Among others, we recall here the work of Beltrametti and Maczyński.[77]
- As we have seen, most proposed and performed experiments deal with spin-1/2 particles. A powerful generalization to particles with spin-1 and-3/2 has been provided by Garg and Mermin.[78]
- We have already seen that CHSH introduced statistical considerations. Important generalizations to non-deterministic HV theories have been provided, among many others, by Stapp.[79]
- Violation of Bell inequalities by mixed quantum states has been shown by Gisin.[80]
- Quantum mechanics violates the bounds on the strength of correlations imposed by Bell inequalities on HV theories. However, this raises the question of whether quantum-mechanical correlations are in turn bound by a weaker constraint. In the following we wish to answer this question.

First, we restate the Bell inequality (16.69) as

$$|\langle \mathbf{a}, \mathbf{b}\rangle - \langle \mathbf{a}, \mathbf{b}'\rangle + \langle \mathbf{a}', \mathbf{b}'\rangle + \langle \mathbf{a}', \mathbf{b}\rangle| \le 2, \tag{16.165}$$

which stems from the fact that $|x+y| \le |x| + |y|$. Since each of the terms in Eq. (16.165) lies between -1 and $+1$, the natural upper bound for the entire expression is 4. If we demand only that the probabilities satisfy the causal communication constraint, i.e. that

[77] See [Beltrametti/Maczyński 1991].
[78] See [Garg/Mermin 1984].
[79] See [Stapp 1980].
[80] See [Gisin 1991].

they do not violate relativistic locality (see previous section), this bound is unchanged, that is,[81]

$$|\langle \mathbf{a}, \mathbf{b} \rangle - \langle \mathbf{a}, \mathbf{b}' \rangle + \langle \mathbf{a}', \mathbf{b}' \rangle + \langle \mathbf{a}', \mathbf{b} \rangle| \leq 4. \quad (16.166)$$

We have already seen that quantum-mechanical correlations violate the much stronger bound imposed by Eq. (16.165). As a consequence of this situation, the natural question arises: do quantum-mechanical correlations fill the gap between 2 and 4 or, in other words, is there an upper bound for quantum-mechanical correlations smaller than 4?

Tsirelson proved that.[82]

Theorem 16.5 (Tsirelson) *According to quantum mechanics the expression on the lhs of* Eq. (16.165) *must be smaller than* $2\sqrt{2}$.

Proof

Let $\hat{O}^a, \hat{O}^{a'}, \hat{O}^b, \hat{O}^{b'}$ be Hermitian operators on a Hilbert space $\mathcal{H}$ satisfying the condition $[\hat{O}^a, \hat{O}^b] = 0$ and so on for the other couples $(a, b'), (a', b), (a', b')$. Moreover, each operator has eigenvalues 1 and -1. In the context of the second Bell inequality, we define the operator

$$\hat{\mathrm{C}} = \hat{O}^a \hat{O}^b + \hat{O}^{a'} \hat{O}^b + \hat{O}^a \hat{O}^{b'} - \hat{O}^{a'} \hat{O}^{b'}. \quad (16.167)$$

We know that the square of each operator is equal to the identity, which implies (see Prob. 16.22)

$$\begin{aligned} 2\sqrt{2} - \hat{\mathrm{C}} &= \frac{1}{\sqrt{2}} \left[(\hat{O}^a)^2 + (\hat{O}^{a'})^2 + (\hat{O}^b)^2 + (\hat{O}^{b'})^2 \right] - \hat{\mathrm{C}} \\ &= \frac{1}{\sqrt{2}} \left(\hat{O}^a - \frac{\hat{O}^b + \hat{O}^{b'}}{\sqrt{2}} \right)^2 + \frac{1}{\sqrt{2}} \left(\hat{O}^{a'} - \frac{\hat{O}^b - \hat{O}^{b'}}{\sqrt{2}} \right)^2. \end{aligned} \quad (16.168)$$

Since the expression in the rhs consists of the sum of squares of Hermitian operators, it is clearly an operator with expectation value greater than or equal to zero. This leads to the conclusion:

$$\langle \hat{\mathrm{C}} \rangle \leq 2\sqrt{2}. \quad (16.169)$$

A similar argument leads to

$$\langle \hat{\mathrm{C}} \rangle \geq -2\sqrt{2}. \quad (16.170)$$

Q.E.D

The importance of Tsirelson result lies in the fact that it proves that quantum mechanics *does not* fill the entire gap between the bounds of Eqs. (16.165)–(16.166). In other

[81] See [Hillery/Yurke 1995].

[82] See [Tsirelson 1980]. Generalizations can be found in [Khalfin/Tsirelson 1992] [Hillery/Yurke 1995].

words, quantum mechanics certainly allows for correlations that are much stronger than those allowed by HV theories. However, there is a wide spectrum of "hyper-correlations" that do not contradict causal communication constraints but are not allowed by quantum mechanics.

16.10 Conclusion

In this chapter we have seen that quantum mechanics forces us to abandon the concept of separability. Since this represents an important classical principle (see Sec. 1.1), this step constitutes a historical breakthrough. The evolution of the debate concerning entanglement and locality, from 1935 to the present day, is paradigmatic and represents a completely new development for the philosophy and history of science. Indeed, the investigation started with the aim of EPR at proving the incompleteness of quantum mechanics, which led to the proposal of the EPR state, which, in turn, after a careful theoretical examination, performed initially by Schrödinger, resulted in one of the most interesting states of the whole history of science. It should also be noted that most of the theoretical investigation had been performed before there was any possibility of experimental tests of the predictions. On the contrary, by finding mathematical constraints like the Bell inequalities, these tests became for the first time theoretically possible.

We have seen the incompatibility between quantum mechanics and local HV theories. This is the reason that forced some supporters of HV theories to go in the direction of a generalized non-local theory.[83] A related question is whether quantum mechanics is also incompatible with non-local HV theories. It is very difficult to provide a sufficiently general answer to this problem, and probably there is none. However, a more specific question may be formulated: is there any incompatibility between quantum mechanics and non-local HV theories that is physically interesting? Leggett investigated a specific class of the latter and found a new inequality that was inconsistent with quantum mechanics.[84] Successive experimental verification found that experimental results also violate these new inequalities.[85]

Summary

The main results of this chapter may be summarized as follows:

- Quantum-mechanical states are *not dispersion-free*, that is, there cannot be a state in which all observables have zero variance.

[83] As outlined in [Aspect 2007], it looks like we are faced with choosing between realism and locality.
[84] See [Leggett 2003].
[85] See [Gröblacher *et al.* 2007].

- Quantum mechanics *violates separability but not locality stricto sensu*. This means that, though the measurement outcomes on a given subsystem are not independent from the measurement outcomes on another subsystem that is entangled with the former, there is no way to manipulate the setting of a given apparatus on a subsystem in order to force determined outcomes on another subsystem that is entangled with it.
- As a consequence, quantum systems may be *correlated though dynamically and communication-independent*.
- *Bell inequalities* and, in a stronger form, the *Tsirelson inequality* set precise bounds that can be violated by quantum (but not by classical) systems.
- There is a contradiction between perfect correlation and the EPR assumptions of *reality and separability*.
- *Experimental tests*, which have become possible after and thanks to Bell's contribution, are consistent with the predictions of quantum mechanics and largely inconsistent with those of local realistic theories.

Problems

16.1 Prove the equivalence of Eqs. (16.3) and (16.7).

16.2 Prove that the singlet state (16.11) is an eigenstate of the operators $\hat{\sigma}_{1z}\hat{\sigma}_{2z}$ and $\hat{\sigma}_{1x}\hat{\sigma}_{2x}$, and find the corresponding eigenvalues.

16.3 Prove that $\hat{H}_I$ (Eq. (16.14)) and $\hat{N}_e$ (Eq. (16.16)) commute.

16.4 Prove Lemma 16.1.

16.5 Prove Eq. (16.29).

16.6 Prove Lemma 16.2.

16.7 Prove Eq. (16.42).
(*Hint*: Study the function $f(x) = x + x^{-1}$ and draw its graph. Is there any value of x for which $|f(x)| < 2$?)

16.8 Prove Eq. (16.51).

16.9 Prove Eq. (16.56).

16.10 Prove the result (16.66).

16.11 Derive the result (16.67).

16.12 Show that the CHSH inequality implies inequality (16.57) as a special case.
(*Hint*: Take advantage of the solution of the next problem.)

16.13 Derive inequality (16.69).

16.14 Prove the CH inequality (16.76).
(*Hint*: Take advantage of the following lemma.[86] Given six real numbers x_1, x_2, y_1, y_2, X and Y such that

$$0 \leq x_1, x_2 \leq X, \quad 0 \leq y_1, y_2 \leq Y, \tag{16.171}$$

[86] For the proof of this lemma see [Clauser/Horne 1974].

the inequality

$$- XY \leq f \leq 0 \tag{16.172}$$

is satisfied by the function $f = x_1 y_1 - x_1 y_2 + x_2 y_1 + x_2 y_2 - Y x_2 - X y_1$.)

16.15 Derive Eq. (16.99).

16.16 Derive Eq. (16.107).

16.17 Derive result (16.132).

16.18 Explicitly compute the eight possible detection probabilities of the type (16.144) and derive Eq. (16.145).

16.19 Explain why there are no two-particle fringes in the GHSZ state.

16.20 Calculate the eigenvalues of the density matrix (16.163).

16.21 Show that the state (16.163) violates Bell's inequality if and only if $w > 1/\sqrt{2}$.

16.22 Derive Eq. (16.168).

Further reading

Bell, John S., *Speakable and Unspeakable in Quantum Mechanics*, Cambridge: Cambridge University Press, 1987, 1994.

Bohm, David, A suggested interpretation of the quantum theory in terms of "hidden variables." *Physical Review*, **85** (1952), 166–93.

Clauser, J. F. and Shimony, A., Bell's theorem: experimental tests and implications. *Reporting Progress Physics*, **41** (1978), 1881–927.

Einstein, A., Podolsky, B. and Rosen, N., Can quantum-mechanical description of physical reality be considered complete? *Physical Review*, **47** (1935), 777–80.

Hillery, M. and Yurke, B., Bell's theorem and beyond. *Quantum Optics*, **7** (1995), 215–27.

17 Entanglement: quantum information and computation

In this chapter we shall deal with the most recent and challenging development of quantum theory, and also one of the most important ones for interpretational, foundational, and even technological issues. This field finds its roots in the observation that quantum states can be viewed as bricks of information in a way that is intrinsically different from classical information. As a consequence, the ability to manipulate quantum states translates immediately into a new form of information processing and exchange. What has been discovered during this conceptual passage is that this type of information processing is in many respects much richer than its classical analogue. This has contributed to the understanding of quantum states as an extension of the classical concept of state and not as a defective reality (see also Subsec. 2.3.4 and Sec. 15.5). The impossibility of knowing the value of all observables at the same time that had been seen as a strong limitation in the early days of quantum mechanics, now turns out to be a manifestation of a different – but not necessarily poorer – resource. On the contrary, we have increasingly discovered that superposition and entanglement are additional informational resources. These resources, for instance, allow for certain particular computations that are much faster on a quantum device than on its classical counterpart, and therefore the former is able to solve problems that cannot be practically solved using classical means. As we shall see, however, quantum devices are incredibly sensitive to decoherence, which destroys this specific ability: the fight against decoherence represents the biggest challenge for quantum technology in the coming years.

In Sec. 17.1 we shall introduce the von Neumann entropy. In Sec. 17.2 we shall see how to deal with entanglement by making use of the concept of information, and in Sec. 17.3 we shall discuss the relationships between measurement and information. In Sec. 17.4 the basic unit of quantum information processing is introduced, the qubit, while in Sec. 17.5 the important informational consequence of quantum non-separability is considered: teleportation. In Sec. 17.6 we shall instead consider an important application of the no-cloning theorem: quantum cryptography. In Sec. 17.7 we consider the basic elements of quantum computation, while in Sec. 17.8 we introduce the fundamental algorithms of quantum computation.

17.1 Information and entropy

The entropy of a state describing a physical system $\mathcal{S}$ is a quantity expressing the randomness of $\mathcal{S}$. It was introduced in 1877 by Boltzmann and can be defined as (see Eq. (1.60))

$$S_B = -k_B \ln w_E, \tag{17.1}$$

where k_B is the Boltzmann constant and w_E represents the total number of different configurations the system. Shannon[1] regarded this randomness as the amount of information that can be obtained about the system. In fact, if $\mathcal{S}$ has a large degree of disorder, a much larger amount of information is needed in order to determine it than the amount of information that is necessary to determine a system having a smaller degree of randomness. The Shannon entropy can be defined as

$$S_S = -\sum \wp_k \ln \wp_k, \tag{17.2}$$

where $\wp_k$ represents the probability that the k-th configuration occurs. Shannon's definition of entropy coincides with the one used in physics when the probabilities $\wp$ represent the canonical distribution. Equation (17.2) is a general formula of wide applicability. Note that entropy and information are to a certain extent complementary concepts because entropy defines the amount of uncertainty of the system before we receive information and information represents our posterior knowledge of the state of the system. In a classical frame, where the uncertainty of a system is only due to subjective ignorance, the acquisition of information normally decreases the entropy of a system. In quantum mechanics, as we shall see, this is not necessarily the case.

Let us define the quantum entropy,[2] the quantum analogue of Eq. (17.2) for a quantum state $\hat{\rho}$ as (see Eq. (5.33))

$$S_{VN}(\hat{\rho}) = -\mathrm{Tr}(\hat{\rho} \ln \hat{\rho}). \tag{17.3}$$

In fact, the density matrix can be seen as the operator which carries maximal information about the state of the system – though, as we know (see Secs. 15.1–15.2), this maximal information cannot be extracted by a measurement on a single system. S_{VN} is referred to as the *von Neumann entropy*.

If we consider an orthonormal basis $\{|b_k\rangle\}$ of eigenvectors of the density operator $\hat{\rho}$ for a system $\mathcal{S}$ such that (see Eq. (5.26))

$$\hat{\rho}\,|\,b_k\rangle = r_k\,|\,b_k\rangle, \tag{17.4}$$

where the r_k's are the eigenvalues of $\hat{\rho}$, we may rewrite Eq. (17.3) as

$$S_{VN}(\hat{\rho}) = -\sum_j r_j \ln r_j, \tag{17.5}$$

which is formally similar to the expression (17.2). For a complete mixture (see Sec. 5.6), as expected, we have the maximal entropy (see Prob. 17.1)

$$S^{(n)}_{\mathrm{Max}} = \ln(n), \tag{17.6}$$

[1] See [Shannon 1948].

[2] See [von Neumann 1927c] [*von Neumann* 1932, 202] [*von Neumann* 1955, 379–81]. The only difference with formulation (5.33) is that here we have taken the Boltzmann constant $k_B = 1$. The reason is that we deal here with pure informational aspects and not with thermodynamical or statistic–mechanics considerations.

where n is the dimension of the system, and here and in the following we drop the subscript VN for the sake of notation. Indeed, for an n-dimensional system, the completely mixed state has the form (see Eq. (5.61))

$$\hat{\hat{\rho}} = \begin{bmatrix} \frac{1}{n} & 0 & \cdots & 0 \\ 0 & \frac{1}{n} & \cdots & 0 \\ \cdots & \cdots & \cdots & \cdots \\ 0 & 0 & \cdots & \frac{1}{n} \end{bmatrix}. \tag{17.7}$$

In the case of pure states, instead, we have $S_{\mathrm{VN}} = 0$ (see Prob. 17.2). This means that pure states are highly ordered states, in the sense we have given to this term in Eqs. (17.1)–(17.2).

The properties of von Neumann entropy are as follows:[3]

- *Non-negativity*:

$$S_{\mathrm{Max}}^{(n)} \geq S(\hat{\rho}) \geq 0, \tag{17.8}$$

where the von Neumann entropy is equal to zero if and only if $\hat{\rho}$ is a pure state and is equal to $S_{\mathrm{Max}}^{(n)}$ only when we have a complete mixture. In other words, as anticipated, S expresses the degree of mixing. This is strictly related to the opposite concept of purity, already defined in Eq. (5.57). Araki and Lieb[4] proved that the entropy of a compound system 1–2, with subsystems 1, 2, satisfies the inequality

$$|S_1 - S_2| \leq S_{12} \leq S_1 + S_2, \tag{17.9}$$

which shows that, if the compound system is in a pure state, then the two subsystems must have equal entropy.
- *Unitary invariance*. Since unitary transformations do not change the spectrum of any observable, and neither therefore of any density matrix $\hat{\rho}$, we have

$$S(\hat{\rho}) = S(\hat{U}\hat{\rho}\hat{U}^{\dagger}), \tag{17.10}$$

for any unitary transformation $\hat{U}$.
- *Concavity*. For all projectors $\hat{P}_j$, we have

$$S\left(\sum_j w_j \hat{P}_j\right) \geq \sum_j w_j S\left(\hat{P}_j\right), \tag{17.11}$$

with $w_j \geq 0$ and $\sum_j w_j = 1$. Concavity means that a mixture of certain pure states has always more entropy than the sum of the corresponding pure states.
- *Subadditivity*. If we have $\hat{\rho} \in \mathcal{H}$, where $\mathcal{H} = \mathcal{H}_1 \otimes \mathcal{H}_2$ and the reduced density operators $\hat{\varrho}_1 = \mathrm{Tr}_2\hat{\rho} \in \mathcal{H}_1, \hat{\varrho}_2 = \mathrm{Tr}_1\hat{\rho} \in \mathcal{H}_2$, then

$$S(\hat{\rho}) \leq S(\hat{\varrho}_1 \otimes \hat{\varrho}_2) = S(\hat{\varrho}_1) + S(\hat{\varrho}_2), \tag{17.12}$$

[3] See [*Lindblad* 1983, 20–21] [Wehrl 1978, 236–37, 242].
[4] See [Araki/Lieb 1970].

where equality holds iff $\hat{\rho} = \hat{\varrho}_1 \otimes \hat{\varrho}_2$. Von Neumann entropy also satisfies a strong subadditivity theorem,[5] i.e.

$$S(\hat{\rho}_{123}) + S(\hat{\rho}_2) \leq S(\hat{\rho}_{12}) + S(\hat{\rho}_{23}). \tag{17.13}$$

In the case of pure states, we introduce a further property. If $\hat{\rho}_{12}$ is the density matrix of a composite system $\mathcal{S}_{12}$ in a pure state, then we have (see Prob. 17.3)

$$S(\hat{\varrho}_1) = S(\hat{\varrho}_2), \tag{17.14}$$

where

$$\hat{\varrho}_1 = \mathrm{Tr}_2\left(\hat{\rho}_{12}\right) \quad \text{and} \quad \hat{\varrho}_2 = \mathrm{Tr}_1\left(\hat{\rho}_{12}\right). \tag{17.15}$$

We now introduce an important concept, that of *relative entropy*, i.e. the entropy of a state $\hat{\rho}_1$ relative to another state $\hat{\rho}_2$[6]

$$S(\hat{\rho}_1 || \hat{\rho}_2) = \mathrm{Tr}\left[\hat{\rho}_1(\ln \hat{\rho}_1 - \ln \hat{\rho}_2)\right]. \tag{17.16}$$

The relative entropy satisfies non-negativity and unitary invariance, and is characterized by a third property, given by the following theorem:

Theorem 17.1 (Lindblad) *For all density matrices $\hat{\rho}, r'$, and for any completely positive active map $\mathcal{T}$* (see Subsecs. 8.1.1 and 9.10.1), *we have*

$$S(\mathcal{T}\hat{\rho}||\mathcal{T}\hat{\rho}') \leq S(\hat{\rho}||\hat{\rho}'). \tag{17.17}$$

The larger $S(\hat{\rho}||\hat{\rho}')$, the more information for discriminating between the two states that can be obtained from an observation. Therefore, the theorem expresses also a general property of the loss of information in the sense that a time evolution obeying a certain form of Markovian equation does not make the two states more easily distinguishable.[7] The concept of relative entropy will turn out to be useful in the context of measures of entanglement.

17.2 Entanglement and information

In Sec. 5.5 we have defined entanglement as a fundamental – and genuinely quantum – feature of microscopic systems. Since then, we have encountered this concept many times throughout the book (see, e.g., Chs. 9, 14, and 16). In this section, we shall provide a more quantitative grounding of this property.

[5] See [Lieb/Ruskai 1973a, Lieb/Ruskai 1973b].
[6] See [Lindblad 1973] [Uhlmann 1977].
[7] See [*Lindblad* 1983, 22–23].

17.2.1 Mutual information

Let us define the entanglement between systems 1 and 2 as

$$\begin{aligned} \mathrm{E}(1,2) &= -[S(1,2) - (S(1) + S(2))] \\ &= S(1) + S(2) - S(1,2), \end{aligned} \tag{17.18}$$

where $S(1,2)$ is the joint (total) entropy of systems 1 and 2, and

$$S(1) = S(\hat{\varrho}_1) \quad \text{and} \quad S(2) = S(\hat{\varrho}_2) \tag{17.19}$$

are the entropies calculated on the reduced density matrices (17.15) of the subsystems 1 and 2, respectively, relative to $\hat{\rho}_{12}$. This reflects the fact that entanglement is a quantum form of mutual information: two entangled systems are correlated because they share an amount of information that is not foreseen classically. Recently, it has been shown[8] that entangled systems can share a potential information that can be exchanged for free in a successive instant of time. We shall discuss below this aspect in connection to teleportation.

It is useful at this stage to mention that, classically, we express the mutual information between two generic random variables J, K as follows:

$$\begin{aligned} I(J:K) &= S(J) - S(J|K) \\ &= S(K) - S(K|J) \\ &= S(J) + S(K) - S(J,K), \end{aligned} \tag{17.20}$$

where

$$S(J|K) = -\sum_{j,k} p(j,k) \ln p(j|k) \tag{17.21}$$

is the conditional entropy between J and K, and j, k span all possible values of J, K, respectively, and

$$S(J,K) = -\sum_{j,k} \wp_{j,k} \ln(\wp_{j,k}) \tag{17.22}$$

is the classical joint entropy of the systems J and K, while $\wp_{j,k}$ is the joint probability distribution of events j, k.

It is also possible to calculate the entanglement relative to given observables[9] – this is very useful for choosing the observables that optimally express the entanglement between subsystems. In this case we can write the actual entropies of observables $\hat{O}_1$ and $\hat{O}_2$ of the system 1 and 2 as

$$S(\hat{O}_1) = -\sum_j \langle j|\hat{\rho}_1|j\rangle \ln\langle j|\hat{\rho}_1|j\rangle \tag{17.23a}$$

$$S(\hat{O}_2) = -\sum_k \langle k|\hat{\rho}_2|k\rangle \ln\langle k|\hat{\rho}_2|k\rangle \tag{17.23b}$$

[8] See [Horodecki *et al.* 2005].

[9] See [Barnett/Phoenix 1989].

where $\{|j\rangle\}$ ($\{|k\rangle\}$) are the eigenstates of the observable $\hat{O}_1$ ($\hat{O}_2$), i.e. the states of the system 1 (2) that will be the output states when measuring the observable $\hat{O}_1$ ($\hat{O}_2$). The joint entropy of the two quantum observables is given by

$$S(\hat{O}_1, \hat{O}_2) = -\sum_j \sum_k \langle j,k|\hat{\rho}|k,j\rangle \ln\langle j,k|\hat{\rho}|k,j\rangle, \tag{17.24}$$

where $\hat{\rho}$ is the density matrix of the global quantum system, and we have used the notation $|j,k\rangle = |j\rangle \otimes |k\rangle$. Then, we may define the entanglement between these two observables as

$$\mathrm{E}(\hat{O}_1, \hat{O}_2) = S(\hat{O}_1) + S(\hat{O}_2) - S(\hat{O}_1, \hat{O}_2). \tag{17.25}$$

It is interesting to note that we always have

$$\mathrm{E}(\hat{O}_1, \hat{O}_2) \leq \mathrm{E}(1,2) \tag{17.26}$$

i.e. that the information contained in the correlation between any two observables pertaining to the two subsystems – given by Eq. (17.25) – cannot exceed the total information content of the correlation between the two systems – given by Eq. (17.18). Moreover, it can be proved[10] that, if the state of the compound system is pure, then

$$\mathrm{E}(\hat{O}_1, \hat{O}_2) \leq \frac{1}{2}\mathrm{E}(1,2). \tag{17.27}$$

17.2.2 More about Bell inequalities

We can also interpret the Bell inequalities in terms of information and entropy. Braunstein and Caves[11] formulated an information-theoretic Bell inequality of the type

$$S(\hat{O}^a|\hat{O}^b) \leq S(\hat{O}^a|\hat{O}^{b'}) + S(\hat{O}^{b'}|\hat{O}^{a'}) + S(\hat{O}^{a'}|\hat{O}^b). \tag{17.28}$$

Inequality (17.28) is deduced (i) from the expression

$$S(\hat{O}^a|\hat{O}^b) \leq S(\hat{O}^a) \leq S(\hat{O}^a, \hat{O}^b), \tag{17.29}$$

where the lhs inequality means that removing a condition never decreases the entropy carried by a quantity, and the rhs inequality means that two observables never carry less entropy than each of them separately; and (ii) from the following generalization of inequality (17.29)

$$\begin{aligned} S(\hat{O}^a, \hat{O}^b) &\leq S(\hat{O}^a, \hat{O}^{a'}, \hat{O}^b, \hat{O}^{b'}) \qquad (17.30)\\ &= S(\hat{O}^a|\hat{O}^{a'} \wedge \hat{O}^b \wedge \hat{O}^{b'}) + S(\hat{O}^{b'}|\hat{O}^{a'} \wedge \hat{O}^b) + S(\hat{O}^{a'}|\hat{O}^b) + S(\hat{O}^b), \end{aligned}$$

[10] See [Barnett/Phoenix 1991].

[11] See [Braunstein/Caves 1988] [Braunstein/Caves 1990].

where the expansion of the rhs is a recursion of the definition of conditional entropy, i.e. [see, also Eq. (9.170)]

$$S(\hat{O}^a, \hat{O}^b) = S(\hat{O}^a|\hat{O}^b) + S(\hat{O}^b). \tag{17.31}$$

Proof

In order to deduce Eq. (17.28) from inequality (17.30) one first needs to substitute the first term of the rhs of expression (17.30) with the first term of the rhs of inequality (17.28) by means of the following application of inequality (17.29):

$$S(\hat{O}^a|\hat{O}^{a'} \wedge \hat{O}^b \wedge \hat{O}^{b'}) \leq S(\hat{O}^a|\hat{O}^{b'}). \tag{17.32}$$

Then, one substitutes the second term of the rhs of expression (17.30) with the second term of the rhs of inequality (17.28) by means of another application of inequality (17.29), i.e.

$$S(\hat{O}^{b'}|\hat{O}^{a'} \wedge \hat{O}^b) \leq S(\hat{O}^{b'}|\hat{O}^{a'}). \tag{17.33}$$

Finally, one leaves the third term of the rhs of inequality (17.30), and one substitutes the lhs term $S(\hat{O}^a, \hat{O}^b)$ with the term $S(\hat{O}^a|\hat{O}^b)$ by applying Bayes' rule (17.31), which also implies the deletion of the fourth term, $S(\hat{O}^b)$, on both the lhs and the rhs of expression (17.30).

Q.E.D

The authors show that, in the context of the Bell's spin-like experiments (see Figs. 16.7 and 16.8), inequality (17.28) is violated by quantum mechanics and that the information deficit increases with increasing spin number s of the involved particles (see Fig. 17.1).

It is possible to arrive at a result formally similar to the previous one by using different conceptual instruments.[12] Recalling the definition (17.20) of mutual information, we now define the informational distance $\delta(\xi, \xi')$ between the variables ξ and ξ' as

$$\begin{aligned} \delta(\xi, \xi') &= S(\xi|\xi') + S(\xi'|\xi) \\ &= S(\xi, \xi') - S(\xi : \xi') \\ &= 2S(\xi, \xi') - S(\xi) - S(\xi'). \end{aligned} \tag{17.34}$$

The informational distance is positive definite and symmetric. It can be proved[13] that the informational distance satisfies the following triangular inequality for the three quantities ξ, ξ', ξ'':

$$\delta(\xi, \xi') + \delta(\xi', \xi'') \geq \delta(\xi, \xi''). \tag{17.35}$$

12 See [Schumacher 1991].
13 See [Schumacher 1990].

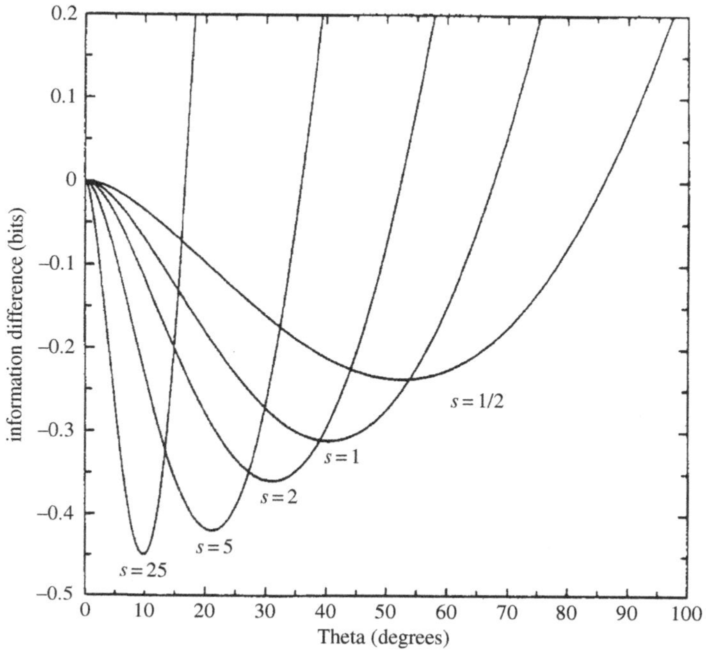

Figure 17.1 Information difference in bits versus angle θ in degrees for different values of the spin number $s = 1/2, 2, 5$, and 25. The maximum information deficit for $s = 1/2$ is -0.2369 bits at 52.31°; for $s = 25$ the maximal deficit equals -0.4493 bits at 9.798°. Adapted from [Braunstein/Caves 1988, 664].

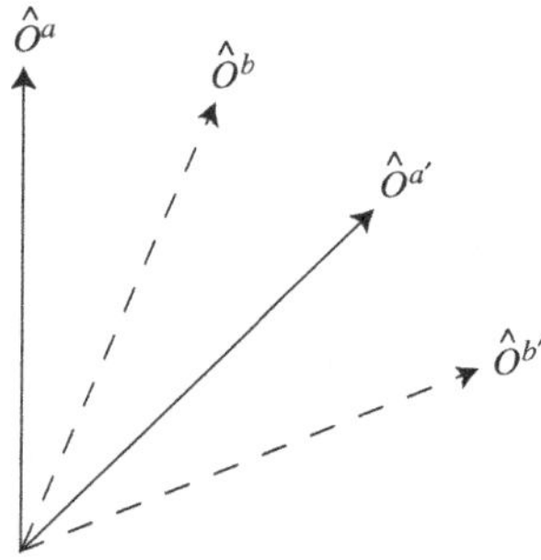

Figure 17.2 Spin measurement axes yielding a violation of the information-distance quadrilateral inequality for a singlet state. The angles between the vectors are all equal to $\pi/8$.

Similarly we can state a quadrilateral inequality, the so-called quadrilateral information-distance Bell inequality for the four Bell-like quantum observables

$$\delta(\hat{O}^a, \hat{O}^b) + \delta(\hat{O}^b, \hat{O}^{a'}) + \delta(\hat{O}^{a'}, \hat{O}^{b'}) \geq \delta(\hat{O}^a, \hat{O}^{b'}). \tag{17.36}$$

If we fix the information distance between $\hat{O}^a$ and $\hat{O}^b$ (separated by angle θ) to be

$$\delta(\hat{O}^a, \hat{O}^b) = 2f\left(\frac{\theta}{2}\right), \tag{17.37}$$

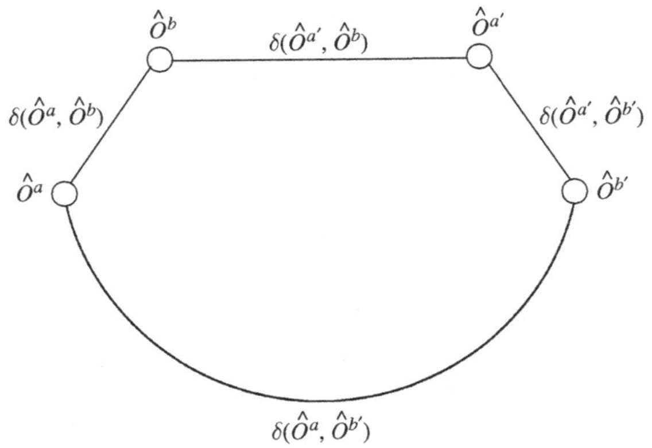

Figure 17.3 **Schematic information-theoretic representation of quantum non-separability. The information distance $\delta(\hat{O}^a, \hat{O}^{b'})$ is greater than is allowed by the classical metric properties of informational distance.**

where $f(\phi) = -\cos^2\phi \ln \cos^2\phi - \sin^2\phi \ln \sin^2\phi$, and take the angle between $\hat{O}^a$ and $\hat{O}^b$, between $\hat{O}^b$ and $\hat{O}^{a'}$, between $\hat{O}^{a'}$ and $\hat{O}^{b'}$ to be equal to $\pi/8$ (see Fig. 17.2), we arrive at the following values:

$$\delta(\hat{O}^a, \hat{O}^b) = \delta(\hat{O}^{a'}, \hat{O}^b) = \delta(\hat{O}^{a'}, \hat{O}^{b'}) = 2f\left(\frac{\pi}{16}\right) \simeq 0.323, \tag{17.38a}$$

$$\delta(\hat{O}^a, \hat{O}^{b'}) = 2f\left(\frac{3\pi}{16}\right) \simeq 1.236. \tag{17.38b}$$

But since $0.323 + 0.323 + 0.323 < 1.236$ it is evident that the Bell inequality (17.36) is violated (see Fig. 17.3).[14]

17.2.3 Vedral and co-workers' measure

As we have seen, and as was expected, entanglement is a quantity that may take on different values. A stringent formulation of entanglement's measure has been given by Knight and co-workers.[15] They established three conditions which such a measure has to fulfill:

- $\mathrm{E}(\hat{\rho}) = \mathrm{E}(\hat{\rho}_1, \hat{\rho}_2) = 0$.
- Local unitary operations leave $\mathrm{E}(\hat{\rho})$ invariant, i.e.,

$$\mathrm{E}(\hat{\rho}) = \mathrm{E}\left(\hat{U}_1 \otimes \hat{U}_2 \hat{\rho} \hat{U}_1^\dagger \otimes \hat{U}_2^\dagger\right). \tag{17.39}$$

- The measure of the entanglement $\mathrm{E}(\hat{\rho})$ cannot increase under local measurement and classical correlation: if we represent such an operation by $\mathcal{T}$, then $\mathrm{E}(\mathcal{T}\hat{\rho}) \leq \mathrm{E}(\hat{\rho})$.

14 In analogy with Eq. (17.20) it is also possible to obtain a violation of mutual-information Bell-like inequalities [Cerf/Adami 1997].

15 See [Vedral *et al.* 1997].

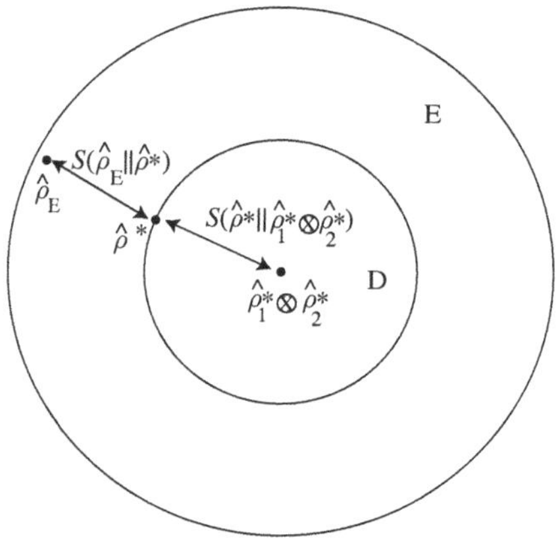

Figure 17.4 The set of all density matrices of a compound system is represented by E + D. The subset D of all disentangled states is represented by the inner circle. An entangled density matrix $\hat\rho_E$ belongs to E, and $\hat\rho^*$ is the disentangled density operator which minimizes the relative entropy $S(\hat\rho_E \parallel \hat\rho_D)$, thus representing the amount of quantum correlations in $\hat\rho_E$. A product density matrix $\hat\rho_1^* \otimes \hat\rho_2^*$ is obtained by tracing $\hat\rho^*$ over $\mathcal{S}_1, \mathcal{S}_2$. The relative entropy $S(\hat\rho^* \parallel \hat\rho_1^* \otimes \hat\rho_2^*)$ represents the classical part of correlations in state $\hat\rho_E$ (see also Subsec. 16.8.2).

The reason for the first requirement is evident (separable states contain no entanglement); the reason for the second requirement is that local unitary transformations only represent local changes of basis and leave quantum correlations unchanged; and the reason for the third condition is that each increase in correlations achieved by $\mathcal{T}$ is classical in nature, and hence entanglement is not increased. Moreover, since each form of operation is local, correlations cannot be increased by this means.

Let us now consider the set of all density matrices of a compound system containing two subsystems $\mathcal{S}_1, \mathcal{S}_2$ (see Fig. 17.4). We take D to be the subset of all disentangled states, and E to be the subset of all entangled ones.[16]

Next we define the entanglement of a density operator $\hat\rho_E$ in terms of the relative entropy (17.16) as

$$E(\hat\rho_E) = \min S(\hat\rho_E \parallel \hat\rho_D), \qquad (17.40)$$

where $\hat\rho_D$ is an arbitrary density operator $\in$ D.[17] The difficulty here is that the relative entropy is not symmetric, as should be the case for a proper distance. However, it appears reasonable to define quantum entanglement as a distance from (relative to) a disentangled state. Our first requirement is satisfied for $\hat\rho_D = \hat\rho_E$, the second one is automatically satisfied, because D is invariant under local unitary transformations (see Prob. 17.4), and the third one is satisfied by Th. 17.1 – where the operation $\mathcal{T}$ is defined by Th. 9.1 (p. 329).

16 Note that the sets E + D and D are convex but not E.
17 See [Vedral *et al.* 1997].

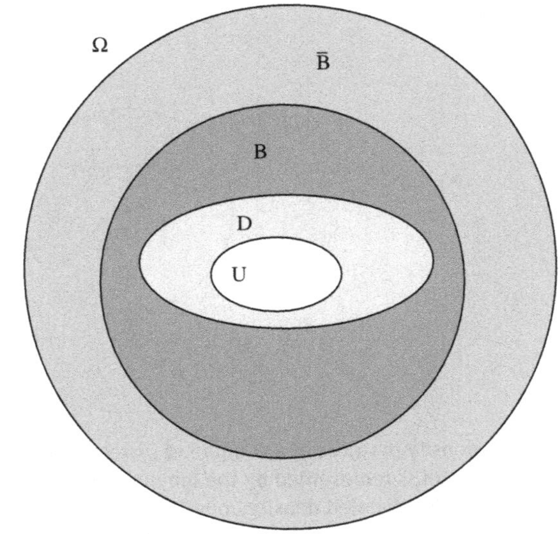

Figure 17.5 Representation of all density matrices. The set of all possible density matrices of a given compound system is denoted by $\Omega = D + E$ (see Fig. 17.4). U is the set of all uncorrelated density matrices, i.e. of the type $\hat{\rho} = \hat{\rho}_1 \otimes \hat{\rho}_2$. The set D also comprehends classically correlated density operators. The set B consists of all density matrices that satisfy Bell inequalities (see also Fig. 16.23), whereas $\overline{B}$ is the set of all quantum density operators violating Bell inequalities (see Prob. 17.5).

Therefore, the amount of entanglement given by Eq. (17.40) can be interpreted as finding a state $\hat{\rho}^*$ in D that is the closest to $\hat{\rho}_E$ under the "measure" $S(\hat{\rho}_E \parallel \hat{\rho}_D)$. The state $\hat{\rho}^*$ approximates the classical correlations of $\hat{\rho}_E$ as closely as possible. In this way we are able to divide the correlations in the quantum-mechanical component $E(\hat{\rho}_E)$ and the classical one $S(\hat{\rho}^* \parallel \hat{\rho}_1^* \otimes \hat{\rho}_2^*)$ (see Fig. 17.5).

17.2.4 Decompression

In order to control and to determine the amount of entanglement a technique of *decompression* of the information has been developed.[18] It is an inverse operation with respect to the purification (see Subsec. 5.5.3): by starting with a number of highly entangled pairs shared by two distant parties, we end up (by local operations) with a greater number of pairs with a lower entanglement. Here we will only sketch the procedure.

Such an operation is a local copying (see Sec. 15.2) of non-local quantum correlations. By this means we can control the entanglement, because if we are able to optimally split the original entanglement of a single pair into two equally entangled pairs (having the same state), we have a means of defining half the entanglement of the original pair, which, for example, can be a maximally entangled state like the singlet state (see Fig. 17.6).

[18] See [Bužek *et al.* 1997].

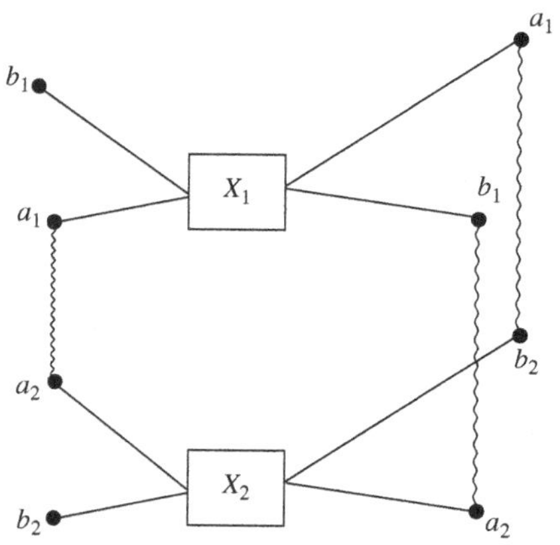

Figure 17.6 Scheme for decompression of information. The input is on the left (entanglement between a_1 and a_2) and the output is on the right (entanglement between a_1 and b_2 and between b_1 and a_2).

17.3 Measurement and information

17.3.1 Information downloading

When measuring, we expect to obtain a mixed state of the object system and the apparatus, which means an increase in entropy if the initial state is a pure state. An increase in entropy can be obtained by discarding the initial potential information that cannot be acquired,[19] which consists exactly of the quantum correlations that exist between the observed system and the apparatus. The information I that can be acquired, instead, has to do with the sum of the partial entropies of the subsystems.

In order to obtain such a result, we assume that the initial total state of the combined system, before the premeasurement, be a product state relative to the two subsystems. The initial entropy is then equal to the sum of the entropies of the subsystems, and this entropy will not change during the measurement (if the two subsystems are isolated). However, during premeasurement, the two subsystems may become correlated and now we can discard the correlation and obtain a final useful piece of information (a function of the sum of the final entropies of the two subsystems) which is greater than the sum of the initial partial entropies. We have already considered this process in general terms (see Sec. 9.4). Now, we wish to consider this process in the context of informational exchange.

Formally, for an initial state $\hat{\rho}_i^{12} = \hat{\rho}_i^1 \otimes \hat{\rho}_i^2$ we have

$$S(\hat{\rho}_i^{12}) = S(\hat{\rho}_i^1) + S(\hat{\rho}_i^2) = S(\hat{\rho}_f^{12}) \leq S(\hat{\varrho}_f^1) + S(\hat{\varrho}_f^2), \tag{17.41}$$

[19] See [*Lindblad* 1983, 27–28, 44–46, 59].

having discarded the part of information due to entanglement, measured by

$$\mathrm{E}(\hat{\rho}_f^{12}) = S(\hat{\varrho}_f^1 \otimes \hat{\varrho}_f^2) - S(\hat{\rho}_f^{12}), \tag{17.42}$$

in agreement with Eq. (17.18). We recall that the quantity $S(\hat{\varrho}_f^1) + S(\hat{\varrho}_f^2)$ represents the degree of disorder of the system in its final state (ideally a complete mixture) and defines the amount of uncertainty of the system about which we may receive a posteriori information, while $\hat{\rho}_f^{12}$ represents the final state of the compound system. We assume here that the two subsystems are open systems so that the correlation (17.42) can be downloaded into a larger system (see also Ch. 14). Hence, as we know, we have to introduce the concept of environment. The a posteriori information that we may gain from a system in a mixed state whose entropy has the form $S(\hat{\varrho}_f^1) + S(\hat{\varrho}_f^2)$ is given by the difference between this entropy and the entropy associated with the final outcome – a much more ordered state $\hat{\rho}_m$ (see Subsecs. 9.10.2 and 14.3.1) – that is,

$$I_m = -\Delta S = S(\hat{\varrho}_f^1 \otimes \hat{\varrho}_f^2) - S(\hat{\rho}_m). \tag{17.43}$$

It is this information gain that represents the true irreversibility in the measurement process (see Sec. 9.5).

In order to gain a better understanding of Eq. (17.41), let us consider[20] the initial density matrix $\hat{\rho}_{\mathcal{SAE}}(t_0)$ for the system $\mathcal{S}$ plus the apparatus $\mathcal{A}$ plus the environment $\mathcal{E}$, with

$$\hat{\rho}_{\mathcal{SAE}}(t_0) = \hat{\rho}_{\mathcal{SA}}(t_0)\hat{\rho}_{\mathcal{E}}(t_0). \tag{17.44}$$

Due to the strong subadditivity property (see Eq. (17.13)), at a later time t we must have

$$S_t(\mathcal{S}, \mathcal{A}, \mathcal{E}) + S_t(\mathcal{A}) \leq S_t(\mathcal{S}, \mathcal{A}) + S_t(\mathcal{A}, \mathcal{E}), \tag{17.45}$$

where $S_t(\mathcal{S}, \mathcal{A}, \mathcal{E})$ is the entropy of the whole system at time t. Since the evolution of the density matrix of the total system is unitary, it follows that

$$S_{t_0}(\mathcal{S}, \mathcal{A}, \mathcal{E}) = S_t(\mathcal{S}, \mathcal{A}, \mathcal{E}). \tag{17.46}$$

Due to the initial state, in which $\mathcal{S}$ and $\mathcal{A}$ are uncoupled with the environment $\mathcal{E}$, we have

$$S_{t_0}(\mathcal{S}, \mathcal{A}, \mathcal{E}) = S_{t_0}(\mathcal{S}, \mathcal{A}) + S_{t_0}(\mathcal{E}). \tag{17.47}$$

Suppose now that $\mathcal{S}$ is a spectator while $\mathcal{A}$ and $\mathcal{E}$ interact for time t. Thus we have

$$S_t(\mathcal{S}) = S_{t_0}(\mathcal{S}) \tag{17.48a}$$

and also

$$S_t(\mathcal{A}, \mathcal{E}) = S_{t_0}(\mathcal{A}, \mathcal{E}). \tag{17.48b}$$

The lack of entanglement or of correlation between $\mathcal{A}$ and $\mathcal{E}$ at t_0 implies that

$$S_{t_0}(\mathcal{A}, \mathcal{E}) = S_{t_0}(\mathcal{A}) + S_{t_0}(\mathcal{E}). \tag{17.49}$$

Combining all these relations, we now show that it is possible to derive

$$\mathrm{E}_t(\mathcal{S}, \mathcal{A}) \leq \mathrm{E}_{t_0}(\mathcal{S}, \mathcal{A}), \tag{17.50}$$

[20] See [Partovi 1989].

where $E(\mathcal{S}, \mathcal{A})$ may be derived from Eq. (17.42) and expresses the complex of quantum correlations between $\mathcal{S}$ and $\mathcal{A}$. To this purpose, let us add $S_t(\mathcal{S})$ to both sides of Eq. (17.45) and rewrite it as

$$S_t(\mathcal{S}) + S_t(\mathcal{A}) - S_t(\mathcal{S}, \mathcal{A}) \leq S_t(\mathcal{S}) + S_t(\mathcal{A}, \mathcal{E}) - S_t(\mathcal{S}, \mathcal{A}, \mathcal{E}). \tag{17.51}$$

By using Eqs. (17.46) – (17.49), we obtain

$$S_t(\mathcal{S}) + S_t(\mathcal{A}) - S_t(\mathcal{S}, \mathcal{A}) \leq S_{t_0}(\mathcal{S}) + S_{t_0}(\mathcal{A}) - S_{t_0}(\mathcal{S}, \mathcal{A}), \tag{17.52}$$

which is Eq. (17.50) in explicit form. In other words, from a measurement we expect a decrease in the quantum correlations between the system and the apparatus, i.e. the entanglement between the system and the apparatus must somehow be reduced. As a consequence, there is an increase in entropy.

17.3.2 Bounds for information

A problem of great interest for measurement problems is the existence of bounds for the information gain. Holevo[21] proved the existence of an upper bound for the information one can extract from a quantum system. Holevo's theorem shows that it is impossible to extract the whole information contained in a quantum system.

Let us first write a *continuous* expression for the mutual information between an input $\boldsymbol{\eta}$ and an output $\mathbf{x}$ which (classically) is given by

$$I_M(\boldsymbol{\eta} : \mathbf{x}) = S(\boldsymbol{\eta}) - S(\boldsymbol{\eta}|\mathbf{x}), \tag{17.53}$$

where

$$S(\boldsymbol{\eta}|\mathbf{x}) = -\int d\mathbf{x}\wp(\mathbf{x}) \int d\boldsymbol{\eta}\wp(\boldsymbol{\eta}|\mathbf{x}) \ln \wp(\boldsymbol{\eta}|\mathbf{x}) \tag{17.54}$$

is the *continuous* expression for the conditional entropy.

Suppose, in a quantum-mechanical context, an effect $\hat{E}(dx)$ (see Sec. 9.10) so that, when it is measured in a state $\hat{\rho}$, the output probability distribution $\wp_{\hat{\rho}}$ of $\hat{x}$ is given by

$$\wp_{\hat{\rho}}(dx) = \mathrm{Tr}[\hat{E}(dx)\hat{\rho}]. \tag{17.55}$$

Suppose now that $\boldsymbol{\eta}$ is a parameter that has a probability distribution $\wp(d\boldsymbol{\eta})$ on the projectors $\hat{P}_\eta$ on the Hilbert space of the system. Hence we write the output conditional distribution as $\wp(dx|\boldsymbol{\eta}) = \wp_{\hat{\rho}_\eta}(dx)$ and define a mixture of projectors $\hat{P}_\eta$

$$\hat{\hat{\rho}} = \int \hat{P}_\eta \wp(d\boldsymbol{\eta}). \tag{17.56}$$

[21] See [Holevo 1973b]. See also [Fuchs/Caves 1995].

Then, we may state the following:

Theorem 17.2 (Holevo) *The information which can be obtained from the considered system is subjected to the following bound:*

$$I_M(\boldsymbol{\eta} : x) \leq S(\hat{\hat{\rho}}) - \int d\boldsymbol{\eta}\wp(d\boldsymbol{\eta})S(\hat{P}_\eta), \tag{17.57}$$

where the equality holds if all the $\hat{P}_\eta$'s commute.

We omit here the proof of the theorem.[22] Since we have $S(\hat{P}_\eta) \geq 0$, inequality (17.57) implies *a fortiori*

$$I_M(\boldsymbol{\eta} : x) \leq S(\hat{\hat{\rho}}), \tag{17.58}$$

which means that, independently of the quantum measurement one can perform, the related information transfer is never larger than the entropy of the mixture one obtains.

Holevo's theorem is a further confirmation of the impossibility of measuring (by repeated measurements) the density matrix of a single system – which represents, as already stated, the maximal amount of information contained in a physical state (see Ch. 15). Once more we face the duality between state and observables, in the sense that we may recover the information content of a state only through the specific perspective of a measurement of a given observable.

17.4 Qubits

Before entering into the details of quantum information processing, we need to define its basic unit: the qubit. A classical bit of information is represented by a system that can be in either of two states, say, 0 or 1. From this simple observation, one may derive the whole binary logic that is at the heart of classical information processing, which employs logical binary operators (AND, NAND, OR, XOR, etc.), instantiated by gates. At the quantum-mechanical level, the most natural candidate for replacing a classical bit is the state of a two-level system, whose basic components may be written as

$$| 0 \rangle = \begin{pmatrix} 1 \\ 0 \end{pmatrix}, \quad | 1 \rangle = \begin{pmatrix} 0 \\ 1 \end{pmatrix}. \tag{17.59}$$

22 See the original article, which is in turn based on a result of Uhlmann [Uhlmann 1977].

This is the so called quantum bit of information or, in short, a *qubit*. The most striking difference between a classical bit and a qubit is, as we know, that any two-level quantum system may be in *any* superposition of these two components, namely

$$| \psi \rangle = c_0 \,| 0 \rangle + c_1 \,| 1 \rangle . \tag{17.60}$$

As we shall see, it is precisely this fact, together with its natural consequences (e.g. entanglement), that makes quantum information processing much richer than its classical counterpart. Generalizing to the n-qubit case, we make use of 2^n basis states of the type

$$| j_1 \rangle_1 \otimes | j_2 \rangle_2 \otimes \ldots \otimes | j_n \rangle_n , \tag{17.61}$$

where each of the j_l may take on the values 0, 1.

Let us consider, in particular, a system composed of two two-level subsytems. In this case, we can store two qubits of information, for instance

$$| \Psi \rangle = \frac{1}{2} \left(| 00 \rangle + | 01 \rangle + | 10 \rangle + | 11 \rangle \right), \tag{17.62}$$

where we have adopted the shorthand notation

$$| jk \rangle = | j \rangle_1 \otimes | k \rangle_2 , \quad j, k = 0, 1. \tag{17.63}$$

When dealing with two or more subsystems, it is very important to distinguish between entangled and separable states (see also Subsec. 5.5.1). For instance, consider the following two states:

$$| \Psi \rangle = c_\alpha \,| 00 \rangle + c_\beta \,| 01 \rangle , \tag{17.64a}$$

$$\left| \Psi' \right\rangle = c'_\alpha \,| 00 \rangle + c'_\beta \,| 11 \rangle . \tag{17.64b}$$

Though they may look quite similar, the latter is entangled while the former is not. In fact, state (17.64a) may be written as

$$| \Psi \rangle = | 0 \rangle_1 \otimes \left(c_\alpha \,| 0 \rangle_2 + c_\beta \,| 1 \rangle_2 \right), \tag{17.65}$$

which clearly shows that it is a product state of subsystems 1 and 2. On the other hand, state (17.64b) cannot be written in such a factorized form.

Sometimes in the following we shall find it convenient to use the notation $\{| \uparrow \rangle , | \downarrow \rangle \}$ instead of $\{| 0 \rangle , | 1 \rangle \}$ to denote the basis states of a qubit, especially when dealing with spin systems.

17.5 Teleportation

In this section we describe one of the most striking examples of the power of quantum information processing: teleportation. By this term, we mean a procedure that is able to

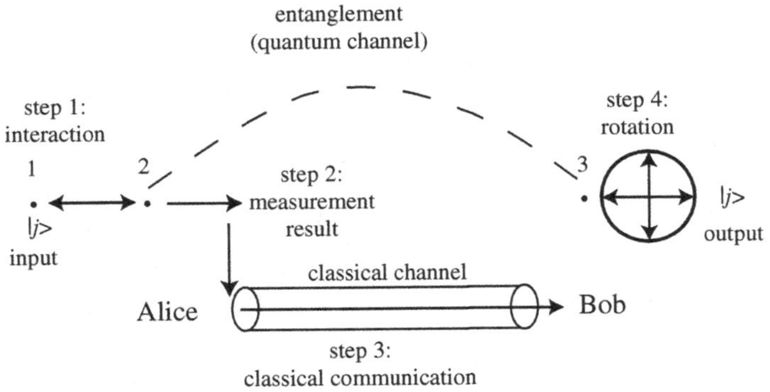

Figure 17.7 Scheme of teleportation. In step 1, Alice makes particle 1 (the qubit to be teleported) interact with particle 2 (which is entangled with particle 3). In step 2 she reads the outcome of the Bell measurement performed on particles 1 and 2. In step 3, she classically communicates this result to Bob (the "owner" of particle 3). In function of this classical piece of information, in step 4 Bob rotates the state of particle 3 and obtains as output the input qubit.

transfer with certainty the state of an input quantum system onto the state of a distant output system of the same type.

As we know, an instantaneous transfer of information is not possible. However, it is possible to "teleport" some information by exploiting EPR correlations and without violating the Einstein locality (the locality *stricto sensu*) (see also Ch. 16).[23] Suppose that "Alice" wishes to give "Bob" some information about a quantum system (a particle labelled "1") prepared in state $|j\rangle_1$ – called the *ancilla* – unknown to her as well as to Bob. For this purpose, Alice allows the ancilla to interact with a particle 2 that is entangled with a particle 3, previously given to Bob (see Fig. 17.7). Now, Alice performs a special kind of measurement on the compound system $1+2$, so that, by classically telling Bob the result of this measurement, Bob can reconstruct the same state of 1 onto 3. In order to express this procedure formally, let particles 2 and 3 be in the EPR state (see Eqs. (6.194) and (16.11))

$$\left|\Psi^-\right\rangle_{23} = \frac{1}{\sqrt{2}}\left(|\uparrow\rangle_2|\downarrow\rangle_3 - |\downarrow\rangle_2|\uparrow\rangle_3\right). \tag{17.66}$$

Now, Alice performs on particles 1 and 2 the measurement of the Bell operator

$$\hat{\mathcal{B}} = \hat{\boldsymbol{\sigma}}_1 \cdot \mathbf{a}\left(\hat{\boldsymbol{\sigma}}_2 \cdot \mathbf{b} + \hat{\boldsymbol{\sigma}}_2 \cdot \mathbf{b}'\right) + \hat{\boldsymbol{\sigma}}_1 \cdot \mathbf{a}'\left(\hat{\boldsymbol{\sigma}}_2 \cdot \mathbf{b} - \hat{\boldsymbol{\sigma}}_2 \cdot \mathbf{b}'\right), \tag{17.67}$$

associated with inequality (16.69) and introduced by Braunstein, Mann, and Revzen,[24] whose eigenbasis is given by (see Subsec. 16.6.3)

23 See [Bennett/Wiesner 1992, Bennett *et al.* 1993].

24 See [Braunstein *et al.* 1992a].

$$|\Psi^-\rangle_{12} = \frac{1}{\sqrt{2}}(|\uparrow\rangle_1|\downarrow\rangle_2 - |\downarrow\rangle_1|\uparrow\rangle_2), \tag{17.68a}$$

$$|\Psi^+\rangle_{12} = \frac{1}{\sqrt{2}}(|\uparrow\rangle_1|\downarrow\rangle_2 + |\downarrow\rangle_1|\uparrow\rangle_2), \tag{17.68b}$$

$$|\Phi^-\rangle_{12} = \frac{1}{\sqrt{2}}(|\uparrow\rangle_1|\uparrow\rangle_2 - |\downarrow\rangle_1|\downarrow\rangle_2), \tag{17.68c}$$

$$|\Phi^+\rangle_{12} = \frac{1}{\sqrt{2}}(|\uparrow\rangle_1|\uparrow\rangle_2 + |\downarrow\rangle_1|\downarrow\rangle_2), \tag{17.68d}$$

which is a complete orthonormal basis spanning the Hilbert space $\mathcal{H} = \mathcal{H}_1 \oplus \mathcal{H}_2$ of particles 1 and 2 (see Prob. 17.6).

If we write the unknown state of the ancilla in the form

$$|j\rangle_1 = c|\uparrow\rangle_1 + c'|\downarrow\rangle_1, \tag{17.69}$$

then the complete state of 1 + 23 before the measurement is given by

$$|\Psi\rangle_{123} = \frac{c}{\sqrt{2}}(|\uparrow\rangle_1|\uparrow\rangle_2|\downarrow\rangle_3 - |\uparrow\rangle_1|\downarrow\rangle_2|\uparrow\rangle_3) + \frac{c'}{\sqrt{2}}(|\downarrow\rangle_1|\uparrow\rangle_2|\downarrow\rangle_3 - |\downarrow\rangle_1|\downarrow\rangle_2|\uparrow\rangle_3). \tag{17.70}$$

The previous equation may be rewritten in terms of basis (17.68) as (see Prob. 17.7)

$$\begin{aligned}|\Psi\rangle_{123} = \frac{1}{2}\Big[&|\Psi^-\rangle_{12}(-c|\uparrow\rangle_3 - c'|\downarrow\rangle_3) + |\Psi^+\rangle_{12}(-c|\uparrow\rangle_3 + c'|\downarrow\rangle_3)\\ &+ |\Phi^-\rangle_{12}(c|\downarrow\rangle_3 + c'|\uparrow\rangle_3) + |\Phi^+\rangle_{12}(c|\downarrow\rangle_3 - c'|\uparrow\rangle_3)\Big].\end{aligned} \tag{17.71}$$

Now, after Alice's measurement, the system 12 is projected into one of the four pure states superposed in Eq. (17.71), depending on the measurement outcome. As a consequence, the four possible output states for particle 3 (Bob's one) are very simply related to the original state $|j\rangle_1$ which Alice wished to teleport:

$$\hat{U}_k|k\rangle_3 = |j\rangle_1, \quad k = 1,2,3,4, \tag{17.72}$$

where (see Prob. 17.8)

$$\hat{U}_1 = \begin{bmatrix} -1 & 0 \\ 0 & -1 \end{bmatrix}; \quad \hat{U}_2 = \begin{bmatrix} -1 & 0 \\ 0 & 1 \end{bmatrix}; \quad \hat{U}_3 = \begin{bmatrix} 0 & 1 \\ 1 & 0 \end{bmatrix}; \quad \hat{U}_4 = \begin{bmatrix} 0 & 1 \\ -1 & 0 \end{bmatrix} \tag{17.73}$$

are simple rotations in the two-dimensional Hilbert space of particle 3 (apart from $\hat{U}_1$, which is minus the identity operator). It is then clear that Bob, depending on Alice's measurement result, can recover with certainty the original state $|j\rangle_1$ with the proper application of one of the above transformations.

In conclusion, teleportation is based on two channels: a *classical channel*, by which Alice communicates the result of her measurement and a *quantum channel* (the EPR pair 2–3), by means of which, using the entanglement, Bob – after having received the classical information, and hence without violating the locality – instantaneously recovers the state $|j\rangle_1$ on particle 3.[25] However, the original state $|j\rangle_1$ is destroyed, in accordance with the

[25] Recently, it has been shown [Meschke *et al.* 2006] that entropy as well as information is subjected to quantization and quantum limits of transmission, both of which are independent from the material constituting the channel. This then raised the interesting connection of this result with the capacity of a quantum channel [Schwab 2006].

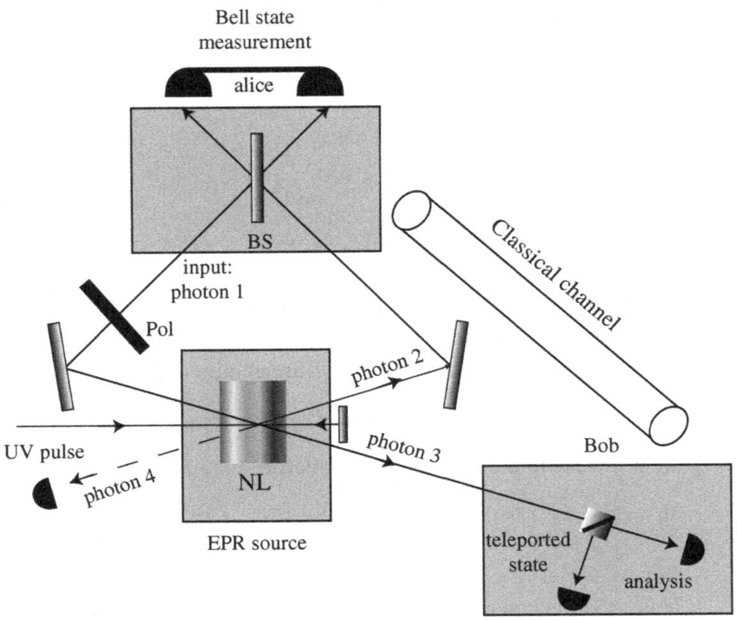

Figure 17.8 Zeilinger and co-workers' scheme of teleportation experiment. A pulse of ultraviolet radiation passing through a non-linear crystal NL produces the EPR photon pair 2–3. After retroflection during its second passage through NL, the radiation creates another pair, 1–4, of which the photon 1 is to be teleported and photon 4 is a trigger indicating that the other photon is under way. Alice looks for coincidence counts of photons 1 and 2, after the BS. Finally, Bob, after receiving the classical bit of information, may retrieve the input state of photon 1 through appropriate detection.

no-cloning theorem (see Sec. 15.2). If we define an *ebit* as the amount of entanglement between a maximally entangled pair of two-state systems (for example two spin-$1/2$ particles in a singlet state), then by teleportation we transmit a qubit by means of a shared ebit and a two-bit piece of classical information. Note that an ebit is a weaker resource than a qubit: in fact the transmission of a qubit can always be used to create one ebit, while the sharing of one ebit or many ebits does not suffice to transmit a qubit (we also need classical information). Teleportation has been experimentally realized, first with photons[26] (see Fig. 17.8), and later also with atoms.[27]

17.6 Quantum cryptography

One of the most important aspects of information technology in our society is how to establish a secure way to exchange secret messages. By secure we mean here that the

26 See [Bouwmeester *et al.* 1997] [Furusawa *et al.* 1998]. See also the theoretical proposal [Vitali *et al.* 2000].
27 See [Barrett *et al.* 2004] [Riebe *et al.* 2004].

message cannot be eavesdropped by extraneous agencies. Cryptography is precisely the technique that allows party A (the sender, traditionally called Alice) to encode a message and party B (the receiver, traditionally called Bob) to decode it using a certain key with a procedure such that the eventual eavesdropper (traditionally called Eve) is not able to break the secrecy of the original message.

This practical problem has roots dating back to ancient societies and their needs for politics, war, and economy. For example, Julius Caesar used to communicate secret messages taking advantage of a simple substitution method: he simply replaced each letter of the latin alphabet in the original message by the letter that follows it alphabetically by three places. In this case, the receiver must simply invert this transformation (i.e. subtract three positions to each letter) to recover the original message. In such case, the key is the translation rule from latin to the Caesar cypher and vice versa. Of course, for this scheme to be reliable, the key must be kept secret and known only to Alice and Bob. Schemes of this type are called *symmetric* or private-key protocols. In these protocols, two obvious requirements should be observed: the key must be as long as the message and must be used only once, both not being satisfied by the Caesar cypher.

In other schemes, the encoding key may be publicly announced by the receiver to the sender, but the decoding key is known only to the former, and this ensures that only the receiver is able to decode the message. These other schemes rely on the enormous difficulty of inverting certain mathematical transformations. For instance, multiplying two large prime numbers is relatively easy but, on the other hand, factoring the result into its prime factors may take a very long time (as we shall see in Sec. 17.8). For this reason, such schemes are called *asymmetric* or public-key protocols.

Unfortunately, it has never been proved that the transformations used in public-key cryptography are truly difficult (i.e. the time required to solve them grows exponentially with the length of the input). The only thing we know is that up to now there are no classical algorithms able to solve them in a polynomial time. Moreover, as we shall see in the next sections, a quantum computer could in principle solve this class of problems. On the other hand, private-key protocols rely on secure communication in order to establish a sufficiently long key, to be used only once. Such a communication is often performed by a human courier but that is, however, not completely reliable. Indeed, there does not exist a classical procedure that allows for a completely safe key sending, since any key can in principle be intercepted without the knowledge of the interested parties. This problem is known as the key-distribution problem.

Things stand in a completely different light when making use of quantum information. As a matter of fact, through *quantum key distribution*, quantum mechanics allows two parties to establish a totally secure private key, by transmitting information in quantum superpositions or entangled states. Essentially, what is used here are some basic quantum principles:

- Any measurement will *perturb* the state of the measured system [Ch. 9].
- It is impossible to *distinguish* with a single measurement two non-orthogonal states (Sec. 15.1).
- It is impossible to *clone* a quantum state (Sec. 15.2).

Table 17.1 An example of sequence transmission in Bennett and Brassard's protocol for quantum key distribution

Alice's random bits	0	1	1	1	0	0	1	0
Alice's random chosen basis	+	+	×	×	×	+	+	×
Photon polarization sent by Alice	→	↑	↗	↗	↘	→	↑	↘
Bob's chosen basis	+	×	+	×	×	×	+	+
Bob's measurement result	→	↘	↑	↗	↘	↗	↑	→
Shared secret key	0	–	–	1	0	–	1	–

The first two facts can lead to the third, especially when we consider Yuen and D'Ariano's theorem (Th. 15.2: p. 549).

Quantum cryptography was inaugurated by a pioneering study of Wiesner, proposed in the 1970s but only published later on.[28] Bennett and Brassard shortly after published a seminal paper that is often known as BB84.[29] In Bennett and Brassard's protocol, Alice and Bob are connected by a quantum communication channel which allows the transmission of quantum states. In addition, they may communicate via a public classical channel. Both these channels may possibly not be secure. In particular, if the quantum channel is represented by the transmission of photons (e.g. in an optical fiber or in free space), they can make use of two possible bases of polarization,

$$+ = \{|\uparrow\rangle, |\rightarrow\rangle\} \quad \text{and} \quad \times = \{|\nearrow\rangle, |\searrow\rangle\}, \tag{17.74}$$

representing, respectively, a vertical–horizontal polarization and a 45°–135° polarization state. First, Alice establishes a one-to-one correspondence between the classical bits (0, 1) that she desires to communicate with each state of the two bases, for instance

$$0 \leftrightarrow |\rightarrow\rangle, \quad |\searrow\rangle, \tag{17.75a}$$

$$1 \leftrightarrow |\uparrow\rangle, \quad |\nearrow\rangle. \tag{17.75b}$$

Then, she sends several bits of information by choosing at random one or the other basis. Bob will measure the photons, choosing again at random one of the two bases. After this exchange and measurement, they publicly tell each other which basis they have used. They will immediately discard photon transmissions where the two bases do not match (on average 50% of the transmitted bits). The bits for which Alice and Bob chose the same basis constitute the shared key. The situation is schematically shown in Tab. 17.1.

After this step, in order to exclude any possibility of eavesdropping, Bob takes a subset of the key and publicly compares his measurement results with Alice's inputs. If they do not match, this means that Eve has intercepted and measured the photons, changing their initial state in a random way (in 50% of the cases, i.e. when Eve choses a different basis from that used by Alice), so that Bob has again a 50% probability of obtaining a result that is a mismatch relative to the input. Therefore, the final mismatch between input and output will be of the order of 25%. When this is the case, Alice and Bob discard

[28] See [Wiesner 1983].
[29] See [Bennett/Brassard 1984].

their sequence and start the procedure again, continuing until they can be sure that no eavesdropping has occurred. In other words, for any single compared bit Alice and Bob have a 25% probability of detecting the presence of an eavesdropper, if any. By increasing the number n of compared bits, Alice and Bob may increase the probability of detecting Eve's presence, i.e. the reliability of their procedure, according to

$$\wp_d = 1 - \left(\frac{3}{4}\right)^n . \tag{17.76}$$

For instance, if Alice and Bob would like to exclude the presence of Eve at a confidence level of 99.99%, they need to compare 33 bits. As said, if they find a mismatch, they should repeat the protocol from the very beginning, over and over again until they find a perfect matching between their respective reference subset. Obviously, the number of bits to be sent must be sufficiently large in order to allow a sufficiently high level of reliability. This makes Eve's attempts at eavesdropping even more difficult. Needless to say, Eve would ideally like to intercept Alice's bit, make a copy of it, and resend it unperturbed to Bob. However, as we know, this is prohibited by the no-cloning theorem.

In the BB84 protocol Alice and Bob exploit the superposition principle to share a secret key. A second very interesting protocol for quantum key distribution is due to Artur Ekert[30] and is based on entanglement. In the Ekert's scheme (sometimes called the EPRBE protocol, from Einstein–Podolsky–Rosen–Bell–Ekert) Alice and Bob make use of two qubits emitted from a common source in a maximally entangled state of the form (17.68d), i.e.

$$| \Phi^+ \rangle = \frac{1}{\sqrt{2}} (| \uparrow, \uparrow \rangle + | \rightarrow, \rightarrow \rangle). \tag{17.77}$$

Again, when Alice and Bob use the same basis, they obtain the same results, providing them with a common key. In this case, moreover, it is possible to make use of Bell's inequality in order to check the security of the protocol: Alice and Bob have a third choice of basis, so that, as they establish a key, they collect enough data to test Bell's inequality (see Secs. 16.4–16.5). They can, in this way, check that the source actually emits the entangled state (17.77) and not just factorized states, ruling out the presence of an eavesdropper.

It should be noted that quantum cryptography may be only exploited to establish and distribute a private key, not to transmit any message. Such a key may, of course, be used with any chosen encoding algorithm to encode and to decode a message, which can be transmitted over standard communication channels.

Concerning physical implementations, to our knowledge the longest distance over which quantum key distribution (following the BB84 protocol) has been realized is 148.7 km.[31] In free space (using entangled photons in the Ekert's scheme), the record distance is 144 km.[32] Quantum cryptography is a very active field of research and has seen some very interesting developments in recent years, such as privacy amplification, information reconciliation,

30 See [Ekert 1991].
31 [Hiskett *et al.* 2006].
32 [Ursin *et al.* 2007].

quantum digital signature, quantum fingerprinting, etc. These, however, go far beyond the scope of this book.

17.7 Elements of quantum computation

The first idea of a quantum computation able to exploit the potentialities represented by the qubit is due to Deutsch,[33] but the first conceptual hint is due to Feynman,[34] who had already proposed a simulation of physical processes on a quantum computer, and to Benioff,[35] who showed that a Turing machine can be simulated by the unitary evolution of a quantum system. A classical *Turing machine* is a recursive device composed of a processing unit in the form of a write/read head and a memory with unlimited storage capacity in the form of an infinite tape divided into cells. Each cell can have a symbol from a finite alphabet. The tape is scanned, one cell at a time, by the read/write head. The head can be in one of a finite set of states. The machine action is made up of discrete steps, and each step is determined by two initial conditions: the current state of the head and the symbol that occupies the cell currently scanned. Given these two conditions, the machine receives a three-part instruction for its next step, which specifies:

- the next state of the head;
- the symbol to write into the scanned cell;
- whether the head has to move (left or right) along the tape or to stop.

In the classical case, a Turing machine is an irreversible device, whereas, if we desire to make use of the specificity of quantum information, we need unitary transformations (see Sec. 17.4 and also below Subsec. 17.7.2), which are reversible. However, as we have already shown at a general level (in Sec. 17.3) and show in the following in the context of quantum computation (see especially Subsec. 17.7.5), an irreversible device can be embedded in a larger device that is reversible.

17.7.1 Gates and circuits

A classical or quantum computer consists wires and logic gates. The *wires* transmit information whereas a *gate* is a transformation of bits or qubits. A logic gate is any function $f : \{0, 1\}^j \rightarrow \{0, 1\}^k$ from a certain number j of input (qu-)bits to a certain number k of output (qu-)bits. When $k = 1$, f is called a *Boolean function* (see Sec. 2.4). Some very simple quantum logic gates are the gate that does nothing, i.e. the identity operator gate (therefore, this turns out to be a wire), the phase addition gate, which acts as

$$\hat{U}_\phi \left(c_0 \left|0\right\rangle + c_1 \left|1\right\rangle\right) = \left(c_0 \left|0\right\rangle + c_1 e^{\imath\phi} \left|1\right\rangle\right), \tag{17.78}$$

[33] See [Deutsch 1985].
[34] See [Feynman 1982].
[35] See [Benioff 1982].

where

$$\hat{U}_\phi = \begin{bmatrix} 1 & 0 \\ 0 & e^{\imath\phi} \end{bmatrix}, \tag{17.79}$$

and the NOT gate, which has the power to change any $|0\rangle$ into $|1\rangle$ and any $|1\rangle$ into $|0\rangle$, i.e.

$$\hat{\sigma}_x (c_0 |0\rangle + c_1 |1\rangle) = (c_0 |1\rangle + c_1 |0\rangle), \tag{17.80}$$

where

$$\hat{\sigma}_x = \begin{bmatrix} 0 & 1 \\ 1 & 0 \end{bmatrix} \tag{17.81}$$

is the unitary NOT gate for a qubit and turns out to be identitical to the Pauli x-spin matrix. The other Pauli matrices (see Eqs. (6.154)) also represent unitary quantum gates.

A more complex example is represented by the so-called *Hadamard gate*, which on a single qubit acts as

$$\hat{U}_H|0\rangle = \frac{1}{\sqrt{2}}(|0\rangle + |1\rangle), \quad \hat{U}_H|1\rangle = \frac{1}{\sqrt{2}}(|0\rangle - |1\rangle), \tag{17.82}$$

where

$$\hat{U}_H = \frac{1}{\sqrt{2}}\begin{bmatrix} 1 & 1 \\ 1 & -1 \end{bmatrix}. \tag{17.83}$$

The input states may also be represented by polarization states of a photon (for instance, vertical and horizontal). Then, the state $|0\rangle$ may be prepared by sending a photon beam through, say, a horizantal polarization filter, and through a vertical polarization filter for preparing $|1\rangle$.

Note that the Hadamard gate transforms a state 0 or 1 in a "halfway" between this state and its negation. However, two subsequent applications of the Hadamard gate are not equal to the NOT gate but to the identity gate, which shows that the Hadamard gate is the inverse of itself.[36] On an initial two-qubit state the Hadamard transform has the form

$$\hat{U}_H^{\otimes 2}|00\rangle = \left(\frac{|0\rangle_1 + |1\rangle_1}{\sqrt{2}}\right)\left(\frac{|0\rangle_2 + |1\rangle_2}{\sqrt{2}}\right) = \frac{1}{2}(|00\rangle + |01\rangle + |10\rangle + |11\rangle). \tag{17.84}$$

In the case of three qubits we have

$$\begin{aligned} \hat{U}_H^{\otimes 3}|000\rangle &= \left(\frac{|0\rangle_1 + |1\rangle_1}{\sqrt{2}}\right)\left(\frac{|0\rangle_2 + |1\rangle_2}{\sqrt{2}}\right)\left(\frac{|0\rangle_3 + |1\rangle_3}{\sqrt{2}}\right) \\ &= 2^{-\frac{3}{2}}(|000\rangle + |001\rangle + |010\rangle + |011\rangle \\ &\quad + |100\rangle + |101\rangle + |110\rangle + |111\rangle), \end{aligned} \tag{17.85}$$

which is the binary codification of the numbers from 0 to 7. In general, we have

$$\hat{U}^{\otimes n}|x\rangle = 2^{-\frac{n}{2}}\sum_y (-1)^{x\cdot y}|y\rangle. \tag{17.86}$$

where $x \cdot y$ is the binary multiplication bit by bit followed by addition modulo 2 (symbolized by $\oplus$), i.e. $0 + 0 = 0$, $0 + 1 = 1 + 0 = 1$, $1 + 1 = 0$.

[36] As for the Pauli matrices, this happens because $\hat{U}_H$ is both Hermitian and unitary.

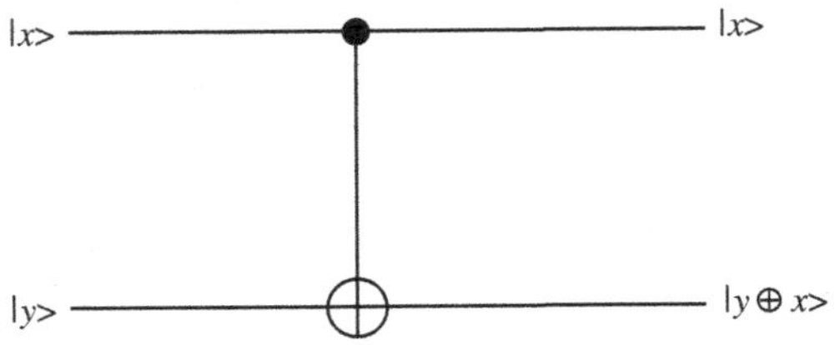

Figure 17.9 Representation of the CNOT gate. The top line represents the control qubit, the bottom line the target qubit.

A controlled not (CNOT) gate is a transformation on a control qubit and on a target qubit such that the target qubit is flipped if and only if the control qubit is 1, i.e.

$$\hat{U}_{\text{CNOT}}\,|00\rangle = |00\rangle\,, \quad \hat{U}_{\text{CNOT}}\,|01\rangle = |01\rangle\,, \tag{17.87a}$$

$$\hat{U}_{\text{CNOT}}\,|10\rangle = |11\rangle\,, \quad \hat{U}_{\text{CNOT}}\,|11\rangle = |10\rangle\,, \tag{17.87b}$$

where the first element of the pair is the control, the second the target qubit, and we have

$$\hat{U}_{\text{CNOT}} = \begin{bmatrix} 1 & 0 & 0 & 0 \\ 0 & 1 & 0 & 0 \\ 0 & 0 & 0 & 1 \\ 0 & 0 & 1 & 0 \end{bmatrix} = |0\rangle\langle 0| \otimes \hat{I} + |1\rangle\langle 1| \otimes \hat{\sigma}_x. \tag{17.88}$$

Deutsch called the CNOT gate the measurement gate, because, if the target qubit is prepared in the 0 state, it can always learn about the state of the control qubit (see also Fig. 17.9). It is also easy to see that we have

$$\hat{U}_{\text{CNOT}}\,(c_0\,|0\rangle + c_1\,|1\rangle)\,|0\rangle = c_0\,|00\rangle + c_1\,|11\rangle\,, \tag{17.89a}$$

$$\hat{U}_{\text{CNOT}}\,(c_0\,|00\rangle + c_1\,|11\rangle) = (c_0\,|0\rangle + c_1\,|1\rangle)\,|0\rangle\,, \tag{17.89b}$$

which shows that the CNOT gate may entangle and disentangle states and that it is the inverse of itself ($\hat{U}^2_{\text{CNOT}} = \hat{I}$). We finally note that it is possible to implement the CNOT gate by means of an interferometer (see Fig. 17.10).[37]

A Mach–Zehnder interferometer can be seen as a quantum circuit of logical wires and gates (see Fig. 17.11), that is, the series beam splitter-phase shifter-beam splitter (with two additional mirrors after the first BS) can be thought of as a succession Hadamard gate–phase gate–Hadamard gate, that is,

[37] See [O'Brien *et al.* 2003].

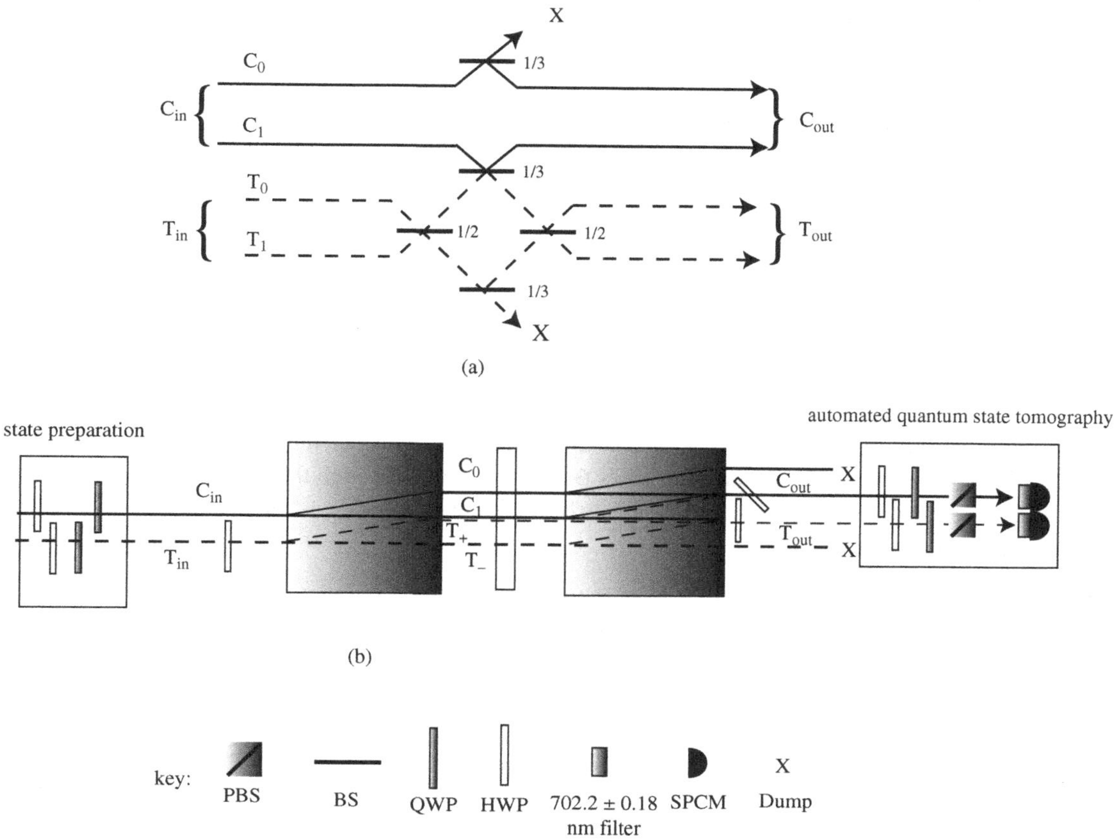

Figure 17.10 Implementation of a CNOT gate by means of a polarization interferometer. (a) Conceptual framework of the experiment by means of an interferomenter and 1/3 and 1/2 beam splitters. A sign change (π phase shift) occurs upon reflection off the lower side of the three lower BSs for the target and off the upper side of the two other BSs for the control. Note that only the control (C) $| 0\rangle$ is entangled with the target (T). (b) A pair of photons is produced on the left by SPDC and collected into single-mode optical fibres (not shown). Then, each fiber is collimated; a half-wave plate (HWP) and a quarter-wave plate (QWP) in each input beam allow preparation of any separable two-qubit state. First, the control and target qubits are split by a polarization beam splitter (PBS) and the target qubit is polarization encoded through a HWP that rotates the polarization by 45°. The operation of all three 1/3 BSs is realized by a single HWP. For further details, see the original article [O'Brien *et al.* 2003].

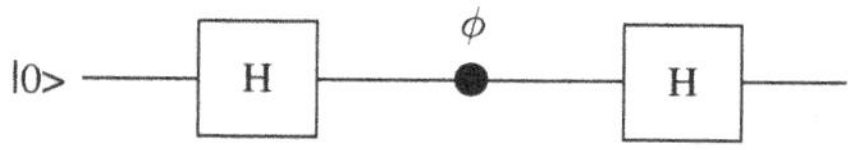

Figure 17.11 The quantum computation device as an equivalent of a Mach–Zehnder interferometer.

$$
\begin{aligned}
|0\rangle &\overset{\hat{U}_H}{\mapsto} \frac{1}{\sqrt{2}}(|0\rangle + |1\rangle) \\
&\overset{\hat{U}_\phi}{\mapsto} \frac{1}{\sqrt{2}}\left(|0\rangle + e^{\imath\phi}|1\rangle\right) \\
&\overset{\hat{U}_H}{\mapsto} \frac{1}{2}\left[\left(1+e^{\imath\phi}\right)|0\rangle + \left(1-e^{\imath\phi}\right)|1\rangle\right] \\
&= \cos\frac{\phi}{2}|0\rangle - \imath\sin\frac{\phi}{2}|1\rangle .
\end{aligned} \tag{17.90}
$$

17.7.2 Gates as unitary transformations

Unitarity is the only constraint on quantum gates. This has the consequence that there are a lot of non-trivial quantum gates that may be built. Though this set is infinite, the general properties of unitary operation may be determined. In fact, any unitary 2×2 matrix $\hat{U}$ may be decomposed as[38]

$$
\hat{U}(\alpha, \beta, \gamma, \delta) = e^{\imath\alpha}\hat{R}_z(\beta)\hat{R}_y(\gamma)\hat{R}_z(\delta) \tag{17.91}
$$

for some real numbers α, β, γ, and δ, and where $\hat{R}_z(\beta)$ is the two-dimensional rotation matrix about the z-axis of an angle β and similarly for the other matrices. The three rotation matrices about the x-, y-, and z-axes are (see Subsec. 6.1.2)

$$
\hat{R}_x(\theta) = \cos\frac{\theta}{2}\hat{I} - \imath\sin\frac{\theta}{2}\hat{\sigma}_x = \begin{bmatrix} \cos\frac{\theta}{2} & -\imath\sin\frac{\theta}{2} \\ -\imath\sin\frac{\theta}{2} & \cos\frac{\theta}{2} \end{bmatrix}, \tag{17.92a}
$$

$$
\hat{R}_y(\theta) = \cos\frac{\theta}{2}\hat{I} - \imath\sin\frac{\theta}{2}\hat{\sigma}_y = \begin{bmatrix} \cos\frac{\theta}{2} & -\sin\frac{\theta}{2} \\ \sin\frac{\theta}{2} & \cos\frac{\theta}{2} \end{bmatrix}, \tag{17.92b}
$$

$$
\hat{R}_z(\theta) = \cos\frac{\theta}{2}\hat{I} - \imath\sin\frac{\theta}{2}\hat{\sigma}_z = \begin{bmatrix} e^{-\imath\frac{\theta}{2}} & 0 \\ 0 & e^{\imath\frac{\theta}{2}} \end{bmatrix}. \tag{17.92c}
$$

In other words, we may write any 2×2 unitary matrix as

$$
\hat{U}(\alpha, \beta, \gamma, \delta) = e^{\imath\alpha}\begin{bmatrix} e^{-\imath\frac{\beta}{2}} & 0 \\ 0 & e^{\imath\frac{\beta}{2}} \end{bmatrix}\begin{bmatrix} \cos\frac{\gamma}{2} & -\sin\frac{\gamma}{2} \\ \sin\frac{\gamma}{2} & \cos\frac{\gamma}{2} \end{bmatrix}\begin{bmatrix} e^{-\imath\frac{\delta}{2}} & 0 \\ 0 & e^{\imath\frac{\delta}{2}} \end{bmatrix}. \tag{17.93}
$$

The reason is that the rows and columns of a unitary matrix are orthonormal. From Eq. (17.93) it follows that there exist real numbers α, β, γ, and δ such that (see Prob. 17.9)

$$
\hat{U}(\alpha, \beta, \gamma, \delta) = \begin{bmatrix} e^{\imath\left(\alpha-\frac{\beta}{2}-\frac{\delta}{2}\right)}\cos\frac{\gamma}{2} & -e^{\imath\left(\alpha-\frac{\beta}{2}+\frac{\delta}{2}\right)}\sin\frac{\gamma}{2} \\ e^{\imath\left(\alpha+\frac{\beta}{2}-\frac{\delta}{2}\right)}\sin\frac{\gamma}{2} & e^{\imath\left(\alpha+\frac{\beta}{2}+\frac{\delta}{2}\right)}\cos\frac{\gamma}{2} \end{bmatrix}. \tag{17.94}
$$

[38] See [*Nielsen/Chuang* 2000, 174–77].

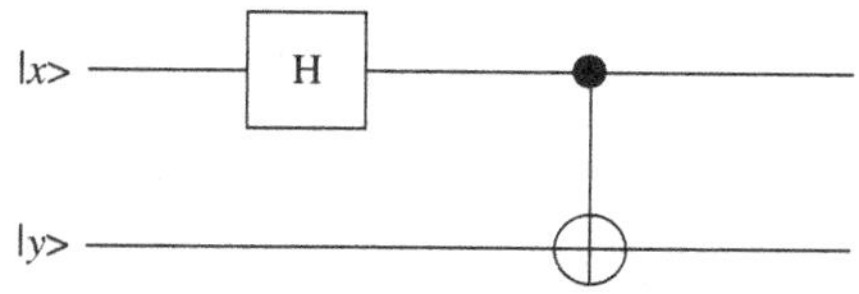

Figure 17.12 Generation of Bell states by means of a Hadamard gate followed by a CNOT gate.

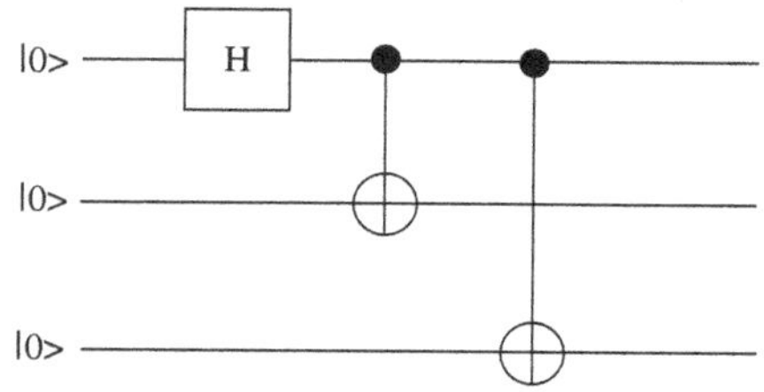

Figure 17.13 Preparation of a GHSZ state.

17.7.3 Producing entanglement

Using the simple device shown in Fig. 17.12, it is possible to produce any of the Bell states. Neglecting normalization, we have

$$| 0\rangle \, | 0\rangle \overset{\hat{U}_H}{\mapsto} (| 0\rangle + | 1\rangle) \, | 0\rangle \overset{\hat{U}_{\text{CNOT}}}{\mapsto} | 00\rangle + | 11\rangle, \tag{17.95a}$$

$$| 0\rangle \, | 1\rangle \overset{\hat{U}_H}{\mapsto} (| 0\rangle + | 1\rangle) \, | 1\rangle \overset{\hat{U}_{\text{CNOT}}}{\mapsto} | 01\rangle + | 10\rangle, \tag{17.95b}$$

$$| 1\rangle \, | 0\rangle \overset{\hat{U}_H}{\mapsto} (| 0\rangle - | 1\rangle) \, | 0\rangle \overset{\hat{U}_{\text{CNOT}}}{\mapsto} | 00\rangle - | 11\rangle, \tag{17.95c}$$

$$| 1\rangle \, | 1\rangle \overset{\hat{U}_H}{\mapsto} (| 0\rangle - | 1\rangle) \, | 1\rangle \overset{\hat{U}_{\text{CNOT}}}{\mapsto} | 01\rangle - | 10\rangle. \tag{17.95d}$$

We immediately recognize that the rhs of the previous equations coincide with Eqs. (17.68). As we know, a very interesting entangled state is represented by the GHSZ state (see Eq. (16.149))

$$| \Psi_{\text{GHZ}}\rangle = \frac{1}{\sqrt{2}} (| 000\rangle \pm | 111\rangle), \tag{17.96}$$

which can be easily prepared using quantum gates through the circuit shown in Fig. 17.13, i.e.

$$\begin{aligned} | 0\rangle \, | 00\rangle \quad &\overset{\hat{U}_H}{\mapsto} \frac{1}{\sqrt{2}} (| 0\rangle + | 1\rangle) \, | 00\rangle \\ &\overset{\hat{U}_{\text{CNOT}}}{\mapsto} \frac{1}{\sqrt{2}} (| 000\rangle + | 110\rangle) \\ &\overset{\hat{U}_{\text{CNOT}}}{\mapsto} \frac{1}{\sqrt{2}} (| 000\rangle + | 111\rangle), \end{aligned} \tag{17.97}$$

where the Hadamard operation is on the first qubit and the targets of the CNOT transformations are, successively, the second and third qubits. This can be generalized to any number of qubits, so that we can create any N-entangled states, since we do not need anything more than two-qubit gates. In fact, it is always possible to decompose a three-qubit gate into two-qubit gates. In general, any multiple qubit logic gate can be decomposed into CNOT and single qubit gates.

A universal set of gates, by which any computation can be performed, is given by the phase, the Hadamard, and the CNOT gates.

17.7.4 Computation and decoherence

Suppose that, in the device shown in Fig. 17.11, during the phase shift, the system interacts with the environment (see Sec. 9.4) in the initial state $|\,\mathcal{E}\rangle$. In this case, we have

$$\begin{aligned}|\,0\rangle\,|\,\mathcal{E}\rangle &\overset{\hat{U}_H}{\mapsto} \frac{1}{\sqrt{2}}(|\,0\rangle + |\,1\rangle)\,|\,\mathcal{E}\rangle \\ &\overset{\hat{U}_\phi}{\mapsto} \frac{1}{\sqrt{2}}\left(|\,0\rangle\,|\,e_0\rangle + e^{\imath\phi}\,|\,1\rangle\,|\,e_1\rangle\right) \\ &\overset{\hat{U}_H}{\mapsto} \frac{1}{2}\left[|\,0\rangle\left(|\,e_0\rangle + e^{\imath\phi}\,|\,e_1\rangle\right) + |\,1\rangle\left(|\,e_0\rangle - e^{\imath\phi}\,|\,e_1\rangle\right)\right],\end{aligned} \tag{17.98}$$

where $|\,e_0\rangle$ and $|\,e_1\rangle$ are two different – but not necessarily orthogonal – states of the environment. If we calculate the probability of obtaining the outcome $|\,0\rangle$, independently from the state of the environment, we obtain

$$\begin{aligned}\wp(|\,0\rangle) &= \frac{1}{4}\left(\langle e_0 \,|\, e_0\rangle + e^{\imath\phi}\langle e_0 \,|\, e_1\rangle + \langle e_1 \,|\, e_1\rangle + e^{\imath\phi}\langle e_1 \,|\, e_0\rangle\right) \\ &= \frac{1}{2}(1 + \cos\phi\,\langle e_0 \,|\, e_1\rangle),\end{aligned} \tag{17.99}$$

which obviously shows less visibility interference than the probability calculated following Eq. (17.90), which is $\wp(|\,0\rangle) = \cos^2(\phi/2)$. In the limit where $\langle e_0 \,|\, e_1\rangle$ goes to zero, the interference is completely destroyed. This shows that decoherence can represent a big problem for quantum computation, because the action of the environment can destroy the coherence that is crucial for the quantum computing device, before the desired result is obtained.

Let us now consider the problem of decoherence in more detail. We may write the interaction of a qubit with the environment as follows:

$$|\,0\rangle\,|\,\mathcal{E}\rangle \mapsto |\,0\rangle\,|\,e_{00}\rangle + |\,1\rangle\,|\,e_{01}\rangle, \tag{17.100a}$$

$$|\,1\rangle\,|\,\mathcal{E}\rangle \mapsto |\,0\rangle\,|\,e_{10}\rangle + |\,1\rangle\,|\,e_{11}\rangle. \tag{17.100b}$$

Equations (17.100) may be written in compact form as

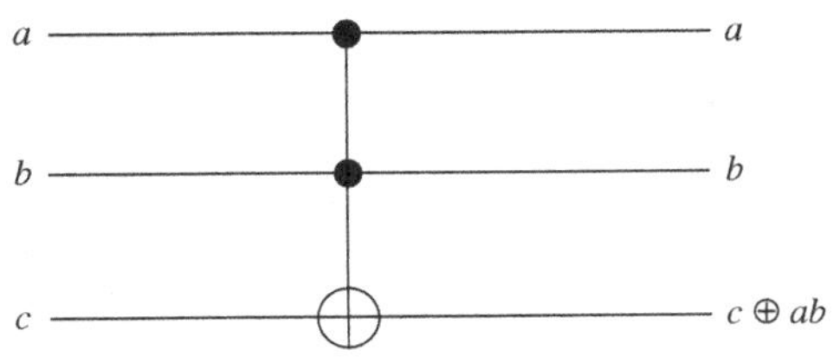

Figure 17.14 **The Toffoli gate.**

$$\begin{bmatrix} |e_{00}\rangle & |e_{10}\rangle \\ |e_{10}\rangle & |e_{11}\rangle \end{bmatrix} \begin{pmatrix} |0\rangle \\ |1\rangle \end{pmatrix}. \tag{17.101}$$

This 2 × 2 matrix can be decomposed in terms of the Pauli matrices as

$$\hat{I}\,|e_0\rangle + \hat{\sigma}_x\,|e_1\rangle + \imath\hat{\sigma}_y\,|e_2\rangle + \hat{\sigma}_z\,|e_3\rangle\,, \tag{17.102}$$

where

$$|e_0\rangle = \frac{|e_{00}\rangle + |e_{11}\rangle}{2}, \quad |e_1\rangle = \frac{|e_{01}\rangle + |e_{10}\rangle}{2}, \tag{17.103a}$$

$$|e_2\rangle = \frac{|e_{01}\rangle - |e_{10}\rangle}{2}, \quad |e_3\rangle = \frac{|e_{00}\rangle - |e_{11}\rangle}{2}, \tag{17.103b}$$

so that we have in general that the action of the environment is represented by the operator

$$\sum_{j=0}^{3} \hat{\sigma}_j\,|e_j\rangle\,, \tag{17.104}$$

where $\hat{\sigma}_0 = \hat{I}$, $\hat{\sigma}_1 = \hat{\sigma}_x$, $\hat{\sigma}_2 = \hat{\sigma}_y$, and $\hat{\sigma}_3 = \hat{\sigma}_z$. The operator $\hat{\sigma}_x$ represents a flip error, the operator $\hat{\sigma}_z$ a phase error, and finally the operator $\hat{\sigma}_y$ a combination of both errors.

17.7.5 Reversibility and irreversibility

As is well known, there are classical gates that are irreversible. Typical examples are the OR and NAND (NOT AND) gates, because we cannot infer the input from the output. These gates are therefore not unitary in quantum computation. However, in classical computation we may replace any gate by an equivalent circuit containing only reversible elements by making use of a reversible gate known as the *Toffoli gate*. Here we have two control bits and a target bit that is flipped if and only if the two control bits are 1 (see Fig. 17.14 and Tab. 17.2). The Toffoli gate can easily be used to implement the NAND gate: it suffices to set the target input to 1. In this case, the target output will be $1 \oplus ab$, which is equivalent to $\neg ab$.

Table 17.2 Toffoli truth table. *ab* are the control bits

Inputs			Outputs		
a	b	c	a	b	$c \oplus ab$
0	0	0	0	0	0
0	0	1	0	0	1
0	1	0	0	1	0
0	1	1	0	1	1
1	0	0	1	0	0
1	0	1	1	0	1
1	1	0	1	1	1
1	1	1	1	1	0

Table 17.3 Fredkin truth table. *c* is the control bit

Inputs			Outputs		
a	b	c	a'	b'	c
0	0	0	0	0	0
0	0	1	0	0	1
0	1	0	0	1	0
0	1	1	1	0	1
1	0	0	1	0	0
1	0	1	0	1	1
1	1	0	1	1	0
1	1	1	1	1	1

The quantum Toffoli gate can be represented by the following 8×8 unitary matrix:

$$\hat{U}_{\text{TOF}} = \begin{bmatrix} 1 & 0 & 0 & 0 & 0 & 0 & 0 & 0 \\ 0 & 1 & 0 & 0 & 0 & 0 & 0 & 0 \\ 0 & 0 & 1 & 0 & 0 & 0 & 0 & 0 \\ 0 & 0 & 0 & 1 & 0 & 0 & 0 & 0 \\ 0 & 0 & 0 & 0 & 1 & 0 & 0 & 0 \\ 0 & 0 & 0 & 0 & 0 & 1 & 0 & 0 \\ 0 & 0 & 0 & 0 & 0 & 0 & 0 & 1 \\ 0 & 0 & 0 & 0 & 0 & 0 & 1 & 0 \end{bmatrix}. \tag{17.105}$$

Another interesting reversible gate is the *Fredkin gate*, whose truth table is shown in Tab. 17.3. If $c = 0$, then a and b remain unchanged. However, if $c = 1$, a and b are swapped. The Fredkin gate also has the property that the number 1 is conserved between the input and the output. It is a universal gate that can be used to simulate the AND, NOT, CROSSOVER, and FANOUT (cloning) gates.

The quantum Fredkin gate can be represented as

$$\hat{U}_{\text{FRED}} = \begin{bmatrix} 1 & 0 & 0 & 0 & 0 & 0 & 0 & 0 \\ 0 & 1 & 0 & 0 & 0 & 0 & 0 & 0 \\ 0 & 0 & 1 & 0 & 0 & 0 & 0 & 0 \\ 0 & 0 & 0 & 1 & 0 & 0 & 0 & 0 \\ 0 & 0 & 0 & 0 & 1 & 0 & 0 & 0 \\ 0 & 0 & 0 & 0 & 0 & 0 & 1 & 0 \\ 0 & 0 & 0 & 0 & 0 & 1 & 0 & 0 \\ 0 & 0 & 0 & 0 & 0 & 0 & 0 & 1 \end{bmatrix}. \tag{17.106}$$

The most interesting point of the previous discussion is that an irreversible computation can be transformed in a reversible one by adding additional inputs and outputs. This is closely related to the problem of measurement. In fact, what from the perspective of the apparatus plus object system seems to be an irreversible process, becomes a reversible one if we take the environment into account, until the information downloaded in the environment can be in principle recovered (see Secs. 9.4 and 17.3).

Another related question is: is the computation energy-free or does it require some energy loss? Landauer[39] showed that only where there is some information erasure is there loss of energy by computing. Since in quantum computation all operations are unitary, there is in principle no need to pump energy into the system.

On the other hand, measurement requires selection which implies energy expenditure and local irreversibility. A recent approach to quantum computation makes use of the techniques of entangling and measuring, and is called for this reason one-way quantum computation.[40]

17.8 Quantum algorithms and error correction

The most fascinating and innovative aspect of quantum computation is that it allows us to solve in an efficient way problems that classical computation cannot. An *algorithm* is a method in which a set of well-defined instructions, applied to a certain initial state, can complete a required task, or solve a given problem. From a completely general viewpoint, let us consider an algorithm which takes an input of M digits and gives an answer after a certain number N of steps (or time) (see Fig. 17.15). One says that an algorithm is *easy* or *efficient* if the time taken to execute it (proportional to N) increases no faster than a polynomial function of the size of the input M (polynomial-time solution). On the contrary, solutions that are not efficient require a time that is exponential in the size of the input. Many operations can be performed using efficient algorithm, either logarithmic (e.g. multiplication) or polynomial (in most of the cases). However, there is a class of operations whose algorithms are not efficient: among them perhaps the most known example is the

39 See [Landauer 1961, Landauer 1996a] and [Bennett 1982].
40 See [Walther *et al.* 2005].

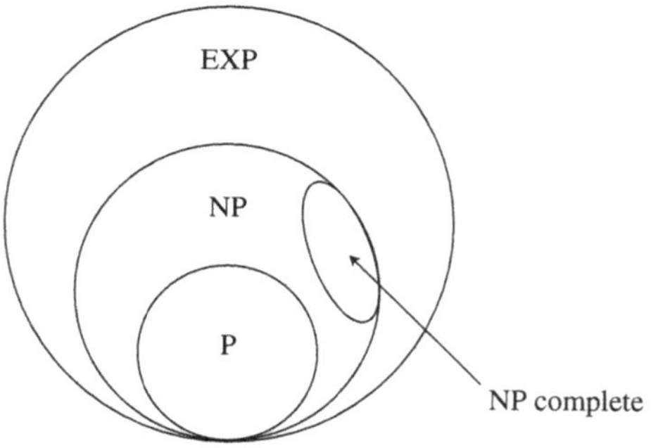

Figure 17.15 Representation of computational complexity. In the "space" of all possible problems (algorithms), P stands for the subset of algorithms that require a polynomial time calculation, NP for algorithms requiring polynomial time on a non-deterministic computer (roughly speaking a computer with an unbound degree of parallelism), EXP for exponential time calculation. Problems are designated "NP-complete" if their solutions can be quickly (polynomially) checked for correctness. Many kinds of optimization problems (e.g. that of the "traveling salesman") belong to the NP complete category. A typical example of problem belonging to EXP is constituted by the game of chess. It is conjectured that P $\neq$ NP, although it is not clear how to attack this conjecture.

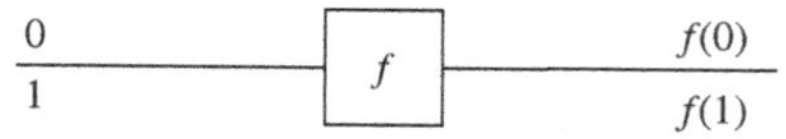

Figure 17.16 The evaluation of a Boolean function of an initial bit.

problem of factoring (see Prob. 17.10). We can pictorially represent the degrees of computational complexity (see Fig. 17.15). The Cook's theorem states that if NP-complete = P, then NP $\subset$ P. This is so because we can map any NP problem to a NP-complete one with a polynomial-time procedure. We consider Deutsch's algorithm first.

17.8.1 Deutsch's algorithm

As we have said, any Boolean function is a function $f(x)$, $x = 0, 1$, that maps from the truth values $\{0, 1\}$ to the truth values $\{0, 1\}$. A function is called *constant* if it always maps to the same value, i.e. $f_1(0) = f_1(1) = 0$ or $f_2(0) = f_2(1) = 1$. A function is called *balanced* if it maps for one half of the x to 0 and for the other half of the x to 1, so that $f_3(0) = 0$ and $f_3(1) = 1$ or $f_4(0) = 1$ and $f_4(1) = 0$. Suppose now that there is a device that can compute a Boolean function and that is allowed to run only once (see Fig. 17.16). *Deutsch's problem* is whether, under these conditions, it is possible to know if the function is constant or balanced. Classically, it is impossible to answer this question by running the device only once. Quantum-mechanically, this can be done. In order to see this, let us first consider the quantum-mechanical evaluation of a function. We may write the following transformation:

$$|x\rangle\,|0\rangle \overset{\hat{U}_f}{\mapsto} |x\rangle\,|f(x)\rangle\,, \tag{17.107}$$

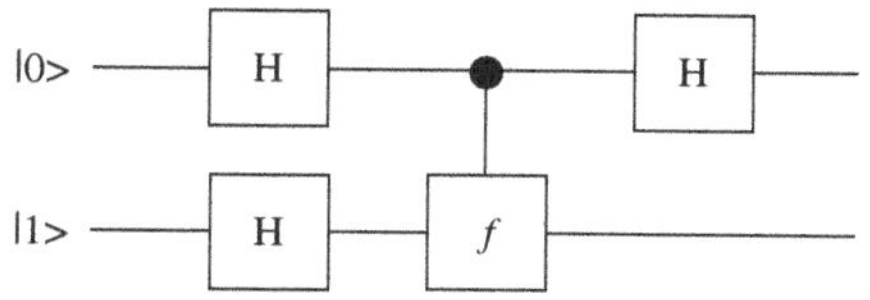

Figure 17.17 Device for solving Deutsch's problem of evaluating if a Boolean function is constant or balanced in a single run.

where the map is unitary. This can be implemented by

$$\sum_x |x\rangle\,|y\rangle \overset{\hat{U}_f}{\mapsto} \sum_x |x\rangle\,|y \oplus f(x)\rangle\,, \tag{17.108}$$

where $|y \oplus f(x)\rangle$ is the computed modulo of the maximum value of the register.

In order to solve Deutsch's problem let us consider the device shown in Fig. 17.17. The input state is transformed by the first two Hadamard gates according to

$$|0\rangle\,|1\rangle \overset{\hat{U}_H,\hat{U}_H}{\mapsto} \hat{U}_H\,|0\rangle\,\hat{U}_H\,|1\rangle = \frac{1}{2}(|0\rangle + |1\rangle)(|0\rangle - |1\rangle), \tag{17.109}$$

where the first two output states represent the x and the second two output states represent the y. It is easy to verify that the Boolean transformation will give (see Prob. 17.11)

$$\hat{U}_H\,|0\rangle\,\hat{U}_H\,|1\rangle \overset{\hat{U}_f}{\mapsto} \pm\frac{1}{2}(|0\rangle + |1\rangle)(|0\rangle - |1\rangle) \quad \text{if} \quad f(0) = f(1), \tag{17.110a}$$

$$\hat{U}_H\,|0\rangle\,\hat{U}_H\,|1\rangle \overset{\hat{U}_f}{\mapsto} \pm\frac{1}{2}(|0\rangle - |1\rangle)(|0\rangle - |1\rangle) \quad \text{if} \quad f(0) \neq f(1). \tag{17.110b}$$

Finally, the final Hadamard gate will give

$$\pm\frac{1}{2}(|0\rangle + |1\rangle)(|0\rangle - |1\rangle) \overset{\hat{U}_H}{\mapsto} \pm|0\rangle\left(\frac{|0\rangle - |1\rangle}{\sqrt{2}}\right), \tag{17.111a}$$

$$\pm\frac{1}{2}(|0\rangle - |1\rangle)(|0\rangle - |1\rangle) \overset{\hat{U}_H}{\mapsto} \pm|1\rangle\left(\frac{|0\rangle - |1\rangle}{\sqrt{2}}\right). \tag{17.111b}$$

This shows that, apart from an irrelevant phase factor, a measurement on the first qubit immediately gives the requested answer: 0 implies that the function is constant while 1 implies that the function is balanced. In fact, the $\hat{U}_f$ gate acts as a CNOT gate. It is interesting to note that the state created by the Boolean evaluation gate is an entangled state. In fact, the third term in the superposition in the final state of Eqs. (17.110) changes the sign according to whether $f(0) = f(1)$ or $f(0) \neq f(1)$.

The previous result may be generalized to n qubits in the following form (see Fig. 17.18) (it is called the Deutsch–Jozsa algorithm). The first two Hadamard gates induce the transformation

$$|0\rangle^{\otimes n}\,|1\rangle \overset{\hat{U}_H^{\otimes n},\hat{U}_H}{\mapsto} \sum_{x\in\{0,1\}^n} \frac{|x\rangle}{\sqrt{2^n}}\left(\frac{|0\rangle - |1\rangle}{\sqrt{2}}\right). \tag{17.112}$$

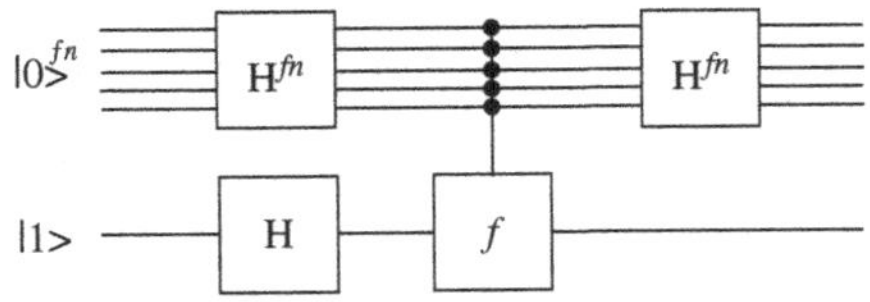

Figure 17.18 Device for solving Deutsch's problem for $n + 1$ input states.

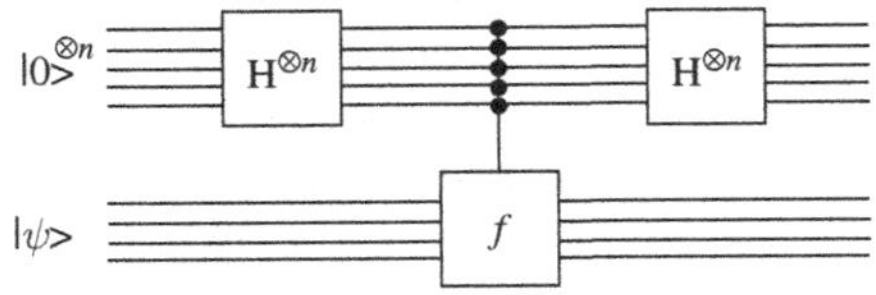

Figure 17.19 Device for solving Deutsch's problem for $n + m$ input states.

Then, the Boolean function gate acts as

$$\sum_{x\in\{0,1\}^n} \frac{|x\rangle}{\sqrt{2^n}} \left(\frac{|0\rangle - |1\rangle}{\sqrt{2}}\right) \overset{\hat{U}_f}{\mapsto} \sum_{x\in\{0,1\}^n} \frac{(-1)^{f(x)}|x\rangle}{\sqrt{2^n}} \left(\frac{|0\rangle - |1\rangle}{\sqrt{2}}\right). \tag{17.113}$$

The action of the final Hadamard gates may be represented by

$$\sum_{x\in\{0,1\}^n} \frac{(-1)^{f(x)}|x\rangle}{\sqrt{2^n}} \left(\frac{|0\rangle - |1\rangle}{\sqrt{2}}\right) \overset{\hat{U}_H^{\otimes n}}{\mapsto} \sum_z \sum_x \frac{(-1)^{x\cdot z + f(x)}|z\rangle}{2^n} \left(\frac{|0\rangle - |1\rangle}{\sqrt{2}}\right), \tag{17.114}$$

where $x \cdot z$ is the bitwise product between x and z, modulo 2.

A further generalization is provided when the input states are $n + m$ with $m > 1$. While the n inputs of the upper register (see Fig. 17.19) transform according to

$$|0\rangle \overset{\hat{U}_H}{\mapsto} \sum_x |x\rangle\,, \tag{17.115}$$

we may always prepare the m inputs of the lower register as

$$|\Psi\rangle = \sum_{y=0}^{2^m-1} e^{\frac{2\pi\imath}{2^m}y}|y\rangle\,. \tag{17.116}$$

After the Boolean gate the state will be

$$\begin{aligned}\sum_x |x\rangle\,|\Psi\rangle &\overset{\hat{U}_f}{\mapsto} \sum_x |x\rangle \sum_y e^{\frac{2\pi\imath}{2^m}y}|y \oplus f(x)\rangle \\ &= \sum_{xy} e^{-\frac{2\pi\imath}{2^m}f(x)} e^{\frac{2\pi\imath}{2^m}(y+f(x))}|x\rangle\,|y \oplus f(x)\rangle \\ &= \sum_x e^{-\frac{2\pi\imath}{2^m}f(x)}|x\rangle \left(\sum_y e^{\frac{2\pi\imath}{2^m}y}\right)|y\rangle\,. \end{aligned}\tag{17.117}$$

It is then clear that only the upper register has changed according to

Box 17.1 Example of factorization

Suppose we wish to factorize the number $N = 15$. We take an arbitrary number $\xi < 15$ which has no common prime factors with N, say $\xi = 7$. Now we define a function

$$f(x) = \xi^x \bmod 15, \tag{17.118}$$

where the x's are positive integers. $f(x)$ is the periodic function of which we wish to find the period, and mathematically it is the rest of the division of ξ^x by N. For $x = 0, 1, 2, 3, 4, 5, 6, \ldots$ we find that $f(x) = 1, 7, 4, 13, 1, 7, 4, \ldots$, i.e. the period of $f(x)$ is $r = 4$. Now we find the largest common divisors of N and $\xi^{\frac{r}{2}} \pm 1 = \{50, 48\}$, which are respectively 5 and 3. These are the desired factors of N.

Quantum-mechanically the solution to this problem is possible thanks to the superposition. In fact, a quantum system can be in a superposition of all: $f(0), f(1), f(2), f(3), f(4), \ldots$. It is evident that, if we perform a measurement, we cannot obtain all the values together. However, if we are able to evaluate the wave function, we can perform a Fourier transform so that we obtain the period r. In fact, the different values $f(0), f(1), f(2), f(3), f(4), \ldots$ are related to r as the fine structure of the interference pattern to the envelope of the same (overall width). This has been realized experimentally with a quantum computer [Vandersypen *et al.* 2001].

$$| x \rangle \mapsto e^{-\frac{2\pi\imath}{2^m} f(x)} | x \rangle \, . \tag{17.119}$$

This means that one can give any phase factor to $| x \rangle$ to the extent of the minimum precision allowed by the size of the second register from 0 to 2π with steps equal to $2\pi/2^m$. If one desires to be more precise, one has to increase the number m of inputs.

This is the power of quantum computation. A quantum computer is allowed to follow different paths simultaneously and then to "amplify" some answers at the expense of some other answers. For this reason, quantum computation is more powerful than randomized computation, where you have (and can add) only probabilities and not probability amplitudes, a fact that allows us to modify the structure of the possible answers. For example, the *primality* problem, where one is asked whether a given (large) integer number is prime or not, is difficult for classical computation but is efficient both for randomized computation and for quantum computation. As we shall see however, factoring is difficult both for classical and randomized computation, whereas it is efficient for quantum computation.

17.8.2 Shor's algorithm

As we have shown, the most well-known problem for which classically there is no known fast algorithm, is that of the factorization of large integer numbers: the number of computational steps increases exponentially with the binary dimension of the input. Note that factoring belongs to NP but not to NP-complete. Fortunately, the problem of factorizing

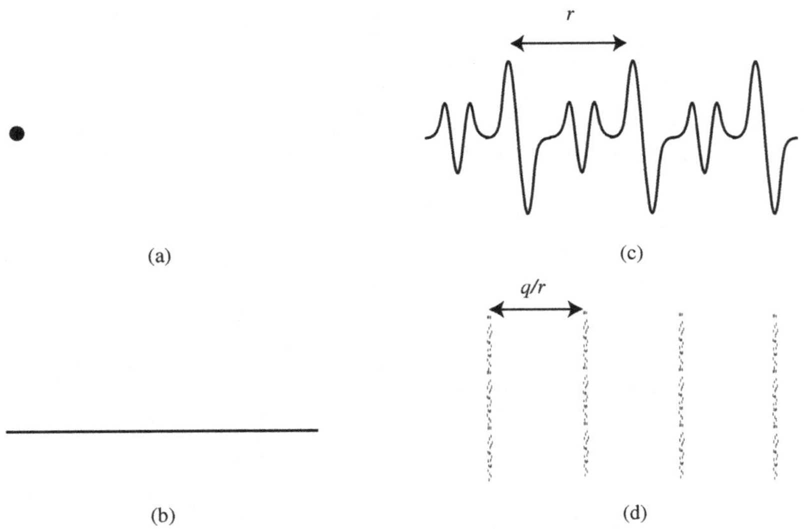

Figure 17.20 **Shor's superfast quantum Fourier sampling uses quantum interference to measure the period r of a periodic function $f(x)$. The period may be exponentially larger than the number of qubits involved in the computation. (a) The computer starts in the state $|x, f(x)\rangle = |0, 0\rangle$. (b) The x-register is put in a superposition of all possible values (see Eq. (17.120)). (c) The value $f(x)$ is computed in the y-register simultaneously for all x values (see Eq. (17.121)). (d) A Fourier transform of the x-register is performed (see Eq. (17.122)). Measuring the Fourier transform of x yields a result $c = \lambda q/r$, from which the period r can be deduced (see Eq. (17.124)).**

a number can be transformed to that of finding the period of a periodic function, and the latter problem can be solved quantum-mechanically.

This is the important result of Shor's theorem:[41] by evaluating a wave function on a superposition of exponentially many arguments, each one representing a value of the requested periodic function of the type (17.118) of Box 17.1, computing a parallel Fourier transform on the superposition, and finally sampling the Fourier power spectrum, one obtains the desired period (see Fig. 17.20).

Let us now generalize the example given in Box 17.1. We choose a number ξ at random and find a period r such that $\xi^r \equiv 1 \bmod N$, where N is the number to be factorized. Now we choose a smooth number q (a number with small prime factors) such that $N^2 < q < 2N^2$ and prepare the input state

$$|\psi_i\rangle = \frac{1}{\sqrt{q}} \sum_{x=0}^{q-1} |x, 0\rangle, \tag{17.120}$$

[41] See [Shor 1994].

from which, with a quantum-computational step, one can obtain

$$|\psi_f\rangle = \frac{1}{\sqrt{q}} \sum_{x=0}^{q-1} |x, \xi^x \bmod N\rangle. \tag{17.121}$$

Now we peform a Fourier transform on this pure state so as to obtain

$$|\tilde{\psi}_f\rangle = \frac{1}{q} \sum_{m=0}^{q-1} \sum_{x=0}^{q-1} e^{\imath 2\pi xm/q} |m, \xi^x \bmod N\rangle, \tag{17.122}$$

where use has been made of the discrete Fourier transform

$$\hat{U}_{\mathrm{DF}} : |x\rangle \mapsto \frac{1}{\sqrt{q}} \sum_{m=0}^{q-1} e^{\imath 2\pi xm/q} |m\rangle. \tag{17.123}$$

It is interesting to note that classically one needs approximately $N \log(N) = n2^n$ steps to Fourier transform $N = 2^n$ discrete inputs, whereas on a quantum computer the Fourier transform takes approximately $\log^2(N) = n^2$ steps. We can now measure both arguments of the superposition (17.122), obtaining a certain value c for m in the first one, and some $\xi^k \bmod N$ as the answer to the second one (k being any number between 0 and r). The probability for such a result will be

$$\wp(c, \xi^k) = \left| \frac{1}{q} \sum_{x=0}^{q-1}{}' e^{\imath 2\pi xc/q} \right|^2, \tag{17.124}$$

where the prime indicates a restricted sum over values which satisfy $\xi^x = \xi^k$. $\wp(c, x^k)$ is periodic in c with period q/r. A measurement gives, with high probability, $c = \lambda q/r$ – where λ is an integer corresponding to one of the peaks shown in Fig. 17.20(c)–(d). But since we know q, we can determine r with few trials.

17.8.3 Grover's algorithm

Grover's algorithm is related to a database search when the database is unstructured.[42] For instance, suppose that you are searching for a telephone number and you do not know to whom it belongs. Then, if N is the length of the database, you need N steps or at least $N/2$ steps (with probability 50%) if you perform a randomized search. Grover's algorithm is a faster quantum-computation solution to this problem. Grover's algorithm may also be described as "inverting a function." Roughly speaking, if we have a function $y = f(x)$ that can be evaluated on a quantum computer, this algorithm allows us to calculate x when y is given. Inverting a function is related to the database searching problem because it is always possible to find a function that produces a particular value of y if x matches a desired entry in the database, and another value of y for other values of x.

[42] See [Grover 1996].

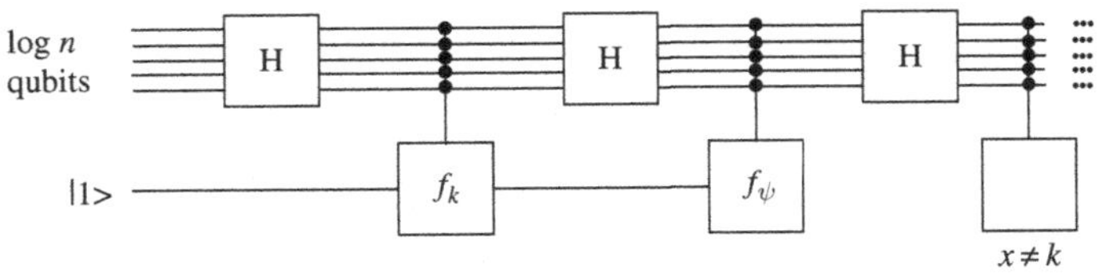

Figure 17.21 Implementation of Grover's algorithm.

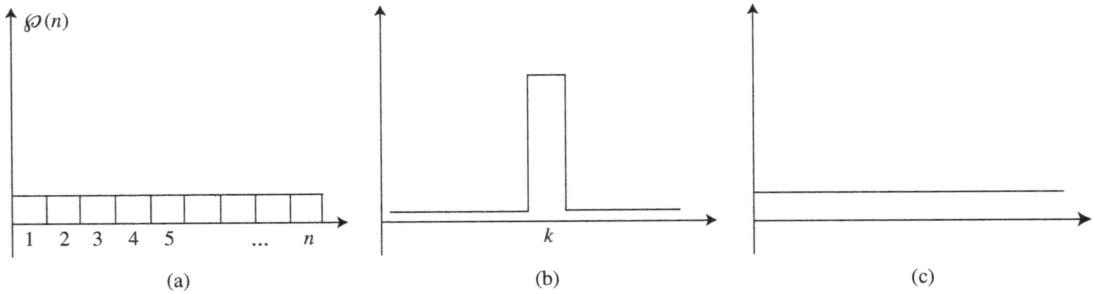

Figure 17.22 Computational steps in Grover's algorithm. (a) The probability distribution is initially uniform. (b) The probability distribution is strongly peaked about the searched state $|k\rangle$. It takes $O(\sqrt{N})$ steps to go from (a) to (b) and $O(\sqrt{N})$ steps to go from (b) to (c). Entanglement is introduced through the Boolean function evaluation.

Without going into solving equations, the steps involved in Grover's algorithm are as follows: we first generate an initial superposition state $|\Psi\rangle$ of all the possible states $|x\rangle$ of the qubits. We then apply two transformations (f_k and f_Ψ in Fig. 17.21) that act as two reflections and whose combined result is to "move" the input state toward the searched state $|k\rangle$. An explicit calculation shows that, repeating $\sqrt{N}$ times the steps above, a final measurement of an observable whose eigenstates are the $|x\rangle$'s, gives with high probability the searched item. The previous steps are pictorially represented in Fig. 17.22.

17.8.4 Quantum computers

As we have said, quantum superpositions can be used to overcome problems for which classical computation based on Boolean codes and gates only gives a solution after a very long computation time. Several physical schemes have been proposed in the last 15 years as possible candidates for quantum computers. Among them, we cite optical (included cavities), nuclear magnetic resonance, ion trap, superconducting, and semiconducting quantum computers.

The biggest difficulty in quantum computation is of technical order, and consists both in maintaining the coherence of the system to be used as processor and to build robust quantum gates to control the operations. An important step under the last respect has been the construction of good CNOT gates.[43]

[43] See [Plantenberg *et al.* 2007].

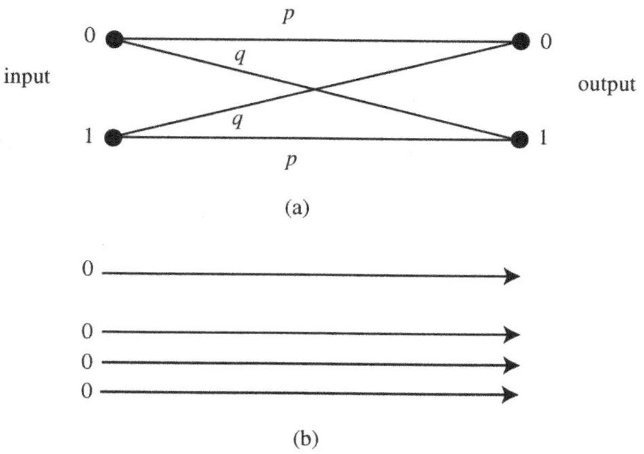

Figure 17.23 Classical error correction. (a) The correct probability p has an output which reproduces the input and the error probability $q = 1 - p$ is shown. (b) Here three qubits are encoded in place of a single one (0) and the output is taken as the majority voting among the outputs.

A new way to consider quantum computation has been proposed by Zeilinger and co-workers.[44] The idea is to make use of an irreversible quantum computation by combining entanglement and measurement. Information is carried by the correlation between the physical qubits, which are subjected to information processing and finally the output is transferred onto physical readout qubits. Now, while entanglement may decrease in the physical qubits as a result of measurement it may increase in the encoded qubits. This gives us a more general lesson and suggests that there are only two ways in which information can be modified:[45]

- by *selecting* it from a pre-existing pool, or
- by *sharing* it.

17.8.5 Quantum error correction

We have considered (in Subsec. 17.7.4) the negative effects of decoherence on quantum computation. The question we want to answer here is: can we do anything about it? The most obvious solution is to isolate the system so that it does not interact with the environment. However, this cannot be done so easily and in general can be done only up to a certain extent. A better idea is inspired by the classical (Shannon) error correction: repeat the input and tolerate noise (see Fig. 17.23). Then, what is the probability that there is no error in the system? In the case of three-qubit encoding of Fig. 17.23(b), this is shown in

[44] See [Walther *et al.* 2005].
[45] See [Auletta 2005].

Table 17.4 Example of classical error correction. The probability of different numbers of errors is shown for the case of three-qubit encoding

Number of errors	Probability
0	$(1-q)^3$
1	$3q\,(1-q)^2$
2	$3q^2\,(1-q)$
3	q^3

Tab. 17.4, where the first two situations are correctly interpreted by the error-correction procedure while the last two are wrong.

The total probability of error, after decoding, will be

$$3q^2\,(1-q)+q^3=3q^2-2q^3\simeq 3q^2, \tag{17.125}$$

which has to be compared with the error probability without correction, which is q. As a consequence, if q is small, there will be a big gain in applying this procedure.

In order to use an analogue protocol in quantum mechanics (where we have qubits instead of bits), we can define a fidelity F of the transmission as

$$F=\langle\psi\,|\hat{\rho}|\,\psi\rangle, \tag{17.126}$$

where $|\,\psi\rangle$ is the input state and $\hat{\rho}$ represents the final state, which, after decoherence, is generally a mixture (see Prob. 17.12). It is clear that, when $F=1$, we have perfect transmission. Then, we may encode a superposition input state as follows:

$$c_0\,|\,0\rangle+c_1\,|\,1\rangle \overset{\text{encoding}}{\mapsto} c_0\,\big|\,\overline{000}\big\rangle+c_1\,\big|\,\overline{111}\big\rangle, \tag{17.127}$$

where

$$\big|\,\overline{0}\big\rangle=\frac{1}{\sqrt{2}}(|\,0\rangle+|\,1\rangle),\quad \big|\,\overline{1}\big\rangle=\frac{1}{\sqrt{2}}(|\,0\rangle-|\,1\rangle). \tag{17.128}$$

This enconding may be performed as shown in Fig. 17.24.

Now, let us consider the decoherence effect (see also Eq. (17.98)):

$$\big|\,\overline{0}\big\rangle\,|\,\mathcal{E}\rangle\mapsto|\,0\rangle\,|\,e_0\rangle+|\,1\rangle\,|\,e_1\rangle=\left(\big|\,\overline{0}\big\rangle+\big|\,\overline{1}\big\rangle\right)|\,e_0\rangle+\left(\big|\,\overline{0}\big\rangle-\big|\,\overline{1}\big\rangle\right)|\,e_1\rangle, \tag{17.129a}$$

$$\big|\,\overline{1}\big\rangle\,|\,\mathcal{E}\rangle\mapsto|\,0\rangle\,|\,e_0\rangle-|\,1\rangle\,|\,e_1\rangle=\left(\big|\,\overline{0}\big\rangle+\big|\,\overline{1}\big\rangle\right)|\,e_0\rangle-\left(\big|\,\overline{0}\big\rangle-\big|\,\overline{1}\big\rangle\right)|\,e_1\rangle. \tag{17.129b}$$

From the first equation, we obtain

$$\big|\,\overline{0}\big\rangle\,(|\,e_0\rangle+|\,e_1\rangle)+\big|\,\overline{1}\big\rangle\,(|\,e_0\rangle-|\,e_1\rangle)=\big|\,\overline{0}\big\rangle\,|\,e_+\rangle+\big|\,\overline{1}\big\rangle\,|\,e_-\rangle. \tag{17.130}$$

Therefore, if only the first qubit decoheres, we can write

$$\left(c_0\,\big|\,\overline{000}\big\rangle+c_1\,\big|\,\overline{111}\big\rangle\right)|\,\mathcal{E}\rangle\mapsto|\,\Psi\rangle, \tag{17.131}$$

where the output state is

$$|\,\Psi\rangle=\left(c_0\,\big|\,\overline{000}\big\rangle+c_1\,\big|\,\overline{111}\big\rangle\right)|\,e_+\rangle+\left(c_0\,\big|\,\overline{100}\big\rangle+c_1\,\big|\,\overline{011}\big\rangle\right)|\,e_-\rangle. \tag{17.132}$$

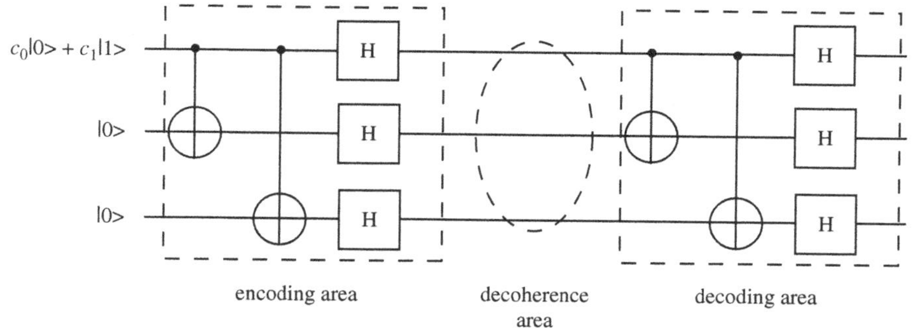

Figure 17.24 Example of quantum circuit for error correction. In the encoding area, the two CNOT gates and the three Hadamard gates contribute to generate an entangled state of the three qubits, which is subject to decoherence in the central area. Finally, the decoding area is crucial in order to invert the initial encoding.

By performing the decoding, we obtain

$$| \Psi \rangle \overset{\text{decoding}}{\mapsto} (c_0 \, | 0 \rangle \, + c_1 \, | 1 \rangle \,) \, | 0 \rangle \, | 0 \rangle \, | e_+ \rangle \, + (c_0 \, | 1 \rangle \, + c_1 \, | 0 \rangle \,) \, | \cdot \rangle \, | \cdot \rangle \, | e_- \rangle . \quad (17.133)$$

It suffices to read the second and third qubits: if one gets $| 00 \rangle$, then one knows that the input has been correctly decoded on the first qubit, otherwise, whatever $| \cdot \rangle$ may be, one knows which unitary operation one has to apply on the first qubit to get back the correct input state. As in classical error correction, the price to pay for this error-correction is the enlargment of the system.

Let us now introduce a time dependence in the environment dynamics, so that we may write the scalar product present in Eq. (17.99) as

$$\langle e_0(t) \, | \, e_1(t) \rangle = \gamma(t). \quad (17.134)$$

We have two possible decays: slow parabolic decay and exponential decay. Whatever the interaction is, $\gamma(t)$ will go to zero in time and quantum error correction will help a little. We will compare two scenarios: one where there is no encoding and decoding (no error correction), and one in which we encode a qubit in three qubits, and then decode them. If we start with a state $| \psi \rangle \, = c_0 \, | 0 \rangle \, + c_1 \, | 1 \rangle$, after a certain time we shall end up with

$$\hat{\rho}(t) = \begin{bmatrix} |c_0|^2 & c_0 c_1^* \gamma(t) \\ c_0^* c_1 \gamma^*(t) & |c_1|^2 \end{bmatrix}, \quad (17.135)$$

whose fidelity is given by $F = \left\langle \psi \left| \hat{\rho}(t) \right| \psi \right\rangle$. In the case of an initial mixed density matrix $\hat{\rho}_0$, a natural extension of the previous expression is

$$F = \text{Tr} \left[\hat{\rho}_0 \hat{\rho}(t) \right]. \quad (17.136)$$

Let us consider the interaction Hamiltonian

$$\hat{H} = \lambda(t) \hat{\sigma}_z \cdot \hat{O}, \quad (17.137)$$

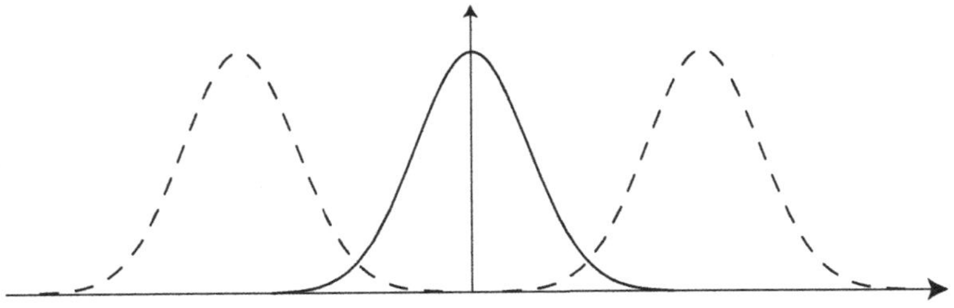

Figure 17.25 **Environmental wave functions and their overlapping as a function of time in quantum computation.**

where $\hat{O}$ may represent, e.g. in the optical case, a quadrature of the field. We have then the following time evolution:

$$\begin{aligned} |\,\psi\rangle\,|\,\mathcal{E}\rangle &\mapsto e^{-\frac{\imath}{\hbar}\int_0^\tau dt\lambda(t)\hat{\sigma}_z\cdot\hat{O}}\,(c_0\,|\,0\rangle\,+c_1\,|\,1\rangle\,)\,|\,\mathcal{E}\rangle \\ &= e^{-\frac{\imath}{\hbar}\Lambda\hat{O}}c_0\,|\,0\rangle\,|\,\mathcal{E}\rangle\,+e^{\frac{\imath}{\hbar}\Lambda\hat{O}}c_1\,|\,1\rangle\,|\,\mathcal{E}\rangle, \end{aligned} \tag{17.138}$$

where

$$\int_0^\tau dt\lambda(t) = \Lambda. \tag{17.139}$$

The final state of Eq. (17.138) is justified by the fact that $\hat{\sigma}_z$ is a constant of motion, i.e.

$$\hat{\sigma}_z\,|\,0\rangle\,=+\,|\,0\rangle\,,\quad \hat{\sigma}_z\,|\,1\rangle\,=-\,|\,1\rangle\,. \tag{17.140}$$

Now, suppose that $\mathcal{E}(x)$ represents the wave function corresponding to the ket $|\,\mathcal{E}\rangle$ in the x representation, i.e.

$$|\,\psi\rangle\,|\,\mathcal{E}\rangle\,\mapsto c_0\,|\,0\rangle\,|\,\mathcal{E}_{x+\Lambda}\rangle\,+c_1\,|\,1\rangle\,|\,\mathcal{E}_{x-\Lambda}\rangle\,. \tag{17.141}$$

We see that the overlap between these two wave functions tends to zero as time passes (see Fig. 17.25), i.e.

$$\langle\mathcal{E}_{x+\Lambda}\,|\,\mathcal{E}_{x-\Lambda}\rangle \rightarrow 0. \tag{17.142}$$

In the Bloch sphere language (see Subsec. 14.3.2)

$$\begin{aligned} \hat{\rho} &= \frac{1}{2}\left(\hat{I}+\mathbf{s}\cdot\hat{\boldsymbol{\sigma}}\right) \\ &= \frac{1}{2}\begin{bmatrix} 1+s_z & s_x+\imath s_y \\ s_x-\imath s_y & 1-s_z \end{bmatrix}, \end{aligned} \tag{17.143}$$

where $\hat{\boldsymbol{\sigma}}$ is the vector of the Pauli matrices, the initial Bloch vector preserves the projection onto the z-axis: in the limit $t \rightarrow \infty$, it will approach a vector pointing up ($|\,0\rangle$), where the off-diagonal elements are zero.

The important point is that, whatever $\gamma(t)$ is, quantum error correction does better when the energy is preserved, i.e. when there is no energy exchange with the environment. This is called *dissipation-free decoherence*.

Now we come back to the two scenarios announced previously. For scenario 1 (with no encoding and decoding) the fidelity is

$$\begin{aligned} F(t) &= \begin{pmatrix} c_0^* & c_1^* \end{pmatrix} \begin{bmatrix} |c_0|^2 & c_0 c_1^* \gamma(t) \\ c_0^* c_1 \gamma^*(t) & |c_1|^2 \end{bmatrix} \begin{pmatrix} c_0 \\ c_1 \end{pmatrix} \\ &= 1 - 2|c_0|^2|c_1|^2 \left[1 - \gamma(t)\right] \\ &= 1 - \frac{1}{2}\sin^2\theta \left[1 - \gamma(t)\right], \end{aligned} \tag{17.144}$$

where

$$c_0 = \cos\frac{\theta}{2}, \quad c_1 = \sin\frac{\theta}{2} e^{\imath\phi}. \tag{17.145}$$

The effect of decoherence depends on the state and it is therefore suitable to average over all possible states. In order to do this, let us perform an integration over the whole solid angle:

$$\begin{aligned} \langle F(t)\rangle &= \frac{1}{2}\int_0^\pi d\theta \left(1 - \frac{1}{2}\sin^2\theta \left[1 - \gamma(t)\right]\right)\sin\theta \\ &= \frac{1}{3}\left[2 + \gamma(t)\right] = \frac{1}{3}\left[2 + e^{-\gamma t}\right] \\ &\simeq 1 - \frac{1}{3}\gamma t, \end{aligned} \tag{17.146}$$

where $\gamma(t) = e^{-\gamma t}$ and the latter approximation is valid for small t.

Now, let us analyze the second scenario (shown in Fig. 17.24). In this case, we can identify which qubits have decohered, and following the perturbation theory we can identify one-qubit, two-qubit, and so on, interactions. If we take this into account, we may build a so-called error-syndrome table that allows to recognize which qubit has decohered and what transformation one has to apply in order to recover the input state, as we have shown previously.

An explicit calculation shows that the average fidelity over all possible configurations and all possible states is given by

$$\langle F(t)\rangle = 1 - \frac{1}{2}\gamma^2 t^2, \tag{17.147}$$

which shows that quantum error correction effectively provides a significant gain with respect to Eq. (17.146).

Summary

- We have introduced the concepts of *von Neumann entropy* – characterized by non-negativity, unitary invariance, concavity, subadditivity – and its relation with information.
- We have discussed the relation between information and entanglement and shown that entanglement is a form of *mutual information* between quantum systems.

- Furthermore, we have considered the problem of *measurement* in informational terms and also considered bounds on information acquisition.
- A relevant issue is represented by *teleportation*, that is, the ability to make use of entanglement for sharing information and – together with a classical communication – for transporting a quantum state between distant locations. This protocol may also have very interesting technological applications.
- Superposition and entanglement are genuinely quantum resources that can be used for establishing and sharing a secret key in symmetric cryptographic protocols. Quantum cryptographic schemes are able to circumvent the known limitations of classical protocols.
- We have then introduced the important concept of *qubit*, which represents a generalization of the classical bit by allowing informational states that are superpositions of a binary code. Quantum *gates* and *circuits* permit effective quantum information processing and computing. Among the quantum gates the CNOT gate is of particular relevance as it may be used to entangle qubits.
- Quantum computation is an interesting form of information processing that allows for solutions to problems that are very time-consuming to deal with in classical computation. In particular, we recall here *Deutsch's algorithm* for evaluating in a single step whether a given function is constant or balanced, *Shor's algorithm*, which allows for the polynomial-time factorization of large integer numbers, and *Grover's algorithm*, which can be used for speeding up database searches.
- Several physical schemes have been proposed to realize quantum computation. All of them have the same fatal enemy, namely decoherence. In order to combat the negative effects of decoherence, a quantum version of *error correction* has been formulated.

Problems

17.1 Prove that, for a quantum system of dimensions n, the maximal entropy is $\ln(n)$.

17.2 Prove Property (17.14) of von Neumann entropy.
(*Hint:* Take advantage of Schmidt decomposition (see Subsec. 5.5.3) and of the fact that $S(\hat{\rho})$ is completely determined by the eigenvalues of $\hat{\rho}$.)

17.3 Prove that in the case of pure states we have $S_{\rm VN} = 0$.

17.4 In the context of Fig. 17.4, prove that the subset D is invariant under local unitary transformations.

17.5 In the context of Figs. 17.4 and 17.5, prove the following relations among the subsets

$$\Omega = \mathrm{D} \cup \mathrm{E}, \quad \mathrm{D} \cap \mathrm{E} = \emptyset, \tag{17.148a}$$

$$\mathrm{U} \subset \mathrm{D}, \quad \mathrm{D} \subset \mathrm{B}, \tag{17.148b}$$

$$\mathrm{B} \cap \overline{\mathrm{B}} = \emptyset, \quad \mathrm{B} \cup \overline{\mathrm{B}} = \Omega, \tag{17.148c}$$

$$\overline{\mathrm{B}} \cup (\mathrm{B}/\mathrm{D}) = \mathrm{E}. \tag{17.148d}$$

17.6 Verify that the states (17.68) are an eigenbasis of the operator (17.67).
(*Hint:* Take advantage of the solution to Prob. 16.8.)

17.7 Show that the states (17.70) and (17.71) are identical.

17.8 Prove that from suitable application of transformations (17.73) to the state of the particle 3 we always obtain the state (17.69).

17.9 Compute explicitly the matrix $\hat{U}(\alpha, \beta, \gamma, \delta)$ in Eq. (17.94).

17.10 Show that factoring is indeed a classical example of non-polynomial algorithm.

17.11 Prove Eqs. (17.110).

17.12 An alternative definition of the distance between two states $\hat{\rho}_1$ and $\hat{\rho}_2$ (see also Probs. 14.15–14.16 and Box 14.1) that is particularly interesting for qubits is given by

$$\mathrm{d}(\hat{\rho}_1, \hat{\rho}_2) = \frac{1}{2}\mathrm{Tr}|\hat{\rho}_1 - \hat{\rho}_2|, \qquad (17.149)$$

and a suitable representation for qubits is provided by the Bloch vector. If $\hat{\rho}_1$ and $\hat{\rho}_2$ have Bloch vectors $\mathbf{r}$ and $\mathbf{s}$ respectively (see Eq. (17.143)),

$$\hat{\rho}_1 = \frac{1}{2}\left(\hat{I} + \mathbf{r}\cdot\hat{\boldsymbol{\sigma}}\right), \quad \hat{\rho}_2 = \frac{1}{2}\left(\hat{I} + \mathbf{s}\cdot\hat{\boldsymbol{\sigma}}\right), \qquad (17.150)$$

compute the distance by making use of this formalism so as to obtain

$$\mathrm{d}(\hat{\rho}_1, \hat{\rho}_2) = \frac{1}{2}|\mathbf{r} - \mathbf{s}|. \qquad (17.151)$$

Further reading

Deutsch, D. and Ekert, A., Quantum computation. *Physics World* (March 1998), 47–51.

Gisin, N., Ribordy, G., Tittel, W., and Zbinden, H., Quantum cryptography. *Review of Modern Physics*, **74** (2002), 145–95.

Nielsen, M. A. and Chuang, I. L., *Quantum Computation and Quantum Information*, Cambridge: Cambridge University Press, 2000.

Pan, J.-W., Bouwmeester, D., Weinfurter, H. and Zeilinger, A., Experimental entanglement swapping: entangling photons that never interacted. *Physics Review Letters*, **80** (1998), 3891.

Wehrl, Alfred, General properties of entropy. *Review of Modern Physics*, **50** (1978), 221–60.

Zeilinger, Anton, Quantum entangled bits step closer to it. *Science*, **289** (2000), 405–406.

Bibliography

Author names in *italics* refer to books.

[*Abramowitz/Stegun* 1964] Abramowitz, M. and Stegun, I. A., *Handbook of Mathematical Functions with Formulas, Graphs, and Mathematical Tables*, US Department of Commerce.

[Agarwal 1971] Agarwal, G. S., Brownian motion of a quantum oscillator. *Physical Review*, **A4**, 739–47.

[Aharonov/Anandan 1987] Aharonov, Y. and Anandan, J., Phase change during a cyclic quantum evolution. *Physical Review Letters*, **58**, 1593–96.

[Aharonov/Bohm 1959] Aharonov, Y. and Bohm, D., Significance of electromagnetic potentials in the quantum theory. *Physical Review*, **115**, 485–91.

[Aharonov/Bohm 1961] —, Time in the quantum theory and the uncertainty relation for time and energy. *Physical Review*, **122**, 1649–58.

[Aharonov/Vaidman 1993] Aharonov, Y. and Vaidman, L., Measurement of the Schrödinger wave of a single particle. *Physics Letters*, **178A**, 38–42.

[Aharonov *et al.* 1993] Aharonov, Y., Anandan, J., and Vaidman, L., Meaning of the wave function. *Physical Review*, **A47**, 4616–26.

[Albert/Loewer 1988] Albert, D. and Loewer, B., Interpreting the Many Worlds Interpretation. *Synthese*, **77**, 195–213.

[Allcock 1969] Allcock, G. R., The time of arrival in quantum mechanics. I–III. *Annals of Physics*, **53**, 253–348.

[Alter/Yamamoto 1998] Alter, O. and Yamamoto, Y., Impossibility of determining the unknown quantum wavefunction of a single system: quantum non-demolition measurements, measurements without entanglement and adiabatic measurements. *Fortschritte der Physik*, **46**, 817–27.

[Anderson *et al.* 1995] Anderson, M. H., Ensher, J. R., Matthews, M. R., Wieman, C. E., and Cornell, E. A., Observation of Bose–Einstein condensation in a dilute atomic vapor. *Science*, **269**, 198–201.

[*Apostol* 1969] Apostol, Tom M., *Calculus*, New York: Wiley, 2nd edn.

[Araki/Lieb 1970] Araki, H. and Lieb, E. H., Entropy inequalities. *Communications in Mathematical Physics*, **18**, 160–70.

[*Arnold* 1978] Arnold, Vladimir I., *Mathematical Methods of Classical Mechanics* (tr. from russ.), New York: Springer: 1978; 2nd edn. 1989.

[Aspect 1976] Aspect, Al., Proposed experiment to test the nonseparability of quantum mechanics. *Physical Review*, **D14**, 1944–45.

[Aspect 2007] —, To be or not to be local. *Nature*, **446**, 866–67.

[Aspect *et al.* 1982a] Aspect, A., Grangier, P., and Roger, G., Experimental realization of Einstein–Podolsky–Rosen–Bohm Gedankenexperiment. *Physical Review Letters*, **48**, 91–94.

[Aspect *et al.* 1982b] Aspect, A., Dalibard, J., and Roger, G., Experimental tests of Bell's inequalities. *Physical Review Letters*, **49**, 1804–7.

[*Atkins/Friedman* 2005] Atkins, P. and Friedman, R., *Molecular Quantum Mechanics*, Oxford: University Press, 4th edn.

[*Auletta* 2000] Auletta, Gennaro, *Foundations and Interpretation of Quantum Mechanics: In the Light of a Critical-Historical Analysis of the Problems and of a Synthesis of the Results*, Singapore: World Scientific, 2000; rev. edn. 2001.

[Auletta 2003] —, Some lessons of quantum mechanics for cognitive science, *Intellectica*, **36–37**, 293–317.

[Auletta 2004] —, Critical examination of the conceptual foundations of classical mechanics in the light of quantum physics. *Epistemologia*, **27**, 55–82.

[Auletta 2005] —, Quantum information as a general paradigm. *Foundations of Physics*, **35**, 787–815.

[Auletta 2006] —, The problem of information, in *Proceedings of the I Workshop on the Relationships Between Science and Philosophy*, edn. G. Auletta. Vatican City: Libreria Editrice Vaticana, pp. 109–127.

[Auletta *et al.* 2008] Auletta, G., Colagé, I., and Torcal, L., Interpretation of quantum mechanics consistent with quantum mechanics. Draft.

[Auletta/Tarozzi 2004a] Auletta, G. and Tarozzi, G., Wavelike correlations versus path detection: another form of complementarity. *Foundations of Physics Letters*, **17**, 889–95.

[Auletta/Tarozzi 2004b] Auletta, G. and Tarozzi, G., On the physical reality of quantum waves. *Foundations of Physics*, **34**, 1675–94.

[Badurek *et al.* 1983] Badurek, G., Rauch, H., and Summhammer, J., Time-dependent superposition of spinors. *Physical Review Letters*, **51**, 1015–18.

[Balian/Bloch 1970] Balian, R. and Bloch, C., Distribution of eigenfrequencies for the wave equation in a finite domain I: three-dimensional. *Annals of Physics*, **60**, 40.

[Balian/Bloch 1971] —, Asymptotic evaluation of the Green's function for large quantum numbers. *Annals of Physics*, **63**, 592.

[Balian/Bloch 1972] —, Distribution of eigenfrequencies for the wave equation in a finite domain. III. Eigenfrequency. *Annals of Physics*, **69**, 76.

[Balian/Bloch 1974] —, Solution of the Schrödinger equation in terms of classical paths. *Annals of Physics*, **85**, 514.

[Balian/Duplantier 1977] Balian, R. and Duplantier, B., Electromagnetic waves near perfect conductors. I. Multiple scattering expansions. Distribution of modes. *Annals of Physics*, **104**: 300.

[Balian/Duplantier 1978] Balian, R. and Duplantier, B., Electromagnetic waves near perfect conductors. II. Casimir effect. *Annals of Physics*, **112**: 165.

[Ballentine 1970] Ballentine, Leslie E., The statistical interpretation of quantum mechanics. *Review of Modern Physics*, **42**, 358–81.

[*Ballentine* 1990] —, *Quantum Mechanics*, Englewood Cliffs: Prentice Hall; Singapore: World Scientific, 1998, 1999.

[Barchielli *et al.* 1982] Barchielli, A., Lanz, L., and Prosperi, G. M., A model for the macroscopic description and continual observations in quantum mechanics. *Nuovo Cimento*, **72B**, 79–12.

[Bardroff *et al.* 1995] Bardroff, P. J., Mayr, E., and Schleich, W. P., Quantum state endoscopy: measurement of the quantum state in a cavity. *Physical Review*, **A51**, 4963–66.

[Bargmann 1954] Bargmann, V., *Annals of Mathematics*, **59**, 1.

[Barnett/Phoenix 1989] Barnett, S. M. and Phoenix, S. J. D., Entropy as a measure of quantum optical correlation. *Physical Review*, **A40**, 2404–409.

[Barnett/Phoenix 1991] —, Information theory, squeezing, and state correlations. *Physical Review*, **A44**, 535–45.

[*Barnett/Radmore* 2002] Barnett, S. M. and Radmore, P. M., *Methods in Theoretical Quantum Optics*, Oxford: Oxford University Press.

[Barrett *et al.* 2004] Barrett, M. D., Chiaverini, J., Schaetz, T., *et al.* Deterministic quantum teleportation of atomic qubits. *Nature*, **429**, 737–39.

[*Barrow et al.* 2004] Barrow, J. D., Davies, P. C. W., and Harper, C. L. (eds.), *Science and Ultimate Reality: Quantum Theory, Cosmology and Complexity*, Cambridge: Cambridge University Press.

[Bayes 1763] Bayes, Thomas, An essay toward solving a problem in the doctrine of chances. *Philosophical Transactions of the Royal Society of London*, **53**, 370–418.

[Bell 1964] Bell, John S., On Einstein Podolsky Rosen paradox. *Physics*, **1**, 195–200; rep. in [*Bell* 1987, 14–21].

[Bell 1966] —, On the problem of hidden variables in quantum mechanics. *Review of Modern Physics*, **38**, 447–52; rep. in [*Bell* 1987, 1–13].

[Bell 1971] —, Introduction to the hidden-variable question, in [*D'Espagnat* 1971, 171–81]; rep. in [*Bell* 1987, 29–39].

[Bell 1981] —, Quantum mechanics for cosmologists, in *Quantum Gravity 2*, G. J. Isham, R. Penrose, D. Sciama (edn.), Oxford: Clarendon; rep. in [*Bell* 1987, 117–38].

[Bell 1981] —, Bertlmann's socks and the nature of reality. *Journal de Physique Supplementa*, **3**, 41–61; rep. in [*Bell* 1987, 139–58].

[*Bell* 1987] —, *Speakable and Unspeakable in Quantum Mechanics*, Cambridge: Cambridge University Press, 1987, 1994.

[*Beltrametti/Cassinelli* 1981] Beltrametti, E. and Cassinelli, G., *The Logic of Quantum Mechanics*, Redwood City: Addison-Wesley.

[Beltrametti/Maczyński 1991] Beltrametti, E. and Maczyński, M. J., On a characterization of classical and nonclassical probabilities. *Journal of Mathematical Physics*, **32**, 1280–86.

[Benioff 1982] Benioff, Paul A., Quantum mechanical models of Turing machines that dissipate no energy. *Physical Review Letters*, **48**, 1581–85.

[Bennett 1982] Bennett, Charles H., The thermodynamics of computation: a review. *International Journal of Theoretical Physics*, **21**, 905–40.

[Bennett 1995] —, Quantum information and computation. *Physics Today*, **48** (Oct.), 24–30.

[Bennett/Brassard 1984] Bennett, C. H. and Brassard, G., Quantum cryptography: Public Key distribution and coin tossing. *Proceedings IEEE International Conference on Computer Systems*, New York: IEEE, 1984, 175–79.

[Bennett/Wiesner 1992] Bennett, C. H. and Wiesner, S. J., Communication via one- and two-particle operators on EPR states. *Physical Review Letters*, **69**, 2881–84.

[Bennett *et al.* 1993] Bennett, C. H., Brassard, G., Crepeau, C., Jozsa, R., Peres, A., and Wootters, W. K., Teleporting an unknown quantum state via dual classical and EPR channels. *Physical Review Letters*, **70**, 1895–99.

[Bergquist *et al.* 1986] Bergquist, J. C., Hulet, R. G., Itano, W. M., and Wineland, D. J., Observation of quantum jumps in a single atom. *Physical Review Letters*, **57**, 1699–1702.

[Berry 1984] Berry, Michael V., Quantal phase factors accompanying adiabatic changes. *Proceedings of the Royal Society of London*, **A392**, 45–57; rep. in [*Shapere/Wilczek* 1989, 124–36].

[Berry 1987] —, The adiabatic phase and the Pancharatnam's phase for polarized light. *Journal of Modern Optics*, **34**, 1401–407; rep. in [*Shapere/Wilczek* 1989, 67–73].

[Berry 1989] —, The quantum phases, five years after, in [*Shapere/Wilczek* 1989, 7–28].

[Bertrand/Bertrand 1987] Bertrand, J. and Bertrand, P., A tomographic approach to Wigner's function. *Foundations of Physics*, **17**, 397–405.

[*Bialynicki-B. et al.* 1992] Bialynicki-Birula, I., Cieplak, M., and Kaminski, J., *Theory of Quanta*, Oxford: Oxford University Press.

[*Bjorken/Drell* 1964] Bjorken, J. D. and Drell, S. D., *Relativistic Quantum Mechanics*, New York: McGraw-Hill.

[*Bjorken/Drell* 1965] —, *Relativistic Quantum Fields*, New York: McGraw-Hill.

[*Blokhintsev* 1965] Blokhintsev, Dimitrii I., *The Philosophy of Quantum Mechanics*, 1965; Engl. trans. Dordrecht: Reidel, 1968.

[Blokhintsev 1976] —, Statistical ensembles in quantum mechanics, in [*Flato et al.* 1976, 147–58]

[*Bocheński* 1970] Bocheński, Innocentius M., *A History of Formal Logic* (Eng. trans.), New York: Chelsea.

[*Bohm* 1951] Bohm, David, *Quantum Theory*, New York: Prentice-Hall.

[Bohm 1952] —, A suggested interpretation of the quantum theory in terms of "hidden variables". *Physical Review*, **85**, 166–93.

[*Bohm* 1979] Bohm, Arno, *Quantum Mechanics: Foundations and Applications*, New York: Springer; 2nd edn. 1986; 3rd edn. 1993.

[*Bohm/Hiley* 1993] Bohm, D. and Hiley, B. J., *The Undivided Universe. An Ontological Interpretation of Quantum Theory*, London: Routledge.

[*Bohm et al.* 2003] Bohm, A., Mostafazadeh, A., Koizumi, H., Niu, Q., and Zwanziger, J., *The Geometric Phase in Quantum Systems*, Berlin: Springer.

[Bohr 1913] Bohr, Niels, On the constitution of atoms and molecules. *Philosophical Magazine*, **26**, 1–25, 476–502, 857–75.

[Bohr 1920] —, Über die Linienspektren der Elemente. *Zeitschrift für Physik*, **2**, 423–69.

[Bohr 1928] —, The quantum postulate and the recent development of atomic theory. *Nature*, **121**, 580–90.

[Bohr 1935a] —, Quantum mechanics and physical reality. *Nature*, **136**, 65.

[Bohr 1935b] —, Can quantum-mechanical description of physical reality be considered complete? *Physical Review*, **48**, 696–702.

[Bohr 1948] —, On the notions of causality and complementarity. *Dialectica*, **1**, 312–19.

[Bohr 1949] —, Discussion with Einstein on epistemological problems in atomic physics, in [*Schilpp* 1949, 201–241].

[Born 1926] Born, Max, Quantenmechanik der Stoßvorgänge. *Zeitschrift für Physik*, **38**, 803–27

[Born 1926] —, Die Adiabatenprinzip in der Quantenmechanik. *Zeitschrift für Physik*, **40**, 167–92.

[Born 1927a] —, Quantenmechanik und Statistik. *Naturwissenschaften*, **15**, 238–42.

[Born 1927b] —, Physical aspects of quantum theory. *Nature*, **119**, 354–57.

[Born/Jordan 1925] Born, M. and Jordan, P., Zur Quantenmechanik. *Zeitschrift für Physik*, **34**, 858–88.

[Born *et al.* 1926] Born, M., Heisenberg, W., and Jordan, P., Zur Quantenmechanik II. *Zeitschrift für Physik*, **35**, 557–615.

[Bose 1924] Bose, Satyendranath, Planck's Gesetz und Lichtquantenhypothese. *Zeitschrift für Physik*, **26**, 178–81.

[Bouwmeester *et al.* 1997] Bouwmeester, D., Pan, J.-W., Mattle, K., Eibl, M., Weinfurter, H., and Zeilinger, A., Experimental Quantum Teleportation. *Nature*, **390**, 575–79.

[Bouwmeester *et al.* 1999] Bouwmeester, D., Pan, J.-W., Daniell, M., Weinfurter, H., and Zeilinger, A., Observation of three-photon Greenberger–Horne–Zeilinger entanglement. *Physical Review Letters*, **82**, 1345–99.

[*Braginsky/Khalili* 1992] Braginsky, V. B. and Khalili, F. Y., *Quantum Measurement*, Cambridge: Cambridge University Press.

[Braginsky *et al.* 1980] Braginsky, V. B., Vorontsov, Y. I., and Thorne, K. S., Quantum nondemolition measurements. *Science*, **209**, 547–57.

[*Brandsen* 1983] Brandsen, B. C. J., *Physics of Atoms and Molecules*, London: Longman.

[Braunstein/Caves 1988] Braunstein, S. L. and Caves, C. M., Information-theoretic Bell inequalities. *Physical Review Letters*, **61**, 662–65.

[Braunstein/Caves 1990] —, Wringing out better Bell inequalities. *Annals of Physics*, **202**, 22–56.

[Braunstein *et al.* 1992a] Braunstein, S. L., Mann, A., and Revzen, M., Maximal violation of Bell inequalities for mixed states. *Physical Review Letters*, **68**, 3259–61.

[Breit/Wigner 1936] Breit, G. and Wigner, E. P., Capture of slow neutrons. *Physical Review*, **49**, 519.

[Brillouin 1926] Brillouin, Léon, La mécanique ondulatoire de Schrödinger : une méthode générale de resolution par approximations successives. *Comptes Rendus à l'Academie des Sciences* **183**: 24–26.

[*Brillouin* 1960] —, *Wave Propagation and Group Velocity*, New York: Academic Press.

[Brune *et al.* 1992] Brune, M., Haroche, S., Raimond, J. M., Davidovich, L., and Zagury, N., Manipulation of photons in a cavity by dispersive atom–field coupling: Quantum-nondemolition measurements and generation of "Schrödinger cat" states. *Physical Review*, **A45**, 5193–214.

[Brune *et al.* 1996] Brune, M., Hagley, E., Dreyer, J., *et al.*, Observing the progressive decoherence of the "meter" in a quantum measurement. *Physical Review Letters*, **77**, 4887–90.

[Busch 2003] Busch, Paul, Quantum states and generalized observables: a simple proof of Gleason's theorem. *Physical Review Letters*, **91**, 120403-1-3.

[*Busch et al.* 1991] Busch, P., Lahti, P., and Mittelstaedt, P., *The Quantum Theory of Measurement*, Berlin: Springer, 1991; 2nd edn. 1996.

[*Busch et al.* 1995] Busch, P., Grabowski, M., and Lahti, P. J., *Operational Quantum Physics*, Berlin: Springer.

[Bužek/Hillery 1996] Bužek, V. and Hillery, M., Quantum copying: beyond the no-cloning theorem. *Physical Review*, **A54**, 1844–52.

[Bužek *et al.* 1996] Bužek, V., Adam, G., and Drobný, G., Quantum state reconstruction and detection of quantum coherences on different observation levels. *Physical Review*, **A54**, 804–820.

[Bužek *et al.* 1997] Bužek, V., Vedral, V., Plenio, M. B., Knight, P. L., and Hillery, M., Broadcasting of entanglement via local copying. *Physical Review*, **A55**, 3327–32.

[*Byron/Fuller* 1969–70] Byron, F. W. Jr. and Fuller, R. W., *Mathematics of Classical and Quantum Physics*, 1969–70; New York: Dover, 1992.

[Cahill/Glauber 1969] Cahill, K. E. and Glauber, R. J., Density operators and quasiprobability distributions. *Physical Review*, **177**, 1882–1902.

[Caldeira/Leggett 1985] Caldeira, A. and Leggett, A. Influence of damping on quantum interference: an exactly soluble model. *Physical Review*, **A31**, 1059–66.

[Cantrell/Scully 1978] Cantrell, C. D. and Scully, M. O., The EPR paradox revisited. *Physics Letters*, **43C**, 499–508.

[*Carmichael* 1993] Carmichael, Howard J., *An Open Systems Approach to Quantum Optics*, Heidelberg: Springer.

[*Carmichael* 1999] Carmichael, Howard J., *Statistical Methods in Quantum Optics*, Berlin: Springer.

[Carruthers/Nieto 1968] Carruthers, P. and Nieto, M. M., Phase and angle variables in quantum mechanics. *Review of Modern Physics*, **40**, 411–40.

[Carmichael/Orozco 2003] Carmichael, H. and Orozco, L. A., Single atom lases orderly light. *Nature*, **425**, 246–47.

[Caser 1992] Caser, Serge, Local vacua, in [*Selleri* 1992, 19–36].

[Caves/Drummond 1994] Caves, C. M. and Drummond, P. D., Quantum limits on bosonic communication rates. *Review of Modern Physics*, **66**, 481–537.

[Caves *et al.* 1980] Caves, C. M., Thorne, K. S., Drever, R. W. P., Sandberg, V. D., and Zimmerman, M., On the measurement of a weak classical force coupled to a quantum-mechanical oscillator, I. *Review of Modern Physics*, **52**, 341–92.

[Cerf/Adami 1997] Cerf, N. and Adami, C., Entropic Bell inequalities. *Physical Review*, **A55**, 3371–74.

[Chiao/Steinberg 1997] Chiao, R. Y. and Steinberg, A. M., Tunneling times and superluminality. *Progress in Optics*, **37**, 345–405.

[Cini 1983] Cini, Marcello, Quantum theory of measurement without wave packet collapse. *Nuovo Cimento*, **73B**, 27–54.

[Cini *et al.* 1979] Cini, M., De Maria, M., Mattioli, G., and Nicoló, F., Wave packet reduction in quantum mechanics: a model of a measuring apparatus. *Foundations of Physics*, **9**, 479–500.

[Cirac/Zoller 1994] Cirac, J. I. and Zoller, P., Preparation of macroscopic superpositions in many-atoms systems. *Physical Review*, **A50**, R2799–802.

[Clauser/Horne 1974] Clauser, J. F. and Horne, M. A., Experimental consequences of objective local theories. *Physical Review*, **D10**, 526–35.

[Clauser/Shimony 1978] Clauser, J. F. and Shimony, A., Bell's theorem: experimental tests and implications. *Reporting Progress Physics*, **41**, 1881–927.

[Clauser *et al.* 1969] Clauser, J. F., Horne, M. A., Shimony, A., and Holt, R. A., Proposed experiment to test local hidden-variable theories. *Physical Review Letters*, **23**, 880–84.

[*Clifton et al.* 2004] Clifton, R., Butterfield, J., and Halvorson, H. (eds.), *Quantum Entanglement: Selected Papers*, Oxford: Oxford University Press.

[Compton 1923] Compton, Arthur H., A quantum theory of the scattering of x-rays by light elements. *Physical Review*, **21**, 483–502.

[Cook/Kimble 1985] Cook, R. J. and Kimble, H. J., Possibility of direct observation of quantum jumps. *Physical Review Letters*, **54**: 1023–26.

[Corbett/Home 2000] Corbett, J. V. and Home, D., Quantum effects involving interplay between unitary dynamics and kinematic entanglement. *Physical Review*, **A62**, 062103.

[Croca 1987] Croca, J. R., Neutron interferometry can prove (or refute) the existence of de Broglie's waves. *Foundations of Physics*, **17**, 971–80.

[Cummings 1965] Cummings, F. W., Stimulated emission of radiation in a single mode. *Physical Review*, **A140**, 1051–56.

[Daneri *et al.* 1962] Daneri, A., Loinger, A., and Prosperi, G. M., Quantum theory of measurement and ergodicity conditions. *Nuclear Physics*, **33**, 297–319; rep. in [*Wheeler/Zurek* 1983, 657–79].

[D'Ariano/Yuen 1996] D'Ariano, G. M. and Yuen, H. P., Impossibility of measuring the wave function of a single quantum system. *Physical Review Letters*, **76**, 2832–35.

[D'Ariano *et al.* 1994] D'Ariano, G. M., Macchiavello, C., and Paris, M. G. A., Detection of the density matrix through optical homodyne tomography without filtered back projection. *Physical Review*, **A50**: 4298–302.

[D'Ariano *et al.* 1995] D'Ariano, G. M., Leonhardt, U., and Paul, H., Homodyne detection of the density matrix of the radiation field. *Physical Review*, **A52**, R1801–804.

[*Davies* 1976] Davies, E. B., *Quantum Theory of Open Systems*, London: Academic Press.

[*Davies* 1989] Davies, Paul C. W. (edn.), *The New Physics*, Cambridge: Cambridge University Press, 1989, 1990, 1992, 1993.

[Davisson/Germer 1927] Davisson, C. J. and Germer, L. H., Diffraction of electrons by a crystal of nickel. *Physical Review*, **30**, 705–740.

[*Davydov* 1976] Davydov, A. S., *Quantum Mechanics*, Oxford: Pergamon, 2nd edn.

[de Broglie 1924] de Broglie, Louis, Sur la définition générale de la correspondance entre onde et mouvement. *Comptes Rendus à l'Academie des Sciences*, **179**, 39–40.

[de Broglie 1925] —, Recherche sur la théorie des quanta. *Annales de Physique*, **3**, 22–138.

[de Broglie 1927a] —, La structure de la matière et du rayonnement et la mécanique ondulatoire. *Comptes Rendus à l'Academie des Sciences*, **184**, 273–74.

[de Broglie 1927b] —, Sur le role des ondes continues y en mécanique ondulatoire. *Comptes Rendus à l'Academie des Sciences*, **185**, 380–82.

[de Broglie 1927c] —, La mécanique ondulatoire et la structure atomique de la matière et du rayonnement. *Journal de Physique*, **8**, 225–41.

[de Broglie 1955] —, Une interprétation nouvelle de la mécanique ondulatoire est-elle possible? *Nuovo Cimento*, **1**, 37–50.

[*de Broglie* 1956] —, *Une tentative d'Interpretation causale et non-linéaire de la mécanique ondulatoire*, Paris: Gauthier-Villars.

[de Muynck *et al.* 1991] de Muynck, W. M., Stoffels, W. W., and Martens, H., Joint measurement of interference and path observables in optics and neutron interferometry. *Physica*, **B175**, 127–32.

[*D'Espagnat* 1971] D'Espagnat, Bernard (edn.), *Foundations of Quantum Mechanics*, New York: Academic Press.

[Deutsch 1983] Deutsch, David, Uncertainty in quantum mechanics. *Physical Review Letters*, **50**, 631–33.

[Deutsch 1985] —, Quantum theory, the Church–Turing principle and the universal quantum computer. *Proceedings of the Royal Society of London*, **A400**, 97–117.

[Deutsch/Ekert 1998] Deutsch, D. and Ekert, A., Quantum computation. *Physics World*, March, 47–51.

[Dewdney *et al.* 1993] Dewdney, C., Hardy, L., and Squires, E., How late measurements of quantum trajectories can fool a detector. *Physics Letters*, **184A**, 6–11.

[DeWitt 1970] DeWitt, Bryce S., Quantum mechanics and reality. *Physics Today*, **23.9**, 30–40.

[Dholakia 2008] Dholakia, Kishan, Against the spread of the light. *Nature*, **451**, 413.

[Dirac 1926a] Dirac, Paul A. M., On quantum algebra. *Proceedings of the Cambridge Philosophical Society*, **23**, 412–18.

[Dirac 1926b] —, On theory of quantum mechanics. *Proceedings of the Royal Society of London*, **A112**, 661.

[*Dirac* 1930] —, *The Principles of Quantum Mechanics*, 4th edn., Oxford: Clarendon, 1958, 1993.

[Dodonov *et al.* 1999] Dodonov, V. V., Man'ko, O. V., Man'ko, V. I. and Wünsche, A., Energy-sensitive and "classical-like" distances between quantum states. *Physica Scripta*, **59**, 81–89.

[Dowling *et al.* 1991] Dowling, J. P., Schleich, W. P., and Wheeler, J. A., Interference in phase space. *Annalen der Physik*, **48**, 423–502.

[Dragoman 1998] Dragoman, D., The Wigner distribution function in optics and optoelectronics. *Progress in Optics*, **37**, 1–56.

[Eberhard 1978] Eberhard, Philippe H., Bell's theorem and the different concepts of locality. *Nuovo Cimento*, **46B**, 392–419.

[Eberhard 1993] —, Background-level and counter efficiencies required for a loop-free EPR experiment. *Physical Review*, **A47**, R747–50.

[*Edmonds* 1957] Edmonds, A. R., *Angular Momentum in Quantum Mechanics*, Princeton: Princeton University Press, 1957, 1960, 1985.

[*Einstein* CP] Einstein, Albert, *The Collected Papers of Albert Einstein*, Princeton: Princeton University Press, 1993.

[Einstein 1905] —, Über einen die Erzeugung und Verwandlung des Lichtes betreffenden heuristischen Gesichtspunkt. *Annalen der Physik*, **17**, 132–48.

[Einstein 1917] —, Zur Quantentheorie der Strahlung. *Physikalische Zeitschrift*, **18**, 121–28.

[Einstein 1949] —, Remarks concerning the essays brought together in this co-operative volume. in [*Schilpp* 1949, 665–88].

[Einstein *et al.* 1935] Einstein, A., Podolsky, B., and Rosen N., Can quantum-mechanical description of physical reality be considered complete? *Physical Review*, **47**, 777–80.

[Ekert 1991] Ekert, Artur K., Quantum cryptography based on Bell's theorem. *Physical Review Letters*, **67**, 661–63.

[Elby/Bub 1994] Elby, A. and Bub, J., Triorthogonal uniqueness theorem and its relevance to the interpretation of quantum mechanics. *Physical Review*, **A49**, 4213–16.

[Elitzur/Vaidman 1993] Elitzur, A. C. and Vaidman, L., Quantum mechanical interaction-free measurements. *Foundations of Physics*, **23**, 987–97.

[Emig *et al.* 2007] Emig, T., Graham, N., Jaffe, R. L., and Kardar, M., *Physical Review Letters*, **99**, 170403.

[Engelmann/Fick 1959] Engelmann, F. and Fick, E., Die Zeit in der Quantenmechanik. *Nuovo Cimento Supplementa*, **12**, 63–72.

[Englert 1996] Englert, Berthold-Georg, Fringe visibility and which-way information: an inequality. *Physical Review Letters*, **77**, 2154–57.

[Englert *et al.* 1993] Englert, B.-G., Scully, M. O., Süssman, G., and Walther, H., Surrealistic Bohm trajectories. *Zeitschrift für Naturforschung*, **47a**, 1175–86.

[Englert *et al.* 1995] Englert, B.-G, Scully, M. O., and Walther, H., Complementarity and uncertainty. *Nature*, **375**, 367–68.

[Everett 1957] Everett, Hugh III, "Relative state" formulation of quantum mechanics. *Review of Modern Physics*, **29**, 454–62.

[Fano 1957] Fano, Ugo, Description of states in quantum mechanics by density matrix and operator techniques. *Review of Modern Physics*, **29**, 74–93.

[*Fano* 1971] Fano, Guido, *Mathematical Methods of Quantum Mechanics*, New York: McGraw-Hill.

[Faraci *et al.* 1974] Faraci, G., Gutkowski, D., Notarigo, S., and Pennisi, A. R., An experimental test of the EPR paradox. *Lettere a Nuovo Cimento*, **9**, 607–611.

[Fermi 1926a] Fermi, Enrico, Sulla quantizzazione del gas perfetto monoatomico. *Rendiconti dell'Accademia Nazionale dei Lincei*, **3**, 145.

[Fermi 1926b] —, Zur Quantelung des idealen einatomigem Gases. *Zeitschrift für Physik*, **36**, 902, Germ. trans. of [Fermi 1926a].

[Fermi 1928] —, Eine statistische Methode zur Bestimmung einiger Eigenschaften des Atoms und ihre Anwendung auf die Theorie des periodischen Systems der Elemente. *Zeitschrift für Physik*, **48**, 73–79.

[Fermi 1949] —, On the origin of the cosmic radiation. *Physical Review*, **75**, 1169–74.

[Ferrero *et al.* 1990] Ferrero, M., Marshall, T. W., and Santos, E., Bell's theorem: local realism versus quantum mechanics. *American Journal of Physics*, **58**, 683–88.

[Feynman 1948] Feynman, Richard P., Space–time approach to non-relativistic quantum mechanics. *Review of Modern Physics*, **20**, 367–87.

[Feynman 1982] —, Simulating physics with computers. *International Journal of Theoretical Physics*, **21**, 467–88.

[Feynman/Hibbs 1965] Feynman, R. P. and Hibbs, A. R., *Quantum Mechanics and Path Integrals*, New York: McGraw-Hill.

[*Feynman et al.* 1965] Feynman, R., Leighton, R. B., and Sands, M., *Lectures on Physics*, Reading (Massachussetts), Addison-Wesley, vol. I: 1963, 1977; vol. II: 1964, 1977; vol. III: 1965, 1966.

[Fick/Engelmann 1963a] Fick, E. and Engelmann, F., Quantentheorie der Zeitmessung, I. *Zeitschrift für Physik*, **175**, 271–82.

[Fick/Engelmann 1963b] Fick, E. and Engelmann, F., Quantentheorie der Zeitmessung, II. *Zeitschrift für Physik*, **178**, 551–62.

[*Fiolhais et al.* 2003] Fiolhais, C. Nogueira, F., and Marques, M. (eds.), *A Primer in Density Functional Theory*, Berlin: Springer-Verlag.

[*Flato et al.* 1976] Flato, M., Maric, Z., Milojevic, A., Sternheimer, D., and Vigier, J.-P. (eds.), *Quantum Mechanics, Determinism, Causality, and Particles*, Dordrecht: Reidel.

[Fleischhauer 2007] Fleischhauer, Michael, Indistinguishable from afar. *Nature*, **445**, 605–606.

[Fock 1932] Fock, V. A., Konfigurationsraum und zweite Quantelung. *Zeitschrift für Physik*, **75**, 622–47.

[Fock/Krylov 1947] Fock, V. A. and Krylov, N., *Journal of Physics (USSR)*, **11**, 112.

[Fokker 1914] Fokker, A. D., The median energy of rotating electrical dipoles in radiation field. *Annalen der Physik*, **43**, 810–20.

[Fortunato *et al.* 1999a] Fortunato, M., Tombesi, P., and Schleich, W. P., Endoscopic tomography and quantum non-demolition. *Physical Review*, **A59**, 718–27.

[Fortunato *et al.* 1999b] Fortunato, M., Raimond, J. M., Tombesi, P., and Vitali, D., Autofeedback scheme for preservation of macroscopic coherence in microwave cavities. *Physical Review*, **A60**, 1687–97.

[Franson/Potocki 1988] Franson, J. D. and Potocki, K. A., Single-photon interference over large distances. *Physical Review*, **A37**, 2511–15.

[Freedman/Clauser 1972] Freedman, S. J. and Clauser, J. F., Experimental test of local hidden-variable theories. *Physical Review Letters*, **28**, 938–41.

[Freedman *et al.* 1976] Freedman, S. J., Holt, R. A., and Papaliolios, C., Experimental status of hidden variable theories, in [*Flato et al.* 1976, 43–59].

[Freyberger/Herkommer 1994] Freyberger, M. and Herkommer, A. M., Probing a quantum state via atomic deflection. *Physical Review Letters*, **72**, 1952–55.

[Freyberger *et al.* 1997] Freyberger, M., Bardroff, P. J., Leichtle, C., Schrade, G., and Schleich, W. P., The Art of measuring quantum states. *Physics World*, **10.11**, 41–45.

[Friedman *et al.* 2000] Friedman, J. R., Patel, V., Chen, W., Tolpygo, S. K., and Lukens, J. E., Quantum superposition of distinct macroscopic states. *Nature*, **406**, 43–46.

[Fry 1995] Fry, Edward S., Bell inequalities and two experimental tests with mercury. *Quantum Optics*, **7**, 229–58.

[Fuchs/Caves 1995] Fuchs, C. A. and Caves, C. M., Bounds for accessible information in quantum mechanics, in [*Greenberger/Zeilinger* 1995, 706–714].

[*Fuchs/Schweigert* 1997] Fuchs, J. and Schweigert, C., *Symmetries, Lie Algebras and Representations*, Cambridge: Cambridge University Press.

[Furusawa *et al.* 1998] Furusawa, A., Sørensen, J. L., Braunstein, S. L., Fuchs, C. A., Kimble, H. J., and Polzik, E. S., Unconditional quantum teleportation. *Science* **282**: 706–709.

[*Gardiner* 1991] Gardiner, Crispin W., *Quantum Noise*, Berlin: Springer.

[Garg/Mermin 1984] Garg, A. and Mermin, N. D., Farka's lemma and the nature of reality: statistical implications of quantum correlations. *Foundations of Physics*, **14**, 1–39.

[Garuccio/Selleri 1984] Garuccio, A. and Selleri, F., Enhanced photon detection in EPR type experiments. *Physics Letters*, **103A**, 99–103.

[Garuccio *et al.* 1982] Garuccio, A., Rapisarda, V., and Vigier, J.-P., Next experimental set-up for the detection of de Broglie waves. *Physics Letters*, **90A**, 17–19.

[Gea-Banacloche 1990] Gea-Banacloche, Julio, Collapse and revival of the state vector in the Jaynes–Cummings model: an example of state preparation by a quantum apparatus. *Physical Review Letters*, **65**, 3385–88.

[Gell-Mann/Hartle 1990] Gell-Mann, M. and Hartle, J. B, Quantum mechanics in the light of quantum cosmology, in [*Zurek* 1990, 425–69].

[Gell-Mann/Hartle 1993] —, Classical equations for quantum systems. *Physical Review*, **D47**, 3345–82.

[Gerlach/Stern 1922a] Gerlach, W. and Stern, O., Der experimentelle Nachweis des magnetischen Moments des Silberatoms. *Zeitschrift für Physik*, **8**, 110–111.

[Gerlach/Stern 1922b] —, Der experimentelle Nachweis der Richtungsquantelung im Magnetfeld. *Zeitschrift für Physik*, **9**, 349–52.

[Gerlach/Stern 1922c] —, Das magnetische Moment des Silberatoms. *Zeitschrift für Physik*, **9**, 353–55.

[Ghirardi *et al.* 1986] Ghirardi, G. C., Rimini, A., and Weber, T., Unified dynamics for microscopic and macroscopic systems. *Physical Review*, **D34**, 470–91.

[Ghosh *et al.* 2003] Ghosh, S., Rosembaum, T. F., Aeppli, G., and Coppersmith, S. N., Entangled quantum state of magnetic dipoles. *Nature*, **425**, 48–51.

[Ginsberg *et al.* 2007] Ginsberg, N. S., Garner, S. R., and Vestergaard Hau, L., Coherent control of optical information with matter wave dynamics. *Nature*, **445**, 623–26.

[Gisin 1984] Gisin, Nicolas, Quantum measurement and stochastic processes. *Physical Review Letters*, **52**, 1657–60.

[Gisin 1991] —, Bell's inequality holds for all non-product states. *Physics Letters*, **154A**, 201–202.

[Gisin *et al.* 2002] Gisin, N., Ribordy, G., Tittel, W., and Zbinden, H., Quantum Cryptography. *Review of Modern Physics*, **74**, 145–95.

[*Giulini* et al. 1996] Giulini, D., Joos, E., Kiefer, C., Kupsch, J., Stamatescu, I.-O., and Zeh, H. D. (eds.), *Decoherence and the Appearance of a Classical World in Quantum Theory*, Berlin: Springer.

[Glauber 1963a] Glauber, Roy J., Photon correlations. *Physical Review Letters*, **10**, 84–86.

[Glauber 1963b] —, The quantum theory of optical coherence. *Physical Review*, **130**, 2529–39.

[Glauber 1963c] —, Coherent and incoherent states of the radiation field. *Physical Review*, **131**, 2766–88.

[Glauber 1966] —, Classical behavior of systems of quantum oscillators. *Physics Letters*, **21**, 650–52.

[Gleason 1957] Gleason, Andrew M., Measures on the closed subspaces of a Hilbert space. *Journal of Mathematics and Mechanics*, **6**, 885–93.

[*Gnedenko* 1969] Gnedenko, B. V., *The Theory of Probability*, Moscow: Mir, 1969, 1988.

[Gödel 1949] Gödel, Kurt, A Remark about the relationship between relativity theory and idealistic philosophy. in [*Schilpp* 1949, 557–62].

[*Goldin* 1982] Goldin, Edwin, *Waves and Photons: An Introduction to Quantum Optics*, New York: J. Wiley.

[*Goldstein* 1950] Goldstein, Herbert, *Classical Mechanics*, Massachusets: Addison-Wesley, 1950, 1965.

[Gordon *et al.* 1954] Gordon, J. P., Zeiger, H. J., and Townes, C. H., Molecular microwave oscillator and new hyperfine structure in the microwave spectrum of NH3. *Physical Review*, **95**, 282–84.

[Gordon *et al.* 1955] —, The masernew type of microwave amplifier, frequency standard, and spectrometer. *Physical Review*, **99**, 1264–74.

[*Gradstein/Ryshik* 1981] Gradstein, I. and Ryshik, I., *Tafeln-Tables*, Moskau: MIR.

[Grangier *et al.* 1986a] Grangier, P., Roger, G., and Aspect, A., Experimental evidence for a photon anticorrelation effect on a beam splitter: a new light on single-photon interferences. *Europhysics Letters*, **1**, 173–79.

[Grangier *et al.* 1986b] —, A new light on single photon interferences, in [*Greenberger* 1986, 98–107].

[*Greenberger* 1986] Greenberger, Daniel M. (edn.), *New Techniques and Ideas in Quantum Measurement Theory*, New York: Academy of Sciences.

[Greenberger/Yasin 1988] Greenberger, D. M. and Yasin, A., Simultaneous wave and particle knowledge in a neutron interferometer. *Physics Letters*, **128A**, 391–94.

[*Greenberger/Zeilinger* 1995] Greenberger, D. M. and Zeilinger, A. (eds.), *Fundamental Problems in Quantum Theory*, New York: Academy of Sciences.

[Greenberger *et al.* 1989] Greenberger, D. M., Horne, M. A., and Zeilinger, A., Going beyond Bell's theorem, in [*Kafatos* 1989, 69–72].

[Greenberger *et al.* 1990] Greenberger, D. M., Horne, M. A., Shimony, A., and Zeilinger, A., Bell's theorem without inequalities. *American Journal of Physics*, **58**, 1131–43.

[*Greiner/Müller* 1984] Greiner, W. and Müller, B., *Quantum Mechanics Symmetries*, Tun: H. Deutsch, 1984; Berlin: Springer, 1989, 1994.

[Griffiths 1984] Griffiths, Robert B., Consistent histories and the interpretation of quantum mechanics. *Journal of Statistical Physics*, **36**, 219–72.

[Gröblacher *et al.* 2007] Gröblacher, S., Paterek, T., Kaltenbaek, R., *et al.*, An experimental test of non-local realism. *Nature*, **446**, 871–75.

[Grot *et al.* 1996] Grot, N., Rovelli, C., and Tate, R. S., Time-of-arrival in quantum mechanics. *Physical Review*, **A54**, 4676–90.

[Grover 1996] Grover, Lov K., A fast quantum mechanical algorithm for database searching. *Proceedings STOC 1996*, Philadelphia: 212–19. ArXiv:quant-ph/9605043v3.

[*Gudder* 1988] Gudder, Stanley P., *Quantum Probability*, Boston: Academic Press.

[Haji-Hassan *et al.* 1987] Haji-Hassan, T., Duncan, A. J., Perrie, W., Beyer, H. J., and Kleinpoppen, H., Experimental investigation of the possibility of enhanced photon detection in EPR type experiments. *Physics Letters*, **123A**, 110–114.

[*Halmos* 1951] Halmos, Paul R., *Introduction to Hilbert Space*, New York: Chelsea, 1951, 2nd edn. 1957.

[Hanbury Brown/Twiss 1956] Hanbury Brown, R. and Twiss, R. Q., Correlation between photons in two coherent beams of light. *Nature*, **177**, 27.

[Hanbury Brown/Twiss 1958] —, Interferometry of the intensity fluctuations in light. II. An experimental test of the theory for partially coherent light. *Proceedings of the Royal Society*, **A243**, 291–319.

[Hardy 1992] Hardy, Lucien, On the existence of empty waves in quantum theory. *Physics Letters*, **167A**, 11–16.

[*Harrison* 2000] Harrison, Walter A., *Applied Quantum Mechanics*, Singapore: World Scientific.

[*Hartle* 2003] Hartle, James B., *Gravitation: An Introduction to Einstein's General Relativity*, B. Cummings.

[Hasegawa *et al.* 2006] Hasegawa, Y., Loidl, R., Badurek, G., Baron, M., and Rauch, H., Quantum Contextuality in a Single–Neutron Optical Experiment. *Physical Review Letters*, **97**, 230401-1–4.

[Heisenberg 1925] Heisenberg, Werner, Über quantentheoretische Umdeutung kinematischer und mechanischer Beziehungen. *Zeitschrift für Physik*, **33**, 879–93.

[Heisenberg 1927] —, Über den anschaulichen Inhalt der quantentheoretischen Kinematik und Mechanik. *Zeitschrift für Physik*, **43**, 172–98.

[*Heisenberg* 1958] —, *Physics and Philosophy*, New York: Harper.

[Hellmuth *et al.* 1986] Hellmuth, T., Zajonc, A. G., and Walther, H., Realizations of "delayed choice" experiments, in [*Greenberger* 1986, 108–14].

[*Helstrom* 1976] Helstrom, Carl W., *Quantum Detection and Estimation Theory*, New York: Academic Press.

[*Herzberg* 1950] Herzberg, Gerhard, *Molecular spectra and molecular structure. I. Spectra of diatomic molecules*, New York: Van Nostrand Reinhold.

[Herzog *et al.* 1995] Herzog, T. J., Kwiat, P. G., Weinfurter, H., and Zeilinger, A., Complementarity and the quantum eraser. *Physical Review Letters*, **75**, 3034–37.

[Hilgevoord/Uffink 1983] Hilgevoord, J. and Uffink, J. B. M., Overall width, mean peak width and the uncertainty principle. *Physics Letters*, **95A**, 474–76.

[Hilgevoord/Uffink 1988] Hilgevoord, J. and Uffink, J. B. M., The mathematical expression of the uncertainty principle, in [*van Der Merwe et al.* 1988, 91–114].

[Hillery/Yurke 1995] Hillery, M. and Yurke, B., Bell's theorem and beyond. *Quantum Optics*, **7**, 215–27.

[Hillery *et al.* 1984] Hillery, M. A., O'Connell, R. F., Scully, M. O., and Wigner, E., Distribution functions in physics: fundamentals. *Physics Reports*, **106**, 121–67.

[*Hirsch/Smale* 1974] Hirsch, M. W. and Smale, S., *Differential Equations, Dynamical Systems, and Linear Algebra*, San Diego: Academic Press.

[Hiskett *et al.* 2006] Hiskett, P. A., Rosenberg, D., Peterson, *et al.*, Long-distance quantum key distribution in optical fiber. *New Journal of Physics*, **8**, 193.

[Holevo 1973a] Holevo, A. S., Statistical decision theory for quantum systems. *Journal Multivariate Analysis*, **3**, 337–94.

[Holevo 1973b] —, Some estimates for information quantity transmitted by quantum communication channels. *Problems Information Transmission*, **9**, 177–83.

[*Holevo* 1982] —, *Probabilistic and Statistical Aspects of Quantum Theory* (Engl. trans.), Amsterdam: North Holland.

[Horodecki *et al.* 1996] Horodecki, R., Horodecki, P., and Horodecki, M., Separability of mixed states: necessary and sufficient conditions. *Physics Letters*, **223A**, 1–8.

[Horodecki *et al.* 2005] Horodecki, M., Oppenheim, J., and Winter, A., Partial quantum information. *Nature*, **436**, 673–76.

[Howard 1985] Howard, Don, Einstein on locality and separability. *Studies in the History and Philosophie of Science*, **16**, 171–201.

[*Huang* 1963] Huang, Kerson, *Statistical Mechanics*, New York: Wiley, 1963, 1987.

[*Hubel* 1988] Hubel, D. H., *Eye, Brain, Vision*, New York: Freeman.

[Husimi 1937] Husimi, Kodi, Studies in the foundations of quantum mechanics. *Proceedings of the Physico-Mathematical Society of Japan*, **19**, 766–89.

[Imamoḡlu 1993] Imamoḡlu, A., Logical reversibility in quantum–nondemolition measurements. *Physical Review*, **A47**, R4577–80.

[Itano *et al.* 1987] Itano, W. H., Bergquist, J. C., Hulet, R. G., and Wineland, D. J., Radiative decay rates in Hg^+ from observations of quantum jumps in a single ion. *Physical Review Letters*, **59**, 2732–35.

[Kimble 1990] Kimble, H. Jeffrey, in *Fundamental Systems in Quantum Optics*, J. Dalibard, J. M. Raymond, J. Zinn-Justin (eds.), Les Houches School Section LIII 1990, Amsterdam: Elsevier.

[*Jackson* 1962] Jackson, John D., *Classical Electrodynamics*, New York: J. Wiley, 1962; 2nd edn. 1975.

[*Jammer* 1966] Jammer, Max, *The Conceptual Development of Quantum Mechanics*, New York: McGraw-Hill, 1966; 2nd edn.: Thomas Pub., 1989.

[*Jammer* 1974] Jammer, Max, *The Philosophy of Quantum Mechanics. The Interpretation of Quantum Mechanics in Historical Perspective*, New York: Wiley, 1974.

[*Jauch* 1968] Jauch, Joseph M., *Foundations of Quantum Mechanics*, Massachussets: Addison-Wesley.

[Jauch *et al.* 1967] Jauch, J. M., Wigner, E. P., and Yanase, M. M., Some comments concerning measurements in quantum mechanics. *Nuovo Cimento*, **48B**, 144–51.

[Jaynes/Cummings 1963] Jaynes, E. T. and Cummings, F. W., Comparison of quantum and semiclassical radiation theories with application to the beam maser. *Proceedings IEEE*, **51**, 89–109.

[Joos/Zeh 1985] Joos, E. and Zeh, H. D., The emergence of classical properties through interaction with the environment. *Zeitschrift für Physik*, **B59**, 223–43.

[Jordan 1983] Jordan, Thomas F., Quantum correlations do not transmit signals. *Physics Letters*, **94A**, 264.

[Josephson 1962] Josephson, Brian D., Possible new effects in superconductive tunnelling. *Physics Letters*, **1**, 251–53.

[Judge 1963] Judge, D., On the uncertainty relation for L_z and ϕ. *Physics Letters*, **5**, 189.

[Judge/Lewis 1963] Judge, D. and Lewis, J. T., On the commutator $[L_z, f(\phi)]$. *Physics Letters*, **5**, 190.

[*Kafatos* 1989] Kafatos, Menas (edn.), *Bell's Theorem, Quantum Theory, and Conceptions of the Universe*, Dordrecht: Kluwer.

[Kastner 1993] Kastner, Marc A., Artificial atoms. *Physics Today*, January, 24–31.

[Khalfin/Tsirelson 1992] Khalfin, L. A. and Tsirelson, B. S., Quantum/classical correspondence in the light of Bell's inequalities. *Foundations of Physics*, **22**, 879–948.

[Kim *et al.* 2000] Kim, Y.-H, Yu, R., Kulik, S. P., Shih, Y. H., and Scully, M. O., A delayed choice quantum eraser. *Physical Review Letters*, **84**, 1–5.

[Kimble 1990] Kimble, H. Jeffrey, Quantum fluctuations in quantum optics squeezing and related phenomena, in *Fundamental Systems in Quantum Optics*, J. Dalibard, J. M. Raymond, J. Zinn-Justin (eds.), Les Houches School Section LIII 1990, Amsterdam: Elzevier.

[*Kittel* 1958] Kittel, Charles, *Elementary Statistical Physics*, 1958; New York: Dover.

[Knöll/Orlowski 1995] Knöll, L. and Orlowski, A., Distance between density operators: applications to the Jaynes–Cummings model. *Physical Review*, **A51**, 1622–30.

[Kochen/Specker 1967] Kochen, S. and Specker, E., The problem of hidden variables in quantum mechanics. *Journal of Mathematics and Mechanics*, **17**, 59–87.

[Kofman *et al.* 2001] Kofman, A. G., Kurizki, G., and Opatrny, T., Zeno and anti-Zeno effects for photon polarization dephasing. *Physical Review*, **A63**, 042108.

[Kramers 1926] Kramers, Hendrik A., Wellenmechanik und halbzahlige Quantisierung. *Zeitschrift für Physik*, **39**, 828–40.

[*Kraus* 1983] Kraus, Karl, *States, Effects and Operations*, Berlin: Springer.

[Kraus 1987] —, Complementary observables and uncertainty relations. *Physical Review*, **D35**, 3070–75.

[Krenn/Zeilinger 1996] Krenn, G. and Zeilinger, A., Entangled entanglement. *Physical Review*, **A54**, 1793–97.

[Kühn *et al.* 1994] Kühn, H., Welsch, D.-G., and Vogel, W., Determination of density matrices from field distributions and quasiprobabilities. *Journal of Modern Optics*, **41**, 1607–613.

[*Kuhn* 1978] Kuhn, Thomas S., *Black-Body Theory and the Quantum Discontinuity. 1894–1912*, Oxford: Clarendon Press.

[Kwiat *et al.* 1994] Kwiat, P. G., Eberhard, P. H., Steinberg, A. M., and Chiao, R. Y., Proposal for a loophole-free Bell inequality experiment. *Physical Review*, **A49**, 3209–20.

[Kwiat *et al.* 1995a] Kwiat, P. G., Weinfurter, H., Herzog, T., Zeilinger, A., and Kasevich, M., Interaction-free measurement. *Physical Review Letters*, **74**, 4763–66.

[Kwiat *et al.* 1995b] Kwiat, P. G., Weinfurter, H., Herzog, T., Zeilinger, A., and Kasevich, M., Experimental realization of interaction-free measurements, in [*Greenberger/Zeilinger* 1995, 383–93].

[Kwiat *et al.* 1996] Kwiat, P., Weinfurter, H., and Zeilinger, A., Quantum seeing in the dark. *Scientific American*, Nov., 72–78.

[Kwiat *et al.* 1999] Kwiat, P. G., Waks, E., White, A. G., Appelbaum, I., and Eberhard, P. H., Ultrabright source of polarization-entangled photons. *Physical Review*, **A60**, R773–76.

[Lamb 1969] Lamb, Willis E. Jr., Nonrelativistic quantum mechanics. *Physics Today*, **22** (April), 23–28.

[*Landau/Lifshitz* 1976a] Landau, Lev D. and Lifshitz, E. M., *Mechanics* (Engl. trans.), vol. I of *The Course of Theoretical Physics*, Oxford: Pergamon.

[*Landau/Lifshitz* 1976b] —, *Quantum Mechanics. Non Relativistic Theory* (Engl. trans.), vol. III of *The Course of Theoretical Physics*, Oxford: Pergamon.

[*Landau/Lifshitz* 1976c] —, *Statistical Physics, Part I* (Engl. trans.), vol. V of *The Course of Theoretical Physics*, Oxford: Pergamon.

[*Landau/Lifshitz* 1976] —, *Statistical Physics, Part II* (Engl. trans.), vol. IX of *The Course of Theoretical Physics*, Oxford: Pergamon.

[Landauer 1961] Landauer, Rolf, Irreversibility and heat generation in the computing process. *IBM Journal of Research and Development*, **5**, 183–91.

[Landauer 1996a] —, Minimal energy requirements in communication. *Science*, **272**, 1914–19.

[Leavens 1995] Leavens, C. R., Bohm trajectory and Feynman path approaches to the "tunneling time problem." *Foundations of Physics*, **25**, 229–68.

[Leggett 1989] Leggett, Anthony J., Low temperature physics, superconductivity and superfluidity, in [*Davies* 1989, 268–88].

[Leggett 2003] —, Nonlocal hidden-variable theories and quantum mechanics: an incompatibility theorem. *Foundations of Physics*, **33**, 1469–93.

[Leibfried *et al.* 2005] Leibfried, D., Knill, E., Seidelin, S., *et al.*, Creation of a six-atom "Schrödinger cat" state. *Nature*, **438**, 639–42.

[Leonhardt 1995] Leonhardt, Ulf, Quantum-state tomography and discrete Wigner function. *Physical Review Letters*, **74**, 4101–105.

[*Leonhardt* 1997] —, *Measuring the Quantum State of Light*, Cambridge: Cambridge University Press.

[Leonhardt/Raymer 1996] Leonhardt, U. and Raymer, M. G., Observation of moving wave packets reveals their quantum state. *Physical Review Letters*, **76**, 1985–89.

[Lévy-Leblond 1976] Lévy-Leblond, Jean-Marc, Who is afraid of nonhermitian operators? A quantum description of angle and phase. *Annals of Physics*, **101**, 319–41.

[Lewenstein 2007] Lewenstein, Maciej, The social life of atoms. *Nature*, **445**, 372–75.

[Lewis 1926] Lewis, G. N., The conservation of photons. *Nature*, **118**, 874–75.

[Lieb 1973] Lieb, Elliott H., The classical limit of quantum spin systems. *Communications in Mathematical Physics*, **31**, 327–340.

[Lieb/Ruskai 1973a] Lieb, E. H. and Ruskai, M. B., Proof of the strong subadditivity of quantum-mechanical entropy. *Journal of Mathematical Physics*, **14**, 1938–41.

[Lieb/Ruskai 1973b] —, A fundamental property of quantum-mechanical entropy. *Physical Review Letters*, **30**, 434–36.

[Lindblad 1973] Lindblad, Göran, Entropy, information and quantum measurement. *Communications in Mathematical Physics*, **33**, 305–322.

[*Lindblad* 1983] —, *Non-Equilibrium Entropy and Irreversibility*, Dordrecht: Reidel.

[Lockwood 1996a] Lockwood, Michael, "Many Minds" interpretations of quantum mechanics. *British Journal for Philosophy of Science*, **47**, 159–88.

[Loinger 2003] Loinger, Angelo, Sull'equazione radiale di Schrödinger dell'atomo di idrogeno. Nota storico-didattica. *Il saggiatore*, **19.1–2**, 70–73.

[*Loudon* 1973] Loudon, Rodney, *The Quantum Theory of Light*, London: Clarendon; 2nd edn. 1983.

[Loudon/Knight 1987] Loudon, R. and Knight, P. L., Squeezed light. *Journal of Modern Optics*, **34**, 709–59.

[Louisell 1963] Louisell, William H., Amplitude and phase uncertainty relations. *Physics Letters*, **7**, 60–61.

[*Louisell* 1973] —, *Quantum Statistical Theory of Radiation*, New York: Wiley.

[Lüders 1951] Lüders, Gerhart, Über die Zustandsänderung durch Meßprozeß. *Annalen der Physik*, **8**, 322–28.

[Mabuchi/Zoller 1996] Mabuchi, H. and Zoller, P., Inversion of quantum jumps in quantum optical systems under continuous observation. *Physical Review Letters*, **76**, 3108–11.

[*Mandel/Wolf* 1995] Mandel, L. and Wolf, E., *Optical Coherence and Quantum Optics*, Cambridge: University Press.

[Mandelstam/Tamm 1945] Mandelstam, L. and Tamm, I. G., The uncertainty relation between energy and time in non-relativistic quantum mechanics. *Journal of Physics (USSR)*, **9**, 249–54.

[*Mandl/Shaw* 1984] Mandl, F. and Shaw, G., *Quantum Field Theory*, Chichester: Wiley.

[*Ma'nko/Dodonov* 2003] Ma'nko, V. I. and Dodonov, V. V., *Theory of Non–Classical States of Light*, CRC Press.

[*Marinari/Parisi* 2008] Marinari, E. and Parisi, G. *Trattatello di Probabilità*, Springer-Verlag, in press.

[Marshall/Santos 1985] Marshall, T. W. and Santos, E., Local realist model for the coincidence rates in atomic-cascade experiments. *Physics Letters*, **107A**, 164–68.

[Marshall *et al.* 1983] Marshall, T. W, Santos, E., and Selleri, F., Local realism has not been refuted by atomic cascade experiments. *Physics Letters*, **98A**, 5–9.

[Martens/De Muynck 1990] Martens, H. and De Muynck, W. M., Nonideal quantum measurements. *Foundations of Physics*, **20**, 255–81.

[*Maslow* 1994] Maslow, V. P., *The Complex WKB Method for Nonlinear Equations 1: Linear Theory*, (trans. from Russian), Basel: Birkhäuser-Verlag.

[Maslow/Fedoriuk 1981] Maslow, V. P. and Fedoriuk, M. V., *Semi-Classical Approximation in Quantum Mechanics* (trans. from Russian), Dordrecht: Reidel.

[*Maxwell* 1873] Maxwell, James C., *A Treatise on Electricity and Magnetism*, Oxford: Clarendon Press.

[McKeever *et al.* 2003] McKeever, J., Boca, A., Boozer, A. D., Buck, J. R., and Kimble, H. J., Experimental realization of a one-atom laser in the regime of strong coupling. *Nature*, **425**, 268–71.

[*McWeeny* 1979] McWeeny, Roy (edn.), *Coulson's Valence*, Oxford: University Press, 3rd edn.

[*Mehra/Rechenberg* 1982–2001] Mehra, J. and Rechenberg, H., *The Historical Development of Quantum Theory*, New York: Springer.

[Mensky 1996] Mensky, Michael B., A note on reversibility of quantum jumps. *Physics Letters*, **222A**, 137–40.

[Mermin 1986] Mermin, N. David, The EPR experiment – thoughts about the "loophole," in [*Greenberger* 1986, 422–27].

[*Merzbacher* 1970] Merzbacher, E., *Quantum Mechanics*, New York: Wiley.

[Meschke *et al.* 2006] Meschke, M., Guichard, W., and Pekola, J. P., Single-mode heat conduction by photons. *Nature*, **444**, 187–90.

[*Messiah* 1958] Messiah, Albert, *Quantum Mechanics* (trans. from French), 1958; Amsterdam: North-Holland, 1961; New York: Dover, 1999.

[Milburn 1986] Milburn, Gerald J., Quantum and classical Liouville dynamics of the anharmonic oscillator. *Physical Review*, **A33**, 674–85.

[Milburn/Walls 1983] Milburn, G. J. and Walls, D. F., Quantum solutions of the damped harmonic oscillator. *American Journal of Physics*, **51**, 1134–36.

[Misra/Sudarshan 1977] Misra, B. and Sudarshan, E. C. G., The Zeno's paradox in quantum theory. *Journal of Mathematical Physics*, **18**, 756–63.

[Mittelstaedt *et al.* 1987] Mittelstaedt, P., Prieur, A., and Schieder, R., Unsharp particle-wave duality in a photon split-beam experiment. *Foundations of Physics*, **17**, 891–903.

[Monroe *et al.* 1996] Monroe, C., Meekhof, D. M., King, B. E., and Wineland, D. J., A "Schrödinger cat" superposition state of an atom. *Science*, **272**, 1131–36.

[*Morrison* 1990] Morrison, Michael A., *Understanding Quantum Physics*, Upper Saddle River, NJ: Prentice Hall.

[Morse 1929] Morse, P. M., Diatomic molecules according to the wave mechanics. II. Vibrational levels. *Physical Review*, **34**, 57–64.

[Naik *et al.* 2006] Naik, A., Buu, O., LaHaye, M. D., *et al.*, Cooling a nanomechanical resonator with quantum back–action. *Nature*, **443**, 193–96.

[*Natterer* 1986] Natterer, F., *The Mathematics of Computerized Tomography*, New York: Wiley.

[Neder *et al.* 2007] Neder, I., Ofek, N., Heiblum, M., Mahalu, D., and Umansky, V., Interference between two indistinguishable electrons from independent sources. *Nature*, **448**, 333–37.

[*Newton* 1704] Newton, Isaac, *Opticks or A Treatise of the Reflections, Refractions, Inflections and Colours of Light*, 1704, 2nd edn. lat. 1706, 3rd edn. 1721, London, 4th edn. 1730; New York, Dover, 1952, 1979.

[Nielsen/Caves 1997] Nielsen, M. A. and Caves, C. M., Reversible quantum operations and their application to teleportation. *Physical Review*, **A55**, 2547–56.

[*Nielsen/Chuang* 2000] Nielsen, M. A. and Chuang, I. L., *Quantum Computation and Quantum Information*, Cambridge: Cambridge University Press, 2000, 2002.

[O'Brien *et al.* 2003] O'Brien, J. L., Pride, G. J., White, A. G., Ralph, T. C., and Branning, D., Demonstration of an all-optical quantum controlled-NOT gate. *Nature*, **426**, 264–67.

[Omnès 1988] Omnès, Roland, Logical reformulation of quantum mechanics. I–III. *Journal of Statistical Physics*, **53**, 893–975.

[Omnès 1989] —, Logical reformulation of quantum mechanics. IV. *Journal of Statistical Physics*, **57**, 357–382.

[Omnès 1990] —, From Hilbert space to common sense: a synthesis of recent progress in the interpretation of quantum mechanics. *Annals of Physics*, **201**, 354–447.

[Ou/Mandel 1988] Ou, Z. Y. and Mandel, L., Violation of Bell's inequality and classical probability in a two-photon correlation experiment. *Physical Review Letters*, **61**, 50–53.

[Ourjoumtsev *et al.* 2007] Ourjoumtsev, A., Jeong, H., Tualle-Brouri, R., and Grangier, P., Generation of optical "Schrödinger cats" from photon number states. *Nature*, **448**, 784–86.

[Pan *et al.* 1998] Pan, J.-W., Bouwmeester, D., Weinfurter, H., and Zeilinger, A., Experimental entanglement swapping: entangling photons that never interacted. *Physical Review Letters*, **80**, 3891–94.

[Pancharatnam 1956] Pancharatnam, S., Generalized theory of interference, and its applications. *Proceedings of Indian Academy of Sciences*, **A44**, 247–62; rep. in [*Shapere/Wilczek* 1989, 51–66].

[*Parisi* 1988] Parisi, Giorgio, *Statistical Field Theory*, Redwood City, CA: Addison-Wesley.

[Parisi 2003] —, Two spaces looking for a geometer. *Bulletin of Symbolic Logic*, **9**, 181–96.

[Parisi 2005a] —, Brownian motion. *Nature*, **433**, 221.

[Parisi 2005b] —, Plancks legacy to statistical mechanics. *ArXiv*, cond–math/0101293.

[Partovi 1989] Partovi, Hossein M., Irreversibility, reduction and entropy increase in quantum measurements. *Physics Letters*, **137A**, 445–50.

[Paul 1962] Paul, Harry, Über quantenmechanische Zeitoperatoren. *Annalen der Physik*, **9**, 252–61.

[Pauli 1925] Pauli, Wolfgang, Über den Zusammenhang des Abschlusses der Elektronengruppen im Atom mit der Komplexstruktur der Spektren. *Zeitschrift für Physik*, **31**, 765–83.

[Pauli 1927] —, Zur Quantenmechanik des magnetischen Elektrons. *Zeitschrift für Physik*, **43**, 601–23.

[*Pauli* 1980] —, *General Principles of Quantum Mechanics*, Berlin, Springer, 1980; Engl. trans. of "Die allgemeinen Prinzipien der Wellenmechanik," first published in Geiger/Scheel (eds.), *Handbuch der Physik*, XXIV, Berlin, Springer, 1933: 83–272 and published again in 1958.

[Paz *et al.* 1993] Paz, J.P., Habib, S., and Zurek, W.H., Reduction of the wave packet: preferred observable and decoherence time scale. *Physical Review*, **D47**, 488–501.

[Pegg *et al.* 2005] Pegg, D.T., Barnett, S.M., Zambrini, R., Franke-Arnold, S., and Padget, M., Minimum uncertainty states of angular momentum and angular position. *New Journal of Physics*, **7**, 62.

[*Peirce CP*] Peirce, Charles S., *The Collected Papers*, vols. I–VI (Charles Hartshorne and Paul Weiss eds.), Cambridge, MA: Harvard University Press, 1931–1935; vols. VII–VIII (Arthur W. Burks edn.), Cambridge, MA: Harvard University Press, 1958.

[Peres 1980] Peres, Asher, Measurement of time by quantum clocks. *American Journal of Physics*, **48**, 552–57.

[*Peres* 1993] —, *Quantum Theory. Concepts and Methods*, Dordrecht: Kluwer.

[Peres 1995] —, Higher order Schmidt decompositions. *Physics Letters*, **A202**, 16–17.

[Peres 1996] —, Separability criterion for density matrices. *Physical Review Letters*, **77**, 1413–15.

[Perreault/Cronin 2005] Perreault, J.D. and Cronin, A.D., Observation of atom wave phase shifts induced by Van Der Waals atom–surface interactions. *Physical Review Letters*, **95**, 133201-1–4. (e-version)

[Perrie *et al.* 1985] Perrie, W., Duncan, A.J., Beyer, H.J., and Kleinpoppen, H., Polarization correlation of the two photons emitted by metastable atomic deuterium: a test of Bell's inequality. *Physical Review Letters*, **54**, 1790–93.

[*Peshkin/Tonomura* 1989] Peshkin, M. and Tonomura, A., *The Aharonov–Bohm Effect*, Berlin: Springer.

[Pfleegor/Mandel 1967] Pfleegor, R.L. and Mandel, L., Interference effects at the single photon level. *Physics Letters*, **24A**, 766–67.

[Philippidis *et al.* 1979] Philippidis, C., Dewdney, C., and Hiley, B.J., Quantum interference and the quantum potential. *Nuovo Cimento*, **52B**, 15–28.

[Phoenix/Barnett 1993] Phoenix, S. J. D. and Barnett, S. M., Non-local interatomic correlations in the micromaser. *Journal of Modern Optics*, **40**, 979–83.

[*Pitowsky* 1989] Pitowsky, Itamar, *Quantum Probability – Quantum Logic*, Berlin: Springer.

[Planck 1900a] Planck, Max, Über die Verbesserung der Wien'schen Spektralgleichung. *Verhandlungen der Deutchen Physikalischen Gesellschaft*, **2**, 202–204.

[Planck 1900b] —, Zur Theorie des Gesetzes der Energieverteilung im Normalspektrum. *Verhandlungen der Deutchen Physikalischen Gesellschaft*, **2**, 237–45.

[Planck 1917] —, Über einen Satz der statistischen Dynamik und seine Erweiterung in der Quantentheorie. *Sitzungberichte der Preussichen Akademie der Wissenschaften: Phys. Math. Kl.*, **1917**, 324–41.

[Plantenberg *et al.* 2007] Plantenberg, J. H., de Groot, P. C., Harmans, C. J. P. M., and Mooij, J. E., Demonstration of controlled-NOT quantum gates on a pair of superconducting quantum bits. *Nature*, **447**, 836–39.

[*Plauger et al.* 2000] Plauger, P. J., Lee, M., Musser, D., and Stepanov, A. A., *C++ Standard Template Library*, Prentice Hall PTR.

[*Poincaré* 1892] Poincaré, Henri, *Theorie Mathématique de la Lumière*, Paris: Gauthiers-Villars.

[*Poincaré* 1902] —, *La Science et l'Hypothèse*, Paris, 1902; Flammarion, 1968.

[Prokhorov 1958] Prokhorov, A. M., Molecular amplifier and generator for submillimeter waves. *Soviet Physics JETP*, **7**, 1140–41; Engl. trans. of *Soviet Physics JETP*, **34**, 1658–59.

[*Prugovečki* 1971] Prugovečki, Eduard, *Quantum Mechanics in Hilbert Space*, 1971; 2nd edn., New York: Academic Press, 1981.

[Prugovečki 1977] —, Information-theoretical aspects of quantum measurement. *International Journal of Theoretical Physics*, **16**, 321–33.

[Quadt 1989] Quadt, Ralf, The nonobjectivity of past events in quantum mechanics. *Foundations of Physics*, **19**, 1027–35.

[Rauch 2000] Rauch, Helmut, Reality in neutron interference experiments. In J. Ellis and D. Amati (eds.), *Quantum Reflections*, Cambridge: Cambridge University Press, pp. 28–68.

[Reid/Walls 1986] Reid, M. D. and Walls, D. F., Violations of classical inequalities in quantum optics. *Physical Review*, **A34**, 1260–76.

[Renninger 1960] Renninger, Mauritius K., Messungen ohne Störung des Meßobjekts. *Zeitschrift für Physik*, **158**, 417–21.

[Riebe *et al.* 2004] Riebe, M., Häffner, H., Roos, C. F., *et al.*, Deterministic quantum teleportation with atoms. *Nature*, **429**, 734–37.

[*Risken* 1996] Risken, Hannes, *The Fokker–Planck Equation: Methods of Solution and Application*, Berlin: Springer.

[Robertson 1929] Robertson, H. P., The uncertainity principle. *Physical Review*, **34**, 163–64.

[Roukes 2006] Roukes, Michael, Observing and the observable. *Nature*, **443**, 154–55.

[Royer 1985] Royer, Antoine, Measurement of the Wigner function. *Physical Review Letters*, **55**, 2745–48.

[Royer 1989] —, Measurement of quantum states and the Wigner function. *Foundations of Physics*, **19**, 3–30.

[Saif *et al.* 1998] Saif, F., Bialynicki-Birula, I., Fortunato, M., and Schleich, W. P., Fermi accelerator in atom optics. *Physical Review*, **A58**, 4779–83.

[Santos 1991] Santos, Emilio, Does quantum mechanics violate the Bell inequalities? *Physical Review Letters*, **66**, 1388–90; Errata: 3227.

[Santos 1992] —, Critical analysis of the empirical tests of local hidden-variable theories. *Physical Review*, **A46**, 3646–56.

[Sauter *et al.* 1986] Sauter, T., Neuhauser, D., Blatt, R., and Toschek, P. E., Observation of quantum jumps. *Physical Review Letters*, **57**, 1696–98.

[*Schiff* 1955] Schiff, Leonard, *Quantum Mechanics*, 1955; 3rd edn., New York: McGraw-Hill, 1968.

[*Schilpp* 1949] Schilpp, Arthur (edn.), *Albert Einstein. Philosopher-Scientist*, La Salle, IL: Open Court; 3rd edn. 1988.

[*Schleich* 2001] Schleich, Wolfgang P., *Quantum Optics in Phase Space*, Berlin: Wiley-VCH.

[*Schleich/Raymer* 1997] Schleich, W. P. and Raymer, M. (eds.), *Quantum State Preparation and Measurement*, *Journal of Modern Optics* **44**, issues 11–12.

[Schmidt 1907] Schmidt, E., Zur Theorie des linearen und nichtlinearen Integralgleichungen. I Teil: Entwicklung willkürlicher Funktionen nach Systemen vorgeschriebener. *Mathematische Annalen*, **63**, 433–76.

[Schrödinger 1926] Schrödinger, Erwin, Quantisierung als Eigenwertproblem. I–II. *Annalen der Physik*, **79**, 361–76 and 489–527.

[Schrödinger 1935] —, Die gegenwärtige Situation in der Quantenmechanick. I–III. *Naturwissenschaften*, **23**, 807–12, 823–28, 844–49.

[Schawlow/Townes 1958] Schawlow, A. L. and Townes, C. H., Infrared and optical masers. *Physical Review*, **112**, 1940–49.

[Schiller *et al.* 1996] Schiller, S., Breitenbach, G., Pereira, S. F., Müller, T., and Mlynek, J., Quantum statistics of the squeezed vacuum by measurement of the density matrix in the number representation. *Physical Review Letters*, **77**, 2933–36.

[Schumacher 1990] Schumacher, Benjamin W., Information from quantum measurements, in [*Zurek* 1990, 29–37].

[Schumacher 1991] —, Information and quantum nonseparability. *Physical Review*, **A44**, 7047–52.

[*Schweber* 1961] Schweber, Silvan S., *An Introduction to Relativistic Quantum Field Theory*, New York: Harper and Row.

[*Scully/Zubairy* 1997] Scully, M. O. and Zubairy, M. S., *Quantum Optics*, Cambridge: Cambridge University Press.

[Scully *et al.* 1978] Scully, M. O., Shea, R., and McCullen, J. D., State reduction in quantum mechanics: a calculation example. *Physics Reports*, **43C**, 485–98.

[Scully *et al.* 1991] Scully, M. O., Englert, B.-G., and Walther, H., Quantum optical tests of complementarity. *Nature*, **351**, 111–16.

[Schwab 2006] Schwab, Keith, Information on heat. *Nature*, **444**, 161.

[*Segrè* 1964] Segrè, Emilio, *Nuclei and Particles*, Benjamin, 1964; 2nd edn., 1977.

[Selleri 1969] Selleri, Franco, On the wave function of quantum mechanics. *Lettere a Nuovo Cimento*, **1**, 908–910.

[Selleri 1982] —, On the direct observability of quantum waves. *Foundations of Physics*, **12**, 1087–112.

[*Selleri* 1992] — (ed.), *Wave–Particle Duality*, New York, Plenum, 1992.

[Shannon 1948] Shannon, Claude E., A mathematical theory of communication. *Bell System Technical Journal*, **27**, 379–423; 623–56.

[*Shapere/Wilczek* 1989] Shapere, A. and Wilczek, F. (eds.), *Geometric Phases in Physics*, Singapore: World Scientific.

[Sherman/Kurizki 1992] Sherman, B. and Kurizki, G., Preparation and detection of macroscopic quantum superpositions by two-photon field-atom interactions. *Physical Review*, **A45**, R7674–77.

[Shih/Alley 1988] Shih, Y. H. and Alley, C. O., New type of EPR–Bohm experiment using pairs of light quanta produced by optical parametric down conversion. *Physical Review Letters*, **61**, 2921–24.

[Shimony 1971] Shimony, Abner, Experimental test of local hidden-variable theories, in [*D'Espagnat* 1971, 182–94]; rep. in [*Shimony* 1993, II, 77–89].

[Shimony 1984] —, Contextual hidden variables theories and Bell's inequalities. *British Journal for the Philosophy of Science*, **35**, 25–45.

[*Shimony* 1993] Shimony, Abner, *Search for a Naturalistic Point of View*, Cambridge: University Press.

[Shor 1994] Shor, Peter W., Algorithms for quantum computation: discrete log and factoring, in *Proc. 35th Annual Symposium on the Foundations of Computer Science*, Los Alamos: IEEE Computer Society Press, p. 1: 124.

[Siviloglou *et al.* 2007] Siviloglou, G. A., Broky, J., Dogariu, A., and Christodoulides, D. N., Observation of accelerating Airy beams. *Physical Review Letters*, **99**, 213901-1–4.

[*Slater* 1951] Slater, John C., *Quantum Theory of Matter*, New York: McGraw-Hill.

[Smithey *et al.* 1993] Smithey, D. T., Beck, M., Raymer, M. G., and Faridani, A., Measurement of the Wigner distribution and the density matrix of a light mode using optical homodyne tomography: application to squeezed states and the vacuum. *Physical Review Letters*, **70**, 1244–47.

[Sommerfeld 1912] Sommerfeld, Arnold, Rapport sur l'application de la théorie de l'élément d'action aux phénomènes moléculaires non périodiques, in *La Théorie du Rayonnement et les Quanta – Rapports et Discussions de la Réunion Tenue à Bruxelles*, P. Langevin and L. de Broglie (eds.), Paris: Gauthier-Villars, pp. 313–72.

[Stapp 1971] Stapp, Henry P., S-matrix interpretation of quantum theory. *Physical Review*, **D3**, 1303–20.

[Stapp 1980] —, Locality and reality. *Foundations of Physics*, **10**, 767–95.

[Stenner *et al.* 2003] Stenner, M. D., Gauthier, D. J., and Neifeld, M. A., The speed of information in a "fast-light" optical medium. *Nature*, **425**, 695–98.

[Storey *et al.* 1994] Storey, E. P., Tan, S. M., Collett, M. J., and Walls, D. F., Path detection and the uncertainty principle. *Nature*, **367**, 626–28.

[*Streater/Wightman* 1978] Streater, R. F. and Wightman, A. S., *PCT, Spin and Statistics, and All That*, 2nd edn. 1978; Princeton: Princeton University Press, 2000.

[Sudarshan 1963] Sudarshan, E. C. G., Equivalence between semiclassical and quantum mechanical descriptions of statistical light beams. *Physical Review Letters*, **10**, 277–79.

[Summhammer *et al.* 1983] Summhammer, J., Badurek, G., Rauch, H., Kischko, U., and Zeilinger A., Direct observation of fermion spin superposition by neutron interferometry. *Physical Review*, **A27**, 2523–32.

[Susskind/Glogower 1964] Susskind, L. and Glogower, J., Quantum mechanical phase and time operator. *Physics*, **1**, 49.

[Tan *et al.* 1991] Tan, S. M., Walls, D. F., and Collett, M. J., Nonlocality of a single photon. *Physical Review Letters*, **66**, 252–55.

[Tarozzi 1985] Tarozzi, Gino, Experimental tests of the properties of the quantum-mechanical wave function. *Lettere a Nuovo Cimento*, **42**, 438–42.

[Tarozzi 1996] —, Quantum measurements and macroscopical reality: epistemological implications of a proposed paradox. *Foundations of Physics*, **26**, 907–917.

[*Taylor/Lay* 1958] Taylor, A. E. and Lay, D. C., *Introduction to Functional Analysis*, New York: Wiley; 2nd edn. 1980; rep. Malabar: Krieger, 1986.

[Thomas 1927] Thomas, L. H., *Proceedings of the Cambridge Philosophical Society*, **23**, 542.

[*Thorne* 1994] Thorne, Kip S., *Black Holes and Time Warps: Einstein's Outrageous Legacy*, New York: W. W. Norton.

[Thorne *et al.* 1978] Thorne, K. S., Drever, R. W. P., Caves, C. M., Zimmermann, M., and Sanberg, V. D., Quantum nondemolition measurements of harmonic oscillators. *Physical Review Letters*, **40**, 667–71.

[*Tinkham* 1996] Thinkham, M., *Introduction to Superconductivity*, New York: McGraw-Hill, 2nd edn.

[Townsend *et al.* 1997] Townsend, C., Ketterle, W., and Stringari, S., Bose-Einstein condensation. *Physics World*, March, 29.

[Tsirelson 1980] Tsirelson, Boris S., Quantum generalizations of Bell's inequality. *Letters in Mathematical Physics*, **4**, 93–100.

[Ueda/Kitagawa 1992] Ueda, M. and Kitagawa, M., Reversibility in quantum measurement processes. *Physical Review Letters*, **68**, 3424–27.

[Uhlmann 1977] Uhlmann, A., Relative entropy and the Wigner–Yanase-Dyson–Lieb concavity in an interpolation theory. *Communications in Mathematical Physics*, **54**, 21–32.

[Uhlenbeck/Goudsmit 1925] Uhlenbeck, G. E. and Goudsmit, S. A., Ersetzung der Hypothese vom unmechanischen Zwang durch eine Forderung bezüglich des inneren Verhaltens jedes einzelnen Elektrons. *Naturwissenschaften*, **47**, 953–54.

[Uhlenbeck/Goudsmit 1926] —, Spinning electrons and the structure of spectra. *Nature*, **117**, 264–65.

[Unruh 1979] Unruh, William G., Quantum nondemolition and gravity-wave detection. *Physical Review*, **D19**, 2888–96.

[Unruh 1994] —, Comment: Reality and measurement of the wave function. *Physical Review*, **A50**, 882–87.

[Ursin *et al.* 2007] Ursin, R., Tiefenbacher, F., Schmitt–Manderbach, *et al.*, Entanglement-based quantum communication over 144 km. *Nature Physics*, **3**, 481–86.

[*van Der Merwe et al.* 1988] van Der Merwe, A., Selleri, F., and Tarozzi, G. (eds.), *Microphysical Reality and Quantum Formalism*, vol. I, Dordrecht: Kluwer.

[Vandersypen *et al.* 2001] Vandersypen, L. M. K., Steffen, M., Breyta, G., Yannoni, C. S., Sherwood, M. K., and Chuang, I. L., Experimental realization of Shor's quantum factoring algorithm using nuclear magnetic resonance. *Nature*, **414**, 883–87.

[van der Wal *et al.* 2000] van der Wal, C. H., ter Haar, A. C. J., Wilhelm, F. K., *et al.*, Quantum superposition of macroscopic persistent-current states. *Science*, **290**, 773–77.

[van Hove 1955] van Hove, Léon, Quantum-mechanical perturbations giving rise to a statistical transport equation. *Physica*, **21**, 517–40.

[van Hove 1957] —, The approach to equilibrium in quantum statistics. *Physica*, **23**, 441–80.

[van Hove 1959] —, The ergodic behaviour of quantum many-body systems. *Physica*, **25**, 268–76; rep. in [*Wheeler/Zurek* 1983, 648–56].

[Varadarajan 1962] Varadarajan, V. S., Probability in physics and a theorem on simultaneous observability. *Communications in Pure and Applied Mathematics*, **15**, 189–217.

[Vedral 2003] Vedral, Vlatko, Entanglement hits the big time. *Nature*, **425**, 28–29.

[Vedral/Plenio 1998] Vedral, V. and Plenio, M. B., Entanglement measures and purification procedures. *Physical Review*, **A57**, 1619–33.

[Vedral *et al.* 1997] Vedral, V., Plenio, M. B., Rippin, M. A., and Knight, P. L., Quantifying entanglement. *Physical Review Letters*, **78**, 2275–79.

[Vedral *et al.* 1997] Vedral, V., Plenio, M. B., Jacobs, K., and Knight, P. L., Statistical inference, distinguishability of quantum states, and quantum entanglement. *Physical Review*, **A56**, 4452–55.

[Vichi *et al.* 1998] Vichi, L., Inguscio, M., Stringari, S., and Tino, G. M., Quantum degeneracy and interaction effects in spin-polarized Fermi–Bose mixtures. *Journal of Physics*, **B31**, L899–L907.

[Vitali *et al.* 2000] Vitali, D., Fortunato, M., and Tombesi, P., Complete Quantum Teleportation with a Kerr Nonlinearity. *Physical Review Letters*, **85**, 445–48.

[Vogel/Risken 1989] Vogel, K. and Risken, H., Determination of quasiprobability distributions in terms of probability distributions for the rotated quadrature phase. *Physical Review*, **A40**, 2847–49.

[Vogel *et al.* 1993] Vogel, K., Akulin, V. M., and Schleich, W. P., Quantum state engineering of the radiation field. *Physical Review Letters*, **71**, 1816–19.

[*von Neumann* CW] von Neumann, John, *Collected Works*, Oxford: Pergamon, 1963.

[von Neumann 1927a] —, Mathematische Begründung der Quantenmechanik. *Göttinger Nachrichten*; rep. in [*von Neumann* CW, I, 151–207].

[von Neumann 1927b] —, Wahrscheinlichkeitstheoretischer Aufbau der Quantemechanik *Göttinger Nachrichten*; rep. in [*von Neumann* CW, I, 208–35].

[von Neumann 1927c] —, Thermodinamik quantmechanischer Gesamtheiten. *Göttinger Nachrichten*; rep. in [*von Neumann* CW, I, 236–54].

[von Neumann 1929] —, Beweis des Ergodensatzes und des H-Theorems in der neuen Mechanik. *Zeitschrift für Physik*, **57**, 30–70; rep. in [*von Neumann* CW, I, 558–98].

[*von Neumann* 1932] —, *Mathematische Grundlagen der Quantenmechanik*, Berlin: Springer, 1932, 1968, 1996.

[*von Neumann* 1955] —, *Mathematical Foundations of Quantum Mechanics*, Princeton, University Press, 1955; Engl. trans. of *Mathematische Grundlagen der Quantenmechanik*, Berlin: Springer, 1932, 1968, 1996.

[Wallentowitz/Vogel 1995] Wallentowitz, S. and Vogel, W., Reconstruction of the quantum mechanical state of a trapped ion. *Physical Review Letters*, **75**, 2932–25.

[Wallentowitz/Vogel 1996] —, Unbalanced homodyning for quantum state measurement. *Physical Review*, **A53**, 4528–33.

[Walls/Milburn 1985] Walls, D. F. and Milburn, G. J., Effect of dissipation on quantum coherence. *Physical Review*, **A31**, 2403–408.

[*Walls/Milburn* 1994] —, *Quantum Optics*, Berlin: Springer.

[Walther *et al.* 2005] Walther, P., Resch, K. J., Rudolph, T., *et al.*, Experimental one-way quantum computing. *Nature*, **434**, 169–76.

[Wang *et al.* 1991a] Wang, L. J., Zou, X. Y., and Mandel, L., Experimental test of the Broglie guided-wave theory for photons. *Physical Review Letters*, **66**, 1111–14.

[Wang *et al.* 1991b] Wang, L. J., Zou, X. Y., and Mandel, L., Induced coherence without induced emission. *Physical Review*, **A44**, 4614–22.

[Wehrl 1978] Wehrl, Alfred, General properties of entropy. *Review of Modern Physics*, **50**, 221–60.

[Weihs 2007] Weihs, Gregor, The truth about reality. *Nature*, **445**, 723–24.

[Wentzel 1926] Wentzel, Gregor, Eine Verallgemeinerun der Quantenbedingungen für die Zwecke der Wellenmechanik. *Zeitschrift für Physik*, **38**, 518–29.

[*Weyl* 1936] Weyl, Hermann, *The Classical Groups: Their Invariants and Representations*, Princeton: Princeton University Press, 1939, 1946; 2nd edn. 1953, 1966, 1997.

[*Weyl* 1950] —, *The Theory of Groups and Quantum Mechanics*, Dover Publications, 1950; Engl. tr. of *Gruppentheorie und Quantenmechanik*, 1928, 1931.

[Wheeler 1978] Wheeler, John A., The "past" and the "delayed-choice" double-slit experiment, in *Mathematical Foundations of Quantum Theory*, A. R. Marlow (edn.), New York: Academic Press, pp. 9–48.

[Wheeler 1983] —, Law without law, in [*Wheeler/Zurek* 1983, 182–213].

[*Wheeler/Zurek* 1983] Wheeler, J. A. and Zurek, W. (eds.), *Quantum Theory and Measurement*, Princeton: Princeton University Press.

[A. White *et al.* 1998] White, A. G., Mitchell, J. R., Nairz, O., and Kwiat, P. G., "Interaction-free" imaging. *Physical Review*, **A58**, 605–13.

[*White* 1934] White, Harvey E., *Introduction to Atomic Spectra*, New York: McGraw-Hill.

[Wiesner 1983] Wiesner, Stephen J., Conjugate coding. *Sigact News*, **15**, 78–88.

[Wigner 1932] Wigner, Eugene P., On the quantum correction for thermodynamic equilibrium. *Physical Review*, **40**, 749–59.

[Wigner 1952] —, Die Messung quantenmechanischer Operatoren. *Zeitschrift für Physik*, **133**, 101–108.

[*Wigner* 1959] —, *Group Theory*, New York: Academic Press.

[Wigner 1961] —, Remarks on the mind–body question, in *The Scientist Speculates*, I. J. Good (ed.), London: Heinemann; rep. in [*Wheeler/Zurek* 1983, 168–81].

[Wootters/Zurek 1979] Wootters, W. K. and Zurek, W. H., Complementarity in the double-slit experiment: quantum nonseparability and a quantitative statement of Bohr's principle. *Physical Review*, **D19**, 473–84.

[Wootters/Zurek 1982] —, A single quantum cannot be cloned. *Nature*, **299**, 802–803.

[Yuen 1976] Yuen, Horace P., Two-photon coherent states of the radiation field. *Physical Review*, **A13**, 2226–43.

[Yuen 1986] Yuen, Horace P., Amplification of quantum states and noiseless photon amplifiers. *Physics Letters*, **113A**, 405–407.

[Yurke/Stoler 1986] Yurke, B. and Stoler, D., Generating quantum mechanical superpositions of macroscopically distinguishable states via amplitude dispersion. *Physical Review Letters*, **57**, 13–16.

[Yurke/Stoler 1992a] —, EPR effects from independent particle sources. *Physical Review Letters*, **68**, 1251–54.

[Yurke/Stoler 1992b] —, Bell's inequality experiments using independent-particle sources. *Physical Review*, **A46**, 2229–34.

[Zeh 1970] Zeh, H. Dieter, On the interpretation of measurement in quantum theory. *Foundations of Physics*, **1**, 69–76; rep. in [*Wheeler/Zurek* 1983, 342–49].

[Zeilinger 2000] Zeilinger, Anton, Quantum entangled bits step closer to It. *Science*, **289**, 405–406.

[Zou *et al.* 1992] Zou, X. Y., Grayson, T. P., Wang, L. J., and Mandel, L., Can an empty de Broglie pilot wave induce coherence?. *Physical Review Letters*, **68**, 3667–69.

[Zucchetti *et al.* 1996] Zucchetti, A., Vogel, W., Tasche, M., and Welsch, D.-G., Direct sampling of density matrices in field-strength bases. *Physical Review*, **A54**, 1678–81.

[Żukowski *et al.* 1993] Żukowski, M., Zeilinger, A., Horne, M. A., and Ekert, A. K., "Event-ready-detectors" Bell experiment via entanglement swapping. *Physical Review Letters*, **71**, 4287–90.

[Żukowski *et al.* 1995] Żukowski, M., Zeilinger, A., and Weinfurter, H., Entangling photons radiated by independent sources, [*Greenberger/Zeilinger* 1995, 91–102].

[Zurek 1981] Zurek, Wojciech H., Pointer basis of quantum apparatus: into what mixture does the wave packet collapse? *Physical Review*, **D24**, 1516–25.

[Zurek 1982] —, Environment-induced superselection rules. *Physical Review*, **D26**, 1862–80.

[Zurek 1986] —, Reduction of the wavepacket: how long does it take?, in *Frontiers of Nonequilibrium Statistical Physics*, G. T. Moore and M. O. Scully (eds.), New York: Plenum Press, pp. 145–51.

[*Zurek* 1990] — (ed.), *Complexity, Entropy and the Physics of Information*, Redwood City, CA: Addison-Wesley.

[Zurek 2007] —, Quantum origin of quantum jumps: breaking of unitary symmetry induced by information transfer in the transition from quantum to classical. *Physical Review*, **A76**, 052110-1–5.

Author index

Subject index